Texts and Monographs in Physics

Albert Einstein (1879–1955), in his study, Berlin 1916.
(ETH Archive, Zurich, Switzerland)

Norbert Straumann

General Relativity

With Applications to Astrophysics

With 111 Figures

Professor Norbert Straumann
Universität Zurich
Institut für Theorestische Physik
Winterthurer Strasse 190
8057 Zurich, Switzerland

Library of Congress Control Number: 2004104635

ISBN 3-540-21924-2 Springer-Verlag Berlin Heidelberg

Cover design: *design & production* GmbH, Heidelberg

Straumann: General Relativity with Application to Astrophysics

First Indian Reprint 2009

ISBN 978-81-8489-210-9

This edition is published by Springer (India) Private Limited,
A part of Springer Science+Business Media, Registered Office: 906-907,
Akash Deep Building, Barakhamba Road, New Delhi – 110 001, India.

Printed in India by Rashtriya Printers, Delhi.

Preface

Physics and mathematics students are as eager as ever to become acquainted with the foundations of general relativity and some of its major applications in astrophysics and cosmology. I hope that this textbook gives a comprehensive and timely introduction to both aspects of this fascinating field, and will turn out to be useful for undergraduate and graduate students.

This book is a complete revision and extension of my previous volume *'General Relativity and Relativistic Astrophysics'* that appeared about twenty years ago in the Springer Series *'Texts and Monographs in Physics'*; however, it cannot be regarded just as a new edition.

In Part I the foundations of general relativity are thoroughly developed. Some of the more advanced topics, such as the section on the initial value problem, can be skipped in a first reading.

Part II is devoted to tests of general relativity and many of its applications. Binary pulsars – our best laboratories for general relativity – are studied in considerable detail. I have included an introduction to gravitational lensing theory, to the extend that the current literature on the subject should become accessible. Much space is devoted to the study of compact objects, especially to black holes. This includes a detailed derivation of the Kerr solution, Israel's proof of his uniqueness theorem, and a derivation of the basic laws of black hole physics. Part II ends with Witten's proof of the positive energy theorem.

All the required differential geometric tools are developed in Part III. Readers who have not yet studied a modern mathematical text on differential geometry should not read this part in linear order. I always indicate in the physics parts which mathematical sections are going to be used at a given point of the discussion. For example, Cartan's powerful calculus of differential forms is not heavily used in the foundational chapters. The mathematical part should also be useful for other fields of physics. Differential geometric and topological tools play an increasingly important role, not only in quantum field theory and string theory, but also in classical disciplines (mechanics, field theory).

A textbook on a field as developed and extensive as general relativity must make painful omissions. When the book was approaching seven hundred pages, I had to give up the original plan to also include cosmology. This is really deplorable, since cosmology is going through a fruitful and exciting period. On the other hand, this omission may not be too bad, since several really good texts on various aspects of cosmology have recently appeared. Some of them are listed in the references. Quantum field theory on curved spacetime backgrounds is not treated at all. Other topics have often been left out because my emphasis is on direct physical applications of the theory. For this reason nothing is said, for instance, about black holes with hair (although I worked on this for several years).

The list of references is not very extensive. Beside some useful general sources on mathematical, physical, and historical subjects, I often quoted relatively recent reviews and articles that may be most convenient for the reader to penetrate deeper into various topics. Pedagogical reasons often have priority in my restrictive selection of references.

The present textbook would presumably not have been written without the enduring help of Thiemo Kessel. He not only typed very carefully the entire manuscript, but also urged me at many places to give further explanations or provide additional information. Since – as an undergraduate student – he was able to understand all the gory details, I am now confident that the book is readable. I thank him for all his help.

I also thank the many students I had over the years from the University of Zurich and the ETH for their interest and questions. This feedback was essential in writing the present book. Many of the exercises posed in the text have been solved in my classes.

Finally, I would like to thank the Tomalla Foundation and the Institute for Theoretical Physics of the University of Zurich for financial support.

Post Script I would be grateful for suggestions of all kind and lists of mistakes. Since I adopted – contrary to my habits – the majority convention for the signature of the metric, I had to change thousands of signs. It is unlikely that no sign errors remained, especially in spinorial equations.

Comments can be sent to me at *<norbert@physik.unizh.ch>*.

Zurich, January 2004 *Norbert Straumann*

Contents

Part I The General Theory of Relativity

Part I

The General Theory of Relativity

Introduction

"I worked horribly strennously, strange that one can endure that."

—A. Einstein

(On a postcard to his friend M. Besso in Nov. 1915)

The discovery of the general theory of relativity (GR) has often been justly praised as one of the greatest intellectual achievements of a single human being. At the ceremonial presentation of Hubacher's bust of A. Einstein in Zürich, W. Pauli said:

"The general theory of relativity then completed and - in contrast to the special theory - worked out by Einstein alone without simultaneous contributions by other researchers, will forever remain the classic example of a theory of perfect beauty in its mathematical structure."

Let us also quote M. Born:

"(The general theory of relativity) seemed and still seems to me at present to be the greatest accomplishment of human thought about nature; it is a most remarkable combination of philosophical depth, physical intuition and mathematical ingenuity. I admire it as a work of art."

The origin of the GR is all the more remarkable when one considers that, aside from a minute advance of the perihelion of Mercury, that remained in the data after all perturbations in the Newtonian theory had been accounted for, no experimental necessity for going beyond the Newtonian theory of gravitation existed. Purely theoretical considerations led to the genesis of GR. The Newtonian law of gravitation as an action-at-a-distance law is not compatible with the special theory of relativity. Einstein and other workers were thus forced to try to develop a relativistic theory of gravitation. It is remarkable that Einstein was soon convinced that gravitation has no place in the framework of special relativity. In his lecture *On the Origins of the General Theory of Relativity* Einstein enlarged on this subject as follows:

"I came a first step closer to a solution of the problem when I tried to treat the law of gravitation within the framework of special relativity theory. Like most authors at that time, I tried to formulate

a field law for gravity, since the introduction of action at a distance was no longer possible, at least in any natural way, due to the elimination of the concept of absolute simultaneity.
The simplest and most natural procedure was to retain the scalar Laplacian gravitational potential and to add a time derivative to the Poisson equation in such a way that the requirements of the special theory would be satisfied. In addition, the law of motion for a point mass in a gravitational field had to be adjusted to the requirements of special relativity. Just how to do this was not so clear, since the inertial mass of a body might depend on the gravitational potential. In fact, this was to be expected in view of the principle of mass-energy equivalence.
However, such considerations led to a result which made me extremely suspicious. According to classical mechanics, the vertical motion of a body in a vertical gravitational field is independent of the horizontal motion. This is connected with the fact that in such a gravitational field the vertical acceleration of a mechanical system, or of its center of mass, is independent of its kinetic energy. Yet, according to the theory which I was investigating, the gravitational acceleration was not independent of the horizontal velocity or of the internal energy of the system.
This in turn was not consistent with the well known experimental fact that all bodies experience the same acceleration in a gravitational field. This law, which can also be formulated as the law of equality of inertial and gravitational mass, now struck me in its deep significance. I wondered to the highest degree about its validity and supposed it to be the key to a deeper understanding of inertia and gravitation. I did not seriously doubt its strict validity even without knowing the result of the beautiful experiment of Eötvös, which - if I remember correctly - I only heard of later. I now gave up my previously described attempt to treat gravitation in the framework of the special theory as inadequate. It obviously did not do justice to precisely the most fundamental property of gravitation."

In the first chapter we shall discuss several arguments to convince the reader that a satisfactory theory of gravitation cannot be formulated within the framework of the special theory of relativity.

After nearly ten years of hard work, Einstein finally completed the general theory of relativity in November 1915. How hard he struggled with the theory is indicated by the following passage in a letter to A. Sommerfeld:

"At present I occupy myself exclusively with the problem of gravitation and now believe that I shall master all difficulties with the help of a friendly mathematician here (Marcel Grossmann). But one thing is certain, in all my life I have never labored nearly as hard, and I have become imbued with great respect for mathematics, the subtler

part of which I had in my simple-mindedness regarded as pure luxury until now. Compared with this problem, the original relativity is child's play."

In GR the rigid spacetime structure of the special theory of relativity (SR) is generalized. Einstein arrived at this generalization on the basis of his *principle of equivalence*, according to which gravitation can be "locally" transformed away in a freely falling, non-rotating system. (In the space age, this has become obvious to everybody.) This means that on an infinitesimal scale, relative to a *locally inertial system*, such as we have just described, special relativity remains valid. The metric field of the SR varies, however, over finite regions of spacetime. Expressed mathematically, spacetime is described by a (differentiable) pseudo-Riemannian manifold. The metric field g has the signature[1] $(-+++)$ and describes not only the metric properties of space and time as well as its causality properties, but also the gravitational field. The physical metric field is thus curved and dynamical, both influencing and being influenced by all other physical processes. In this sense GR unifies geometry *and* gravitation, whence the name *geometrodynamics*, coined by J.A. Wheeler, is indeed appropriate. Einstein had this profound insight some time during the summer 1912.

The theory consists of two main parts. The first describes the action of gravity on other physical processes. In collaboration with M. Grossmann, Einstein soon succeeded in generalizing non-gravitational (matter) laws, such as Maxwell's equations. The coupling of matter variables to gravitational fields is largely determined by the equivalence principle. This will be discussed extensively in Chap. 1. Briefly, this comes about as follows.

It is a mathematical fact that in a Lorentz manifold one can introduce (normal) coordinates such that, at a given point p, the metric takes the normal form $(g_{\mu\nu}(p)) = \text{diag}(-1,1,1,1)$, and in addition its components have vanishing first derivatives: $g_{\mu\nu,\lambda}(p) = 0$. Relative to such a coordinate system, the gravitational field is locally transformed away (up to higher order inhomogeneities). The kinematical structure of GR thus permits locally inertial systems. According to the equivalence principle, the laws of SR (for example, Maxwell's equations) should hold in such systems. This requirement permits one to formulate the non-gravitational laws in the presence of a gravitational field. Mathematically this amounts to the substitution of ordinary derivatives by *covariant* derivatives. The equivalence principle thus prescribes - up to certain ambiguities - the coupling of mechanical, electromagnetic and other systems to external gravitational fields. This corner stone enabled Einstein to derive important results, such as the gravitational redshift of light, long before the theory was completed.

Conversely, the metric field depends on the energy-momentum distribution of all forms of matter. This dependence is expressed in *Einstein's field*

[1] In this case, the pseudo-Riemannian manifold is called a *Lorentz manifold.*

equations - the hard core of the theory - which relates the curvature of spacetime to the energy-momentum tensor of matter. Einstein discovered these highly nonlinear partial differential equations after a long and difficult search, and showed that they are almost uniquely determined by only a few requirements. At the same time, he recognized that his new theory reduces in lowest order to the Newtonian one for the case of quasistationary weak fields and slowly changing matter sources (and test particles). In particular, Newton's $1/r^2$ law is an unavoidable consequence of GR. The new field theory of gravity also described deviations from Newtonian theory, in particular the anomalous precession of the perihelion of Mercury. Einstein once said about his intellectual odyssey:

> *"In the light of present knowledge, these achievements seem to be almost obvious, and every intelligent student grasps them without much trouble. Yet the years of anxious searching in the dark, with their intense longing, their alternations of confidence and exhaustion and the final emergence into the light - only those who have experienced this can understand it."*

The history of Einstein's struggle is by now well documented. Some particularly valuable references are collected in the bibliography.

Contrary to original expectations, GR had little immediate influence on further progress in physics. With the development of quantum mechanics, the theory of matter and electromagnetism took a direction which had little connection with the ideas of GR. (Einstein himself, and some other important physicists, did not participate in these developments and clung to classical field concepts.) Nothing would indicate that such phenomena would not be compatible with special relativity. This was one of the reasons that progress in GR slowed. Another was that for a long time no really novel observations were made for which Newtonian theory would be inadequate.

The situation changed radically with an incredible chain of important astronomical discoveries in the 1960s and 1970s which brought GR and relativistic astrophysics to the forefront of present day research in physics. We now know that objects having extremely strong gravitational field exist in the Universe. Catastrophic events, such as stellar collapse or explosions in the centers of active galaxies give rise to not only strong, but also rapidly varying, gravitational fields. This is where GR finds its proper applications. For sufficiently massive objects, gravitation dominates at some point over all other interactions, due to its universally attractive and long range character. Not even the most repulsive nuclear forces can always prevent the final collapse to a black hole. So-called horizons appear in the spacetime geometry, behind which matter disappears and thus, for all practical purposes of physics and astrophysics, ceases to exist. For such dramatic events GR must be used in its full ramifications, and is no longer merely a small correction to the Newtonian theory. The observational evidence for black holes in some X-ray binary systems, as well as for supermassive black holes in galactic centers is still indirect, but has become overwhelming. We are undoubtedly in the middle of a truly Golden Age of astrophysics and cosmology.

As a result of these astronomical observations, theorists have performed relevant investigations on the stability of gravitating systems, gravitational collapse, the physics of black holes, the emission of gravitational radiation, and other applied fields of GR have been created and are further developed. Soon we shall have a gravitational wave astronomy, allowing us to study highly dynamical strong field processes, like the coalescence of black holes. Here Einstein's equations come into play in their full glory. Surely, gravitation wave searches will again strengthen the interplay of theory and experiment. Cosmology is another observationally driven, rapidly expanding area where GR plays a crucial role.

It is now widely recognized that GR is a non-Abelian gauge theory of a special type, in that it has a common geometrical structure with the gauge theories of particle physics[2]. In all these theories spacetime-dependent invariance principles specify the dynamics. In most unifications attempts beyond the Standard Model of particle physics, gravity is an essential and unavoidable part. Also for this reason, particle physicists should nowadays have a good knowledge of GR and its differential geometric methods.

[2] For a historically oriented account of this, see [51].

1 Physics in External Gravitational Fields

"I was made aware of these (works by Ricci and Levi-Cività) by my friend Grossmann in Zürich, when I put to him the problem to investigate generally covariant tensors, whose components depend only on the derivatives of the coefficients of the quadratic fundamental form."

—A. Einstein (1955)

We already emphasized in the introduction that the principle of equivalence is one of the foundation pillars of the general theory of relativity. It leads naturally to the kinematical framework of general relativity and determines, suitable interpreted, the coupling of physical systems to external gravitational fields. This will be discussed in detail in the present chapter.

1.1 Characteristic Properties of Gravitation

Among the known fundamental interactions only the electromagnetic and gravitational are of long range, thus permitting a classical description in the macroscopic limit. While there exists a highly successful quantum electrodynamics, a (unified) quantum description of gravity remains a fundamental theoretical task.

Strength of the Gravitational Interaction

Gravity is by far the weakest of the four fundamental interactions. If we compare for instance the gravitational and electrostatic force between two protons, we find in obvious notation

$$\frac{G\,m_p^2}{r^2} = 0.8 \times 10^{-36}\frac{e^2}{r^2}\,. \qquad (1.1)$$

The tiny ratio of the two forces reflects the fact that the **Planck mass**

$$M_{Pl} = \left(\frac{\hbar c}{G}\right)^{1/2} = 1.2 \times 10^{19}\,\frac{GeV}{c^2} \simeq 10^{-5} g \qquad (1.2)$$

is huge in comparison to known mass scales of particle physics. The numerical factor on the right in equation (1.1) is equal to $\alpha^{-1} m_p^2/M_{Pl}^2$, where $\alpha = e^2/\hbar c \simeq 1/137$ is the fine structure constant. Quite generally, gravitational effects in atomic physics are suppressed in comparison to electromagnetic ones by factors of the order $\alpha^n (m/M_{Pl})^2$, where $m = m_e, m_p, \ldots$ and $n = 0, \pm 1, \cdots$. There is thus no chance to measure gravitational effects on the atomic scale. Gravity only becomes important for astronomical bodies. For sufficiently large masses it sooner or later predominates over all other interactions and will lead to the catastrophic collapse to a black hole. One can show (see Chap. 6) that this is always the case for stars having a mass greater than about

$$\frac{M_{Pl}^3}{m_N^2} \simeq 2M_\odot , \tag{1.3}$$

where m_N is the nucleon mass. Gravity wins because it is not only long range, but also universally attractive. (By comparison, the electromagnetic forces cancel to a large extent due to the alternating signs of the charges, and the exclusion principle for the electrons.) In addition, not only matter, but also antimatter, and every other form of energy acts as a source for gravitational fields. At the same time, gravity also acts on all forms of energy.

Universality of Free Fall

Since the time of Galilei, we learned with increasing precision that all test bodies fall at the same rate. This means that for an appropriate choice of units, the inertial mass is equal to the gravitational mass. Newton established that the "weight" of a body (its response to gravity) is proportional to the "quantity of matter" in it already to better than a part in 1000. He achieved this with two pendulums, each 11 feet long ending in a wooden box. One was a reference; into the other he put successively "gold, silver, lead, glass, common salt, wood, water and wheat". Careful observations showed that the times of swing are independent of the material. In Newtons words:

> *"And by experiments made with the greatest accuracy, I have always found the quantity of matter in bodies to be proportional to their weight."*

In Newton's theory of gravity there is no explanation for this remarkable fact. A violation would not upset the conceptual basis of the theory. As we have seen in the introduction, Einstein was profoundly astonished by this fact[1]. The equality of the inertial and gravitational masses has been experimentally established with an accuracy of one part in 10^{12} (For a review, see

[1] In popular lectures which have only recently been published [54], H. Hertz said about inertial and gravitational mass:

> *"And in reality we do have two properties before us, two most fundamental properties of matter, which must be thought as being completely in-*

[74]). This remarkable fact suggests the validity of the following **universality of free fall**, also called the *Weak Equivalence Principle*:

Weak Equivalence Principle (WEP). *The motion of a test body in a gravitational field is independent of its mass and composition (at least when one neglects interactions of spin or of a quadrupole moment with field gradients).*

For the Newtonian theory, universality is of course a consequence of the equality of inertial and gravitational masses. We postulate that it holds generally, in particular also for large velocities and strong fields.

Equivalence Principle

The equality of inertial and gravitational masses provides experimental support for a stronger version of the principle of equivalence.

Einstein's Equivalence Principle (EEP). *In an arbitrary gravitational field no local non-gravitational experiment can distinguish a freely falling non-rotating system (local inertial system) from a uniformly moving system in the absence of a gravitational field.*

Briefly, we may say that gravity can be locally transformed away. Today of course, this is a well known fact to anyone who has watched space flight on television.

Remarks.

1. The EEP implies (among other things) that inertia and gravity cannot be (uniquely) separated.
2. The formulation of the EEP is somewhat vague, since it is not entirely clear what is meant by a local experiment. At this point, the principle is thus of a heuristic nature. We shall soon translate it into a mathematical requirement.
3. The EEP is even for test bodies stronger than the universality of free fall, as can be seen from the following example. Consider a fictitious world in which, by a suitable choice of units, the electric charge is equal to the mass of the particles and in which there are no negative charges. In a classical framework, there are no objections to such a theory, and by definition, the universality property is satisfied. However, the principle of equivalence is not satisfied. Consider a homogeneous magnetic field.

---

dependent of each other, but in our experience, and only in our experience, appear to be exactly equal. This correspondence must mean much more than being just a miracle . . . . We must clearly realize, that the proportionality between mass and inertia must have a deeper explanation and cannot be considered as of little importance, just as in the case of the equality of the velocities of electrical and optical waves."

Since the radii and axes of the spiral motion are arbitrary, there is no transformation to an accelerated frame of reference which can remove the effect of the magnetic field on all particles at the same time.

4. We do not discuss here the so-called *strong equivalence principle (SEP)* which includes self-gravitating bodies and experiments involving gravitational forces (e.g., Cavendish experiments). Interested readers are referred to [74] and [6].

Gravitational Red- and Blueshifts

An almost immediate consequence of the EEP is the gravitational redshift (or blueshift) effect. (Originally, Einstein regarded this as a crucial test of GR.) Following Einstein, we consider an elevator cabin in a static gravitational

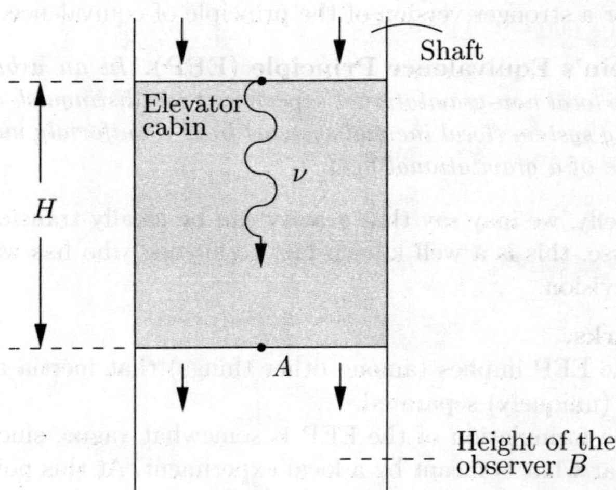

Fig. 1.1. EEP and blueshift.

field. For simplicity, we consider a homogeneous field of strength (acceleration) g, but the result (1.4) below also applies for inhomogeneous fields; this is obvious if the height H of the elevator is taken to be infinitesimal. Suppose the elevator cabin is dropped from rest at time $t = 0$, and that at the same time a photon of frequency ν is emitted from its ceiling toward the floor (see Fig. 1.1). The EEP implies two things:

a) The light arrives at a point A of the floor at time $t = H/c$;
b) no frequency shift is observed in the freely falling cabin.

Consider beside A an observer B at rest in the shaft at the same height as the point A of the floor when the photon arrives there. Clearly, B moves relative to A with velocity $v \simeq gt$ (neglecting higher order terms in t). Therefore, B

sees the light Doppler shifted to the blue by the amount (in first order)[2]

$$z := \frac{\Delta\nu}{\nu} \simeq \frac{v}{c} \simeq \frac{gH}{c^2}.$$

If we write this as

$$z = \frac{\Delta\phi}{c^2}, \tag{1.4}$$

where $\Delta\phi$ is the difference in the Newtonian potential between the receiver and the emitter at rest at different heights, the formula also holds for inhomogeneous gravitational fields to first order in $\Delta\phi/c^2$. (The exact general relativistic formula will be derived in Sect. 1.9.)

Since the early 1960s the consequence (1.4) of the EEP has been tested with increasing accuracy. The most precise result so far was achieved with a rocket experiment that brought a hydrogen-maser clock to an altitude of about $10,000\,km$. The data confirmed the prediction (1.4) to an accuracy of 2×10^{-4}. Gravitational redshift effects are routinely taken into account for Earth-orbiting clocks, such as for the Global Positioning System (GPS). For further details see [74].

At the time when Einstein formulated his principle of equivalence in 1907, the prediction (1.4) could not be directly verified. Einstein was able to convince himself of its validity indirectly, since (1.4) is also a consequence of the conservation of energy. To see this, consider two points A and B, with

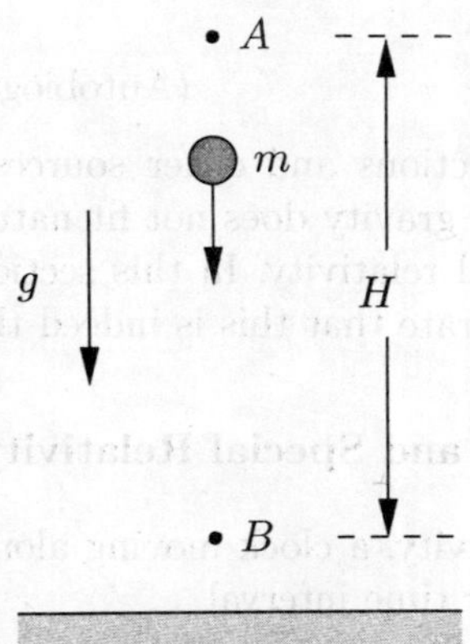

Fig. 1.2. Conservation of energy.

separation H in a homogeneous gravitational field (see Fig. 1.2). Let a mass m fall with initial velocity zero from A to B. According to the Newtonian

[2] B is, of course, not an inertial observer. It is, however, reasonable to assume that B makes the same measurements as a freely falling (inertial) observer B' momentarily at rest relative to B.

theory, it has the kinetic energy $m\,g\,H$ at point B. Now let us assume that at B the entire energy of the falling body (rest energy plus kinetic energy) is annihilated to a photon, which subsequently returns to the point A. If the photon did not interact with the gravitational field, we could convert it back to the mass m and gain the energy $m\,g\,H$ in each cycle of such process. In order to preserve the conservation of energy, the photon must experience a redshift. Its energy must satisfy

$$E_{lower} = E_{upper} + m\,g\,H = m\,c^2 + m\,g\,H = E_{upper}\left(1 + \frac{g\,H}{c^2}\right).$$

For the wavelengths we then have (h is Planck's constant)

$$1 + z = \frac{\lambda_{upper}}{\lambda_{lower}} = \frac{h\,\nu_{lower}}{h\,\nu_{upper}} = \frac{E_{lower}}{E_{upper}} = 1 + \frac{g\,H}{c^2},$$

in perfect agreement with (1.4).

1.2 Special Relativity and Gravitation

> *"That the special theory of relativity is only the first step of a necessary development became completely clear to me only in my efforts to represent gravitation in the framework of this theory."*
>
> —A. Einstein
>
> (Autobiographical Notes, 1949)

From several later recollections and other sources we know that Einstein recognized very early that gravity does not fit naturally into the framework of special relativity special relativity. In this section, we shall discuss some arguments which demonstrate that this is indeed the case.

Gravitational Redshift and Special Relativity

According to special relativity, a clock moving along the timelike world line $x^\mu(\lambda)$ measures the proper time interval

$$\Delta\tau = \int_{\lambda_1}^{\lambda_2} \sqrt{-\eta_{\mu\nu}\frac{dx^\mu}{d\lambda}\frac{dx^\nu}{d\lambda}}\,d\lambda, \tag{1.5}$$

where $\eta_{\mu\nu}$ is the Minkowski metric. In the presence of a gravitational field, (1.5) can no longer be valid, as is shown by the following argument.

Consider a redshift experiment in the Earth's gravitational field and assume that a special relativistic theory of gravity exists, which need not be further specified here. For such an experiment we may neglect all masses other than that of the Earth and regard the Earth as being at rest relative

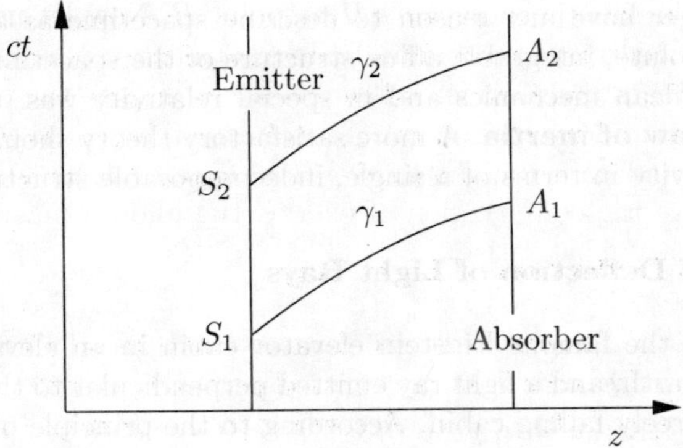

Fig. 1.3. Redshift in the Earth field.

to some inertial system. In a spacetime diagram (height z above the Earth's surface versus time), the Earth's surface, the emitter and the absorber all move along world lines of constant z (see Fig. 1.3).

The transmitter is supposed to emit at a fixed frequency from S_1 to S_2. The photons registered by the absorber move along world lines γ_1 and γ_2, that are not necessarily straight lines at an angle of 45°, due to a possible interaction with the gravitational field, but must be *parallel*, since we are dealing with a static situation. Thus, if the flat Minkowski geometry holds and the time measurement is given by (1.5), it follows that the time difference between S_1 and S_2 must be equal to the time difference between A_1 and A_2. Thus, there would be no redshift. This shows that at the very least (1.5) is no longer valid. The argument does not exclude the possibility that the metric $g_{\mu\nu}$ might be proportional to $\eta_{\mu\nu}$. (This possibility will be rejected below.)

Global Inertial Systems Cannot Be Realized in the Presence of Gravitational Fields

In Newtonian-Galilean mechanics and in special relativity, the law of inertia distinguishes a special class of equivalent frames of reference (inertial systems). Due to the universality of gravitation, only the free fall of electrically neutral test bodies can be regarded as particularly distinguished motion in the presence of gravitational fields. Such bodies experience, however, relative accelerations. There is no operational procedure to uniquely separate inertia and gravitation. In spite of this, the fiction of a linear affine Galilee spacetime (with a flat affine connection) is maintained in the traditional presentation of Newton's theory[3], and gravity is put on the side of the forces. But since the concept of an inertial system cannot be defined operationally, we are deprived of an essential foundation of the special theory of relativity.

[3] A more satisfactory formulation was given by E. Cartan, [52] and K. Friederichs, [53] (see also Exercise 2.2)

We no longer have any reason to describe spacetime as a linear affine space. The absolute, integrable affine structure of the spacetime manifold in Newtonian-Galilean mechanics and in special relativity was, after all, suggested by the **law of inertia**. A more satisfactory theory should account for inertia and gravity in terms of a single, indecomposable structure.

Gravitational Deflection of Light Rays

Consider again the famous Einstein elevator cabin in an elevator shaft attached to the Earth, and a light ray emitted perpendicular to the direction of motion of the freely falling cabin. According to the principle of equivalence, the light ray propagates along a straight line inside the cabin relative to the cabin. Since the elevator is accelerated relative to the Earth, one expects that the light ray propagates along a parabolic path relative to the Earth. This consequence of the EEP holds, a priori, only *locally.* It does not necessarily imply bending of light rays from a distant source traversing the gravitational field of a massive body and arriving at a distant observer. Indeed, we shall see later that it is possible to construct a theory which satisfies the principle of equivalence, but in which there is no deflection of light. (For a detailed discussion of how this comes about, see [73].) At any rate, the deflection of light is an experimental fact (the precise magnitude of the effect does not concern us at the moment).

It is therefore not possible to describe the gravitational field (as in the Einstein-Fokker theory, discussed in Sect. 2.2) in terms of a **conformally flat metric**, i.e., by a metric field proportional to the Minkowski metric: $g_{\mu\nu}(x) = \phi(x)\eta_{\mu\nu}$, where $\phi(x)$ plays the role of the gravitational potential. Indeed, for such a metric the light cones are the same as in the Minkowski spacetime; hence, there is no light deflection.

Theories of Gravity in Flat Spacetime

> *"I see the most essential thing in the overcoming of the inertial system, a thing that acts upon all processes, but undergoes no reaction. This concept is, in principle, no better than that of the center of the universe in Aristotelian physics."*
>
> —A. Einstein (1954)

In spite of these arguments one may ask, how far one gets with a theory of gravity in Minkowski spacetime, following the pattern of well understood field theories, such as electrodynamics. Attempts along these lines have a long tradition, and are quite instructive. Readers with some background in special relativistic (classical) field theory should find the following illuminating. One may, however, jump directly to the conclusion at the end of this subsection (p. 20).

Scalar Theory. Let us first try a scalar theory. This simplest possibility was studied originally by Einstein, von Laue, and others, but was mainly developed by G. Nordstrøm, [49].

The field equation for the scalar field φ, generalizing the Newtonian potential, in the limit of *weak* fields (linear field equation) is unique:

$$\Box\varphi = -4\pi G T . \tag{1.6}$$

Here, T denotes the trace of the energy-momentum tensor $T^{\mu\nu}$ of matter. For a Newtonian situation this reduces to the Poisson equation.

We formulate the equation of motion of a test particle in terms of a Lagrangian. For weak fields this is again unique:

$$L(x^\mu, \dot{x}^\mu) = -\sqrt{-\eta_{\mu\nu}\dot{x}^\mu\dot{x}^\nu}\,(1+\varphi) , \tag{1.7}$$

because only for this the Newtonian limit for weak static fields and small velocities of the test bodies comes out right:

$$L(\boldsymbol{x}, \dot{\boldsymbol{x}}) \approx \frac{1}{2}\dot{\boldsymbol{x}}^2 - \varphi + \text{const} .$$

The basic equations (1.6) and (1.7) imply a perihelion motion of the planets, but this comes out wrong, even the sign is incorrect. One finds $(-1/6)$ times the value of general relativity (see Exercise 1.3). In spite of this failure we add some further instructive remarks.

First, we want to emphasize that the interaction is necessarily *attractive*, independent of the matter content. To show this, we start from the general form of the Lagrangian density for the scalar theory

$$\mathcal{L} = -\frac{1}{2}\partial_\mu\phi\,\partial^\mu\phi + g\,T\cdot\phi + \mathcal{L}_{mat} , \tag{1.8}$$

where ϕ is proportional to φ and g is a coupling constant. Note first that only g^2 is significant: Setting $\tilde{\phi} = g\phi$, we have

$$\mathcal{L} = -\frac{1}{2g^2}\partial_\mu\tilde{\phi}\,\partial^\mu\tilde{\phi} + T\cdot\tilde{\phi} + \mathcal{L}_{mat} ,$$

involving only g^2. Next, it has to be emphasized that it is not allowed to replace g^2 by $-g^2$, otherwise the field energy of the gravitational field would be negative. (This "solution" of the energy problem does not work.) Finally, we consider the field energy for *static* sources.

The total (canonical) energy-momentum tensor

$$T^\mu_{\ \nu} = \frac{\partial\mathcal{L}}{\partial\phi_{,\mu}}\phi_{,\nu} + \ldots - \delta^\mu_{\ \nu}\mathcal{L}$$

gives for the ϕ-contribution

$$(T_\phi)_{\mu\nu} = \partial_\mu\phi\,\partial_\nu\phi - \frac{1}{2}\eta_{\mu\nu}\partial_\lambda\phi\,\partial^\lambda\phi + \eta_{\mu\nu}g\,T\,\phi\,.$$

For the corresponding total energy we find

$$\begin{aligned} E &= \int (T_\phi)_{00}\,d^3x = \frac{1}{2}\int ((\boldsymbol{\nabla}\phi)^2 - 2g\,T\,\phi)d^3x \\ &= \frac{1}{2}\int (\phi(-\Delta\phi) - 2g\,T\,\phi)d^3x = \frac{1}{2}g\int T\,\phi\,d^3x\,. \end{aligned} \tag{1.9}$$

Since $\Delta\phi = -g\,T$, we have

$$\phi(\boldsymbol{x}) = \frac{g}{4\pi}\int \frac{T(\boldsymbol{x}')}{|\boldsymbol{x}-\boldsymbol{x}'|}\,d^3x'\,.$$

Inserting this in (1.9) gives finally

$$E = -\frac{g^2}{4\pi}\frac{1}{2}\int \frac{T(\boldsymbol{x})T(\boldsymbol{x}')}{|\boldsymbol{x}-\boldsymbol{x}'|}\,d^3x\,d^3x'\,,$$

showing that indeed the interaction is attractive.

This can also be worked out in quantum field theory by computing the effective potential corresponding to the one-particle exchange diagram with the interaction Lagrangian $\mathcal{L}_{int} = g\,\bar{\psi}\psi\,\phi_{m=0}$. One finds

$$V_{eff} = -\frac{g^2}{4\pi}\frac{1}{|\boldsymbol{x}-\boldsymbol{x}'|}$$

both for fermion-fermion and fermion-antifermion interactions. The same result is found for the exchange of massless spin-2 particles, while for spin-1 we obtain *repulsion* between particles, and attraction between particles and antiparticles (see Exercise 1.4).

The scalar theory predicted that there is *no* light deflection, simply because the trace of the electromagnetic energy-momentum tensor vanishes. For this reason Einstein urged in 1913 astronomers (E. Freundlich in Potsdam) to measure the light deflection during the solar eclipse the coming year in the Crimea. Shortly before the event the first world war broke out. Over night Freundlich and his German colleagues were captured as prisoners of war and it took another five years before the light deflection was observed. For further discussion of the scalar theory we refer to [75], and references therein.

Tensor (spin-2) Theory. We are led to study the spin-2 option. (There are no consistent higher spin equations with interaction.) This means that we try to describe the gravitational field by a symmetric tensor field $h_{\mu\nu}$. Such a field has 10 components. On the other hand, we learned from Wigner that in the massless case there are only *two* degrees of freedom. How do we achieve the truncation from 10 to 2?

Recall first the situation in the *massive case*. There we can require that the trace $h = h^\mu_\mu$ vanishes, and then the field $h_{\mu\nu}$ transforms with respect to the homogeneous Lorentz group irreducibly as $D^{(1,1)}$ (in standard notation). With respect to the subgroup of rotations this reduces to the reducible representation

$$D^1 \otimes D^1 = D^2 \oplus D^1 \oplus D^0 .$$

The corresponding unwanted spin-1 and spin-0 components are then eliminated by imposing 4 subsidery conditions

$$\partial_\mu h^\mu{}_\nu = 0 .$$

The remaining 5 degrees of freedom describe (after quantization) massive spin-2 particles (W. Pauli and M. Fierz, [76]; see, e.g., the classical book [17]).

In the *massless case* we have to declare certain classes of fields as physically equivalent, by imposing - as in electrodynamics - a **gauge invariance**. The gauge transformations are

$$h_{\mu\nu} \longrightarrow h_{\mu\nu} + \partial_\mu \xi_\nu + \partial_\nu \xi_\mu , \tag{1.10}$$

where ξ_μ is an arbitrary vector field.

Let us first consider the *free* spin-2 theory which is unique (W. Pauli and M. Fierz)

$$\mathcal{L} = -\frac{1}{4} h_{\mu\nu,\sigma} h^{\mu\nu,\sigma} + \frac{1}{2} h_{\mu\nu,\sigma} h^{\sigma\nu,\mu} + \frac{1}{4} h_{,\sigma} h^{,\sigma} - \frac{1}{2} h_{,\sigma} h^{\nu\sigma}{}_{,\nu} . \tag{1.11}$$

Let $G_{\mu\nu}$ denote the Euler-Lagrange derivative of $\mathcal{L}$,

$$\begin{aligned} G_{\mu\nu} = {} & \frac{1}{2}\partial^\sigma \partial_\sigma h_{\mu\nu} + \partial_\mu \partial_\nu h - \partial_\nu \partial^\sigma h_{\mu\sigma} - \partial_\mu \partial^\sigma h_{\sigma\nu} \\ & + \eta_{\mu\nu} \left(\partial^\alpha \partial^\beta h_{\alpha\beta} - \partial^\sigma \partial_\sigma h \right) . \end{aligned} \tag{1.12}$$

The free field equations

$$G_{\mu\nu} = 0 \tag{1.13}$$

are identical to the linearized Einstein equations (as shown in Sect. 4.1) and describe, for instance, the propagation of weak gravitational fields.

The gauge invariance of $\mathcal{L}$ (modulo a divergence) implies the identity

$$\partial_\nu G^{\mu\nu} \equiv 0 , \qquad \text{“linearized Bianchi identity”} . \tag{1.14}$$

This should be regarded in analogy to the identity $\partial_\mu(\Box A^\mu - \partial^\mu \partial_\nu A^\nu) \equiv 0$ for the left-hand side of Maxwell's equations.

Let us now introduce couplings to matter. The simplest possibility is the linear coupling

$$\mathcal{L}_{int} = -\frac{1}{2} \kappa h_{\mu\nu} T^{\mu\nu} , \tag{1.15}$$

leading to the field equation

$$G^{\mu\nu} = \frac{\kappa}{2} T^{\mu\nu}. \tag{1.16}$$

This can, however, not yet be the final equation, but only an approximation for weak fields. Indeed, the identity (1.14) implies $\partial_\nu T^{\mu\nu} = 0$ which is unacceptable (in contrast to the charge conservation of electrodynamics). For instance, the motion of a fluid would then not at all be affected by the gravitational field. Clearly, we must introduce a *back-reaction* on matter. Why not just add to $T^{\mu\nu}$ in (1.16) the energy-momentum tensor ${}^{(2)}t^{\mu\nu}$ which corresponds to the **Pauli-Fierz Lagrangian** (1.11)? But this modified equation cannot be derived from a Lagrangian and is still not consistent, but only the second step of an iteration process

$$\mathcal{L}_{free} \longrightarrow {}^{(2)}t^{\mu\nu} \longrightarrow \mathcal{L}_{cubic} \longrightarrow {}^{(3)}t^{\mu\nu} \longrightarrow \ldots ?$$

The sequence of arrows has the following meaning: A Lagrangian which gives the quadratic terms ${}^{(2)}t^{\mu\nu}$ in

$$G^{\mu\nu} = \frac{\kappa}{2}\left(T^{\mu\nu} + {}^{(2)}t^{\mu\nu} + {}^{(3)}t^{\mu\nu} + \ldots\right) \tag{1.17}$$

must be cubic in $h_{\mu\nu}$, and in turn leads to cubic terms ${}^{(3)}t^{\mu\nu}$ of the gravitational energy-momentum tensor. To produce these in the field equation (1.17), we need quartic terms in $h_{\mu\nu}$, etc.. This is an infinite process. By a clever reorganization it stops already after the second step, and one arrives at field equations which are equivalent to Einstein's equations (see [77]). The physical metric of GR is given in terms of $\phi^{\mu\nu} = h^{\mu\nu} - \frac{1}{2}\eta^{\mu\nu}h$ by

$$\sqrt{-g}\, g^{\mu\nu} = \eta^{\mu\nu} - \phi^{\mu\nu}, \tag{1.18}$$

where $g := \det(g_{\mu\nu})$.

At this point one can reinterprete the theory geometrically. Thereby the flat metric disappears completely and one arrives in a pedestrian way to GR. This approach is further discussed in [75]. There it is also shown that $g_{\mu\nu}$ is really the *physical* metric.

Conclusion. *The consequent development of the theory shows that it is possible to eliminate the flat Minkowski metric, leading to a description in terms of a curved metric which has a direct physical meaning. The originally postulated Lorentz invariance turns out to be physically meaningless and plays no useful role. The flat Minkowski spacetime becomes a kind of unobservable ether. The conclusion is inevitable that spacetime is a pseudo-Riemannian (Lorentzian) manifold, whereby the metric is a dynamical field, subjected to field equations.*

* * *

Exercises.

Exercise 1.1. Consider a homogeneous electric field in the z-direction and a charged particle with $e = m$. Show that a particle which originally at rest moves faster in the vertical direction than a particle which was originally moving horizontally.

Exercise 1.2. Consider a self gravitating body (star) moving freely in the neighborhood of a black hole. Estimate at which distance D the star is disrupted by relative forces due to inhomogeneities of the gravitational field (tidal force **tidal forces**).

Solution. Relative gravitational accelerations in Newtonian theory are determined by the second derivative of the Newtonian potential. A satellite with mass M and radius R at distance r from a compact body (neutron star, black hole) of mass M_c experiences a tidal force at the surface (relative to the center) of magnitude

$$\left|\frac{d}{dr}\left(\frac{GM_c}{r^2}\right)R\right| .$$

Once this becomes larger than the gravitational acceleration of its own field at the surface, the satellite will be disrupted. The critical distance D is thus estimated to be

$$D \simeq \left(\frac{2M_c}{M}\right)^{1/3} R .$$

Let us introduce the average mass density $\bar{\rho}$ of the satellite by $M = \frac{4\pi}{3}R^3\bar{\rho}$, then

$$D \simeq \left(\frac{3}{2\pi}\right)^{1/3}\left(\frac{M_c}{\bar{\rho}}\right)^{1/3}$$

Put in the numbers for $M_c \simeq 10^8 M_\odot$ and the parameters of the sun for the satellite. Compare D with the *Schwarzschild radius* $R_s = 2GM_c/c^2$ for M_c.

Exercise 1.3. Determine the perihelion motion for Nordstrøm's theory of gravity (basic equations (1.6) and (1.7)). Compare the result with that of GR, derived in Sect. 3.3. Even the sign turn out to be wrong.

Exercise 1.4. Show that a vector theory of gravity, similar to electrodynamics, leads necessarily to *repulsion*.

1.3 Spacetime as a Lorentzian Manifold

"Either, therefore, the reality which underlies space must form a discrete manifold, or we must seek the ground of its metric relations (measure conditions) outside it, in binding forces which act upon it."
—B. Riemann (1854)

The discussion of Sect. 1.2 has shown that in the presence of gravitational fields the spacetime description of SR has to be generalized. According to the EEP, special relativity remains, however, valid in "infinitesimal" regions. This suggests that the metric properties of spacetime have to be described by a symmetric tensor field $g_{\mu\nu}(p)$ for which it is not possible to find coordinate systems such that $g_{\mu\nu}(p) = \text{diag}(-1, 1, 1, 1)$ in finite regions of spacetime. This should only be possible when no true gravitational fields are present. We therefore postulate: *The mathematical model for* ***spacetime*** *(i.e., the set of all elementary events) in the presence of gravitational fields is a pseudo-Riemannian manifold M, whose metric g has the same signature as the Minkowski metric. The pair (M, g) is called a* ***Lorentz manifold*** *and g is called a* ***Lorentzian metric****.*

Remark. At this point, readers who are not yet familiar with (pseudo-) Riemannian geometry should study the following sections of the differential geometric part at the end of the book: All of Chaps. 9 and 10, and Sects. 13.1-13.6. These form a self-contained subset and suffice for most of the basic material covered in the first two chapters (at least for a first reading). References to the differential geometric Part III will be indicated by DG.

As for Minkowski spacetime, the metric g determines, beside the metric properties, also the causal relationships, as we shall see soon. At the same time, we also interpret the metric field as the **gravitational potential**. In the present chapter our goal is to describe how it influences non-gravitational systems and processes.

Among the metric properties, the generalization of (1.5) of the **proper time** interval for a timelike curve $x^\mu(\lambda)$ (i.e., a curve with timelike tangent vectors) is

$$\Delta\tau = \int_{\lambda_1}^{\lambda_2} \sqrt{-g_{\mu\nu}\frac{dx^\mu}{d\lambda}\frac{dx^\nu}{d\lambda}}\, d\lambda\,. \tag{1.19}$$

A good (atomic) clock, moving along $x^\mu(\lambda)$, measures this proper time.

The coupling of the metric to non-gravitational systems should satisfy two principles. First, the basic equations must have *intrinsic* meaning in (M, g). In other words, they should be expressible in terms of the intrinsic calculus on Lorentz manifolds, developed in DG. Equivalently, the basic laws should not distinguish any coordinate system. All charts of any atlas, belonging to the differential structure, are on the same footing. One also says that the physical laws have to be *covariant* with respect to smooth coordinate transformations (or are *generally covariant*). Let us formulate this property more precisely:

Definition. *A system of equations is* ***covariant with respect to the group*** *$\mathcal{G}(M)$ of (germs of) smooth coordinate transformations, provided that for any element of $\mathcal{G}(M)$ the quantities appearing in the equations can be transformed to new quantities in such a way that*

(i) the assignment preserves the group structure of $\mathcal{G}(M)$;
(ii) both the original and the transformed quantities satisfy the same system of equations.

Only generally covariant laws have an intrinsic meaning in the Lorentz manifold. If a suitable calculus is used, these can be formulated in a coordinate-free manner. The general covariance is at this point a matter of course. It should, however, not be confused with general **invariance**. The difference of the two concepts will be clarified later when we shall consider the coupled dynamical system of metric plus matter variables. It will turn out that general invariance is, like gauge invariance, a powerful symmetry principle (see Sect. 2.5).

From DG, Sect. 13.3, we know that in a neighborhood of every point p a coordinate system exists, such that

$$g_{\mu\nu}(p) = \eta_{\mu\nu} \quad \text{and} \quad g_{\mu\nu,\lambda}(p) = 0\,, \tag{1.20}$$

where $(\eta_{\mu\nu}) = \text{diag}(-1, 1, 1, 1)$. Such coordinates are said to be inertial or normal at p, and are interpreted as **locally inertial systems**. We also say that such a coordinate system is *locally inertial with origin p*. The metric g describes the behavior of clocks and measuring sticks in such locally inertial systems, exactly as in special relativity. Relative to such a system, the usual laws of electrodynamics, mechanics, etc. in the special relativistic form are locally valid. The form of these laws for an arbitrary system is to a large extend determined by the following two requirements (we shall discuss possible ambiguities in Sect. 1.4.3):

a) Aside from the metric and its derivatives, the laws should contain only quantities which are also present in the special theory of relativity[4].
b) The laws must be generally covariant and reduce to the special relativistic form at the origin of a locally inertial coordinate system.

These requirements provide a *mathematical* formulation of Einstein's Equivalence Principle. We shall soon arrive at a more handy prescription.

1.4 Non-gravitational Laws in External Gravitational Fields

We now apply this mathematical formulation of the EEP and discuss possible ambiguities at the end of this section. Familiarity with the concept of

[4] It is not permitted to introduce in addition to $g_{\mu\nu}$ other "external" (absolute) elements such as a flat metric which is independent of g.

covariant differentiation (DG, Sects. 13.1-13.6) will be assumed. We add here some remarks about notation.

Remarks (Coordinate-free Versus Abstract Index Notation). Modern mathematical texts on differential geometry usually make use of indices for vectors, tensors, etc. only for their components relative to a local coordinate system or a **frame**, i.e. a local basis of vector fields. If indices for the objects themselves are totally avoided, computations can, however, quickly become very cumbersome, especially when higher rank tensors with all sorts of contractions are involved. For this reason, relativists usually prefer what they call the **abstract index notation**. This has *nothing* to do with coordinates or frames. For instance, instead of saying: "... let Ric be the Ricci tensor, u a vector field, and consider $Ric(u,u)$ ...", one says: "... let $R_{\mu\nu}$ be the Ricci tensor, u^μ a vector field, and consider $R_{\mu\nu}\, u^\mu u^\nu$, ..." In particular, the number and positions of indices specify which type of tensor is considered, and repeated indices indicate the type of contraction that is performed.

Some authors (e.g., R. Wald in his text book [7]) distinguish this abstract meaning of indices from the usual component indices by using different alphabets. We do not want to adopt this convention, since it should be clear from the context whether the indices can be interpreted abstractly, or whether they refer to special coordinates or frames which are adapted to a particular spacetime (with certain symmetries, distinguished submanifolds, ...).

When dealing, for example, with differential forms we often avoid abstract indices and follow the habits of mathematicians. We hope that the reader will not be disturbed by our notational flexibility. After having studied the present subsection, he should be familiar with our habits.

1.4.1 Motion of a Test Body in a Gravitational Field

What is the equation of motion of a freely falling test particle? Let $\gamma(\tau)$ be its timelike world line, parameterized by the proper time τ. According to (1.19) its tangent vector $\dot\gamma$ (four-velocity) satisfies

$$g_{\mu\nu}\frac{dx^\mu}{d\tau}\frac{dx^\nu}{d\tau} = -1 \quad \text{or} \quad g(\dot\gamma,\dot\gamma) =: \langle\dot\gamma,\dot\gamma\rangle = -1 \,. \tag{1.21}$$

Consider some arbitrary point p along the orbit $\gamma(\tau)$, and introduce coordinates which are locally inertial at p. The weak equivalence principle implies

$$\left.\frac{d^2x^\mu}{d\tau^2}\right|_p = 0 \,. \tag{1.22}$$

Since the Christoffel symbols $\Gamma^\mu{}_{\alpha\beta}$ vanish at p, we can write this as

$$\frac{d^2x^\mu}{d\tau^2} + \Gamma^\mu{}_{\alpha\beta}\frac{dx^\alpha}{d\tau}\frac{dx^\beta}{d\tau} = 0 \tag{1.23}$$

at p. This is just the geodesic equation (DG, Sect. 13.3), which is generally covariant. Therefore, equation (1.23) holds in any coordinate system. Moreover, since the point p is arbitrary, it is valid along the entire orbit of the test body. In coordinate-free notation it is equivalent to the statement that $\dot{\gamma}$ is autoparallel along γ

$$\nabla_{\dot{\gamma}}\dot{\gamma} = 0 \,. \tag{1.24}$$

Note that (1.21) and (1.23) are compatible. This follows with the Ricci identity from

$$\frac{d}{d\tau}\langle\dot{\gamma},\dot{\gamma}\rangle = \nabla_{\dot{\gamma}}\langle\dot{\gamma},\dot{\gamma}\rangle = 2\langle\nabla_{\dot{\gamma}}\dot{\gamma},\dot{\gamma}\rangle = 0 \,.$$

The geodesic equation (1.23) is the Euler-Lagrange equation for the variational principle

$$\delta\int\sqrt{-g_{\mu\nu}\frac{dx^\mu}{d\lambda}\frac{dx^\nu}{d\lambda}}\,d\lambda = 0 \tag{1.25}$$

(see the solution of Exercise 1.5).

The basic equation (1.23) can be regarded as the generalization of the Galilean law of inertia in the presence of a gravitational field. It is a great triumph that the universality of the inertial/gravitational mass ratio is automatic. It is natural to regard the connection coefficients ${\Gamma^\mu}_{\alpha\beta}$ as the gravitational-inertial **field strength** relative to the coordinates $\{x^\mu\}$.

World Lines of Light Rays

Using the same arguments, the following equations for the world line $\gamma(\lambda)$ of a light ray, parameterized by an affine parameter λ, are obtained

$$\frac{d^2x^\mu}{d\lambda^2} + {\Gamma^\mu}_{\alpha\beta}\frac{dx^\alpha}{d\lambda}\frac{dx^\beta}{d\lambda} = 0 \,, \tag{1.26a}$$

$$g_{\mu\nu}\frac{dx^\mu}{d\lambda}\frac{dx^\nu}{d\lambda} = 0 \,. \tag{1.26b}$$

In other words, the world lines of light rays are null geodesics. (See also Exercise 1.7.) Later (in Sect. 1.8) we shall derive these equations from Maxwell's equations in the eikonal approximation.

At each spacetime point $p \in M$ we can consider, as in Minkowski spacetime, the past and future **null cones** in the tangent space T_pM. These are tangent to past and future **light cones**, generated by light rays ending up in p, respectively emanating from p. The set of all these light cones describes the **causal structure** of spacetime.

Relativists are used to draw spacetime diagrams. A typical example that illustrates some of the basic concepts we have introduced so far is shown in Fig. 1.4.

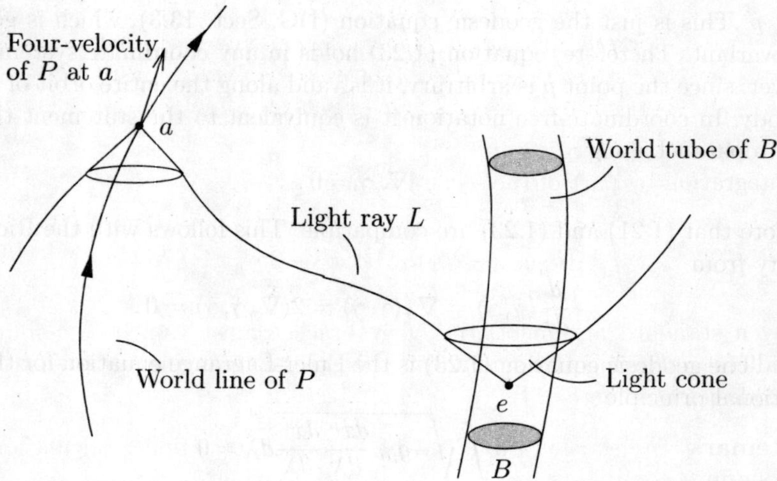

Fig. 1.4. Spacetime diagram representing a particle P, a body B and a light ray L emitted at $e \in B$ and absorbed at a point of P's worldline.

* * *

Exercises.

Exercise 1.5. Let

$$\begin{aligned} \gamma : [a, b] &\longrightarrow M \\ \tau &\longmapsto \gamma(\tau) \,, \end{aligned}$$

be a smooth timelike curve (at least of class C^2). Show that when γ minimizes the **distance**

$$L(\gamma) = \int_a^b \sqrt{-\langle \dot{\gamma}, \dot{\gamma} \rangle} \, d\tau \tag{1.27}$$

for fixed endpoints p and q, then γ is a geodesic if τ is the proper time.

Solution. Note first that $L(\gamma)$ is independent of the parameterization. For simplicity, we assume that the minimizing curve γ lies in the domain U of a chart with associated coordinates $\{x^\mu\}$. Consider a family $\{\gamma_\varepsilon\}$, $-\alpha < \varepsilon < \alpha$, of smooth ($C^2$) curves from p to q $\big(\gamma_\varepsilon \subset U, \ \varepsilon \in (-\alpha, \alpha)\big)$, defined by the coordinates $x^\mu(\tau, \varepsilon) = x^\mu(\tau) + \varepsilon \, \xi^\mu(\tau)$, with $\xi^\mu(a) = \xi^\mu(b) = 0$. We use the notation $\cdot = \partial/\partial\tau$ and $' = \partial/\partial\varepsilon$. Since the quantity

$$L(\gamma_\varepsilon) = \int_a^b \left[-g_{\mu\nu}\big(x(\tau, \varepsilon)\big) \dot{x}^\mu(\tau, \varepsilon) \dot{x}^\nu(\tau, \varepsilon)\right]^{1/2} d\tau$$

attains a minimum at $\varepsilon = 0$, we have

$$L'(\gamma_\varepsilon)|_{\varepsilon=0} = -\frac{1}{2}\int_a^b \Big[\partial_\lambda g_{\mu\nu}(\gamma(\tau))\xi^\lambda \dot{x}^\mu(\tau)\dot{x}^\nu(\tau)$$

$$+2g_{\mu\nu}(\gamma(\tau))\dot{\xi}^\mu(\tau)\dot{x}^\nu(\tau)\Big]\,d\tau = 0\,.$$

Integration by part of the second term gives

$$\int_a^b \left[(\partial_\lambda g_{\mu\nu} - 2\partial_\mu g_{\lambda\nu})\dot{x}^\mu \dot{x}^\nu - 2g_{\lambda\nu}\ddot{x}^\nu\right]\xi^\lambda d\tau = 0\,.$$

By a standard argument, the curly bracket must vanish. Renaming some indices and using the symmetry of $g_{\mu\nu}$, one finds that $x^\mu(\tau)$ satisfies the geodesic equation (1.23).

Remark. We show in DG, Sect. 14.4 how this variational calculation can be done in a coordinate-free manner. For the second derivative, see, e.g., [39], Chap. 10.

Exercise 1.6. Beside $L(\gamma)$ one can consider the **energy functional**

$$E(\gamma) = \frac{1}{2}\int \langle\dot{\gamma},\dot{\gamma}\rangle\,d\tau\,, \tag{1.28}$$

which depends on the parameterization. From the solution of Exercise 1.5 it should be obvious, that minimization of $E(\gamma)$ again leads to the geodesic equation (we shall often use this for practical calculations). Show by a direct calculation that the geodesic equation implies

$$\frac{d}{d\tau}\Big(g_{\mu\nu}(x(\tau))\dot{x}^\mu \dot{x}^\nu\Big) = 0\,,$$

whence the parameterization is proportional to proper time. It should be clear that the variational principle for the energy functional also applies to null geodesics.

Exercise 1.7. Consider a conformal change of the metric

$$g \longmapsto \tilde{g} = \mathrm{e}^{2\phi} g\,. \tag{1.29}$$

Show that a null geodesic for g is also a null geodesic for $\tilde{g}$.

Hints. In transforming the geodesic equation one has to carry out a reparameterization $\lambda \longmapsto \tilde{\lambda}$ such that $d\tilde{\lambda}/d\lambda = \mathrm{e}^{2\phi}$. The relation between the Christoffel symbols for the two metrics is

$$\tilde{\Gamma}^\mu{}_{\alpha\beta} = \Gamma^\mu{}_{\alpha\beta} + \delta^\mu{}_\alpha\phi_{,\beta} + \delta^\mu{}_\beta\phi_{,\alpha} - g_{\alpha\beta}g^{\mu\nu}\phi_{,\nu}\,. \tag{1.30}$$

1.4.2 Energy and Momentum "Conservation" in the Presence of an External Gravitational Field

According to the special theory of relativity, the energy-momentum tensor $T^{\mu\nu}$ of a closed system satisfies, as a result of translation invariance, the conservation law

$$T^{\mu\nu}{}_{,\nu} = 0\,.$$

In the presence of a gravitational field, we define a corresponding tensor field on (M, g) such that it reduces to the special relativistic form at the origin of a locally inertial system.

Example (Energy-momentum Tensor for an Ideal Fluid). In SR the form of the energy-momentum tensor is established as follows. A fluid is by definition **ideal** if a comoving observer sees the fluid around him as isotropic. So let us consider a **local rest frame**, that is an inertial system such that the fluid is at rest at some particular spacetime point. Relative to this system (indicated by a tilde) the energy-momentum tensor has at that point the form

$$\tilde{T}^{00} = \rho\,, \quad \tilde{T}^{0i} = \tilde{T}^{i0} = 0\,, \quad \tilde{T}^{ij} = p\,\delta_{ij}\,, \tag{1.31}$$

where ρ is the **proper energy density** and p is the **pressure** of the fluid ($c = 1$). The four-velocity u^μ of the fluid has in the local rest frame the value $\tilde{u}^0 = 1$, $\tilde{u}^i = 0$, and hence we can write (1.31) as

$$\tilde{T}^{\mu\nu} = (\rho + p)\tilde{u}^\mu \tilde{u}^\nu + p\,\eta^{\mu\nu}\,.$$

Since this is a tensor equation it holds in any inertial system.

In the presence of gravitational fields our general prescription leads uniquely to

$$T^{\mu\nu} = (\rho + p)u^\mu u^\nu + p\,g^{\mu\nu}, \tag{1.32}$$

with the normalization

$$g_{\mu\nu}u^\mu u^\nu = -1\,. \tag{1.33}$$

of the four-velocity field. For an introduction to special relativistic fluid dynamics, see for instance [16].

Remark. We shall discuss in Sect. 2.3.3 a general method of constructing the energy-momentum tensor in the framework of the Lagrangian formalism.

At the origin $p \in M$ of a locally inertial system we have, by the EEP, $T^{\mu\nu}{}_{,\nu} = 0$ at p. We may just as well write $T^{\mu\nu}{}_{;\nu} = 0$ at p, where the semicolon denotes the covariant derivative of the tensor field. This equation is generally covariant, and hence is valid in any coordinate system. We thus arrive at

$$T^{\mu\nu}{}_{;\nu} = 0\,. \tag{1.34}$$

Conclusion. *From this consideration, we conclude quite generally that the physical laws of special relativity are changed in the presence of a gravitational field simply by the substitution of covariant derivatives for ordinary derivatives (comma⟶semicolon rule). This is an expression of the principle of equivalence. (Possible ambiguities for higher order derivatives are discussed at the end of this section.)*

In this manner the coupling of the gravitational field to physical systems is determined in an extremely simple manner.

We may write (1.34) as follows: General calculational rules give (DG, Eq. (13.24))

$$T^{\mu\nu}{}_{;\sigma} = T^{\mu\nu}{}_{,\sigma} + \Gamma^{\mu}_{\sigma\lambda} T^{\lambda\nu} + \Gamma^{\nu}_{\sigma\lambda} T^{\mu\lambda} .$$

Hence,

$$T^{\mu\nu}{}_{;\nu} = T^{\mu\nu}{}_{,\nu} + \Gamma^{\mu}_{\nu\lambda} T^{\lambda\nu} + \Gamma^{\nu}_{\nu\lambda} T^{\mu\lambda} .$$

Now we have[5]

$$\Gamma^{\nu}_{\nu\lambda} = \frac{1}{\sqrt{-g}} \partial_\lambda (\sqrt{-g}) , \tag{1.35}$$

where g is the determinant of $(g_{\mu\nu})$. Hence, (1.34) is equivalent to

$$\frac{1}{\sqrt{-g}} \partial_\lambda (\sqrt{-g}\, T^{\mu\nu}) + \Gamma^{\mu}{}_{\nu\lambda} T^{\lambda\nu} = 0 . \tag{1.36}$$

Because of the second term in (1.36), this is *no longer* a conservation law. We cannot form any constants of the motion from (1.36). This should also not be expected, since the system under consideration can exchange energy and momentum with the gravitational field.

Equations (1.34) (or (1.36)) and (1.32) provide the basic hydrodynamic equations for an ideal fluid in the presence of a gravitational field (see the exercises below).

Remark. In the derivation of the field equations for the gravitational field, (1.34) will play an important role.

* * *

---

[5] From linear algebra we know (Cramer's rule) that $g\, g^{\mu\nu}$ is the cofactor (minor) of $g_{\mu\nu}$, hence $\partial_\alpha g = \frac{\partial g}{\partial g_{\mu\nu}} \partial_\alpha g_{\mu\nu} = g\, g^{\mu\nu} \partial_\alpha g_{\mu\nu}$. This gives

$$\begin{aligned} \Gamma^{\nu}_{\nu\alpha} &= g^{\mu\nu} \frac{1}{2} (\partial_\alpha g_{\mu\nu} + \partial_\nu g_{\mu\alpha} - \partial_\mu g_{\nu\alpha}) = \frac{1}{2} g^{\mu\nu} \partial_\alpha g_{\mu\nu} \\ &= \frac{1}{2g} \partial_\alpha g = \frac{1}{\sqrt{-g}} \partial_\alpha (\sqrt{-g}) . \end{aligned}$$

Exercises.

Exercise 1.8. Contract Eq. (1.34) with u^μ and show that the stress-energy tensor (1.32) for a perfect fluid leads to

$$\nabla_u \rho = -(\rho + p)\nabla \cdot u \,. \tag{1.37}$$

Exercise 1.9. Contract Eq. (1.34) with the "projection tensor"

$$h_{\mu\nu} = g_{\mu\nu} + u_\mu u_\nu \tag{1.38}$$

and derive the following general relativistic **Euler equation** for a perfect fluid:

$$(\rho + p)\nabla_u u = -\mathrm{grad}\, p - (\nabla_u p)u \,. \tag{1.39}$$

The **gradient** of a function f is the vector field $\mathrm{grad}\, f := (df)^\sharp$.

1.4.3 Electrodynamics

We assume that the reader is familiar with the four-dimensional tensor formulation of electrodynamics in SR. The basic dynamical object is the antisymmetric electromagnetic field tensor $F_{\mu\nu}$, which unifies the electric and magnetic fields as follows:

$$(F_{\mu\nu}) = \begin{pmatrix} 0 & -E_1 & -E_2 & -E_3 \\ E_1 & 0 & B_3 & -B_2 \\ E_2 & -B_3 & 0 & B_1 \\ E_3 & B_2 & -B_1 & 0 \end{pmatrix}.$$

In the language of differential forms (DG, Sect. 12) $F_{\mu\nu}$ can be regarded as the components of the 2-form

$$\begin{aligned} F &= \frac{1}{2} F_{\mu\nu}\, dx^\mu \wedge dx^\nu \\ &= (E_1 dx^1 + E_2 dx^2 + E_3 dx^3) \wedge dx^0 \\ &\quad + B_1\, dx^2 \wedge dx^3 + B_2\, dx^3 \wedge dx^1 + B_3\, dx^1 \wedge dx^2 \,, \end{aligned} \tag{1.40}$$

sometimes called the **Faraday form**. The homogeneous Maxwell equations are

$$\partial_\lambda F_{\mu\nu} + \partial_\mu F_{\nu\lambda} + \partial_\nu F_{\lambda\mu} = 0 \,, \tag{1.41}$$

expressing that the Faraday 2-form is closed:

$$dF = 0 \,. \tag{1.42}$$

Obviously, this law makes no use of a metric.

If $j^\mu = (\rho, \boldsymbol{J})$ denotes the current four-vector, the inhomogeneous Maxwell equations are ($c = 1$),

$$\partial_\nu F^{\mu\nu} = 4\pi j^\mu . \tag{1.43}$$

With the calculus of differential forms this can be written as

$$\delta F = -4\pi J , \tag{1.44}$$

where δ is the codifferential[6] and J denotes the current 1-form

$$J = j_\mu dx^\mu . \tag{1.45}$$

The generalization of these fundamental equations to Einstein's gravity theory is simple: We have to define $F_{\mu\nu}$ and j^μ such that they transform as tensor fields, and have the same meaning as in SR in locally inertial systems. Secondly, we must apply the $\partial_\mu \longrightarrow \nabla_\mu$ rule. Maxwell's equations in GR are thus

$$\nabla_\lambda F_{\mu\nu} + \nabla_\mu F_{\nu\lambda} + \nabla_\nu F_{\lambda\mu} = 0 , \tag{1.46a}$$
$$\nabla_\nu F^{\mu\nu} = 4\pi j^\mu , \tag{1.46b}$$

with

$$F^{\mu\nu} = g^{\mu\alpha} g^{\nu\beta} F_{\alpha\beta} . \tag{1.47}$$

Because of (1.42) the metric should drop out in (1.46a). The reader may verify explicitly that the following identity

$$\nabla_\lambda F_{\mu\nu} + \ldots = \partial_\lambda F_{\mu\nu} + \ldots$$

holds for any antisymmetric tensor field $F_{\mu\nu}$. A more general statement is derived in DG, Sect. 13.4 (Eq. (13.26)).

As expected, the inhomogeneous equations (1.46b) imply (covariant) current conservation

$$\nabla_\mu j^\mu = 0 . \tag{1.48}$$

This follows from the identity $\nabla_\mu \nabla_\nu F^{\mu\nu} \equiv 0$. A simple way to show this is to note that for an antisymmetric tensor field $F^{\mu\nu}$ and a vector field j^μ the following identities hold (see Exercise 1.10)

$$\nabla_\nu F^{\mu\nu} \equiv \frac{1}{\sqrt{-g}} \partial_\nu (\sqrt{-g}\, F^{\mu\nu}) , \tag{1.49}$$

$$\nabla_\mu j^\mu \equiv \frac{1}{\sqrt{-g}} \partial_\mu (\sqrt{-g}\, j^\mu) . \tag{1.50}$$

These can also be used to rewrite (1.46b) and (1.48) as

$$\frac{1}{\sqrt{-g}} \partial_\nu (\sqrt{-g}\, F^{\mu\nu}) = 4\pi j^\mu , \tag{1.51}$$

$$\partial_\mu (\sqrt{-g}\, j^\mu) = 0 . \tag{1.52}$$

[6] Note the sign convention for δ adopted in DG, Sect. 12.6.3, which is not universally used.

In terms of differential forms, things are again much more concise. Due to the identity $\delta \circ \delta = 0$, equation (1.44) implies immediately $\delta J = 0$, and this is equivalent to (1.48) (see DG, Exercise 13.7). Because of Gauss' Theorem (DG, Theorem 12.12), the vanishing of the divergence of J implies an integral conservation law (conservation of electric charge).

The energy-momentum tensor of the electromagnetic field can be read off from the expression in SR

$$T^{\mu\nu} = \frac{1}{4\pi}\left[F^{\mu\alpha}F^{\nu}{}_{\alpha} - \frac{1}{4}g^{\mu\nu}F_{\alpha\beta}F^{\alpha\beta}\right]. \tag{1.53}$$

Note that its trace vanishes.

The Lorentz equation of motion for a charged test mass becomes in GR

$$m\left(\frac{d^2x^\mu}{d\tau^2} + \Gamma^\mu{}_{\alpha\beta}\frac{dx^\alpha}{d\tau}\frac{dx^\beta}{d\tau}\right) = e\,F^\mu{}_\nu\frac{dx^\nu}{d\tau}. \tag{1.54}$$

The homogeneous Maxwell equation (1.42) allows us also in GR to introduce vector potentials, at least locally. By Poincaré's Lemma (DG, Sect. 12.4), F is locally exact

$$F = d\,A. \tag{1.55}$$

In components, with $A = A_\mu dx^\mu$, we have

$$F_{\mu\nu} = \partial_\mu A_\nu - \partial_\nu A_\mu \quad (\equiv \nabla_\mu A_\nu - \nabla_\nu A_\mu). \tag{1.56}$$

As in SR there is a **gauge freedom**

$$A \longrightarrow A + d\chi \quad \text{or} \quad A_\mu \longrightarrow A_\mu + \partial_\mu\chi, \tag{1.57}$$

where χ is any smooth function. This can be used to impose gauge conditions, for instance the **Lorentz condition**

$$\nabla_\mu A^\mu = 0 \qquad (\text{or } \delta A = 0). \tag{1.58}$$

We stay in this class if χ in (1.57) is restricted to satisfy

$$\Box\chi := \nabla_\mu\nabla^\mu\chi = 0. \tag{1.59}$$

In terms of the four-potential A^μ we can write the inhomogeneous Maxwell equations (1.46b) as

$$\nabla_\nu\nabla^\nu A^\mu - \nabla_\nu\nabla^\mu A^\nu = -4\pi j^\mu. \tag{1.60}$$

Let us impose the Lorentz condition. We can use this in the second term with the help of the Ricci identity for the commutator of two covariant derivatives (DG, Eq. (13.89))

$$(\nabla_\mu\nabla_\nu - \nabla_\nu\nabla_\mu)A^\alpha = R^\alpha{}_{\beta\mu\nu}A^\beta. \tag{1.61}$$

This leads to

$$\nabla_\nu \nabla^\nu A^\mu - R^\mu_{\ \nu} A^\nu = -4\pi j^\mu . \qquad (1.62)$$

Note that in SR (1.62) reduces to the inhomogeneous wave equation $\partial_\nu \partial^\nu A^\mu = -4\pi j^\mu$. If we would substitute here covariant derivatives, we would miss the curvature term in (1.62). This example illustrates possible *ambiguities* in applying the $\partial \longrightarrow \nabla$ rule to second order differential equations, because covariant derivatives do not commute. In passing, we mention that without the curvature term we would, however, loose gauge invariance (see Exercise 1.11).

Let us finally derive a wave equation for F in vacuum ($J = 0$). With the calculus of exterior forms this is extremely simple: From $dF = 0$ and $\delta F = 0$ we deduce

$$\Box F = 0 , \qquad (1.63)$$

where

$$\Box = \delta \circ d + d \circ \delta . \qquad (1.64)$$

In Exercise 1.13 the reader is asked to write this in terms of covariant derivatives, with the result (1.66).

* * *

Exercises.

Exercise 1.10. Derive the identities (1.49) and (1.50).

Exercise 1.11. Show that the curvature term in (1.60) is needed in order to maintain gauge invariance within the Lorentz gauge class.

Exercise 1.12. Use the Ricci identity (1.59), as well as $\nabla_\mu \nabla_\nu f = \nabla_\mu \nabla_\nu f$ for functions f, to derive the following Ricci identity for covariant vector fields

$$\omega_{\alpha;\mu\nu} - \omega_{\alpha;\nu\mu} = R^\lambda_{\ \alpha\mu\nu} \omega_\lambda , \qquad (1.65)$$

and its generalization for arbitrary tensor fields.

Exercise 1.13. As an application of the last exercise apply ∇^λ on the homogeneous Maxwell equations in the form (1.46a) and use the vacuum Maxwell equations $\nabla_\nu F^{\mu\nu} = 0$ to show that

$$F_{\mu\nu;\lambda}^{\ \ \ \ \ ;\lambda} + (R^\sigma_{\ \mu} F_{\nu\sigma} - R^\sigma_{\ \nu} F_{\mu\sigma}) + R_{\alpha\beta\mu\nu} F^{\alpha\beta} = 0 . \qquad (1.66)$$

Exercise 1.14. Show that Maxwell's vacuum equations are invariant under conformal changes $g \longrightarrow e^{2\phi} g$ of the metric.

Exercise 1.15. Show that the equation

$$\Box \psi + \frac{1}{6} R \psi = 0 \qquad (1.67)$$

for a scalar field ψ is invariant under conformal changes of the metric for a suitable transformation law for ψ.

1.5 The Newtonian Limit

For any generalization of a successful physical theory it is crucial to guarantee that the old theory is preserved within certain limits. The Newtonian theory should be an excellent approximation for slowly varying weak gravitational fields and small velocities of material bodies. At this point we can check only part of this requirement, because the dynamical equation for the metric field is not yet known to us.

We consider a test particle moving slowly in a quasi-stationary weak gravitational field. For weak fields, there are coordinate systems which are nearly Lorentzian. This means that

$$g_{\mu\nu} = \eta_{\mu\nu} + h_{\mu\nu}\,, \qquad |h_{\mu\nu}| \ll 1\,. \tag{1.68}$$

For a slowly moving particle (in comparison with the speed of light) we have $dx^0/d\tau \simeq 1$ and we neglect $dx^i/d\tau$ $(i = 1,2,3)$ in comparison to $dx^0/d\tau$ in the geodesic equation (1.23). We then obtain

$$\frac{d^2x^i}{dt^2} \simeq \frac{d^2x^i}{d\tau^2} = -\Gamma^i{}_{\alpha\beta}\frac{dx^\alpha}{d\tau}\frac{dx^\beta}{d\tau} \simeq -\Gamma^i{}_{00}\,. \tag{1.69}$$

Thus only the components $\Gamma^i{}_{00}$ appear in the equation of motion. To first order in $h_{\mu\nu}$ these are given by

$$\Gamma^i{}_{00} \simeq -\frac{1}{2}h_{00,i} + h_{0i,0}\,. \tag{1.70}$$

For quasi-stationary fields we can neglect the last term, $\Gamma^i{}_{00} \simeq -\frac{1}{2}h_{00,i}$, obtaining

$$\frac{d^2x^i}{dt^2} \simeq \frac{1}{2}\partial_i h_{00}\,. \tag{1.71}$$

This agrees with the **Newtonian equation of motion**

$$\frac{d^2\boldsymbol{x}}{dt^2} = -\boldsymbol{\nabla}\phi\,, \tag{1.72}$$

where ϕ is the Newtonian potential, if we set $h_{00} \simeq -2\phi + \text{const}$. For an isolated system ϕ and h_{00} should vanish at infinity. So we arrive at the important relation

$$g_{00} \simeq -1 - 2\phi\,. \tag{1.73}$$

Note that we only obtain information on the component g_{00} for a Newtonian situation. However, this does not mean that the other components of $h_{\mu\nu}$ must be small in comparison to h_{00}. The almost Newtonian approximation of the other components will be determined in Sect. 4.2. (In this connection an interesting remark is made in Exercise 1.17.) The following table shows that for most situations the correction in (1.73) is indeed very small.

Table 1.1. Numerical illustration of Eq. (1.73).

ϕ/c^2	*On the surface of*
10^{-9}	the Earth
10^{-6}	the Sun
10^{-4}	a white dwarf
10^{-1}	a neutron star
10^{-39}	a proton

Remark. In the Newtonian limit, the Poisson equation for ϕ will follow from Einstein's field equation.

* * *

Exercises.

Exercise 1.16. Use (1.73) to derive the Newtonian limit of the basic equation for an ideal fluid.

Exercise 1.17. The result (1.73) might suggest that the metric for a Newtonian situation is approximately

$$g = -(1+2\phi)dt^2 + dx^2 + dy^2 + dz^2 \,.$$

Compute for this metric the deflection of light by the sun.

Remark. It turns out that the deflection angle is only *half*[7] of the value of GR that will be derived in Sect. 3.4. The reason for this famous factor 2 is that the correct Newtonian approximation will be found to be

$$g = -(1+2\phi)dt^2 + (1-2\phi)\left(dx^2 + dy^2 + dz^2\right) . \qquad (1.74)$$

Thus, the spatial part of the metric is non-Euclidean. This result will be derived in Chap. 4.2.

1.6 The Redshift in a Stationary Gravitational Field

The derivation of the gravitational redshift in this section is a bit pedestrian, but instructive. A more elegant treatment will be given in Sect. 1.9.

We consider a clock in an arbitrary gravitational field which moves along an arbitrary timelike world line (not necessarily in free fall). According to

[7] Einstein got this result in 1911 during his time in Prague. He obtained the correct value only after he found his final vacuum equation in November 1915.

the principle of equivalence, the clock rate is unaffected by the gravitational field when one observes it from a locally inertial system. Let Δt be the time between "ticks" of clocks *at rest* in some inertial system in the absence of a gravitational field. In the locally inertial system $\{\xi^\mu\}$ under consideration, we then have for the coordinate intervals $d\xi^\mu$ between two ticks

$$\Delta t = \sqrt{-\eta_{\mu\nu}\, d\xi^\mu d\xi^\nu}\,.$$

In an arbitrary coordinate system $\{x^\mu\}$ we obviously have

$$\Delta t = \sqrt{-g_{\mu\nu}\, dx^\mu dx^\nu}\,.$$

Hence,

$$\frac{dt}{\Delta t} = \left(-g_{\mu\nu}\frac{dx^\mu}{dt}\frac{dx^\nu}{dt}\right)^{-1/2}, \qquad (1.75)$$

where $dt = dx^0$ denotes the time interval between two ticks relative to the system $\{x^\mu\}$. If the clock is at rest relative to this system, i.e. $dx^i/dt = 0$, we have in particular

$$\frac{dt}{\Delta t} = \frac{1}{\sqrt{-g_{00}}}\,. \qquad (1.76)$$

This is true for any clock. For this reason, we cannot verify (1.75) or (1.76) locally. However, we can compare the time dilations at two different points with each other. For this purpose, we specialize the discussion to the case of a **stationary** field. By this we mean that we can choose the coordinates x^μ such that the $g_{\mu\nu}$ are independent of t. Now consider two clocks at rest at the points 1 and 2. (One can convince oneself that the clocks are at rest in any other coordinate system in which the $g_{\mu\nu}$ are independent of time. The concept "at rest" has an intrinsic meaning for stationary fields. See Sect. 1.9 for a geometrical discussion.)

Let a periodic wave be emitted at point 2. Since the field is stationary, the time (relative to our chosen coordinate system) which a wave crest needs to move from point 2 to point 1 is constant[8]. The time between the arrival of successive crests (or troughs) at point 1 is thus equal to the time dt_2 between their emission at point 2, which is, according to (1.76)

$$dt_2 = \Delta t \frac{1}{\sqrt{-g_{00}(x_2)}}\,.$$

[8] From $g_{\mu\nu}\, dx^\mu dx^\nu = 0$ along the light rays, we have

$$dt = \frac{1}{g_{00}} \cdot \left(-g_{i0}\, dx^i - \sqrt{(g_{i0}g_{j0} - g_{ij}g_{00})dx^i\, dx^j}\right).$$

The time interval being discussed is equal to the integral of the right hand side from 2 to 1, and is thus constant.

If on the other hand, we consider the same atomic transition at point 1, then, according to (1.76) the time dt_1 between two wave crests, as observed at point 1 is

$$dt_1 = \Delta t \frac{1}{\sqrt{-g_{00}(x_1)}} .$$

For a given atomic transition, the ratio of frequencies observed at point 1 for light emitted at the points 2 and 1, respectively, is equal to

$$\frac{\nu_2}{\nu_1} = \sqrt{\frac{g_{00}(x_2)}{g_{00}(x_1)}} . \tag{1.77}$$

For weak fields, $g_{00} \simeq -1 - 2\phi$ with $|\phi| \ll 1$, we have

$$\frac{\Delta\nu}{\nu} = \frac{\nu_2}{\nu_1} - 1 \simeq \phi(x_2) - \phi(x_1) , \tag{1.78}$$

in agreement with our previous result in Sect. 1.1. The experimental situation was already discussed there.

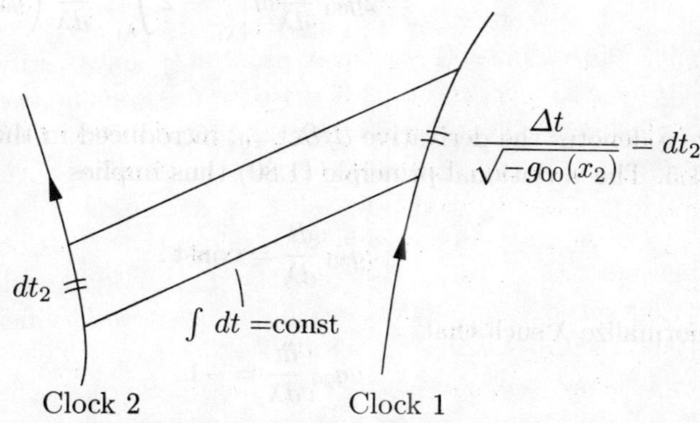

Fig. 1.5. Gravitational redshift in a stationary field.

1.7 Fermat's Principle for Static Gravitational Fields

In the following we shall study in more detail light rays in a **static** gravitational field. A characteristic property of a static field is that in suitable coordinates the metric splits as

$$ds^2 = g_{00}(\boldsymbol{x})\, dt^2 + g_{ik}(\boldsymbol{x})\, dx^i dx^k . \tag{1.79}$$

Thus there are no off-diagonal elements g_{0i} and the $g_{\mu\nu}$ are independent of time. We shall give an intrinsic definition of a static field in Sect. 1.9.

If λ is an affine parameter, the paths $x^\mu(\lambda)$ of light rays can be characterized by the variational principle (using standard notation)

$$\delta \int_{\lambda_1}^{\lambda_2} g_{\mu\nu} \frac{dx^\mu}{d\lambda}\frac{dx^\nu}{d\lambda} d\lambda = 0 , \tag{1.80}$$

where the endpoints of the path are held fixed. In addition, we have (see Exercise 1.6)

$$g_{\mu\nu} \frac{dx^\mu}{d\lambda}\frac{dx^\nu}{d\lambda} = 0 . \tag{1.81}$$

Consider now a static spacetime with a metric of the form (1.79). If we vary only $t(\lambda)$, we have

$$\begin{aligned} \delta \int_{\lambda_1}^{\lambda_2} g_{\mu\nu} \frac{dx^\mu}{d\lambda}\frac{dx^\nu}{d\lambda} d\lambda &= \int_{\lambda_1}^{\lambda_2} 2 g_{00} \frac{dt}{d\lambda} \delta\left(\frac{dt}{d\lambda}\right) d\lambda \\ &= \int_{\lambda_1}^{\lambda_2} 2 g_{00} \frac{dt}{d\lambda}\frac{d}{d\lambda}(\delta t) d\lambda \\ &= 2 g_{00} \frac{dt}{d\lambda}\delta t\bigg|_{\lambda_1}^{\lambda_2} - 2\int_{\lambda_1}^{\lambda_2} \frac{d}{d\lambda}\left(g_{00}\frac{dt}{d\lambda}\right) \delta t \, d\lambda , \end{aligned} \tag{1.82}$$

where δ denotes the derivative $\partial/\partial\varepsilon|_{\varepsilon=0}$, introduced in the solution of Exercise 1.5. The variational principle (1.80) thus implies

$$g_{00} \frac{dt}{d\lambda} = \text{const} .$$

We normalize λ such that

$$g_{00} \frac{dt}{d\lambda} = -1 . \tag{1.83}$$

Now consider a general variation of the path $x^\mu(\lambda)$, for which only the *spatial* endpoints $x^i(\lambda)$ are held fixed, while the condition $\delta t = 0$ at the endpoints is dropped. For such variations, we find from (1.82) and (1.83), making use of the variational principle (1.80),

$$\delta \int_{\lambda_1}^{\lambda_2} g_{\mu\nu} \frac{dx^\mu}{d\lambda}\frac{dx^\nu}{d\lambda} d\lambda = -2\,\delta t\bigg|_{\lambda_1}^{\lambda_2} = -2\,\delta \int_{\lambda_1}^{\lambda_2} dt . \tag{1.84}$$

If the varied orbit is also traversed at the speed of light (just as the original path), the left-hand side of (1.84) is equal to zero and for the varied light-like curves we have also

$$\sqrt{-g_{00}}\, dt = d\sigma , \tag{1.85}$$

where $d\sigma^2 = g_{ik}\, dx^i dx^k$ is the 3-dimensional Riemannian metric of the spatial sections. We thus have

$$\delta \int_{\lambda_1}^{\lambda_2} dt = 0 = \delta \int_{\lambda_1}^{\lambda_2} \frac{1}{\sqrt{-g_{00}}}\, d\sigma\,. \tag{1.86}$$

This is **Fermat's principle of least time**. The second equality in (1.86) determines the spatial path of the light ray. The time has been completely eliminated in this formulation: Equation (1.86) is valid for an arbitrary portion of the path of the light ray, for any variation such that the ends are held fixed. A comparison with Fermat's principle in optics shows that the role of the index of refraction has been taken over by $(g_{00})^{-1/2}$.

Equation (1.86) states that the path of a light ray is a geodesic in the spatial sections for what is often called the **Fermat metric**

$$g_F = g_{ik}^F \, dx^i dx^k, \tag{1.87}$$

where $g_{ik}^F = g_{ik}/(-g_{00})$. We thus have the variational principle for the spatial path γ of a light ray

$$\delta \int g_F(\dot\gamma, \dot\gamma)\, d\lambda = 0\,, \tag{1.88}$$

where the spatial endpoints are kept fixed. Instead of the energy functional for g_F we can, of course, also use the length functional. This result is useful for calculating the propagation of light rays in gravitational fields. In many situations it suffices to use the almost Newtonian approximation (1.74) for the metric. Fermat metric is then

$$g_F = \frac{1-2\phi}{1+2\phi}\, d\boldsymbol{x}^2\,, \tag{1.89}$$

with $d\boldsymbol{x}^2 = (dx^1)^2 + (dx^2)^2 + (dx^3)^2$. Fermat's principle becomes

$$\delta \int (1-2\phi)|\dot{\boldsymbol{x}}^2(\lambda)|\, d\lambda = 0\,,$$

where $|\dot{\boldsymbol{x}}|$ denotes the Euclidean norm of $d\boldsymbol{x}/d\lambda$. This agrees with the Fermat principle of geometrical optics

$$\delta \int n(\boldsymbol{x}(\lambda))|\dot{\boldsymbol{x}}(\lambda)|\, d\lambda = 0$$

for the refraction index

$$n = 1 - 2\phi\,. \tag{1.90}$$

This can be used as the starting point for much of gravitational lensing theory, an important branch of present day astronomy. Section 4.8 will be devoted to this topic.

* * *

Exercise 1.18. Consider a stationary source-free electromagnetic field $F_{\mu\nu}$ in a static gravitational field g with metric (1.79). Show that the time independent scalar potential φ satisfies the Laplace equation

$$\Delta(g_F)\varphi = 0 , \tag{1.91}$$

where $\Delta(g_F)$ is the 3-dimensional Laplace operator for the Fermat metric g_F.

Hints. Use the conformal invariance of Maxwell's equations and work with the metric g/g_{00}.

1.8 Geometric Optics in Gravitational Fields

In most instances gravitational fields vary even over macroscopic distances so little that the propagation of light and radio waves can be described in the geometric optics limit (**ray optics**). We shall derive in this section the laws of geometric optics in the presence of gravitational fields from Maxwell's equations (see also the corresponding discussion in books on optics). In addition to the geodesic equation for light rays, we shall find a simple propagation law for the polarization vector.

The following characteristic lengths are important for our analysis:

1. The wavelength λ.
2. A typical length L over which the amplitude, polarization and wavelength of the wave vary significantly (for example the radius of curvature of a wave front).
3. A typical "radius of curvature" for the geometry; more precisely, take

$$R = \left| \begin{matrix} \text{typical component of the Riemannian tensor} \\ \text{in a typical local inertial system} \end{matrix} \right|^{-1/2} .$$

The region of validity for geometric optics is

$$\lambda \ll L \qquad \text{and} \qquad \lambda \ll R . \tag{1.92}$$

Consider a wave which is highly monochromatic in regions having a size smaller than L (more general cases can be treated via Fourier analysis). Now separate the four-vector potential A_μ into a rapidly varying real phase ψ and a slowly varying complex amplitude $\mathcal{A}_\mu$ (eikonal ansatz)

$$A_\mu = \mathrm{Re}\{\mathcal{A}_\mu \, \mathrm{e}^{\mathrm{i}\psi}\} .$$

It is convenient to introduce the small parameter $\varepsilon = \lambda/\min(L, R)$. We may expand $\mathcal{A}_\mu = a_\mu + \varepsilon b_\mu + \dots$, where $a_\mu, b_\mu, \dots$ are independent of λ. Since $\psi \propto \lambda^{-1}$, we replace ψ by ψ/ε. We thus seek solutions of the form

$$A_\mu = \mathrm{Re}\{(a_\mu + \varepsilon b_\mu + \ldots)\mathrm{e}^{\frac{\mathrm{i}\psi}{\varepsilon}}\}. \tag{1.93}$$

In the following let $k_\mu = \partial_\mu \psi$ be the wave number, $a = \sqrt{(a_\mu \bar{a}^\mu)}$ the scalar amplitude and $f_\mu = a_\mu / a$ the polarization vector, where f_μ is a complex unit vector. By definition, **light rays** are integral curves of the vector field k^μ and are thus perpendicular to the surfaces of constant phase ψ, in other words perpendicular to the **wave fronts**.

Now insert the geometric-optics ansatz (1.93) into Maxwell's equations. In vacuum, these are given (see Sect. 1.4.3)

$$A^{\nu;\mu}{}_{;\nu} - A^{\mu;\nu}{}_{;\nu} = 0. \tag{1.94}$$

We use the Ricci identity

$$A^{\nu;\mu}{}_{;\nu} = A^{\mu;\nu}{}_{;\nu} + R^\mu_{\ \nu} A^\nu \tag{1.95}$$

and impose the Lorentz gauge condition

$$A^\nu{}_{;\nu} = 0. \tag{1.96}$$

Equation (1.94) then takes the form

$$A^{\mu;\nu}{}_{;\nu} - R^\mu_{\ \nu} A^\nu = 0. \tag{1.97}$$

If we now insert (1.93) into the Lorentz condition, we obtain

$$0 = A^\nu{}_{;\nu} = \mathrm{Re}\left\{\left(\mathrm{i}\frac{k_\mu}{\varepsilon}(a^\mu + \varepsilon b^\mu + \ldots) + (a^\mu + \varepsilon b^\mu + \ldots)_{;\mu}\right)\mathrm{e}^{\frac{\mathrm{i}\psi}{\varepsilon}}\right\}. \tag{1.98}$$

From the leading term, it follows that $k_\mu a^\mu = 0$, or equivalently

$$k_\mu f^\mu = 0. \tag{1.99}$$

Thus, the polarization vector is perpendicular to the wave vector. The next order in (1.98) leads to $k_\mu b^\mu = \mathrm{i}\, a^\mu{}_{;\mu}$. Now substitute (1.93) in (1.97) to obtain

$$\begin{aligned} 0 &= -A^{\mu;\nu}{}_{;\nu} + R^\mu_{\ \nu} A^\nu \\ &= \mathrm{Re}\left\{\left(\frac{1}{\varepsilon^2} k^\nu k_\nu (a^\mu + \varepsilon b^\mu + \ldots) - 2\frac{\mathrm{i}}{\varepsilon} k^\nu (a^\mu + \varepsilon b^\mu + \ldots)_{;\nu} \right.\right. \\ &\quad \left.\left. - \frac{\mathrm{i}}{\varepsilon} k^\nu{}_{;\nu}(a^\mu + \varepsilon b^\mu + \ldots) - (a^\mu + \ldots)_{;\nu}^{\ \ \nu} + R^\mu_{\ \nu}(a^\nu + \ldots)\right)\mathrm{e}^{\frac{\mathrm{i}\psi}{\varepsilon}}\right\}. \end{aligned} \tag{1.100}$$

This gives, in order ε^{-2}, $k^\nu k_\nu a^\mu = 0$, which is equivalent to

$$k^\nu k_\nu = 0, \tag{1.101}$$

telling us that the wave vector is null. Using $k_\mu = \partial_\mu \psi$ we obtain the general relativistic **eikonal equation**

$$g^{\mu\nu}\,\partial_\mu\psi\,\partial_\nu\psi = 0\,. \tag{1.102}$$

The terms of order ε^{-1} give

$$k^\nu k_\nu b^\mu - 2\,\mathrm{i}\left(k^\nu a^\mu{}_{;\nu} + \frac{1}{2}k^\nu{}_{;\nu}a^\mu\right) = 0\,.$$

With (1.101), this implies

$$k^\nu a^\mu{}_{;\nu} = -\frac{1}{2}k^\nu{}_{;\nu}a^\mu\,. \tag{1.103}$$

As a consequence of these equations, we obtain the geodesic law for the propagation of light rays: Equation (1.101) implies

$$0 = (k^\nu k_\nu)_{;\mu} = 2\,k^\nu k_{\nu;\mu}\,.$$

Now $k_\nu = \psi_{,\nu}$ and since $\psi_{;\nu;\mu} = \psi_{;\mu;\nu}$ we obtain, after interchanging indices,

$$k^\nu k_{\mu;\nu} = 0\,, \qquad (\nabla_k k = 0)\,. \tag{1.104}$$

We have thus demonstrated that, as a consequence of Maxwell's equations, the paths of light rays are *null geodesics*.

Now consider the amplitude $a^\mu = a\,f^\mu$. From (1.103) we have

$$\begin{aligned} 2\,ak^\nu a_{,\nu} &= 2\,ak^\nu a_{;\nu} = k^\nu(a^2)_{;\nu} = k^\nu(a_\mu\bar{a}^\mu)_{;\nu} \\ &= \bar{a}^\mu k^\nu a_{\mu;\nu} + a_\mu k^\nu \bar{a}^\mu{}_{;\nu} \overset{(1.103)}{=} -\frac{1}{2}k^\nu{}_{;\nu}\left(\bar{a}^\mu a_\mu + a_\mu\bar{a}^\mu\right), \end{aligned}$$

so that

$$k^\nu a_{,\nu} = -\frac{1}{2}k^\nu{}_{;\nu}\,a\,. \tag{1.105}$$

This can be regarded as a propagation law for the scalar amplitude. If we now insert $a^\mu = a\,f^\mu$ into (1.103) we obtain

$$\begin{aligned} 0 &= k^\nu(a\,f^\mu)_{;\nu} + \frac{1}{2}k^\nu{}_{;\nu}a\,f^\mu \\ &= a\,k^\nu f^\mu_{;\nu} + f^\mu\left(k^\nu a_{;\nu} + \frac{1}{2}k^\nu{}_{;\nu}a\right) \overset{(1.105)}{=} a\,k^\nu f^\mu_{;\nu} \end{aligned}$$

or

$$k^\nu f^\mu_{;\nu} = 0\,, \qquad (\nabla_k f = 0)\,. \tag{1.106}$$

We thus see that the polarization vector f^μ is perpendicular to the light rays and is parallel-propagated along them.

Remark. The gauge condition (1.99) is consistent with the other equations: Since the vectors k^μ and f^μ are parallel transported along the rays, one must specify the condition $k_\mu f^\mu = 0$ at only one point. For the same reason, the equations $f_\mu \bar{f}^\mu = 1$ and $k_\mu k^\mu = 0$ are preserved.

Equation (1.105) can be rewritten as follows. After multiplying by a, we have

$$(k^\nu \nabla_\nu) a^2 + a^2 \nabla_\nu k^\nu = 0$$

or

$$(a^2 k^\mu)_{;\mu} = 0 \,, \tag{1.107}$$

thus $a^2 k^\mu$ is a conserved "current".

Quantum mechanically this has the meaning of a conservation law for the number of photons. Of course, the photon number is not in general conserved; it is an **adiabatic invariant**, in other words, a quantity which varies very slowly for $R \gg \lambda$, in comparison to the photon frequency.

Let us consider the eikonal equation (1.102) for the almost Newtonian metric (1.74)

$$-(1-2\phi)(\partial_t \psi)^2 + (1+2\phi)(\boldsymbol{\nabla}\psi)^2 = 0 \,.$$

Since ϕ is time independent we set

$$\psi(\boldsymbol{x}, t) = S(\boldsymbol{x}) - \omega t \tag{1.108}$$

and obtain (up to higher orders in ϕ)

$$(\boldsymbol{\nabla} S)^2 = n^2 \omega^2 \,, \qquad n = 1 - 2\phi \,. \tag{1.109}$$

This has the standard form of the eikonal equation in ray optics with refraction index n. The connection between n and the Newtonian potential ϕ was already found earlier with the help of Fermat's principle.

* * *

Exercises.

Exercise 1.19. Consider light rays, i.e., integral curves $x^\mu(\lambda)$ of $\nabla^\mu \psi$. Derive from the eikonal equation (1.102) that $\dot{x}^\mu$ is an autoparallel null vector.

Exercise 1.20. Show that the energy-momentum tensor, averaged over a wavelength, is

$$\langle T^{\mu\nu} \rangle = \frac{1}{8\pi} a^2 k^\mu k^\nu \,.$$

In particular, the energy flux is

$$\langle T^{0j} \rangle = \langle T^{00} \rangle n^j \,,$$

where $n^j = k^j / k^0$. The Equations (1.104) and (1.107) imply

$$8\pi \nabla_\nu \langle T^{\mu\nu} \rangle = \nabla_\nu (a^2 k^\mu k^\nu) = \nabla_\nu (a^2 k^\nu) k^\mu + a^2 k^\nu \nabla_\nu k^\mu = 0 \,.$$

1.9 Stationary and Static Spacetimes

For this section the reader should be familiar with parts of DG covered in Chaps. 10 and 11.

In Sect. 1.5 we defined somewhat naively a gravitational field to be stationary if there exist coordinates $\{x^\mu\}$ for which the components $g_{\mu\nu}$ of the metric tensor are independent of $t = x^0$. We translate this definition into an intrinsic property of the spacetime (M, g). Let $K = \partial/\partial x^0$, i.e., $K^\mu = \delta^\mu_0$. The Lie derivative $L_K g$ of the metric tensor is then

$$\begin{aligned}(L_K g)_{\mu\nu} &= K^\lambda g_{\mu\nu,\lambda} + g_{\lambda\nu} K^\lambda{}_{,\mu} + g_{\mu\lambda} K^\lambda{}_{,\nu} \\ &= 0 + 0 + 0\,, \end{aligned} \tag{1.110}$$

so that

$$L_K g = 0\,. \tag{1.111}$$

A vector field K which satisfies (1.110) is a **Killing field** or an **infinitesimal isometry**. This leads us to the

Definition 1.1. *A spacetime (M, g) is **stationary** if there exists a timelike Killing field K.*

This means that observers moving with the flow of the Killing field K recognize no changes (see DG, Theorem 11.11). It may be, as in the case for black holes, that a Killing field is only timelike in some open region of M. We then say that this part of spacetime is stationary.

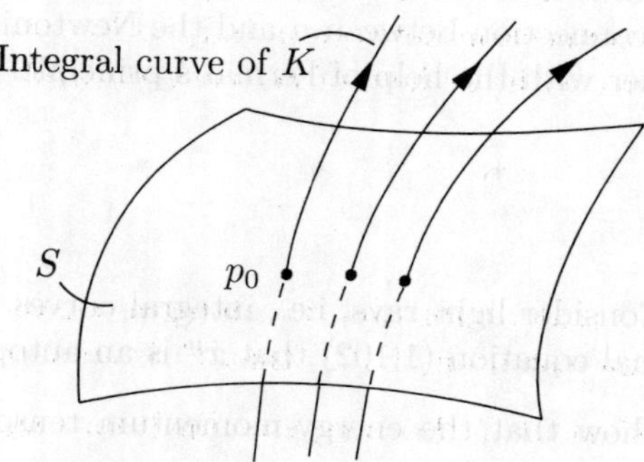

Fig. 1.6. Adapted (stationary) coordinates.

Let us conversely show that Definition 1.1 implies the existence of local coordinates for which the $g_{\mu\nu}$ are independent of time. Choose a spacelike hypersurface S of M and consider the integral curves of K passing through S (see Fig. 1.6). In S we choose arbitrary coordinates and introduce local coordinates of M as follows: If $p = \phi_t(p_0)$, where $p_0 \in S$ and ϕ_t is the flow of K, then the (Lagrange-)coordinates of p are $(t, x^1(p_0), x^2(p_0), x^3(p_0))$. In terms of these coordinates, we have $K = \partial/\partial x^0$, and $L_K g = 0$ implies (using (1.110))

$$g_{\mu\nu,0} + 0 + 0 = 0\,.$$

We call such coordinates to be **adapted** to the Killing field.

Static fields are special cases of stationary fields. The following heuristic consideration will lead us to their proper definition. We choose adapted coordinates and assume that $g_{0i} = 0$ for $i = 1, 2, 3$. Then the Killing field is orthogonal to the spatial sections $\{t = \text{const.}\}$. The 1-form ω corresponding to K ($\omega = K^\flat$, $\omega_\mu = K_\mu = g_{\mu\nu}K^\nu$) is then

$$\omega = g_{00}\, dt = \langle K, K\rangle\, dt\,. \tag{1.112}$$

This implies trivially the **Frobenius condition**

$$\omega \wedge d\omega = 0\,. \tag{1.113}$$

Conversely, let us assume that the Frobenius condition holds for a stationary spacetime with Killing field K. We apply the interior product i_K to (1.113) and use Cartan's formula $L_K = d \circ i_K + i_K \circ d$:

$$0 = i_K(\omega \wedge d\omega) = \underbrace{(i_K\omega)}_{\omega(K) = \langle K,K\rangle} d\omega - \omega \wedge \underbrace{i_K d\omega}_{L_K\omega - d\langle K,K\rangle}\,.$$

We expect that $L_K\omega = 0$; indeed, for any vector field X we have

$$(L_K\omega)(X) = K\big(\omega(X)\big) - \omega\big([K, X]\big) = K\langle K, X\rangle - \langle K, [K, X]\rangle\,.$$

On the other hand

$$0 = (L_K g)(K, X) = K\langle K, X\rangle - \langle [K, K], X\rangle - \langle K, [K, X]\rangle\,,$$

and the right-hand sides of both equations agree.

Using the abbreviation $V := \langle K, K\rangle \neq 0$ we thus arrive at

$$V d\omega + \omega \wedge dV = 0 \qquad \text{or} \qquad d(\omega/V) = 0\,. \tag{1.114}$$

Together with the Poincaré Lemma we see that locally $\omega = V\, df$ for a function f. We use this function as our time coordinate t,

$$\omega = \langle K, K\rangle\, dt\,. \tag{1.115}$$

K is perpendicular to the spacelike sections $\{t = \text{const.}\}$. Indeed, for a tangential vector field X to such a section, $\langle K, X\rangle = \omega(X) = V\, dt(X) = V(Xt) = 0$. In adapted coordinates we have therefore $K = \partial_t$ and $g_{0i} = \langle \partial_t, \partial_i\rangle = \langle K, \partial_i\rangle = 0$.

Summarizing, if the Frobenius condition (1.113) for the timelike Killing field is satisfied, the metric splits locally as

$$g = g_{00}(\boldsymbol{x})\, dt^2 + g_{ik}(\boldsymbol{x})\, dx^i dx^k\,, \tag{1.116}$$

and

$$g_{00} = \langle K, K\rangle\,, \tag{1.117}$$

where $K = \partial/\partial t$. This leads us to the

Definition 1.2. *A stationary spacetime (M, g) with timelike Killing field is* ***static****, if $\omega = K^\flat$ satisfies the Frobenius condition $\omega \wedge d\omega = 0$, whence locally $\omega = \langle K, K\rangle dt$ for an adapted time coordinate t, which is unique up to an additive constant.*

The flow of K maps the hypersurfaces $\{t = \text{const.}\}$ isometrically onto each other. An *observer at rest* moves along integral curves of K.

Killing Equation

According to DG, Eq. (13.101) we have for any vector field X and its associated 1-form $\alpha = X^\flat$ the identity

$$\nabla\alpha = \frac{1}{2}(L_X g - d\alpha). \tag{1.118}$$

For the special case $X = K$ and $\alpha = \omega$ this gives

$$\nabla\omega = -\frac{1}{2}\, d\omega\,. \tag{1.119}$$

In components, this is equivalent to the **Killing equation**

$$K_{\mu;\nu} + K_{\nu;\mu} = 0\,. \tag{1.120}$$

A shorter derivation in terms of local coordinates goes as follows. From (1.110) we obtain for a Killing field

$$K^\lambda g_{\mu\nu,\lambda} + g_{\lambda\nu} K^\lambda{}_{,\mu} + g_{\mu\lambda} K^\lambda{}_{,\nu} = 0\,.$$

Now introduce, for a given point p, normal coordinates with origin p. At this point the last equation reduces to $K_{\mu,\nu} + K_{\nu,\mu} = 0$ or, equivalently, to (1.120). But (1.120) is generally invariant and so holds in any coordinate system.

The reader may wonder, how one might obtain (1.114) in terms of local coordinates. We want to demonstrate that such a derivation can be faster. We write the Frobenius condition (1.113) in components

$$K_\mu K_{\nu,\lambda} + K_\nu K_{\lambda,\mu} + K_\lambda K_{\mu,\nu} = 0\,.$$

The left hand side does not change if partial derivatives are replaced by covariant derivatives. If then multiply the resulting equation by K^λ and use the Killing equation (1.120), we obtain

$$-K_\mu(K^\lambda K_\lambda)_{;\nu} + K_\nu(K^\lambda K_\lambda)_{;\mu} + K^\lambda K_\lambda(K_{\mu;\nu} - K_{\nu;\mu}) = 0\,.$$

This implies

$$[K_\nu/\langle K,K\rangle]_{,\mu} - [K_\mu/\langle K,K\rangle]_{,\nu} = 0\,,$$

and this is equivalent to (1.114).

The Redshift Revisited

The discussion of the gravitational redshift in Sect. 1.6 was mathematically a bit ugly. Below we give two derivations which are more satisfactory, mathematically.

First Derivation. In the eikonal approximation (see Sect. 1.8) we have for the electromagnetic field tensor

$$F_{\mu\nu} = \mathrm{Re}\big(f_{\mu\nu}\mathrm{e}^{\mathrm{i}\psi}\big)\,,$$

where $f_{\mu\nu}$ is a slowly varying amplitude. Light rays are integral curves of the vector field $k^\mu = \psi^{,\mu}$ and are null geodesics (see Sect. 1.8). Since $k^\mu\psi_{,\mu} = 0$, the light rays propagate along surfaces of constant phase, i.e. wave fronts.

Consider now the world lines, parameterized by proper time, of a transmitter and an observer, as well as two light rays which connect the two (see Fig. 1.7). Let the corresponding phases be $\psi = \psi_0$ and $\psi = \psi_0 + \Delta\psi$. We

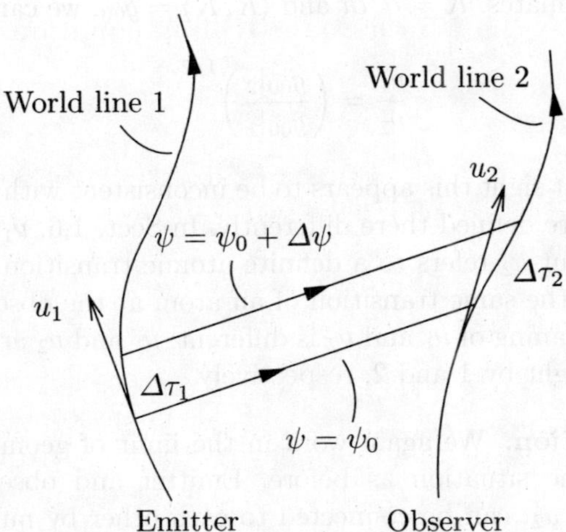

Fig. 1.7. Redshift (first derivation).

denote the interval of proper time between the events at which the two light rays intersect the world line i ($i = 1, 2$) by $\Delta\tau_i$. The four-velocities of the emitter and observer are denoted by u_1^μ and u_2^μ, respectively. Obviously,

$$u_1^\mu(\partial_\mu\psi)_1\Delta\tau_1 = \Delta\psi = u_2^\mu(\partial_\mu\psi)_2\Delta\tau_2\,. \tag{1.121}$$

If ν_1 and ν_2 are the frequencies assigned to the light by 1 and 2, respectively, then (1.121) gives

$$\frac{\nu_1}{\nu_2} = \frac{\Delta\tau_2}{\Delta\tau_1} = \frac{\langle k, u_1\rangle}{\langle k, u_2\rangle}. \tag{1.122}$$

This equation gives the combined effects of Doppler and gravitational redshifts (and is also useful in SR).

Now, we specialize (1.122) to a *stationary* spacetime with Killing field K. For an observer at rest (along an integral curve of K) with four-velocity u, we have

$$K = \left(-\langle K, K\rangle\right)^{1/2} u\,. \tag{1.123}$$

Furthermore, we note that $\langle k, K\rangle$ is *constant* along a light ray, since

$$\nabla_k\langle k, K\rangle = \langle\nabla_k k, K\rangle + \langle k, \nabla_k K\rangle = 0\,.$$

Note that the last term is equal to $k^\alpha k^\beta K_{\alpha;\beta}$ and vanishes as a result of the Killing equation; alternatively, due to (1.119) it is proportional to $d\omega(k,k) = 0$.

If both emitter and observer are at rest, we obtain from (1.122) and (1.123)

$$\frac{\nu_1}{\nu_2} = \left(\frac{\langle K, K\rangle_2}{\langle K, K\rangle_1}\right)^{1/2}. \tag{1.124}$$

In adapted coordinates, $K = \partial/\partial t$ and $\langle K, K\rangle = g_{00}$, we can write (1.124) as

$$\frac{\nu_1}{\nu_2} = \left(\frac{g_{00}|_2}{g_{00}|_1}\right)^{1/2}. \tag{1.125}$$

Remark. At first sight this appears to be inconsistent with (1.77). However, the frequencies are defined there differently. In Sect. 1.6, ν_1 and ν_2 are both measured at 1, but ν_2 refers to a definite atomic transition at 2, while ν_1 is the frequency of the same transition of an atom at the observers position 1. In (1.125) the meaning of ν_1 and ν_2 is different: ν_1 and ν_2 are the frequencies assigned to the light by 1 and 2, respectively.

Second Derivation. We again work in the limit of geometric optics, and consider the same situation as before. Emitter and observer, with four-velocities u_1 and u_2, can be connected to each other by null geodesics with tangent vectors k (see Fig. 1.8). We assume that for a finite τ_1-interval, null geodesics exist which are received by the observer. This family of null

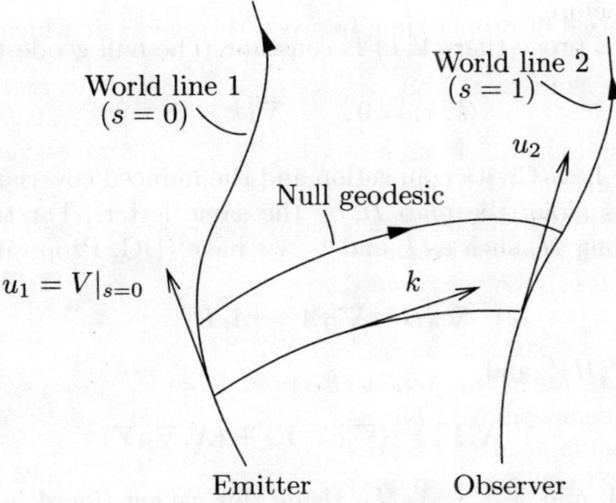

Fig. 1.8. Redshift (second derivation).

geodesics can be parameterized by the emission time τ_1 or by the observer time τ_2, and defines a function $\tau_2(\tau_1)$. The frequency ratio r is clearly the derivative of this function,

$$r = \frac{d\tau_2}{d\tau_1} \,. \tag{1.126}$$

We can parameterize the null geodesics by an affine parameters s, such that $s = 0$ along the world line 1 and $s = 1$ along 2. In what follows we parameterize the 1-parameter family of null geodesics by $(s, \tau_1) \longmapsto H(s, \tau_1)$. With this set up, we are in a situation that has been studied generally in DG, Sect. 14.4. We use the concepts and results which have been developed there (including Sect. 14.2 and Sect. 14.3).

k is a tangential vector field along the map H

$$k = TH \circ \frac{\partial}{\partial s} \,. \tag{1.127}$$

Beside this we also use the field of tangent vectors for curves of constant s

$$V = TH \circ \frac{\partial}{\partial \tau_1} \,. \tag{1.128}$$

Note that

$$V|_{s=0} = u_1 \,, \qquad V|_{s=1} = r\, u_2 \,, \tag{1.129}$$

because $\tau_1 \longmapsto H(s = 1, \tau_1)$ is the world line 2, parameterized by τ_1.

We shall show below that $\langle V, k\rangle$ is constant along a null geodesic. Using this, the ratio r can easily be computed. We obtain the previous result (1.122) from

$$\langle V, k_1\rangle_1 = \langle u_1, k\rangle = \langle V, k\rangle_2 = r\langle u_2, k\rangle \,.$$

The rest is as before.

It remains to prove that $\langle V, k\rangle$ is constant. The null geodesics satisfy

$$\langle k, k\rangle = 0\,, \qquad \nabla_k k = 0\,. \tag{1.130}$$

(We denote the Levi-Civita connection and the induced covariant derivatives for vector fields along the map H by the same letter.) For the tangential vector fields along H, such as k and V, we have (DG, Proposition 14.4 and 14.5)

$$\nabla_A B - \nabla_B A = [A, B]\,, \tag{1.131}$$

where A,$B \in \mathcal{X}(H)^T$ and

$$A'\langle X, Y\rangle = \langle \nabla_A X, Y\rangle + \langle X, \nabla_A Y\rangle \tag{1.132}$$

for $A = TH \circ A'$ and X,$Y \in \mathcal{X}(H)$. Using this we get (for $A' = \partial/\partial s$)

$$\frac{\partial}{\partial s}\langle V, k\rangle = \langle \nabla_k V, k\rangle + \langle V, \nabla_k k\rangle = \langle \nabla_k V, k\rangle\,.$$

From (1.127) and (1.128) we see that $[V, k] = 0$ (see DG, Eq. (14.8)). Hence,

$$\frac{\partial}{\partial s}\langle V, k\rangle = \langle \nabla_k V, k\rangle = \langle \nabla_V k, k\rangle = \frac{1}{2}\frac{\partial}{\partial \tau_1}\langle k, k\rangle = 0\,.$$

1.10 Spin Precession and Fermi Transport

Suppose that an observer moves along a timelike world line in a gravitational field (not necessarily in free fall). One might, for example, consider an astronaut in a space capsule. For practical reasons he will choose a coordinate system in which all apparatus attached to his capsule is at rest. What is the equation of motion of a freely falling test body in this coordinate system? More specifically, the following questions arise:

1. How should the observer orient his space ship so that "Coriolis forces" do not appear?
2. How does one describe the motion of a gyroscope? One might expect that it will not rotate relative to the frame of references, provided the latter is chosen such that Coriolis forces are absent.
3. It is possible to find a spatial frame of reference for an observer at rest in a stationary field, which one might call Copernican? What is the equation of motion of a spinning top in such a frame of reference? Under what conditions will it not rotate relative to the Copernican frame?

1.10.1 Spin Precession in a Gravitational Field

By **spin** we mean either the polarization vector of a particle (i.e., the expectation value of the spin operator for a particle in a particular quantum mechanical state) or the intrinsic angular momentum of a rigid body, such as a gyroscope.

In both cases this is initially defined only relative to a local inertial system in which the body is at rest (its *local rest system*). In this system the spin is described by a three vector $\boldsymbol{S}$. For a gyroscope or for an elementary particle, the equivalence principle implies that in the local rest system, in the absence of external forces,

$$\frac{d}{dt}\boldsymbol{S}(t) = 0\,. \tag{1.133}$$

(We assume that the interaction of the gyroscope's quadrupole moment with inhomogeneities of the gravitational field can be neglected; this effect is studied in Exercise 1.22 at the end of this section). We now define a four-vector S which reduces to $(0, \boldsymbol{S})$ in the local rest system. This last requirement can be expressed invariantly as

$$\langle S, u\rangle = 0\,, \tag{1.134}$$

where u is the four-velocity.

We shall now rewrite (1.133) in a covariant form. For this we consider $\nabla_u S$. In the local rest system (indicated by R) we have

$$(\nabla_u S)_R = \left(\frac{dS^0}{dt}, \frac{d}{dt}\boldsymbol{S}\right) = \left(\frac{dS^0}{dt}, \boldsymbol{0}\right). \tag{1.135}$$

It follows from Equation (1.134) that

$$\langle \nabla_u S, u\rangle = -\langle S, \nabla_u u\rangle = -\langle S, a\rangle\,, \tag{1.136}$$

where $a = \nabla_u u$ is the acceleration. Hence,

$$\langle \nabla_u S, u\rangle = -\left.\frac{dS^0}{dt}\right|_R = -\langle S, a\rangle\,. \tag{1.137}$$

From (1.135) and (1.137) we then have

$$(\nabla_u S)_R = \left(\langle S, a\rangle, \boldsymbol{0}\right) = \left(\langle S, a\rangle u\right)_R\,. \tag{1.138}$$

The desired covariant equation is thus

$$\nabla_u S = \langle S, a\rangle\, u\,. \tag{1.139}$$

Equation (1.134) is consistent with (1.139). Indeed from (1.139) we find

$$\langle u, \nabla_u S\rangle = \langle S, a\rangle\langle u, u\rangle = -\langle S, a\rangle = -\langle S, \nabla_u u\rangle\,,$$

so that $\nabla_u\langle u, S\rangle = 0$.

Thomas Precession

For Minkowski spacetime (1.139) reduces to

$$\dot{S} = \langle S, \dot{u}\rangle\, u\,, \tag{1.140}$$

where the dot means differentiation with respect to proper time. One can easily derive the Thomas precession from this equation.

Let $x(\tau)$ denote the path of a particle. The instantaneous rest system (at time τ) is obtained form the laboratory system via the special Lorentz transformation $\Lambda(\boldsymbol{\beta})$, where $\boldsymbol{\beta} = \boldsymbol{v}/c$ and $\boldsymbol{v}$ the 3-velocity. With respect to this family of instantaneous rest systems S has the form $S = (0, \boldsymbol{S}(t))$, where t is the time in the laboratory frame. We obtain the equation of motion for $\boldsymbol{S}(t)$ easily form (1.140). Since S is a four vector, we have, in the laboratory frame, with standard notation of SR

$$S = \left(\gamma\boldsymbol{\beta}\cdot\boldsymbol{S}, \boldsymbol{S} + \boldsymbol{\beta}\frac{\gamma^2}{\gamma+1}\boldsymbol{\beta}\cdot\boldsymbol{S}\right). \tag{1.141}$$

In addition,

$$u = (\gamma, \gamma\boldsymbol{\beta})\,, \qquad \dot{u} = (\dot{\gamma}, \dot{\gamma}\boldsymbol{\beta} + \gamma\dot{\boldsymbol{\beta}})\,. \tag{1.142}$$

Hence,

$$\begin{aligned}\langle S, \dot{u}\rangle &= -\dot{\gamma}\gamma\boldsymbol{\beta}\cdot\boldsymbol{S} + (\dot{\gamma}\boldsymbol{\beta} + \gamma\dot{\boldsymbol{\beta}})\left(\boldsymbol{S} + \boldsymbol{\beta}\frac{\gamma^2}{\gamma+1}\boldsymbol{\beta}\cdot\boldsymbol{S}\right)\\ &= \gamma\left(\dot{\boldsymbol{\beta}}\cdot\boldsymbol{S} + \frac{\gamma^2}{\gamma+1}\dot{\boldsymbol{\beta}}\cdot\boldsymbol{\beta}\,\boldsymbol{\beta}\cdot\boldsymbol{S}\right).\end{aligned} \tag{1.143}$$

From (1.140) and (1.143) we then obtain

$$(\gamma\boldsymbol{\beta}\cdot\boldsymbol{S})^{\cdot} = \gamma^2\left(\dot{\boldsymbol{\beta}}\cdot\boldsymbol{S} + \frac{\gamma^2}{\gamma+1}\dot{\boldsymbol{\beta}}\cdot\boldsymbol{\beta}\,\boldsymbol{\beta}\cdot\boldsymbol{S}\right),$$

$$\left(\boldsymbol{S} + \boldsymbol{\beta}\frac{\gamma^2}{\gamma+1}\boldsymbol{\beta}\cdot\boldsymbol{S}\right)^{\cdot} = \boldsymbol{\beta}\gamma^2\left(\dot{\boldsymbol{\beta}}\cdot\boldsymbol{S} + \frac{\gamma^2}{\gamma+1}\dot{\boldsymbol{\beta}}\cdot\boldsymbol{\beta}\,\boldsymbol{\beta}\cdot\boldsymbol{S}\right).$$

After some rearrangements, one finds

$$\dot{\boldsymbol{S}} = \boldsymbol{S}\times\boldsymbol{\omega}_T\,, \tag{1.144}$$

where $\boldsymbol{\omega}_T = \frac{\gamma-1}{\beta^2}\boldsymbol{\beta}\times\dot{\boldsymbol{\beta}}$. This is the well-known expression of the Thomas precession.

1.10.2 Fermi Transport

Definition. *Let $\gamma(s)$, with s the proper time, be a timelike curve with tangent vector $u = \dot{\gamma}$ satisfying $\langle u, u\rangle = -1$. The* ***Fermi derivative*** $\boldsymbol{F}_u$ *of a vector field X along γ is defined by*

$$\boldsymbol{F}_u X = \nabla_u X - \langle X, a\rangle u + \langle X, u\rangle a\,, \tag{1.145}$$

where $a = \nabla_u u$.

Since $\langle S, u\rangle = 0$ we may write (1.139) in the form

$$\boldsymbol{F}_u S = 0\,. \tag{1.146}$$

It is easy to show that the Fermi derivative (1.145) has the following important properties:

1. $\boldsymbol{F}_u = \nabla_u$ if γ is a geodesic;
2. $\boldsymbol{F}_u u = 0$;
3. If $\boldsymbol{F}_u X = \boldsymbol{F}_u Y = 0$ for vector fields X, Y along γ, then $\langle X, Y\rangle$ is constant along γ;
4. If $\langle X, u\rangle = 0$ along γ, then

$$\boldsymbol{F}_u X = (\nabla_u X)_\perp\,. \tag{1.147}$$

Here $\perp$ denotes the projection perpendicular to u.

These properties show that the Fermi derivative is a natural generalization of ∇_u.

We say that a vector field X is **Fermi transported** along γ if $\boldsymbol{F}_{\dot\gamma} X = 0$. Since this equation is linear in X, Fermi transport defines (analogously to parallel transport) a two parameter family of isomorphisms

$$\tau^{\boldsymbol{F}}_{t,s} : T_{\gamma(s)}(M) \longrightarrow T_{\gamma(t)}(M)\,.$$

One can show that

$$\boldsymbol{F}_{\dot\gamma} X\big(\gamma(t)\big) = \left.\frac{d}{ds}\right|_{s=t} \tau^{\boldsymbol{F}}_{t,s} X\big(\gamma(s)\big)\,.$$

The proof is similar to that of Theorem 13.1 in DG.

As in the case of the covariant derivative (see DG, Sect. 13.4), the Fermi derivative can be extended to arbitrary tensor fields such that the following properties hold:

1. $\boldsymbol{F}_u$ transforms a tensor field of type (r, s) into another tensor field of the same type;
2. $\boldsymbol{F}_u$ commutes with contractions;
3. $\boldsymbol{F}_u(S \otimes T) = (\boldsymbol{F}_u S) \otimes T + S \otimes (\boldsymbol{F}_u T)$;
4. $\boldsymbol{F}_u f = df/ds$, when f is a function;
5. $\tau^{\boldsymbol{F}}_{t,s}$ induces linear isomorphisms

$$T_{\gamma(s)}(M)^r_s \longrightarrow T_{\gamma(t)}(M)^r_s\,.$$

We now consider the world line $\gamma(\tau)$ of an accelerated observer (τ is the proper time). Let $u = \dot{\gamma}$ and let $\{e_i\}$, with $i = 1, 2, 3$, be an arbitrary orthonormal frame along γ perpendicular to $e_0 := \dot{\gamma} = u$. We then have

$$\langle e_\mu, e_\nu \rangle = \eta_{\mu\nu} ,$$

where $(\eta_{\mu\nu}) = \mathrm{diag}(-1, 1, 1, 1)$. For the acceleration $a := \nabla_u u$ it follows from $\langle u, u \rangle = -1$ that $\langle a, u \rangle = 0$. We set

$$\omega_{ij} = \langle \nabla_u e_i, e_j \rangle = -\omega_{ji} . \tag{1.148}$$

If $e^\mu = \eta^{\mu\nu} e_\nu$ and $\omega_i^{\ j} := \langle \nabla_u e_i, e^j \rangle$, we have

$$\begin{aligned} \nabla_u e_i &= \langle \nabla_u e_i, e^\alpha \rangle e_\alpha \\ &= -\langle \nabla_u e_i, u \rangle u + \langle \nabla_u e_i, e^j \rangle e_j \\ &= \langle e_i, \nabla_u u \rangle u + \omega_i^{\ j} e_j , \end{aligned}$$

so that

$$\nabla_u e_i = \langle e_i, a \rangle u + \omega_i^{\ j} e_j . \tag{1.149}$$

Adding a vanishing term, this can be rewritten as $\nabla_u e_i = -\langle e_i, u \rangle a + \langle e_i, a \rangle u + \omega_i^{\ j} e_j$.

Let

$$(\omega_\alpha^{\ \beta}) = \begin{pmatrix} 0 & 0 \\ 0 & \omega_i^{\ j} \end{pmatrix}$$

Then $\nabla_u e_\alpha = -\langle e_\alpha, u \rangle a + \langle e_\alpha, a \rangle u + \omega_\alpha^{\ \beta} e_\beta$, since for $\alpha = 0$ the right-hand side is equal to $-\langle u, u \rangle a + \langle u, a \rangle u = a = \nabla_u u$. Using (1.145) this can be written in the form

$$\boldsymbol{F}_u e_\alpha = \omega_\alpha^{\ \beta} e_\beta . \tag{1.150}$$

$\omega_\alpha^{\ \beta}$ thus describes the deviation from Fermi transport. For a spinning top we have $\boldsymbol{F}_u S = 0$ and $\langle S, u \rangle = 0$. If we write $S = S^i e_i$, then

$$0 = \boldsymbol{F}_u S = \frac{dS^i}{d\tau} e_i + S^j \boldsymbol{F}_u e_j = \frac{dS^i}{d\tau} e_i + S^j \omega_j^{\ i} e_i ,$$

hence,

$$\frac{dS^i}{d\tau} = \omega_j^{\ i} S^j . \tag{1.151}$$

Thus the top precesses relative to the frame $\{e_i\}$ with angular velocity $\boldsymbol{\Omega}$, where

$$\omega_{ij} = \varepsilon_{ijk} \Omega^k . \tag{1.152}$$

We may write (1.151) in three-dimensional vector notation

$$\frac{d\boldsymbol{S}}{d\tau} = \boldsymbol{S} \times \boldsymbol{\Omega} . \tag{1.153}$$

If the frame $\{e_i\}$ is Fermi transported along γ, then clearly $\boldsymbol{\Omega} = 0$. We shall evaluate (1.148) for the angular velocity in a number of instances. A first example is given in Sect. 1.10.4.

1.10.3 The Physical Difference Between Static and Stationary Fields

We consider now an observer at rest in a stationary spacetime with timelike Killing field K. The observer thus moves along an integral curve $\gamma(\tau)$ of K. His four-velocity u is

$$u = \left(-\langle K, K\rangle\right)^{-1/2} K\,. \tag{1.154}$$

We now choose and orthonormal triad $\{e_i\}$ along γ which is **Lie-transported**

$$L_K e_i = 0\,, \tag{1.155}$$

for $i = 1, 2, 3$. Note that the e_i remain perpendicular to K and hence to u. Indeed, it follows from

$$0 = L_K g(X, Y) = K\langle X, Y\rangle - \langle L_K X, Y\rangle - \langle X, L_K Y\rangle$$

that the orthogonality of the vector fields X and Y is preserved when $L_K X = L_K Y = 0$. Also note that K itself is Lie-transported: $L_K K = [K, K] = 0$.

The $\{e_i\}$ can then be interpreted as "axes at rest" and define what one may call a "Copernican system" (see Fig. 1.9). We are interested in the change of the spin relative to this system. Our starting point is (1.148) or, making use of (1.154)

$$\omega_{ij} = \left(-\langle K, K\rangle\right)^{-1/2} \langle e_j, \nabla_K e_i\rangle\,. \tag{1.156}$$

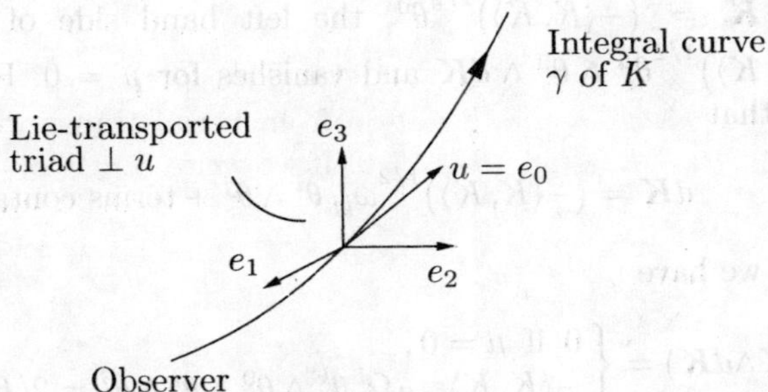

Fig. 1.9. Spin precession in stationary fields relative to a "Copernican system".

Now

$$0 = T(K, e_i) = \nabla_K e_i - \nabla_{e_i} K - [K, e_i]\,,$$

and $[K, e_i] = L_K e_i = 0$. Hence (1.156) implies

$$\omega_{ij} = \left(-\langle K, K\rangle\right)^{-1/2} \langle e_j, \nabla_{e_i} K\rangle = \left(-\langle K, K\rangle\right)^{-1/2} \nabla \boldsymbol{K}(e_j, e_i)\,,$$

where $\boldsymbol{K} = K^\flat$. Since ω_{ij} is antisymmetric

$$\omega_{ij} = -\big(-\langle K, K\rangle\big)^{-1/2}\frac{1}{2}\big(\nabla\boldsymbol{K}(e_i, e_j) - \nabla\boldsymbol{K}(e_j, e_i)\big)$$

or, since any one-form φ satisfies $\nabla\varphi(X, Y) - \nabla\varphi(Y, X) = -d\varphi(X, Y)$, we have also

$$\omega_{ij} = \frac{1}{2}\big(-\langle K, K\rangle\big)^{-1/2} d\boldsymbol{K}(e_i, e_j)\,. \tag{1.157}$$

We shall show below that

$$\omega_{ij} = 0\,, \quad \text{if and only if } \boldsymbol{K}\wedge d\boldsymbol{K} = 0\,. \tag{1.158}$$

From this it follows that a *Copernican system does not rotate if and only if the stationary field is static.*

The one-form $*(\boldsymbol{K}\wedge d\boldsymbol{K})$ can be regarded as a measure of the "absolute" rotation, because the vector $\Omega = \Omega^k e_k$ can be expressed in the form

$$\boldsymbol{\Omega} = \frac{1}{2}\langle K, K\rangle^{-1} *(\boldsymbol{K}\wedge d\boldsymbol{K})\,, \tag{1.159}$$

where $\boldsymbol{\Omega}$ denotes the one-form corresponding to Ω.

It remains to derive Equation (1.158). Let $\{\theta^\mu\}$ denote the dual basis of $\{e_\mu\}$, $e_0 = u$. Form well-known properties of the $*$-operation (see DG, Sect. 12.6.2), we have

$$\theta^\mu \wedge (\boldsymbol{K}\wedge d\boldsymbol{K}) = \eta\,\langle\theta^\mu, *(\boldsymbol{K}\wedge d\boldsymbol{K})\rangle\,. \tag{1.160}$$

Since $\boldsymbol{K} = \big(-\langle K, K\rangle\big)^{1/2}\theta^0$, the left hand side of (1.160) is equal to $\big(-\langle K, K\rangle\big)^{1/2}\theta^\mu\wedge\theta^0\wedge d\boldsymbol{K}$ and vanishes for $\mu = 0$. From (1.157) we conclude that

$$d\boldsymbol{K} = \big(-\langle K, K\rangle\big)^{1/2}\omega_{ij}\,\theta^i\wedge\theta^j + \text{terms containing } \theta^0\,.$$

Hence we have

$$\theta^\mu\wedge(\boldsymbol{K}\wedge d\boldsymbol{K}) = \begin{cases} 0 \text{ if } \mu = 0\,, \\ -\langle K, K\rangle\varepsilon_{ijl}\Omega^l\,\theta^k\wedge\theta^0\wedge\theta^i\wedge\theta^j = 2\langle K, K\rangle\,\eta\,\Omega^k \text{ if } \mu = k\,, \end{cases}$$

where we used $\theta^k\wedge\theta^0\wedge\theta^i\wedge\theta^j = -\varepsilon_{ijk}\,\eta$. From this and (1.160) we get

$$\big(\theta^\mu, *(\boldsymbol{K}\wedge d\boldsymbol{K})\big) = \begin{cases} 0 & \text{if } \mu = 0\,, \\ 2\,\langle K, K\rangle\Omega^k & \text{if } \mu = k\,. \end{cases}$$

The left-hand side of this expression is equal to the contravariant components of $*(\boldsymbol{K}\wedge d\boldsymbol{K})$ and hence (1.159) follows. Obviously (1.159) implies (1.158).

1.10.4 Spin Rotation in a Stationary Field

The spin rotation relative to the Copernican system is given by (1.159). We now write this in terms of adapted coordinates, with $K = \partial/\partial t$ and $g_{\mu\nu}$ independent of $t = x^0$. Then

$$\begin{aligned} \boldsymbol{K} &= g_{00}\, dt + g_{0i}\, dx^i\,, \\ d\boldsymbol{K} &= g_{00,k}\, dx^k \wedge dt + g_{0i,k}\, dx^k \wedge dx^i, \\ \boldsymbol{K} \wedge d\boldsymbol{K} &= (g_{00} g_{0i,j} - g_{0i} g_{00,j})\, dt \wedge dx^j \wedge dx^i + g_{0k} g_{0i,j}\, dx^k \wedge dx^j \wedge dx^i\,. \end{aligned}$$

From DG, Exercise 12.7, we then have

$$\begin{aligned} *(\boldsymbol{K} \wedge d\boldsymbol{K}) &= g_{00}^2 \left(\frac{g_{0i}}{g_{00}}\right)_{,j} *(dt \wedge dx^j \wedge dx^i) + g_{0k} g_{0i,j} *(dx^k \wedge dx^j \wedge dx^i) \\ &= \frac{g_{00}^2}{\sqrt{-g}} \varepsilon_{ijl} \left(\frac{g_{0i}}{g_{00}}\right)_{,j} \left[(g_{lk} dx^k + g_{l0} dx^0) - \left(g_{l0} dx^0 + \frac{g_{l0} g_{k0}}{g_{00}} dx^k \right) \right], \end{aligned}$$

where we used $*(dt \wedge dx^j \wedge dx^i) = \eta^{0jil} g_{l\mu}\, dx^\mu = -\frac{1}{\sqrt{-g}} \varepsilon_{jil}\, g_{l\mu}\, dx^\mu$ in the last step. We thus obtain

$$\boldsymbol{\Omega} = \frac{g_{00}}{2\sqrt{-g}} \varepsilon_{ijl} \left(\frac{g_{0i}}{g_{00}}\right)_{,j} \left(g_{lk} - \frac{g_{l0} g_{k0}}{g_{00}} \right) dx^k \qquad (1.161)$$

and it follows immediately that

$$\Omega = \frac{g_{00}}{2\sqrt{-g}} \varepsilon_{ijl} \left(\frac{g_{0i}}{g_{00}}\right)_{,j} \left(\partial_k - \frac{g_{k0}}{g_{00}} \partial_0 \right). \qquad (1.162)$$

We shall later apply this equation to the field outside of a rotating star or black hole. At sufficiently large distances we have

$$g_{00} \simeq -1\,, \qquad g_{ij} \simeq \delta_{ij}\,, \qquad \frac{g_{0k}}{g_{00}} \ll 1\,,$$

and a good approximation to (1.162) is given by

$$\Omega \simeq -\frac{1}{2}\, \varepsilon_{ijk}\, g_{0i,j}\, \partial_k\,. \qquad (1.163)$$

Since $e_k \cong \partial_k$, the gyroscope rotates relative to the Copernican frame with angular velocity

$$\boldsymbol{\Omega} \simeq -\frac{1}{2} \boldsymbol{\nabla} \times \boldsymbol{g}\,, \qquad (1.164)$$

where $\boldsymbol{g} := (g_{01}, g_{02}, g_{03})$.

We have shown that in a stationary (but not static) field, a gyroscope rotates relative to the Copernican system (relative to the "fixed stars") with angular velocity (1.162). In a weak field this can be approximated by (1.164). This means that *the rotation of a star drags along the local inertial system* (**Lense-Thirring effect**). We shall discuss possible experimental tests of this effect later.

1.10.5 Adapted Coordinate Systems for Accelerated Observers

We again consider the world line $\gamma(\tau)$ of an (accelerated) observer. Let $u = \dot{\gamma}$ and let $\{e_i\}$ be an arbitrary orthonormal frame along γ which is perpendicular to $e_0 = u$. As before, $a = \nabla_u u$. Now construct a local coordinate system as follows: at every point on $\gamma(\tau)$ consider spacelike geodesics $\alpha(s)$ perpendicular to u, with proper length s as affine parameter. Thus $\alpha(0) = \gamma(\tau)$; let $n = \dot{\alpha}(0) \perp u$. In order to distinguish the various geodesics, we denote the geodesic through $\gamma(\tau)$ in the direction n with affine parameter s by $\alpha(s, n, \tau)$ (see Fig. 1.10). We have

$$n = \left(\frac{\partial}{\partial s}\right)_{\alpha(0,n,\tau)} , \qquad \langle n, n \rangle = 1 . \tag{1.165}$$

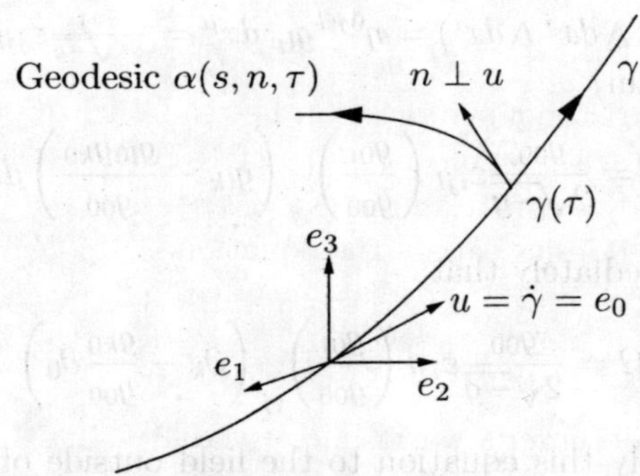

Fig. 1.10. Adapted coordinates for accelerated observer.

Each point $p \in M$ in the vicinity of the observer's world line lies on precisely one of these geodesics. If $p = \alpha(s, n, \tau)$ and $n = n^j e_j$ we assign the following coordinates to p

$$(x^0(p), \ldots, x^3(p)) = (\tau, s\, n^1, s\, n^2, s\, n^3) . \tag{1.166}$$

This means

$$\begin{aligned} x^0\big(\alpha(s,n,\tau)\big) &= \tau \\ x^j\big(\alpha(s,n,\tau)\big) &= s\, n^j = s\, n_j = s \langle n, e_j \rangle . \end{aligned} \tag{1.167}$$

Calculation of the Christoffel Symbols Along $\gamma(\tau)$. Along the observer's world line, we have by construction

$$\left.\frac{\partial}{\partial x^\alpha}\right|_\gamma = e_\alpha , \tag{1.168}$$

and hence $g_{\alpha\beta} = \langle \partial_\alpha, \partial_\beta \rangle = \eta_{\alpha\beta}$ along $\gamma(\tau)$. If $\Gamma^\alpha{}_{\beta\gamma}$ denote the Christoffel symbols relative to the tetrad $\{e_\alpha\}$ (see DG, Sect. 13.7), then $\nabla_u e_\alpha = \nabla_{e_0} e_\alpha = \Gamma^\beta{}_{0\alpha} e_\beta$ and thus

$$\langle e_\beta, \nabla_u e_\alpha \rangle = \eta_{\beta\gamma} \Gamma^\gamma{}_{0\alpha} . \tag{1.169}$$

In particular,

$$\begin{aligned} \Gamma^0{}_{00} &= -\langle u, \nabla_u u \rangle = 0 , \\ \Gamma^j{}_{00} &= \langle e_j, \nabla_u u \rangle = \langle e_j, a \rangle = a^j , \\ \Gamma^0{}_{0j} &= -\langle u, \nabla_u e_j \rangle \overset{(1.149)}{=} \langle e_j, a \rangle = a^j . \end{aligned}$$

If we make use of $\omega_{ij} = \langle \nabla_u e_i, e_j \rangle = -\omega_{ji}$, introduced in (1.148), and the angular velocity Ω^i,

$$\omega_{jk} = \varepsilon_{ijk} \Omega^i , \tag{1.170}$$

then

$$\Gamma^j{}_{0k} = -\varepsilon_{ijk} \Omega^i . \tag{1.171}$$

The remaining Christoffel symbols can be read off from the equation for the geodesics $s \longmapsto \alpha(s, n, \tau)$. According to (1.167) the coordinates x^μ for these geodesics are $x^0(s) =$ const. and $x^j(s) = s\, n^j$, hence $d^2 x^\alpha / ds^2 = 0$. On the other hand, the geodesics satisfy the equation

$$0 = \frac{d^2 x^\alpha}{ds^2} + \Gamma^\alpha{}_{\beta\gamma} \frac{dx^\beta}{ds} \frac{dx^\gamma}{ds} = \Gamma^\alpha{}_{jk} n^j n^k .$$

Hence, along the observer's world line γ we have

$$\Gamma^\alpha{}_{jk} = 0 . \tag{1.172}$$

The partial derivatives of the metric coefficients can be determined from the Christoffel symbols. The general relation is

$$0 = g_{\alpha\beta;\gamma} = g_{\alpha\beta,\gamma} - \Gamma^\mu{}_{\alpha\gamma} g_{\mu\beta} - \Gamma^\mu{}_{\beta\gamma} g_{\alpha\mu} .$$

If we substitute $g_{\alpha\beta} = \eta_{\alpha\beta}$ along $\gamma(\tau)$ and our previously derived results for the Christoffel symbols, we find

$$\begin{aligned} g_{\alpha\beta,0} &= 0 , \qquad & g_{ik,l} &= 0 , \\ g_{00,j} &= -2a^j , \qquad & g_{0j,k} &= -\varepsilon_{jkl} \Omega^l . \end{aligned}$$

These relations, together with $g_{\alpha\beta} = \eta_{\alpha\beta}$ along $\gamma(\tau)$, imply that the metric near γ is given by

$$\begin{aligned} g = &-(1 + 2\boldsymbol{a} \cdot \boldsymbol{x})(dx^0)^2 - 2\, \varepsilon_{jkl}\, x^k \Omega^l\, dx^0 dx^j \\ &+ \delta_{jk}\, dx^j dx^k + O(|\boldsymbol{x}|^2)\, dx^\alpha dx^\beta . \end{aligned} \tag{1.173}$$

From this we see that

1. The acceleration leads to the additional term
$$\delta g_{00} = -2\boldsymbol{a}\cdot\boldsymbol{x}\,. \tag{1.174}$$

2. Since the observer's coordinates axes rotate ($\Omega^i \neq 0$), the metric has the "non-diagonal" term
$$g_{0j} = -\varepsilon_{jkl}\, x^k \Omega^l = -(\boldsymbol{x}\times\boldsymbol{\Omega})^j\,. \tag{1.175}$$

3. The lowest order corrections are not affected by the curvature. The curvature shows itself in second order.
4. If $a = \nabla_u u = 0$ and $\boldsymbol{\Omega} = 0$ (no acceleration and no rotation) we have a local inertial system along $\gamma(\tau)$.

Motion of a Test Body

Suppose that the observer, whose world line is $\gamma(\tau)$ (an astronaut in a capsule, for example), observes a nearby freely falling body. This obeys the equation of motion
$$\frac{d^2x^\alpha}{d\tau^2} + \Gamma^\alpha{}_{\beta\gamma}\frac{dx^\beta}{d\tau}\frac{dx^\gamma}{d\tau} = 0\,.$$
We now replace the proper time τ of the test body by the coordinate time t, using
$$\frac{d}{d\tau} = \left(\frac{dt}{d\tau}\right)\frac{d}{dt} =: \gamma\frac{d}{dt}\,.$$
Since $dx^\alpha/d\tau = \gamma(1, dx^k/dt)$, we obtain
$$\frac{d^2x^\alpha}{dt^2} + \frac{1}{\gamma}\frac{d\gamma}{dt}\frac{dx^\alpha}{dt} + \Gamma^\alpha{}_{\beta\gamma}\frac{dx^\beta}{dt}\frac{dx^\gamma}{dt} = 0\,. \tag{1.176}$$
For $\alpha = 0$, this becomes
$$\frac{1}{\gamma}\frac{d\gamma}{dt} + \Gamma^0{}_{\beta\gamma}\frac{dx^\beta}{dt}\frac{dx^\gamma}{dt} = 0\,.$$
Substitution of this into (1.176) for $\alpha = j$ gives
$$\frac{d^2x^j}{dt^2} + \left(-\frac{dx^j}{dt}\Gamma^0{}_{\beta\gamma} + \Gamma^j{}_{\beta\gamma}\right)\frac{dx^\beta}{dt}\frac{dx^\gamma}{dt} = 0\,. \tag{1.177}$$

We work this out relative to the coordinate system introduced above for small velocities of the test body. To first order in $v^j = dx^j/dt$ the equation of motion is
$$\frac{dv^j}{dt} - v^j\,\Gamma^0{}_{00} + \Gamma^j{}_{00} + 2\,\Gamma^j{}_{k0}v^k = 0\,. \tag{1.178}$$
Since the particle is falling in the vicinity of the observer, the spatial coordinates (1.167) are small. To first order in x^k and v^k we have

$$\frac{dv^j}{dt} = v^j \Gamma^0{}_{00}\Big|_{\boldsymbol{x}=0} - \Gamma^j{}_{00}\Big|_{\boldsymbol{x}=0} - x^k \Gamma^j{}_{00,k}\Big|_{\boldsymbol{x}=0} - 2\, v^k \Gamma^j{}_{k0}\Big|_{\boldsymbol{x}=0} .$$

The Christoffel symbols along γ have already been determined. If we insert these, we find

$$\frac{dv^j}{dt} = -a^j - 2\,\varepsilon_{jik}\Omega^i v^k - x^k \Gamma^j{}_{00,k}\Big|_{\boldsymbol{x}=0} .$$

The quantity $\Gamma^j{}_{00,k}$ is obtained from the Riemann tensor (see DG, Eq. (13.31))

$$R^\alpha{}_{\beta\gamma\delta} = \Gamma^\alpha{}_{\beta\delta,\gamma} - \Gamma^\alpha{}_{\beta\gamma,\delta} + \Gamma^\alpha{}_{\gamma\mu}\Gamma^\mu{}_{\beta\delta} - \Gamma^\alpha{}_{\delta\mu}\Gamma^\mu{}_{\beta\gamma} ,$$

so that

$$\Gamma^j{}_{00,k} = R^j{}_{0k0} + \Gamma^j{}_{0k,0} - \Gamma^j{}_{k\mu}\Gamma^\mu{}_{00} + \Gamma^j{}_{0\mu}\Gamma^\mu{}_{0k} .$$

For $\boldsymbol{x} = 0$, we have

$$\begin{aligned}
\Gamma^j{}_{0k,0} &= -\varepsilon_{jkm}\Omega^m{}_{,0} , \\
\Gamma^j{}_{k\mu}\Gamma^\mu{}_{00} &= 0 , \\
\Gamma^j{}_{0\mu}\Gamma^\mu{}_{0k} &= \Gamma^j{}_{00}\Gamma^0{}_{0k} + \Gamma^j{}_{0m}\Gamma^m{}_{0k} = a^j a^k + \varepsilon_{mjn}\,\Omega^n \varepsilon_{kml}\,\Omega^l .
\end{aligned}$$

Hence,

$$x^k \Gamma^j{}_{00,k} = x^k \Big(R^j{}_{0k0} - \varepsilon_{jkm}\,\Omega^m{}_{,0} + a^j a^k + \varepsilon_{mjn}\,\Omega^n \varepsilon_{kml}\,\Omega^l \Big) .$$

Expressed in three-dimensional vector notation, we end up with

$$\dot{\boldsymbol{v}} = -\boldsymbol{a}(1 + \boldsymbol{a}\cdot\boldsymbol{x}) - 2\,\boldsymbol{\Omega}\times\boldsymbol{v} - \boldsymbol{\Omega}\times(\boldsymbol{\Omega}\times\boldsymbol{x}) - \dot{\boldsymbol{\Omega}}\times\boldsymbol{x} + \boldsymbol{f} , \tag{1.179}$$

where

$$f^j := R^j{}_{00k} x^k . \tag{1.180}$$

The first term in (1.179) is the usual "inertial acceleration", including the relativistic correction $(1 + \boldsymbol{a}\cdot\boldsymbol{x})$, which is a result of (1.174). The terms containing $\boldsymbol{\Omega}$ are well-known from classical mechanics. The force $R^j{}_{00k}x^k$ is a consequence of the inhomogeneity of the gravitational field.

If the frame $\{e_i\}$ is Fermi transported ($\boldsymbol{\Omega} = 0$), the Coriolis force (third term) vanishes. If the observer is freely falling, then also $\boldsymbol{a} = 0$ and only the **tidal force** f^j remains. This cannot be transformed away. We shall discuss this in more detail at the beginning of the next chapter.

* * *

Exercises.

Exercise 1.21. Carry out the rearrangements leading to the Thomas precession (1.144).

Exercise 1.22. A non spherical body in an inhomogeneous gravitational field experiences a torque which results in a time dependence of the spin four vector. Suppose that the center of mass is freely falling along a geodesic with four-velocity u. Show that S satisfies the equation of motion

$$\nabla_u S^\rho = \eta^{\rho\beta\alpha\mu}\, u_\mu u^\sigma u^\lambda\, t_{\beta\nu}\, R^\nu{}_{\sigma\alpha\lambda},$$

where $t_{\beta\nu}$ is the "reduced quadrupole moment tensor"

$$t_{ij} = \int \rho \left(x^i x^j - \frac{1}{3}\delta_{ij} \right) d^3x$$

in the rest frame of the center of mass and $t^{\alpha\beta}u_\beta = 0$. It is assumed that the Riemann tensor is determined by an external field which is nearly constant over distances comparable to the size of the test body.

Exercise 1.23. Show that

$$H = \frac{1}{2} g^{\mu\nu}(\pi_\mu - eA_\mu)(\pi_\nu - eA_\nu)$$

is the Hamiltonian which describes the motion of a charged particle with charge e in a gravitational field (π_μ is the *canonical momentum*).

Exercise 1.24 (Maxwell Equations in a Static Spacetime). Let g be a static spacetime. In adapted coordinates

$$g = -\alpha^2\, dt^2 + g_{ik}\, dx^i dx^k\,, \tag{1.181}$$

where α (the **lapse**) and g_{ik} are independent of t. We introduce an orthonormal tetrad of 1-forms: $\theta^0 = \alpha\, dt$, $\{\theta^i\}$ an orthonormal triad for the spatial metric. Relative to this basis we decompose the Faraday 2-form as in SR

$$F = E \wedge \theta^0 + B\,, \tag{1.182}$$

where E is the electric 1-form $E = E_i\,\theta^i$ and B the magnetic 2-form $B = \frac{1}{2}B_{ij}\,\theta^i \wedge \theta^j$.

1. Show that the homogeneous Maxwell equation $dF = 0$ splits as

$$\boldsymbol{d}B = 0\,, \quad \boldsymbol{d}(\alpha E) + \partial_t B = 0\,, \tag{1.183}$$

where $\boldsymbol{d}$ is the 3-dimensional (spatial) exterior derivative.

2. Show that
$$*F = -H \wedge \theta^0 + D\,, \tag{1.184}$$
where
$$H = *B\,, \qquad D = *E\,. \tag{1.185}$$
The symbol $*$ denotes the Hodge dual for the spatial metric $g_{ik}\, dx^i dx^k$.
3. Decompose the current 1-form J as
$$J = \rho\,\theta^0 + j_k\,\theta^k \equiv \rho\,\theta^0 + j$$
and show that the inhomogeneous Maxwell equation $\delta F = -J$ splits as
$$\boldsymbol{\delta} E = 4\pi\rho\,, \quad \boldsymbol{\delta}(\alpha B) = \boldsymbol{\partial}_t E + 4\pi\alpha j,, \tag{1.186}$$
where $\boldsymbol{\delta}$ is the spatial codifferential.
4. Deduce the continuity equation
$$\partial_t\rho + \boldsymbol{\delta}(\alpha j) = 0\,. \tag{1.187}$$

2 Einstein's Field Equations

"There is something else that I have learned form the theory of gravitation: No collection of empirical facts, however comprehensive, can ever lead to the setting up of such complicated equations. A theory can be tested by experience, but there is no way from experience to the formulation of a theory. Equations of such complexity as are the equations of the gravitational field can be found only through the discovery of a logically simple mathematical condition which determines the equations completely or (at least) almost completely. Once one has those sufficiently strong formal conditions, one requires only little knowledge of facts for the setting up of a theory; in the case of the equations of gravitation it is the four-dimensionality and the symmetric tensor as expression for the structure of space which, together with the invariance concerning the continuous transformation-group, determine the equation almost completely."

—A. Einstein

(Autobiographical Notes, 1949)

In the previous chapter we examined the kinematical framework of the general theory of relativity and the effect of gravitational fields on physical systems. The hard core of the theory, however, consists of **Einstein's field equations**, which relate the metric field to matter. After a discussion of the physical meaning of the curvature tensor, we shall first give a simple physical motivation for the field equations and will then show that they are determined by only a few natural requirements[1].

2.1 Physical Meaning of the Curvature Tensor

In a local inertial system the "field strengths" of the gravitational field (the Christoffel symbols) can be transformed away. Due to its tensor character,

---

[1] For an excellent historical account of Einstein's struggle which culminated in the final form of his gravitational field equations, presented on November 25 (1915), we refer to A. Pais, [55]. Since the publication of this master piece new documents have been discovered, which clarified what happened during the crucial weeks in November 1915 (see, e.g., [56])

this is not possible for the curvature. Physically, the curvature describes the "tidal forces" as we shall see in the following. This was indicated already in the discussion of (1.179).

Consider a family of timelike geodesics, having the property that in a sufficiently small open region of the Lorentz manifold (M, g) precisely one geodesic passes through every point. Such a collection is called a **congruence of timelike geodesics**. It might represent dust or a swarm of freely falling bodies. The tangent field to this set of curves, with proper time s as curve parameter, is denoted by u. Note that $\langle u, u\rangle = -1$.

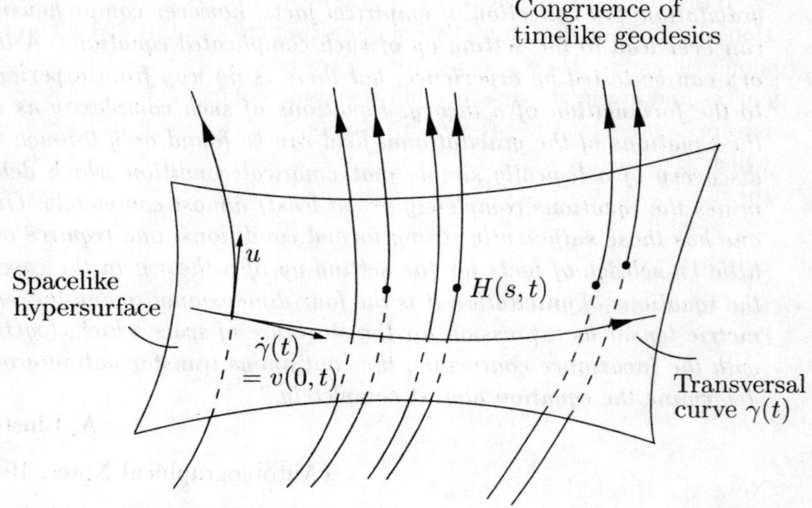

Fig. 2.1. Geodesic congruence.

Now let $\gamma(t)$ be a curve transversal to the congruence being investigated. This means that the tangent vector $\dot{\gamma}$ is never parallel to u in the region under consideration, as indicated in Fig. 2.1. Imagine that every point on the curve $\gamma(t)$ moves a distance s along the geodesics which passes through that point. Let the resulting point be $H(s,t)$. This defines a map $H : Q \longrightarrow M$ from some open region Q of $\mathbb{R}^2$ into M. For each t, the curve $s \longmapsto H(s,t)$ is a timelike geodesic with tangent vectors $u(H(s,t))$. Such 1-parameter variations of geodesics have been studied in DG, Sect. 14.4, using general results developed in Sects. 14.3 and 14.2. With the concept introduced there, we can say that u defines a vector field $u \circ H$ along the map H which is tangential

$$u \circ H = TH \circ \frac{\partial}{\partial s} \quad \text{or} \quad u\big(H(s,t)\big) = T_{(s,t)}H \cdot \frac{\partial}{\partial s}\,. \tag{2.1}$$

(In other words, u and $\partial/\partial s$ are H-related.) We shall use for $u \circ H$ the same letter u. Let v be the tangential vector field along the map H belonging to $\partial/\partial t$

$$v = TH \circ \frac{\partial}{\partial t}. \tag{2.2}$$

The vectors $v(s,t)$ are tangent to the curves $t \longmapsto H(s,t)$ (s fixed), and represent the separation of points which are moved with the same proper time along the neighboring geodesics of the congruence (beginning at arbitrary starting points).

Since $\partial/\partial s$ and $\partial/\partial t$ commute, the tangential fields u and v also commute, i.e., $[u,v]=0$, which is equivalent to the statement that v is invariant under the flow of the four-velocity field u; we also say that v is Lie-transported[2].

If one adds a multiple of u to v, the resulting vector represents the separation of points on the same two neighboring curves, but at different distances along the curve. However, we are interested only in the distance between the curves and not in the distance between particular points on these curves. This is represented by the projection of v on the subspace of the tangent space which is perpendicular to u. Thus the relevant (infinitesimal) separation vector n is

$$n = v + \langle v, u\rangle\, u. \tag{2.3}$$

Since $\langle u,u\rangle = -1$, n is indeed perpendicular to u. We now show that n is also Lie-transported. We have

$$L_u n = [u,n] = [u,v] + [u, \langle v,u\rangle\, u] = \left(u\langle v,u\rangle\right) u\,,$$

$$u\,\langle v,u\rangle = \frac{\partial}{\partial s}\langle v,u\rangle\,.$$

Now, the Ricci identity and $\langle u,u\rangle = -1$ imply[3]

$$0 = \frac{\partial}{\partial t}\langle u,u\rangle = 2\,\langle \nabla_v u, u\rangle\,.$$

Furthermore, since $\nabla_v u = \nabla_u v$ (because u and v commute and the torsion vanishes), it follows that

$$\frac{\partial}{\partial s}\langle u,v\rangle = \langle \nabla_u u, v\rangle + \langle u, \nabla_u v\rangle = \langle u, \nabla_v u\rangle = 0\,. \tag{2.4}$$

This shows that indeed

$$L_u n = 0\,. \tag{2.5}$$

Next, we consider

$$\nabla_u^2 v = \nabla_u \nabla_u v = \nabla_u(\nabla_v u) = (\nabla_u\nabla_v - \nabla_v\nabla_u)u\,.$$

[2] The reader is invited to generalize the proof of Theorem 11.11 (given in DG, Sect. 14.1) to vector fields along maps.

[3] To simplify notation, we use for the induced covariant derivative of fields along the map H the same letter ∇ (instead of $\tilde{\nabla}$ in the quoted sections of DG).

Due to $[u, v] = 0$, we obtain

$$\nabla_u^2 v = R(u, v)\, u \,. \tag{2.6}$$

This is called the **Jacobi equation** for the field v; see also DG, Sect. 14.4.

We now show that n also satisfies this equation. From (2.4) it follows that

$$\nabla_u n = \nabla_u v + \big(u\langle v, u\rangle\big)\, u + \langle v, u\rangle \nabla_u u = \nabla_u v \,. \tag{2.7}$$

Furthermore,

$$R(u, n)\, u = R(u, v)\, u + \langle v, u\rangle R(u, u)\, u = R(u, v)\, u \,. \tag{2.8}$$

From (2.6), (2.7) and (2.8) we see that n satisfies the Jacobi equation

$$\nabla_u^2 n = R(u, n)\, u \,. \tag{2.9}$$

The Jacobi field n is everywhere perpendicular to u. Physicists usually call this the **equation for geodesic deviation**. For a given u the right hand side defines at each point $p \in M$ a linear map $n \longmapsto R(u, n)\, u$ of the subspace of T_pM perpendicular to u. We might call this the tidal force operator. Its trace is equal to $-\mathrm{Ric}(u, u)$, as is shown below.

Now let $\{e_j\}$, for $j = 1, 2, 3$, be an orthonormal frame perpendicular to u, which is parallel transported along an arbitrary geodesic of the congruence (according to the discussion in Sect. 1.10, this is a *non-rotating* frame of reference). Since n is orthogonal to u, it has the expansion

$$n = n^i\, e_i \,. \tag{2.10}$$

Note that

$$\nabla_u n = (u\, n^i) e_i + n^i \nabla_u e_i = \frac{dn^i}{ds} e_i \,.$$

The triad $\{e_j\}$, together with $e_0 = u$ form a tetrad along the chosen geodesic. From (2.9) we obtain

$$\frac{d^2 n^i}{ds^2} e_i = n^j R(e_0, e_j)\, e_0 = n^j R^i{}_{00j} e_i$$

or

$$\frac{d^2 n^i}{ds^2} = R^i{}_{00j} n^j \,. \tag{2.11}$$

If we set $K_{ij} = R^i{}_{00j}$, Equation (2.11) can be written in matrix form

$$\frac{d^2}{ds^2} \boldsymbol{n} = K\, \boldsymbol{n} \,, \tag{2.12}$$

where $\boldsymbol{n} = (n^1, n^2, n^3)$ and $K = (K_{ij})$. Note that $K = K^T$ and

$$\operatorname{Tr} K = R^i{}_{00i} = R^\mu{}_{00\mu} = -R_{00} = -R_{\mu\nu}\, u^\mu u^\nu \,. \tag{2.13}$$

Equation (2.11) (or (2.12)) describes the relative acceleration ("tidal forces") of neighboring freely falling test bodies.

Comparison with Newtonian Theory

The paths of two neighboring test bodies, denoted by $x^i(t)$ and $x^i(t) + n^i(t)$, satisfy the equation of motion

$$\ddot{x}^i(t) = -\left(\frac{\partial \phi}{\partial x^i}\right)_{\boldsymbol{x}(t)},$$

$$\ddot{x}^i(t) + \ddot{n}^i(t) = -\left(\frac{\partial \phi}{\partial x^i}\right)_{\boldsymbol{x}(t)+\boldsymbol{n}(t)}.$$

If we take the difference of these two equations and retain only terms linear in $\boldsymbol{n}$, we obtain

$$\ddot{n}^i(t) = -\frac{\partial^2 \phi}{\partial x^i \partial x^j} n^j(t). \tag{2.14}$$

Thus we again find (2.12), but with

$$K = -\frac{\partial^2 \phi}{\partial x^i \partial x^j}. \tag{2.15}$$

Note that $\operatorname{Tr} K = -\Delta\phi$, whence in matter-free space $\operatorname{Tr} K = 0$. As we shall see, this is also true in GR. We thus have the correspondence

$$R^i{}_{0j0} \longleftrightarrow \frac{\partial^2 \phi}{\partial x^i \partial x^j}. \tag{2.16}$$

In particular,

$$R_{\mu\nu}\, u^\mu u^\nu \longleftrightarrow \Delta\phi. \tag{2.17}$$

Summary. ***Variations*** *of the gravitational "field strengths" are described by the Riemann tensor. The tensor character of this quantity implies that such inhomogeneities, unlike the field strengths* $\Gamma^\mu{}_{\alpha\beta}$*, can not be transformed away.* ***Relative accelerations*** *or* ***tidal forces*** *of freely falling test bodies are described by the Riemann tensor, via the equation of geodesic deviation (2.9) or (2.11) for an adapted tetrad.*

These remarks are also important for understanding the effect of a gravitational wave on a mechanical detector (see Sect. 4.3).

* * *

Exercises.

Exercise 2.1. Use the results of Sect. 1.5 to show that in the Newtonian limit the components $R^i{}_{0j0}$ of the curvature tensor are given approximately by $\partial_i\partial_j\phi$.

Exercise 2.2. Write the Newtonian equations of motion for a potential as a geodesic equation in a four dimensional affine space. Compute the Christoffel symbols and the Riemann tensor, and show that the affine connection is not metric.

Exercise 2.3. Generalize the considerations of this section to an arbitrary congruence of timelike curves (replace parallel transport everywhere by Fermi transport). Derive, in particular, the following generalization of Equation (2.9)

$$\boldsymbol{F}_u^2 n = R(u,n)\,u + (\nabla_n a)_\perp + \langle a, n\rangle\, a\,,$$

where $a = \nabla_u u$ and $\perp$ denotes the projection orthogonal to u. (Important consequences of this equation, in particular the *Landau-Raychaudhuri equation*, are studied, for instance, in [8].)

Exercise 2.4. Consider a static Lorentz manifold (M, g). We known (see Sect. 1.9) that this is locally isometric to a direct product manifold $\mathbb{R} \times \Sigma$ with metric

$$g = -\varphi^2\, dt^2 + h\,, \tag{2.18}$$

where h is a Riemannian metric on Σ and φ is a smooth function on Σ. (More accurately, one should write in (2.18) $\pi^* h$ instead of h, where π denotes the projection from $\mathbb{R}\times\Sigma$ to the second factor.) Show that the function φ satisfies the equation

$$\frac{1}{\varphi}\bar{\Delta}\varphi = R_{\mu\nu}\, u^\mu u^\nu\,, \tag{2.19}$$

where $\bar{\Delta}$ denotes the Laplacian on (Σ, h), and u^μ is the four-velocity field

$$u = \frac{1}{\varphi}\frac{\partial}{\partial t}\,. \tag{2.20}$$

Solution. We first give some hints for an elementary derivation, and then show how the result can be obtained in a straightforward manner with the help of Cartan's structure equations (see DG, Sect. 13.7).

Hints. Use the Killing equation for the Killing field $K = \partial/\partial t$ and $d(\frac{K^\flat}{\varphi^2}) = 0$ to show first that $\varphi\, u_{\alpha;\beta} = -u_\beta\, \varphi_{,\alpha}$. This implies that $a = \nabla_u u$ is given by $a_\alpha = -(\frac{1}{\varphi})\varphi_{,\alpha}$ and that $u_{\alpha;\beta} = a_\alpha u_\beta$. Then express the divergence of a_α in two different ways.

We introduce an orthonormal tetrad $\{\theta^\mu\}$ on $\mathbb{R}\times\Sigma$ consisting of $\theta^0 = \varphi\, dt$ and an orthonormal triad $\{\theta^i\}$, for $i = 1, 2, 3$, of (Σ, h). The connection forms of (Σ, h) relative to $\{\theta^i\}$ are denoted by $\bar{\omega}^i{}_j$ and those of (M, g) relative to $\{\theta^\mu\}$ by $\omega^\mu{}_\nu$. Bars always refer to quantities on (Σ, h). (We identify, for instance, θ^i with the pull-back $\pi^*\theta^i$.)

First, we determine the connection forms $\omega^\mu{}_\nu$. Comparison of

$$d\theta^0 = d\varphi \wedge dt = \frac{1}{\varphi} d\varphi \wedge \theta^0 \equiv \frac{1}{\varphi} \varphi_{,i}\, \theta^i \wedge \theta^0 \tag{2.21}$$

with the first structure equation (see DG, Eq. (13.70))

$$d\theta^0 = -\omega^0{}_i \wedge \theta^i \tag{2.22}$$

invites us to guess that

$$\omega^0{}_i = \frac{1}{\varphi} \varphi_{,i}\, \theta^0 , \qquad \omega^i{}_j = \bar{\omega}^i{}_j . \tag{2.23}$$

This does not only satisfy (2.22), but also the other components of the first structure equations

$$d\theta^i + \omega^i{}_0 \wedge \theta^0 + \omega^i{}_j \wedge \theta^j = 0 ,$$

thanks to the first structure equations on (Σ, h).

Next, we compute the curvature forms. We have

$$\begin{aligned} \Omega^0_i &= d\omega^0{}_i + \omega^0{}_j \wedge \omega^j{}_i \\ &= d\left(\frac{1}{\varphi}\varphi_{,i}\right) \wedge \theta^0 + \frac{1}{\varphi}\varphi_{,i}\frac{1}{\varphi}\varphi_{,j}\, \theta^j \wedge \theta^0 + \frac{1}{\varphi}\varphi_{,j}\, \theta^0 \wedge \bar{\omega}^j{}_i \\ &= \frac{1}{\varphi}(d\varphi_{,i} - \bar{\omega}^j{}_i \varphi_{,j}) \wedge \theta^0 . \end{aligned}$$

The expression in the bracket of the last line is equal to the covariant derivative $\bar{\nabla}$ of $\varphi_{,i}$, i.e., equal to $\bar{\nabla}_j \bar{\nabla}_i \varphi \wedge \theta^j$. Thus, we have

$$\Omega^0_i = \frac{1}{2} R^0{}_{i\alpha\beta}\, \theta^\alpha \wedge \theta^\beta = \frac{1}{\varphi} \bar{\nabla}_j \bar{\nabla}_i \varphi\, \theta^j \wedge \theta^0 . \tag{2.24}$$

From this we obtain

$$\begin{aligned} R^0{}_{ij0} &= R^i{}_{0j0} = \frac{1}{\varphi} \bar{\nabla}_j \bar{\nabla}_i \varphi , \\ R^0{}_{ijk} &= R^i{}_{0jk} = 0 , \end{aligned} \tag{2.25}$$

and hence for the Ricci tensor

$$R_{00} = \frac{1}{\varphi} \bar{\Delta}\varphi , \qquad R_{0i} = 0 . \tag{2.26}$$

But $R_{00} = R_{\mu\nu}\, u^\mu u^\nu$ (relative to the base $\{\theta^\mu\}$ and its dual $\{e_\nu\}$, $u = e_0$). Therefore, the first equation of (2.26) agrees with the claim (2.19).

For later use, we also work out the other components of the Ricci tensor. From Equation (2.23) we obtain

$$\Omega^i{}_j = d\omega^i{}_j + \omega^i{}_0 \wedge \omega^0{}_j + \omega^i{}_k \wedge \omega^k{}_j = d\bar{\omega}^i{}_j + \bar{\omega}^i{}_k \wedge \bar{\omega}^k{}_j = \bar{\Omega}^i{}_j . \tag{2.27}$$

This implies in particular $R^i{}_{jkl} = \bar{R}^i{}_{jkl}$. Hence, with (2.25),

$$R_{ij} = R^\lambda{}_{i\lambda j} = R^0{}_{i0j} + R^k{}_{ikj} = -\frac{1}{\varphi}\bar{\nabla}_j\bar{\nabla}_i\varphi + \bar{R}^k{}_{ikj}$$

or

$$R_{ij} = -\frac{1}{\varphi}\bar{\nabla}_j\bar{\nabla}_i\varphi + \bar{R}_{ij}\,. \tag{2.28}$$

From (2.26) and (2.28) we obtain for the Riemann scalar

$$R = -2\frac{1}{\varphi}\bar{\Delta}\varphi + \bar{R}\,. \tag{2.29}$$

Relative to any basis which is adapted to the split (2.18), the Einstein tensor is found to be

$$\begin{aligned} G_{ij} &= \bar{G}_{ij} - \frac{1}{\varphi}(\bar{\nabla}_j\bar{\nabla}_i\varphi - h_{ij}\bar{\Delta})\,, \\ G_{0i} &= 0\,, \\ G_{\mu\nu}\,u^\mu u^\nu &= \frac{1}{2}\bar{R}\,. \end{aligned} \tag{2.30}$$

2.2 The Gravitational Field Equations

"You will be convinced of the general theory of relativity once you have studied it. Therefore I am not going to defend it with a single word."

—A. Einstein

(On a postcard to A. Sommerfeld, Feb. 8, 1916)

In the absence of true gravitational fields one can always find a coordinate system in which $g_{\mu\nu} = \eta_{\mu\nu}$ everywhere. This is the case when the Riemann tensor vanishes (modulo global questions, as discussed in Sect. 13.9 of DG). One might call this fact the "*zeroth law of gravitation*". In the presence of gravitational fields, spacetime will become curved. The field equations must describe the dependence of the curvature of the metric field on the energy-momentum distributions in terms of partial differential equations.

Heuristic "Derivation" of the Field Equations

As a starting point for the arguments which will lead to the field equations we recall from Sect. 1.5 that for a weak, quasi-stationary field generated by a non-relativistic mass distribution, the component g_{00} of the metric tensor (relative to a suitable coordinate system which is nearly Lorentzian on a global scale) is related to the Newtonian potential by

$$g_{00} \simeq -(1 + 2\phi) . \tag{2.31}$$

The Newtonian potential ϕ satisfies the **Poisson equation**

$$\Delta\phi = 4\pi G \rho . \tag{2.32}$$

We have seen that the left-hand side of (2.32) is equal to minus the trace of the tidal tensor (2.15). According to (2.17) this corresponds to $R_{\mu\nu}\, u^\mu u^\nu$ in GR for a freely falling observer with four-velocity u^μ. On the other hand, ρ in the Poisson equation is approximatively equal to $T_{\mu\nu}\, u^\mu u^\nu$ and $T = T^\mu{}_\mu$ under Newtonian conditions. So the GR-analogue of Poisson's equation is

$$R_{\mu\nu}\, u^\mu u^\nu = 4\pi G\Big(\lambda T_{\mu\nu} - (1-\lambda)\, g_{\mu\nu} T\Big) u^\mu u^\nu ,$$

where λ is a constant to be determined. Since this should hold for all observers, we are led to the equation

$$R_{\mu\nu} = 4\pi G\Big(\lambda T_{\mu\nu} - (1-\lambda)\, g_{\mu\nu} T\Big) .$$

Using the trace of this equation one finds that it is equivalent to

$$G_{\mu\nu} = 4\pi G\Big(\lambda T_{\mu\nu} - \frac{1}{2}(2-\lambda)\, g_{\mu\nu} T\Big) , \tag{2.33}$$

where $G_{\mu\nu}$ is the Einstein tensor (see DG, Eq. (13.53))

$$G_{\mu\nu} = R_{\mu\nu} - \frac{1}{2} g_{\mu\nu} R , \qquad R = R^\mu{}_\mu . \tag{2.34}$$

From Sect. 1.4 we know that the energy-momentum tensor must satisfy $\nabla_\nu T^{\mu\nu} = 0$. If we use this in (2.34) and also the reduced Bianchi identity

$$\nabla_\nu G^{\mu\nu} = 0 , \tag{2.35}$$

we conclude that for $\lambda \neq 2$ the trace T would have to be constant. This unphysical condition is avoided for $\lambda = 2$. We are thus naturally led to **Einstein's field equations**

$$G_{\mu\nu} = 8\pi G\left(T_{\mu\nu} - \frac{1}{2} g_{\mu\nu} T\right) . \tag{2.36}$$

This final form is contained in Einstein's conclusive paper [58] submitted on 25 November 1915[4].

[4] The story that D. Hilbert found the definite equations a few days earlier is now known to be wrong (see [61]). Archival material reveals that Hilbert revised the proofs of his paper [62] in a major way, *after* he had seen Einstein's publication from November 25. The theory he originally submitted was not generally covariant and contained no explicit field equations. It is indeed strange that Hilbert made no indications of his revisions, and kept the submission date 20 November 1915. This is documented in [61].

Remark. Let us consider a static solution of the Einstein equations for an ideal fluid at rest. Using the notation in Exercise 2.4 the fluid four-velocity is $u = \frac{1}{\varphi}\partial_t$, and we obtain as an exact relation the following GR-version of Poisson's equation

$$\frac{1}{\varphi}\bar{\Delta}\varphi = 4\pi G(\rho + 3p)\,. \tag{2.37}$$

Note the combination $\rho + 3p$ as the source of the potential φ.

The Question of Uniqueness

> *"A chief attraction of the theory lies in its logical completeness. If a single one of the conclusions drawn from it proves wrong, it must be given up; to modify it without destroying the whole structure seems to be impossible ."* —A. Einstein (1919)

Now that we have motivated the field equations with simple physical arguments, we must ask how uniquely they are determined. The following investigation will show that one has very little freedom.

It is natural to postulate that the ten potentials $g_{\mu\nu}$ satisfy equations of the form

$$\mathcal{D}_{\mu\nu}[g] = T_{\mu\nu}\,, \tag{2.38}$$

where $\mathcal{D}_{\mu\nu}[g]$ is a tensor field constructed pointwise from $g_{\mu\nu}$ and its first and second derivatives. Furthermore, this tensor valued local functional should satisfy the identity

$$\nabla_\nu \mathcal{D}^{\mu\nu} = 0\,, \tag{2.39}$$

in order to guarantee that $\nabla_\nu T^{\mu\nu} = 0$ is a consequence of the field equations.

Remarks.

1. The requirement that the field equations contain derivatives of $g_{\mu\nu}$ only up to second order is certainly reasonable. If this were not the case, one would have to specify for the **Cauchy problem** the initial values not only for the metric and its first derivative, but also higher derivatives on a spacelike surface, in order to determine the development of the metric.
2. Since $\nabla_\nu T^{\mu\nu} = 0$ is also a consequence of the non-gravitational laws (see Sect. 1.4), the coupled system of equations contains *four* **identities**. As we shall later discuss in more detail, general invariance of the theory implies this. Indeed, general invariance is a kind of gauge invariance: Solutions of the coupled dynamical system - consisting of gravitational and matter equations - that can be transformed into each other by a diffeomorphism, are physically equivalent. Since the diffeomorphism group can (locally) be described by four smooth functions, the existence of four identities in the coupled system is necessary. This is completely analogous to the situation in gauge field theories. (More on this in Sect. 2.3.5.) On the linearized level, this was already discussed in Sect. 1.2 (see, in particular, Eqs. (1.10) and (1.14)).

It is very satisfying that in four spacetime dimensions the following theorem holds:

Theorem 2.1 (D. Lovelock). *A tensor $\mathcal{D}_{\mu\nu}[g]$ with the required properties is in four dimensions a linear combination of the metric and the Einstein tensor*

$$\mathcal{D}_{\mu\nu}[g] = a\,G_{\mu\nu} + b\,g_{\mu\nu}\,, \tag{2.40}$$

where $a, b \in \mathbb{R}$.

Remarks. This theorem was established relatively late (see [78]). Previously one had made the additional assumption that the tensor $\mathcal{D}_{\mu\nu}[g]$ was linear in the *second* derivatives. It is remarkable that it is unnecessary to postulate linearity in the second derivatives only in four dimensions[5]. It is also by no means trivial that it is not necessary to require the symmetry of $\mathcal{D}_{\mu\nu}$. The theorem thus shows that a gravitational field satisfying (2.38) cannot be coupled to a non-symmetric energy-momentum tensor.

We first prove the following

Theorem 2.2. *Let $K[g]$ be a functional which assigns to every smooth pseudo-Riemannian metric field g a tensor field K (of arbitrary type) such that the components of $K[g]$ at the point $p \in M$ depend smoothly on g_{ij} and its derivatives $g_{ij,k}$, $g_{ij,kl}$ at p. Then K is determined pointwise by g and the curvature tensor by elementary operations of tensor algebra.*

Proof. In normal coordinates (see Sect. 13.3 of DG) the geodesics are straight lines $x^i(s) = s\,a^i$. Hence,

$$\Gamma^k{}_{ij}(s\,a^m)a^i\,a^j = 0\,, \tag{2.41}$$

with arbitrary constants a^i. Since Γ is symmetric, we have $\Gamma^k{}_{ij}(0) = 0$, implying $g_{ij,k} = 0$. Furthermore, at $x = 0$ we can bring g_{ij} in normal form: $g_{ij}(0) = \varepsilon_i\,\delta_{ij}$, with $\varepsilon_i = \pm 1$. In a neighborhood of $x = 0$ we may then expand in powers of x^i

$$g_{ik} = \varepsilon_i\,\delta_{ij} + \frac{1}{2}\beta_{ik,rs}\,x^r x^s + \dots\,.$$

Thus, if $\Gamma_{ikr} := g_{ij}\Gamma^j{}_{kr}$,

$$\begin{aligned}\Gamma_{ikr} &= \frac{1}{2}\left(g_{ik,r} + g_{ir,k} - g_{kr,i}\right)\\ &= \frac{1}{2}\left(\beta_{ik,rs}\,x^s + \beta_{ir,ks}\,x^s - \beta_{kr,is}\,x^s\right) + \dots\,.\end{aligned} \tag{2.42}$$

From (2.41) we have $\Gamma^k{}_{ij}(x)x^i x^j = 0$ and hence

[5] In higher dimensions additional terms are possible. We shall state Lovelock's general result at the end of Sect. 2.6.

$$\Gamma_{ikr}(x)x^k x^r = 0\,. \tag{2.43}$$

If we insert (2.42) into (2.43) we obtain

$$(\beta_{ik,rs} + \beta_{ir,ks} - \beta_{kr,is})x^k x^r x^s = 0\,. \tag{2.44}$$

We shall use the abbreviation

$$\beta_{ik,rs} = \left(\frac{ik}{rs}\right) = \left.\frac{\partial^2 g_{ik}}{\partial x^r \partial x^s}\right|_{x=0}\,. \tag{2.45}$$

The x^k in (2.44) are arbitrary; hence the contribution to the coefficients which is symmetric in (k, r, s) must vanish. In other words, the following cyclic sum must vanish

$$\sum_{(k,r,s)} \left[\left(\frac{ik}{rs}\right) + \left(\frac{ir}{ks}\right) - \left(\frac{kr}{is}\right)\right] = 0$$

or

$$2\left[\left(\frac{ik}{rs}\right) + \left(\frac{ir}{sk}\right) + \left(\frac{is}{kr}\right)\right] - \left(\frac{rs}{ik}\right) - \left(\frac{ks}{ir}\right) - \left(\frac{kr}{is}\right) = 0\,. \tag{2.46}$$

The negative terms are obtained from the positive ones by interchanging numerators and denominators. Instead of i, one can distinguish k or r

$$2\left[\left(\frac{ik}{rs}\right) + \left(\frac{kr}{si}\right) + \left(\frac{ks}{ir}\right)\right] - \left(\frac{rs}{ik}\right) - \left(\frac{is}{kr}\right) - \left(\frac{ir}{ks}\right) = 0,$$

$$2\left[\left(\frac{rk}{is}\right) + \left(\frac{ir}{sk}\right) + \left(\frac{rs}{ki}\right)\right] - \left(\frac{is}{rk}\right) - \left(\frac{ks}{ir}\right) - \left(\frac{ki}{rs}\right) = 0\,.$$

If we add these three equations, we obtain

$$\left(\frac{ik}{rs}\right) + \left(\frac{ri}{ks}\right) + \left(\frac{kr}{is}\right) = 0\,. \tag{2.47}$$

Finally, interchange i and s in (2.46); the result is

$$2\left[\left(\frac{sk}{ri}\right) + \left(\frac{sr}{ik}\right) + \left(\frac{si}{kr}\right)\right] - \underbrace{\left(\frac{ri}{sk}\right) - \left(\frac{ki}{sr}\right) - \left(\frac{kr}{is}\right)}_{=0} = 0\,.$$

If we now again interchange i and s, we obtain

$$\left(\frac{ik}{rs}\right) + \left(\frac{ir}{sk}\right) + \left(\frac{is}{kr}\right) = 0\,. \tag{2.48}$$

A comparison with (2.47) yields

$$\begin{pmatrix} i\,s \\ \hline k\,r \end{pmatrix} = \begin{pmatrix} k\,r \\ \hline i\,s \end{pmatrix} . \tag{2.49}$$

From the general expression for the Riemann tensor (see DG, Eq.(13.31)), one obtains in normal coordinates

$$R_{iklm} = \frac{1}{2}\left[\begin{pmatrix} i\,m \\ \hline k\,l \end{pmatrix} + \begin{pmatrix} k\,l \\ \hline m\,i \end{pmatrix} - \begin{pmatrix} i\,l \\ \hline k\,m \end{pmatrix} - \begin{pmatrix} k\,m \\ \hline i\,l \end{pmatrix}\right] \stackrel{(2.49)}{=} \begin{pmatrix} i\,m \\ \hline k\,l \end{pmatrix} - \begin{pmatrix} i\,l \\ \hline k\,m \end{pmatrix}$$

Using this and (2.48) one obtains

$$\begin{aligned} R_{iklm} + R_{lkim} &= \begin{pmatrix} i\,m \\ \hline k\,l \end{pmatrix} - \begin{pmatrix} i\,l \\ \hline k\,m \end{pmatrix} + \begin{pmatrix} l\,m \\ \hline k\,i \end{pmatrix} - \begin{pmatrix} i\,l \\ \hline k\,m \end{pmatrix} \\ &\stackrel{(2.49)}{=} \begin{pmatrix} i\,m \\ \hline k\,l \end{pmatrix} - \begin{pmatrix} k\,i \\ \hline l\,m \end{pmatrix} - 2\begin{pmatrix} i\,l \\ \hline k\,m \end{pmatrix} \stackrel{(2.48)}{=} -3\begin{pmatrix} i\,l \\ \hline k\,m \end{pmatrix} . \end{aligned}$$

Hence, in normal coordinates

$$\begin{aligned} g_{ik} &= \varepsilon_i \delta_{ij} - \frac{1}{6}\left(R_{irks} + R_{kris}\right)x^r x^s + \dots , \\ g_{ik} &= \varepsilon_i\,\delta_{ij} - \frac{1}{3} R_{irks} x^r x^s + \dots . \end{aligned} \tag{2.50}$$

Thus R_{irks} expressed in terms of normal coordinates determines all the second derivatives of g_{ik} at p. The first derivatives vanish. Since R_{irks} is a tensor, the statement of the theorem follows immediately. □

We now prove Theorem 2.1 under the *additional* assumption that $\mathcal{D}_{\mu\nu}[g]$ is *linear* in the second derivatives. (For the generalization we refer once more to [78]. Unfortunately, the proof of Lovelock's theorem is too complicated to be reproduced here.)

Proof (Partial proof of Theorem 2.1). Since the $R^{\mu}{}_{\alpha\beta\gamma}$ are linear in the second derivatives of the $g_{\mu\nu}$, Theorem 2.2 implies that $\mathcal{D}_{\mu\nu}[g]$ must have the following form

$$\mathcal{D}_{\mu\nu}[g] = c_1\,R_{\mu\nu} + c_2\,R\,g_{\mu\nu} + c_3\,g_{\mu\nu} ,$$

with constants c_1, c_2 and c_3. As a result of the contracted Bianchi identity $G^{\mu\nu}{}_{;\nu} = 0$, we get the restriction

$$\mathcal{D}^{\mu\nu}{}_{;\nu} = \left(c_2 + \frac{1}{2}c_1\right)\left(g_{\mu\nu}R\right)_{;\nu} = \left(c_2 + \frac{1}{2}c_1\right) g_{\mu\nu} R_{,\nu} = 0 .$$

Hence $c_2 + \frac{1}{2}c_1 = 0$, and $\mathcal{D}_{\mu\nu}$ must be of the form (2.40). □

Theorem 2.1 implies that the field equations having the structure (2.38), with the required properties for $\mathcal{D}_{\mu\nu}[g]$, must be of the form

$$G_{\mu\nu} = \kappa\, T_{\mu\nu} - \Lambda\, g_{\mu\nu}\,, \tag{2.51}$$

where Λ and κ are two constants. The field equations (2.51) are highly *non-linear* partial differential equations, even in vacuum. This must be so, since every form of energy, including the "energy of the gravitational field", is a source of gravitational fields. This non-linearity makes a detailed analysis of the field equations extremely difficult. Fortunately, one has found a number of physically relevant exact solutions. Beside this, qualitative insight has been gained, for example, by the study of the initial value problem (see Sect. 2.8). As in any other field of physics, approximation methods allow us to treat relevant problems, such as binaries of compact objects (see Chapters 4 and 5). And, of course, numerical methods are playing an increasingly important role, especially for the investigation of highly dynamical processes.

Remarks.

1. The energy-momentum tensor $T^{\mu\nu}$ must be obtained from a theory of matter (just as is the case for the current density j^μ in Maxwell's equations). We shall discuss the construction of $T^{\mu\nu}$ (and j^μ) in the framework of a Lagrangian field theory in Sect. 2.3. In practical applications one often uses simple phenomenological models (such as ideal or viscous fluids).
2. We shall obtain the field equations in the next section from a variational principle.

Newtonian Limit, Interpretation of the Constants Λ and κ

The field equations (2.51) contain two constants Λ and κ. In order to interprete them, we consider the Newtonian limit. We need

$$R_{\mu\nu} = \partial_\lambda \Gamma^\lambda{}_{\nu\mu} - \partial_\nu \Gamma^\lambda{}_{\lambda\mu} + \Gamma^\sigma{}_{\nu\mu}\Gamma^\lambda{}_{\lambda\sigma} - \Gamma^\sigma{}_{\lambda\mu}\Gamma^\lambda{}_{\nu\sigma}\,.$$

As in Sect. 1.5, we can introduce a nearly Lorentzian system for weak, quasi-stationary fields, in which

$$g_{\mu\nu} = \eta_{\mu\nu} + h_{\mu\nu}\,, \qquad |h_{\mu\nu}| \ll 1 \tag{2.52}$$

and $h_{\mu\nu}$ is almost independent of time. We may neglect the quadratic terms $\Gamma\Gamma$ of the Ricci tensor,

$$R_{\mu\nu} \simeq \partial_\lambda \Gamma^\lambda{}_{\mu\nu} - \partial_\nu \Gamma^\lambda{}_{\lambda\mu}\,.$$

Since the field is quasi-stationary, we have $R_{00} \simeq \partial_l \Gamma^l{}_{00}$ and $\Gamma^l{}_{00} \simeq \frac{1}{2}\partial_l g_{00}$. In the Newtonian limit, we have $g_{00} \simeq -(1+2\phi)$. We thus find, as already suggested by (2.17),

$$R_{00} \simeq \Delta\phi \,. \tag{2.53}$$

Equation (2.51) is equivalent to

$$R_{\mu\nu} = 8\pi G \left(T_{\mu\nu} - \frac{1}{2} g_{\mu\nu} T \right) + \Lambda g_{\mu\nu} \,, \tag{2.54}$$

with $\kappa = 8\pi G$. For non-relativistic matter $|T_{ij}| \ll |T_{00}|$, and hence we obtain for $\Lambda = 0$ the Poisson equation (2.32) for the $(0,0)$ component of (2.54), using Equation (2.53). A small non-zero Λ in (2.51) leads to

$$\Delta\phi = 4\pi G \rho - \Lambda \,. \tag{2.55}$$

Note that instead of (2.37), we obtain now as an *exact* equation for $\varphi = \sqrt{-g_{00}} \simeq 1 + \phi$

$$\frac{1}{\varphi} \bar{\Delta} \varphi = 4\pi G (\rho + 3\,p) - \Lambda \,. \tag{2.56}$$

On the Cosmological Constant Λ

We shall now say more on the mysterious Λ-term in (2.51). This term was introduced by Einstein when he applied GR for the first time to cosmology. In his influential paper [63] of 1917 he proposed a **static** cosmological model which is **spatially closed**. His assumption that the Universe is quasistatic was not unreasonable at the time, because the relative velocities of stars as observed were small. The completely novel assumption that space is closed had to do with Machian ideas about inertia[6].

More precisely, Einstein's cosmological model (M, g) is

$$M = \mathbb{R} \times S_a^3 \,, \qquad g = -dt^2 + h \,, \tag{2.57}$$

where S_a^3 is the 3-sphere with radius a and h its standard metric. The Ricci and Einstein tensors for the metric (2.57) can be obtained by specializing the results of Exercise 2.4. We need only the time-time components, and obtain

$$R_{00} = 0 \,, \qquad G_{00} = \frac{1}{2} \bar{R} \,, \tag{2.58}$$

---

[6] In these early years, Einstein believed that GR should satisfy what he named Mach's principle, in the sense that the metric field should be determined uniquely by the energy-momentum tensor (see Einstein's paper [63]). His intention was to eliminate all vestiges of absolute space. For a detailed discussion we refer to the last part of his Princeton lectures [66]. Only later Einstein came to realize that the metric field is not an epiphenomenon of matter, but has an independent existence, and his enthusiasm for Mach's principle decreased. In a letter to F. Pirani he wrote in 1954: "As a matter of fact, one should no longer speak of Mach's principle at all." For more on this, see [55].

where $\bar{R}$ is the Riemann scalar of S_a^3. For this space of constant curvature one finds (see Exercise 2.8) $\bar{R} = \frac{6}{a^2}$, thus

$$G_{00} = \frac{3}{a^2} . \tag{2.59}$$

With these geometrical results we go into Einstein's field equation. Because of the symmetries of the model, the energy-momentum tensor must have the following diagonal form

$$(T_{\mu\nu}) = \mathrm{diag}(\rho, p, p, p) . \tag{2.60}$$

From (2.36) we obtain

$$\frac{3}{a^2} = 8\pi G\rho + \Lambda . \tag{2.61}$$

On the other hand, (2.54) gives

$$0 = 4\pi G(\rho + 3p) - \Lambda . \tag{2.62}$$

Einstein first tried to solve his original field equations from November 1915 (without the Λ-term). Then, however, (2.62) cannot be satisfied for $p \geq 0$ (e.g., for dust). It was this failure that led him to introduce the Λ-term as a possibility which is compatible with the principles of GR. Indeed, as we have seen, nothing forbids the presence of this term. With its inclusion the two equations (2.61) and (2.62) can be satisfied and lead for dust ($p = 0$) to the two basic relations

$$4\pi G\rho = \frac{1}{a^2} = \Lambda \tag{2.63}$$

of Einstein's cosmological model. For obvious reasons, the Λ-term is often called the **cosmological term**. Because of recent exciting cosmological observations, the **cosmological constant** Λ has again become a central issue, both for cosmology and particle physics. Before indicating that there is a deep mystery connected with the cosmological constant, we rewrite Einstein's equations in the following suggestive manner

$$G_{\mu\nu} = 8\pi G\left(T_{\mu\nu} + T_{\mu\nu}^{\Lambda}\right) , \tag{2.64}$$

with

$$T_{\mu\nu}^{\Lambda} = \frac{\Lambda}{8\pi G} g_{\mu\nu} .$$

This has the form of the energy-momentum tensor of an ideal fluid with energy density ρ_Λ and pressure p_Λ given by

$$\rho_\Lambda = \frac{\Lambda}{8\pi G}, \qquad p_\Lambda = -\rho_\Lambda . \tag{2.65}$$

Also a positive Λ corresponds to a positive ρ_Λ, it has a repulsive effect. This becomes manifest if we write (2.56) as

$$\frac{1}{\varphi}\Delta\varphi = 4\pi G\Big((\rho + 3\,p) + \underbrace{(\rho_\Lambda + 3\,p_\Lambda)}_{-2\,\rho_\Lambda}\Big)\,. \tag{2.66}$$

Since the energy density and the pressure appear in the combination $\rho + 3\,p$, we understand that a positive ρ_Λ leads to a repulsion. In the Newtonian limit we have $\varphi \simeq 1 + \phi$ and $p \ll \rho$, hence we obtain the modified Poisson equation

$$\Delta\phi = 4\pi G(\rho - 2\,\rho_\Lambda)\,. \tag{2.67}$$

Without gravity, we do not care about the absolute energy of the vacuum, because only energy *differences* matter. However, when we consider the coupling to gravity, the vacuum energy ρ_{vac} acts like a cosmological constants. In order to see this, first consider the vacuum expectation value of the energy-momentum tensor operator in Minkowski spacetime. Since the vacuum state is Lorentz invariant, this expectation value is an invariant symmetric tensor, hence proportional to the metric tensor. For a curved metric we expect that this is still the case, up to higher curvature terms

$$\langle T_{\mu\nu}\rangle_{vac} = g_{\mu\nu}\rho_{vac} + \text{higher curvature terms}\,. \tag{2.68}$$

The *effective* cosmological constant, which controls the large scale behavior of the Universe, is given by

$$\Lambda = 8\pi G\rho_{vac} + \Lambda_0\,, \tag{2.69}$$

where Λ_0 is a bare cosmological constant in Einstein's field equations.

We know from astronomical observations that ρ_Λ can not be larger than about the so-called critical density

$$\begin{aligned}\rho_{crit} &= \frac{3\,H_0^2}{8\pi G} = 1.88\times 10^{-9} h_0^2\, g\, cm^{-3}\\ &= 8\times 10^{-47} h_0^2\, GeV^4\,,\end{aligned} \tag{2.70}$$

where H_0 is the Hubble parameter and $h_0 = H_0/(100\, km\, s^{-1}\, Mpc^{-1})$. Observationally, $h_0 \simeq 0.6 - 0.7$.

It is complete mystery as to why the two terms in (2.69) should almost exactly cancel. This is the famous **Λ-problem**. It is true that we are unable to calculate the vacuum energy density in quantum field theories, like the Standard Model of particle physics. But we can attempt to make what appear to be reasonable order-of-magnitude estimates for the various contributions. These expectations turn out to be in gigantic conflict with the facts. For a detailed discussion of this and extensive references, see [79].

The Einstein-Fokker Theory

One can obtain field equations different from those of Einstein if one limits a priori the degrees of freedom of the metric tensor (beyond the requirement that the signature be Lorentzian). For example, one might require that the metric be **conformally flat**, i.e., that g has the form

$$g = \phi^2(x)\eta\,, \tag{2.71}$$

where η is a flat metric, so that the corresponding Riemann tensor vanishes, and ϕ is a function. According to DG, Sect. 13.10 conformal flatness is locally equivalent to the vanishing of the Weyl tensor. Since the metric (2.71) contains only one scalar degree of freedom $\phi(x)$, we need a scalar equation. The only possibility which satisfies the requirement that ϕ appears linearly in the second derivatives is

$$R = 24\pi GT\,, \tag{2.72}$$

where $T = T^{\mu}{}_{\mu}$ and R is the scalar Riemann curvature. The constants have been chosen such that in the Newtonian limit Equation (2.72) reduces to the Poisson equation. This scalar theory was considered by Einstein and Fokker; in the linearized approximation it agrees with the original Nordstrøm theory (see Sect. 1.2). As we emphasized previously, there is no bending of light rays in such a theory. In addition, the precession of the perihelion of Mercury is not correctly predicted.

However, this scalar theory *does* satisfy the principle of equivalence. Thus we have an example to demonstrate that the equivalence principle alone (without additional assumptions) does not predict the bending of light rays.

* * *

Exercises.

Exercise 2.5. Choose a coordinate system such that (2.71) has the form

$$g_{\mu\nu} = \phi^2\eta_{\mu\nu}\,, \qquad (\eta_{\mu\nu}) = \mathrm{diag}(-1,+1,+1,+1) \tag{2.73}$$

and show that in this case

$$R = -6\phi^{-3}\eta^{\mu\nu}\partial_\mu\partial_\nu\phi\,. \tag{2.74}$$

Equation (2.72) then reads

$$\Box\phi = -4\pi GT\phi^3\,, \tag{2.75}$$

where $\Box := \eta^{\mu\nu}\partial_\mu\partial_\nu$, and this reduces indeed to the Poisson equation in the Newtonian limit.

Exercise 2.6. Consider a static, spherically symmetric solution for a point source subject to the boundary condition $g_{\mu\nu} \longrightarrow \eta_{\mu\nu}$ for $r \longrightarrow \infty$.

Exercise 2.7. Write down the differential equations for planetary orbits (geodesics for $g_{\mu\nu}$). Show that the precession of the periastron is given by

$$\Delta\varphi = -\pi\left(\frac{4\pi GM}{L}\right)^2 ,$$

where M is the central mass (solar mass) and $L = r^2\dot{\varphi}$ is the constant in the law of equal areas (L is proportional to the orbital angular momentum). We shall see in Sect. 3.3 that this result is equal to $(-1/6)$ times the value in GR.

Exercise 2.8. Introduce for the n-sphere of radius a $S_a^n = \{x \in \mathbb{R}^{n+1} : |x| = a\}$ coordinates by stereographic projection (see Fig. 2.2). Show that after a rescaling of the coordinates x^i, for $i = 1, \ldots, n$, the induced metric of S^n is

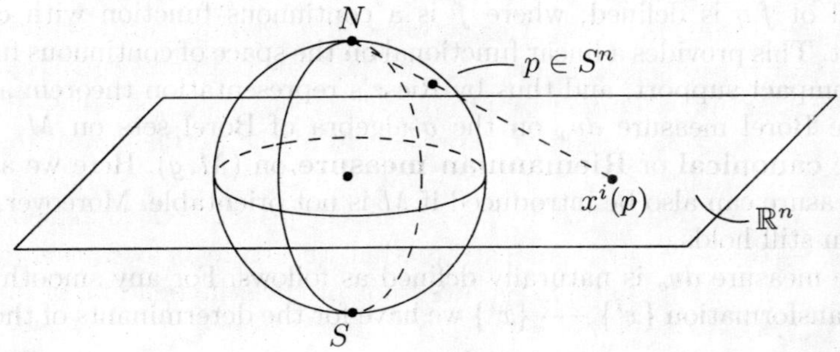

Fig. 2.2. Stereographic projection from the north pole.

$$g = \frac{1}{\psi^2}\sum_{i=1}^{n}(dx^i)^2 , \qquad \psi(x) = 1 + \frac{k}{4}\sum_{i=1}^{n}(x^i)^2 , \tag{2.76}$$

with $k = \frac{1}{a^2}$. As in Exercise 2.5 show that for such a conformally flat metric the Riemann scalar is given by

$$R = 2(n-1)\psi\Delta\psi - n(n-1)\big(\nabla\psi\big)^2 \tag{2.77}$$

(the right-hand side is formed with the Euclidean metric), and conclude from this that for the particular ψ in (2.76) this becomes

$$R = k\,n(n-1) . \tag{2.78}$$

We used this for the Einstein universe ($n = 3$): $R = \frac{6}{a^2}$. Energetic students may compute the Riemann tensor, with the result

$$R_{ijkl} = k\big(g_{ik}g_{jl} - g_{il}g_{jk}\big) . \tag{2.79}$$

2.3 Lagrangian Formalism

In this section we shall obtain Einstein's field equations from an action principle. This will enable us to give a general formula for the energy-momentum tensor in the framework of a Lagrangian field theory, and to show that its covariant divergence vanishes as a consequence of the equations for the matter fields. In this connection we shall discuss the necessity for four identities in the coupled dynamical system, and point out the close analogy with electrodynamics. We begin with a mathematical supplement.

Canonical Measure on a Pseudo-Riemannian Manifold

In DG, Sect. 12.6.1 the canonical (Riemannian) volume form η for an oriented pseudo-Riemannian manifold (M, g) is introduced, and in Sect. 12.7.2 the integral of $f\,\eta$ is defined, where f is a continuous function with compact support. This provides a linear functional on the space of continuous functions with compact support, and thus by Riesz's representation theorem a unique positive Borel measure dv_g on the σ-algebra of Borel sets on M. We call dv_g the **canonical** or **Riemannian measure** on (M, g). Here we add that this measure can also be introduced if M is not orientable. Moreover, Gauss' theorem still holds.

The measure dv_g is naturally defined as follows. For any smooth coordinate transformation $\{x^i\} \longrightarrow \{\bar{x}^i\}$ we have for the determinants of the metric tensor

$$\sqrt{|\bar{g}(\bar{x})|} = \sqrt{|g(x)|}\left|\left(\det\frac{\partial x^k}{\partial \bar{x}^l}\right)\right| . \tag{2.80}$$

From this, and the transformation formula for integrals, it follows that for a continuous function f with compact support in a coordinate patch $U \subset M$, the integral

$$\int_M f\,dv_g := \int_U f(x)\sqrt{|g(x)|}\,d^4x \tag{2.81}$$

is *independent* of the coordinate system. If the support of f is not in the domain of a chart, we define the integral with the help of a partition of unity, exactly as for the integration of n-forms (see DG, Sect. 12.7.1). Again, this provides a linear functional of such functions and thus a measure dv_g on M. One often says that dv_g is the measure *associated* to the metric g. (Usually the index g will be dropped if no confusion can arise.) In local coordinates the Riemannian measure is given by

$$dv_g = \sqrt{|g(x)|}\,d^4x\,. \tag{2.82}$$

For a proof of the divergence theorem for non-orientable manifolds, see [45].

2.3.1 The Einstein-Hilbert Action

According to Lovelock's theorem 2.1 the most general Lagrangian density $\mathcal{L}[g]$, whose Euler-Lagrange derivative contains no higher than second order derivatives of the metric field, is in *four* dimensions proportional to the Riemann scalar $R[g]$, apart from a additive constant (leading to a cosmological term). The additional freedom in higher dimensions will be mentioned in Sect. 2.6.

For this reason we now study variations of the **Einstein-Hilbert action** functional

$$S_D[g] = \int_D R[g]\, dv_g\,, \tag{2.83}$$

where D is a compact region with smooth boundary ∂D. We write everything for a Lorentz manifold, but the formalism is independent of the number of dimensions. Assume, for simplicity, that D is contained in a coordinate patch, but the final result (Eq. (2.84) below) is independent of this. Variational derivatives ($\partial/\partial\varepsilon|_{\varepsilon=0}$ for a 1-parameter family of metrics g_ε, $-\alpha < \varepsilon < \alpha$) will be denoted by the letter δ (not to be confused with the codifferential).

It was shown by Einstein and Hilbert[7] that the functional derivative of (2.83) is proportional to the Einstein tensor. More precisely, we show that the first variation of (2.83) is given by

$$\delta S_D[g] = \int_D G_{\mu\nu}\, \delta g^{\mu\nu}\, dv + \int_D \operatorname{div} W\, dv\,, \tag{2.84}$$

where W is a vector field defined later.

We have

$$\begin{aligned} \delta \int_D R\, dv &= \int_D \delta\left(g^{\mu\nu} R_{\mu\nu} \sqrt{-g}\right) d^4x \\ &= \int_D \delta R_{\mu\nu}\, g^{\mu\nu} \sqrt{-g}\, d^4x + \int_D R_{\mu\nu} \delta\left(g^{\mu\nu} \sqrt{-g}\right) d^4x\,. \end{aligned} \tag{2.85}$$

We first consider $\delta R_{\mu\nu}$. Recall

$$R_{\mu\nu} = \partial_\alpha \Gamma^\alpha{}_{\mu\nu} - \partial_\nu \Gamma^\alpha{}_{\mu\alpha} + \Gamma^\rho{}_{\mu\nu} \Gamma^\alpha{}_{\rho\alpha} - \Gamma^\rho{}_{\mu\alpha} \Gamma^\alpha{}_{\rho\nu}\,.$$

Things are enormously simplified by using a normal system of coordinates with origin $p \in D$. Then we have at p:

$$\delta R_{\mu\nu} = \partial_\alpha(\delta\Gamma^\alpha{}_{\mu\nu}) - \partial_\nu(\delta\Gamma^\alpha{}_{\mu\alpha})\,.$$

---

[7] In the light of recent historical studies it is no more justified to associate the action principle entirely to Hilbert. For a detailed discussion and references we refer to [80]. Beside Einstein and Hilbert, also H.A. Lorentz published a series of papers on general relativity based on a variational principle.

Now, the variations $\delta\Gamma^{\alpha}{}_{\mu\nu}$ of the Christoffel symbols transform as tensors, as is obvious from the transformation law DG, Equation (13.4). At p we can therefore write

$$\delta R_{\mu\nu} = \nabla_{\alpha}(\delta\Gamma^{\alpha}{}_{\mu\nu}) - \nabla_{\nu}(\delta\Gamma^{\alpha}{}_{\mu\alpha}). \tag{2.86}$$

Since this is a tensor equation, known as the **Palatini identity**, it holds everywhere in every coordinate system. It implies

$$g^{\mu\nu}\,\delta R_{\mu\nu} = W^{\alpha}{}_{;\alpha}\,, \tag{2.87}$$

where W^{α} is the vector field

$$W^{\alpha} = g^{\mu\nu}\,\delta\Gamma^{\alpha}{}_{\mu\nu} - g^{\alpha\nu}\,\delta\Gamma^{\mu}{}_{\mu\nu}\,. \tag{2.88}$$

Thus the first term on the right of (2.85) is the volume integral of $\operatorname{div} W$.

For the second term we need $\delta\sqrt{-g}$. With Cramer's rule (footnote in Sect. 1.4.2) we get, using that $g^{\alpha\beta}\delta g_{\alpha\beta} = -g_{\alpha\beta}\,\delta g^{\alpha\beta}$,

$$\begin{aligned}\delta(\sqrt{-g}) &= -\frac{1}{2}\frac{1}{\sqrt{-g}}\delta g = -\frac{1}{2}\frac{1}{\sqrt{-g}} g\, g^{\alpha\beta}\,\delta g_{\alpha\beta} \\ &= -\frac{1}{2}\sqrt{-g}\, g_{\alpha\beta}\delta g^{\alpha\beta}\,. \end{aligned} \tag{2.89}$$

If we insert this in (2.85) we obtain indeed (2.84). Clearly, this holds without the assumption that D is in the domain of a chart.

For later use we rewrite (2.89) as

$$\delta(dv) = -\frac{1}{2} g_{\alpha\beta}\,\delta g^{\alpha\beta}\,dv\,. \tag{2.90}$$

If the variations $\delta g_{\mu\nu}$ vanish outside a region contained in D, only the first term in (2.84) survives

$$\delta\int_D R\,dv = \int_D G_{\mu\nu}\delta g^{\mu\nu}\,dv\,. \tag{2.91}$$

In addition, (2.90) gives

$$\delta\int_D dv = -\frac{1}{2}\int_D g_{\mu\nu}\delta g^{\mu\nu}\,dv\,.$$

So Einstein's vacuum equations, with a cosmological term, are obtained from the action principle

$$\delta\int(R - 2\Lambda)\,dv = 0\,. \tag{2.92}$$

Boundary Term. We want to write (2.88) a bit more explicitely. Using again normal coordinates with origin p, we see that in any coordinate system we have everywhere

$$\delta \Gamma^{\mu}{}_{\alpha\beta} = \frac{1}{2} g^{\mu\nu} \Big((\delta g_{\nu\alpha})_{;\beta} + (\delta g_{\nu\beta})_{;\alpha} - (\delta g_{\alpha\beta})_{;\nu} \Big) . \tag{2.93}$$

This gives

$$W^{\alpha}{}_{;\alpha} = (\delta g^{\alpha\beta})_{;\alpha\beta} - (g^{\alpha\mu} \delta g_{\alpha\mu})^{;\beta}{}_{;\beta} . \tag{2.94}$$

We shall later interprete the boundary term in (2.84) geometrically in terms of the second fundamental form (extrinsic curvature) of ∂D (see Sect. 2.6).

2.3.2 Reduced Bianchi Identity and General Invariance

We can use the variational equation (2.91) to obtain another derivation of the reduced Bianchi identity (this presentation assumes familiarity with DG, Chap. 12). For the moment we shall keep the discussion more general, so that we can make use of it later in another context.

For this purpose we consider a 4-form $\Omega[\psi]$, which is a functional of certain (tensor) fields ψ. We assume that it is generally invariant in the following sense

$$\varphi^{*}(\Omega[\psi]) = \Omega[\varphi^{*}(\psi)] , \tag{2.95}$$

for all $\varphi \in \mathrm{Diff}(M)$. In particular, (2.95) is valid for the flow ϕ_t of a vector field X. Hence, it follows by differentiating with respect to t for $t = 0$ that

$$L_X(\Omega[\psi]) = \left.\frac{d}{dt}\right|_{t=0} \Omega[\phi_t^{*}(\psi)] .$$

We integrate this equation over a region D which has a smooth boundary and compact closure. From Cartan's formula (see DG, Sect. 12.5)

$$L_X = d \circ i_X + i_X \circ d ,$$

and Stokes' theorem we find

$$\int_D L_X \Omega = \int_D d\, i_X \Omega = \int_{\partial D} i_X \Omega = 0 ,$$

when X vanishes on the boundary ∂D. Thus, we have

$$\int_D \left.\frac{d}{dt}\right|_{t=0} \Omega[\phi_t^{*}(\psi)] = 0 . \tag{2.96}$$

In particular, we obtain from (2.91) and (2.96), for $\delta g_{\mu\nu} = (L_X g)_{\mu\nu}$ and $\Omega = (R\eta)[g]$,

$$\int_D G^{\mu\nu} (L_X g)_{\mu\nu}\, \eta = 0 . \tag{2.97}$$

(Recall that $(L_X g)_{\mu\nu}$ is a special variational derivative, determined by a local flow of the vector field X.)

The components of the Lie derivative $L_X g$ are (see DG, Sect. 11)

$$(L_X g)_{\mu\nu} = X^\lambda g_{\mu\nu,\lambda} + g_{\lambda\nu} X^\lambda{}_{,\mu} + g_{\mu\lambda} X^\lambda{}_{,\nu} . \tag{2.98}$$

In a normal system with origin in p this reduces to

$$(L_X g)_{\mu\nu} = X_{\mu,\nu} + X_{\nu,\mu} .$$

Thus, in an arbitrary coordinate system we have the useful expression

$$(L_X g)_{\mu\nu} = X_{\mu;\nu} + X_{\nu;\mu} . \tag{2.99}$$

(Of course, the indices can be interpreted abstractly.) This equation also follows from (1.118). Hence,

$$\int_D G^{\mu\nu}(X_{\mu;\nu} + X_{\nu;\mu})\eta = 0$$

or, using the symmetry of $G^{\mu\nu}$,

$$\int_D \left(G^{\mu\nu} X_\mu\right)_{;\nu} \eta - \int_D G^{\mu\nu}{}_{;\nu} X_\mu \, \eta = 0 .$$

According to Gauss' theorem the first term vanishes. Since D and the values of X_μ in D are arbitrary, the contracted Bianchi identity

$$G^{\mu\nu}{}_{;\nu} = 0 , \qquad \nabla_\nu G^{\mu\nu} = 0$$

must hold.

The contracted Bianchi identity shows that the vacuum field equations $G_{\mu\nu} = 0$ are not independent. That this must be the case becomes apparent from the following considerations. Suppose that g is a vacuum solution. For every diffeomorphism ϕ the diffeomorphically transformed field $\phi^*(g)$ must also be a vacuum solution, since g and $\phi^*(g)$ are physically equivalent. The field equations can thus only determine equivalence classes of diffeomorphic metric fields. Since Diff(M) has four "degrees of freedom", we expect four identities.

Equivalently, one may argue as follows: For every solution of the $g_{\mu\nu}$ as functions of particular coordinates, one can find a coordinate transformation such that four of the $g_{\mu\nu}$ can be made to take on arbitrary values. Hence only six of the ten field equations for the $g_{\mu\nu}$ can be independent.

A similar situation exists in electrodynamics. The fields A_μ are physically equivalent within a "gauge class". The function χ in the gauge transformation

$$A_\mu \longrightarrow A_\mu + \partial_\mu \chi$$

is arbitrary and for this reason an identity must hold in the field equations for the A_μ. In vacuum, these become

$$\Box A_\mu - \partial_\mu \partial_\nu A^\nu = 0 .$$

The left-hand side identically satisfies

$$\partial^\mu (\Box A_\mu - \partial_\mu \partial_\nu A^\nu) = 0 ,$$

and this is the identity we are looking for. This can be extended to non-Abelian gauge theories. We shall generalize this discussion to the coupled matter-field system further on.

2.3.3 Energy-Momentum Tensor in a Lagrangian Field Theory

In a Lagrangian field theory one can give a general formula for the energy-momentum tensor.

Let $\mathcal{L}$ be the Lagrangian density for a set of "matter fields" ψ_A, for $A = 1, \ldots, N$ (we include the electromagnetic field among the fields ψ_A). For simplicity, we consider only tensor fields (spinor fields in GR will be introduced in Chap. 8). If we assume that we know $\mathcal{L}$ from local physics in flat space, the principle of equivalence prescribes the form of $\mathcal{L}$ in the presence of a gravitational field. We must simply write $g_{\mu\nu}$ in place of $\eta_{\mu\nu}$ and replace ordinary derivatives by covariant ones, as we discussed in Sect. 1.4. Thus (dropping the index A) we have

$$\mathcal{L} = \mathcal{L}(\psi, \nabla\psi, g) . \tag{2.100}$$

For example, the Lagrangian density for the electromagnetic field is

$$\mathcal{L} = -\frac{1}{16\pi} F_{\mu\nu} F_{\sigma\rho} g^{\mu\sigma} g^{\nu\rho} . \tag{2.101}$$

The matter field equations are obtained by means of a variational principle

$$\delta \int_D \mathcal{L}\eta = 0 ,$$

where the fields ψ are varied inside of D such that the variations vanish on the boundary. As usual, δ means differentiation with respect to the parameter of a 1-parameter family of field variations. Now, we have

$$\delta \int_D \mathcal{L}\eta = \int_D (\delta\mathcal{L})\eta = \int_D \left(\frac{\partial \mathcal{L}}{\partial \psi} \delta\psi + \frac{\partial \mathcal{L}}{\partial (\nabla\psi)} \delta(\nabla\psi) \right) \eta . \tag{2.102}$$

The covariant derivative ∇ commutes with the variational derivative δ. This is obvious when one considers coordinate expressions for the various quantities, since δ commutes with partial differentiation. Hence, we obtain

$$\frac{\partial \mathcal{L}}{\partial(\nabla\psi)}\delta(\nabla\psi) = \nabla\cdot\left(\frac{\partial \mathcal{L}}{\partial(\nabla\psi)}\delta\psi\right) - \left(\nabla\cdot\frac{\partial \mathcal{L}}{\partial(\nabla\psi)}\right)\delta\psi\,. \tag{2.103}$$

The first term on the right-hand side is the divergence of a vector field (this becomes clear when one inserts indices). Hence, by Gauss' theorem it does not contribute to (2.102). From (2.102) we thus obtain

$$\int_D \left(\frac{\partial \mathcal{L}}{\partial\psi} - \nabla\cdot\frac{\partial \mathcal{L}}{\partial(\nabla\psi)}\right)\delta\psi\,\eta = 0\,,$$

which implies, with a standard argument, the **Euler-Lagrange equations** for the matter fields

$$\frac{\partial \mathcal{L}}{\partial\psi_A} - \nabla_\mu\cdot\frac{\partial \mathcal{L}}{\partial(\nabla_\mu\psi_A)} = 0 \tag{2.104}$$

In order to obtain an expression for the energy-momentum tensor, we consider variations of the action integral, which are induced by variations in the metric. The Lagrangian $\mathcal{L}$ depends on the metric both explicitly and implicitly, through the covariant derivatives of the matter fields. In addition, the 4-form η is an invariant functional of the metric g. Hence, we have

$$\delta\int_D \mathcal{L}\eta = \int_D\left\{\left(\frac{\partial \mathcal{L}}{\partial(\nabla_\lambda\psi)}\delta(\nabla_\lambda\psi) + \frac{\partial \mathcal{L}}{\partial g_{\mu\nu}}\delta g_{\mu\nu}\right)\eta + \mathcal{L}\delta\eta\right\}\,. \tag{2.105}$$

We first consider the term $\delta\eta$. In local coordinates,

$$\eta = \sqrt{-g}\,dx^0\wedge\ldots\wedge dx^3\,,$$

and thus

$$\delta\eta = -\left(2\sqrt{-g}\right)^{-1}\delta g\,dx^0\wedge\ldots\wedge dx^3\,.$$

Since $\delta g = g\,g^{\mu\nu}\delta g_{\mu\nu}$, it follows that

$$\delta\eta = \frac{1}{2}\sqrt{-g}\,g^{\mu\nu}\delta g_{\mu\nu}\,dx^0\wedge\ldots\wedge dx^3$$

or, in agreement with (2.90),

$$\delta\eta = \frac{1}{2}g^{\mu\nu}\delta g_{\mu\nu}\,\eta\,. \tag{2.106}$$

Even if $\delta\psi = 0$, we have $\delta(\nabla_\lambda\psi)\neq 0$, since the Christoffel symbols vary. For the tensor $\delta\Gamma^\mu{}_{\alpha\beta}$ we use (2.93). With this equation it is possible to express $\delta(\nabla_\lambda\psi)$ in terms of $\delta g_{\alpha\beta;\gamma}$. After integration by parts, one can bring the variation (2.105) of the action integral to the form

$$\delta\int_D \mathcal{L}\,\eta = \frac{1}{2}\int_D T^{\mu\nu}\delta g_{\mu\nu}\,\eta\,, \tag{2.107}$$

whereby *only* g is varied. We identify the tensor $T^{\mu\nu}$ in (2.107) with the **energy-momentum tensor**. Note that $T^{\mu\nu}$ is automatically symmetric. As a first justification, we note that its covariant divergence vanishes.

The proof of $T^{\mu\nu}{}_{;\nu} = 0$ is very similar to the derivation of the Bianchi identity in Sect. 2.2. We start with (2.96), which is valid for every invariant functional $\mathcal{L}[\psi, g]\,\eta$, i.e., $\varphi^*\big(\mathcal{L}[\psi, g]\big) = \mathcal{L}[\varphi^*\psi, \varphi^* g]$ for all $\varphi \in \mathrm{Diff}(M)$. That is,

$$\int_D \frac{d}{dt}\bigg|_{t=0} \big(\mathcal{L}\,\eta[\phi_t^*\psi, \phi_t^* g]\big) = 0\,. \tag{2.108}$$

As usual, ϕ_t denotes the flow of a vector field X, which vanishes on the boundary of D. The variations of the fields ψ do not contribute, due to the matter equations (2.104). Hence, using (2.107), we have

$$\int_D T^{\mu\nu}(L_X g)_{\mu\nu}\,\eta = 0\,. \tag{2.109}$$

Exactly as in the proof of the contracted Bianchi identity (see Sect. 2.3.2), we obtain

$$T^{\mu\nu}{}_{;\nu} = 0, \qquad \nabla_\nu T^{\mu\nu} = 0\,. \tag{2.110}$$

For familiar systems, the definition (2.107) leads to the usual expressions for the energy-momentum tensor. We show this for the electromagnetic field. In this case, the Lagrangian is given by (2.101). If we vary only the metric g and use (2.106), we obtain

$$\begin{aligned}\delta\int_D \mathcal{L}\,\eta &= \int_D \big(\delta\mathcal{L}\,\eta + \mathcal{L}\,\delta\eta\big) = \int_D \Big(\delta\mathcal{L} + \frac{1}{2}\mathcal{L} g^{\mu\nu}\delta g_{\mu\nu}\Big)\eta \\ &= \int_D \Big(-\frac{1}{8\pi}F_{\mu\nu}F_{\sigma\rho}g^{\mu\sigma}\,\delta g^{\nu\rho} + \frac{1}{2}\mathcal{L} g^{\mu\nu}\delta g_{\mu\nu}\Big)\,\eta\,.\end{aligned}$$

However, the equation $g^{\nu\beta}g_{\alpha\beta} = \delta^\nu{}_\alpha$ implies that

$$\delta g^{\nu\beta}g_{\alpha\beta} + g^{\nu\beta}\delta g_{\alpha\beta} = 0\,,$$

so that

$$\delta g^{\nu\rho} = -g^{\nu\beta}g^{\alpha\rho}\delta g_{\alpha\beta}\,.$$

Hence, we obtain for the variation of the action integral

$$\delta\int_D \mathcal{L}\eta = \frac{1}{8\pi}\int_D \Big(F_{\mu\nu}F_{\sigma\rho}g^{\mu\sigma}g^{\nu\beta}g^{\alpha\rho}\delta g_{\alpha\beta} - \frac{1}{4}F_{\mu\nu}F^{\mu\nu}g^{\alpha\beta}\delta g_{\alpha\beta}\Big)\,\eta\,.$$

Thus we obtain from (2.107)

$$T^{\alpha\beta} = \frac{1}{4\pi}\Big(F^{\alpha\sigma}F^\beta{}_\sigma - \frac{1}{4}F_{\mu\nu}F^{\mu\nu}g^{\alpha\beta}\Big)\,,$$

and

$$T_{\alpha\beta} = \frac{1}{4\pi}\left(F_{\alpha\mu}F_{\beta\nu}g^{\mu\nu} - \frac{1}{4}g_{\alpha\beta}F_{\mu\nu}F^{\mu\nu}\right) . \tag{2.111}$$

This agrees with (1.53).

Remark. In SR the energy-momentum tensor is usually derived from the translation invariance. This results in the conserved canonical energy-momentum tensor, which is in general not symmetric, but which can always be symmetrized. One could also use the above procedure in SR: replace *formally* $\eta_{\mu\nu}$ by a curved metric $g_{\mu\nu}$, use (2.107) and then set $g_{\mu\nu}$ equal to the flat metric at the end. The resulting tensor is then divergenceless (by (2.110)) and symmetric. The relationship of the two procedures is discussed in [85].

2.3.4 Analogy with Electrodynamics

The definition of $T^{\mu\nu}$ is similar to the definition of the current density j^μ in electrodynamics. We decompose the full Lagrangian $\mathcal{L}$ for matter into a part $\mathcal{L}_F$ for the electromagnetic field and a term $\mathcal{L}_M$, which contains the interaction between the charged fields and the electromagnetic field. As before,

$$\mathcal{L}_F = -\frac{1}{16\pi}F_{\mu\nu}F^{\mu\nu} ,$$

where $F_{\mu\nu} = A_{\nu,\mu} - A_{\mu,\nu}$. By varying only the A_μ, such that the variation vanishes on the boundary of D, we obtain

$$\begin{aligned} \delta\int_D \mathcal{L}_F\eta &= \frac{1}{8\pi}\int_D F^{\mu\nu}\delta(A_{\mu,\nu} - A_{\nu,\mu})\,\eta \\ &= \frac{1}{8\pi}\int_D F^{\mu\nu}\delta(A_{\mu;\nu} - A_{\nu;\mu})\,\eta \\ &= \frac{1}{4\pi}\int_D F^{\mu\nu}\,\delta A_{\mu;\nu}\eta \\ &= -\frac{1}{4\pi}\int_D F^{\mu\nu}{}_{;\nu}\delta A_\mu\eta . \end{aligned} \tag{2.112}$$

We define the current density j^μ by

$$\delta\int_D \mathcal{L}_M\eta = \int_D j^\mu\delta A_\mu\,\eta , \tag{2.113}$$

where only the A_μ are varied.

Maxwell's equations follow from Hamilton's variational principle

$$\delta\int_D \mathcal{L}\eta = 0, \quad \mathcal{L} = \mathcal{L}_F + \mathcal{L}_M , \tag{2.114}$$

together with (2.112) and (2.113). Equation (2.113) is analogous to (2.107). The divergence of the electromagnetic current vanishes as a consequence of the gauge invariance of the action $\int_D \mathcal{L}_M\eta$. Let us show this in detail.

If we subject all the fields to a gauge transformation with gauge function $s\,\chi(x)$ and denote the derivative with respect to s at $s=0$ by δ_χ, then the $\delta_\chi\psi$ of the matter fields (with exception of the A_μ) do not contribute to the variational equation

$$\delta_\chi \int_D \mathcal{L}_M \eta = 0\,,$$

as a result of the field equations for matter. Using (2.113) and $\delta_\chi A_\mu = \chi_{,\mu}$, we obtain

$$0 = \delta_\chi \int_D \mathcal{L}_M \eta = \int_D j^\mu \chi_{,\mu} \eta\,.$$

If χ vanishes on the boundary of D, we obtain, after an integration by parts,

$$\begin{aligned} 0 &= \int_D j^\mu \chi_{,\mu} \eta = \int_D j^\mu \chi_{;\mu} \eta = \int_D (j^\mu \chi)_{;\mu} \eta - \int_D j^\mu{}_{;\mu} \chi\, \eta \\ &= -\int_D j^\mu{}_{;\mu} \chi\, \eta\,. \end{aligned}$$

Hence, the covariant divergence of the electromagnetic current vanishes

$$j^\mu{}_{;\mu} = 0\,. \tag{2.115}$$

This equation is thus a consequence of Maxwell's equations on the one hand, because of the identity $F^{\mu\nu}{}_{;\nu;\mu} = 0$. On the other hand it also follows from the matter equations (the Euler-Lagrange equations for $\mathcal{L}_M$). This means that the coupled system of Maxwell's equations and matter equations is not independent. This must be the case, since the transformed fields $\{A^\chi_\mu, \psi^\chi\}$, obtained from a solution $\{A_\mu, \psi\}$ of the coupled equations for matter and the electromagnetic field, is a equivalent solution. (These remarks are easily generalized to Yang-Mills theories.)

2.3.5 Meaning of the Equation $\nabla \cdot T = 0$

In general relativity the equation $T^{\mu\nu}{}_{;\nu} = 0$ plays a completely analogous role as (2.115) in electrodynamics. For every solution $\{g, \psi\}$ of the coupled system of gravitational and matter equations (we now include the electromagnetic field among the matter fields), and every $\varphi \in \text{Diff}(M)$, $\{\varphi^*(g), \varphi^*(\psi)\}$ is a physically equivalent solution. We expect thus four identities, since Diff(M) has four degrees of freedom. (Instead of "active" diffeomorphisms, we could just as well consider "passive" coordinate transformations.) These four identities are a result of the fact that (as was first emphasized by Hilbert) $\nabla \cdot T = 0$ is a consequence of both the matter equations and the gravitational field equations. The role of the gauge group of electrodynamics is now taken over by the group Diff(M), which of course is non-Abelian.

The Equations of Motion and $\nabla \cdot T = 0$

If we take $T^{\mu\nu}$ to be the energy-momentum tensor of an ideal fluid with pressure $p = 0$ (incoherent dust), so that

$$T^{\mu\nu} = \rho u^\mu u^\nu \,, \tag{2.116}$$

and assume that the quantity of fluid is conserved, then we have

$$\int_{\phi_t(D)} \rho\eta = \int_D \rho\eta \,,$$

where ϕ_t is the flow of u. Since,

$$\int_{\phi_t(D)} \rho\eta = \int_D \phi_t^*(\rho\eta)$$

and D is arbitrary, the conservation law is equivalent to

$$L_u(\rho\eta) = 0 \,. \tag{2.117}$$

But, we have

$$L_u(\rho\eta) = (L_u\rho)\,\eta + \rho\, L_u\eta = (\nabla_u\rho + \rho\,\nabla\cdot u)\eta = \nabla\cdot(\rho u)\eta \,.$$

Hence, the conservation law can be written as

$$(\rho u^\mu)_{;\mu} = 0 \,. \tag{2.118}$$

It follows from $T^{\mu\nu}{}_{;\nu} = 0$ that

$$(\rho u^\nu)_{;\nu} u^\mu + \rho u^\nu u^\mu{}_{;\nu} = 0 \,,$$

whence

$$\nabla_u u = 0 \,. \tag{2.119}$$

This says that the integral curves of u (streamlines) are geodesics. This law of motion is thus a consequence of the field equations. In the exercises we shall see other examples for this.

2.3.6 Variational Principle for the Coupled System

Einstein's field equations and the matter equations follow from the variational principle

$$\delta \int_D \left(\frac{R}{16\pi G} + \mathcal{L} \right) \eta = 0 \,, \tag{2.120}$$

since according to (2.91) and (2.107), variation of g only results in

$$\delta \int_D R\eta = -\int_D G^{\mu\nu}\delta g_{\mu\nu}\,\eta\,,$$

$$\delta \int_D \mathcal{L}\eta = \frac{1}{2}\int_D T^{\mu\nu}\delta g_{\mu\nu}\,\eta\,.$$

Hence, we obtain

$$G_{\mu\nu} = 8\pi G\, T_{\mu\nu}\,.$$

* * *

Exercises.

Exercise 2.9. Consider a neutral scalar field with Lagrangian density

$$\mathcal{L} = \frac{1}{2}\varphi_{;\mu}\varphi^{;\mu} - U(\varphi)\,. \tag{2.121}$$

1. Derive the Euler-Lagrange equation for φ.
2. Show that the energy-momentum tensor, defined by (2.107) becomes

$$T_{\mu\nu} = \varphi_{;\mu}\varphi_{;\nu} - g_{\mu\nu}\mathcal{L}\,. \tag{2.122}$$

3. Show that Einstein's field equations imply the Euler-Lagrange equation for φ in points where $d\varphi \neq 0$.

Exercise 2.10. Investigate the last point of the previous exercise for the electromagnetic field.

Exercise 2.11. Decompose the Einstein-Hilbert density as

$$R\sqrt{-g} = \mathcal{A}(g, \partial g) + \text{divergence}\,, \tag{2.123}$$

such that $\mathcal{A}(g, \partial g)$ is only a function of $g_{\mu\nu}$ and its first derivatives.

Solution. We start from the original expression

$$\mathcal{R} := \sqrt{-g}\, g^{\mu\nu}\, R_{\mu\nu} = \sqrt{-g}\, g^{\mu\nu}\left(\Gamma^{\alpha}{}_{\mu\nu,\alpha} - \Gamma^{\alpha}{}_{\mu\alpha,\nu} + \Gamma^{\beta}{}_{\mu\nu}\Gamma^{\alpha}{}_{\beta\alpha} - \Gamma^{\beta}{}_{\mu\alpha}\Gamma^{\alpha}{}_{\beta\nu}\right). \tag{2.124}$$

The first two terms can be rewritten as

$$\begin{aligned}\sqrt{-g}\, g^{\mu\nu}\left(\Gamma^{\alpha}{}_{\mu\nu,\alpha} - \Gamma^{\alpha}{}_{\mu\alpha,\nu}\right) &= \left(\sqrt{-g}\, g^{\mu\nu}\Gamma^{\alpha}{}_{\mu\nu}\right)_{,\alpha} - \left(\sqrt{-g}\, g^{\mu\nu}\Gamma^{\alpha}{}_{\mu\alpha}\right)_{,\nu} \\ &\quad + \Gamma^{\alpha}{}_{\mu\alpha}\left(\sqrt{-g}\, g^{\mu\nu}\right)_{,\nu} - \Gamma^{\alpha}{}_{\mu\nu}\left(\sqrt{-g}\, g^{\mu\nu}\right)_{,\alpha}\,.\end{aligned} \tag{2.125}$$

This already solves the problem, but we can simplify the expression for $\mathcal{A}(g, \partial g)$. We use the familiar formula

$$(\sqrt{-g})_{,\nu} = \sqrt{-g}\,\Gamma^{\beta}{}_{\nu\beta}$$

and take the derivatives of $g^{\mu\nu}$ from

$$0 = g^{\mu\nu}{}_{;\alpha} = g^{\mu\nu}{}_{,\alpha} + \Gamma^{\mu}{}_{\alpha\beta} g^{\beta\nu} + \Gamma^{\nu}{}_{\alpha\beta} g^{\mu\beta} ,$$

to write the last two terms in (2.125) as

$$\Gamma^{\alpha}{}_{\mu\alpha}(\sqrt{-g}\, g^{\mu\nu})_{,\nu} - \Gamma^{\alpha}{}_{\mu\nu}(\sqrt{-g}\, g^{\mu\nu})_{,\alpha} = g^{\mu\nu}(\Gamma^{\alpha}{}_{\beta\nu}\Gamma^{\beta}{}_{\alpha\mu} - \Gamma^{\alpha}{}_{\mu\nu}\Gamma^{\beta}{}_{\alpha\beta})\sqrt{-g} =: 2\mathcal{A} .$$

If we use this in (2.125) and (2.124), the combination $\mathcal{A}$ shows up once more, and we finally obtain

$$\mathcal{R} = \mathcal{A} + (\sqrt{-g}\, g^{\mu\nu}\Gamma^{\alpha}{}_{\mu\nu})_{,\alpha} - (\sqrt{-g}\, g^{\mu\nu}\Gamma^{\alpha}{}_{\mu\alpha})_{,\nu} , \tag{2.126}$$

with

$$\mathcal{A} = \frac{1}{2} g^{\mu\nu}(\Gamma^{\alpha}{}_{\beta\nu}\Gamma^{\beta}{}_{\alpha\mu} - \Gamma^{\alpha}{}_{\mu\nu}\Gamma^{\beta}{}_{\alpha\beta})\sqrt{-g} . \tag{2.127}$$

2.4 Non-localizability of the Gravitational Energy

In SR the conservation laws for energy and momentum of a closed system are a consequence of the invariance with respect to translations in time and space. Except for special solutions, translations do not act as isometries on a Lorentz manifold and for this reason a general conservation law for energy and momentum does *not* exist in GR. This has been disturbing to many people, but one simple has to get used to this fact. There is no "energy-momentum tensor for the gravitational field". Independently of any formal arguments, Einstein's equivalence principle tells us directly that there is no way to localize the energy of the gravitational field: The "gravitational field" (the connection $\Gamma^{\mu}{}_{\alpha\beta}$) can be locally transformed away. But if there is no field, there is locally no energy and no momentum. This is closely analogous to the situation with regard to charge conservation in non-Abelian gauge theories.

However, it is still possible to define the *total* energy and *total* momentum of an isolated system with an asymptotically flat geometry. In a cosmological context this is, however, in general not possible. (We shall discuss this in detail in Sect. 2.7.)

At this point, we shall merely show the following: If a Killing field K exists, i.e., a field K such that $L_K g = 0$, then a conservation law follows from $\nabla \cdot T = 0$ for the vector field

$$P^{\mu} = T^{\mu\nu} K_{\nu} . \tag{2.128}$$

Because $T^{\mu\nu}$ is symmetric we obtain with the Killing equation (1.120)

$$P^{\mu}{}_{;\mu} = T^{\mu\nu}{}_{;\mu} K_{\nu} + T^{\mu\nu} K_{\nu;\mu} = \frac{1}{2} T^{\mu\nu}(K_{\mu;\nu} + K_{\nu;\mu}) = 0 . \tag{2.129}$$

If D is a region having a smooth boundary ∂D and compact closure $\bar{D}$, then Gauss' theorem (see DG, Theorem 12.12) implies

$$\int_{\partial D} P^\mu \, d\sigma_\mu = \int_D P^\mu{}_{;\mu} \, dv = 0 \,, \tag{2.130}$$

where dv is the measure corresponding to g.

In SR one has ten Killing fields, corresponding to the ten-dimensional Lie algebra of the inhomogeneous Lorentz group. The corresponding quantities (2.128) are precisely the ten conserved densities. In a Lorentz system, the ten independent Killing fields are

$$^{(\alpha)}T = \frac{\partial}{\partial x^\alpha} \,, \tag{2.131}$$

which generate the translations, and

$$^{(\alpha\beta)}M = \eta_{\alpha\gamma} \, x^\gamma \frac{\partial}{\partial x^\beta} - \eta_{\beta\gamma} \, x^\gamma \frac{\partial}{\partial x^\alpha} \,, \tag{2.132}$$

which generate the homogeneous Lorentz transformations. One easily verifies that the Killing equation is satisfied for (2.131) and (2.132). The corresponding conserved quantities (2.128) are $T^{\mu\alpha}$ for (2.131) and $T^{\mu\beta}x^\alpha - T^{\mu\alpha}x^\beta$ for (2.132), in other words the well-known angular momentum density.

2.5 On Covariance and Invariance

> *"Few words have been abused by physicists more than relativity, symmetry, covariance, invariance and gauge or coordinate transformation."*
>
> —A. Trautman (1972)

Confusions around these concepts began already in 1917 when E. Kretschmann first pointed out that the principle of general covariance has no physical content whatever. By appropriate reformulations of a theory, the covariance group can always be enlarged. Well known examples are the Lagrangian and Hamiltonian formulations of classical mechanics. In the former the covariance group is the diffeomorphism group of configuration space, while in Hamilton's theory it becomes the larger symplectic group of phase space (the cotangent bundle of configuration space). It is also clear that the laws of SR can be made generally covariant. Einstein entirely concurred with Kretschmann's view, but emphasized the 'heuristic value' of the covariance principle. This was the beginning of one of the most confusing chapters of the history of GR with strange discussions over decades. (See [64] for an authorative account.)

Unfortunately, it took a long time until a clear distinction between the notions of **covariance** and **invariance** was made. In a textbook this was the first time done by J.L. Anderson, [9]. Below we give an abbreviated version of this discussion.

The hart of Anderson's clarification is a distinction between **absolute** and **dynamical objects**. Before we attempt to give general definitions on the basis of field equations, the following examples should be helpful and illustrate the main points.

We consider three "theories" for a symmetric tensor field g of signature $(-1,1,1,1)$ on $\mathbb{R}^4$, regarded as a differentiable manifold.

$$\begin{array}{lll} (a) & R_{\mu\nu\alpha\beta}(g) = 0\,; & \\ (b) & R(g) = 0\,, & C_{\mu\nu\alpha\beta}(g) = 0\,; \\ (c) & R_{\mu\nu}(g) = 0\,. & \end{array}$$

In all three cases the group $\mathrm{Diff}(\mathbb{R}^4)$ is a covariance group, because the Riemann tensor, its contractions, and the Weyl tensor satisfy the covariance property

$$\mathrm{Riem}(\varphi^* g) = \varphi^*\big(\mathrm{Riem}(g)\big)\,, \text{ etc.}. \tag{2.133}$$

The "theory" (a) just says that the metric field g is the Minkowski metric. It has no dynamical degrees of freedom, and is the prototype of an absolute object: Each solution of the "field equation" (a) is of the form

$$g = \varphi_*\eta, \qquad \varphi \in \mathrm{Diff}(\mathbb{R}^4), \tag{2.134}$$

where η is the standard metric

$$\eta = \eta_{\mu\nu}\, dx^\mu dx^\nu\,, \qquad (\eta_{\mu\nu}) = \mathrm{diag}(-1,1,1,1)\,. \tag{2.135}$$

The theory (b) is just the Einstein-Fokker theory, discussed at the end of Sect. 2.2. The vanishing of the Weyl tensor implies that g is conformally flat. In other words, g is of the form

$$g = \varphi_*(\phi^2\eta)\,, \qquad \varphi \in \mathrm{Diff}(\mathbb{R}^4)\,, \tag{2.136}$$

where ϕ is a scalar field. This is the only dynamical degree of freedom. The first equation in (b) implies that ϕ satisfies the field equation

$$\Box_\eta \phi = 0\,. \tag{2.137}$$

The conformal structure in this theory is absolute: The object $\tilde{g}_{\mu\nu} = g_{\mu\nu}/(-g)^{1/4}$ is an absolute tensor density, in that it is diffeomorphic (as a tensor density) to $\eta_{\mu\nu}$ (as in (2.135)).

Finally, (c) is Einstein's vacuum theory of the gravitational field. In this theory, g is a fully dynamical object. We cannot point out any absolute elements, apart of the Lorentzian signature.

Which subgroups of the common covariance group $\mathrm{Diff}(\mathbb{R}^4)$ of the three theories should we consider as invariance group? With Anderson we adopt the definition that the invariance group consists of those elements of $\mathcal{G}$ which leave the absolute objects invariant. For (a) this is the inhomogeneous Lorentz group, for (b) it is the conformal group, which is a finite dimensional Lie group (see [31], p.352), and for (c) the invariance group agrees with the covariance group. In particular, the Einstein-Fokker theory is generally covariant (as emphasized in the original paper), however, *not* generally invariant. In GR, on the other hand, the metric is entirely dynamical, and for this reason, "**general relativity**" is an appropriate naming.

Given a theory with a set of fields of various types and field equations, we can formalize the concepts of absolute and dynamical objects, and give general definitions of the invariance and covariance groups. The examples discussed above suggest how this has to be done. First we recall what we mean by a covariance group.

Definition 2.1. *Given a set of fields on a differentiable manifold M, collectively denoted by $\mathcal{F}(x)$, and corresponding field equations*

$$\mathcal{D}[\mathcal{F}] = 0\,, \tag{2.138}$$

we say that $\mathcal{G} \subset Diff(M)$ is a ***covariance group****, if to each $\varphi \in \mathcal{G}$ there is a map $\mathcal{F} \longrightarrow \varphi_* \mathcal{F}$, such that*

(i) $\varphi \longmapsto \varphi_*$ is a realization of $\mathcal{G}$: $id_* = id$, $(\varphi_1 \circ \varphi_2)_* = \varphi_{1*} \circ \varphi_{2*}$ for $\varphi_1, \varphi_2 \in \mathcal{G}$;

(ii) With $\mathcal{F}$ also $\varphi_*\mathcal{F}$ satisfies the field equations (2.138).

The action of $\mathcal{G}$ on the fields defines an equivalence relation.

Definition 2.2. *Two solutions $\mathcal{F}_1$ and $\mathcal{F}_2$ of (2.138) are* ***equivalent*** *relative to the covariance group $\mathcal{G}$, if there exists a $\varphi \in \mathcal{G}$ such that $\varphi_*\mathcal{F}_1 = \mathcal{F}_2$.*

It is obvious that this defines an equivalence relation. The corresponding equivalence classes are called $\mathcal{G}$-classes.

Next, we isolate in an intrinsic fashion the absolute elements.

Definition 2.3. *Assume that a subset $\mathcal{A} \subset \mathcal{F}$ of fields has the following property: Each $\mathcal{G}$-class contains all solutions of the fields in $\mathcal{A}$. (There is only one $\mathcal{G}$-class for $\mathcal{A}$.) Then the fields in $\mathcal{A}$ are* ***absolute objects****. The non-absolute elements are* ***dynamical****.*

Absolute elements[8] can usually brought in some charts into a standard form, given in terms of a set of fixed numerical constants (like the flat metric in SR, Eq.(2.135)).

At this point, we can say what we mean by invariance or symmetry groups.

[8] It may happen that by an unfortunate choice of the independent fields of the theory, a field is absolute according to our definition, while this is not the case physically. An example, mentioned in [65], would be a nowhere-vanishing vector field on a Lorentz manifold. Such vector fields are for instance used in dust models. The problem disappears, however, if one chooses as independent vector field the product of the density of the dust and its four-velocity field.

Definition 2.4. *The subgroup*

$$Inv = \{\varphi \in \mathcal{G} : \varphi_* A = A, A \in \mathcal{A}\}$$

*is an **invariance (symmetry) group** of the system (2.138).*

The reader is invited to illustrate these concepts with other examples. For a more detailed discussion we refer also to [65]. Note that gauge covariance and gauge invariance can be distinguished in a similar fashion.

* * *

Exercise 2.12. Consider a metric field g and a 2-form F satisfying the coupled system

$$\mathrm{Riem}(g) = 0\,,$$
$$dF = 0\,, \qquad \delta_g F = 0\,.$$

What are the largest covariance and invariance groups?

2.6 The Tetrad Formalism

In this section we shall formulate Einstein's field equations in terms of differential forms. This has certain advantages, also for practical calculations. We shall use Sects. 13.7 and 13.8 of DG.

In the following, $\{\theta^\alpha\}$ denotes a (local) basis of 1-forms and $\{e_\alpha\}$ the corresponding dual basis of (local) vector fields. We shall often assume these bases to be orthonormal, so that

$$g = -\theta^0 \otimes \theta^0 + \sum_{i=1}^{3} \theta^i \otimes \theta^i\,. \tag{2.139}$$

Under a change of basis

$$\bar{\theta}^\alpha(x) = A^\alpha{}_\beta(x)\theta^\beta(x) \tag{2.140}$$

the connection forms $\omega = (\omega^\alpha{}_\beta)$ transform inhomogeneously

$$\bar{\omega} = A\omega A^{-1} - dA\,A^{-1}\,, \tag{2.141}$$

while the curvature $\Omega = (\Omega^\alpha{}_\beta)$ is a tensor-valued 2-form

$$\bar{\Omega} = A\Omega A^{-1}\,. \tag{2.142}$$

With respect to an orthonormal basis we have (see Eq. (13.69) of DG)

$$\omega_{\alpha\beta} + \omega_{\beta\alpha} = 0 \,, \tag{2.143}$$

so that $\omega(x) \in so(1,3)$, the Lie algebra of the Lorentz group. For orthonormal tetrad fields $\{\theta^\alpha\}$, (2.140) is a spacetime dependent Lorentz transformation

$$\bar{\theta}(x) = \Lambda(x)\theta(x) \,, \tag{2.144a}$$

$$\bar{\omega} = \Lambda(x)\omega(x)\Lambda^{-1}(x) - d\Lambda(x)\Lambda^{-1}(x) \,, \tag{2.144b}$$

where $\Lambda(x) \in L_+^\uparrow$ (if orientations are preserved). For any given point x_0 there is always a $\Lambda(x)$ such that $\bar{\omega}(x_0) = 0$; according to (2.144b), $\Lambda(x)$ must satisfy the equation $\Lambda^{-1}(x_0)d\Lambda(x_0) = \omega(x_0)$. One can easily convince oneself that this is always possible, since $\omega(x_0) \in so(1,3)$. If $\omega(x_0) = 0$, then $\theta^\alpha(x)$ describes a **local inertial frame** with origin x_0. At the point x_0, the exterior covariant derivative D coincides with the exterior derivative d.

Obviously, the metric (2.139) is invariant under the transformation (2.144a). In analogy to the gauge transformations of electrodynamics (and Yang-Mills theories) the local Lorentz transformations (2.144a) are also called **gauge transformations**. General relativity is invariant with respect to such transformations and is thus a (special) **non-Abelian gauge theory**. The tetrad fields $\{\theta^\alpha\}$ can be regarded as the potentials of the gravitational field, since by (2.139), variations of the θ^α induce variations in the metric.

The 4-form $*R = R\eta$ in the Einstein-Hilbert variational principle can be represented in the tetrad formalism as

$$*R = \eta^{\mu\nu} \wedge \Omega_{\mu\nu} \,, \tag{2.145}$$

where $\eta^{\mu\nu} = *(\theta^\mu \wedge \theta^\nu)$. Indices are raised or lowered with the metric coefficients $g_{\mu\nu}$, appearing in $g = g_{\mu\nu}\, \theta^\mu \otimes \theta^\nu$.

We prove (2.145). Clearly

$$\eta_{\alpha\beta} \wedge \Omega^{\alpha\beta} = \frac{1}{2}\eta_{\alpha\beta} R^{\alpha\beta}{}_{\mu\nu} \wedge \theta^\mu \wedge \theta^\nu \,.$$

From DG, Exercise 12.7 of Sect. 12.6.2, we have

$$*(\theta^\mu \wedge \theta^\nu) = \frac{1}{2}\eta_{\alpha\beta\sigma\rho}\, g^{\alpha\mu} g^{\beta\nu} \theta^\sigma \wedge \theta^\rho \,,$$

which implies that

$$\eta_{\alpha\beta} = \frac{1}{2}\eta_{\alpha\beta\sigma\rho}\, \theta^\sigma \wedge \theta^\rho \,. \tag{2.146}$$

Hence, we obtain

$$\eta_{\alpha\beta} \wedge \theta^\mu \wedge \theta^\nu = \frac{1}{2}\eta_{\alpha\beta\sigma\rho}\, \theta^\sigma \wedge \theta^\rho \wedge \theta^\mu \wedge \theta^\nu = (\delta^\mu{}_\alpha \delta^\nu{}_\beta - \delta^\mu{}_\beta \delta^\nu{}_\alpha)\eta \,,$$

and thus

$$\eta_{\alpha\beta} \wedge \Omega^{\alpha\beta} = \frac{1}{2}(\delta^\mu{}_\alpha \delta^\nu{}_\beta - \delta^\mu{}_\beta \delta^\nu{}_\alpha) R^{\alpha\beta}{}_{\mu\nu}\, \eta = R\eta = *R \,.$$

Variation of Tetrad Fields

Under a variation $\delta\theta^\alpha$ of the orthonormal tetrad fields $*R$ varies as

$$\delta(\eta_{\alpha\beta} \wedge \Omega^{\alpha\beta}) = \delta\theta^\alpha \wedge (\eta_{\alpha\beta\gamma} \wedge \Omega^{\beta\gamma}) + \text{exact differential}\,, \tag{2.147}$$

where $\eta_{\alpha\beta\gamma} = *(\theta^\alpha \wedge \theta^\beta \wedge \theta^\gamma)$.

We want to prove (2.147). As for (2.146), one finds that

$$\eta_{\alpha\beta\gamma} = \eta_{\alpha\beta\gamma\delta}\,\theta^\delta\,. \tag{2.148}$$

Now, we have

$$\delta(\eta_{\alpha\beta} \wedge \Omega^{\alpha\beta}) = \delta\eta_{\alpha\beta} \wedge \Omega^{\alpha\beta} + \eta_{\alpha\beta} \wedge \delta\Omega^{\alpha\beta}$$

From (2.146) and (2.148) we obtain

$$\delta\eta_{\alpha\beta} = \frac{1}{2}\delta(\eta_{\alpha\beta\gamma\delta}\,\theta^\gamma \wedge \theta^\delta) = \delta\theta^\gamma \wedge \eta_{\alpha\beta\gamma}\,.$$

The second structure equation gives

$$\delta\Omega^{\alpha\beta} = d\delta\omega^{\alpha\beta} + \delta\omega^\alpha{}_\lambda \wedge \omega^{\lambda\beta} + \omega^\alpha{}_\lambda \wedge \delta\omega^{\lambda\beta}\,.$$

Hence,

$$\begin{aligned}\delta(\eta_{\alpha\beta} \wedge \Omega^{\alpha\beta}) &= \delta\theta^\lambda \wedge (\eta_{\alpha\beta\gamma} \wedge \Omega^{\alpha\beta}) + d(\eta_{\alpha\beta} \wedge \delta\omega^{\alpha\beta}) \\ &\quad -d\eta_{\alpha\beta} \wedge \delta\omega^{\alpha\beta} + \eta_{\alpha\beta} \wedge \left(\delta\omega^\alpha{}_\lambda \wedge \omega^{\lambda\beta} + \omega^\alpha{}_\lambda \wedge \delta\omega^{\lambda\beta}\right).\end{aligned}$$

The last line is equal to $\delta\omega^{\alpha\beta} \wedge D\,\eta_{\alpha\beta}$. However,

$$D\,\eta_{\alpha\beta} = 0 \tag{2.149}$$

for the Levi-Cività connection, as we shall show below. We thus obtain

$$\delta(\eta_{\alpha\beta} \wedge \Omega^{\alpha\beta}) = \delta\theta^\lambda \wedge (\eta_{\alpha\beta\gamma} \wedge \Omega^{\alpha\beta}) + d(\eta_{\alpha\beta} \wedge \delta\omega^{\alpha\beta})\,, \tag{2.150}$$

i.e. (2.147).

It remains to show that $D\,\eta_{\alpha\beta} = 0$. Relative to an orthonormal system the expression $\eta_{\alpha\beta\gamma\delta}$ is constant. Since orthonormality is preserved under parallel transport (for a metric connection) we conclude that

$$D\eta_{\alpha\beta\gamma\delta} = 0\,. \tag{2.151}$$

Together with (2.146) and the first structure equation, this implies

$$\begin{aligned}D\eta_{\alpha\beta} &= \frac{1}{2}D\big(\eta_{\alpha\beta\mu\nu}\,\theta^\mu \wedge \theta^\nu\big) = \eta_{\alpha\beta\mu\nu} D\theta^\mu \wedge \theta^\nu \\ &= \eta_{\alpha\beta\mu\nu}\Theta^\mu \wedge \theta^\nu = \Theta^\mu \wedge \eta_{\alpha\beta\mu}\,,\end{aligned}$$

thus (2.149) for the Levi-Cività connection.

In the same manner, one can show that

$$D\eta_{\alpha\beta\gamma} = 0\,. \tag{2.152}$$

Clearly, both (2.149) and (2.152) hold for any frame.

The Einstein-Hilbert Action

As in (2.147) we split off an exact differential from $*R$. If we use in

$$*R = \Omega^\alpha_\beta \wedge \eta_\alpha^{\ \beta}$$

for Ω^α_β the second structure equation, we get

$$*R = d(\omega^\alpha{}_\beta \wedge \eta_\alpha^{\ \beta}) + \omega^\alpha{}_\beta \wedge d\eta_\alpha^{\ \beta} + \omega^\alpha{}_\sigma \wedge \omega^\sigma{}_\beta \wedge \eta_\alpha^{\ \beta}$$

or with (2.149), i.e.,

$$d\eta_\alpha^{\ \beta} - \omega^\sigma{}_\alpha \wedge \eta_\sigma^{\ \beta} + \omega^\beta{}_\sigma \wedge \eta_\alpha^{\ \sigma} = 0,$$

we obtain

$$*R = d(\omega^\alpha{}_\beta \wedge \eta_\alpha^{\ \beta}) + \mathcal{L}_{EH}\,, \tag{2.153}$$

where

$$\mathcal{L}_{EH} = \omega^\alpha_{\ \beta} \wedge \omega^\sigma{}_\alpha \wedge \eta_\sigma^{\ \beta}\,. \tag{2.154}$$

The splitting (2.153) and (2.154) is equivalent to the one we derived earlier in Exercise 2.11 of Sect. 2.3. This can easily by verified by using a coordinate basis $\{dx^\mu\}$ for $\{\theta^\mu\}$ (see Exercise 2.13).

For the Einstein-Hilbert action we obtain from (2.153)

$$\int_D *R = \int_D \mathcal{L}_{EH} + \int_{\partial D} \omega^\alpha{}_\beta \wedge \eta_\alpha^{\ \beta}\,. \tag{2.155}$$

The *boundary term* can be expressed geometrically, using the results of DG, Appendix A about hypersurfaces. We adapt (locally) the orthonormal basis to ∂D and use the Gauss formula (A.3), as well as (A.8a):

$$\omega^i{}_j = \bar{\omega}^i{}_j\,, \qquad \omega^0{}_i = -K_{ij}\,\theta^j \ \text{ on } \partial D \tag{2.156}$$

(bars indicate objects on ∂D). Thus

$$\omega^\alpha{}_\beta \wedge \eta_\alpha^{\ \beta} = \bar{\omega}^i{}_j \wedge \eta_i^{\ j} - 2K_{ij}\,\theta^j \wedge \eta_0^{\ i} \ \text{ on } \partial D\,.$$

Now, one can easily see (use Eq. (2.166) below) that

$$\theta^j \wedge \eta_{0i}|_{\partial D} = \delta^j_{\ i}\mathrm{Vol}_{\partial D}\,. \tag{2.157}$$

Hence, we obtain

$$\omega^\alpha{}_\beta \wedge \eta_\alpha^{\ \beta}|_{\partial D} = \bar{\omega}^i{}_j \wedge \eta_i^{\ j} - 2\,\mathrm{Tr}(\bar{K})\mathrm{Vol}_{\partial D}\,. \tag{2.158}$$

This gives

$$\int_D *R = \int_D \mathcal{L}_{EH} - 2\int_{\partial D} \mathrm{Tr}(\bar{K})\mathrm{Vol}_{\partial D} + \int_{\partial D} \bar{\omega}^i{}_j \wedge \eta_i^{\ j}\,. \tag{2.159}$$

The last term in this equation involves only the induced metric on ∂D, and thus gives no contribution to the variation of the action for variations of the metric which vanish on ∂D. However, the boundary integral over the trace of the extrinsic curvature has to be kept. (Examples will be considered later.)

The total Lagrangian density for Einstein's field equations is

$$\mathcal{L} = \frac{1}{16\pi G} * R + \mathcal{L}_{mat} . \tag{2.160}$$

Under variation of the θ^α let

$$\delta\mathcal{L}_{mat} = \delta\theta^\alpha \wedge *T_\alpha , \tag{2.161}$$

where $*T_\alpha$ are the 3-forms of energy and momentum for matter. Together with (2.150) the variation of the total Lagrangian is

$$\delta\mathcal{L} = \delta\theta^\alpha \wedge \left(\frac{1}{16\pi G} \eta_{\alpha\beta\gamma} \wedge \Omega^{\beta\gamma} + *T_\alpha \right) + \text{exact differential} ,$$

and the field equations become

$$-\frac{1}{2} \eta_{\alpha\beta\gamma} \wedge \Omega^{\beta\gamma} = 8\pi G * T_\alpha . \tag{2.162}$$

Since the $\eta_{\alpha\beta\gamma}$ and $\Omega^{\beta\gamma}$ transforms as tensors, (2.162) holds not only for orthonormal tetrads. T_α is related to the energy-momentum tensor $T_{\alpha\beta}$ by

$$T_\alpha = T_{\alpha\beta}\, \theta^\beta . \tag{2.163}$$

With this identification, the Eqs. (2.162) are in fact equivalent to the field equations in the classical form

$$R_{\mu\nu} - \frac{1}{2} g_{\mu\nu} R = 8\pi G\, T_{\mu\nu} . \tag{2.164}$$

We show this by explicit calculation. Using (2.163) and

$$\eta^\alpha := *\theta^\alpha , \tag{2.165a}$$

$$\eta_\alpha = \frac{1}{3!} \eta_{\alpha\beta\gamma\delta}\, \theta^\beta \wedge \theta^\gamma \wedge \theta^\delta = \frac{1}{3} \theta^\beta \wedge \eta_{\alpha\beta} , \tag{2.165b}$$

Equation (2.162) can be written in the form

$$-\frac{1}{4} \eta_{\alpha\mu\nu} \wedge \theta^\sigma \wedge \theta^\rho\, R^{\mu\nu}{}_{\sigma\rho} = 8\pi G\, T_{\alpha\beta}\, \eta^\beta .$$

In this we use the middle two of the following easily proved identities:

$$\begin{aligned}
&\theta^\beta \wedge \eta_\alpha = \delta^\beta{}_\alpha \eta , \\
&\theta^\gamma \wedge \eta_{\alpha\beta} = \delta^\gamma{}_\beta \eta_\alpha - \delta^\gamma{}_\alpha \eta_\beta , \\
&\theta^\delta \wedge \eta_{\alpha\beta\gamma} = \delta^\delta{}_\gamma \eta_{\alpha\beta} + \delta^\delta{}_\beta \eta_{\gamma\alpha} + \delta^\delta{}_\alpha \eta_{\beta\gamma} , \\
&\theta^\varepsilon \wedge \eta_{\alpha\beta\gamma\delta} = \delta^\varepsilon{}_\delta \eta_{\alpha\beta\gamma} - \delta^\varepsilon{}_\gamma \eta_{\delta\alpha\beta} + \delta^\varepsilon{}_\beta \eta_{\gamma\delta\alpha} - \delta^\varepsilon{}_\alpha \eta_{\beta\gamma\delta} ,
\end{aligned} \tag{2.166}$$

and obtain for the left-hand side

$$
\begin{aligned}
&-\frac{1}{4}R^{\mu\nu}{}_{\sigma\rho}\Big(\delta^\rho{}_\nu(\delta^\sigma{}_\mu\eta_\alpha - \delta^\sigma{}_\alpha\eta_\mu) + \delta^\rho{}_\mu(\delta^\sigma{}_\alpha\eta_\nu - \delta^\sigma{}_\nu\eta_\alpha) + \delta^\rho{}_\alpha(\delta^\sigma{}_\nu\eta_\mu - \delta^\sigma{}_\mu\eta_\nu)\Big)\\
&= -\frac{1}{2}R^{\mu\nu}{}_{\mu\nu}\,\eta_\alpha + R^{\mu\nu}{}_{\alpha\nu}\eta_\mu\\
&= R^{\beta\nu}{}_{\alpha\nu}\eta_\beta - \frac{1}{2}R^{\mu\nu}{}_{\mu\nu}\delta^\beta{}_\alpha\eta_\beta\\
&= \left(R^\beta{}_\alpha - \frac{1}{2}\delta^\beta{}_\alpha R\right)\eta_\beta\,.
\end{aligned}
$$

This proves (2.164).

It follows from (2.152) and the second Bianchi identity $D\Omega = 0$ that the absolute exterior differential of the left-hand side of (2.162) vanishes. Hence Einstein's field equations imply

$$D * T_\alpha = 0\,. \tag{2.167}$$

This equation is equivalent to $\nabla \cdot T = 0$. Indeed, we have

$$D(T_{\alpha\beta}\eta^\beta) = (D\,T_{\alpha\beta}) \wedge \eta^\beta = T_{\alpha\beta;\gamma}\,\theta^\gamma \wedge \eta^\beta = T_\alpha{}^\beta{}_{;\beta}\,\eta\,.$$

Consequences of the Invariance Properties of the Lagrangian $\mathcal{L}$

Let us rewrite (2.147) in the form

$$-\frac{1}{2}\delta * R = \delta\theta^\alpha \wedge *G_\alpha + \text{exact differential}\,, \tag{2.168}$$

with $*G_\alpha = -\frac{1}{2}\eta_{\alpha\beta\gamma} \wedge \Omega^{\beta\gamma}$. According to the derivation of (2.164),

$$G_\alpha = G_{\alpha\beta}\,\theta^\beta\,, \tag{2.169}$$

where, as usual, $G_{\alpha\beta} = R_{\alpha\beta} - \frac{1}{2}g_{\alpha\beta}R$ is the Einstein tensor. We now consider two special types of variations $\delta\theta^\alpha$.

Local Lorentz Invariance. For an infinitesimal Lorentz transformation in (2.144a) we have

$$\delta\theta^\alpha(x) = \lambda^\alpha_\beta(x)\theta^\beta(x)\,,$$

where $(\lambda^\alpha_\beta(x)) \in so(1,3)$. Since $*R$ is invariant under such transformations, it follows from (2.168) that

$$0 = \lambda_{\alpha\beta}(\theta^\beta \wedge *G^\alpha - \theta^\alpha \wedge *G^\beta) + \text{exact differential}\,.$$

If we integrate this over a region D such that the support of λ is compact and contained in D, it follows, with $\lambda_{\alpha\beta} + \lambda_{\beta\alpha} = 0$, that

$$\theta^\alpha \wedge *G^\beta = \theta^\beta \wedge *G^\alpha \tag{2.170}$$

or, using (2.169),

$$\theta^\alpha \wedge \eta_\gamma \, G^{\beta\gamma} = \theta^\beta \wedge \eta_\gamma \, G^{\alpha\gamma} \, .$$

With $\theta^\alpha \wedge \eta_\gamma = \delta^\alpha_{\ \gamma}\eta$, we get

$$G^{\alpha\beta} = G^{\beta\alpha} \, . \tag{2.171}$$

In the same manner, the local Lorentz invariance of $\mathcal{L}_{mat}$, together with the equations for matter, imply

$$T_{\alpha\beta} = T_{\beta\alpha} \, . \tag{2.172}$$

Invariance under Diff(M). We now produce the variations by Lie derivatives L_X, with compact support of X contained in D. This time we have, as a result of the invariance of $*R$ under Diff(M),

$$L_X\theta^\alpha \wedge *G_\alpha + \text{exact differential} = 0 \, , \tag{2.173}$$

or

$$\int_D L_X\theta^\alpha \wedge *G_\alpha = 0 \, . \tag{2.174}$$

Since $L_X\theta^\alpha = d\, i_X\theta^\alpha + i_X d\theta^\alpha$ with $d\theta^\alpha = -\omega^\alpha_{\ \beta} \wedge \theta^\beta$, we have

$$\begin{aligned} L_X\theta^\alpha \wedge *G_\alpha &= d(i_X\theta^\alpha \wedge *G_\alpha) - i_X\theta^\alpha \wedge d * G_\alpha - \underbrace{i_X(\omega^\alpha_{\ \beta} \wedge \theta^\beta)}_{(i_X\omega^\alpha_{\ \beta}) \wedge \theta^\beta - \omega^\alpha_{\ \beta} \wedge i_X\theta^\beta} \wedge * G_\alpha \\ &= -i_X\theta^\alpha \wedge \left(d * G_\alpha - \omega^\beta_{\ \alpha} \wedge *G_\alpha\right) - (i_X\omega^\alpha_{\ \beta}) \wedge \theta^\beta \wedge *G_\alpha \\ &\quad + \text{exact differential} \, . \end{aligned}$$

The second term vanishes due to (2.170) and the antisymmetry of $\omega_{\alpha\beta}$. According to this, we have

$$L_X\theta^\alpha \wedge *G_\alpha = -\underbrace{(i_X\theta^\alpha)}_{X^\alpha} D * G_\alpha + \text{exact differential} \, , \tag{2.175}$$

and from (2.174) we conclude

$$\int_D X^\alpha D * G_\alpha = 0 \, . \tag{2.176}$$

We again obtain the contracted Bianchi identity

$$D * G_\alpha = 0 \, . \tag{2.177}$$

In an identical manner, the invariance of $\mathcal{L}_{mat}$, together with the equations for matter, imply

$$D * T_\alpha = 0 \, . \tag{2.178}$$

There are well-known results in a new form.

Example (Energy-momentum Tensor for the Electromagnetic Field). The Lagrangian of the electromagnetic field is

$$\mathcal{L}_{em} = -\frac{1}{8\pi} F \wedge *F \,. \tag{2.179}$$

We compute its variation with respect to simultaneous variations of F and θ^α. We have

$$\delta(F \wedge *F) = \delta F \wedge *F + F \wedge \delta * F \,. \tag{2.180}$$

The second term of (2.180) can be computed as follows: From DG, Exercise 12.2 of Sect. 12.6.2, we have

$$\theta^\alpha \wedge \theta^\beta \wedge *F = F \wedge *(\theta^\alpha \wedge \theta^\beta) = F \wedge \eta^{\alpha\beta}$$

and hence

$$\delta(\theta^\alpha \wedge \theta^\beta) \wedge *F + (\theta^\alpha \wedge \theta^\beta) \wedge \delta * F = \delta F \wedge \eta^{\alpha\beta} + F \wedge \delta\eta^{\alpha\beta} \,.$$

Contracting with $\frac{1}{2}F_{\alpha\beta}$ results in

$$\underbrace{\frac{1}{2}F_{\alpha\beta}\delta(\theta^\alpha \wedge \theta^\beta) \wedge * F}_{\delta\theta^\alpha \wedge F_{\alpha\beta}\,\theta^\beta} + F \wedge \delta * F = \delta F \wedge *F + \frac{1}{2}F \wedge \underbrace{\delta\eta^{\alpha\beta}}_{\delta\theta^\gamma \wedge \eta^{\alpha\beta}{}_\gamma}\, F_{\alpha\beta} \,.$$

If we insert this into (2.180), we obtain

$$\delta\left(-\frac{1}{2}F \wedge *F\right) = -\delta F \wedge *F + \frac{1}{2}\delta\theta^\alpha \wedge \left[F_{\alpha\beta}\,\theta^\beta \wedge *F - \frac{1}{2}F \wedge F^{\mu\nu}\eta_{\alpha\mu\nu}\right] . \tag{2.181}$$

In the square bracket, use $F_{\alpha\beta}\,\theta^\beta = i_\alpha F$, with $i_\alpha := i_{e_\alpha}$, as well as $F^{\mu\nu}\eta_{\alpha\mu\nu} = 2i_\alpha * F$, which is obtained from

$$i_\alpha * F = i_\alpha\left(\frac{1}{2}F^{\mu\nu}\eta_{\mu\nu}\right) = \frac{1}{4}F^{\mu\nu}\eta_{\mu\nu\rho\sigma}\, i_\alpha(\theta^\rho \wedge \theta^\sigma) = \frac{1}{2}F^{\mu\nu}\eta_{\mu\nu\alpha} \,.$$

This gives

$$\delta\left(-\frac{1}{8\pi}F \wedge *F\right) = -\frac{1}{4\pi}\delta F \wedge *F + \delta\theta^\alpha \wedge *T^{elm}_\alpha \,, \tag{2.182}$$

where

$$*T^{elm}_\alpha = \frac{1}{8\pi}\Big(i_\alpha F \wedge *F - F \wedge i_\alpha * F\Big) \,. \tag{2.183}$$

We now use (2.166) to rewrite the square bracket in (2.181) in the form

$$[\ldots] = F_{\alpha\beta}\,\theta^\beta \wedge *F - \frac{1}{2} F \wedge F^{\mu\nu}\eta_{\mu\nu\alpha}$$

$$= \frac{1}{2} F_{\alpha\beta}F^{\sigma\rho} \underbrace{\theta^\beta \wedge \eta_{\sigma\rho}}_{\delta^\beta{}_\rho \eta_\sigma - \delta^\beta{}_\sigma \eta_\rho}$$

$$- \frac{1}{4} F_{\sigma\rho}F^{\mu\nu} \underbrace{\theta^\sigma \wedge \theta^\rho \wedge \eta_{\alpha\mu\nu}}_{\delta^\rho{}_\nu(\delta^\sigma{}_\mu \eta_\alpha - \delta^\sigma{}_\alpha \eta_\mu) + \delta^\rho{}_\mu(\delta^\sigma{}_\alpha \eta_\nu - \delta^\sigma{}_\nu \eta_\alpha) + \delta^\rho{}_\alpha(\delta^\sigma{}_\nu \eta_\mu - \delta^\sigma{}_\mu \eta_\nu)}$$

$$= -\frac{1}{2} F_{\mu\nu}F^{\mu\nu}\eta_\alpha + 2F_{\alpha\nu}F^{\mu\nu}\eta_\mu \,.$$

We thus also have

$$T^{elm}_\alpha = -\frac{1}{4\pi}\Big(F_{\alpha\nu}F_\mu{}^\nu \theta^\mu + \frac{1}{4} F_{\mu\nu}F^{\mu\nu}\theta_\alpha\Big) = T^{elm}_{\alpha\beta}\theta^\beta \,, \tag{2.184}$$

where $T^{elm}_{\alpha\beta}$ agrees with (2.111).

Lovelock's Theorem in Higher Dimensions

We mention at this point the generalization of Lovelock's theorem to dimensions > 4. This is of some relevance for Kaluza-Klein theories.

In *four* dimensions we know that the most general Lagrangian density, leading to second order field equations, is

$$\mathcal{L} = a\,\eta + b\,\Omega^{\alpha\beta} \wedge \eta_{\alpha\beta} \,,$$

where $a, b \in \mathbb{R}$. In n dimensions this generalizes to

$$\mathcal{L} = \sum_{m=0}^{[n/2]-1} \lambda_m \Omega^{\alpha_1\alpha_2} \wedge \ldots \wedge \Omega^{\alpha_{2m-1}\alpha_{2m}} \wedge \eta_{\alpha_1\ldots\alpha_{2m}} \,, \tag{2.185}$$

with

$$\eta_{\alpha_1\ldots\alpha_m} = \frac{1}{(n-m)!}\eta_{\alpha_1\ldots\alpha_n}\theta^{\alpha_{m+1}} \wedge \ldots \wedge \theta^{\alpha_n} \,. \tag{2.186}$$

Note that $m = 0$ gives a cosmological term. Each of the remaining ones becomes in dimension m a so-called **Euler form** (up to a numerical factor) and determines the **Euler characteristic class**. Since we shall not make use of this, we illustrate this point only with an example.

The Einstein-Hilbert Lagrangian density $\Omega^{\alpha\beta} \wedge \eta_{\alpha\beta}$ becomes in two dimensions for an orthonormal 2-bein $\{\theta^\alpha\}$

$$\Omega^{\alpha\beta}\eta_{\alpha\beta} = 2\Omega^{01}\eta_{01} = 2\Omega^{01} = R\,\eta$$

or, using the second structure equation,

$$R\,\eta = \Omega^{\alpha\beta}\eta_{\alpha\beta} = 2\,d\omega^0{}_1 .$$

This is locally exact. Its integral has a topological meaning. According to the **Gauss-Bonnet theorem** the integral of $R\,\eta$ over a compact two-dimensional manifold is 4π times the **Euler characteristic** (R is twice the **Gaussian curvature**). For all this see, e.g., [42].

The situation is similar in higher dimensions. The characteristic classes are, for instance, treated in S. Kobayashi and K. Nomizu, [35]. One might call the terms in (2.185) for $m > 0$ the "dimensionally continued Euler forms".

The result (2.185) shows that Einstein's equations in higher dimensions are not as unique as in four dimensions.

* * *

Exercises.

Exercise 2.13. Show that (2.154) agrees with (2.127) of Exercise 2.11.

Hints. Use $\omega^\alpha_{\ \beta} = \Gamma^\alpha_{\ \sigma\beta}\theta^\sigma$ and (2.146).

Exercise 2.14. Show that the Einstein Lagrangian density can be written relative to an orthonormal basis as follows

$$\frac{1}{2}\eta_{\alpha\beta}\wedge\Omega^{\alpha\beta} = -\frac{1}{2}(d\theta^\alpha\wedge\theta^\beta)\wedge*(d\theta_\beta\wedge\theta_\alpha) + \frac{1}{4}(d\theta^\alpha\wedge\theta_\alpha)\wedge*(d\theta^\beta\wedge\theta_\beta) + \text{exact differential}. \tag{2.187}$$

Hints. Set $d\theta^\alpha = \frac{1}{2}F^\alpha_{\ \beta\gamma}\,\theta^\beta\wedge\theta^\gamma$ and express both sides in terms of $F^\alpha_{\ \beta\gamma}$.

Exercise 2.15. In the Euclidean path integral approach to quantum gravity, one is interested in solutions of the (classical) *Euclidean* Einstein equations. (Of particular interest are finite action solutions, called **gravitational instantons** because of the close analogy to the Yang-Mills instantons.) A variety of solutions with selfdual curvature are known in the Euclidean case.

Rewrite the Euclidean vacuum field equations in terms of the "dual" curvature forms

$$\tilde{\Omega}_{\alpha\beta} = \frac{1}{2}\eta_{\alpha\beta\gamma\delta}\,\Omega^{\gamma\delta}$$

and show that they are automatically satisfied if the curvature is (anti-) selfdual $\tilde{\Omega}_{\alpha\beta} = \pm\Omega_{\alpha\beta}$. What follows from $*\tilde{\Omega}_{\alpha\beta} = \pm\Omega_{\alpha\beta}$?

Exercise 2.16. Equation (2.113) can be written in the form

$$\delta\int_D \mathcal{L}_M\eta = \int_D *J\wedge\delta A .$$

Use this and (2.182) to derive Maxwell's equations from the variational principle

$$\delta \int_D \left(-\frac{1}{8\pi} F \wedge *F + \mathcal{L}_M \eta \right) = 0 .$$

Exercise 2.17. Express the term for $m = 2$ in (2.185) explicitly in terms of the Riemann tensor.

The result is

$$\Omega^{\alpha\beta} \wedge \Omega^{\gamma\delta} \eta_{\alpha\beta\gamma\delta} =: \mathcal{L}_{GB}\, \eta ,$$

where the **Gauss-Bonnet scalar** $\mathcal{L}_{GB}$ is given by

$$\mathcal{L}_{GB} = R_{\alpha\beta\gamma\delta}\, R^{\alpha\beta\gamma\delta} - 4\, R_{\alpha\beta} R^{\alpha\beta} + R^2 .$$

Energetic readers may derive its varitational derivative, which is proportional to

$$H_{\mu\nu} = (R\, R_{\mu\nu} - 2 R_{\mu\nu} R^{\lambda}_{\ \nu} - 2\, R^{\sigma\rho} R_{\mu\sigma\nu\rho} + R_{\mu}^{\ \sigma\rho\lambda} R_{\nu\sigma\rho\lambda}) - \frac{1}{4}\, g_{\mu\nu} \mathcal{L}_{GB} .$$

2.7 Energy, Momentum, and Angular Momentum for Isolated Systems

Far away from an isolated system (like a neutron star), where spacetime is nearly flat, one expects that mass, momentum, and angular momentum can be defined in terms of the asymptotically weak gravitational field, and that these quantities are conserved. Surprisingly, this is an elusive subject in GR, worrying relativists up to the present day.

In this section we derive the so-called **Arnowitt-Deser-Misner (ADM)** expressions for energy, momentum, and angular momentum at *spatial* infinity. These concepts are not very useful for discussing gravitational radiation. For this one has to define mass and momentum at *null* infinity, called **Bondi mass** and **momentum**. This construction, introduced in the early 1960s, will described much later (see Sect. 5.1).

We first rewrite the field equations in the form of a continuity equation, from which differential "conservation laws" follow immediately. Our starting point is (2.162) in the form

$$-\frac{1}{2} \Omega_{\beta\gamma} \wedge \eta^{\beta\gamma}{}_{\alpha} = 8\pi G * T_{\alpha} . \tag{2.188}$$

We rewrite the left-hand side with the second structure equation[9]

[9] Using $dg_{\alpha\beta} = \omega_{\alpha\beta} + \omega_{\beta\alpha}$ one easily derives from

$$\Omega^{\alpha}_{\ \beta} = d\omega^{\alpha}{}_{\beta} + \omega^{\alpha}{}_{\lambda} \wedge \omega^{\lambda}{}_{\beta}$$

that

$$\Omega_{\beta\gamma} = d\omega_{\beta\gamma} - \omega_{\alpha\beta} \wedge \omega^{\sigma}{}_{\gamma}. \tag{2.189}$$

The contribution of the first term is proportional to

$$d\omega_{\beta\gamma} \wedge \eta^{\beta\gamma}{}_{\alpha} = d(\omega_{\beta\gamma} \wedge \eta^{\beta\gamma}{}_{\alpha}) + \omega_{\beta\gamma} \wedge d\eta^{\beta\gamma}{}_{\alpha}\,. \tag{2.190}$$

According to (2.152) $D\eta^{\beta\gamma}{}_{\alpha} = 0$, so that

$$d\eta^{\beta\gamma}{}_{\alpha} + \omega^{\beta}{}_{\sigma} \wedge \eta^{\sigma\gamma}{}_{\alpha} + \omega^{\gamma}{}_{\sigma} \wedge \eta^{\beta\sigma}{}_{\alpha} - \omega^{\sigma}{}_{\alpha} \wedge \eta^{\beta\gamma}{}_{\sigma} = 0\,.$$

If this is used in (2.190), the result is

$$d\omega_{\beta\gamma} \wedge \eta^{\beta\gamma}{}_{\alpha} = d(\omega_{\beta\gamma} \wedge \eta^{\beta\gamma}{}_{\alpha}) + \omega_{\beta\gamma} \wedge \left(-\omega^{\beta}{}_{\sigma} \wedge \eta^{\sigma\gamma}{}_{\alpha} - \omega^{\gamma}{}_{\sigma} \wedge \eta^{\beta\sigma}{}_{\alpha} + \omega^{\sigma}{}_{\alpha} \wedge \eta^{\beta\gamma}{}_{\sigma}\right).$$

If we add the contribution of the second term of (2.189) to this, we obtain the identity

$$-\frac{1}{2}\Omega_{\beta\gamma} \wedge \eta^{\beta\gamma}{}_{\alpha} = -\frac{1}{2}d(\omega_{\beta\gamma} \wedge \eta^{\beta\gamma}{}_{\alpha}) - 8\pi G * t_{\alpha}\,, \tag{2.191}$$

where[10]

$$*t_{\alpha} = \frac{1}{16\pi G}\omega_{\beta\gamma} \wedge \left(\omega_{\sigma\alpha} \wedge \eta^{\beta\gamma\sigma} - \omega^{\gamma}{}_{\sigma} \wedge \eta^{\beta\gamma}{}_{\alpha}\right). \tag{2.192}$$

This quantity is closely related to Einstein's 'energy-momentum' complex which he introduced in 1916.

Using (2.191) in (2.188) we obtain Einstein's field equations in the form

$$-\frac{1}{2}d\left(\omega_{\beta\gamma} \wedge \eta^{\beta\gamma}{}_{\alpha}\right) = 8\pi G * (T_{\alpha} + t_{\alpha})\,. \tag{2.193}$$

This implies the 'conservation laws'

$$d(*T_{\alpha} + *t_{\alpha}) = 0\,. \tag{2.194}$$

$$\Omega_{\beta\gamma} = d\omega_{\beta\gamma} - \omega_{\alpha\beta} \wedge \omega^{\sigma}{}_{\gamma}\,.$$

In a similar manner, one obtains from the first structure equation

$$d\theta_{\beta} - \omega^{\sigma}{}_{\beta} \wedge \theta_{\sigma} = 0\,.$$

[10] **Remark (for Readers Familiar with Connections in Principle Fibre Bundles).** The identity (2.191) and its derivation can be interpreted globally on the frame bundle, when ω is regarded as an $so(1,3)$-valued connection form and θ as the canonical $\mathbb{R}^n$-valued 1-form (soldering form). In standard terminology (see [35]), $*t_{\alpha}$ is a pseudo-tensorial 3-form on the principle frame bundle.

At first sight one might interprete $*t_\alpha$ as energy and momentum 3-forms of the gravitational field. However, we know from the discussion in Sect. 2.4 that it is meaningless to talk about such concepts. This is reflected here by the fact that $*t_\alpha$ does not transform as a tensor with respect to gauge transformations. When $\omega_{\beta\gamma}(x) = 0$ (which can always be made to hold at some given point x), then $*t_\alpha = 0$. Conversely, $*t_\alpha$ vanishes even in flat space only with respect to global Lorentz systems. A physical meaning can at best be assigned to integrals of $*t_\alpha$ over spacelike hypersurfaces. For isolated systems with asymptotically flat geometry, this is in fact the case, when one chooses the reference frame $\{\theta^\alpha\}$ to be asymptotically Lorentzian[11].

In order to be able to define the total angular momentum in such cases, we need a conservation law of the form (2.194) so that $t_{\alpha\beta}$ (in $t_\alpha = t_{\alpha\beta}\theta^\beta$) is *symmetric* with respect to a natural basis. Unfortunately, this is not the case for (2.192). For this reason, we rewrite the Einstein equations in still another form[12].

As a starting point we again take (2.162), and insert

$$\eta^{\alpha\beta\gamma} = \eta^{\alpha\beta\gamma\delta}\theta_\delta \, . \qquad (2.195)$$

Together with (2.189) this gives (the basis is not necessarily orthonormal)

$$-\frac{1}{2}\eta^{\alpha\beta\gamma\delta}\theta_\delta \wedge \left(d\omega_{\beta\gamma} - \omega_{\alpha\beta}\wedge\omega^\sigma{}_\gamma\right) = 8\pi G * T^\alpha \, .$$

In the first term, we perform a "partial integration" and use the first structure equation (see the footnote on p. 110)

$$d(\omega_{\beta\gamma}\wedge\theta_\delta) = \theta_\delta \wedge d\omega_{\beta\gamma} - \omega_{\beta\gamma}\wedge \underbrace{d\theta_\delta}_{\omega_{\sigma\delta}\wedge\theta^\sigma} \, .$$

Hence,

$$-\frac{1}{2}\eta^{\alpha\beta\gamma\delta} d(\omega_{\beta\gamma}\wedge\theta_\delta) = 8\pi G * (T^\alpha + t^\alpha_{LL}) \, , \qquad (2.196)$$

where the right-hand side now contains the so called **Landau-Lifshitz 3-forms**. They are given by

$$*t^\alpha_{LL} = -\frac{1}{16\pi G}\eta^{\alpha\beta\gamma\delta}\left(\omega_{\alpha\beta}\wedge\omega^\sigma{}_\gamma\wedge\theta_\delta - \omega_{\beta\gamma}\wedge\omega_{\sigma\delta}\wedge\theta^\sigma\right) . \qquad (2.197)$$

We now multiply (2.196) by $\sqrt{-g}$; it follows from

$$\eta^{\alpha\beta\gamma\delta} = -\frac{1}{\sqrt{-g}}\varepsilon_{\alpha\beta\gamma\delta} \, ,$$

[11] The use of pseudotensors appeared to many researchers in GR to violate the whole spirit of this generally invariant theory, and criticism of Einstein's conservation law was widespread (see, e.g., the discussions of Einstein with F. Klein, [60]). Einstein defended his point of view in detail in [59].

[12] This discussion follows partly [30], Sect. 4.2.11.

that

$$-d(\sqrt{-g}\,\eta^{\alpha\beta\gamma\delta}\omega_{\beta\gamma}\wedge\theta_\delta) = 16\pi G\sqrt{-g}\,(*T^\alpha + *t^\alpha_{LL})$$

or

$$-d(\sqrt{-g}\,\omega^{\beta\gamma}\wedge\eta^\alpha{}_{\beta\gamma}) = 16\pi G\sqrt{-g}\,(*T^\alpha + *t^\alpha_{LL})\,. \tag{2.198}$$

From this we obtain the differential conservation law

$$d\Big(\sqrt{-g}(*T^\alpha + *t^\alpha_{LL})\Big) = 0\,. \tag{2.199}$$

In a coordinate basis $\theta^\alpha = dx^\alpha$, $t^{\alpha\beta}$ corresponding to (2.197) is now symmetric (see Exercise 2.17)

$$dx^\rho \wedge *t^\alpha_{LL} = dx^\alpha \wedge *t^\rho_{LL}\,. \tag{2.200}$$

As before, the Landau-Lifshitz 3-forms (2.197) do not transform as a tensor under gauge transformations. In the following we use the notation

$$\tau^\alpha = T^\alpha + t^\alpha_{LL}\,. \tag{2.201}$$

From (2.199), i.e., $d(\sqrt{-g} * \tau^\alpha) = 0$ and the symmetry of τ^α,

$$dx^\alpha \wedge *\tau^\beta = dx^\beta \wedge *\tau^\alpha\,, \tag{2.202}$$

we obtain

$$d(\sqrt{-g} * M^{\alpha\beta}) = 0\,, \tag{2.203}$$

where

$$*M^{\alpha\beta} = x^\alpha * \tau^\beta - x^\beta * \tau^\alpha\,. \tag{2.204}$$

In fact, we have

$$d(\sqrt{-g} * M^{\alpha\beta}) = dx^\alpha \wedge *\tau^\beta - dx^\beta \wedge *\tau^\alpha = 0\,.$$

Interpretation

We consider an isolated system with asymptotically flat geometry. All coordinate systems will be assumed to be asymptotically Lorentzian. Consider a spacelike hypersurface Σ which is non-singular and asymptotically flat (precisions will follow). We interpret

$$P^\alpha = \int_\Sigma \sqrt{-g} * \tau^\alpha \tag{2.205}$$

as the total (ADM) four-momentum and

$$J^{\alpha\beta} = \int_\Sigma \sqrt{-g} * M^{\alpha\beta} \tag{2.206}$$

as the total (ADM) angular momentum of the isolated system. These quantities can be decomposed according to (2.201) into contributions from matter and from the gravitational field. The total four momentum P^α and the total angular momentum $J^{\alpha\beta}$ are constant in time if the gravitational fields fall off sufficiently fast at spacelike infinity. (This will be described in detail later.)

We can use the field equations to express P^α and $J^{\alpha\beta}$ in terms of two-dimensional flux integrals. These will be considered as the appropriate definitions of energy, momentum, and angular momentum at spacelike infinity.

If we integrate (2.198) over a three-dimensional spacelike region $D_3 \subset \Sigma$, we obtain

$$16\pi G \int_{D_3} \sqrt{-g} * \tau^\alpha = -\int_{\partial D_3} \sqrt{-g}\, \omega^{\beta\gamma} \wedge \eta^\alpha{}_{\beta\gamma} .$$

In the limit when D_3 becomes all of Σ we obtain

$$P^\alpha = -\frac{1}{16\pi G} \int_{S^2_\infty} \sqrt{-g}\, \omega^{\beta\gamma} \wedge \eta^\alpha{}_{\beta\gamma} . \tag{2.207}$$

The region of integration must be extended over a "surface at infinity". Note that one obtains the same expression for the four-momentum from (2.193). We shall see that (2.207) agrees with the original ADM four-momentum.

We now write the total angular momentum also as a flux integral. If we use the field equations (in the form (2.198)) in (2.204), we obtain

$$\begin{aligned} 16\pi G \sqrt{-g} * M^{\rho\alpha} &= x^\rho dh^\alpha - x^\alpha dh^\rho \\ &= d(x^\rho h^\alpha - x^\alpha h^\rho) - (dx^\rho \wedge h^\alpha - dx^\alpha \wedge h^\rho) , \end{aligned} \tag{2.208}$$

where

$$h^\alpha := -\sqrt{-g}\, \omega^{\beta\gamma} \wedge \eta^\alpha{}_{\beta\gamma} . \tag{2.209}$$

The last term on the right-hand side can also be written as an exact differential. We have

$$\begin{aligned} dx^\rho \wedge h^\alpha - dx^\alpha \wedge h^\rho &= \sqrt{-g}\, \omega^{\beta\gamma} \wedge \underbrace{dx^\rho \wedge \eta^\alpha{}_{\beta\gamma}}_{\delta^\rho{}_\gamma \eta^\alpha{}_\beta + \delta^\rho{}_\beta \eta_\gamma{}^\alpha + g^{\rho\alpha}\eta_{\beta\gamma}} - (\alpha \longleftrightarrow \rho) \\ &= \sqrt{-g}\Big(\omega^{\beta\rho} \wedge \eta^\alpha_\beta + \omega^{\rho\beta} \wedge \eta_\beta^\alpha - (\alpha \longleftrightarrow \rho)\Big) \\ &= \sqrt{-g}\Big(\omega_\beta{}^\rho \wedge \eta^{\alpha\beta} + \omega^\rho{}_\beta \wedge \eta^{\beta\alpha} - \\ &\quad -\omega_\beta{}^\alpha \wedge \eta^{\rho\beta} + \omega^\alpha{}_\beta \wedge \eta^{\beta\rho}\Big) . \end{aligned}$$

We now use $D\eta^{\rho\alpha} = 0$, which means that

$$d\eta^{\rho\alpha} + \omega^\rho{}_\beta \wedge \eta^{\beta\alpha} + \omega^\alpha{}_\beta \wedge \eta^{\eta\beta} = 0 ,$$

and obtain

$$dx^\rho \wedge h^\alpha - dx^\alpha \wedge h^\rho = \sqrt{-g}\Big(\omega_\beta{}^\rho \wedge \eta^{\alpha\beta} - (\rho \longleftrightarrow \alpha) - d\eta^{\rho\alpha}\Big). \tag{2.210}$$

However, if we insert (using $\omega^\mu{}_\nu = \Gamma^\mu{}_{\alpha\nu} dx^\alpha$)

$$\omega_\beta{}^\rho \wedge \eta^{\alpha\beta} = \Gamma_{\beta\mu}{}^\rho dx^\mu \wedge \eta^{\alpha\beta} = \Gamma_\beta{}^{\beta\rho}\eta^\alpha - \Gamma_\beta{}^{\alpha\rho}\eta^\beta$$

in (2.210), we obtain

$$dx^\rho \wedge h^\alpha - dx^\alpha \wedge h^\rho = \sqrt{-g}\Big(\Gamma^\beta{}_\beta{}^\rho \eta^\alpha - \Gamma^\beta{}_\beta{}^\alpha \eta^\rho - d\eta^{\rho\alpha}\Big).$$

Since

$$\Gamma^\beta{}_\beta{}^\rho = \frac{1}{\sqrt{-g}} g^{\mu\rho}\partial_\mu \sqrt{-g}\,,$$

we finally obtain

$$\begin{aligned} dx^\rho \wedge h^\alpha - dx^\alpha \wedge h^\rho &= -\sqrt{-g}\, d\eta^{\rho\alpha} + \partial_\mu \sqrt{-g}\underbrace{\left(g^{\mu\rho}\eta^\alpha - g^{\mu\alpha}\eta^\rho\right)}_{dx^\mu \wedge \eta^{\alpha\rho}} \\ &= -\sqrt{-g}\, d\eta^{\rho\alpha} - d\sqrt{-g} \wedge \eta^{\rho\alpha}\,, \end{aligned}$$

so that

$$dx^\rho \wedge h^\alpha - dx^\alpha \wedge h^\rho = -d\big(\sqrt{-g}\,\eta^{\rho\alpha}\big). \tag{2.211}$$

If we make use of this in (2.208), we obtain for the total angular momentum (2.206)

$$J^{\rho\alpha} = \frac{1}{16\pi G}\int_{S^2_\infty} \Big((x^\rho h^\alpha - x^\alpha h^\rho) + \sqrt{-g}\,\eta^{\rho\alpha}\Big)$$

or

$$J^{\rho\alpha} = \frac{1}{16\pi G}\int_{S^2_\infty} \sqrt{-g}\Big((x^\rho \eta^\alpha{}_{\beta\gamma} - x^\alpha \eta^\rho{}_{\beta\gamma}) \wedge \omega^{\beta\gamma} + \eta^{\rho\alpha}\Big). \tag{2.212}$$

Again, regard this as our primary definition of angular momentum at spacelike infinity.

P^α and $J^{\alpha\beta}$ are gauge invariant in the following sense: Under a transformation

$$\begin{aligned} \theta(x) &\longrightarrow A(x)\theta(x)\,, \\ \omega(x) &\longrightarrow A(x)\omega(x)A^{-1}(x) - dA(x)A^{-1}(x)\,, \end{aligned} \tag{2.213}$$

which reduces asymptotically to the identity, the flux integrals (2.207) and (2.212) remain invariant. To see this we note that the homogeneous contributions to (2.213) obviously do not change the flux integrals. The inhomogeneous term gives an additional surface integral of an exact differential, which vanishes by Stokes' theorem.

Thus P^α and $J^{\alpha\beta}$ transform tensorially under every transformation which leaves the flat metric asymptotically invariant, since every such transformation can be represented as the product of a Lorentz transformation (with respect to which P^α and $J^{\alpha\beta}$ transform as tensors) and a transformation which reduces to the identity asymptotically.

In order to establish the connection with presentations found in other texts (for example, [10]), we use the result of Exercise 2.19. This allows us to write the field equation (2.198) in the form

$$H^{\mu\alpha\nu\beta}{}_{,\alpha\beta} = 16\pi G(-g)(T^{\mu\nu} + t^{\mu\nu}_{LL}). \tag{2.214}$$

The expression (2.197) for $t^{\mu\nu}_{LL}$ can be written explicitly in terms of the metric. The result is (see [10])

$$\begin{aligned}(-g)t^{\alpha\beta}_{LL} = \frac{1}{16\pi G}\Big\{&\tilde{g}^{\alpha\beta}{}_{,\lambda}\tilde{g}^{\lambda\mu}{}_{,\mu} - \tilde{g}^{\alpha\lambda}{}_{,\lambda}\tilde{g}^{\beta\mu}{}_{,\mu} \\ &+\frac{1}{2}g^{\alpha\beta}g_{\lambda\mu}\tilde{g}^{\lambda\nu}{}_{,\rho}\tilde{g}^{\rho\mu}{}_{,\nu} - \big(g^{\alpha\lambda}g_{\mu\nu}\tilde{g}^{\beta\nu}{}_{,\rho}\tilde{g}^{\mu\rho}{}_{,\lambda} \\ &+g^{\beta\lambda}g_{\mu\nu}\tilde{g}^{\alpha\nu}{}_{,\rho}\tilde{g}^{\mu\rho}{}_{,\lambda}\big) + g_{\mu\lambda}g^{\nu\rho}\tilde{g}^{\alpha\lambda}{}_{,\nu}\tilde{g}^{\beta\mu}{}_{,\rho} \\ &+\frac{1}{8}\big(2g^{\alpha\lambda}g^{\beta\mu} - g^{\alpha\beta}g^{\lambda\mu}\big)\big(2g_{\nu\rho}g_{\sigma\tau} - g_{\rho\sigma}g_{\nu\tau}\big)\tilde{g}^{\nu\tau}{}_{,\lambda}\tilde{g}^{\rho\sigma}{}_{,\mu}\Big\}.\end{aligned} \tag{2.215}$$

This complicated expression is quadratic in the $\tilde{g}^{\alpha\beta}{}_{,\mu}$, where $\tilde{g}^{\alpha\beta} := \sqrt{-g}\,g^{\alpha\beta}$.

Sketch of Derivation of (2.215). Clearly, $t^{\alpha\beta}$ is a linear combination of products of two Christoffel symbols. Next, we note that $\Gamma^{\mu\alpha\beta} := g^{\alpha\sigma}g^{\beta\rho}\Gamma^\mu{}_{\sigma\rho}$ can be written as

$$\Gamma^{\mu\alpha\beta} = \frac{1}{2}\Big(g^{\mu\rho}g^{\alpha\beta}{}_{,\rho} - g^{\alpha\rho}g^{\beta\mu}{}_{,\rho} - g^{\beta\rho}g^{\alpha\mu}{}_{,\rho}\Big).$$

If we use here

$$\tilde{g}^{\alpha\beta}{}_{,\nu} = \sqrt{-g}\big(g^{\alpha\beta}{}_{,\nu} + g^{\alpha\beta}y_\nu\big),$$

where

$$y_\nu = (\ln\sqrt{-g})_{,\nu} = \Gamma^\nu{}_{\nu\lambda} = \frac{1}{2\sqrt{-g}}g_{\sigma\rho}\tilde{g}^{\rho\sigma}{}_{,\nu},$$

we obtain

$$\Gamma^{\mu\alpha\beta} = \Pi^{\mu\alpha\beta} + \Lambda^{\mu\alpha\beta}$$

with

$$\Pi^{\mu\alpha\beta} = \frac{1}{2g}\Big(\tilde{g}^{\alpha\rho}\tilde{g}^{\mu\beta}{}_{,\rho} + \tilde{g}^{\beta\rho}\tilde{g}^{\mu\alpha}{}_{,\rho} - \tilde{g}^{\mu\rho}\tilde{g}^{\alpha\beta}{}_{,\rho}\Big),$$

$$\Lambda^{\mu\alpha\beta} = \frac{1}{2}\Big(g^{\mu\beta}y^\alpha + g^{\mu\alpha}y^\beta - g^{\alpha\beta}y^\mu\Big).$$

This shows that $t^{\alpha\beta}$ is a sum of terms quadratic in $\tilde{g}^{\alpha\beta}{}_{,\mu}$, with coefficients depending only of the metric tensor (and its derivatives). The tedious, but straightforward part of the calculation is to collect the many terms.

It is actually much simpler to derive the result by making use of Appendix B in [4], see also Exercise 2.20.

ADM Expressions for Energy and Momentum

Before rewriting expression (2.207) solely in terms of data on Σ, we state more precisely what we mean by **asymptotic flatness** of the hypersurface Σ, which we assume to be complete[13] to exclude singularities.

We assume that there is a compact set $\mathcal{C} \subset \Sigma$ such that the complement $\Sigma \setminus \mathcal{C}$ is diffeomorphic to the complement of a contractable compact set in $\mathbb{R}^3$. (More generally, $\Sigma \setminus \mathcal{C}$ could be a disjoint union of a finite number of such sets.) Under this diffeomorphism the metric on $\Sigma \setminus \mathcal{C}$ should be of the form

$$g_{ij} = \delta_{ij} + h_{ij} \tag{2.216}$$

in the standards coordinates of $\mathbb{R}^3$, where $h_{ij} = O(1/r)$, $\partial_k h_{ij} = O(1/r^2)$ and $\partial_l \partial_k h_{ij} = O(1/r^3)$. Furthermore, the second fundamental form K_{ij} of $\Sigma \subset M$ should satisfy $K_{ij} = O(1/r^2)$ and $\partial_k K_{ij} = O(1/r^3)$.

Now, we rewrite (2.207), working with coordinates for which the asymptotic conditions stated above are satisfied. If $\doteq$ denotes 'asymptotic equality', we have

$$\omega_{\alpha\beta} \doteq \frac{1}{2}\left(g_{\alpha\gamma,\beta} + g_{\alpha\beta,\gamma} - g_{\beta\gamma,\alpha}\right) dx^\gamma \,. \tag{2.217}$$

Consider first the energy

$$\begin{aligned} P^0 &= \frac{1}{16\pi G} \varepsilon^{0ij}{}_l \int_{S^2_\infty} g_{jk,i}\, dx^k \wedge dx^l \\ &= \frac{1}{16\pi G} \int_{S^2_\infty} \varepsilon_{ijl}\, g_{jk,i}\, dx^k \wedge dx^l \,. \end{aligned}$$

Here, we use $dx^k \wedge dx^l = \varepsilon_{kls} N^s \, dS$, where N^s is the outward normal in Σ

[13] A pseudo-Riemannian manifold is **geodesically complete** if every maximal geodesic is defined on the entire real line. For Riemannian manifolds there is another notion of completeness. Introduce the Riemannian distance $d(p, q)$ between two points as the infimum of $L(\gamma)$ for all piecewise smooth curve segments from p to q. This makes the manifold into a metric space whose topology coincides with the original one. According to an important theorem of Hopf and Rinow, a Riemannian manifold is **complete** as a metric space *if and only if* it is geodesically complete. (Another equivalent statement is that any closed bounded subset is compact.) For such a space each pair of points can be connected by a geodesic. For proofs see, e.g., [41].

of the surface S^2_∞ 'at infinity' and dS its volume form[14]. Then we find the standard ADM expression

$$P^0 = \frac{1}{16\pi G}\int_{S^2_\infty}(\partial_j g_{ij} - \partial_i g_{jj})N^i\,dS\,. \tag{2.218}$$

For the momentum we have in a first step

$$\begin{aligned} P^r &= -\frac{1}{16\pi G}\varepsilon^{r\alpha\beta}{}_s\int_{S^2_\infty}\omega_{\alpha\beta}\wedge dx^s \\ &= -\frac{1}{16\pi G}\varepsilon^{r0j}{}_s\,2\int_{S^2_\infty}\omega_{0j}\wedge dx^s\,. \end{aligned}$$

Here, we can introduce the second fundamental form, using DG, (A.8a), $\omega_{0j}|_\Sigma = K_{ij}\,dx^j$. Then

$$P^r = -\frac{1}{8\pi G}\varepsilon_{rjs}\int_{S^2_\infty}K_{ij}\,dx^i\wedge dx^s$$

or, as before,

$$P_i = \frac{1}{8\pi G}\int_{S^2_\infty}(K_{ij} - \delta_{ij}K^l{}_l)N^j\,dS\,. \tag{2.219}$$

This is the standard ADM expression. We emphasize that (2.218) and (2.219) involve only the asymptotics of the induced metric and of the extrinsic curvature on Σ. The integrals have to be understood, of course, as limits over two-surfaces S^2_R for $R \longrightarrow \infty$.

Positive Energy Theorem

What about the sign of the ADM energy? To answer this question we have to make an general assumption about the energy-momentum tensor T which is satisfied for "normal" matter.

Dominant Energy Condition. *For each point $p \in \Sigma$ and for each timelike vector ξ^μ at p, $T_{\mu\nu}\xi^\mu\xi^\nu \geq 0$ and $-T^\mu{}_\nu\,\xi^\nu$ is a future-directed non-spacelike vector for all future-directed timelike vectors ξ^μ.*

Under this assumption it has been shown that the ADM four-momentum P^μ is a future directed timelike or null vector. Furthermore, $P^\mu = 0$ if and only if spacetime is flat in a neighborhood of Σ.

This **positive energy theorem** says roughly: One cannot construct an object out of "ordinary" matter, i.e., matter with positive local energy density, whose total energy (including gravitational contributions) is negative.

[14] Consider a 3-dimensional Riemannian manifold (N, h) with volume form $\Omega = \sqrt{h}\,dx^1\wedge dx^2\wedge dx^3$. If one writes the equation DG, (12.51),

$$i_X\Omega = \langle X, N\rangle dS\,,$$

in terms of coordinates, one easily finds $dx^j \wedge dx^k = N_i\,\eta^{ijk}dS$.

The first proof of this important theorem was given by R. Schoen and S.T. Yau, [84]. Later, E. Witten found a much simpler proof which makes crucial use of spinor fields (see [223]). Witten showed that it is possible to write with this tool the four-momentum in such a way that the stated properties can be read off. We shall give this proof in Chap. 8, where more will be said about the significance of the positive energy theorem.

* * *

Exercises.

Exercise 2.18. Use the fact that $\Gamma^{\mu}{}_{\alpha\beta} = \Gamma^{\mu}{}_{\beta\alpha}$ in a coordinate basis to verify (2.200).

Exercise 2.19. Show that the left-hand side of the field equations (2.198) can be written in the form

$$-d\left(\sqrt{-g}\,\omega^{\alpha\beta} \wedge \eta^{\mu}{}_{\alpha\beta}\right) = \frac{1}{\sqrt{-g}} H^{\mu\alpha\nu\beta}{}_{,\alpha\beta}\eta_{\nu}\,, \tag{2.220}$$

where

$$H^{\mu\alpha\nu\beta} := \tilde{g}^{\mu\nu}\tilde{g}^{\alpha\beta} - \tilde{g}^{\alpha\nu}\tilde{g}^{\beta\mu} \tag{2.221}$$

is the so called Landau-Lifshitz “super-potential”.

Solution. First note that

$$\begin{aligned} -\sqrt{-g}\,\omega^{\alpha\beta} \wedge \eta^{\mu}{}_{\alpha\beta} &= -(-g)\left(\omega^{\alpha\beta} \wedge dx^{\lambda}\right) g^{\mu\gamma}\,\varepsilon_{\gamma\alpha\beta\lambda} \\ &= -(-g) g^{\mu\gamma} g^{\alpha\tau} g^{\beta\rho} \\ &\quad \times \frac{1}{2}\Big(g_{\sigma\tau,\rho} + g_{\tau\rho,\sigma} - g_{\rho\sigma,\tau}\Big) dx^{\sigma} \wedge dx^{\lambda}\,\varepsilon_{\gamma\alpha\beta\lambda}\,. \end{aligned}$$

Hence, we have

$$\begin{aligned} d\left(\sqrt{-g}\,\omega^{\alpha\beta} \wedge \eta^{\mu}{}_{\alpha\beta}\right) = &-\frac{1}{2}\varepsilon_{\gamma\alpha\beta\lambda}\Big\{(-g) g^{\mu\gamma} g^{\alpha\tau} g^{\beta\rho} \\ &\times \left(g_{\sigma\tau,\rho} + g_{\tau\rho,\sigma} - g_{\rho\sigma,\tau}\right)\Big\}_{,\kappa}\; dx^{\kappa} \wedge dx^{\sigma} \wedge dx^{\lambda}\,. \end{aligned}$$

Due to symmetry, the second term does not contribute. If we denote the left-hand side of the last equation by $(\sqrt{-g})^{-1} H^{\mu\nu}\eta_{\nu}$, we have

$$\frac{1}{\sqrt{-g}} H^{\mu\nu}\eta = -\frac{1}{2}\varepsilon_{\gamma\alpha\beta\lambda}\Big\{\cdots\cdots\Big\}_{,\kappa}\; \underbrace{dx^{\nu} \wedge dx^{\kappa} \wedge dx^{\sigma} \wedge dx^{\lambda}}_{\frac{1}{\sqrt{-g}}\varepsilon_{\nu\kappa\sigma\lambda}\,\eta}$$

or (see DG, Eq. (12.40))

$$H^{\mu\nu} = -\frac{1}{2} 3! \delta^{\nu}_{[\gamma} \delta^{\kappa}_{\alpha} \delta^{\sigma}_{\beta]} \left\{ (-g) g^{\mu\gamma} g^{\alpha\tau} g^{\beta\rho} (g_{\tau\sigma,\rho} - g_{\rho\sigma,\tau}) \right\}_{,\kappa} .$$

Since the last factor is antisymmetric in ρ and τ, it is no longer necessary to antisymmetrize in α and β. Hence only a cyclic sum over (γ, α, β) remains in the expression for $H^{\mu\nu}$

$$H^{\mu\nu} = - \sum_{(\gamma,\alpha,\beta)} \delta^{\nu}_{\gamma} \delta^{\kappa}_{\alpha} \delta^{\sigma}_{\beta} \left\{ \ldots\ldots \right\}_{,\kappa} .$$

The first term in the sum is

$$-\left\{ (-g) g^{\mu\nu} g^{\alpha\tau} g^{\beta\rho} (g_{\beta\tau,\rho} - g_{\rho\beta,\tau}) \right\}_{,\alpha} . \tag{2.222}$$

We now use

$$g^{\beta\rho} g_{\beta\rho,\tau} = \frac{1}{g} g_{,\tau}$$

and

$$g^{\alpha\tau} g^{\beta\rho} g_{\beta\tau,\rho} = -g^{\alpha\rho}{}_{,\rho} .$$

Hence (2.222) is equal to

$$\left(g^{\mu\nu} (-g\, g^{\alpha\beta})_{,\beta} \right)_{,\alpha} .$$

In an analogous manner one can simplify the other terms in the cyclic sum and one easily finds

$$H^{\mu\nu} = H^{\mu\alpha\nu\beta}{}_{,\alpha\beta} ,$$

which is what we wanted to show.

Exercise 2.20. Prove the identity

$$(-g) G^{\mu\nu} = \frac{1}{2} H^{\mu\alpha\nu\beta}{}_{,\alpha\beta} - 8\pi G (-g) t^{\mu\nu}_{LL} , \tag{2.223}$$

with the expressions (2.215) for $t^{\mu\nu}_{LL}$, using the tools developed in Appendix B of [4]. More precisely, start from (B.87) of this reference, and use for the various quantities appearing in this equation for $G^{\mu\nu}$ the formulas (B.08), (B.37), (B.43), (B.58), (B.65) and (B.95).

2.8 The Initial Value Problem of General Relativity

"Mathematicians are like theologians: we regard existence as the prime attribute of what we study. But unlike most theologians, we need not always rely upon faith alone."

—L.C. Evans[15] (1998)

[15] Quote from the textbook [46] by L.C. Evans on partial differential equations. We highly recommend this well written clear graduate text to all those who are eager to read a detailed modern introduction to this vast field of mathematics.

A study of the **Cauchy problem** in GR provides a deeper understanding of the structure of Einstein's field equations. We give an introduction to the subject but discuss only the local problem in some detail. Moreover, we shall avoid technicalities (use of Sobolev spaces and all that). Our aim is to show that the local problem is conceptually very similar to the initial value problem in electrodynamics, although much more difficult because of non-linearities.

In this section we shall make use of the fundamental equations for hypersurfaces of pseudo-Riemannian manifolds, developed in DG, Appendix A. Standard reviews of the subject are [81], [82] and [8]. Global aspects are treated in the recent monograph [83]. A useful source will be the proceedings of the Cargèse summer school (2002) on the Cauchy problem in GR (see [96]).

2.8.1 Nature of the Problem

Recall that any spacelike hypersurface Σ of a Lorentz manifold (M, g) inherits a Riemannian metric $\bar{g}$ (the induced metric or first fundamental form) and the second fundamental form $\bar{K}$. The equations of Gauss and Codazzi-Mainardi show that certain components of the Einstein tensor for g are determined entirely by $\bar{g}$ and $\bar{K}$. If Einstein's vacuum equations hold (we shall discuss only this case) these components have to vanish. Conversely, these equations must be imposed as constraints for the initial data of the Cauchy problem. These remarks lead us to some basic definitions.

An **initial data** set is a triple $(\mathcal{S}, h, K)$, where $(\mathcal{S}, h)$ is a three-dimensional Riemannian manifold and K a covariant 2-tensor on $\mathcal{S}$. A **development** of $(\mathcal{S}, h, K)$ is a triple (M, g, σ), where (M, g) is a Lorentz manifold and σ a diffeomorphism of $\mathcal{S}$ onto a spacelike hypersurface Σ in M, such that the induced metric and the second fundamental form for Σ are the images of h and K under the diffeomorphism $\sigma : \mathcal{S} \longrightarrow \Sigma$. A different development (M', g', σ') is an **extension** of (M, g, σ), provided there exists an injective differentiable map $\phi : M \longrightarrow M'$ that transforms g into g' and respects the images of $\mathcal{S}$ pointwise: $\sigma' = \phi \circ \sigma$ (see Fig. 2.3). A map which satisfies the

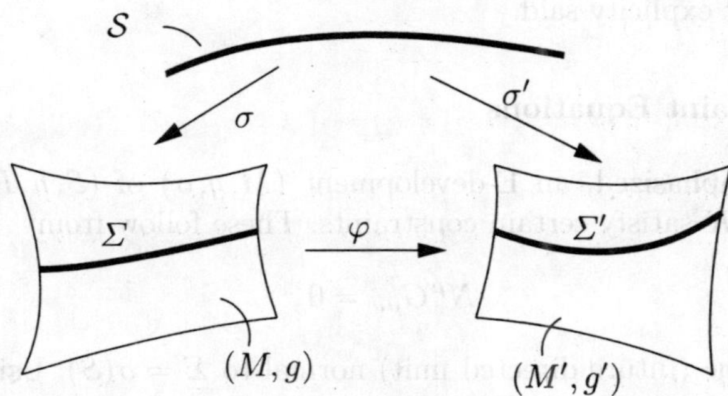

Fig. 2.3. Extension of a development ($\sigma' = \phi \circ \sigma$).

latter property will be called **admissable**. Two developments are **equivalent** if each one is an extension of the other, i.e., if there exists an admissable diffeomorphism $\phi : M \longrightarrow M'$ with $\phi_* g = g'$. A special case would be two extensions (M, g, σ) and (M, g', σ) with a diffeomorphism $\phi : M \longrightarrow M$, which leaves $\sigma(\mathcal{S})$ pointwise invariant and transforms g into g', i.e., $g' = \phi_* g$.

The Cauchy problem concerns what we shall call **E-developments**, namely the existence and uniqueness (modulo equivalence) of a development (M, g, σ) for initial data $(\mathcal{S}, h, K)$ such that g satisfies Einstein's vacuum equations. Of special interest are developments for which $\Sigma = \sigma(\mathcal{S})$ is a **Cauchy surface** of (M, g). By this we mean that Σ is a spacelike hypersurface with the property that any non-spacelike curve (**causal curve**) intersects Σ exactly once (see Fig. 2.4).

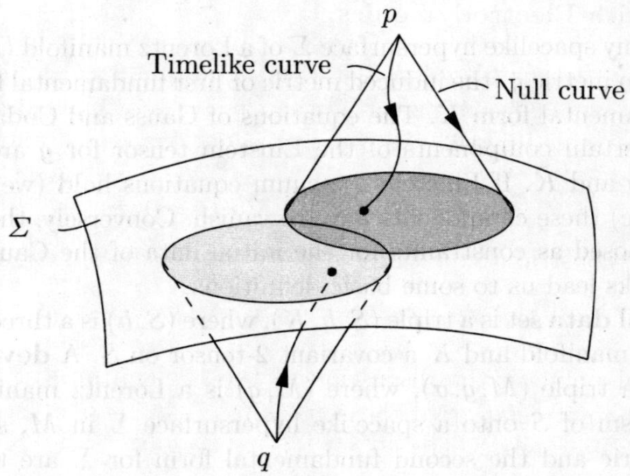

Fig. 2.4. Cauchy hypersurface Σ. Past and future directed causal curves intersect Σ exactly once.

From now on it is *always* assumed that $\sigma(\mathcal{S})$ is a *Cauchy surface*, if the contrary is not explicity said.

2.8.2 Constraint Equations

As already emphasized, an E-development (M, g, σ) of $(\mathcal{S}, h, K)$ can only exist if h and K satisfy certain constraints. These follow from

$$N^\mu G_{\mu\nu} = 0 \,, \tag{2.224}$$

where N^μ is the (future directed unit) normal to $\Sigma = \sigma(\mathcal{S})$. Using DG, Eq. (A.21), we can write $G_{\mu\nu} N^\mu N^\nu = 0$ as the following equation on $\mathcal{S}$

$$R + (\mathrm{Tr} K)^2 - \mathrm{Tr}(K \cdot K) = 0 \,, \tag{2.225}$$

where R is the Riemann scalar for the induced metric. The other components of (2.224) give, according to (A.19),

$$\nabla \operatorname{Tr}(K) - \nabla \cdot K = 0 . \tag{2.226}$$

The **constraint equations** (2.225) and (2.226), called **Hamiltonian constraint** and **momentum constraint**, respectively, are tensor equations on $\mathcal{S}$. They are non-linear elliptic partial differential equations for the initial data h and K.

There is an extended literature on these difficult constraint equations. Before we discuss special types of solutions, it may be helpful to recall the role of constraints in the Cauchy problem for electrodynamics in Minkowski spacetime.

Analogy with Electrodynamics

For simplicity, let us also consider the vacuum equations. As initial data at $t = 0$ we cannot freely prescribe the electric and magnetic fields. These must be constrained by the two Maxwell equations that do not contain time derivatives:

$$\boldsymbol{\nabla} \cdot \boldsymbol{E} = 0 , \qquad \boldsymbol{\nabla} \cdot \boldsymbol{B} = 0 . \tag{2.227}$$

Beside these constraint equations we have the dynamical equations

$$\partial_t \boldsymbol{E} - \boldsymbol{\nabla} \wedge \boldsymbol{B} = 0 , \qquad \partial_t \boldsymbol{B} + \boldsymbol{\nabla} \wedge \boldsymbol{B} = 0 . \tag{2.228}$$

A crucial point in the discussion of the Cauchy problem is that these guarantee the propagation of the constraints (2.224); moreover, the Cauchy problem is well-posed (see Exercise 2.21). By this one means that (a) the problem has a solution, (b) the solution is unique, and (c) the solution depends continuously on the data given in the problem. Something similar happens in GR, as we shall see later.

* * *

For illustration, we discuss two special types of solutions of the constraint equations which make use of conformal techniques.

First we choose $K = 0$, so that $\sigma(\mathcal{S})$ is totally geodesic (see DG, Appendix A). We are then left with the geometrical problem of finding a Riemannian metric h on $\mathcal{S}$ whose curvature scalar R vanishes[16]. For this we make the following ansatz: Let $\tilde{h}$ be any Riemannian metric on $\mathcal{S}$ and set

$$h = \varphi^4 \tilde{h} . \tag{2.229}$$

[16] For a compact $\mathcal{S}$ this is a special case of the **Yamabe problem**: Show that on a compact Riemannian manifold of dimension ≥ 3 there always exists a metric with constant scalar curvature. This problem is solved, but for complete non-compact manifolds there are only a few results.

A straightforward calculation (see Exercise 2.22) shows that $R = 0$ is then equivalent to

$$\tilde{\Delta}\varphi - \frac{1}{8}\tilde{R}\varphi = 0\,, \tag{2.230}$$

where the tildes refer to the auxiliary metric $\tilde{h}$. In the special case when $\tilde{h}$ is chosen to be flat, this reduces to Laplace's equation in $\mathbb{R}^3$.

Next, we generalize this to $K \neq 0$, but require that the mean curvature $\mathrm{Tr}(K)$ vanishes ($\sigma(\mathcal{S})$ is a maximal surface). Assume that we have solved the relatively simple problem

$$\tilde{\nabla}\cdot\tilde{K} = 0 \tag{2.231}$$

for a symmetric traceless tensor $\tilde{K}$. If we then set

$$h = \varphi^4\tilde{h}\,, \qquad K = \frac{1}{\varphi^2}\tilde{K}$$

it turns out (see Exercise 2.22) that

$$\nabla\cdot K = 0 \qquad (\mathrm{Tr}\,K = 0)\,, \tag{2.232}$$

and that the Hamiltonian constraint is satisfied if φ is a solution of the equation

$$\tilde{\Delta}\varphi - \frac{1}{8}\tilde{R}\varphi + \frac{1}{8}\varphi^{-7}\mathrm{Tr}(\tilde{K}\cdot\tilde{K}) = 0\,. \tag{2.233}$$

This equation has local solutions.

Propagation of Constraints

Let us assume that the six dynamical equations $R_{ij} = 0$, for $i = 1, 2, 3$, are satisfied. Then all $G^\alpha{}_\beta$ can be expressed linearly and homogeneously in terms of $G^0{}_\mu$. To see this note that

$$G^0{}_0 = \frac{1}{2}g^{00}R_{00}\,, \qquad G^0{}_i = g^{00}R_{0i}\,,$$

and use this in

$$G^i{}_j = g^{i0}R_{0j} - \frac{1}{2}\delta^i{}_j(g^{00}R_{00} + 2g^{0k}R_{0k})\,,$$
$$G^i{}_0 = g^{i0}R_{00} + g^{ik}R_{0k}\,.$$

If we use this fact in the contracted Bianchi identity

$$0 = \nabla_\alpha G^\alpha{}_\beta = \partial_0 G^0{}_\beta + \partial_i G^i{}_\beta - \Gamma^\sigma{}_{\beta\alpha}G^\alpha{}_\sigma + \Gamma^\alpha{}_{\alpha\sigma}G^\sigma{}_\beta\,,$$

we obtain a first order linear homogeneous system of partial differential equations (PDE) for $G^0{}_\mu$. Thus, if these components vanish at $t = 0$, they also for later times $0 \le t < t_0$, at least for analytic data. Without analycity we can not refer to a known uniqueness theorem for this type of equations. Indeed, there are simple counter examples for uniqueness of such linear systems of PDE[17]. However, the propagation of the constraints can rigorously been proved with the existence and uniqueness theorems discussed below.

2.8.3 Local Existence and Uniqueness Theorems

The classical approach to establish local existence and uniqueness theorems in GR makes crucial use of the fact that for harmonic gauge conditions the Einstein equations become a strictly hyperbolic system for which mathematical theorems can be applied. A second equally important fact is that it suffices to solve the Cauchy problem for this reduced equation, because it guarantees that the harmonic gauge condition propagates. If it is imposed on the initial surface – together with the constraint equations – it holds also for later times. Hence, the Einstein equations are also satisfied.

Analogy with Electrodynamics

Since the situation in electrodynamics is conceptually very analogous, we consider this first. This time we work with the gauge potential A_μ, in terms of which Maxwell's equations are

$$G_\nu := \partial^\mu(\partial_\mu A_\nu - \partial_\nu A_\mu) = 0\,. \tag{2.234}$$

A first observation is that $G_0 = 0$ contains no second time derivatives:

$$\partial^i(\partial_i A_0 - \partial_t A_i) = 0\,. \tag{2.235}$$

This equation says that the divergence of the electric field vanishes. It has to be imposed on the initial data A_μ, $\partial_t A_\mu$ at $t = 0$. The remaining three spatial components of Maxwell's equations (2.234) do not give rise to a well-posed initial value problem for the four potentials A_μ. Clearly, this is due to gauge invariance.

The standard procedure is to restrict the gauge freedom by imposing a gauge condition. Because the analogue of the harmonic gauge condition in GR is the Lorentz gauge condition

$$\Gamma := \partial^\mu A_\mu = 0\,, \tag{2.236}$$

---

[17] For example, there exist functions $a(t,x)$ and $u(t,x)$ in $C^\infty(\mathbb{R}^2, \mathbb{C})$ with supports in $\{(x,t) \in \mathbb{R}^2 : t > 0\}$, such that

$$\partial_t u + a\,\partial_x u = 0\,,$$

but do not vanish everywhere. Such an u and $u \equiv 0$ are both solutions with the initial condition $u = 0$.

we impose this and get from (2.234) the 'reduced equation'

$$^{(L)}G_\mu := \Box A_\mu = 0\,. \tag{2.237}$$

For an arbitrary gauge we have the identity

$$G_\mu = {}^{(L)}G_\mu - \partial_\mu \Gamma\,. \tag{2.238}$$

Equation (2.237) has an initial value formulation with an existence and uniqueness theorem, given the initial data $(A_\mu, \partial_t A_\mu)$.

If we impose on these data the Lorentz condition it propagates as a result of (2.237), if the constraint (2.235), i.e., $G_0 = 0$ is satisfied for the initial data. Indeed, (2.237) implies that Γ satisfies the wave equation. Beside $\Gamma = 0$ for $t = 0$, we also have $\partial_t \Gamma = 0$ at $t = 0$, because this is equivalent to (2.235) if Equation (2.237) is satisfied:

$$\begin{aligned}\partial_t \Gamma &= \partial_t^2 A_0 + \partial^i \partial_t A_i \\ &= -\partial^i \partial_i A_0 + \partial^i \partial_t A_i = -\boldsymbol{\nabla} \cdot \boldsymbol{E}\,.\end{aligned}$$

Summarizing, a solution of the 'reduced' hyperbolic equation (2.237) for which the initial data $(A_\mu, \partial_t A_\mu)$ satisfy the constraint $\boldsymbol{\nabla} \cdot \boldsymbol{E} = 0$ and the Lorentz condition $\Gamma = 0$, solves the Maxwell equations and the Lorentz condition. Furthermore, one obtains in this way all solutions of Maxwell's equations, up to gauge transformations. This is clear from the formulation of the Cauchy problem in terms of $\boldsymbol{E}$ and $\boldsymbol{B}$, as described earlier (see Exercise 2.21). In terms of the gauge potential we can argue as follows.

Given a solution A_μ of Maxwell's equations (2.234) in any gauge, we can find a gauge transformation

$$A_\mu \longrightarrow {}^{(L)}A_\mu = A_\mu + \partial_\mu \chi$$

such that ${}^{(L)}A_\mu$ satisfies the Lorentz condition, by solving the inhomogeneous wave equation

$$\Box \chi = -\partial^\mu A_\mu\,.$$

${}^{(L)}A_\mu$ solves the reduced Equation (2.237), and satisfies initially the constraint (2.235) as well as the Lorentz condition $\Gamma = 0$, and is thus one of the solutions considered above.

Harmonic Gauge Condition

In GR we proceed along the same line of reasoning. We begin by discussing the harmonic gauge condition. (This was used from the very beginning of GR. We now know that Einstein introduced it already in 1912 in his Zürich research notes on the linearized level. In 1916 he used it for the full theory.)

The **harmonic gauge condition** is defined as the requirement that the coordinate functions x^μ are harmonic:

$$\Box x^\mu = \frac{1}{\sqrt{-g}}\partial_\nu(\sqrt{-g}\,g^{\mu\nu}) = 0\,. \tag{2.239}$$

This can be written differently. Consider

$$\Gamma^\alpha := g^{\mu\nu}\Gamma^\alpha{}_{\mu\nu}\,. \tag{2.240}$$

Inserting the explicit expressions for the Christoffel symbols, one easily finds

$$\Gamma^\alpha = -\partial_\mu g^{\mu\alpha} - \frac{1}{2}g^{\alpha\beta}\,g^{\mu\nu}\partial_\beta g_{\mu\nu}\,, \tag{2.241}$$

thus

$$\Gamma^\alpha = -\Box x^\alpha\,. \tag{2.242}$$

Therefore, (2.239) is equivalent to the gauge condition

$$\Gamma^\alpha = 0\,. \tag{2.243}$$

Field Equations in Harmonic Gauge

From now on we work on an open set $M \subset \mathbb{R}^4$ since our considerations are local. The initial surface Σ is taken to be $\{x^0 = 0\}$.

As a preparation we write out explicitly the second order derivatives of the Ricci tensor

$$R_{\mu\nu} = \partial_\alpha\Gamma^\alpha{}_{\mu\nu} - \partial_\mu\Gamma^\alpha{}_{\nu\alpha} + \Gamma^\alpha{}_{\mu\nu}\Gamma^\beta{}_{\alpha\beta} - \Gamma^\beta{}_{\mu\alpha}\Gamma^\alpha{}_{\nu\beta}\,. \tag{2.244}$$

These are contained in the first two terms and one easily verifies that they can be collected as

$$-\frac{1}{2}g^{\alpha\beta}\partial_\alpha\partial_\beta g_{\mu\nu} + \frac{1}{2}\left(g_{\alpha\mu}\partial_\nu\Gamma^\alpha + g_{\alpha\nu}\partial_\mu\Gamma^\alpha\right).$$

So, in an arbitrary coordinate system

$$R_{\mu\nu} = {}^{(h)}R_{\mu\nu} + \frac{1}{2}\left(g_{\alpha\mu}\partial_\nu\Gamma^\alpha + g_{\alpha\nu}\partial_\mu\Gamma^\alpha\right), \tag{2.245}$$

with

$${}^{(h)}R_{\mu\nu} = -\frac{1}{2}g^{\alpha\beta}\partial_\alpha\partial_\beta g_{\mu\nu} + H_{\mu\nu}(g,\partial g)\,. \tag{2.246}$$

Here, $H_{\mu\nu}(g,\partial g)$ is a rational expression of $g_{\mu\nu}$ and $\partial_\alpha g_{\mu\nu}$ with denominator $\det(g_{\mu\nu}) \neq 0$. (By Cramer's formula $g^{\mu\nu}$ is a rational combination of $g_{\mu\nu}$ with denominator $\det(g_{\mu\nu})$.) Clearly, $R_{\mu\nu} ={}^{(h)} R_{\mu\nu}$ in a harmonic gauge $\Gamma^\alpha = 0$.

For the Einstein tensor we obtain from (2.245)

$$G_{\mu\nu} = {}^{(h)}G_{\mu\nu} + \frac{1}{2}\left(g_{\mu\alpha}\partial_\nu\Gamma^\alpha + g_{\nu\alpha}\partial_\mu\Gamma^\alpha\right) - \frac{1}{2}g_{\mu\nu}\partial_\alpha\Gamma^\alpha\,, \tag{2.247}$$

with

$$^{(h)}G_{\mu\nu} = {}^{(h)}R_{\mu\nu} - \frac{1}{2}g_{\mu\nu}\,{}^{(h)}R, \quad {}^{(h)}R = {}^{(h)}R^{\lambda}_{\ \lambda}\,. \tag{2.248}$$

The highest derivative operator $-\frac{1}{2}g^{\alpha\beta}\partial_\alpha\partial_\beta$ in (2.246) acts the same way on each component of the system $g_{\mu\nu}$ (no mixing of highest order derivatives). The **reduced system**, i.e., the field equations in harmonic gauge,

$$^{(h)}R_{\mu\nu} = 0 \tag{2.249}$$

is, in standard terminology (see [46]), a quasilinear hyperbolic system. The highest derivative in (2.249) is the same as for the wave equation of a scalar field. It is of crucial importance that such a system has for smooth initial data $g_{\mu\nu}(0,x^i)$ and $\partial_0 g_{\mu\nu}(0,x^i)$, locally a unique smooth solution. For the relevant theorems[18], see [81]. The domain of dependence is the same as for the wave equation with metric $g_{\mu\nu}$.

Next, we show that it suffices to solve the well-posed Cauchy problem for the reduced system (2.249), because it will turn out that for special initial conditions the gauge condition $\Gamma^\alpha = 0$ is then automatically satisfied. For this end we first derive a linear wave equation for Γ^α. This is obtained by inserting (2.247) into the Bianchi identity. For solutions of (2.249) we find

$$\begin{aligned} 0 = \nabla^\mu G_{\mu\nu} &= \frac{1}{2}g^{\rho\mu}\nabla_\rho\Big(g_{\mu\alpha}\partial_\nu\Gamma^\alpha + g_{\nu\alpha}\partial_\mu\Gamma^\alpha - g_{\mu\nu}\partial_\alpha\Gamma^\alpha\Big) \\ &= \frac{1}{2}g_{\alpha\nu}g^{\rho\mu}\partial_\rho\partial_\mu\Gamma^\alpha + \text{lower order terms in } \Gamma^\alpha\,, \end{aligned}$$

thus,

$$g^{\mu\nu}\partial_\mu\partial_\nu\Gamma^\alpha + A^{\alpha\nu}_\mu(g,\partial g)\partial_\nu\Gamma^\mu = 0\,. \tag{2.250}$$

For this system of linear hyperbolic PDE one again has an existence and uniqueness theorem (see [81]). Hence $\Gamma^\alpha(0,x^i) = 0$ and $\partial_0\Gamma^\alpha(0,x^i) = 0$ imply $\Gamma^\alpha(x^\mu) = 0$. The next point is to show that the constraint equations $G_{\mu 0} = 0$, that necessarily have to be imposed at $t = 0$, imply $\partial_0\Gamma^\alpha(0,x^i) = 0$, while $\Gamma^\alpha(0,x^i) = 0$ is compatible with them.

That the harmonic condition at $t = 0$ is compatible with the constraints is not surprising, because the latter have a geometrical meaning, while the former is a coordinate condition. In Exercise 2.23 we shall verify that the derivatives $\partial_0 g_{0\mu}(0,x^i)$ can indeed be chosen such that $\Gamma^\alpha(0,x^i) = 0$.

From (2.247) we see that for a solution of the reduced system (2.249) the constraint equations imply that at $t = 0$

$$\big(g_{0\alpha}\partial_\nu\Gamma^\alpha + g_{\nu\alpha}\partial_0\Gamma^\alpha\big) - g_{0\nu}\partial_\alpha\Gamma^\alpha = 0\,. \tag{2.251}$$

---

[18] Physicists usually do not care about such theorems, because they take the result for granted. As a warning, we mention that that there are even *linear* PDE without singular points that have no solution anywhere. A famous example was constructed by H. Lewy that also highlights the importance of analycity in the Cauchy-Kovalevskaya theorem (see Chap. 8 of [47]).

Let us work this out for Gaussian normal coordinates[19] relative to $t = 0$, for which (see DG, Eq.(A.23))

$$g_{00}(0, x^k) = -1\,, \quad g_{0j}(0, x^k) = 0\,, \quad \partial_0 g_{ij}(0, x^k) = -2K_{ij}\,. \tag{2.252}$$

Imposing also $\Gamma^\alpha(0, x^k) = 0$, Equation (2.251) becomes $g_{\nu\alpha}\partial_0\Gamma^\alpha = 0$, thus indeed $\partial_0\Gamma^\alpha(0, x^i) = 0$.

Let us summarize what has been shown so far:

Theorem 2.3. *Let $g^0_{\mu\nu}$ and $k^0_{\mu\nu}$ be smooth initial conditions at $t = 0$ in a bounded open domain of $\mathbb{R}^3$ for a Lorentz metric and its first time derivative. Suppose that $\Gamma^\alpha(g^0_{\mu\nu}, k^0_{\mu\nu})$ and $G_{0\mu}(g^0_{\mu\nu}, k^0_{\mu\nu})$ vanish. Then there is locally a unique Lorentz metric $g_{\mu\nu}$ such that*

(i) ${}^{(h)}R_{\mu\nu} = 0$;
(ii) $\left(g_{\mu\nu}(0, x^i), \partial_0 g_{\mu\nu}(0, x^i)\right) = (g^0_{\mu\nu}, k^0_{\mu\nu})$.

Moreover, this metric also satisfies the vacuum equations $R_{\mu\nu} = 0$, as well as $\Gamma^\alpha = 0$.

Remarks.

1. One can also show that $g_{\mu\nu}$ depends continuously on $(g^0_{\mu\nu}, k^0_{\mu\nu})$ relative to some natural topology.
2. We can choose the coordinates such that (2.252) holds at $t = 0$.

Next we prove that we obtain in this way locally *all* solutions of Einstein's vacuum equations.

Theorem 2.4. *Any solution of Einstein's vacuum equation is locally diffeomorphic to a solution described in Theorem 2.3. More specifically, let g be a solution on an open region $U \subset \mathbb{R}^4$ and $\Sigma = U \cap \{x^0 = 0\}$. (After an eventual coordinate shift Σ is non empty.) Then there exists a diffeomorphism φ from a strip $|x^0| < \varepsilon$ in U preserving the points with $x^0 = 0$, such that the standard coordinate functions x^μ of $\mathbb{R}^4$ are harmonic for the metric $\tilde{g} := \varphi_* g$, and $\tilde{g}_{00} = -1$, $\tilde{g}_{0i} = 0$ for $x^0 = 0$.*

Proof. The construction of φ proceeds as follows. Let f^α be four functions on such a strip which are harmonic relative to the metric g,

$$\Box_g f^\alpha = \delta_g \circ df^\alpha = 0\,, \tag{2.253}$$

where δ_g is the codifferential corresponding to g. We solve these wave equations for the initial data

$$f^i(0, x^j) = x^i\,, \qquad f^0(0, x^j) = 0\,,$$

and

$$\partial_0 f^\alpha(0, x^j) = \xi^\alpha\,, \tag{2.254}$$

[19] Gaussian normal coordinates are introduced in Sect. 2.9.3.

where

$$g_{\alpha i}\xi^\alpha = 0\,, \qquad g_{\alpha\beta}\xi^\alpha\xi^\beta = -1\,. \tag{2.255}$$

Clearly, the differentials df^α are linearly independent for a sufficiently small strip U'.

Now, we define $\varphi : U' \longrightarrow V'$ by $x^\mu \circ \varphi = f^\mu$, where x^μ are the standard coordinate functions of $\mathbb{R}^4$ (in other words, $\varphi(x)^\mu = f^\mu(x)$). For showing that

$$\Box_{\varphi_* g}(x^\mu) = 0\,, \tag{2.256}$$

we use the general fact that for a differential form ω, and a diffeomorphism $\varphi : M \longrightarrow M$, one has (exercise)

$$\delta_{\varphi_* g}(\varphi_*\omega) = \varphi_*(\delta_g\omega)\,, \tag{2.257}$$

and thus

$$\Box_{\varphi_* g}(\varphi_*\omega) = \varphi_*(\Box_g\omega)\,. \tag{2.258}$$

With this we have

$$\begin{aligned}\Box_{\varphi_* g}(x^\mu) &= \Box_{\varphi_* g}\big((\varphi_* \circ \varphi^*)x^\mu\big) = \Box_{\varphi_* g}\big(\varphi_*(x^\mu \circ \varphi)\big) \\ &= \Box_{\varphi_* g}(\varphi_* f^\mu) = \varphi_*(\Box_g f^\mu) = 0\,.\end{aligned}$$

This shows that $\varphi_* g$ belongs to the solutions described in Theorem 2.3. By construction $\varphi|\mathcal{S} = id$. □

Remark. After a suitable transformation of the original metric g, we can always assume that (2.252) holds on Σ. Then φ preserves Σ pointwise and by (2.254), (2.255) its induced metric and second fundamental form.

Theorem 2.4 and the uniqueness statement in Theorem 2.3 imply the

Corollary 2.1. *E-developments of $(\mathcal{S}, h, k)$ are locally unique: Two E-developments are extensions of a common development.*

Proof. If $(U_1, g^{(1)})$ and $(U_2, g^{(2)})$, $U_i \subset \mathbb{R}^4$, are two such developments, we can choose local coordinates such that both $\Sigma_i \cap U_i$ are contained in $\{x^0 = 0\}$. For each of the E-developments construct a diffeomorphism as in the remark above. The images of $g^{(1)}$ and $g^{(2)}$ agree by the uniqueness result of Theorem 2.3. This implies that there is an isometry between the two E-developments, if U_1 and U_2 are appropriately reduced, which respects the initial surfaces Σ_1 and Σ_2. □

Let us translate these results into geometrical statements.

Theorem 2.5 (Local Existence and Uniqueness). *Let $(\mathcal{S}, h, K)$ be an initial data set which satisfies the Hamiltonian and momentum constraints (2.225) and (2.226). Then there exists an E-development (M, g, σ). Two such E-developments are extensions of the same E-development.*

Proof. By Theorem 2.3 we can construct an atlas of $\mathcal{S}$ with domains $\{V_i\}$ and E-developments $(U_i, g^{(i)}, \sigma_i)$ of (V_i, h, K). On an intersection $U_i \cap U_j$ the Corollary implies that there is an isometry ψ_{ij} that takes $g^{(i)}$ into $g^{(j)}$, if U_i and U_j are suitable reduced. Therefore, we can patch the local solutions together: Consider the disjoint union of the $(U_i, g^{(i)})$ and identify points $p_i \in U_i$ and $p_j \in U_j$ when $p_j = \psi_{ij}(p_i)$. This proves the existence part.

The uniqueness follows from the local uniqueness (Corollary) by making use of an atlas for $\mathcal{S}$. We leave the details to the reader. □

The existence of a maximal development now follows with Zorn's lemma.

Theorem 2.6 (Global Uniqueness). *Under the hypothesis of Theorem 2.5 there is a maximal E-development which is unique, up to isometries.*

Proof. See, e.g., [82]. □

Note that for uniqueness it is important that the hypersurface Σ is a Cauchy surface. (Recall that we included this property in the definition of a development.) A spacetime which admits a Cauchy surface Σ is **globally hyperbolic**. One can show that it is then diffeomorphic to the product $\mathbb{R} \times \Sigma$.

It is nice to have this general existence theorem, but it does not tell us much about the global aspects of solutions of Einstein's field equation. A major result in this direction was established by D. Christodoulou and S. Klainermann in 1993, [86]. This work, which gave a proof of the global stability of Minkowski spacetime, is revisited in the recent monograph [83].

2.8.4 Characteristics of Einstein's Field Equations

The notion of characteristic surfaces is closely linked with the Cauchy problem of PDE. Let us explain this concept for the wave equation

$$\Box_g \psi = 0\,, \tag{2.259}$$

whose explicit form is

$$g^{\mu\nu}\partial_\mu\partial_\nu\psi - \Gamma^\alpha\partial_\alpha\psi = 0\,. \tag{2.260}$$

The Cauchy problem for this equation is to find a solution for which the value of ψ and its normal derivative on a hypersurface Σ are given.

This problem does not have a unique solution if Σ is chosen such that (2.260) does not determine the second derivatives of ψ on Σ for given Cauchy data. In this case, the hypersurface Σ is said to be a **characteristic surface**, or a **characteristic** of the PDE (2.260). The second derivative can be discontinuous across a characteristic surface. For this reason, a (moving) wave front must be a characteristic.

It is not difficult to show that Σ is a *non-characteristic* for the PDE (2.260), provided

$$g_{\mu\nu}N^{\mu}N^{\nu} \neq 0\,, \tag{2.261}$$

where N^{μ} denotes the unit normal to Σ. If Σ is a non-characteristic one can show that for a smooth solution ψ satisfying the Cauchy conditions, all partial derivatives of ψ on Σ are determined. (The details of the calculations for general quasi linear PDE can be found in [46].)

If Σ is described as $\{S = 0\}$ for a function S, with $dS \neq 0$ on Σ, then (2.261) becomes

$$g^{\mu\nu}\partial_{\mu}S\partial_{\nu}S \neq 0\,. \tag{2.262}$$

In other words, characteristic surfaces are null hypersurfaces; S satisfies the general relativistic eikonal equation (Hamilton-Jacobi equation)

$$g^{\mu\nu}\partial_{\mu}S\partial_{\nu}S = 0\,, \tag{2.263}$$

which we encountered in the study of the geometrical optics limit in Sect. 1.8. These surfaces describe for instance the wave fronts in gravitational lensing.

What are the characteristic surfaces of Einstein's vacuum equations? With our previous results this can easily be answered. We know that the field equations are equivalent to the pair (2.249) and (2.250). For both of them the highest derivative operator is the same as for the wave equation (2.260). Therefore, the characteristic surfaces in GR are, as expected, *null hypersurfaces*.

* * *

Exercises.

Exercise 2.21. Show that the following formulation of the initial value problem for Maxwell's vacuum equations has a unique answer.

1. Initial conditions: $\boldsymbol{E}(\boldsymbol{x},0)$ and $\boldsymbol{B}(\boldsymbol{x},0)$ satisfying the constraints (2.227).
2. Dynamical equations: the first order PDE (2.228).

Hints.

1. Show that the constraints (2.227) propagate.
2. Conclude that $\boldsymbol{E}$ and $\boldsymbol{B}$ satisfy the wave equations and that beside $\boldsymbol{E}$ and $\boldsymbol{B}$ also $\partial_t\boldsymbol{E}$ and $\partial_t\boldsymbol{B}$ are known for $t = 0$. Hence $\boldsymbol{E}$ and $\boldsymbol{B}$ are known for all times.
3. Prove that these fields satisfy the dynamical equations (2.228) by showing that the left-hand sides of (2.228) satisfy all Maxwell equations (by construction) and vanish at $t = 0$. Therefore, they vanish at all times, because there is only the trivial solution of all Maxwell equations with $\boldsymbol{E}(\boldsymbol{x},0) = \boldsymbol{B}(\boldsymbol{x},0) = 0$ (use 2.).

Exercise 2.22.

1. Consider a Riemannian manifold (M, g) of dimension n and the metric $g' = e^f g$ conformal to g. Show that for the corresponding Ricci tensors and scalar curvatures one has the relations

$$R'_{kj} - R_{kj} = -\frac{n-2}{2}\nabla_k\nabla_j f + \frac{1}{2}\Delta f\, g_{ik}$$
$$+\frac{n-2}{4}\nabla_k f\nabla_j f - \frac{n-2}{4}\nabla^l f\nabla_l f\, g_{jk}\,, \tag{2.264}$$

and

$$R'e^f - R = (n-1)\Delta f - \frac{(n-2)(n-1)}{4}\nabla^l f\nabla_l f\,. \tag{2.265}$$

2. Let T^{ij} be a symmetric and traceless tensor whose covariant divergence vanishes, i.e., $\nabla_j T^{ij} = 0$. Show that $\nabla'_j T'^{ij} = 0$, where ∇' is the covariant derivative for g', and

$$T'^{ij} = e^{sf}T^{ij}\,, \qquad s = -\frac{n}{2} - 1\,. \tag{2.266}$$

Solution (of 2.). The relation between the Christoffel symbols corresponding to g' and g is

$$\Gamma'^i_{\ jk} - \Gamma^i_{\ jk} = \frac{1}{2}g^{il}\left(g_{lk}\partial_j f + g_{lj}\partial_k f - g_{jk}\partial_l f\right). \tag{2.267}$$

For both parts of the exercise the computation is simplified by using coordinates normal at the point p for g ($g_{ij}(p) = \delta_{ij}$, $\Gamma^i_{\ jk}(p) = 0$). According to (1.36) we have

$$\nabla'_j T'^{ij} = \frac{1}{\sqrt{g'}}\partial_j\left(\sqrt{g'}T'^{ij}\right) + \Gamma'^i_{\ jk}T'^{jk}\,.$$

In p this reduces to (use $\partial_j T^{ij} = 0$)

$$\nabla'_j T'^{ij} = \left(s + \frac{n}{2}\right)e^{sf}\partial_j f\, T^{ij} + e^{sf}\partial_j f\, T^{ij}$$
$$= 0 \quad \text{for} \quad s + \frac{n}{2} = -1\,.$$

Exercise 2.23. Use the expression (2.241) for Γ^α and show that in Gaussian coordinates (2.252) the harmonic constraint $\Gamma^\alpha(0, x^k) = 0$ can be satisfied by a proper choice of $\partial_0 g_{0\alpha}(0, x^k)$.

Solution. On $\{x^0 = 0\}$ we have in Gaussian coordinates

$$0 = g_{\alpha\lambda}\Gamma^\lambda = g^{\mu\alpha}\partial_\mu g_{\alpha\lambda} - \frac{1}{2}g^{\mu\nu}\partial_\lambda g_{\mu\nu}$$
$$= g^{00}\partial_0 g_{0\lambda} + g^{ij}\partial_i g_{j\lambda} - \frac{1}{2}\left(g^{00}\partial_\lambda g_{00} + g^{ij}\partial_\lambda g_{ij}\right).$$

This gives for $\lambda = 0$

$$g^{00}\partial_0 g_{00} - \frac{1}{2}\left(g^{00}\partial_0 g_{00} + g^{ij}\partial_0 g_{ij}\right) = 0\,,$$

and for $\lambda = k$ we obtain

$$g^{00}\partial_0 g_{0k} + g^{ij}\partial_i g_{jk} - \frac{1}{2}\partial_k g_{ij} = 0 \,.$$

This shows that for a given g_{ij} and K_{ij} these equations can (uniquely) be solved for $\partial_0 g_{0\mu}$.

2.9 General Relativity in 3 + 1 Formulation

The '**3+1**' or **dynamical formulation** of GR plays an important role in basic theoretical investigations and is also very useful in applications, including numerical relativity.

2.9.1 Generalities

We assume that a spacetime (M, g) admits a slicing by a 1-parameter family Σ_t of spacelike hypersurfaces. More precisely, we assume that there is a diffeomorphism $\phi : M \longrightarrow \Sigma \times I$, $I \subset \mathbb{R}$, such that the manifolds $\Sigma_t = \phi^{-1}(\Sigma \times \{t\})$ are spacelike and the curves $\phi^{-1}(\{x\} \times I)$ are timelike. These curves will be called **preferred timelike orbits**. Their tangent vectors define a vector field ∂_t on M. This can be decomposed into normal and parallel components relative to the slicing (see Fig. 2.5):

$$\partial_t = \alpha n + \bar{\beta} \,. \tag{2.268}$$

Here n is the future directed unit normal field with $\langle n, n\rangle = -1$ and $\bar{\beta}$ is tangent to the slices Σ_t. It is common practice to call α the **lapse function** and $\bar{\beta}$ the **shift vector field**. A coordinate system $\{x^i\}$ on Σ induces natural coordinates on M: $\phi^{-1}(m, t)$ has coordinates (x^i, t) if $m \in \Sigma$ has coordinates x^i. The preferred timelike curves have constant spatial coordinates.

Let us set $\bar{\beta} = \beta^i \partial_i$. From $\langle n, \partial_i\rangle = 0$ and (2.268) we find

$$\langle \partial_t, \partial_t\rangle = -(\alpha^2 - \beta^i\beta_i) \,, \quad \langle \partial_t, \partial_i\rangle = \beta_i \,, \tag{2.269}$$

where $\beta_i = g_{ij}\beta^j$ and $g_{ij} = \langle \partial_i, \partial_j\rangle$ are the components of the **induced metric** on Σ_t. Thus in what we could also call comoving coordinates the metric takes the form

$$g = -(\alpha^2 - \beta^i\beta_i)\,dt^2 + 2\beta_i\,dx^i dt + g_{ij}\,dx^i dx^j \tag{2.270}$$

or

$$g = -\alpha^2 dt^2 + g_{ij}\big(dx^i + \beta^i dt\big)\big(dx^j + \beta^j dt\big) \,. \tag{2.271}$$

This shows that dt is orthogonal to $dx^i + \beta^i dt$, for $i = 1, 2, 3$.

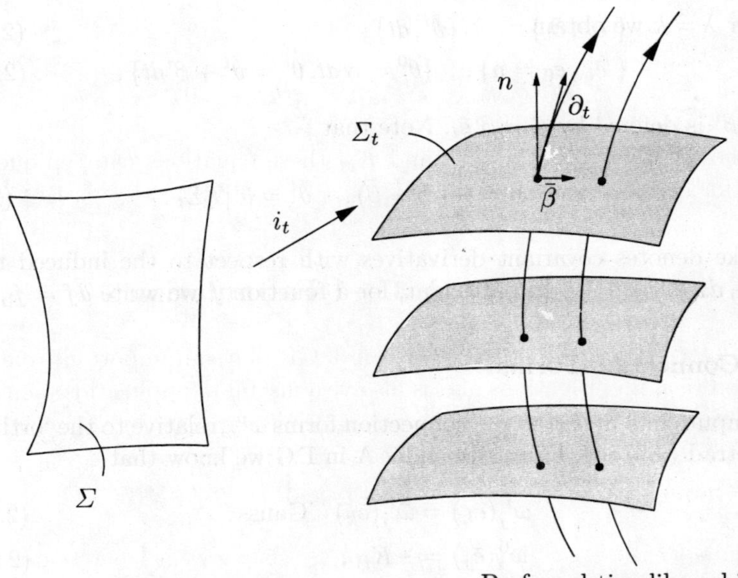

Fig. 2.5. Lapse function and shift vector field.

We call a tensor field S on M **tangential** when it can be regarded as a family of tensor fields on Σ_t, or as a t-dependent tensor field on the 'absolute' space Σ. Relative to a comoving coordinate system it has the form

$$S = S^{i_1 \cdots i_r}_{j_1 \cdots j_s} \left(\partial_{i_1} \otimes \cdots \otimes \partial_{i_r}\right) \otimes \left(dx^{j_1} \otimes \cdots \otimes dx^{j_s}\right). \tag{2.272}$$

(For each t the values of S are in tensor products of $T\Sigma_t$ and $T^*\Sigma_t$.) We shall denote them by bars (except for dx^i and ∂_i). As an often occurring example, let us consider a tangential p-form $\bar{\omega}$ and its exterior derivative $d\bar{\omega}$. We can decompose this as

$$d\bar{\omega} = \bar{d}\bar{\omega} + dt \wedge \partial_t \bar{\omega}, \tag{2.273}$$

where $\bar{d}$ denotes the exterior derivative in Σ_t ($\bar{d} = dx^i \wedge \partial_i$ in comoving coordinates), and $\partial_t \bar{\omega}$ is the partial time derivative. Both $\bar{d}\bar{\omega}$ and $\partial_t \bar{\omega}$ are, of course, tangential. Similarly, other differential operators (covariant derivative, Lie derivative, etc.) can be decomposed.

Beside the dual pair $\{\partial_\mu\}$ and $\{dx^\mu\}$ for comoving coordinates, we also use the dual pair

$$\{\partial_i, n\} \qquad \text{and} \qquad \{dx^i + \beta^i dt, \alpha\, dt\}. \tag{2.274}$$

For our computations it is often convenient to use instead of $\{\partial_i\}$ an orthonormal triad of tangential vector fields $\{\bar{e}_i\}$, together with the dual basis $\{\bar{\vartheta}^i\}$ instead of dx^i. Then we have the following two dual pairs which will be constantly used

$$
\begin{aligned}
&\{\,\bar{e}_i\,,\partial_t\}\,, \quad \{\bar{\vartheta}^i, dt\}\,; && (2.275a)\\
&\{\,\bar{e}_i\,,e_0 = n\}\,, \quad \{\theta^0 = \alpha\, dt, \theta^i = \bar{\vartheta}^i + \beta^i dt\}\,, && (2.275b)
\end{aligned}
$$

where β^i is defined by $\bar{\beta} = \beta^i \bar{e}_i$. Note that

$$
e_0 = n = \frac{1}{\alpha}(\partial_t - \bar{\beta})\,, \quad \bar{\vartheta}^i = \theta^i | T\Sigma_t\,. \tag{2.276}
$$

A stroke denotes covariant derivatives with respect to the induced metric $\bar{g}_t = g_{ij}\, dx^i dx^j$ on Σ_t. In particular, for a function f we write $\bar{d}f = f_{|i}\bar{\vartheta}^i$.

2.9.2 Connection Forms

We compute in a first step the connection forms $\omega^\mu{}_\nu$ relative to the 'orthonormal' tetrad (2.275b). From Appendix A in DG we know that

$$
\begin{aligned}
\omega^i{}_j(\bar{e}_k) &= \bar{\omega}^i{}_j(\bar{e}_k) \quad \text{Gauss}\,, && (2.277a)\\
\omega^0{}_i(\bar{e}_j) &= -K_{ij}\,, && (2.277b)
\end{aligned}
$$

where $\bar{\omega}^i{}_j$ are the connection forms for $(\Sigma_t, \bar{g}_t)$ relative to the dual orthonormal pairs $\{\bar{e}_i\}$, $\{\bar{\vartheta}^i\}$, and K_{ij} are the components of the **second fundamental form** or **extrinsic curvature**

$$
\bar{K} = K_{ij}\, \bar{\vartheta}^i \otimes \bar{\vartheta}^j\,. \tag{2.278}
$$

Next, we compute $\omega^0{}_i(e_0)$. We have, using the first structure equation,

$$
\begin{aligned}
d\theta^0 &= d(\alpha dt) = \bar{d}\alpha \wedge dt = \frac{1}{\alpha}\alpha_{|j}\bar{\vartheta}^j \wedge \theta^0\\
&= \frac{1}{\alpha}\alpha_{|j}\theta^j \wedge \theta^0 = -\omega^0{}_i \wedge \theta^i\,.
\end{aligned}
$$

Thus, we obtain

$$
i_{e_0} d\theta^0 = -\omega^0{}_i(e_0)\theta^i = -\frac{1}{\alpha}\alpha_{|i}\theta^i\,,
$$

whence

$$
\omega^0{}_i(e_0) = \omega^i{}_0(e_0) = \frac{1}{\alpha}\alpha_{|i}\,. \tag{2.279}
$$

From (2.277b) and (2.279) we obtain

$$
\omega^0{}_i = -K_{ij}\theta^j + \frac{1}{\alpha}\alpha_{|i}\theta^0\,. \tag{2.280}
$$

Finally, we need $\omega^i{}_j(e_0)$. The first structure equation and (2.280) give

$$
i_{e_0} d\theta^i = -i_{e_0}\left(\omega^i{}_0 \wedge \theta^0 + \omega^i{}_j \wedge \theta^j\right) = -\left(K^i{}_j + \omega^i{}_j(e_0)\right)\theta^j\,. \tag{2.281}
$$

Hence, we have

$$K^i{}_j + \omega^i{}_j(e_0) = -i_{\bar{e}_j} i_{e_0} d\theta^i \,. \tag{2.282}$$

The right-hand side can be worked out as follows

$$\begin{aligned} i_{e_o} d\theta^i &= i_{e_0} d(\bar{\vartheta}^i + \beta^i dt) = i_{e_0} \big(\underbrace{\bar{d}\bar{\vartheta}^i}_{-\bar{\omega}^i{}_k \wedge \bar{\vartheta}^k} + dt \wedge \partial_t \bar{\vartheta}^i + \bar{d}\beta^i \wedge dt \big) \\ &= \frac{1}{\alpha} i_{\bar{\beta}} (\bar{\omega}^i{}_k \wedge \bar{\vartheta}^k) + \frac{1}{\alpha} \partial_t \bar{\vartheta}^i - \frac{1}{\alpha} \bar{d}\beta^i \,, \end{aligned}$$

and

$$\begin{aligned} i_{\bar{e}_j} i_{e_0} d\theta^i &= \frac{1}{\alpha} \Big[(\bar{\omega}^i{}_k \wedge \bar{\vartheta}^k)(\bar{\beta}, \bar{e}_j) + \partial_t \bar{\vartheta}^i(\bar{e}_j) - \bar{d}\beta^i(\bar{e}_j) \Big] \\ &= \frac{1}{\alpha} \Big[\bar{\omega}^i{}_j(\bar{\beta}) - \beta^i{}_{|j} \Big] + \frac{1}{\alpha} \partial_t \bar{\vartheta}^i(\bar{e}_j) \,. \end{aligned}$$

Let us set

$$\partial_t \bar{\vartheta}^i = c^i{}_j \bar{\vartheta}^j \,, \tag{2.283}$$

giving

$$(\partial_t \bar{g})_{ij} = c_{ij} + c_{ji} \,. \tag{2.284}$$

We then obtain from (2.282)

$$K_{ij} + \omega_{ij}(e_0) = -\frac{1}{\alpha} \Big[\bar{\omega}_{ij}(\bar{\beta}) - \beta_{i|j} + c_{ij} \Big] \,. \tag{2.285}$$

The symmetric and skew symmetric parts of this equation give

$$\omega_{ij}(e_0) = -\frac{1}{\alpha} \bar{\omega}_{ij}(\bar{\beta}) + \frac{1}{2\alpha} (\beta_{i|j} - \beta_{j|i}) - \frac{1}{2\alpha} (c_{ij} - c_{ji}) \,, \tag{2.286a}$$

$$K_{ij} = \frac{1}{2\alpha} \Big[(\beta_{i|j} + \beta_{j|i}) - (c_{ij} + c_{ji}) \Big] \,. \tag{2.286b}$$

Equation (2.286a) solves our problem. The last equation provides an interesting relation. In fact, from (2.284) and

$$(\bar{L}_{\bar{\beta}} \bar{g})_{ij} = \beta_{i|j} + \beta_{j|i} \tag{2.287}$$

we get

$$\bar{K} = -\frac{1}{2\alpha} (\partial_t - \bar{L}_{\bar{\beta}}) \bar{g} \,. \tag{2.288}$$

Later we shall need an explicit expression for $\bar{L}_{\bar{\beta}} \bar{K}$. In

$$\bar{L}_{\bar{\beta}} \bar{K} = (\bar{L}_{\bar{\beta}} K_{ij}) \bar{\vartheta}^i \otimes \bar{\vartheta}^j + K_{ij} \Big(\bar{L}_{\bar{\beta}} \bar{\vartheta}^i \otimes \bar{\vartheta}^j + \bar{\vartheta}^i \otimes \bar{L}_{\bar{\beta}} \bar{\vartheta}^j \Big)$$

we use

$$\begin{aligned} \bar{L}_{\bar{\beta}} \bar{\vartheta}^i &= \bar{d} i_{\bar{\beta}} \bar{\vartheta}^i + i_{\bar{\beta}} \bar{d} \bar{\vartheta}^i = \bar{d}\beta^i - i_{\bar{\beta}} (\bar{\omega}^i{}_k \wedge \bar{\vartheta}^k) \\ &= \bar{D}\beta^i - \bar{\omega}^i{}_k(\bar{\beta}) \bar{\vartheta}^k = \big(\beta^i{}_{|k} - \bar{\omega}^i{}_k(\bar{\beta}) \big) \bar{\vartheta}^k \,. \end{aligned} \tag{2.289}$$

This gives

$$(\bar{L}_{\bar{\beta}}\bar{K})_{ij} = \bar{L}_{\bar{\beta}}K_{ij} + K_{sj}\big(\beta^s{}_{|i} - \bar{\omega}^s{}_i(\bar{\beta})\big) + K_{si}\big(\beta^s{}_{|j} - \bar{\omega}^s{}_j(\bar{\beta})\big)\,. \tag{2.290}$$

In what follows we shall encounter $(\partial_t - \bar{L}_{\bar{\beta}})\bar{K}$. Now, we have

$$\partial_t\bar{K} = \Big(\partial_t K_{ij} + K_{sj}c^s{}_i + K_{is}c^s{}_j\Big)\bar{\vartheta}^i \otimes \bar{\vartheta}^j\,, \tag{2.291}$$

thus

$$(\partial_t\bar{K} - \bar{L}_{\bar{\beta}}\bar{K})_{ij} = (\partial_t - \bar{L}_{\bar{\beta}})K_{ij} + \Big(K_{sj}\big(c^s{}_i - \beta^s{}_{|i} + \bar{\omega}^s{}_i(\bar{\beta})\big) + (i \longleftrightarrow j)\Big)\,.$$

With (2.285) this becomes

$$(\partial_t\bar{K} - \bar{L}_{\bar{\beta}}\bar{K})_{ij} = (\partial_t - \bar{L}_{\bar{\beta}})K_{ij} - 2\alpha(\bar{K}\cdot\bar{K})_{ij} - \alpha\Big(K_{sj}\omega^s{}_i(e_0) + K_{is}\omega^s{}_j(e_0)\Big)\,, \tag{2.292}$$

where $(\bar{K}\cdot\bar{K})_{ij} = K_{is}K^s{}_j$. This formula will turn out to be useful.

2.9.3 Curvature Forms, Einstein and Ricci Tensors

The Gauss and Codazzi-Mainardi equations, derived in DG, Appendix A, provide the tangential parts of the curvature forms

$$\Omega^i{}_j = \bar{\Omega}^i{}_j + K^i{}_r K^j{}_s\,\bar{\vartheta}^r \wedge \bar{\vartheta}^s \qquad \text{on } T\Sigma_t\,, \tag{2.293a}$$

$$\Omega^0{}_i = K_{ij|k}\,\bar{\vartheta}^j \wedge \bar{\vartheta}^k \qquad \text{on } T\Sigma_t\,. \tag{2.293b}$$

These provide already some components of the Einstein and Ricci tensors (see DG, (A.19) and (A.21))

$$G_{00} = \frac{1}{2}\Big(\bar{R} + (K^i{}_i)^2 - K^i{}_j K^j{}_i\Big) \quad \text{(Gauss)} \tag{2.294a}$$

$$G_{0i} = R_{0i} = \bar{\nabla}_i K^j{}_j - \bar{\nabla}_j K^j{}_i \qquad \text{(Codazzi-Mainardi)}\,. \tag{2.294b}$$

For the other components we need all of $\Omega^0{}_i$. With the second structure equation this can be worked out as follows:

$$\begin{aligned}\Omega^i{}_0 &= d\omega^i{}_0 + \omega^i{}_j \wedge \omega^j{}_0 \\ &\overset{(2.280)}{=} -d(K^i{}_j\theta^j) + d\left(\frac{1}{\alpha}\alpha^{|i}\theta^0\right) + \omega^i{}_j \wedge \left(-K^j{}_l\theta^l + \frac{1}{\alpha}\alpha^{|j}\theta^0\right)\,.\end{aligned} \tag{2.295}$$

The first term on the right is, using (2.280) once more,

$$-dK^i{}_j \wedge \theta^j + K^i{}_j(\omega^i{}_l \wedge \theta^l + \omega^j{}_0 \wedge \theta^0)$$
$$= -dK^i{}_j \wedge \theta^j + K^i{}_j(\omega^i{}_l \wedge \theta^l - K^j{}_l\theta^l \wedge \theta^0)\,.$$

We also need

$$\omega^i{}_j \frac{1}{\alpha}\alpha^{|j} \wedge \theta^0 = \omega^i{}_j(\bar{e}_k)\frac{1}{\alpha}\alpha^{|j}\,\theta^k \wedge \theta^0\,,$$

and

$$d\left(\frac{1}{\alpha}\alpha^{|i}\theta^0\right) = d\left(\frac{1}{\alpha}\alpha^{|i}\right) \wedge \theta^0 + \frac{1}{\alpha}\alpha^{|i}\,\underbrace{d\theta^0}_{\frac{1}{\alpha}\alpha_{|j}\,\theta^j \wedge \theta^0}$$
$$= \left(\frac{1}{\alpha}\alpha^{|i}\right)_{,j}\theta^j \wedge \theta^0 + \frac{1}{\alpha}\alpha^{|i}\frac{1}{\alpha}\alpha^{|j}\,\theta^j \wedge \theta^0$$
$$= \frac{1}{\alpha}\alpha_{,ij}\,\theta^j \wedge \theta^0\,.$$

Therefore, the second and last term on the right in (2.295) give together $\frac{1}{\alpha}\alpha^{|i}{}_{|j}\,\theta^j \wedge \theta^0$. Thus we obtain

$$\Omega^i{}_0 = \frac{1}{\alpha}\alpha^{|i}{}_{|j}\,\theta^j \wedge \theta^0 - dK^i{}_j \wedge \theta^j$$
$$+ (K^i{}_j\omega^j{}_l - \omega^i{}_j K^j{}_l) \wedge \theta^l - K^i{}_j K^j{}_l\,\theta^l \wedge \theta^0\,. \tag{2.296}$$

As a check one can restrict this to $T\Sigma_t$ to get the Codazzi-Mainardi equation.

The Ricci tensor is obtained from (see DG, (A.17))

$$R_{\beta\sigma} = \Omega^\alpha{}_\beta(e_\alpha, e_\sigma)\,. \tag{2.297}$$

In particular, using that ω_{ij} is skew symmetric and K_{ij} symmetric, we have

$$R_{00} = \Omega^i{}_0(\bar{e}_i, e_0) = \frac{1}{\alpha}\bar{\Delta}\alpha + dK^i{}_i(e_0) - K^i{}_j K^j{}_i$$
$$= \frac{1}{\alpha}\bar{\Delta}\alpha - K^i{}_j K^j{}_i + \frac{1}{\alpha}(\partial_t - \bar{L}_\beta)K^i{}_i\,.$$

Thus, if $H := K^i{}_i$ we obtain

$$R_{00} = \frac{1}{\alpha}\bar{\Delta}\alpha - K^i{}_j K^j{}_i + \frac{1}{\alpha}(\partial_t - \bar{L}_\beta)H\,. \tag{2.298}$$

Finally, we need

$$R_{ij} = \Omega^0{}_i(e_0, \bar{e}_j) + \Omega^k{}_i(\bar{e}_k, \bar{e}_j)\,. \tag{2.299}$$

According to (2.293a) the second term is

$$\Omega^k{}_i(\bar{e}_k, \bar{e}_j) = \bar{R}_{ij} + H\,K_{ij} - K_{il}K^l{}_j\,. \tag{2.300}$$

For the first term we use (2.296)

$$\Omega^0{}_i(e_0,\bar{e}_j) = \Omega^i{}_0(e_0,\bar{e}_j) = -\frac{1}{\alpha}\alpha_{|ij} - dK_{ij}(e_0) + K_{is}\left(\omega^s{}_j(e_0) + K^s{}_j\right) - \omega^i{}_s(e_0)K^s{}_j\,.$$

This gives

$$R_{ij} = \bar{R}_{ij} + H\,K_{ij} - \frac{1}{\alpha}(\partial_t - \bar{L}_{\bar\beta})K_{ij} + K_{is}\,\omega^s{}_j(e_0) + K_{js}\,\omega^s{}_i(e_0)\,.$$

Now we can use (2.292) for the last three terms, and obtain

$$R_{ij} = \bar{R}_{ij} + H\,K_{ij} - 2(\bar{K}\cdot\bar{K})_{ij} - \frac{1}{\alpha}\left(\partial_t\bar{K} - \bar{L}_{\beta}\bar{K}\right)_{ij} - \frac{1}{\alpha}\alpha_{|ij}\,. \tag{2.301}$$

For the Riemann scalar one obtains[20]

$$R = \bar{R} + H^2 + (\bar{K}\cdot\bar{K})^i{}_i - \frac{2}{\alpha}(\partial_t - \bar{L}_{\bar\beta})H - \frac{2}{\alpha}\bar{\Delta}\alpha\,. \tag{2.302}$$

For later use we collect the results. All components are given relative to the orthonormal tetrad (2.275b):

$$\theta^0 = \alpha\,dt\,, \qquad \theta^i = \bar{\vartheta}^i + \beta^i dt\,.$$

We found

$$R_{00} = \frac{1}{\alpha}\bar{\Delta}\alpha - K^i{}_jK^j{}_i + \frac{1}{\alpha}(\partial_t - \bar{L}_{\bar\beta})H\,, \tag{2.303a}$$

$$R_{0i} = G_{0i} = \bar{\nabla}_iH - \bar{\nabla}_jK^j{}_i\,, \tag{2.303b}$$

$$R_{ij} = \bar{R}_{ij} + H\,K_{ij} - 2K_{il}K^l{}_j - \frac{1}{\alpha}\left(\partial_t\bar{K} - \bar{L}_{\beta}\bar{K}\right)_{ij} - \frac{1}{\alpha}\alpha_{|ij}\,, \tag{2.303c}$$

$$G_{00} = \frac{1}{2}\left(\bar{R} + H^2 - K^i{}_jK^j{}_i\right); \tag{2.303d}$$

$$R = \bar{R} + H^2 - 2K_{ij}K^{ij} - \frac{2}{\alpha}(\partial_t - \bar{L}_{\bar\beta})H - \frac{2}{\alpha}\bar{\Delta}\alpha\,. \tag{2.303e}$$

The right-hand sides of all these equations contain only tangential objects, i.e., time dependent fields on Σ and their derivatives. With obvious changes they hold for arbitrary comoving coordinates relative to the basis $\{dt, dx^i + \beta^i dt\}$, where $\beta = \beta^i\partial_i$.

Let us consider the vacuum field equations without the cosmological term. The four equations $G_{0\mu} = 0$ do not contain time derivatives and give the constraint equations for $\bar{g}$ and $\bar{K}$ on Σ

[20] Note that $(\partial_t\bar{K})^i{}_i \neq \partial_t K^i{}_i$; we find from (2.292)

$$(\partial_t - \bar{L}_{\bar\beta}K)^i{}_i = (\partial_t - L_{\bar\beta})H - 2\alpha(\bar{K}\cdot\bar{K})^i{}_i\,.$$

$$\bar{R} + H^2 - K_{ij}K^{ij} = 0 \qquad \text{(energy constraint)}, \tag{2.304a}$$
$$\bar{\nabla}_i H - \bar{\nabla}_j K^j_{\ i} = 0 \qquad \text{(momentum constraint)}. \tag{2.304b}$$

These elliptic equations have already been discussed in Sect. 2.8 in connection with the Cauchy problem.

The remaining six equations $R_{ij} = 0$ involve time derivatives

$$\partial_t \bar{K} = \bar{L}_{\bar{\beta}}\bar{K} - \overline{\text{Hess}}(\alpha) + \alpha\Big(\text{Ric}(\bar{g}) - 2\bar{K}\cdot\bar{K} + H\,\bar{K}\Big), \tag{2.305}$$

where $\overline{\text{Hess}}(\alpha)$ denotes the **Hessian** of α, whose components are $\alpha_{|ij}$. Beside, we have the identity (2.288) for the time derivative of $\bar{g}$

$$\partial_t \bar{g} = -2\alpha\bar{K} + \bar{L}_{\bar{\beta}}\bar{g}\,. \tag{2.306}$$

The system (2.305) and (2.306) are the **dynamical equations** for $\bar{g}$ and $\bar{K}$. They contain the lapse function α and the shift vector β which have to be fixed by *four* **gauge conditions**.

The simplest choice $\alpha = 1$ and $\bar{\beta} = 0$ corresponds to Gaussian normal coordinates that are often useful.

Gaussian Normal Coordinates

Close to a spacelike (or timelike) hypersurface $\Sigma \subset M$ we can introduce so called **Gaussian normal coordinates** as follows. Let N be as before a field of unit normals on Σ near a point $p \in \Sigma$. Construct through each point $q \in \Sigma$ the geodesic through q tangent to N and assign to a point along such a geodesic as time coordinate t the proper time ($t = 0$ on Σ), and as spatial coordinates those of the point q of a coordinate system of Σ in a neighborhood of p. For sufficiently small $|t|$ we obtain in this way a coordinate system of M in a neighborhood of p (see Fig. 2.6).

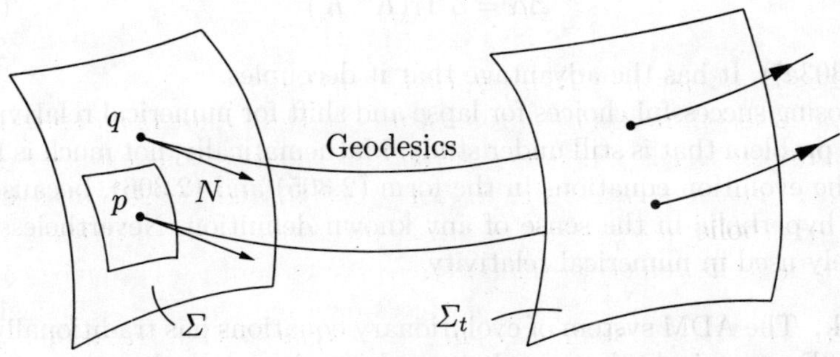

Fig. 2.6. Construction of Gaussian normal coordinates.

We claim that the hypersurfaces Σ_t near Σ with constant t are orthogonal to the geodesics leaving Σ orthogonally (**Gauss's Lemma**). We emphasize that this is a local result; distinct geodesics can meet at some distance from Σ (coordinate singularity).

For a proof of Gauss's Lemma, consider a 1-parameter family c_ε of our congruence of timelike geodesics, with constant coordinates x^2, x^3 and $x^1 = x^1_{(0)} + \varepsilon$, and apply the first variation formula of DG, Sect. 14.3. If Y denotes the variation vector $\partial/\partial\varepsilon|_{\varepsilon=0}$ along the geodesic $c = c_{\varepsilon=0}$, this reads

$$L'(0) = \langle Y, \dot{c} \rangle|_0^{t_0} - \int_0^{t_0} \langle Y, \nabla_{\dot{c}}\dot{c} \rangle \, dt \, .$$

Since the length is constant (equal to t_0), and Y at $c(0)$ is tangent to Σ, thus orthogonal to $\dot{c}(0)$, we obtain with $\nabla_{\dot{c}}\dot{c} = 0$

$$\langle Y, \dot{c} \rangle = 0$$

for all t. Hence, every $\dot{c}$ is orthogonal to $\partial/\partial x^1$, and similarly to all $\partial/\partial x^i$. This proves the lemma.

Since $\langle \partial_t, \partial_t \rangle = -1$ and $\langle \partial_t, \partial_i \rangle = 0$, the metric has in Gaussian normal coordinates the form

$$g = -dt^2 + g_{ik}(x^\mu) \, dx^i dx^k \, , \tag{2.307}$$

thus $\alpha = 1$ and $\bar{\beta} = 0$. We emphasize once more that this slicing usually develops caustics (coordinate singularities).

Maximal Slicing

A traditional condition for the lapse results by requiring **maximal slicing**: $H \equiv \mathrm{Tr}(\bar{K}) = 0$. For vacuum solutions this condition, originally proposed by Lichnerowicz, implies the elliptic equation

$$\bar{\Delta}\alpha = \alpha \, \mathrm{Tr}(K \cdot K) \tag{2.308}$$

(see (2.303a)). It has the advantage that it decouples.

Imposing successful choices for lapse and shift for numerical relativity is a difficult problem that is still under study. Mathematically, not much is known about the evolution equations in the form (2.305) and (2.306), because they are not hyperbolic in the sense of any known definition. Nevertheless, they are widely used in numerical relativity.

Remark. The ADM system of evolutionary equations has traditionally been used in 3D numerical relativity. It turned out, however, that this system is not ideal. Severe instabilities often show up in numerical simulations. There is no good understanding of this, but it is true that the ADM evolutionary equations do not satisfy any known hyperbolicity condition. This situation prompted attempts for developing alternative formulations of Einstein's equations. Among these there are various hyperbolicity formulations (see [90]–[95]), whose mathematical structure is more appealing, but did not lead to substantial improvements in numerical integrations.

Recently, it turned out that a simple way of rewriting the traditional ADM form (2.305), (2.306) of the evolutionary equations leads to remarkably stable numerical evolution. In this formulation (see [87]) one separates out the conformal degrees of freedom of the 3-metric $\bar{g}$ and the trace-free part of the second fundamental form. In addition, the quantities (2.240) for $\bar{g}$ are promoted to independent variables, for which a first order evolutionary equation is derived. (For details, the reader is referred to [88].) That this 'conformal approach' leads to *more stable evolutions* has been demonstrated in a number of simulations (see, e.g., [89]).

* * *

Exercises.

Exercise 2.24. Consider the volume form $\mathrm{vol} = \bar{\vartheta}^1 \wedge \bar{\vartheta}^2 \wedge \bar{\vartheta}^3$ of the 3-metric $\bar{g}$. Show that

$$\partial_t \mathrm{vol} = \left(\bar{\nabla} \cdot \bar{\beta} - \alpha\, \mathrm{Tr}\bar{K}\right) \mathrm{vol}\,.$$

Exercise 2.25. Sometimes the dynamical equation (2.305) is written in terms of the contravariant tensor K^{ij}. This is easily done once one has the identity

$$(\partial_t - L_\beta K)^\sharp = (\partial_t - L_\beta)K^\sharp - 4\alpha K^\sharp \cdot K^\sharp\,.$$

Derive this equation.

Part II

Applications of General Relativity

3 The Schwarzschild Solution and Classical Tests of General Relativity

> *"Imagine my joy at the feasibility of general covariance and the result that the equations give the perihelion motion of Mercury correctly. For a few days I was beside myself with joyous excitement."*
>
> —A. Einstein
>
> (To P. Ehrenfest, Jan. 17, 1916)

The solution of the field equations, which describes the field outside of a spherically symmetric mass distribution, was found by Karl Schwarzschild only two months after Einstein published his field equations. Schwarzschild performed this work under rather unusual conditions. In the spring and summer of 1915 he was assigned to the eastern front. There he came down with an infectious disease and in the fall of 1915 he returned seriously ill to Germany. He died only a few months later, on May 11, 1916. In this short time, he wrote two significant papers, in spite of his illness. One of these dealt with the Stark effect in the Bohr-Sommerfeld theory, and the other solved the Einstein field equations for a static, spherically symmetric field. From this solution he derived the precession of the perihelion of Mercury and the bending of light rays near the sun. Einstein had calculated these effects previously by solving the field equations in post-Newtonian approximation.

3.1 Derivation of the Schwarzschild Solution

The Schwarzschild solution is the unique static, spherically symmetric vacuum spacetime and describes the field outside a spherically symmetric body. It is the most important exact solution of Einstein's field equations.

We know that a *static* spacetime (M, g) has locally the form (see Sect. 1.9)

$$M = \mathbb{R} \times \Sigma\,, \qquad g = -\varphi^2\, dt^2 + h\,, \tag{3.1}$$

where h is a Riemannian metric on Σ and φ is a smooth function on Σ. The timelike Killing field K with respect to which (M, g) is static is ∂_t. We assume that this is the *only* static Killing field and thus that there is a *distinguished* time. Note also that $\varphi^2 = -\langle K, K\rangle$.

We say that a Lorentz manifold is **spherically symmetric**, provided it admits the group $SO(3)$ as an isometry group, in such a way that the group orbits, i.e., the subsets on which the group acts transitively, are two-dimensional spacelike surfaces. If (3.1) is also spherically symmetric, the action of $SO(3)$ is only non-trivial on Σ, i.e., the time t is preserved. This follows from the fact that $K = \partial_t$ is invariant under $SO(3)$ (because of the uniqueness assumption). Indeed, this implies that K is perpendicular to the orbits of $SO(3)$. (A non-vanishing orthogonal projection of K on an orbit would also be invariant, but such an invariant vector can not exist, as is easy to show.) Another way to see that t remains invariant is to note that the 1-form $\omega = \langle K, K\rangle dt$ belonging to K (see Eq. (1.112)) is also invariant. Since $\langle K, K\rangle$ is obviously invariant we conclude that dt remains invariant under the $SO(3)$ action. But since $SO(3)$ is connected this implies that t is preserved.

We now consider the isometric action of $SO(3)$ on (Σ, h). An orbit is a homogeneous two-dimensional Riemannian manifold $\mathcal{S}$ and therefore has constant Gaussian curvature k and is complete. It is known[1] that the standard simply connected surface $M(k)$ (2-sphere of radius $1/\sqrt{k}$ for $k > 0$, hyperbolic plane of curvature $k < 0$, Euclidean plane for $k = 0$) is a **Riemannian covering** of $\mathcal{S}$. By this one means that there is a covering map which is also a local isometry. Since the isometry group of $\mathcal{S}$ includes $SO(3)$, k must be positive and there are only two possibilities for $\mathcal{S}$: Either $\mathcal{S}$ is the 2-sphere of radius $1/\sqrt{k}$ or the projective plane of radius $1/\sqrt{k}$ (obtained from the 2-sphere by identifying antipodal points). Considering - for good reasons - only the first possibility we have a foliation of (Σ, h) by invariant 2-spheres, characterized by the surface A. The **radial Schwarzschild coordinate** r is defined by $A = 4\pi r^2$. We assume that $dr \neq 0$.

Let us now consider the family of geodesics orthogonal to some particular 2-sphere (see Fig. 3.1), for which we introduce polar coordinates and transport these with the family of geodesics. We leave it as an exercise to show that this family is orthogonal to all $SO(3)$ orbits. (Consider, for instance, the scalar product of a Killing field ξ for the $SO(3)$ action with the tangent field u for the geodesic congruence, and compute $u\langle \xi, u\rangle$.) With this construction the metric h takes the form

$$h = \mathrm{e}^{2b(r)} dr^2 + r^2 (d\vartheta^2 + \sin^2\vartheta \, d\varphi^2) \, . \tag{3.2}$$

Clearly, $\langle K, K\rangle$ also depends only on r. Setting $-\langle K, K\rangle = \mathrm{e}^{2a(r)}$, Equation (3.1) takes the form $\Sigma = I \times S^2$, where I is an interval (R, ∞) of $\mathbb{R}^+$, and

$$g = -\mathrm{e}^{2a(r)} dt^2 + \mathrm{e}^{2b(r)} dr^2 + r^2 (d\vartheta^2 + \sin^2\vartheta \, d\varphi^2) \, . \tag{3.3}$$

The **Schwarzschild coordinates** $(t, r, \vartheta, \varphi)$ are well adapted to the symmetries of the spacetime manifold.

The functions $a(r)$ and $b(r)$ approach zero asymptotically (g is asymptotically flat). We must now insert the ansatz (3.3) into the field equations. For this, it is necessary to compute the Ricci tensor (or the Einstein tensor) corresponding to the metric (3.3). This is accomplished most quickly with the

[1] For detailed proofs see Chap.8 of [42].

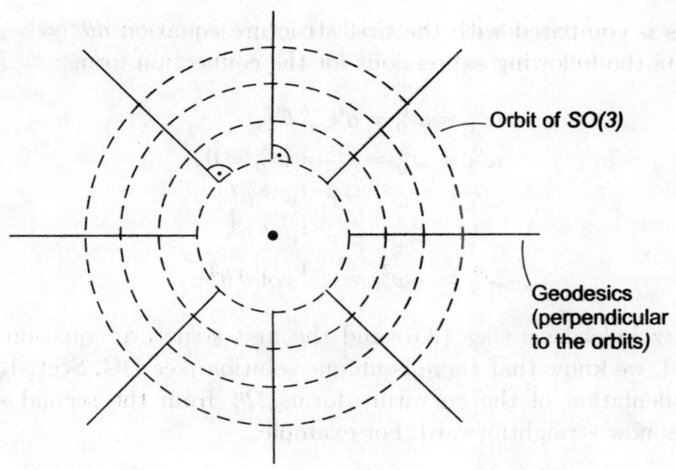

Fig. 3.1. Construction of Schwarzschild coordinates.

help of the Cartan calculus. The traditional computation using Christoffel symbols is, for instance, given in detail in [11].

We choose the following basis of 1-forms:

$$\theta^0 = \mathrm{e}^a \, dt \,, \; \theta^1 = \mathrm{e}^b \, dr \,, \; \theta^2 = r \, d\vartheta \,, \; \theta^3 = r \, \sin\vartheta \, d\varphi \,. \tag{3.4}$$

The metric (3.3) then reads

$$g = g_{\mu\nu} \, \theta^\mu \otimes \theta^\nu \,, \quad (g_{\mu\nu}) = \mathrm{diag}(-1, 1, 1, 1) \,. \tag{3.5}$$

Thus, the basis $\{\theta^\alpha\}$ is orthonormal. Hence, the connection forms $\omega^\mu{}_\nu$ satisfy (see Sect. 2.6)

$$\omega_{\mu\nu} + \omega_{\nu\mu} = 0 \,, \quad \omega_{\mu\nu} := g_{\mu\lambda} \omega^\lambda{}_\nu \,. \tag{3.6}$$

In order to determine these from the first structure equation, we compute the exterior derivatives of θ^μ. With $a' = da/dr$, etc., we have

$$\begin{aligned} d\theta^0 &= a' \mathrm{e}^a \, dr \wedge dt \,, \\ d\theta^1 &= 0 \,, \\ d\theta^2 &= dr \wedge d\vartheta \,, \\ d\theta^3 &= \sin\vartheta \, dr \wedge d\varphi + r \cos\vartheta \, d\vartheta \wedge d\varphi \,. \end{aligned}$$

Expressing the right-hand sides in terms of the basis $\theta^\sigma \wedge \theta^\rho$, we obtain

$$\begin{aligned} d\theta^0 &= a' \mathrm{e}^{-b} \, \theta^1 \wedge \theta^0 \,, \\ d\theta^1 &= 0 \,, \\ d\theta^2 &= r^{-1} \mathrm{e}^{-b} \, \theta^1 \wedge \theta^2 \,, \\ d\theta^3 &= r^{-1} \big(\mathrm{e}^{-b} \, \theta^1 \wedge \theta^3 + \cot\vartheta \, \theta^2 \wedge \theta^3 \big) \,. \end{aligned} \tag{3.7}$$

When this is compared with the first structure equation $d\theta^\alpha = -\omega^\alpha_{\ \beta} \wedge \theta^\beta$, one expects the following expressions for the connection forms:

$$\begin{aligned}
\omega^0_{\ 1} &= \omega^1_{\ 0} = a' e^{-b}\,\theta^0 , \\
\omega^0_{\ 2} &= \omega^2_{\ 0} = \omega^0_{\ 3} = \omega^3_{\ 0} = 0 , \\
\omega^2_{\ 1} &= -\omega^1_{\ 2} = r^{-1} e^{-b}\,\theta^2 , \\
\omega^3_{\ 1} &= -\omega^1_{\ 3} = r^{-1} e^{-b}\,\theta^3 , \\
\omega^3_{\ 2} &= -\omega^2_{\ 3} = r^{-1} \cot\vartheta\,\theta^3 .
\end{aligned} \tag{3.8}$$

This ansatz indeed satisfies (3.6) and the first structure equation. On the other hand, we know that there is unique solution (see DG, Sect. 13.7).

The calculation of the curvature forms $\Omega^\mu_{\ \nu}$ from the second structure equation is now straightforward. For example

$$\begin{aligned}
\Omega^0_{\ 1} &= d\omega^0_{\ 1} + \omega^0_{\ k} \wedge \omega^k_{\ 1} = d\omega^0_{\ 1} = d(a' e^{-b}\,\theta^0) \\
&= (a' e^{-b})'\, dr \wedge \theta^0 + a' e^{-b}\, d\theta^0 \\
&= (a' e^{-b})' e^{-b}\,\theta^1 \wedge \theta^0 + (a' e^{-b})^2\,\theta^1 \wedge \theta^0 , \\
\Omega^0_{\ 1} &= -e^{-2b}(a'^2 - a'b' + a'')\,\theta^0 \wedge \theta^1 ; \\
\Omega^0_{\ 2} &= d\omega^0_{\ 2} + \omega^0_{\ k} \wedge \omega^k_{\ 2} = \omega^0_{\ 1} \wedge \omega^1_{\ 2} = -e^{-2b}\frac{a'}{r}\,\theta^0 \wedge \theta^2 .
\end{aligned}$$

The other components are found similarly. We summarize the results for later use (note that $\Omega^\mu_{\ \nu}$ is proportional to $\theta^\mu \wedge \theta^\nu$; the indices 2 and 3 are equivalent)

$$\begin{aligned}
\Omega^0_{\ 1} &= e^{-2b}(a'b' - a'' - a'^2)\,\theta^0 \wedge \theta^1 , \\
\Omega^0_{\ 2} &= -\frac{a' e^{-2b}}{r}\,\theta^0 \wedge \theta^2 , \\
\Omega^0_{\ 3} &= -\frac{a' e^{-2b}}{r}\,\theta^0 \wedge \theta^3 , \\
\Omega^1_{\ 2} &= \frac{b' e^{-2b}}{r}\,\theta^1 \wedge \theta^2 , \\
\Omega^1_{\ 3} &= \frac{b' e^{-2b}}{r}\,\theta^1 \wedge \theta^3 , \\
\Omega^2_{\ 3} &= \frac{1 - e^{-2b}}{r^2}\,\theta^2 \wedge \theta^3 ,
\end{aligned} \tag{3.9}$$

and $\Omega_{\mu\nu} := g_{\mu\lambda}\Omega^\lambda_{\ \nu} = -\Omega_{\nu\mu}$. From this, one reads off the components of the Riemann tensor with respect to the basis θ^α. The Ricci tensor can conveniently be obtained from the formula

$$R_{\beta\sigma} = \Omega^\alpha_{\ \beta}(e_\alpha, e_\sigma) , \tag{3.10}$$

that is easily verified. For the Einstein tensor one then finds

$$
\begin{aligned}
G^0{}_0 &= -\frac{1}{r^2} + \mathrm{e}^{-2b}\left(\frac{1}{r^2} - \frac{2b'}{r}\right), \\
G^1{}_1 &= -\frac{1}{r^2} + \mathrm{e}^{-2b}\left(\frac{1}{r^2} + \frac{2a'}{r}\right), \\
G^2{}_2 = G^3{}_3 &= \mathrm{e}^{-2b}\left(a'^2 - a'b' + a'' + \frac{a' - b'}{r}\right), \\
G_{\mu\nu} &= 0 \text{ for all other components}. \qquad (3.11)
\end{aligned}
$$

We now solve the vacuum equations. First, note that $G_{00} + G_{11} = 0$ implies $a' + b' = 0$, and hence $a + b = 0$ since $a, b \longrightarrow 0$ asymptotically. The equation $G_{00} = 0$ reads

$$\mathrm{e}^{-2b}\left(\frac{2b'}{r} - \frac{1}{r^2}\right) + \frac{1}{r^2} = 0 .$$

This is equivalent to $(r\,\mathrm{e}^{-2b})' = 1$, which implies that $\mathrm{e}^{-2b} = 1 - 2m/r$, where m is an integration constant. We thus obtain the **Schwarzschild solution**

$$g = -\left(1 - \frac{2m}{r}\right)dt^2 + \frac{dr^2}{1 - 2m/r} + r^2\left(d\vartheta^2 + \sin^2\vartheta\, d\varphi^2\right). \qquad (3.12)$$

The Birkhoff Theorem

Let us allow the functions a and b in (3.3) to be a priori time dependent. In Exercise 3.1 we show that then the components of the Einstein tensor with respect to the basis (3.4) are given by ($\dot{a} = da/dt$, etc.)

$$
\begin{aligned}
G^0{}_0 &= -\frac{1}{r^2} + \mathrm{e}^{-2b}\left(\frac{1}{r^2} - \frac{2b'}{r}\right), \\
G^1{}_1 &= -\frac{1}{r^2} + \mathrm{e}^{-2b}\left(\frac{1}{r^2} + \frac{2a'}{r}\right), \\
G^2{}_2 = G^3{}_3 &= \mathrm{e}^{-2b}\left(a'^2 - a'b' + a'' + \frac{a' - b'}{r}\right) \\
&\quad + \mathrm{e}^{-2a}(-\dot{b}^2 + \dot{a}\,\dot{b} - \ddot{b}), \\
G^1{}_0 &= \frac{2\dot{b}}{r}\mathrm{e}^{-a-b}, \\
G^\mu{}_\nu &= 0 \text{ for all other components}. \qquad (3.13)
\end{aligned}
$$

We now solve the vacuum equations for these expressions. The equation $G^1{}_0 = 0$ implies that b is independent of time and hence it follows from $G_{00} = 0$ that b has the same form as in the static case. From $G_{00} + G_{11} = 0$ we again obtain the relation $a' + b' = 0$. Now, however, we can only conclude that

$$a = -b + f(t) .$$

The other vacuum equations are then all satisfied and the metric is

$$g = -\mathrm{e}^{2f(t)}\left(1 - \frac{2m}{r}\right) dt^2 + \frac{dr^2}{1 - 2m/r} + r^2\left(d\vartheta^2 + \sin^2\vartheta\, d\varphi^2\right).$$

If we introduce the new time coordinate

$$t' = \int \mathrm{e}^{f(t)}\, dt\,,$$

we again obtain the Schwarzschild metric (3.12). For $r > 2m$, a spherically symmetric vacuum field is thus *necessarily* static (**Birkhoff theorem**). This should not be too surprising because of the spin-2 nature of the gravitational field. The Birkhoff theorem is closely related to an analogous fact in electrodynamics: The electromagnetic field outside a time dependent spherical charge distribution is time independent, and equal to the Coulomb field of the total charge.

The physical meaning of the integration constant m in (3.12) becomes clear by comparison with the Newtonian limit at large distances. In this region we must have $g_{00} \simeq -(1 + 2\phi)$, with $\phi = -G\,M/r$. Hence,

$$m = \frac{G\,M}{c^2}\,. \tag{3.14}$$

We shall now show that the integration constant M is also equal to the total energy P^0. For this purpose, we write (3.12) in nearly Lorentzian coordinates. Let

$$\rho = \frac{1}{2}\left(r - m + (r^2 - 2m\,r)^{1/2}\right)\,,$$

whence

$$r = \rho\left(1 + \frac{m}{2\rho}\right)^2\,. \tag{3.15}$$

Substitution into (3.12) results in

$$g = -\left(\frac{1 - m/2\rho}{1 + m/2\rho}\right)^2 dt^2 + \left(1 + \frac{m}{2\rho}\right)^4 \left(d\rho^2 + \rho^2\, d\vartheta^2 + \rho^2 \sin^2\vartheta\, d\varphi^2\right)\,. \tag{3.16}$$

If we now set

$$x^1 = \rho\sin\vartheta\cos\varphi\,,\quad x^2 = \rho\sin\vartheta\sin\varphi\,,\quad x^3 = \rho\cos\vartheta\,,$$

the Schwarzschild metric takes the form

$$g = -h^2(|\boldsymbol{x}|)\, dt^2 + f^2(|\boldsymbol{x}|)\, d\boldsymbol{x}^2\,, \tag{3.17}$$

with

$$h(r) = \frac{1 - m/2r}{1 + m/2r}\,,\quad f(r) = \left(1 + \frac{m}{2r}\right)^2\,. \tag{3.18}$$

Note that the spatial part in (3.17) is conformally flat. With respect to the orthonormal tetrad $\{\theta^\alpha\} = \{h\,dt, f\,dx^i\}$ one finds the following connection forms

$$\omega^{0j} = \frac{h'}{f}\frac{x^j}{r}dt\,, \quad \omega^{jk} = \frac{f'}{fr}(x^j dx^k - x^k dx^j)\,. \tag{3.19}$$

We now compute P^0 from Equation (2.207):

$$\begin{aligned} P^0 &= -\frac{1}{16\pi G}\int_{S^2_\infty} \omega^{jk}\wedge\eta^0{}_{jk} = -\frac{1}{16\pi G}\varepsilon_{0jkl}\int_{S^2_\infty}\omega^{jk}\wedge\theta^l \\ &= \frac{1}{8\pi G}\varepsilon_{0jkl}\int_{S^2_\infty}\frac{f'}{r}x^k\,dx^j\wedge dx^l\,. \end{aligned}$$

Let us integrate over the surface S^2_∞ of a large sphere. Then,

$$\varepsilon_{0jkl}x^k\,dx^k\wedge dx^l = -x^k\varepsilon_{kjl}\,dx^j\wedge dx^l = -r\,2r^2 d\Omega\,,$$

where $d\Omega$ denotes the solid angle element. This gives, as expected,

$$P^0 = -\frac{1}{4\pi G}\lim_{R\to\infty}\int_{r=R} f' r^2\,d\Omega = M\,.$$

It is easy to verify that $P^i = 0$. Furthermore, one obtains a vanishing angular momentum from (2.212).

The Schwarzschild solution (3.12) has an apparent singularity at

$$r = R_S := 2\frac{GM}{c^2}\,, \tag{3.20}$$

where R_S is the so-called **Schwarzschild radius**. Schwarzschild himself was quite disturbed by this "singularity". For this reason, he subsequently investigated the solution of Einstein's field equations for a static, spherically symmetric mass distribution having constant energy density. He showed that the radius of such a configuration must be bigger than $9\,R_S/8$. Schwarzschild was extremely satisfied by this result, since it showed that the singularity is not relevant (for the special case being considered). However, somewhat later, in 1923, Birkhoff proved that a spherically symmetric vacuum solution of Einstein's equations is necessarily *static* for $r > R_S$. Hence, the exterior field for a non-static, spherically symmetric mass distribution is necessarily the Schwarzschild solution for $r > R_S$. Clearly, the lower bound $9R_S/8$ is no longer valid for a non-static situation. Hence it is necessary to investigate in more detail what is going on in the vicinity of the Schwarzschild sphere $r = R_S$. We shall do this in Sect. 3.7. It will turn out that the spacetime geometry is not singular there. Merely the Schwarzschild coordinates fail to properly cover the spacetime region for $r \longrightarrow R_s$ ("coordinate singularity"). The Schwarzschild sphere has nevertheless physical significance (as a horizon). When $r < R_S$, the solution is no longer static. It took remarkably long until all this was clearly understood. Lemaître was the first who recognized in 1933 the fictitious character of the "Schwarzschild singularity".

Geometric Meaning of the Spatial Part of the Schwarzschild Metric

We shall now give a geometrical illustration of the spatial part of the metric (3.12). Consider the two dimensional submanifold $\{\vartheta = \frac{\pi}{2}, t = \text{const}\}$ and represent this as surface of rotation in three-dimensional Euclidean space E^3. This submanifold has the induced metric

$$\hat{g} = \frac{dr^2}{1 - 2m/r} + r^2\, d\varphi^2\,. \tag{3.21}$$

On the other hand, a surface of rotation in E^3 has the induced metric

$$\hat{g} = z'^2\, dr^2 + \left(dr^2 + r^2\, d\varphi^2\right) = \left(1 + z'^2\right) dr^2 + r^2\, d\varphi^2\,,$$

where $z(r)$ describes the surface. (We use cylindrical coordinates.) If we require this to agree with (3.21), then

$$\frac{dz}{dr} = \left(\frac{2m}{r - 2m}\right)^{1/2}.$$

Integration gives $z = \left(8m(r - 2m)\right)^{1/2} + \text{const}$. If we set the integration constant equal to zero, we obtain a paraboloid of revolution

$$z^2 = 8m(r - 2m)\,. \tag{3.22}$$

* * *

Exercises.

Exercise 3.1. Derive the formulas (3.13).

Exercise 3.2. Consider a spherical cavity inside a spherically symmetric non-rotating matter distribution. Show that there the metric is flat. This remark justifies certain Newtonian considerations in cosmology.

Exercise 3.3. Generalize the Schwarzschild solution for Einstein's equations with a cosmological constant.

Exercise 3.4. Determine the solution representing spacetime outside a spherically symmetric *charged* body carrying an electric charge (but no spin and magnetic dipole moment). The result is the **Reissner-Nordstrøm solution** $(c = 1)$

$$g = -\left(1 - \frac{2m}{r} + \frac{Ge^2}{r^2}\right) dt^2 + \left(1 - \frac{2m}{r} + \frac{Ge^2}{r^2}\right)^{-1} dr^2 + r^2\left(d\vartheta^2 + \sin^2\vartheta\, d\varphi^2\right),$$

where m/G represents the gravitational mass and e the electric charge of the body.

Solution. Since the fields are static and spherically symmetric we can use for the metric the ansatz (3.3). It is again convenient to work with the orthonormal tetrad (3.4). Relative to this the Einstein tensor is given in (3.11). We expect that the electromagnetic field has only a radial electric component. Hence we make the ansatz

$$F = \frac{e^2}{r^2}\,\theta^0 \wedge \theta^1 \,. \tag{3.23}$$

For this the components of the energy-momentum tensor (1.53) are readily found to be

$$T^0{}_0 = T^1{}_1 = -T^2{}_2 = -T^3{}_3 = \frac{e^2}{8\pi r^4}\,, \tag{3.24}$$

all other $T_{\mu\nu} = 0$. Hence, the equations $G^0_{\ 0} = 8\pi G\,T^0{}_0$ and $G^1{}_1 = 8\pi G\,T^1{}_1$ become

$$\frac{1}{r^2} - \mathrm{e}^{-2b}\left(\frac{1}{r^2} - \frac{2b'}{r}\right) = \frac{Ge^2}{r^4}\,, \tag{3.25a}$$

$$\frac{1}{r^2} - \mathrm{e}^{-2b}\left(\frac{1}{r^2} + \frac{2a'}{r}\right) = \frac{Ge^2}{r^4}\,. \tag{3.25b}$$

By subtraction we see that again $a+b = 0$. Furthermore, (3.25a) is equivalent to

$$(r\mathrm{e}^{-2b})' = 1 - \frac{e^2\,G}{r^2}\,.$$

The metric functions are thus given by

$$\mathrm{e}^{2a} = \mathrm{e}^{-2b} = 1 - \frac{2m}{r} + \frac{e^2\,G}{r^2}\,, \tag{3.26}$$

where m has the same interpretation as for the Schwarzschild solution. The other components of Einstein's equation are also satisfied.

It remains to show that Maxwell's vacuum equations are also fulfilled. For this we first note that $F = dA$, $A = \frac{e}{r}\,dt$, hence $dF = 0$. Since $*F = -\frac{e}{r}\,\theta^2 \wedge \theta^3 = d(e\cos\varphi\, d\varphi)$, we have also $d * F = 0$.

For later use we write the result in the form

$$g = -\frac{\Delta}{r^2}\,dt^2 + \frac{r^2}{\Delta}\,dr^2 + r^2\left(d\vartheta^2 + \sin^2\vartheta\, d\varphi^2\right),$$

$$F = \frac{e}{r^2}\,dt \wedge dr\,,$$

where

$$\Delta = r^2 - 2mr + e^2 G\,. \tag{3.27}$$

The apparent singularities will be resolved later (see Sect. 3.9.1).

3.2 Equation of Motion in a Schwarzschild Field

We consider a test body in a Schwarzschild field. We know (see Sect. 1.4.1) that its geodesic equation of motion is the Euler-Lagrange equation for the Lagrangian $\mathcal{L} = \frac{1}{2} g_{\mu\nu}\, \dot{x}^\mu \dot{x}^\nu$, which for the Schwarzschild metric (3.12) is given by

$$2\,\mathcal{L} = -\left(1 - \frac{2m}{r}\right) \dot{t}^2 + \frac{\dot{r}^2}{1 - 2m/r} + r^2 (\dot{\vartheta}^2 + \sin^2\vartheta\, \dot{\varphi}^2)\,, \tag{3.28}$$

where the dot denotes differentiation with respect to proper time. By definition of the proper time we have along the orbit $x^\mu(\tau)$

$$2\,\mathcal{L} = -1\,. \tag{3.29}$$

We consider first the ϑ-Euler-Lagrange equation

$$(r^2\dot{\vartheta})^{\cdot} = r^2 \sin\vartheta \cos\vartheta\, \dot{\varphi}^2\,.$$

This implies that if the motion of the test body is initially in the equatorial plane $\vartheta = \frac{\pi}{2}$ (and hence $\dot{\vartheta} = 0$), then $\vartheta \equiv \frac{\pi}{2}$. We may therefore take $\vartheta = \frac{\pi}{2}$ without loss of generality, and thus

$$2\,\mathcal{L} = -\left(1 - \frac{2m}{r}\right) \dot{t}^2 + \frac{\dot{r}^2}{1 - 2m/r} + r^2 \dot{\varphi}^2\,. \tag{3.30}$$

The variables φ and t are cyclic. Hence, we have

$$\frac{\partial \mathcal{L}}{\partial \dot{\varphi}} = r^2 \dot{\varphi} = \text{const} =: L\,, \tag{3.31a}$$

$$-\frac{\partial \mathcal{L}}{\partial \dot{t}} = \dot{t}\left(1 - \frac{2m}{r}\right) = \text{const} =: E\,. \tag{3.31b}$$

Inserting (3.31a) and (3.31b) into (3.29) gives

$$\left(1 - \frac{2m}{r}\right)^{-1} E^2 - \left(1 - \frac{2m}{r}\right)^{-1} \dot{r}^2 - \frac{L^2}{r^2} = 1\,. \tag{3.32}$$

From this we obtain

$$\dot{r}^2 + V(r) = E^2\,, \tag{3.33}$$

with the **effective potential**

$$V(r) = \left(1 - \frac{2m}{r}\right)\left(1 + \frac{L^2}{r^2}\right)\,. \tag{3.34}$$

* * *

The conservation laws (3.31a) and (3.31b) are based on the following general fact. Let $\gamma(\tau)$ be a geodesic with tangent vector u and let ξ be a Killing field. Then

$$\langle u, \xi \rangle = \text{const} \tag{3.35}$$

along the geodesic γ. In fact, using the Ricci identity and the Killing equation we find

$$\begin{aligned} u\langle u, \xi \rangle &= \langle \nabla_u u, \xi \rangle + \langle u, \nabla_u \xi \rangle \\ &= u^\mu u^\nu \xi_{\mu;\nu} = \frac{1}{2} u^\mu u^\nu (\xi_{\mu;\nu} + \xi_{\nu;\mu}) = 0 \,. \end{aligned}$$

For the Schwarzschild metric, $\partial/\partial t$ and $\partial/\partial\varphi$ are Killing fields. The corresponding conservation laws (3.35) agree with (3.31a) and (3.31b) along γ:

$$\begin{aligned} \left\langle u, \frac{\partial}{\partial t} \right\rangle &= u^t \left\langle \frac{\partial}{\partial t}, \frac{\partial}{\partial t} \right\rangle = g_{00}\, u^t = \left(1 - \frac{2m}{r}\right) \dot{t} = \text{const}\,, \\ \left\langle u, \frac{\partial}{\partial \varphi} \right\rangle &= u^\varphi \left\langle \frac{\partial}{\partial \varphi}, \frac{\partial}{\partial \varphi} \right\rangle = g_{\varphi\varphi}\, u^\varphi = r^2 \dot{\varphi} = \text{const}\,. \end{aligned}$$

* * *

In the following, we are primarily interested in the orbit $r(\varphi)$. Use $r' = \dot{r}/\dot{\varphi}$, where the prime denotes differentiation with respect to φ, and (3.31a) to get $\dot{r} = r'\dot{\varphi} = r'\, L/r^2$. Inserting this in (3.33) gives

$$r'^2\, \frac{L^2}{r^4} = E^2 - V(r)\,.$$

As in the Kepler problem we work with $u = 1/r$ and thus $r' = -u'/u^2$. In terms of this variable, we have

$$L^2 u'^2 = E^2 - (1 - 2m\, u)(1 + L^2 u^2)$$

or

$$u'^2 + u^2 = \frac{E^2 - 1}{L^2} + \frac{2m}{L^2}\, u + 2m\, u^3\,. \tag{3.36}$$

Differentiating (3.36) with respect to φ gives

$$2u'\, u'' + 2u\, u' = \frac{2m}{L^2}\, u' + 6m\, u' u^2\,.$$

Hence either $u' = 0$ (circular motion) or u satisfies the simple differential equation

$$u'' + u = \frac{m}{L^2} + 3m\, u^2\,. \tag{3.37}$$

At this point it is instructive to make a comparison with Newtonian theory, in which the Lagrangian for a test particle in a central gravitational potential $\phi(r)$ is given by

$$\mathcal{L} = \frac{1}{2} \left[\left(\frac{dr}{dt}\right)^2 + r^2 \left(\frac{d\varphi}{dt}\right)^2 \right] - \phi(r)\,.$$

Since φ is cyclic, we have

$$r^2 \frac{d\varphi}{dt} =: L = \text{const}\,,$$

and the radial Euler-Lagrange equation is

$$\frac{d^2 r}{dt^2} = r\left(\frac{d\varphi}{dt}\right)^2 - \frac{d\phi}{dr}\,.$$

Now, we can write

$$\begin{aligned} \frac{dr}{dt} &= \frac{dr}{d\varphi}\frac{d\varphi}{dt} = r' \frac{L}{r^2} = -L\,u'\,,\\ \frac{d^2 r}{dt^2} &= -L\,u'' \frac{d\varphi}{dt} = -L^2\,u'' u^2\,. \end{aligned}$$

After some rearrangement we obtain the differential equation

$$u'' + u = \frac{1}{L^2}\frac{d\phi}{dr}\frac{1}{u^2}\,. \tag{3.38}$$

For a Newtonian potential $\phi = -\frac{GM}{r}$ this becomes

$$u'' + u = \frac{GM}{L^2}\,. \tag{3.39}$$

Equation (3.37) contains the additional term $3m\,u^2$. For our planet system this "perturbation" is small, since

$$\begin{aligned} \frac{3m\,u^2}{m/L^2} &= 3\,u^2\,L^2 = 3\frac{1}{r^2}(r^2\dot{\varphi})^2 \simeq 3\left(r\frac{d\varphi}{dt}\right)^2 \frac{1}{c^2}\\ &\simeq 3\,v_\perp^2/c^2 \simeq 7.7\times 10^{-8} \quad \text{for Mercury}\,, \end{aligned}$$

where $v_\perp$ is the velocity component perpendicular to the radius vector.

According to (3.38), we may regard (3.37) as a Newtonian equation of motion for the potential

$$\phi(r) = -\frac{GM}{r} - m\frac{L^2}{r^3}\,. \tag{3.40}$$

It is instructive to have a closer look at the effective potential $V(r)$ in (3.34). As a function of r/m it depends in an interesting way on the ratio L/m. The limiting values are always $V(r) \longrightarrow -\infty$ for $r \longrightarrow 0$ and $V(r) \longrightarrow 1$ for $r \longrightarrow \infty$. When $L/m < 2\sqrt{3}$ the potential has no critical points (see Fig. 3.2), whence any incoming particle with $E^2 < 1$ crashes directly toward $r = 2m$. Other cases are discussed in Exercise 3.5 (see also Fig. 3.3).

* * *

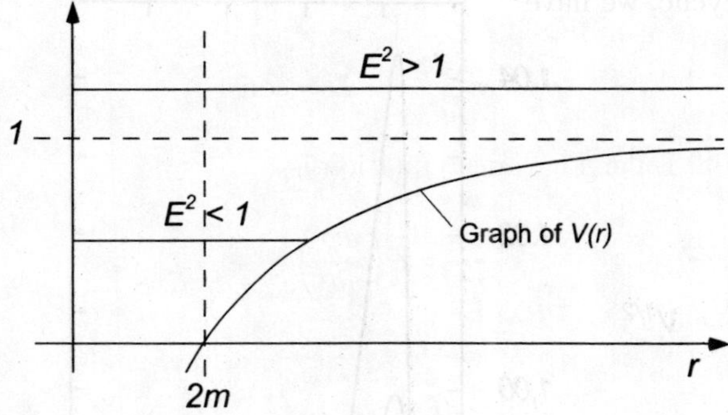

Fig. 3.2. The effective potential for $L/m < 2\sqrt{3}$.

Exercises.

Exercise 3.5. Show that the effective potential (3.34) has the following properties:

1. For $L/m < 2\sqrt{3}$ any incoming particle falls toward $r = 2m$.
2. The most tightly bound, stable circular orbit is at $r = 6m$ with $L/m = 2\sqrt{3}$ and has fractional binding energy of $1 - \sqrt{8/9}$.
3. Any particle with $E \geq 1$ will be pulled into $r = 2m$ if $2\sqrt{3} < L/m < 4$.

3.3 Perihelion Advance

We shall now examine the orbit equation (3.37), treating the term $3m\,u^2$ as a small perturbation. In the Newtonian approximation, the orbit is a Kepler ellipse

$$u(\varphi) = \frac{1}{p}\,(1 + e\cos\varphi)\ , \tag{3.41}$$

where e is the eccentricity and

$$p = a(1 - e^2) = \frac{L^2}{m}\ . \tag{3.42}$$

We now insert this into the perturbation term and obtain from (3.37) to a first approximation

$$u'' + u = m\frac{1}{L^2} + 3m^3\frac{1}{L^4}(1 + e\cos\varphi)^2\ . \tag{3.43}$$

The solution of (3.43) is easily found by noting that the three differential equations

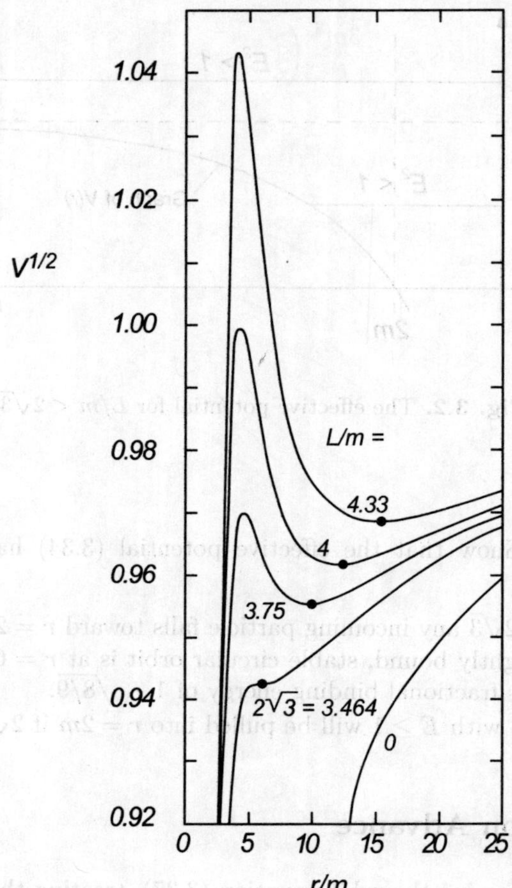

Fig. 3.3. The effective potential.

$$v'' + v = \begin{cases} A \\ A\cos\varphi \\ A\cos^2\varphi\,, \end{cases}$$

with $A = \text{const}$, have the particular solutions

$$v(\varphi) = \begin{cases} A \\ \\ \dfrac{1}{2}A\,\varphi\sin\varphi \\ \\ \dfrac{1}{2}A - \dfrac{1}{6}A\cos 2\varphi\,. \end{cases} \tag{3.44}$$

The general relativistic correction term in (3.43) leads to the following modification of the Kepler ellipse (3.41)

$$u(\varphi) = \frac{m}{L^2}(1 + e\cos\varphi) + \frac{3m^3}{L^4}\left(1 + \frac{e^2}{2} - \frac{e^2}{6}\cos 2\varphi + e\,\varphi\sin\varphi\right) \quad (3.45)$$

with perihelion at $\varphi = 0$ (see Fig. 3.4). The important part of the small correction term is the *secular* contribution proportional to $\varphi\sin\varphi$. The next

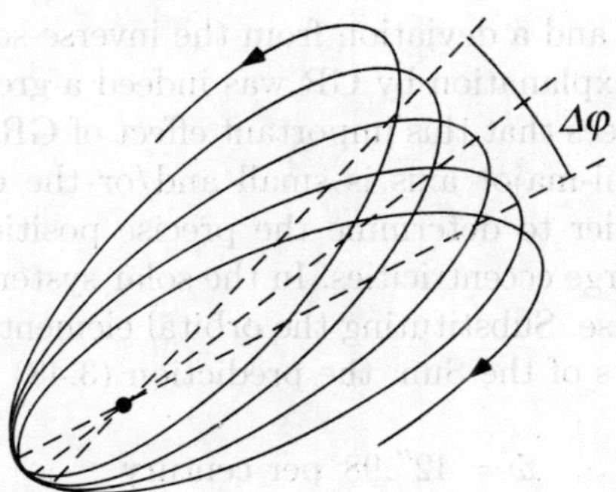

Fig. 3.4. Planetary orbit in GR.

perihelion will appear near $\varphi = 2\pi$ where $u' = 0$. From

$$u' = -\frac{m\,e}{L^2}\sin\varphi + \frac{3m^3\,e}{L^4}\left(\frac{e}{3}\sin 2\varphi + \sin\varphi + \varphi\cos\varphi\right) \quad (3.46)$$

we see that the **perihelion advance** (or **anomaly**) $\Delta\varphi$ per revolution is approximately determined by the first and the last term of this expression:

$$-\sin\Delta\varphi + \frac{3m^2}{L^2}(2\pi + \Delta\varphi)\cos\Delta\varphi \simeq 0\,.$$

Since $\Delta\varphi$ is small we obtain

$$\Delta\varphi \simeq \tan\Delta\varphi \simeq \frac{6\pi m^2}{L^2}\,. \quad (3.47)$$

Using (3.42) this can be written as (recall (3.20))

$$\Delta\varphi = 3\pi\frac{R_S}{a(1-e^2)}\,. \quad (3.48)$$

This famous result was presented by Einstein in a lecture delivered to the Prussian Academy of Sciences in Berlin on 18 November 1915, before he had the definite field equations with matter (see [58]). For the perihelion motion of Mercury he obtained 43 seconds of arc per century, in very satisfactory agreement with observations. Einstein later told his former collaborator A. Fokker that when he saw the result he was so excited that he had heart palpitations.

The perihelion shift of Mercury's orbit was an unsolved problem in celestial mechanics since the announcement by Le Verrier in 1859 that, after the perturbing effect of the other planets on Mercury's orbit had been accounted for, there remained in the data an unexplained advance in its perihelion. Various *ad hoc* proposals were made to account for this excess, such as the existence of a new planet Vulcan near the Sun, a ring of planetoids, a solar quadrupole moment, and a deviation from the inverse-square law of gravitation. – The natural explanation by GR was indeed a great triumph.

From (3.48) one sees that this important effect of GR is most pronounced when the orbit's semi-major axis is small and/or the eccentricity is large. In addition, it is easier to determine the precise position of the perihelion observationally for large eccentricities. In the solar system, *Mercury* provides the most favorable case. Substituting the orbital elements for this planet and the physical constants of the Sun, the prediction (3.48) gives for the rate of perihelion shift

$$\dot{\omega} = 42''.98 \ \text{per century}\,. \tag{3.49}$$

On the other hand, the measured perihelion shift of Mercury is known accurately. The excess shift, after the perturbing effects of the other planets have been accounted for, is known to about 0.1 percent from radar observations of Mercury since 1966. Within this uncertainty theory and observation agree.

Before we take a closer look at the Newtonian perturbations, it should be mentioned that the equinoctial precession of the celestial coordinates used in planetary observations gives an apparent advance of about 5000″ per century. The perturbing effects of the various planets on Mercury's perihelion precession are listed in Table 3.1. Note that the observational accuracy of 0.1 percent corresponds on Mercury's orbit to about 300 meters.

Table 3.1. Newtonian perturbations on Mercury.

Perturbing planet	*Perihelion shift/century*
Venus	277".86±0".27
Earth	90".04±0".08
Mars	2".54
Jupiter	153".58
Saturn	7".30
Uranus	0".14
Neptune	0".04
Sum	531".5±0".3

An additional Newtonian perturbation is caused by the solar quadrupole moment. Let us derive the contribution of the solar oblateness to the perihelion shift. The Newtonian potential exterior to a mass distribution having the matter density $\rho(\boldsymbol{x})$ is

$$\phi(\boldsymbol{x}) = -G \int \frac{\rho(\boldsymbol{x}')}{|\boldsymbol{x}-\boldsymbol{x}'|} d^3x' .$$

For $r > r'$ we use the well-known formula

$$\frac{1}{|\boldsymbol{x}-\boldsymbol{x}'|} = 4\pi \sum_{l=0}^{\infty} \sum_{m=-l}^{+l} \frac{1}{2l+1} \left(\frac{r'}{r}\right)^l \frac{1}{r} Y^*_{lm}(\hat{\boldsymbol{x}}') Y_{lm}(\hat{\boldsymbol{x}}) ,$$

to expand ϕ in terms of multipoles

$$\phi(\boldsymbol{x}) = -4\pi G \sum_{l,m} \frac{Q^*_{lm}}{2l+1} \frac{1}{r^{l+1}} Y_{lm}(\hat{\boldsymbol{x}}) , \tag{3.50}$$

where the

$$Q_{lm} := \int \rho(\boldsymbol{x}') r'^l \, Y_{lm}(\hat{\boldsymbol{x}}') \, d^3x' \tag{3.51}$$

are the multipole moments. Suppose that $\rho(\boldsymbol{x})$ is azimuthally symmetric, and symmetric under reflections at the (x,y)-plane. Then $Q_{lm} = 0$ for $m \neq 0$. The monopole contribution (with $Y_{00} = 1/\sqrt{4\pi}$) is equal to $-GM_\odot/r$. Due to mirror symmetry, the dipole contribution vanishes:

$$\int \rho(\boldsymbol{x}) \underbrace{r' Y_{10}(\hat{\boldsymbol{x}}')}_{\propto z'} \, d^3x' = 0 .$$

The remaining terms give

$$\phi = -\frac{GM_\odot}{r} \left\{ 1 - \sum_{l=2}^{\infty} J_l \left(\frac{R_\odot}{r}\right)^l P_l(\cos\vartheta) \right\} , \tag{3.52}$$

with the *reduced moments*

$$J_l = \frac{-1}{M_\odot R^l_\odot} \int \rho(\boldsymbol{x}') r'^l P_l(\cos\vartheta') \, d^3x' . \tag{3.53}$$

It suffices to keep the quadrupole term $l = 2$. For $\vartheta = \frac{\pi}{2}$ we have

$$\phi = -\frac{GM_\odot}{r} - \frac{1}{2} \frac{GM_\odot J_2 R^2_\odot}{r^3} . \tag{3.54}$$

Note that J_2 is positive for an oblate Sun. This has the same form as (3.40), provided we make the substitution $m \longrightarrow \frac{1}{2} GM_\odot L^{-2} J_2 R^2_\odot$ in the second term. Hence, we obtain from (3.48), using (3.42),

$$\Delta\varphi_{quadrupole} = \frac{6\pi}{a(1-e^2)}\frac{1}{2}J_2\frac{GM_\odot R_\odot^2}{L^2}$$
$$= \frac{6\pi m}{a(1-e^2)}\frac{1}{2}J_2\frac{R_\odot^2}{(GM_\odot/c^2)a(1-e^2)},$$

which means that

$$\Delta\varphi_{quadrupole} = \frac{1}{2}J_2\frac{R_\odot^2}{(GM_\odot/c^2)a(1-e^2)}\Delta\varphi_{Einstein}\,. \tag{3.55}$$

From this one sees that $\Delta\varphi_{quadrupole}$ and $\Delta\varphi_{Einstein}$ depend differently on a and e.

For Mercury one finds in seconds of arc per century

$$\dot{\omega}_{Einstein} + \dot{\omega}_{quadrupole} = 42.98[1 + 3\times 10^{-4}(J_2/10^{-7})]\,. \tag{3.56}$$

Assuming that the Sun rotates uniformly with its observed surface angular velocity one estimates $J_2 \sim 10^{-7}$. This agrees with values inferred from rotating solar models that produce the observed normal modes of solar oscillations (helioseismology). From this we conclude that the solar oblateness effect on the perihelion shift of Mercury is smaller than the observational error. The perihelion anomaly is almost entirely given by GR.

This and other solar system tests of GR discussed below involve weak fields. With the discovery of the binary pulsar PSR 1913+16 by Taylor and Hulse in 1974 strong field effects and the influence of gravitational radiation reaction can be tested. This will be discussed in detail later in Sect. 5.8. At this point we only mention that the periastron advance[2] of this remarkable close binary system of compact objects is per day as large as for Mercury in a century. The reader is encouraged to read the Nobel Prize lecture [97], [98] of Hulse and Taylor.

3.4 Deflection of Light

"This is the most important result obtained in connection with the theory of gravitation since Newton's day."

—J.J. Thomson (6 November 1919)

For light rays in a Schwarzschild field, (3.29) is replaced by $\mathcal{L} = 0$. Instead of (3.32), one then obtains

$$\frac{1}{(1-2m/r)}E^2 - \frac{1}{(1-2m/r)}\dot{r}^2 - \frac{L^2}{r^2} = 0\,, \tag{3.57}$$

[2] Literally, the term perihelion shift should be restricted to orbits around the Sun, while the more general term is periastron.

and (3.36) becomes

$$L^2 u'^2 = E^2 - (1 - 2m\,u)L^2\,u^2\,.$$

The orbit equation for light rays is thus

$$u'^2 + u^2 = \frac{E^2}{L^2} + 2m\,u^3\,. \tag{3.58}$$

Differentiation results in

$$u'' + u = 3m\,u^2\,. \tag{3.59}$$

(Compare this with the orbit equation for a test body (3.37).) For the solar system the right-hand side is very small:

$$3m\,u^2/u = 3R_S/2r \le R_S/R_\odot \sim 10^{-6}\,.$$

If we neglect this, the light path is straight (see Fig. 3.5):

$$u = b^{-1}\sin\varphi\,, \tag{3.60}$$

where b is the impact parameter. Inserting this into the right-hand side of (3.59) gives in first order perturbation theory

$$u'' + u = 3m\,b^{-2}(1 - \cos^2\varphi)\,. \tag{3.61}$$

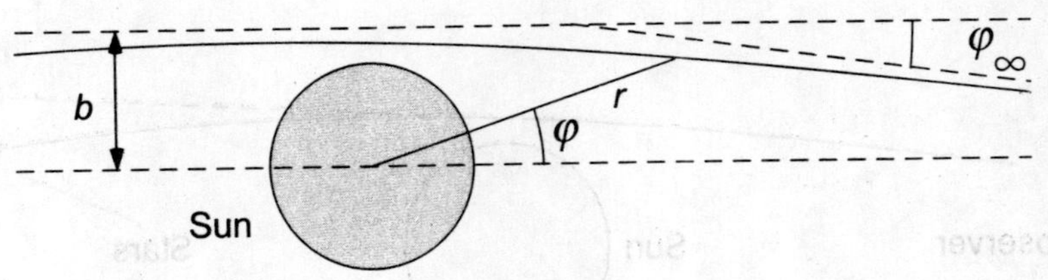

Fig. 3.5. Path of a light ray in the field of the Sun.

This has the particular solution

$$u_1 = \frac{3m}{2b^2}\left(1 + \frac{1}{3}\cos 2\varphi\right),$$

giving to first order

$$u = b^{-1}\sin\varphi + \frac{3m}{2b^2}\left(1 + \frac{1}{3}\cos 2\varphi\right). \tag{3.62}$$

For large r (or small u), $|\varphi|$ is either small or close to π. For small φ we take $\sin\varphi \simeq \varphi$ and $\cos\varphi \simeq 1$. In the limit $u \longrightarrow 0$, φ approaches φ_∞, with

$$\varphi_\infty = -\frac{2m}{b} .$$

The **total deflection** δ is equal to $2|\varphi_\infty|$; thus

$$\delta = \frac{4m}{b} = 4\frac{GM}{c^2 b} = 2\frac{R_S}{b} . \tag{3.63}$$

For the sun we obtain Einstein's famous prediction

$$\delta = 1.75'' R_\odot / b . \tag{3.64}$$

This result follows already from the linearized theory (see Chap. 4). Thus, the bending of light rays does *not* depend on the non-linearity of GR; by contrast, the precession of the perihelion does depend on this, as will be shown in Exercise 3.7.

Historically, this prediction of Einstein was first tested during the solar eclipse of March 29, 1919. Eddington and Dyson organized two expeditions, one to the Brazilian city Sobral, and the other to the island Principe in the gulf of Guinea.

The effect of the deflection of light is observed as an apparent outward shift in the position of the stars during the eclipse. This shift can be determined by photographing the stars in the vicinity of the sun during the eclipse and later the same stars at night. Then one compares the two photographs (see Fig. 3.6).

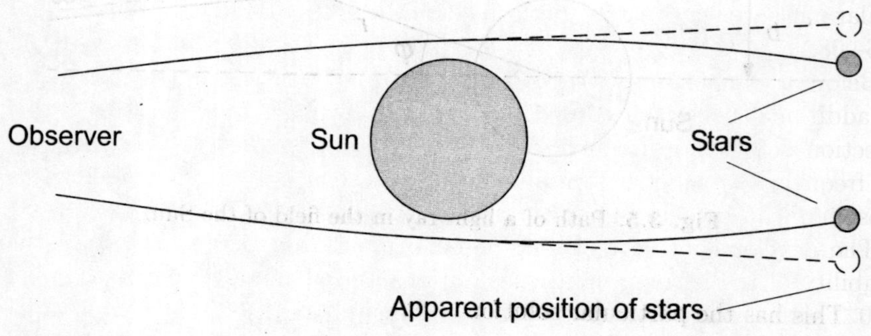

Fig. 3.6. Deflection of light.

Sir F.W. Dyson, the Astronomer Royal, had realized that the expected eclipse should be particularly suitable, since the Sun would be standing before the stellar cluster of the Hyades. The final result of the two teams were

$$\delta = \begin{cases} 1.98'' \pm 0.12'' & \text{Sobral} \\ 1.60'' \pm 0.3'' & \text{Principe}\,. \end{cases}$$

Based in these, Eddington announced that "there can be no doubt that the values confirm Einstein's prediction. A very definite result has been obtained that light is deflected in accordance with Einstein's law of gravitation."

The news made headline in most newspapers, and Einstein quickly rose to international fame. On 27 September 1919 he wrote on a postcard to his mother: "... joyous news today. H.A. Lorentz telegraphed that the English expeditions have actually measured the deflection of starlight from the Sun."

The history around this event is complicated. For good reasons, controversies surrounded the reliability of the results at the time. On the other hand, J.J. Thomson, Chair of the special joint meeting of the Royal Astronomical Society and the Royal Society of London on 6 November 1919, stated: "This is the most important result obtained in connection with the theory of gravitation since Newton's day." For a detailed discussion of the experiments and references to the original literature see [99].

We have mentioned in Sect. 1.2 that in 1914, a German expedition, led by E. Freundlich and funded by Krupp, was sent to the Crimea to photograph the solar eclipse on 21 August. Due to the outbreak of the First World War, the scientists were imprisoned, and their instruments were confiscated by Russian authorities.

Later, numerous measurements of the bending of light have been performed during solar eclipses. The results show considerable scatter and the accuracies were low. After 1969, substantial improvements have been made using radio astronomy. Every year the quasar 3C 279 is eclipsed by the sun on October 8, and thus the deflection of radio waves emitted from this quasar, relative to those from the quasar 3C 273, which is 10° away, can be measured. Similar measurements can be performed on the group 0111+02, 0119+11 and 0116+08.

Before a comparison with GR can be made, it is necessary to correct for the additional deflection caused by the solar corona. For radio waves, this deflection depends on frequency as ω^{-2} (see Exercise 3.6). By observing at two frequencies, the contributions due to the solar corona and the Earth's ionosphere can be determined very precisely.

The development of very-long-baseline interferometry (VLBI) have the capability of measuring angular separations and changes in angles as small as 10^{-4} seconds of arc. A series of transcontinental and intercontinental VLBI quasar and radio galaxy observations (made primarily to monitor the Earth's rotation) was sensitive to the deflection of light over almost the entire celestial sphere. (Note that at 90° from the Sun, the deflection is still 4 milliarcseconds.) A recent analysis of VLBI data [100] yielded for the ratio of the observed deflection to Einstein's prediction the value 0.99997 ± 0.00016.

It has become traditional to parameterize the static, spherically symmetric vacuum field independent of theory. In "isotropic" coordinates, one then has, in place of (3.16) and (3.17)

$$g = -\left[1 - 2\frac{m}{r} + 2\beta\left(\frac{m}{r}\right)^2 + \dots\right] dt^2 + \left[1 + 2\gamma\frac{m}{r} + \dots\right] d\boldsymbol{x}^2 , \tag{3.65}$$

where β and γ are the so-called **Eddington-Robertson parameters**. According to (3.17), we have in GR

$$\beta = \gamma = 1 . \tag{3.66}$$

For a systematic presentation of the so-called Parameterized Post-Newtonian (PPN) formalism and an analysis of solar system tests within this framework, we refer to [6]. The result of Exercise 3.7 shows that the precession of the perihelion is sensitive to non-linearities of the theory (the parameter β), while the light deflection is proportional to $\frac{1}{2}(1+\gamma)$. The measurements discussed so far restrict the Eddington-Robertson parameters as follows:

$$|\gamma - 1| < 3 \times 10^{-4} , \quad |\beta - 1| < 3 \times 10^{-3} . \tag{3.67}$$

Measurements of $(1+\gamma)/2$ are shown in Fig. 3.7.

* * *

Exercises.

Exercise 3.6 (Deflection by the Solar Corona). Compute the deflection of radio waves by the solar corona for the electron density (for $r/R_\odot > 2.5$)

$$n_e(r) = \frac{A}{(r/R_\odot)^6} + \frac{B}{(r/R_\odot)^2} , \tag{3.68}$$

where $A = 10^8$ electrons/cm^3 and $B = 10^6$ electrons/cm^3, which should by fairly realistic.

Hints. The dispersion relation for transverse waves in a plasma (see, e.g., J.D. Jackson, [18]) is given by

$$\omega^2 = c^2 k^2 + \omega_p^2 . \tag{3.69}$$

Here, ω_p is the plasma frequency, which is related to n_e by

$$\omega_p^2 = \frac{4\pi n_e e^2}{m} . \tag{3.70}$$

The corresponding index of refraction is

$$n = c\frac{k}{\omega} = \sqrt{1 - \frac{\omega_p^2}{\omega^2}} . \tag{3.71}$$

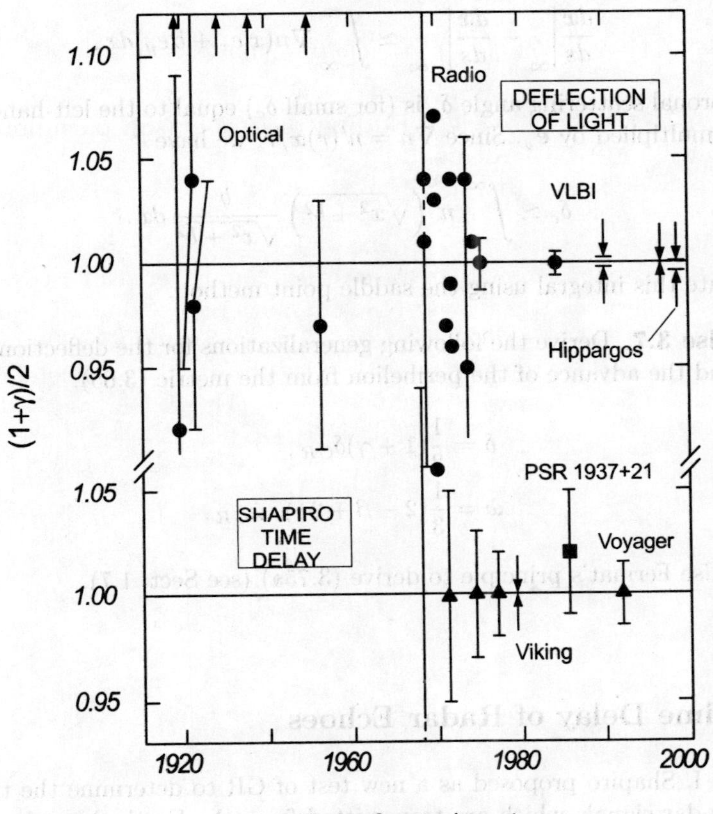

Fig. 3.7. Measurements of the coefficient $(1+\gamma)/2$ from light deflection (upper part) and time delay measurements (discussed in Sect. 3.5). Its GR value is unity. The arrows denote anomalously large values from early eclipse expeditions. The Shapiro time delay measurements using Viking spacecraft yielded an agreement with GR to 0.1 percent and VLBI light deflection measurements have reached 0.02 percent. Hipparcos denotes the optical astrometry satellite, which has reached 0.1 percent. (From [74].)

Let s denote the arc length of the light ray $\boldsymbol{x}(s)$. In any optics text one finds that $\boldsymbol{x}(s)$ satisfies the differential equation (ray equation)

$$\frac{d}{ds}\left(n\frac{d\boldsymbol{x}}{ds}\right) = \nabla n\,. \tag{3.72}$$

(This is also derived in Sect. 4.8.1). Since the deflection is small, we may calculate it to sufficient accuracy by integrating (3.72) along an unperturbed trajectory $y = b$, $-\infty < x < \infty$.

Since $n = 1$ for $r \longrightarrow \infty$, the difference in the direction of the asymptotes is given by

$$\left.\frac{d\boldsymbol{x}}{ds}\right|_{\infty} - \left.\frac{d\boldsymbol{x}}{ds}\right|_{-\infty} \simeq \int_{-\infty}^{\infty} \nabla n(x\,\boldsymbol{e}_x + b\,\boldsymbol{e}_y)dx\,. \tag{3.73}$$

The coronal scattering angle δ_c is (for small δ_c) equal to the left-hand side of (3.73) multiplied by $\boldsymbol{e}_y$. Since $\nabla n = n'(r)\boldsymbol{x}/r$, we have

$$\delta_c \simeq \int_{-\infty}^{\infty} n'\left(\sqrt{x^2+b^2}\right)\frac{b}{\sqrt{x^2+b^2}}\,dx\,. \tag{3.74}$$

Compute this integral using the saddle point method.

Exercise 3.7. Derive the following generalizations for the deflection of light rays and the advance of the perihelion from the metric (3.65):

$$\delta = \frac{1}{2}(1+\gamma)\delta_{GR}\,, \tag{3.75a}$$

$$\dot{\omega} = \frac{1}{3}(2-\beta+2\gamma)\cdot\dot{\omega}_{GR}\,. \tag{3.75b}$$

Hint. Use Fermat's principle to derive (3.75a) (see Sect. 1.7).

3.5 Time Delay of Radar Echoes

In 1964 I. Shapiro proposed as a new test of GR to determine the time delay of radar signals which are transmitted from the Earth through a region near the Sun to another planet or satellite and then reflected back to the Earth. Suppose that a radar signal is transmitted from a point 1 with coordinates $(r_1, \vartheta = \frac{\pi}{2}, \varphi_1)$ to another point 2 with $(r_2, \vartheta = \frac{\pi}{2}, \varphi_2)$, as indicated in Fig. 3.8.

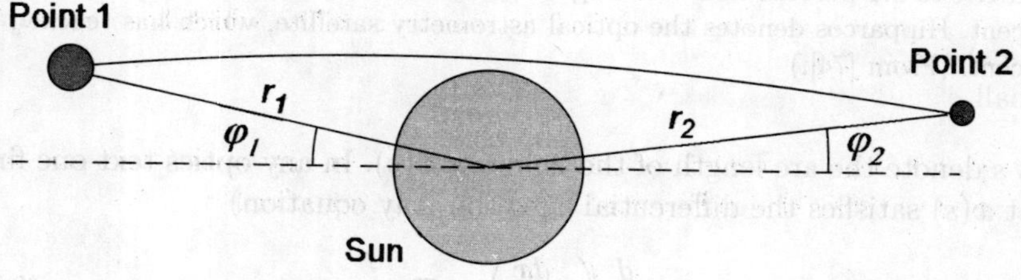

Fig. 3.8. Time delay of radar signals.

We first compute the Schwarzschild coordinate travel time t_{12} required. From (3.57) we have

$$\dot{r}^2 = E^2 - \left(1 - \frac{2m}{r}\right)\frac{L^2}{r^2}. \tag{3.76}$$

Using (3.31b) we can express $\dot{r}$ as

$$\dot{r} = \left(\frac{dr}{dt}\right)\dot{t} = \frac{dr}{dt}\frac{E}{1-2m/r}.$$

Inserting this into (3.76) gives

$$\left(1 - \frac{2m}{r}\right)^{-3}\left(\frac{dr}{dt}\right)^2 = \left(1 - \frac{2m}{r}\right)^{-1} - \left(\frac{L}{E}\right)^2\frac{1}{r^2}. \tag{3.77}$$

At the distance $r = r_0$ of closest approach to the Sun, dr/dt vanishes, so that

$$\left(\frac{L}{E}\right)^2 = \frac{r_0^2}{(1-2m/r_0)}.$$

If we insert this in (3.77), we obtain

$$\left(1 - \frac{2m}{r}\right)^{-3}\left(\frac{dr}{dt}\right)^2 + \left(\frac{r_0}{r}\right)^2\left(1 - \frac{2m}{r_0}\right)^{-1} - \left(1 - \frac{2m}{r}\right)^{-1} = 0. \tag{3.78}$$

According to this equation, the coordinate time which the light requires to from r_0 to r (or the reverse) is

$$t(r, r_0) = \int_{r_0}^{r}\frac{dr}{1-2m/r}\left(1 - \frac{1-2m/r}{1-2m/r_0}\left(\frac{r_0}{r}\right)^2\right)^{-1/2}. \tag{3.79}$$

Hence, for $|\varphi_1 - \varphi_2| > \pi/2$,

$$t_{12} = t(r_1, r_0) + t(r_2, r_0). \tag{3.80}$$

We shall treat the quantity $2m/r$ appearing in the integrand of (3.79) as small, thus obtaining

$$t(r, r_0) \simeq \int_{r_0}^{r}\left\{1 + \frac{2m}{r}\left(1 - \left(\frac{2m}{r_0} - \frac{2m}{r}\right)\left(\frac{r_0}{r}\right)^2\right)^{-1/2}\right\}dr,$$

or

$$t(r, r_0) \simeq \int_{r_0}^{r}\left(1 - \frac{r_0^2}{r^2}\right)^{-1/2}\left(1 + \frac{2m}{r} + \frac{m r_0}{r(r+r_0)}\right)dr.$$

Elementary integration gives

$$t(r, r_0) \simeq \sqrt{r^2 - r_0^2} + 2m \ln \left(\frac{r + \sqrt{r^2 - r_0^2}}{r_0} \right) + m \left(\frac{r - r_0}{r + r_0} \right)^{1/2} . \tag{3.81}$$

For the circuit from point 1 to point 2 and back, it is natural to introduce the following **Shapiro delay in coordinate time**:

$$\begin{aligned} \Delta t := {} & 2\Big(t(r_1, r_0) + t(r_2, r_0) - \sqrt{r_1^2 - r_0^2} - \sqrt{r_2^2 - r_0^2} \Big) \\ \simeq {} & 4m \ln \left(\frac{(r_1 + \sqrt{r_1^2 - r_0^2})(r_2 + \sqrt{r_2^2 - r_0^2})}{r_0^2} \right) \\ & + 2m \left(\sqrt{\frac{r_1 - r_0}{r_1 + r_0}} + \sqrt{\frac{r_2 - r_0}{r_2 + r_0}} \right) . \end{aligned} \tag{3.82}$$

Although this time is not observable, it gives an idea of the magnitude of the general relativistic time delay.

For a round trip from the Earth to Mars and back, we get (for $r_0 \ll r_1, r_2$):

$$\Delta t \simeq 4m \left(\ln \frac{4 r_1 r_2}{r_0^2} + 1 \right) , \tag{3.83a}$$

$$(\Delta t)_{max} \simeq 240 \mu s . \tag{3.83b}$$

To give an idea of the experimental possibilities, we mention that the error in the time measurement of a circuit during the Viking mission was only about $10 ns$.

We emphasize once more that, the coordinate time delay (3.82) cannot be directly measured. For this, it would be necessary to know $(r_1^2 - r_0^2)^{1/2} + (r_2^2 - r_0^2)^{1/2}$ extremely precisely (i.e., to better than $1 km$ for a 1% measurement). Now the various radial coordinates in general use (see, e.g., (3.14)) already differ by an amount of the order of $GM_\odot/c^2 \simeq 1.5 km$. For this reason, the theoretical analysis and practical evaluation of the data from a time delay experiment is much more complicated. One must carry out the measurements of the circuit times over a period of several months and fit the data to a complicated model which describes the motion of the transmitter and receiver very precisely. The post-Newtonian corrections to the equations of motion must also be taken into account[3]. In every reasonably realistic model, there is a set of parameters which are not known well enough. It is necessary to include all these "uninteresting" parameters, along with the relativistic parameter γ (see (3.65)), in a simultaneous fit.

The first ("passive") experiments were performed with radar echoes from Mercury and Venus. The rather complicated topography of these planets limited the precision attainable in these experiments. In addition, the reflected signals were very weak. The result of an early measurement is shown in Fig. 3.9.

[3] A similar, but even more demanding analysis will be carried out in Sect. 5.8, in connection with binary pulsar data. The Shapiro time delay will there also play an important role.

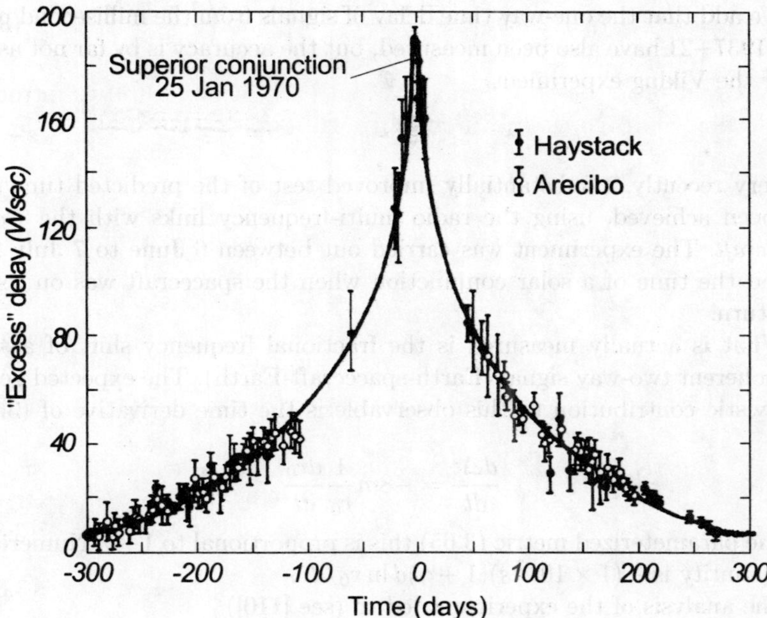

Fig. 3.9. Earth-Venus time-delay measurements. The solid line gives the theoretical prediction. (From [101].)

The ("active") radio observations from satellites, in which the reflected signals are shifted in frequency by a transponder, suffer from the difficulty that the satellites are subject to strong, non-gravitational perturbations, such as radiation pressure, loss of gas, etc.

In all the experiments, time delays due to the solar corona provide an additional source of error. In spite of all these difficulties, the *Viking mission* (see [102]) provided a very precise determination of γ:

$$\gamma = 1 \pm 0.001\,, \tag{3.84}$$

confirming the GR prediction $\gamma = 1$. An important factor in attaining such an amazing precision is that in the Viking mission two transponders landed on Mars, while two others continued to orbit the planet. The orbiting transponders transmitted in both the S-band and the X-band (unfortunately, the two "landers" could only transmit at S-band frequencies). It was thus possible to determine the rather large time delays (up to 100 μs) due to the solar corona fairly precisely (since these are proportional to ν^{-2}). Since two of the transponders are fixed on Mars, the determination of the orbits is more precise than was possible with Mariner 6, 7 an 9. One also discovered that the corona varies considerably with time, and hence that the time delays cannot by determined sufficiently precisely with a parameterized theoretical model.

We add that the one-way time delay of signals from the millisecond pulsar PSR 1937+21 have also been measured, but the accuracy is by far not as high as for the Viking experiment.

* * *

Very recently, a substantially improved test of the predicted time delay has been achieved, using the radio multi-frequency links with the *Cassini spacecraft*. The experiment was carried out between 6 June to 7 July 2002, around the time of a solar conjunction when the spacecraft was on its way to Saturn.

What is actually measured is the fractional frequency shift of a stable and coherent two-way signal (Earth-spacecraft-Earth). The expected general relativistic contribution to this observable is the time derivative of (3.83a), i.e.,

$$\frac{d\Delta t}{dt} = -8m\frac{1}{r_0}\frac{dr_0}{dt}.$$

For the parameterized metric (3.65) this is proportional to $1+\gamma$. Numerically, this quantity is $-(1 \times 10^{-5}s)(1+\gamma)d\ln r_0/dr$.

The analysis of the experiment led to (see [110])

$$\gamma = 1 + (2.1 \pm 2.3) \times 10^{-5}.$$

This is a significant improvement in comparison to (3.84).

* * *

Exercise 3.8. Use the electron density (3.68) to compute the time delay caused by the solar corona.

Hints. The group velocity corresponding to the dispersion relation $\omega = (c^2k^2 + \omega_p^2)^{1/2}$ is

$$v_g = \frac{\partial\omega}{\partial k} = \frac{kc^2}{\sqrt{c^2k^2 + \omega_p^2}} \simeq \left(1 - \frac{1}{2}\frac{\omega_p^2}{\omega^2}\right)c.$$

For the computation of the time delay, one may use the unperturbed straight line path for the light ray. This can be seen with Fermat's principle, according to which the actual path of the ray minimizes the transit time (hence the error is of second order in the perturbation).

3.6 Geodetic Precession

Suppose that a gyroscope with spin S moves along a geodesic. Then the four-vector S is parallel transported (see Sect. 1.10.1). As a special case, we consider a circular motion in the plane $\vartheta = \frac{\pi}{2}$ in the Schwarzschild field and compute the corresponding spin precession. The spin precession for general motion will be studied in post-Newtonian approximation in Sect. 5.6.

In the following, we use the tetrad (3.4) and its dual basis, denoted by $\{e_\alpha\}$. The four-velocity u and the spin S of the gyroscope moving along a geodesic satisfy the equations

$$\langle S, u\rangle = 0\,, \quad \nabla_u S = 0\,, \quad \nabla_u u = 0\,. \tag{3.85}$$

The components of $\nabla_u S$ are given by (see DG, (13.60))

$$(\nabla_u S)^\mu = \dot{S}^\mu + \omega^\mu{}_\beta(u) S^\beta = 0\,, \tag{3.86}$$

where a dot again denotes differentiation with respect to proper time. For circular motion with $\vartheta = \frac{\pi}{2}$, we obviously have $u^1 = u^2 = 0$. If we now use the connection forms (3.8) with $a = -b$, we obtain

$$\dot{S}^0 = -\omega^0{}_\beta(u) S^\beta = -\omega^0{}_1(u) S^1 = -a'\,\mathrm{e}^{-b}\theta^0(u) S^1 = b'\,\mathrm{e}^{-b} u^0 S^1\,.$$

The other equations are derived similarly. The result is

$$\begin{aligned}
\dot{S}^0 &= b'\,\mathrm{e}^{-b} u^0 S^1\,,\\
\dot{S}^1 &= b'\,\mathrm{e}^{-b} u^0 S^0 + \frac{1}{r}\mathrm{e}^{-b} u^3 S^3\,,\\
\dot{S}^2 &= 0\,,\\
\dot{S}^3 &= -\frac{1}{r}\mathrm{e}^{-b} u^3 S^1\,.
\end{aligned} \tag{3.87}$$

From $\nabla_u u = 0$ we obtain in place of the second equation in (3.87)

$$0 = b'\mathrm{e}^{-b}(u^0)^2 + \frac{1}{r}\mathrm{e}^{-b}(u^3)^2\,,$$

which implies that

$$\left(\frac{u^0}{u^3}\right)^2 = -\frac{1}{b'r}\,. \tag{3.88}$$

The following normalized vectors are perpendicular to u:

$$\bar{e}_1 = e_1\,, \quad \bar{e}_2 = e_2\,, \quad \bar{e}_3 = u^3\, e_0 + u^0\, e_3\,. \tag{3.89}$$

When expanded in terms of these, S has only spatial components: $S = \bar{S}^i \bar{e}_i$. Obviously, we have

$$S^0 = u^3\,\bar{S}^3\,, \quad S^1 = \bar{S}^1\,, \quad S^2 = \bar{S}^2\,, \quad S^3 = u^0\,\bar{S}^3\,. \tag{3.90}$$

We now write (3.87) in terms of the $\bar{S}^i$, with the result (dropping the bars)

$$\dot{S}^1 = -b'\mathrm{e}^{-b}\frac{u^0}{u^3} S^3\,, \quad \dot{S}^2 = 0\,, \quad S^3 = b'\mathrm{e}^{-b}\frac{u^0}{u^3} S^1\,. \tag{3.91}$$

In obtaining (3.91), we used (3.88) and $(u^0)^2 - (u^3)^2 = 1$.

We now replace the $\dot{S}^i$ in (3.91) by derivatives with respect to coordinate time. Note that (3.4) implies $u^0 = \mathrm{e}^{-b}\dot{t}$, hence $dS^i/dt = \dot{S}^i/\dot{t} = \dot{S}^i \mathrm{e}^{-b}/u^0$, and therefore

$$\frac{dS^1}{dt} = -\frac{b'}{u^3}\mathrm{e}^{-2b}S^3\,, \qquad \frac{dS^2}{dt} = 0\,, \qquad \frac{dS^3}{dt} = \frac{b'}{u^3}\mathrm{e}^{-2b}S^1\,. \tag{3.92}$$

From (3.4) we have $u^3 = r\dot{\varphi}$, thus

$$\omega := \frac{d\varphi}{dt} = \frac{\dot{\varphi}}{\dot{t}} = \frac{1}{r}\frac{u^3}{u^0}\mathrm{e}^{-b}$$

or, together with (3.88),

$$\omega^2 = \left(\frac{u^3}{u^0}\right)^2 \frac{1}{r^2}\mathrm{e}^{-2b} = -\frac{1}{r}b'\mathrm{e}^{-2b} = \frac{1}{2r}(\mathrm{e}^{-2b})'\,.$$

Using $\mathrm{e}^{-2b} = 1 - 2m/r$ we obtain

$$\omega^2 = \frac{m}{r^3}\,. \tag{3.93}$$

This is the relativistic version of Kepler's third law.

The orbital angular frequency Ω in (3.92) satisfies

$$\begin{aligned}\Omega^2 &= \frac{(b')^2\mathrm{e}^{-4b}}{(u^3)^2} = (b')^2\mathrm{e}^{-4b}\left(-1 - \frac{1}{b'r}\right) = -b'\left(b' + \frac{1}{r}\right)\mathrm{e}^{-4b}\\ &= \omega^2\mathrm{e}^{-2b}(rb'+1) = \omega^2\mathrm{e}^{-2b}\left(-\frac{m}{r}\frac{1}{1-2m/r} + 1\right)\\ &= \omega^2\mathrm{e}^{-2b}\left(\frac{1-3m/r}{1-2m/r}\right),\end{aligned}$$

thus $\Omega^2 = e^2\,\omega^2$, with

$$e^2 = 1 - \frac{3m}{r}\,. \tag{3.94}$$

We write (3.92) in vector notation

$$\frac{d}{dt}\boldsymbol{S} = \boldsymbol{\Omega} \times \boldsymbol{S}\,, \qquad \boldsymbol{\Omega} = (0, e\,\omega, 0)\,. \tag{3.95}$$

In the Newtonian limit $\boldsymbol{\Omega} = (0, \omega, 0)$. Equation (3.95) shows that in the three-dimensional space oriented toward the center and perpendicular to the direction of motion, $\boldsymbol{S}$ precesses retrograde about an axis perpendicular to the plane of the orbit, with frequency $e\,\omega < \omega$. After one complete orbit, the projection of $\boldsymbol{S}$ onto the plane of the orbit has *advanced* by an angle

$$2\pi(1-e) =: \omega_s\frac{2\pi}{\omega} \tag{3.96}$$

(the reader should convince himself of the sign).

The precession frequency $\omega_s = \omega(1-e)$ is given by

$$\omega_s = \left(\frac{m}{r^3}\right)^{1/2}\left[1-\left(1-\frac{3m}{r}\right)^{1/2}\right] \simeq \left(\frac{GM}{r^3}\right)^{1/2}\frac{3GM}{2r},$$

or

$$\omega_s \simeq \frac{3}{2}\frac{(GM)^{3/2}}{r^{5/2}}. \tag{3.97}$$

The square root in the exact expression for ω_s is always well defined, since there are no circular orbits for $r < 3m$.

Taking the Earth as the central mass, one obtains

$$\omega_s \simeq 8.4\left(\frac{R_\oplus}{r}\right)^{5/2} \quad arcs/yr. \tag{3.98}$$

In addition to the geodetic precession (3.98), there is a further small contribution from GR which is due to the Earth's rotation. This **Lense-Thirring precession** will be discussed in Sects. 4.2 and 5.6 (see also Sect. 1.10.4). The Stanford Gyroscope Experiment has as one of its goals the measurement of these effects.

Remarks.

1. The Earth is a natural gyroscope. After one "sidereal year" (i.e., $\Delta\varphi = 2\pi$) the projection of the Earth's axis on the ecliptic has advanced by the angle

   $$2\pi(1-e) = 0.019''.$$

   The sidereal year is characterized by the return of the Sun to the same location with respect to the fixed stars. The pole axis has then moved with respect to the fixed stars. However, this effect is masked by other perturbations of the Earth's axis.
2. The Earth-Moon system can be considered as a "gyroscope", with its axis perpendicular to the orbital plane. The predicted precession, first calculated by de Sitter in 1916, is about 2 arc-seconds per century. Using Lunar laser ranging data over many years, this effect has been measured to about 0.7 percent (see [103]).
3. Another natural gyroscope is the binary pulsar 1913+16. As we shall see in Sect. 5.8, its general relativistic precession amounts to 1.21 *deg* per year, leading to a change in the pulse profile. This has recently been detected with relatively low precision (see [104]).

* * *

Exercise 3.9.
1. Determine the azimuthal velocity of a circular geodesic orbit in the Schwarzschild field. How large is this on the last stable circular orbit? The result is
$$v^{\varphi} = \frac{1}{r} \frac{L}{\sqrt{1 + L^2/r^2}} .$$
This becomes equal to 1/2 for the last stable circular orbit, and approaches 1 for the smallest (unstable) circular orbit.
2. Calculate also the radial acceleration for a non-geodesic circular orbit. Show that this becomes *positive* for $r < 3m$. This anti-intuitive result has led to lots of discussions.

Hints. Use the tools developed in Sect. 3.6.

3.7 Schwarzschild Black Holes

In Chap. 6 we shall show that, in spite of the uncertainties in the equation of state, the mass of non-rotating, cold (neutron) stars is bounded by a few solar masses. Here we mention only a few important results (see also [106]).

Let ρ denote the mass-energy density, and assume that the equation of state is known for $\rho \leq \rho_*$, where ρ_* is not significantly larger than the nuclear matter density $\rho_0 = 2.8 \times 10^{14} g/cm^3$. The mass of a non-rotating neutron star is bounded by
$$M/M_{\odot} \leq 6.75(\rho_0/\rho_*)^{1/2}, \tag{3.99}$$
assuming only that the equation of state above ρ_* obeys $\rho > 0$ and $\partial p/\partial \rho > 0$.

If in addition, one assumes that the speed of sound is less than the speed of light, so that $\partial p/\partial \rho < c^2$, the bound (3.99) can be tightened:
$$M/M_{\odot} \leq 4.0(\rho_0/\rho_*)^{1/2} . \tag{3.100}$$
"Realistic" equations of state give typical values of M_{max} around $2M_{\odot}$. For rapidly rotating neutron stars this limit increases.

Of course, a supercritical mass can exist in a temporary equilibrium, if pressure is built up by means of thermonuclear processes. Sooner or later, however, the nuclear energy sources will be exhausted, and thus a permanent equilibrium is no more possible, unless sufficient mass is "blown off" (for example in a supernova explosion). Otherwise, the star will collapse to a black hole.

In this section, we shall consider qualitatively the spherically symmetric collapse and the detailed properties of spherically symmetric black holes. Many of the qualitatively important aspects of the realistic collapse to a black hole are already encountered in this simple situation. The spherically symmetric collapse is particularly simple, since we know the gravitational field outside of the star. In this region, it depends only on the total mass of the collapsing object, and not on any dynamical details of the implosion, as we have already shown for $r > 2m$ in Sect. 3.1 (Birkhoff theorem). In the following section, we shall extend this result to the case $r \leq 2m$.

3.7.1 The Kruskal Continuation of the Schwarzschild Solution

The Schwarzschild solution

$$g = -\left(1 - \frac{2m}{r}\right) dt^2 + \frac{dr^2}{1 - 2m/r} + r^2(d\vartheta^2 + \sin^2\vartheta d\varphi^2) \tag{3.101}$$

has an apparent singularity at $r = 2m$. We shall see that (3.101) becomes singular at $r = 2m$ only because the chosen coordinate system loses its applicability[4] at $r = 2m$. A first hint of this behavior comes from the observation that, with respect to the orthonormal basis (3.4), the Riemann tensor (3.10) is finite at $r = 2m$. A typical component is

$$R^1{}_{212} = R^1{}_{313} = \mathrm{e}^{-2b}\frac{b'}{r} \sim \frac{1}{r^3}\,.$$

Hence at $r = 2m$, the tidal forces remain finite.

Before we continue the Schwarzschild solution beyond $r = 2m$, let us consider timelike radial geodesics in the vicinity of the horizon. For $L = 0$, we obtain from (3.33) and (3.34) the following equation of motion for a test body in a Schwarzschild field:

$$\dot{r}^2 = 2m/r + E^2 - 1\,. \tag{3.102}$$

Suppose that a radially falling particle is at rest at $r = R$, so that $2m/R = 1 - E^2$, with $E < 1$. From (3.102) we have

$$d\tau = \left(\frac{2m}{r} - \frac{2m}{R}\right)^{-1/2} dr\,. \tag{3.103}$$

One can easily find the parametric representation of the motion (cycloid)

$$r = \frac{R}{2}(1 + \cos\eta)\,, \tag{3.104a}$$

$$\tau = \left(\frac{R^3}{8m}\right)^{1/2} (\eta + \sin\eta)\,. \tag{3.104b}$$

Nothing unusual happens at $r = 2m$. For $\eta = 0$, we have $r = R$ and $\tau = 0$. The proper time for free fall to $r = 0$ (i.e., $\eta = \pi$) is $\tau = (\pi/2)(R^3/2m)^{1/2}$.

[4] It is easy to construct coordinate singularities. For example, consider in $\mathbb{R}^2$ the metric $ds^2 = dx^2 + dy^2$ and introduce in place of x the coordinate $\xi = x^3/3$. The transformed metric is given by

$$ds^2 = (3\xi)^{-4/3} d\xi^2 + dy^2\,,$$

which is "singular" at $\xi = 0$.

On the other hand, consider r as a function of the *coordinate* time t. Together with (3.31b) we obtain

$$\dot{r} = \frac{dr}{dt}\dot{t} = \frac{dr}{dt}\frac{E}{1-2m/r}. \tag{3.105}$$

Let us use the radial coordinate

$$r^* = r + 2m\ln\left(\frac{r}{2m} - 1\right) \tag{3.106}$$

Since

$$\frac{dr^*}{dt} = \frac{1}{1-2m/r}\frac{dr}{dt},$$

it follows from (3.105) that

$$\dot{r} = E\frac{dr^*}{dt}. \tag{3.107}$$

If we now insert this in (3.102), we obtain

$$\left(E\frac{dr^*}{dt}\right)^2 = E^2 - 1 + \frac{2m}{r}. \tag{3.108}$$

Let us look for an approximate solution in the vicinity of $r = 2m$. For $r \downarrow 2m$, we have $r^* \to -\infty$ and the right hand side of (3.108) approaches E^2. Hence, for $r \simeq 2m$,

$$\frac{dr^*}{dt} \simeq -1, \quad \text{and} \quad r^* \simeq -t + \text{const}. \tag{3.109}$$

Thus, we obtain, using (3.106),

$$r \simeq 2m + \text{const}\ e^{-t/2m}, \tag{3.110}$$

and this shows that the test body arrives at the Schwarzschild radius only after *infinitely long* coordinate time.

It is also possible to express t in terms of the parameter η. From (3.31b), (3.104a) and (3.104b), we have

$$dt = \frac{E\,d\tau}{1-2m/r} = E\frac{d\tau}{d\eta}\frac{1}{1-2m/r}d\eta$$

$$= \sqrt{(1-2m/R)}\left(\frac{R^3}{8m}\right)^{1/2}\frac{1+\cos\eta}{1-4m(R(1+\cos\eta))^{-1}}d\eta.$$

With the help of integration tables, one finds

$$\frac{t}{2m} = \ln\left|\frac{(R/2m-1)^{1/2} + \tan\eta/2}{(R/2m-1)^{1/2} - \tan\eta/2}\right|$$

$$+\left(\frac{R}{2m}-1\right)^{1/2}\left(\eta + \frac{R}{4m}(\eta + \sin\eta)\right). \tag{3.111}$$

Here, the constant of integration was chosen such that $t = 0$ for $\eta = 0$ ($r = R$). For $\tan\eta/2 \to (R/2m-1)^{1/2}$, we have $r \to 2m$ and $t \to \infty$ (see Fig. 3.10).

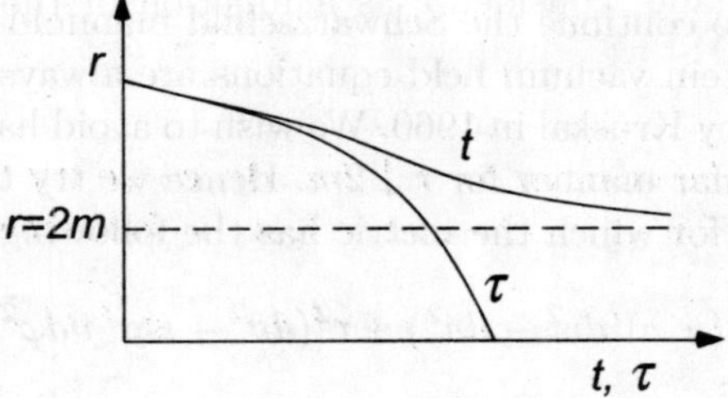

Fig. 3.10. Radial motion of a test body as a function of proper time and coordinate time, respectively.

Next, we consider radial *null* directions for the metric (3.101) (for $r > 2m$ and $r < 2m$). For these we have $ds = 0$ and

$$\frac{dr}{dt} = \pm\left(1 - \frac{2m}{r}\right). \tag{3.112}$$

As $r \downarrow 2m$, the opening angle of the light cones becomes increasingly narrow, as shown in Fig. 3.11.

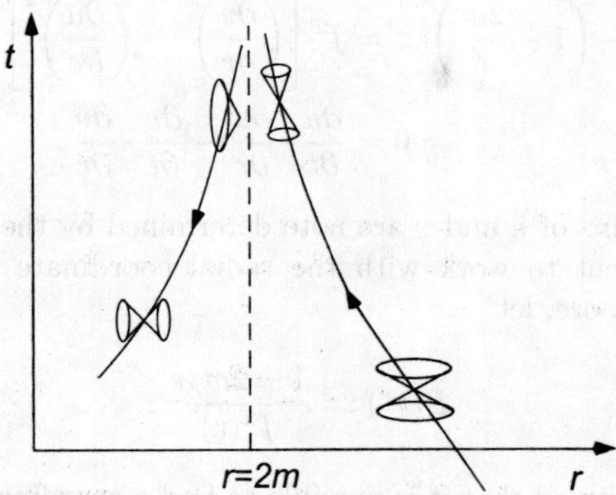

Fig. 3.11. Light cones in Schwarzschild coordinates.

The previous discussion indicates that the simultaneous use of the coordinates r and t is limited. By using the proper time, it is possible to describe events which only occur after $t = \infty$.

When $r > 2m$, t is a distinguished timelike coordinate. The Schwarzschild solution is *static* and the corresponding Killing field K is just $K = \partial/\partial t$ (the corresponding 1-form $K^\flat$ is given by $K^\flat = \langle K, K\rangle dt$). Since the coordinate t is adapted to the Killing field K, it is uniquely defined, up to an additive constant.

We now wish to continue the Schwarzschild manifold ($r > 2m$) in such a way that the Einstein vacuum field equations are always satisfied. This was done most simply by Kruskal in 1960. We wish to avoid having the light cones contract in a singular manner for $r \downarrow 2m$. Hence we try to transform to new coordinates (u, v), for which the metric has the following form

$$g = -f^2(u,v)(dv^2 - du^2) + r^2(d\vartheta^2 + \sin^2\vartheta d\varphi^2)\,. \tag{3.113}$$

For radially emitted light rays we then have $(du/dv)^2 = 1$ when $f^2 \neq 0$. The metrics of the two-dimensional submanifolds $\{\vartheta, \varphi = \text{const}\}$ are conformally equivalent to the 2-dimensional Minkowski metric $dv^2 - du^2$.

We can easily write down the differential equations for the desired transformation (ϑ and φ are unchanged). From

$$g_{\alpha\beta} = \frac{\partial x'^\mu}{\partial x^\alpha}\frac{\partial x'^\nu}{\partial x^\beta} g'_{\mu\nu}$$

it follows that

$$1 - \frac{2m}{r} = f^2\left[\left(\frac{\partial v}{\partial t}\right)^2 - \left(\frac{\partial u}{\partial t}\right)^2\right],$$

$$-\left(1 - \frac{2m}{r}\right)^{-1} = f^2\left[\left(\frac{\partial v}{\partial r}\right)^2 - \left(\frac{\partial u}{\partial r}\right)^2\right],$$

$$0 = \frac{\partial u}{\partial t}\cdot\frac{\partial u}{\partial r} - \frac{\partial v}{\partial t}\cdot\frac{\partial v}{\partial r}\,.$$

Note that the signs of u and v are note determined by these equations.

It is convenient to work with the radial coordinate r^*, introduced in (3.106). Furthermore, let

$$F(r^*) := \frac{1 - 2m/r}{f^2(r)}\,.$$

Here we have assumed that it is possible to find a coordinate transformation such that the function f depends only on r. The transformation equations can now be rewritten as:

$$\left(\frac{\partial v}{\partial t}\right)^2 - \left(\frac{\partial u}{\partial t}\right)^2 = F(r^*)\,, \tag{3.114a}$$

$$\left(\frac{\partial v}{\partial r^*}\right)^2 - \left(\frac{\partial u}{\partial r^*}\right)^2 = -F(r^*)\,, \tag{3.114b}$$

$$\frac{\partial u}{\partial t}\cdot\frac{\partial u}{\partial r^*} = \frac{\partial v}{\partial t}\cdot\frac{\partial v}{\partial r^*}\,. \tag{3.114c}$$

The linear combinations (3.114a)+(3.114b) $\pm$ 2×(3.114c) give

$$\left(\frac{\partial v}{\partial t} + \frac{\partial v}{\partial r^*}\right)^2 = \left(\frac{\partial u}{\partial t} + \frac{\partial u}{\partial r^*}\right)^2, \tag{3.115a}$$

$$\left(\frac{\partial v}{\partial t} - \frac{\partial v}{\partial r^*}\right)^2 = \left(\frac{\partial u}{\partial t} - \frac{\partial u}{\partial r^*}\right)^2. \tag{3.115b}$$

When extracting square roots, we choose the positive roots in the first equation and the negative roots in the second (this avoids vanishing of the Jacobi determinant). We then obtain

$$\frac{\partial v}{\partial t} = \frac{\partial u}{\partial r^*}, \quad \frac{\partial v}{\partial r^*} = \frac{\partial u}{\partial t}. \tag{3.116}$$

This gives

$$\frac{\partial^2 u}{\partial t^2} - \frac{\partial^2 u}{\partial r^{*2}} = 0, \quad \frac{\partial^2 v}{\partial t^2} - \frac{\partial^2 v}{\partial r^{*2}} = 0. \tag{3.117}$$

The general solution of these wave equations is

$$v = h(r^* + t) + g(r^* - t), \tag{3.118a}$$

$$u = h(r^* + t) - g(r^* - t). \tag{3.118b}$$

We now insert this into (3.114a)-(3.114c). Equation (3.114c) is satisfied identically, while (3.114a) and (3.114b) lead to the additional condition:

$$-4h'(r^* + t)g'(r^* - t) = F(r^*). \tag{3.119}$$

For the moment, we assume that $r > 2m$ and hence that $F(r^*) > 0$. Differentiation of (3.119) with respect to r^*, respectively t, leads to

$$\frac{F'(r^*)}{F(r^*)} = \frac{h''(r^* + t)}{h'(r^* + t)} + \frac{g''(r^* - t)}{g'(r^* - t)}, \tag{3.120a}$$

$$0 = \frac{h''(r^* + t)}{h'(r^* + t)} - \frac{g''(r^* - t)}{g'(r^* - t)}. \tag{3.120b}$$

Adding these gives

$$(\ln F(r^*))' = 2(\ln h')'(r^* + t). \tag{3.121}$$

We may regard r^* and $y = r^* + t$ as independent variables. In (3.121), both sides must then be equal to the same constant 2η, say. Introducing suitable constants of integration, we then obtain

$$h(y) = \frac{1}{2}\mathrm{e}^{\eta y}, \quad F(r^*) = \eta^2 \mathrm{e}^{2\eta r^*}. \tag{3.122}$$

From (3.120b) we also find

$$g(y) = -\frac{1}{2}\mathrm{e}^{\eta y}. \tag{3.123}$$

The relative sign of h and g is determined by $F > 0$ and (3.119).

Summarizing, we have

$$u = h(r^* + t) - g(r^* - t) = \frac{1}{2}\mathrm{e}^{\eta(r^*+t)} + \frac{1}{2}\mathrm{e}^{\eta(r^*-t)}$$

$$= \mathrm{e}^{\eta r^*}\cosh(\eta t) = \left(\frac{r}{2m} - 1\right)^{2m\,\eta} \mathrm{e}^{\eta r}\cosh(\eta t),$$

thus

$$u = \left(\frac{r}{2m} - 1\right)^{2m\,\eta} \mathrm{e}^{\eta r}\cosh(\eta t).$$

Similarly,

$$v = \left(\frac{r}{2m} - 1\right)^{2m\,\eta} \mathrm{e}^{\eta r}\sinh(\eta t).$$

Furthermore, we have

$$f^2 = \frac{2m}{\eta^2 r}\left(\frac{r}{2m} - 1\right)^{1-4m\,\eta} \mathrm{e}^{-2\eta\, r}.$$

We now choose η such that $f^2 \neq 0$ for $r = 2m$. This requires $\eta = 1/4m$. In this manner, we are led to the **Kruskal transformation**:

$$u = \sqrt{\frac{r}{2m} - 1}\, \mathrm{e}^{r/4m} \cosh\left(\frac{t}{4m}\right), \tag{3.124a}$$

$$v = \sqrt{\frac{r}{2m} - 1}\, \mathrm{e}^{r/4m} \sinh\left(\frac{t}{4m}\right). \tag{3.124b}$$

By construction, the metric expressed in terms of these new coordinates has the form (3.113), with

$$f^2 = \frac{32m^3}{r}\mathrm{e}^{-r/2m}. \tag{3.125}$$

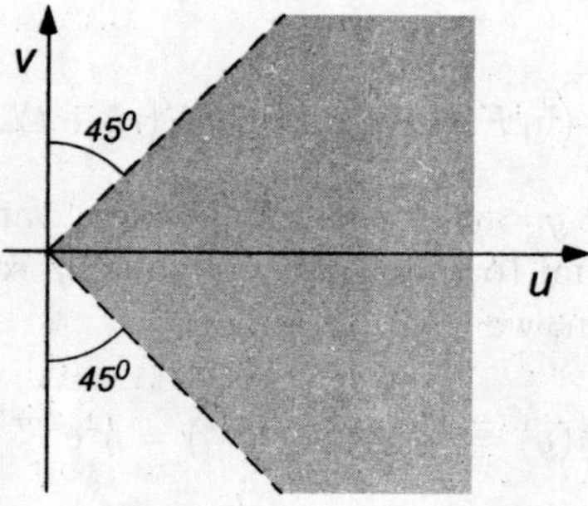

Fig. 3.12. Schwarzschild quadrant in Kruskal coordinates.

Discussion

1. So far we just made a change of coordinates. In the (u, v)-plane, the Schwarzschild region $r > 2m$ corresponds to the shaded quadrant $u > |v|$ of Fig. 3.12.

The lines r = const in the (r, t)-plane corresponds to hyperbolas. This follows from

$$u^2 - v^2 = \left(\frac{r}{2m} - 1\right) e^{r/2m} ,$$
$$\frac{v}{u} = \tanh\left(\frac{t}{4m}\right) . \tag{3.126}$$

In the limit $r \to 2m$ these approach the lines at 45° in Fig. 3.12. According to (3.126), the lines t = const correspond to radial lines through the origin, as shown in Fig. 3.13.

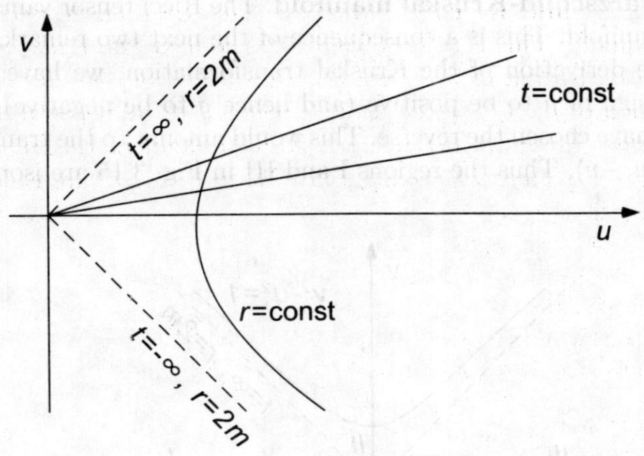

Fig. 3.13. Lines of constant t and r in the Kruskal plane.

2. However, the transformed metric

$$g = -f^2(u, v)(dv^2 - du^2) + r^2(u, v)d\Omega^2 , \tag{3.127}$$

in which r is implicitly defined as a function of u and v by (3.126), is regular in a larger region than the quadrant $u > |v|$.

The region in which (3.127) is non-singular is bounded by the hyperbolas $v^2 - u^2 = 1$, where according to (3.126) we have $r = 0$, and hence by (3.125) f^2 becomes singular. For $v^2 - u^2 < 1$, r is a uniquely defined function of u and v, since the right-hand side of the first equation in (3.126) is a monotonic function when $r > 0$; its derivative with respect to r is proportional to $re^{r/2m}$,

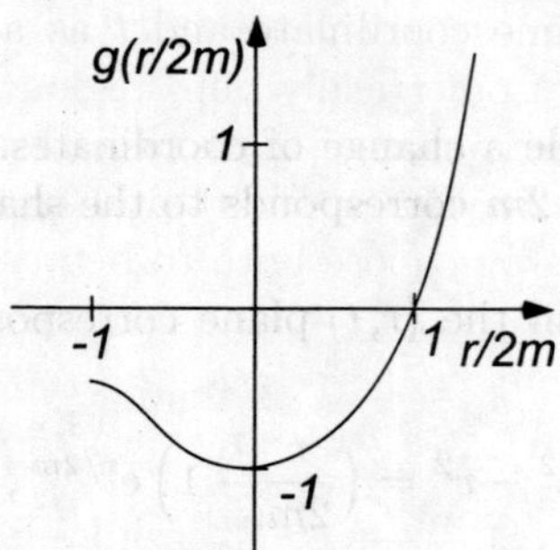

Fig. 3.14. Graph of the function g.

thus positive for $r > 0$. If we write (3.126) as $u^2 - v^2 = g(r/2m)$, the graph of the function g looks as in Fig. 3.14. We have thus isometrically embedded the original Schwarzschild manifold in a larger Lorentz manifold, which is known as the **Schwarzschild-Kruskal manifold**. The Ricci tensor vanishes on the extended manifold. This is a consequence of the next two remarks.

3. In the derivation of the Kruskal transformation, we have arbitrarily chosen the sign of h to be positive (and hence g to be negative). We could just as well have chosen the reverse. This would amount to the transformation $(u, v) \to (-u, -v)$. Thus the regions I and III in Fig. 3.15 are isometric.

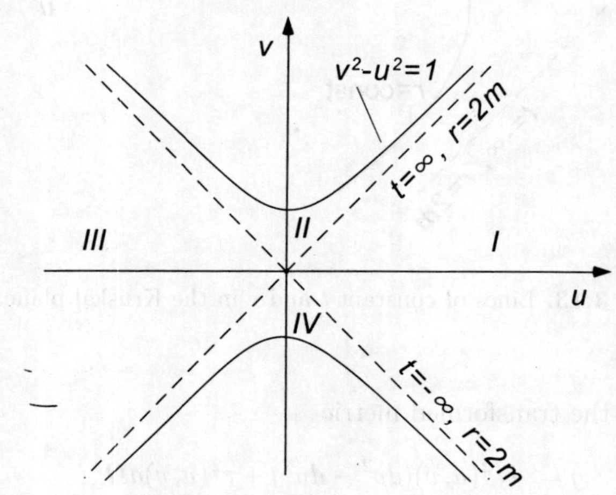

Fig. 3.15. The four quadrants of the Kruskal region.

4. We now consider the original Schwarzschild solution (3.101) formally for $r < 2m$. This also satisfies the vacuum equations, as is easily shown by direct computation (see also the Appendix 3.10 to this chapter). In this

region, r behaves as a time coordinate and t as a space coordinate. This manifold can be mapped isometrically onto region II of Fig. 3.15. In order to see this, consider the derivation of the Kruskal transformation once more. For $0 < r < 2m$, we now have $F(r^*) < 0$, and hence the relative sign of g and h must be positive. If we choose both to be positive, we obtain

$$u = \sqrt{1 - \frac{r}{2m}}\, e^{r/4m} \sinh\left(\frac{t}{4m}\right) ,$$
$$v = \sqrt{1 - \frac{r}{2m}}\, e^{r/4m} \cosh\left(\frac{t}{4m}\right) . \tag{3.128}$$

The transformed metric again has the form (3.127), and the function f is also given in terms of r by (3.125).

The image of $(0 < r < 2m)$ under the transformation (3.128) is precisely the region II. The inverse of (3.128) is given by

$$v^2 - u^2 = \left(1 - \frac{r}{2m}\right) e^{r/2m} , \quad \frac{u}{v} = \tanh\left(\frac{t}{4m}\right) . \tag{3.129}$$

Taking g and h both negative is equivalent to the transformation $(u, v) \to (-u, -v)$, so that the regions II and IV are mutually isometric.

5. Along the hyperbolas $v^2 - u^2 = 1$, the metric is *truly singular*. For example, here the invariant $R_{\alpha\beta\gamma\delta}R^{\alpha\beta\gamma\delta}$ diverges. The Kruskal extension is **maximal**, which means that every geodesic can either be extended to arbitrary large values of the affine parameter, or it runs into the singularity $v^2 - u^2 = 1$ for a finite value of this parameter. The manifold is geodesically incomplete see the footnote on p. 117.

6. The causal relationships in the Schwarzschild-Kruskal manifold are quite obvious. Radial light rays are lines at 45°, as in the Minkowski diagram. Observers in I and III (where the metric is static) can receive signals from IV and send them to II. However, any particle which enters region II must run into the singularity at $r = 0$ within a finite proper time. A particle in region IV must have come from the singularity at a finite previous proper time. There is no causal connection between regions I and III.

The singularity in the future is shielded from distant observers in regions I and III by an **event horizon**. By definition, this is the boundary of the region which is causally connected with distant observers in regions I and III, and is given by the surfaces $r = 2m$. Signals which are sent from IV to I (or III) will already have reached I at $t = -\infty$. It is presumably not possible to observe such light.

7. In Sect. 3.7.2 we shall see that the complete Kruskal extension is not relevant in astrophysics.

8. For a geometric visualization of the Schwarzschild-Kruskal manifold, we represent the two-dimensional surface $\{v = 0, \vartheta = \frac{\pi}{2}\}$ as a surface of rotation in the three-dimensional space (see Fig. 3.16). The upper part of the

embedding diagram corresponds to $u > 0$ (I) and the lower part to $u < 0$ (III). The reader should consider how the funnel-like structure ("Schwarzschild throat", "Einstein-Rosen bridge") changes for slices with increasing v.

9. The Schwarzschild-Kruskal manifold is static in regions I and III, but is *dynamical* in regions II and IV. In these regions, the Killing field $K = \partial/\partial t$ becomes *spacelike*. (At the horizon, K is null: $\langle K, K \rangle = 1 - 2m/r = 0$.) There are no observers at rest in regions II and IV.

In Appendix 3.10 to this chapter, we shall prove a generalization of Birkhoff's theorem (see Theorem 3.1).

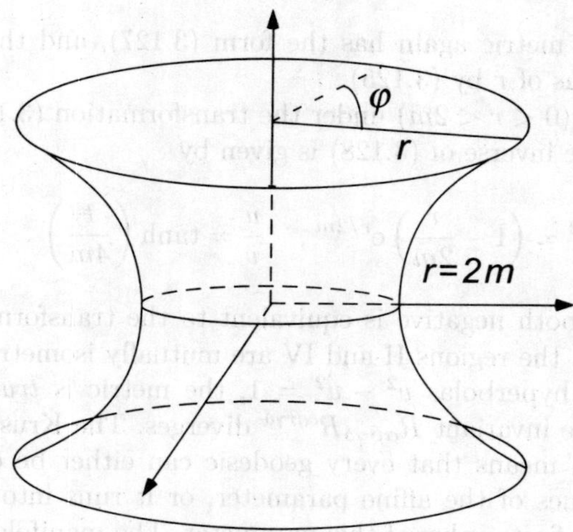

Fig. 3.16. The "Einstein-Rosen bridge".

Eddington-Finkelstein Coordinates

For spherically symmetric gravitational collapse, only part of the regions I and II of the Kruskal diagram are relevant. For this part we shall introduce other, frequently used coordinates, which are originally due to Eddington and were rediscovered by Finkelstein. They are related to the original Schwarzschild coordinates (t, r) in (3.101) by a change of time, but leaving (r, ϑ, φ) unchanged:

$$\begin{aligned} r &= r', \quad \vartheta = \vartheta', \quad \varphi = \varphi', \\ t &= t' - 2m \ln\left(\frac{r}{2m} - 1\right) \quad \text{for } r > 2m, \\ t &= t' - 2m \ln\left(1 - \frac{r}{2m}\right) \quad \text{for } r < 2m. \end{aligned} \tag{3.130}$$

The coordinates r and t' are related to the Kruskal coordinates by

$$u = \frac{1}{2}e^{r/4m}\left(e^{t'/4m} + \frac{r-2m}{2m}e^{-t'/4m}\right),$$
$$v = \frac{1}{2}e^{r/4m}\left(e^{t'/4m} - \frac{r-2m}{2m}e^{-t'/4m}\right). \tag{3.131}$$

From these relations one immediately finds

$$\frac{r-2m}{2m}e^{r/2m} = u^2 - v^2\,, \quad e^{t'/2m} = \frac{r-2m}{2m}\frac{u+v}{u-v}\,. \tag{3.132}$$

With the help of these equations, one can readily convince oneself that the transformation (3.131) is regular for $u > -v$ (i.e., in the region $I \cup II$). Inserting (3.130) into (3.101) gives

$$g = -\left(1 - \frac{2m}{r}\right)dt'^2 + \left(1 - \frac{2m}{r}\right)dr^2 + \frac{4m}{r}dt'\,dr + r^2 d\Omega^2\,. \tag{3.133}$$

This form is valid in the region $I \cup II$. The metric coefficients expressed in terms of the Eddington-Finkelstein coordinates are independent of t'; the price one pays is the existence of non-diagonal terms, i.e., a non-trivial shift.

We determine the light cones for (3.133). For radially propagating light rays the condition $ds^2 = 0$ leads to

$$(dt' + dr)\left[\left(1 - \frac{2m}{r}\right)dt' - \left(1 + \frac{2m}{r}\right)dr\right] = 0\,. \tag{3.134}$$

Hence the radial null directions are determined by

$$\frac{dr}{dt'} = -1\,, \quad \frac{dr}{dt'} = \frac{r-2m}{r+2m}\,. \tag{3.135}$$

The radial null geodesics corresponding to the first of these equations are given by

$$t' + r = \text{const.}$$

These are the straight lines in Fig. 3.17. The second equation of (3.135) shows that the tangents to the null geodesics of the other family have the following properties:

$$\frac{dr}{dt'} \to -1 \quad \text{for } r \to 0\,, \quad \frac{dr}{dt'} \to 0 \quad \text{for } r \to 2m\,. \tag{3.136}$$

The second of these properties implies that the geodesics do not cross the surface $r = 2m$. This is indicated in Fig. 3.17. One can easily show that non-radial null geodesics and also timelike directions have dr/dt' between the two values in (3.135). Hence the light cones have the structure shown in Fig. 3.17.

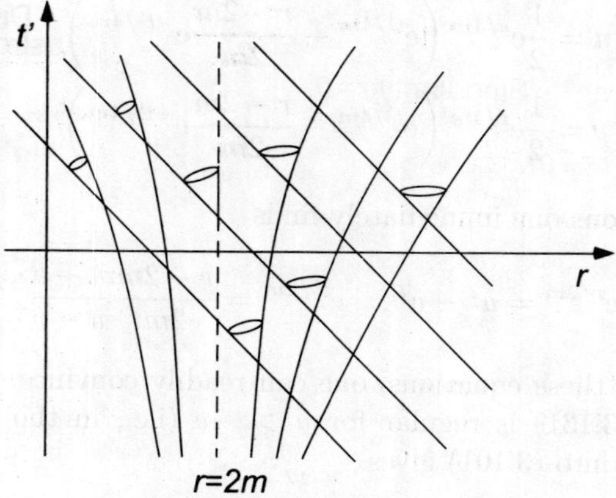

Fig. 3.17. Light cones in Eddington-Finkelstein coordinates.

3.7.2 Spherically Symmetric Collapse to a Black Hole

We now describe qualitatively the spherically symmetric, catastrophic collapse of a supercritical mass. Since, according to the generalized Birkhoff Theorem 3.1, the exterior field is a region of the Kruskal manifold, the collapse is most easily visualized in Kruskal coordinates as in Fig. 3.18. In Fig. 3.19 we show the same process in Eddington-Finkelstein coordinates. From these two figures we may draw the following conclusions:

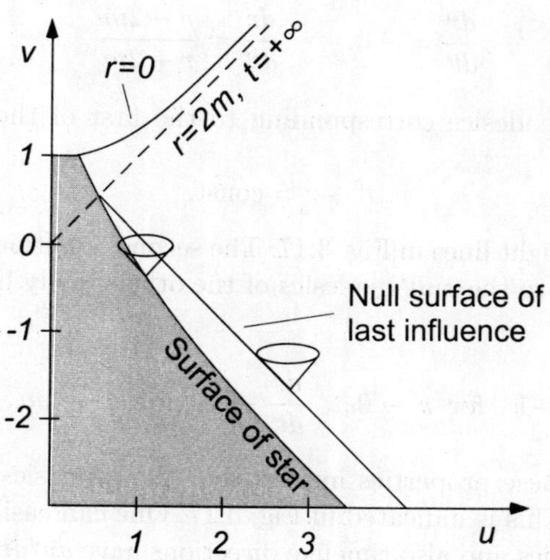

Fig. 3.18. Spherically symmetric collapse in Kruskal coordinates.

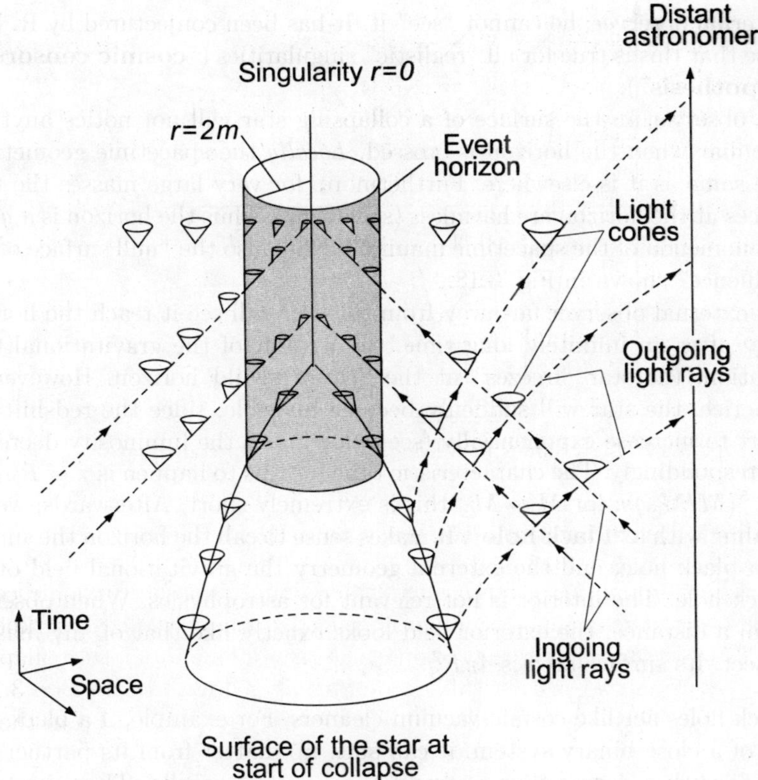

Fig. 3.19. Spacetime diagram of a collapsing star in Eddington-Finkelstein coordinates.

1. Equilibrium is no longer possible when the stellar radius becomes smaller than the radius of the horizon, since the world lines of the stellar surface must lie inside the light cones. Collapse to a singularity cannot be avoided. At some point in the vicinity of the singularity, GR will probably no longer be valid, since quantum effects will become important.
2. If a signal is emitted from inside the horizon, it will not reach a distant observer. The stellar matter is literally cut off from the outside world. Also light rays will approach the singularity. Thus, the horizon is the boundary of the region which is causally connected to a distant observer. It is defined generally by this property. Thus, the event horizon acts like a one-way membrane through which energy and information can pass to the interior, but not to the exterior. The existence of horizons, or causal boundaries, in our universe is a remarkable consequence of GR. The singularity is behind the horizon, and hence has no causal connection to an

external observer; he cannot "see" it. It has been conjectured by R. Penrose that this is true for all "realistic" singularities ("**cosmic censorship hypothesis**").

3. An observer on the surface of a collapsing star will not notice anything peculiar when the horizon is crossed. *Locally* the spacetime geometry is the same as it is elsewhere. Furthermore, for very large masses the tidal forces at the horizon are harmless (see below). Thus the horizon is a *global* phenomenon of the spacetime manifold. Note also the "null surface of last influence" shown in Fig. 3.18.
4. An external observer far away from the star will see it reach the horizon only after an infinitely long time. As a result of the gravitational time dilation, the star "freezes" at the Schwarzschild horizon. However, in practice, the star will suddenly become invisible, since the redshift will start to increase exponentially (see below), and the luminosity decreases correspondingly. The characteristic time for this to happen is $\tau \sim R_S/c \simeq 10^{-5}(M/M_{\odot})s$; for $M \sim M_{\odot}$ this is extremely short. Afterwards, we are dealing with a "**black hole**". It makes sense to call the horizon the surface of a black hole, and the external geometry the gravitational field of the black hole. The interior is not relevant for astrophysics. When observed from a distance, the exterior field looks exactly like that of any massive object. Its surface area is $4\pi R_S^2$.

Black holes act like cosmic vacuum cleaners. For example, if a black hole is part of a close binary system, it can suck up matter from its partner and heat it to such a degree that a strong X-ray source results. There is strong reason to believe that a number of X-ray sources arise from such an accretion mechanism. We shall examine this evidence more closely in Chap. 7.6.

Astrophysicists are by now convinced that gigantic black holes, with perhaps $10^9 M_{\odot}$, exist in the center of many galaxies. The accretion of matter by these black holes would provide a relatively natural explanation for the huge amounts of energy which are set free in a rather small volume as observed in active galactic nuclei.

It now appears likely that the central core of most galaxies contains a black hole of at least $10^6 M_{\odot}$. The best evidence for supermassive black holes has recently been established for our galaxy (see [109] and [214]). Ten years of high resolution astrometric imaging made it possible to determine the orbit for the star ($S2$) currently closest to the compact massive black hole candidate Sagittarius A^* ($Sgr\ A^*$). The measured points and the projection of the best fitting Kepler orbit are shown in Fig. 3.20. The orbital period is 15.2 years, and the pericentre distance of the highly elliptic Kepler orbit is only 17 light hours. The orbital elements require an enclosed very compact mass of $(3.7 \pm 1.5) \times 10^6$ solar masses. This means that the pericentre distance is about 2100 times the Schwarzschild radius. For further details of this remarkable findings, see [109]. The future infrared interferometry will offer the possibility

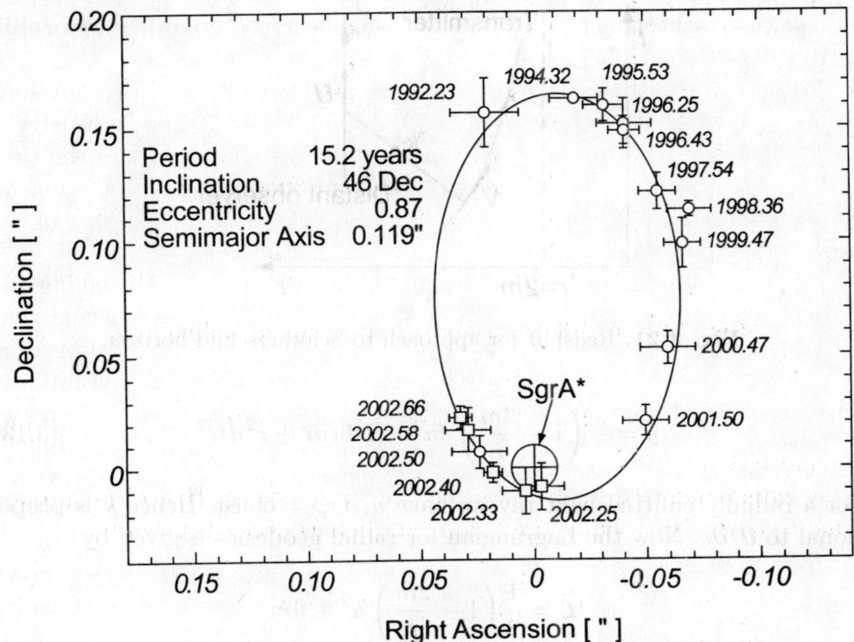

Fig. 3.20. Orbit of the star $S2$ around $Sgr\ A^*$. Beside the data points (with 1 sigma errors) the best-fitting Kepler orbit is shown as a continuous curve. Its main parameters are listed adjacent to the orbit. (From [109].)

to explore relativistic motions at 10-100 Schwarzschild radii from the central dark mass, and tighten its interpretation as a black hole.

The current evidence for both stellar-mass and supermassive black holes will be presented in Sect. 7.6. This has become overwhelming, but is still indirect.

Redshift for a Distant Observer

Suppose that a transmitter approaches the Schwarzschild horizon radially with four-velocity V. The emitted signals with frequency ω_e are received by a distant observer at rest (with four-velocity U) at the frequency ω_0 (see Fig. 3.21). If k is the wave vector, then, according to (1.122), the redshift is given by

$$1 + z = \frac{\omega_e}{\omega_0} = \frac{\langle k, V \rangle}{\langle k, U \rangle} . \tag{3.137}$$

We use the "retarded time"

$$u = t - r^* , \tag{3.138}$$

where r^* is given by (3.106). In terms of the coordinates u, r, ϑ and φ the metric becomes

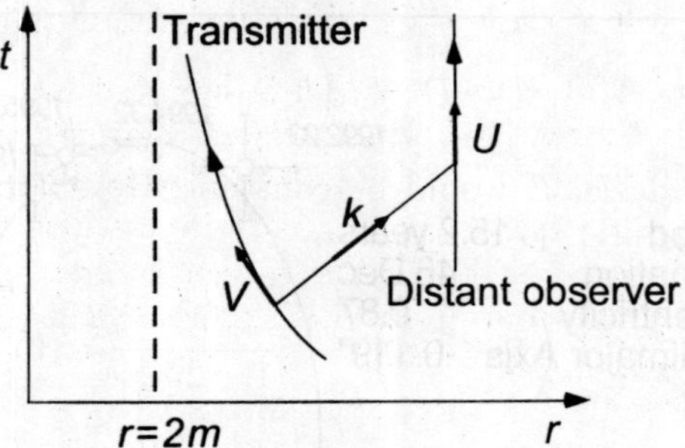

Fig. 3.21. Redshift for approach to Schwarzschild horizon.

$$g = -\left(1 - \frac{2m}{r}\right) du^2 - 2\,du\,dr + r^2 d\Omega^2\,. \tag{3.139}$$

For a radially emitted light ray we have $u, \vartheta, \varphi = \text{const}$. Hence k is proportional to $\partial/\partial r$. Now the Lagrangian for radial geodesics is given by

$$\mathcal{L} = \frac{1}{2}\left(1 - \frac{2m}{r}\right) \dot{u}^2 + \dot{u}\dot{r}\,.$$

Since u is cyclic the following conservation law holds:

$$p_u := \frac{\partial \mathcal{L}}{\partial \dot{u}} = \left(1 - \frac{2m}{r}\right) \dot{u} + \dot{r} = \text{const}\,. \tag{3.140}$$

This shows that for radial light rays ($\dot{u} = 0$), we have $\dot{r} = p_u = \text{const}$. Hence we have, in terms of the coordinates u, r, ϑ and φ,

$$k = \text{const}\,\frac{\partial}{\partial r}\,. \tag{3.141}$$

Together with (3.137) this gives

$$1 + z = \frac{\langle V, \partial/\partial r\rangle}{\langle U, \partial/\partial r\rangle}\,. \tag{3.142}$$

The four-velocity V of the transmitter can be expanded as

$$V = \dot{u}\frac{\partial}{\partial u} + \dot{r}\frac{\partial}{\partial r}\,,$$

and hence $\langle V, \partial/\partial r\rangle = \dot{u}\, g_{ur} = -\dot{u}$. Similarly, we have for a distant observer, $\langle U, \partial/\partial r\rangle = -\dot{t} \simeq -1$, and thus

$$1 + z = \dot{u} = \dot{t} - \left(1 - \frac{2m}{r}\right)^{-1} \dot{r}\,. \tag{3.143}$$

We set

$$E := \dot{t}\left(1 - \frac{2m}{r}\right). \tag{3.144}$$

For a radial geodesic, E would be constant. However, we shall not assume free fall. Since $2\mathcal{L} = 1$, i.e.,

$$\left(1 - \frac{2m}{r}\right)\dot{t}^2 - \frac{\dot{r}^2}{1 - 2m/r} = 1\,, \tag{3.145}$$

we have

$$\dot{r}^2 + \left(1 - \frac{2m}{r}\right) = E^2\,. \tag{3.146}$$

It then follows from (3.143) and (3.144) that

$$1 + z = \frac{E - \dot{r}}{1 - 2m/r} = \left(1 - \frac{2m}{r}\right)^{-1}\left[E + \left(E^2 - 1 + \frac{2m}{r}\right)^{1/2}\right]. \tag{3.147}$$

From $U \simeq \partial/\partial u$ we see that u is the proper time of the observer, that we also denote by τ_0. Keeping (3.147) in mind, we are interested in the radial coordinate r_e of the transmitter as a function of τ_0. From (3.143) we have

$$\frac{dr_e}{d\tau_0} = \frac{dr_e}{du} = \left(\frac{\dot{r}}{\dot{u}}\right)_e = \frac{\dot{r}}{1 + z}\,.$$

If we set (3.147) into this expression and use (3.146), we obtain for $r = r_e$

$$\frac{dr}{d\tau_0} = -\left(1 - \frac{2m}{r}\right)\left(E^2 - 1 + \frac{2m}{r}\right)^{1/2}\left[E + \left(E^2 - 1 + \frac{2m}{r}\right)^{1/2}\right]^{-1}. \tag{3.148}$$

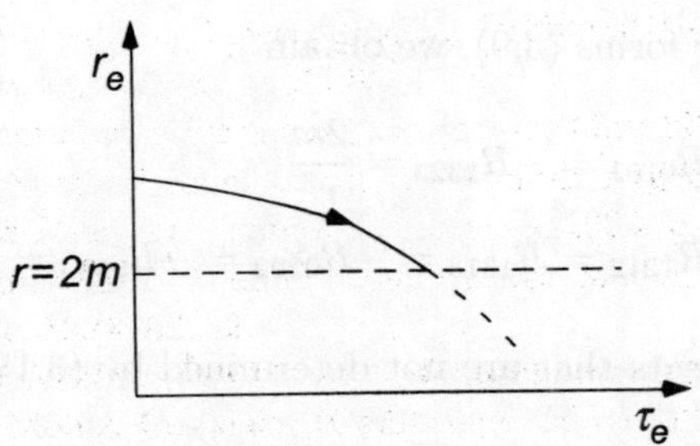

Fig. 3.22. Qualitative behavior of $r_e(\tau_e)$.

Now $r_e(\tau_e)$ has qualitatively the form shown in Fig. 3.22. After a finite proper time the transmitter reaches the Schwarzschild horizon (e.g., in free fall). From (3.146) we see that for $r \simeq 2m$, E is finite, non-zero, and slowly varying. It then follows from (3.148) that in the vicinity of $r = 2m$

$$\frac{dr}{d\tau_0} \simeq -\frac{1}{2}\left(1 - \frac{2m}{r}\right) \simeq -\frac{r-2m}{4m}$$

or

$$r - 2m \simeq \text{const } e^{-\tau_0/4m} . \tag{3.149}$$

With (3.147) we obtain the asymptotic expression

$$1 + z \simeq \frac{4m\,E}{r - 2m} . \tag{3.150}$$

Inserting (3.149) into (3.150) gives the previously mentioned result

$$1 + z \propto e^{\tau_0/4m} , \tag{3.151}$$

which shows that the redshift increases exponentially in the vicinity of $r = 2m$.

Fate of an Observer on the Surface of the Star

As the stellar radius decreases, an observer on the surface of the star experiences constantly increasing tidal forces. Since his feets are attracted more strongly than his head, he experiences a longitudinal stress. At the same time, he is compressed from the sides (lateral stress). Let us calculate these forces.

As a first step, we compute the components of the Riemann tensor relative to the basis (3.4). If we insert

$$e^{2a} = e^{-2b} = \left(1 - \frac{2m}{r}\right) \tag{3.152}$$

into the curvature forms (3.9), we obtain

$$\begin{aligned} R_{0101} &= -R_{2323} = \frac{2m}{r^3} , \\ R_{1212} &= R_{1313} = -R_{0202} = -R_{0303} = \frac{m}{r^3} . \end{aligned} \tag{3.153}$$

All other components that are not determined by (3.153) and by symmetry properties vanish.

The falling observer with four-velocity U will use his rest system as a frame of reference. For $r > 2m$, this is reached by the special Lorentz transformation in the radial direction, which transforms e_0 to the four-velocity U. Remarkably, the components of the Riemann tensor with respect to this new system are unchanged. A Lorentz "boost" has the form

$$\Lambda = \left(\begin{array}{cc|c} \cosh\alpha & \sinh\alpha & 0 \\ \sinh\alpha & \cosh\alpha & 0 \\ \hline 0 & 0 & 1 \end{array}\right).$$

From (3.153), we then obtain, for example,

$$\begin{aligned} R'_{0101} &= \Lambda_0^{\ \mu}\Lambda_1^{\ \nu}\Lambda_0^{\ \sigma}\Lambda_1^{\ \rho}R_{\mu\nu\sigma\rho} \\ &= R_{0101}(\cosh^4\alpha - 2\cosh^2\alpha\sinh^2\alpha + \sinh^4\alpha) = R_{0101}\,. \end{aligned}$$

The invariance of the other components is shown similarly. This invariance is the result of the particular structure of the Schwarzschild geometry.

We now follow the observer in quadrant II. For this, we need some results from the Appendix 3.10 to this chapter. According to (3.214), the metric in quadrant II has the form (3.204) with

$$e^{2b} = e^{-2a} = (2m/t) - 1\,, \qquad R = t\,. \tag{3.154}$$

If one inserts these metric functions into the curvature forms (3.208) relative to the basis (3.205), one obtains the same expressions as we had for $r > 2m$, except that r and t are interchanged. If we denote the timelike variable t in (3.154) by r, then the expressions (3.153) are always valid for the local rest system.

The equation for geodesic deviation implies that the distance between two freely falling test bodies increases according to (2.11) as

$$\ddot{n}^i = R^i_{\ 00j}n^j\,, \tag{3.155}$$

where the dot denotes differentiation with to the proper time. Using (3.153) we thus obtain for our problem

$$\ddot{n}^1 = \frac{2m}{r^3}n^1\,, \quad \ddot{n}^2 = -\frac{m}{r^3}n^2\,, \quad \ddot{n}^3 = -\frac{m}{r^3}n^3\,. \tag{3.156}$$

Let us consider a square prism having mass μ and height l with equal length and width w, as in Fig. 3.23.

We now ask what forces must be present in the body in order to avoid having its various parts move along diverging (or converging) geodesics. From (3.156) we see that the 1-,2- and 3-directions define the principal axes of the stress tensor (the 2- and 3-directions are equivalent). The longitudinal stress is calculated as follows: According to (3.156) an element $d\mu$ of the body at a height h above the center of mass (the distance is measured along e_1) would be accelerated relative to the center of mass with $a = 2m\,r^{-3}h$, if it were free to move. In order to prevent this, a force

$$dF = a\,d\mu = \frac{2m}{r^3}h\,d\mu$$

must act on the element. This force contributes to the stress through the horizontal plane through the center of mass. The total force through this plane is

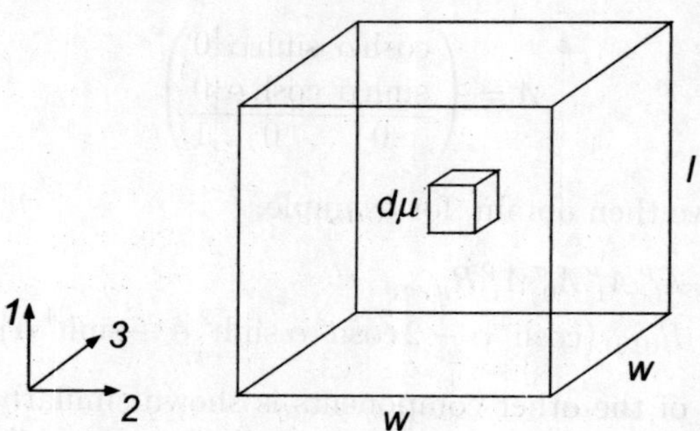

Fig. 3.23. Freely falling square prism.

$$F = \int_0^{l/2} \frac{2m\,h}{r^3}\frac{\mu}{l\,w^2} w^2\, dh$$

or

$$F = \frac{1}{4}\frac{\mu m l}{r^3}.$$

The longitudinal stress T^l is given by $T^l = -F/w^2$, so that

$$T^l = -\frac{1}{4}\frac{\mu m l}{r^3 w^2}. \tag{3.157}$$

Similarly, using (3.156), one finds for the lateral stress $T_\perp$:

$$T_\perp = +\frac{1}{8}\frac{\mu m}{l\, r^3} \tag{3.158}$$

As a numerical example, take $\mu = 75\ kg$, $l = 1.8\ m$ and $w = 0.2\ m$. We find

$$\begin{aligned} T^l &\simeq -1.1 \times 10^{14} M/M_\odot (r/1\ km)^{-3} N/m^2\,, \\ T_\perp &\simeq +0.7 \times 10^{12} M/M_\odot (r/1\ km)^{-3} N/m^2\,. \end{aligned} \tag{3.159}$$

For comparison, recall that 1 atmosphere is $1.013 \times 10^5 N/m^2$.

When the mass of the black hole is very large, the tidal force are even at the Schwarzschild horizon completely harmless.

After the body is finally torn apart, its constituents (electrons and nucleons) move along geodesics. For $r < 2m$, these have the qualitative form indicated in Fig. 3.24. The spatial coordinate t is nearly constant as $r \to 0$ (the reader should provide an analytic discussion). Thus, when the upper (lower) end has the radial coordinate t_2 (t_1), the length L of the body approaches

$$L = g_{tt}(r)^{1/2}(t_2 - t_1) \simeq \left(\frac{2m}{r}\right)^{1/2}(t_2 - t_1) \propto r^{-1/2} \propto (\tau_{\text{collapse}} - \tau)^{-1/3} \to \infty$$

as $r \to 0$. Here, τ is the proper time

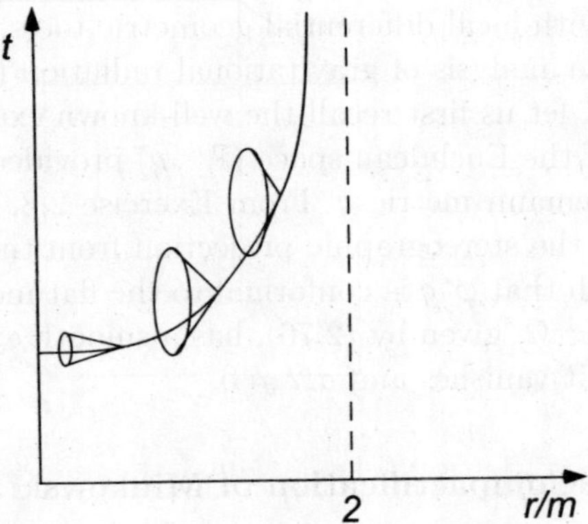

Fig. 3.24. Qualitative form of geodesics for $r < 2m$.

$$\tau = -\int_{\infty}^{r} \left(\frac{2m}{r} - 1\right)^{-1/2} dr\,,$$

and $\tau_{\text{collapse}} = \tau|_{r=0}$. For $r \simeq 0$, we have $\tau_{\text{collapse}} - \tau \propto r^{3/2}$. For the surface area A we find

$$A = \left(g_{\vartheta\vartheta}(r) g_{\varphi\varphi}(r)\right)^{1/2} \Delta\vartheta\, \Delta\varphi \propto r^2 \propto (\tau_{\text{collapse}} - \tau)^{4/3} \to 0\,.$$

The volume V behaves according to

$$V = A\,L \propto r^{3/2} \propto (\tau_{\text{collapse}} - \tau) \to 0\,.$$

It is thus not advisable to fall into a black hole.

Stability of the Schwarzschild Black Hole

The Schwarzschild solution is only physically relevant if it is stable against small perturbations. Wether this is the case is a very difficult unsolved problem. Some insight can be expected from a *linear* stability analysis. It turns out (see [107] and [108]) that the frequencies ω_n of the normal modes in the time dependent factors $\exp(-i\omega_n t)$ are all real, because the ω_n^2 can be shown to be the eigenvalues of a self-adjoint positive operator of the Schrödinger type. (The radial parts of the normal modes are the corresponding eigenfunctions.)

From this linear stability we can, however, not conclude anything about the stability of the non-linear problem (as is well known from simple examples of ordinary non-linear differential equations).

3.8 Penrose Diagram for Kruskal Spacetime

In this section, we shall construct a 'conformal compactification' of the Kruskal spacetime. Such constructions make it possible to discuss the behavior at infinity with local differential geometric tools, a point of view that is important for the analysis of gravitational radiation (see Sect. 5.1).

For illustration, let us first recall the well-known example of a conformal compactification of the Euclidean space $(\mathbb{R}^n, \eta)$ provided by the n-sphere S^n and its natural Riemann metric g: From Exercise 2.8, Eq. (2.76) it follows that the inverse of the stereographic projection from the north pole is a diffeomorphism φ such that $\varphi^* g$ is conformal to the flat metric η. Furthermore, the conformal factor Ω, given by (2.76), has a smooth extension to all of S^n. On the north pole it vanishes and $d\Omega \neq 0$.

3.8.1 Conformal Compactification of Minkowski Spacetime

As a more relevant example for what follows, we now construct a conformal compactification of Minkowski spacetime. In polar coordinates the Minkowski metric is

$$g = -dt^2 + dr^2 + r^2(d\vartheta^2 + \sin^2\vartheta\, d\varphi^2) \,. \tag{3.160}$$

It will be useful to introduce the null coordinates $u = t - r$ and $v = t + r$, in terms of which we have

$$g = -du\, dv + \frac{1}{4}(v-u)^2(d\vartheta^2 + \sin^2\vartheta\, d\varphi^2) \,. \tag{3.161}$$

Note that the hypersurfaces $\{u = \text{const}\}$ and $\{v = \text{const}\}$ are null surfaces, i.e., $\langle du, du\rangle = 0$ and $\langle dv, dv\rangle = 0$. Both coordinates u, v range over the complete real line, subject only to the restriction $v \geq u$. This unbounded range can be mapped diffeomorphically to a bounded region, for instance by the map $(u, v) \longmapsto (U, V)$ given by

$$u = \tan U \,, \qquad v = \tan V \,, \tag{3.162}$$

with $U, V \in \left(-\frac{\pi}{2}, \frac{\pi}{2}\right)$ and the restriction $V - U \geq 0$. Using

$$du\, dv = \frac{1}{\cos^2 U \cos^2 V}\, dU\, dV$$

and

$$(v-u)^2 = \frac{1}{\cos^2 U \cos^2 V} \sin^2(U - V) \,,$$

the metric becomes in the coordinates $(U, V, \vartheta, \varphi)$

$$g = \frac{1}{4\cos^2 U \cos^2 V}\, \tilde{g} \,, \tag{3.163}$$

where

$$\tilde{g} = -4\, dU\, dV + \sin^2(U - V)(d\vartheta^2 + \sin^2\vartheta\, d\varphi^2) \,. \tag{3.164}$$

Thus g and $\tilde{g}$ are conformally related as

$$\tilde{g} = \Omega^2 g, \quad \Omega^2 = 4\cos^2 U \cos^2 V. \tag{3.165}$$

In contrast to g, the metric $\tilde{g}$ is also defined when U or/and V are equal to $\pm\pi/2$. In other words, $\tilde{g}$ is a smooth metric on the compact manifold $\tilde{M} = \left[-\frac{\pi}{2}, \frac{\pi}{2}\right]^2 \times S^2$. The Lorentz manifold $(\tilde{M}, \tilde{g})$ is actually a compact submanifold of the Einstein universe, discussed in Sect. 2.2. This becomes obvious by introducing the coordinates $\tilde{t} = U + V$ and $\chi = V - U$, in terms of which

$$\tilde{g} = -d\tilde{t}^2 + d\chi^2 + \sin^2\chi(d\vartheta^2 + \sin^2\vartheta\, d\varphi^2)\,. \tag{3.166}$$

Here, the spatial part is the metric of S^3 in polar coordinates $(\chi, \vartheta, \varphi)$.

The compact manifold $\tilde{M}$ is the shaded part of the **Einstein cylinder** $\mathbb{R} \times S^3$, shown in Fig. 3.25. The image of Minkowski spacetime is the open part of this compact subset. Its boundary, called **conformal infinity of Minkowski spacetime**, consists of the following pieces: The two null-hypersurfaces $\mathscr{I}^+$ and $\mathscr{I}^-$, given by $V = \pi/2$, $|U| < \pi/2$ for $\mathscr{I}^+$ and $U = -\pi/2$, $|V| < \pi/2$ for $\mathscr{I}^-$, represent *future* and *past* **null infinity**, respectively. This is where null-geodesics end up. The points $i^\pm$, given by $U = V = \pm\pi/2$, represent *future* and *past* **timelike infinity**, where timelike geodesics end. Finally the point i^0, corresponding to $U = -\pi/2$, $V = \pi/2$ represents **spacelike infinity**. This is the end of all spacelike geodesics.

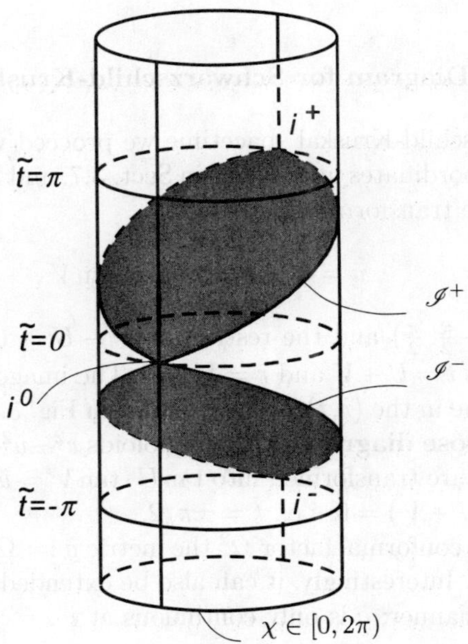

Fig. 3.25. Conformal compactification of Minkowski spacetime. The open part of the shaded region of the Einstein cylinder is conformally isometric to Minkowski spacetime. The various pieces of the conformal boundary are also shown.

Note that the conformal factor vanishes on the conformal boundary of Minkowski spacetime, consisting of $\mathscr{I}^{\pm}$, $i^{\pm}$ and i^0, but $d\Omega \neq 0$ there. This part is fixed by the Minkowski metric. Any further extension, like the part $(-\pi, \pi) \times S^3$ of the Einstein cylinder, is arbitrary.

The conformal diagram in the $(\tilde{t}, \chi)$ coordinates is shown in Fig. 3.26. The interior of the triangle corresponds to Minkowski spacetime.

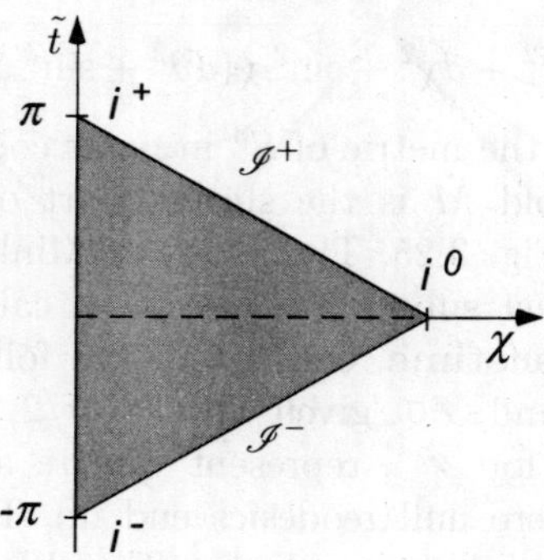

Fig. 3.26. Conformal diagram of Minkowski spacetime. The various pieces of the conformal boundary are also shown.

3.8.2 Penrose Diagram for Schwarzschild-Kruskal Spacetime

For the Schwarzschild-Kruskal spacetime we proceed very similarly. Let u, v be the Kruskal coordinates introduced in Sect. 3.7, and let $\tilde{u} = v - u$, $\tilde{v} = v + u$. We use again the transformation (3.162),

$$\tilde{u} = \tan U\,, \qquad \tilde{v} = \tan V\,,$$

with $(U, V) \in \left(-\frac{\pi}{2}, \frac{\pi}{2}\right)$ and the restriction $V - U \geq 0$. In addition, we use the combinations $\tilde{t} = U + V$ and $\tilde{r} = V - U$. The image of the Schwarzschild-Kruskal spacetime in the $(\tilde{r}, \tilde{t})$-plane is shown in Fig. 3.27. Note, for example, that in this **Penrose diagram** the hyperboloids $v^2 - u^2 = 1$, where the metric (3.113) diverges, are transformed into $\tan U \cdot \tan V = \tilde{u}\,\tilde{v} = v^2 - u^2 = 1$. This means that $\cos(U + V) = 0$, i.e., $\tilde{t} = \pm\pi/2$.

For a suitable conformal factor Ω^2 the metric $\tilde{g} := \Omega^2 g$ has again a smooth extension to $\mathscr{I}^{\pm}$. Interestingly, it can also be extended to i^0, however, not in a differentiable manner; $\tilde{g}$ is only continuous at i^0.

* * *

Exercise 3.10. Sketch the lines $\{t = \text{const}\}$ and $\{r = \text{const}\}$ in the triangle of Fig. 3.26.

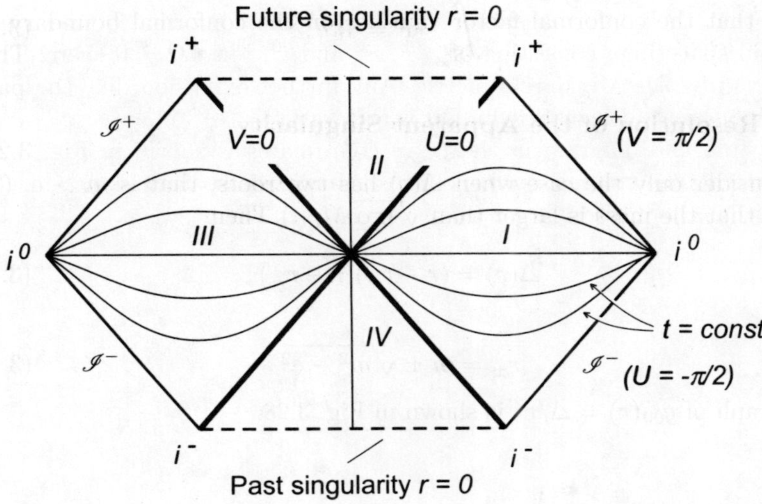

Fig. 3.27. Conformal picture (Penrose diagram) of the Schwarzschild-Kruskal spacetime.

3.9 Charged Spherically Symmetric Black Holes

The generalization to *electrically charged* black holes is of theoretical interest, because some new qualitative features will show up. In particular, we shall encounter two horizons, an external event horizon and an internal "Cauchy horizon". This we shall also find for uncharged, but rotating black holes (see Chap. 6). Astrophysically, charged black holes are of little interest, since macroscopic bodies do not possess sizable net electric charges.

In Exercise 3.4 we derived the Reissner-Nordstrøm solution as the unique two parameter family of static, spherically symmetric solutions of the coupled Einstein and Maxwell equations. In Schwarzschild coordinates the metric and the Faraday form F are given by $(G = 1)$

$$g = -\frac{\Delta}{r^2}\,dt^2 + \frac{r^2}{\Delta}\,dr^2 + r^2(d\vartheta^2 + \sin^2\vartheta\,d\varphi^2)\,, \tag{3.167a}$$

$$F = \frac{e}{r^2}\,dt \wedge dr\,, \tag{3.167b}$$

where

$$\Delta = r^2 - 2m\,r + e^2\,. \tag{3.168}$$

Recall that m is the mass parameter and e the charge, measured far away. Indeed, far away F reduces to the Coulomb field of strength e. In other words, the electric flux is asymptotically[5]

[5] We have $F = E_1\,\theta^1\wedge\theta^0$ with $E_1 = e/r^2$ and thus $*F = E_1\,\theta^2\wedge\theta^3 = e\,d\vartheta\wedge\sin\vartheta\,d\varphi$.

$$\int_{S^2_\infty} *F = 4\pi e\,. \tag{3.169}$$

3.9.1 Resolution of the Apparent Singularity

We consider only the case when $\Delta(r)$ has two roots, that is $m > e$. (This means that the mass is larger than $\sqrt{4\pi\alpha}\, M_{Pl}$.) Then

$$\Delta(r) = (r - r_+)(r - r_-)\,, \tag{3.170}$$

with

$$r_\pm = m \pm \sqrt{m^2 - e^2}\,. \tag{3.171}$$

The graph of $g_{00}(r) = \Delta/r^2$ is shown in Fig. 3.28.

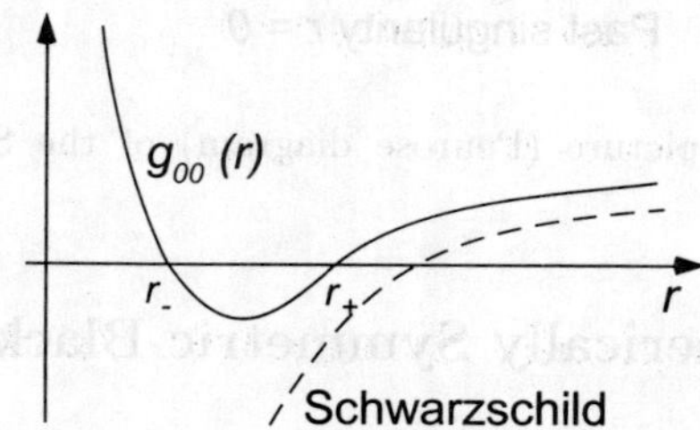

Fig. 3.28. Graph of the metric coefficient $g_{00}(r) = \Delta/r^2$ for the Reissner-Nordstrøm solution. The dotted line shows this quantity for the Schwarzschild solution.

In the following discussion we distinguish the three regions:

$$0 < r < r_- \text{ (region A)}\,, \quad r_- < r < r_+ \text{ (region B)}\,, \quad r > r_+ \text{ (region C)}\,.$$

It will turn out that $\{r = r_+\}$ is an event horizon, while $\{r = r_-\}$ is a 'horizon' of a different kind.

The following radial coordinates r_* generalizes (3.106) for the Schwarzschild solution:

$$r_* := \int \frac{r^2}{\Delta(r)} dr\,. \tag{3.172}$$

Since

$$\frac{r^2}{\Delta} = 1 + \frac{r_+^2}{r_+ - r_-}\left(\frac{1}{r - r_+}\right) - \frac{r_-^2}{r_+ - r_-}\left(\frac{1}{r - r_-}\right) \tag{3.173}$$

we have

$$r_* = r + \frac{r_+^2}{r_+ - r_-}\ln|r - r_+| - \frac{r_-^2}{r_+ - r_-}\ln|r - r_-|\,. \tag{3.174}$$

Next, we generalize the Eddington coordinates. The new time coordinate $\tilde{t}$ is again defined by $t - r =: \tilde{t} - r_*$ or with (3.174)

$$t = \tilde{t} - \frac{r_+^2}{r_+ - r_-} \ln |r - r_+| + \frac{r_-^2}{r_+ - r_-} \ln |r - r_-| . \tag{3.175}$$

From this definition and (3.172) we obtain

$$dt = d\tilde{t} + \left(1 - \frac{r^2}{\Delta}\right) dr . \tag{3.176}$$

Introducing the function

$$f = 1 - \frac{\Delta}{r^2} = \frac{2m}{r} - \frac{e^2}{r} \qquad (g_{00} = 1 - f) \tag{3.177}$$

the transformed metric can be written as

$$g = -(1 - f)\, d\tilde{t}^2 + 2f\, d\tilde{t}\, dr + (1 + f)\, dr^2 + r^2\, d\Omega^2 , \tag{3.178}$$

where $d\Omega^2$ is the metric of the 2-sphere. One sees that this metric is *regular* for all positive r. At the point $r = 0$ we have again a true singularity, as can be seen by calculating the invariant $R_{\alpha\beta\gamma\delta}R^{\alpha\beta\gamma\delta}$.

For a better understanding of (3.178) we investigate the radial null geodesics. For $d\vartheta = d\varphi = 0$ we obtain

$$(d\tilde{t} + dr)\left[(1 - f)d\tilde{t} - (1 + f)dr\right] = 0 . \tag{3.179}$$

Thus one family of null geodesics is

$$\tilde{t} + r = \text{const} , \tag{3.180}$$

and the second family satisfies

$$\frac{d\tilde{t}}{dr} = \frac{1 + f}{1 - f} . \tag{3.181}$$

Numerator and denominator on the right are plotted in Fig. 3.29. At $r_\pm$ the slope (3.181) becomes vertical. The two families of radial null geodesics are shown in Fig. 3.30.

From the structure of the light cones we see that the surface $\{r = r_+\}$ is an event horizon (one way membrane), similar to the one we found for the uncharged Schwarzschild solution. Moreover, one sees that a particle crossing this horizon inwards has to move towards the center of symmetry until it crosses or it reaches asymptotically the surface $\{r = r_-\}$. Note that this is a null hypersurface (characteristic surface).

Somewhat surprisingly, in the region A the light cones are no more inclined towards the center, whence test particles do not need to fall towards the singularity. The following analysis will actually show that *neutral* test particles cannot reach the singularity.

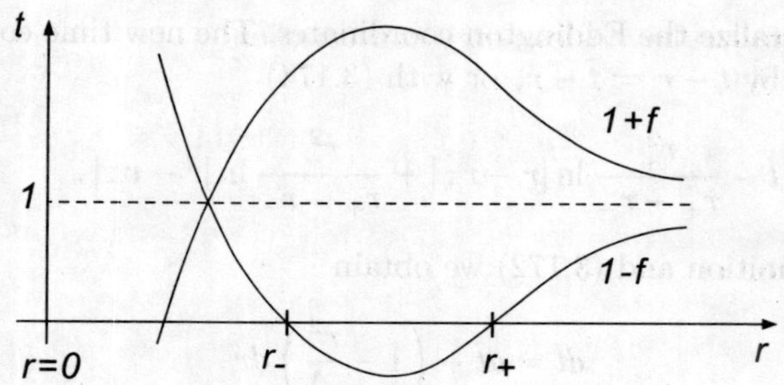

Fig. 3.29. Graphs of $(1+f)$ and $(1-f)$ in Eq. (3.181).

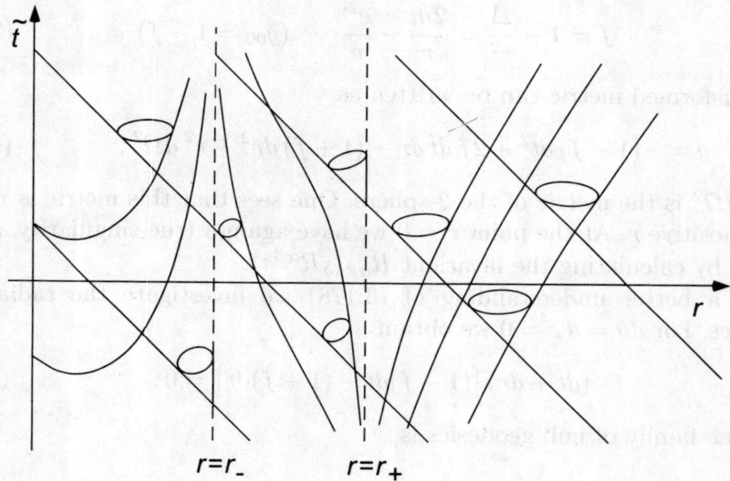

Fig. 3.30. Radial null geodesics and light cones for the metric (3.178).

3.9.2 Timelike Radial Geodesics

Neutral test particles move along timelike geodesics of the metric (3.178). We limit the discussion to radial orbits, but the results can be easily generalized to non-radial timelike geodesics.

The Lagrange function for this problem is

$$\mathcal{L} = \frac{1}{2}\left[(1-f)\left(\frac{d\tilde{t}}{d\tau}\right)^2 - 2f\,\frac{d\tilde{t}}{d\tau}\frac{dr}{d\tau} - (1-f)\left(\frac{dr}{d\tau}\right)^2\right], \tag{3.182}$$

where τ is the proper time. Thus

$$(1-f)\left(\frac{d\tilde{t}}{d\tau}\right)^2 - 2f\,\frac{d\tilde{t}}{d\tau}\frac{dr}{d\tau} - (1-f)\left(\frac{dr}{dt}\right)^2 = 1\,. \tag{3.183}$$

Since $\tilde{t}$ is cyclic, we have the energy conservation

$$(1-f)\frac{d\tilde{t}}{d\tau} - f\,\frac{dr}{d\tau} = \text{const} =: E\,. \tag{3.184}$$

Adopting the notation $\dot{r} = dr/d\tilde{t}$, Eq. (3.183) can be written as

$$\left((1-f) - 2f\,\dot{r} - (1+f)\dot{r}^2\right)\left(\frac{d\tilde{t}}{d\tau}\right)^2 = 1,$$

and (3.184) gives

$$\left((1-f) - f\,\dot{r}\right)\frac{d\tilde{t}}{d\tau} = E\,.$$

Eliminating $d\tilde{t}/d\tau$ from the last two equations leads to the differential equation

$$\left(\dot{r} + \frac{f\,(E^2-(1-f))}{E^2(1+f)+f^2}\right)^2 = \frac{E^2\,(E^2-(1-f))}{E^2(1+f)+f^2} \tag{3.185}$$

for $r(\tilde{t})$. With the abbreviation

$$F := E^2 - (1-f) \tag{3.186}$$

we can rewrite this as

$$\left(\dot{r} + \frac{f\,F}{1+(1+f)F}\right)^2 = \frac{F\,E^2}{(1+(1+f)F)^2}\,. \tag{3.187}$$

This equation requires $F \geq 0$. The allowed region for the radial motion can be seen from Fig. 3.31. For $E^2 < 1$ there are two intersections A' and B' with $F = 0$ (note that $g_{00} \longrightarrow 1$ for $r \longrightarrow \infty$). At these points $\dot{r} = 0$ according to (3.187): $\dot{r} = 0$ for $r = r_{A'}$ or $r = r_{B'}$. The allowed region is $r_{A'} \leq r \leq r_{B'}$; the neutral test particle oscillates between A' and B'.

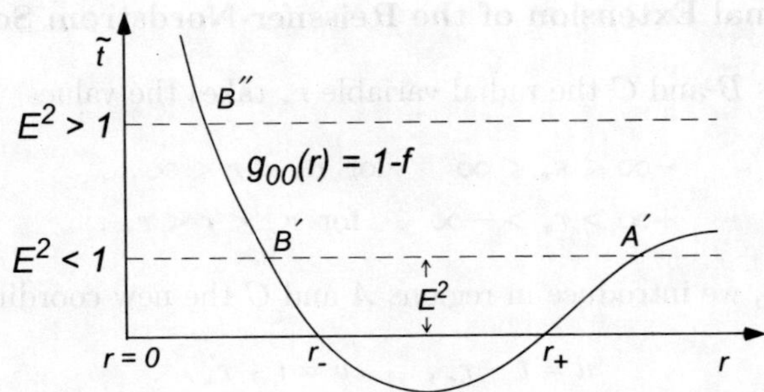

Fig. 3.31. Restrictions for neutral test particle in radial motion.

When $E^2 > 1$ there is only a minimal distance, that approaches $r = 0$ with increasing energy. Is there an intuitive reason for the existence of the non-vanishing minimum distance? Perhaps one can argue as follows: The electrostatic energy density, proportional to $1/r^4$ (see Exercise 3.4), leads to an infinitely large electrostatic energy. Since, however, the total energy has the finite value m, there will be necessarily an infinite amount of negative, non-electrostatic energy. It is because of this that at sufficiently small distances the gravitational force becomes repulsive. (This remark demonstrates that the Reissner-Nordstrøm solution is not very physical.)

Let us discuss the motion of the test particle a bit more closely. From (3.187) we obtain

$$\dot{r} = \frac{-fF \pm E\sqrt{F}}{1 + (1+f)F}\,, \tag{3.188}$$

where the double sign corresponds to the two possibilities $\dot{r} > 0$ or $\dot{r} < 0$. For the negative sign $\dot{r}$ is everywhere negative. In particular, at $r = r_\pm$ $(f = 1)$ we find

$$\dot{r}|_{r_\pm} = -\frac{2E^2}{1+2E^2} < 0\,. \tag{3.189}$$

On the other hand, for the plus sign we get

$$\dot{r}|_{r_\pm} = 0\,. \tag{3.190}$$

From this it follows that, after the particle has passed the point B' in Fig. 3.31, it cannot cross the surface $\{r = r_-\}$, but reaches it asymptotically for $\tilde{t} \longrightarrow \infty$. This conclusion can also be drawn from the orientation of the light cones at r_- (see Fig. 3.30). The motion of the particle in the $(r, \tilde{t})$-plane looks therefore as shown in Fig. 3.32. One finds that it reaches the point B_∞ with coordinates $r = r_-$, $\tilde{t} = \infty$ in a *finite* proper time. Timelike geodesics passing through B_∞ are thus incomplete. Below, we shall extend the manifold, using Kruskal-type coordinates, such that these geodesics do not terminate after a finite affine length.

3.9.3 Maximal Extension of the Reissner-Nordstrøm Solution

In the regions B and C the radial variable r_* takes the values

$$\begin{aligned} -\infty < r_* < \infty &\qquad \text{for } r_+ < r < \infty\,, \\ +\infty > r_* > -\infty &\qquad \text{for } r_- < r < r_+\,. \end{aligned} \tag{3.191}$$

In a first step, we introduce in regions A and C the new coordinates

$$u = t - r_*\,, \qquad v = t + r_*\,. \tag{3.192}$$

Since

$$dt = \frac{1}{2}(du + dv)\,, \quad dr = \frac{\Delta}{r^2}dr_* = \frac{\Delta}{r^2}\frac{1}{2}(dv - du)\,,$$

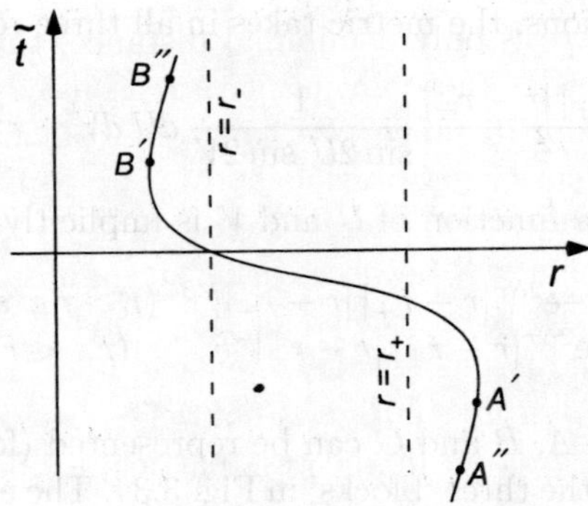

Fig. 3.32. Typical orbit of a neutral test particle.

the metric (3.167a) becomes

$$g = \frac{-\Delta}{r^2}\, du\, dv + r^2\, d\Omega^2 . \tag{3.193}$$

In region B we choose instead

$$u = r_* + t , \qquad v = r_* - t , \tag{3.194}$$

and obtain

$$g = -\frac{|\Delta|}{r^2}\, du\, dv + r^2\, d\Omega^2 .$$

For piecing together the three regions in a smooth fashion and for the maximal extension, we introduce further coordinate transformations. In the regions A and C, we set

$$\begin{aligned} \tan U &= -\mathrm{e}^{-\alpha u} = \mathrm{e}^{-\alpha(t-r_*)} \\ &= -\mathrm{e}^{-\alpha t}\left(\mathrm{e}^{\alpha r}|r-r_+|^{1/2}|r-r_-|^{-\beta/2}\right) , \end{aligned} \tag{3.195}$$

$$\begin{aligned} \tan V &= \mathrm{e}^{\alpha v} = \mathrm{e}^{\alpha(t+r_*)} \\ &= \mathrm{e}^{\alpha t}\left(\mathrm{e}^{\alpha r}|r-r_+|^{1/2}|r-r_-|^{-\beta/2}\right) , \end{aligned} \tag{3.196}$$

where

$$\alpha = \frac{r_+ - r_-}{2\, r_+^2}\ (>0) , \quad \beta = \frac{r_-^2}{r_+^2}\ (<1) . \tag{3.197}$$

In region B, we change some signs:

$$\begin{aligned} \tan U &= \mathrm{e}^{\alpha u} = \mathrm{e}^{\alpha(t+r_*)} \\ &= \mathrm{e}^{\alpha t}\left(\mathrm{e}^{\alpha r}|r-r_+|^{1/2}|r-r_-|^{-\beta/2}\right) , \end{aligned} \tag{3.198}$$

$$\begin{aligned} \tan V &= \mathrm{e}^{\alpha v} = \mathrm{e}^{\alpha(-t+r_*)} \\ &= \mathrm{e}^{-\alpha t}\left(\mathrm{e}^{\alpha r}|r-r_+|^{1/2}|r-r_-|^{-\beta/2}\right) . \end{aligned} \tag{3.199}$$

With these substitutions, the metric takes in all three regions the same form

$$g = \frac{4}{\alpha^2}\frac{|r-r_+|\,|r-r_-|}{r^2}\frac{1}{\sin 2U\,\sin 2V}\,dU\,dV + r^2\,d\Omega^2\,. \tag{3.200}$$

The coordinate r as a function of U and V is implicitly given by

$$\tan U\,\tan V = \begin{cases} -e^{2\alpha r}|r-r_+|\,|r-r_-|^{-\beta} & (r>r_+ \text{ and } 0<r<r_-) \\ e^{2\alpha r}|r-r_+|\,|r-r_-|^{-\beta} & (r_-<r<r_+)\,. \end{cases} \tag{3.201}$$

The three regions A, B and C can be represented (for fixed values of the polar angles ϑ, φ) by the three 'blocks' in Fig. 3.34. The edges are identified as indicated. By this procedure we do not yet obtain a 'geodesically complete' manifold, as the discussion in the last section shows. As for the Kruskal extension, we can obtain three other regions A', B' and C' by applying the transformation $(u,v) \longmapsto (-u,-v)$. Now we can construct a maximal analytic extension by piecing together copies of the six blocks so that an over-lapping edge is covered by either a $u(r)$- or a $v(r)$-system of coordinates (with the exception of the corner of each block). This results in the 'ladder' shown in Fig. 3.33 which extends indefinitely in both directions. Then all geodesics, except those which terminate at the singularity, have infinite affine lengths, both in the past and in the future directions. The metric (3.200) is *everywhere analytic*, except for $r=r_-$, since $|r-r_-|$ appears in (3.201) with a negative power. There the metric is, however, at least C^2.

Since U and V are null coordinates, the properties of the maximally extended spacetime are transparent. For example, it is clear that the spacetime region external to r_+ (in C) is very similar to the region of the Schwarzschild spacetime external to $r=2M=R_S$. Any observer who crosses the surface $\{r=r_+\}$, following a future-directed timelike orbit, is forever lost to an external observer (in C). The surface $\{r=r_+\}$ is an **event horizon**. However, after crossing this horizon, infinitely many possibilities are left open. There is, for instance, the possibility to move into a region A' (as indicated in Fig. 3.33). At the instant when the surface $\{r=r_-\}$ is crossed, something very dramatic happens. A hypothetical observer will witness, in an infinitely blue-shifted flash, a panorama of the entire history of the external world. (This is discussed in detail in [105].) Should he survive this and enter the region A', the observers future is no longer determined by his past history. Indeed, the region A' is outside the domain of dependence of the spacelike slice $\mathcal{S}$ extending the regions C and C' shown in Fig. 3.33. For this reason the surface $\{r=r_-\}$ between B and A' is called a **Cauchy horizon**.

As indicated in Fig. 3.33, an observer entering B has also the option to enter the region A. There is no danger to fall into the singularity. On the contrary he will cross the surface $\{r=r_-\}$ between A and B', and eventually emerge into a new asymptotically flat universe (C or C'). For all this to be possible it is crucial that the singularity $\{r=0\}$ is timelike, while it is spacelike for the Schwarzschild-Kruskal spacetime.

There is, of course, the question whether all this is just mathematics. The maximal extension of the Reissner-Nordstrøm solution is probably not physically realized. (Presumably, the Cauchy horizon is unstable). However, it is a wonderful source for science fiction writers.

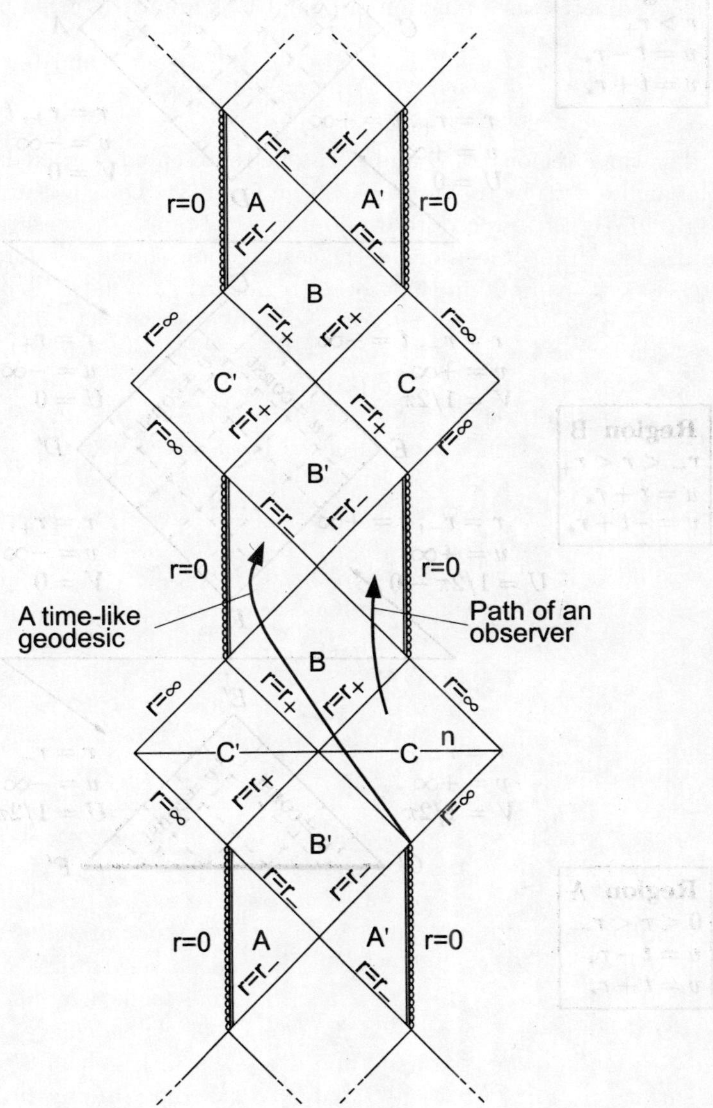

Fig. 3.33. Maximal analytic extension of the Reissner-Nordstrøm solution. Possibilities for timelike geodesics starting in the asymptotic region C.

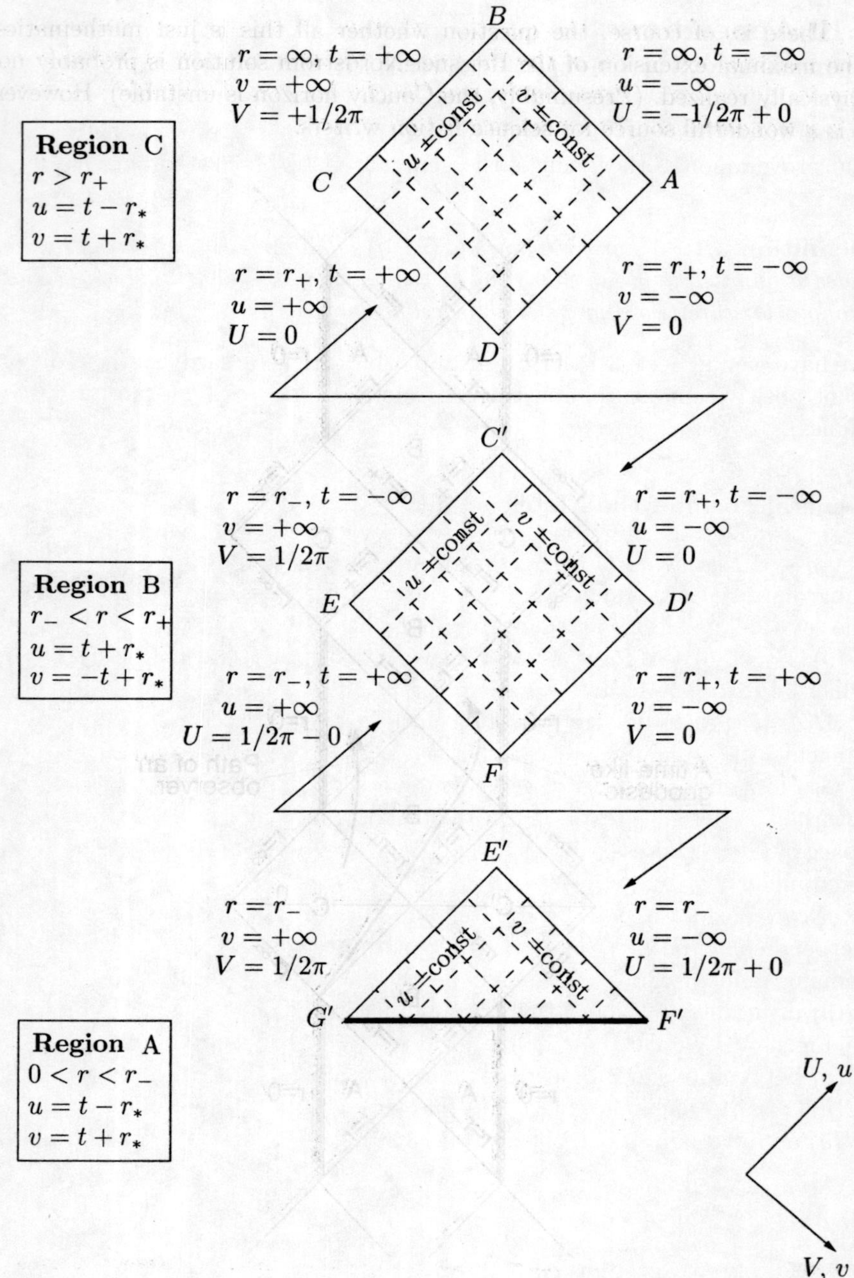

Fig. 3.34. Piecing together the different regions in the Reissner-Nordstrøm spacetime. (Adapted from [105].)

3.10 Appendix: Spherically Symmetric Gravitational Fields

In this appendix, we consider non-stationary but spherically symmetric fields and prove among other things, the generalized Birkhoff theorem. We recall the

Definition 3.1. *A Lorentz manifold (M, g) is **spherically symmetric** provided it admits the group $SO(3)$ as an isometry group, in such a way that the group orbits are two-dimensional spacelike surfaces.*

We have seen in Sect. 3.1 that the group orbits are necessarily surfaces of constant positive curvature, and thus the metric 2-spheres (ignoring projective planes).

General Form of the Metric

Let $q \in M$ and let $\Omega(q)$ be the orbit through q. The (one-dimensional) subgroup of $SO(3)$ which leaves q invariant is the **stabilizer** G_q. We denote the set of all geodesics passing through q which are perpendicular to $\Omega(q)$ by $N(q)$. Locally these form a two dimensional surface (also denoted by $N(q)$) which is invariant under G_q. A vector $v \in T_q(M)$ which is perpendicular to $\Omega(q)$ is transformed into another such vector by G_q, for if $u \in T_q(M)$ is tangent to $\Omega(q)$, and if $g \in G_q$, then $\langle u, g \cdot v\rangle = \langle g^{-1} \cdot u, v\rangle = 0$, since $g^{-1} \cdot u$ is also tangent to $\Omega(q)$. From this it follows that the vector v remains even invariant under G_q (a rotation in the orthogonal complement to the tangent space of $\Omega(q)$ is necessarily trivial). This implies that $N(q)$ remains *pointwise* fixed under G_q.

Now consider a point $p \in N(q)$ and the directions (vectors in $T_p(M)$) perpendicular to $N(q)$. The group G_q permutes these directions, since $N(q)$ remains pointwise invariant under G_q. On the other hand, G_q leaves the orbit $\Omega(p)$ invariant, and hence this must be perpendicular to $N(q)$, as indicated in Fig. 3.35. We can now define locally a bijective mapping $\phi : \Omega(q) \longrightarrow \Omega(p)$ between two group orbits as follows: ϕ maps the point q into the point $\Omega(p) \cap N(q)$. This mapping is *G_q-equivariant*, since q, as well as $\Omega(p)$ and $N(q)$ are invariant under G_q (see Fig. 3.36).

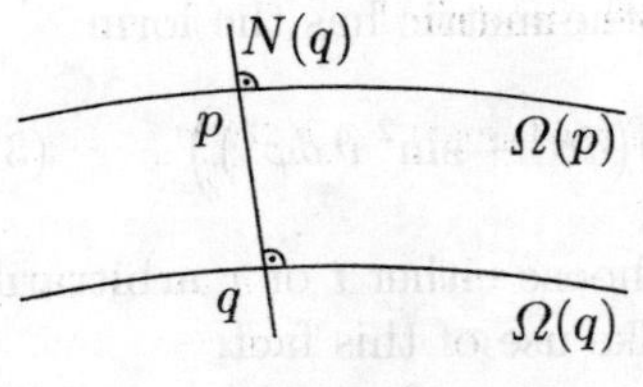

Fig. 3.35.

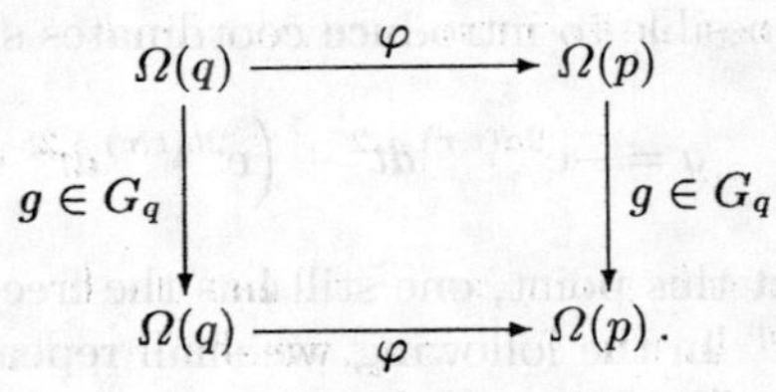

Fig. 3.36.

Therefore vectors of $T_q\Omega(q)$ having the same length are transformed by ϕ into vectors of $T_{\phi(q)}\Omega(p)$ having also the same length. Furthermore, since all points in $\Omega(q)$ are equivalent (with respect to the group action), one obtains the same magnification factors for all points in $\Omega(q)$; this means that the mapping ϕ is **conformal.**

With these remarks we can introduce local coordinates $(t, r, \vartheta, \varphi)$ such that the group orbits be surfaces of constant t and r, and the surfaces N orthogonal to these are the surfaces $\{\vartheta, \varphi = \text{const}\}$. Since ϕ is conformal, and since the orbits are metric 2-spheres, the spacetime metric has the form

$$g = \tilde{g} + R^2(t,r)\big(d\vartheta^2 + \sin\vartheta\, d\varphi^2\big)\,,$$

where $\tilde{g}$ is an indefinite metric in t and r.

Expressed more abstractly, (M, g) is, at least locally, a **warped product**

$$M = \tilde{M} \times_R S^2\,, \tag{3.202}$$

where $\tilde{M}$ is a 2-dimensional Lorentz manifold. This means that M is the product manifold $\tilde{M} \times S^2$, and the metric has the form

$$g = \tilde{g} + R^2\hat{g}\,, \tag{3.203}$$

where $\tilde{g}$ and $\hat{g}$ denote a Lorentz metric on $\tilde{M}$ and the standard metric on S^2, respectively, and R is a function on $\tilde{M}$, called the **warped function**. In the Appendix B of DG, we discuss general warped products of two pseudo-Riemannian manifolds and compute, in particular, the Ricci and Einstein tensors in terms of those of the factors and the warping function. This turns out to be useful in many instances.

We can choose the coordinates r and t on $\tilde{M}$ such that $\tilde{g}$ is diagonal:

$$\tilde{g} = -e^{2a(t,r)}dt^2 + e^{2b(t,r)}dr^2\,.$$

Indeed, if we start from the general form

$$\tilde{g} = -\alpha^2 dt^2 - 2\alpha\beta\, dt dr + \gamma^2\, dr^2\,,$$

we can use the fact that $\alpha\, dt + \beta\, dr$ has an integrating factor: $\alpha\, dt + \beta\, dr = A\, dt'$ (see Exercise C.1 in DG, Appendix C). In terms of the coordinates (t', r) the metric $\tilde{g}$ has thus the form $\tilde{g} = -A^2\, dt'^2 + B^2\, dr^2$. Hence, it is always possible to introduce coordinates such that the metric has the form

$$g = -e^{2a(t,r)}dt^2 + \Big(e^{2b(t,r)}dr^2 + R^2(t,r)\big(d\vartheta^2 + \sin^2\vartheta\, d\varphi^2\big)\Big)\,. \tag{3.204}$$

At this point, one still has the freedom to choose either t or r arbitrarily on $\tilde{M}$. In the following, we shall repeatedly make use of this fact.

We will need the connection forms and curvature forms corresponding to the metric (3.204), as well as the Ricci and Einstein tensors not only in this appendix, but also in later chapters. Let us compute these relative to the orthonormal basis

$$\theta^0 = e^a\, dt\,, \quad \theta^1 = e^b\, dr\,, \quad \theta^2 = R\, d\vartheta\,, \quad \theta^3 = R \sin\vartheta\, d\varphi\,. \tag{3.205}$$

Exterior differentiation of the θ^μ results immediately in (the prime denotes $\partial/\partial r$ and the dot denotes $\partial/\partial t$):

$$\begin{aligned}
d\theta^0 &= a' e^{-b}\, \theta^1 \wedge \theta^0\,, \\
d\theta^1 &= \dot{b}\, e^{-a}\, \theta^0 \wedge \theta^1\,, \\
d\theta^2 &= \frac{\dot{R}}{R} e^{-a}\, \theta^0 \wedge \theta^2 + \frac{R'}{R} e^{-b}\, \theta^1 \wedge \theta^2\,, \\
d\theta^3 &= \frac{\dot{R}}{R} e^{-a}\, \theta^0 \wedge \theta^3 + \frac{R'}{R} e^{-b}\, \theta^1 \wedge \theta^3 + \frac{1}{R} \cot\vartheta\, \theta^2 \wedge \theta^3\,.
\end{aligned} \tag{3.206}$$

Comparison with the first structure equation leads to

$$\begin{aligned}
\omega^0{}_1 &= \omega^1{}_0 = a' e^{-b} \theta^0 + \dot{b}\, e^{-a}\, \theta^1\,, \\
\omega^0{}_2 &= \omega^2{}_0 = \frac{\dot{R}}{R} e^{-a}\, \theta^2\,, \\
\omega^0{}_3 &= \omega^3{}_0 = \frac{\dot{R}}{R} e^{-a}\, \theta^3\,, \\
\omega^2{}_1 &= -\omega^1{}_2 = \frac{R'}{R} e^{-b}\, \theta^2\,, \\
\omega^3{}_1 &= -\omega^1{}_3 = \frac{R'}{R} e^{-b}\, \theta^3\,, \\
\omega^3{}_2 &= -\omega^2{}_3 = \frac{1}{R} \cot\vartheta\, \theta^3\,.
\end{aligned} \tag{3.207}$$

Verify that the first structure equations are indeed satisfied.

The curvature forms are obtained from the second structure equations by a straightforward calculation. The result is

$$\begin{aligned}
\Omega^0{}_1 &= E\, \theta^0 \wedge \theta^1\,, \\
\Omega^0{}_2 &= \tilde{E}\, \theta^0 \wedge \theta^2 + H\, \theta^1 \wedge \theta^2\,, \\
\Omega^0{}_3 &= \tilde{E}\, \theta^0 \wedge \theta^3 + H\, \theta^1 \wedge \theta^3\,, \\
\Omega^1{}_2 &= \tilde{F}\, \theta^1 \wedge \theta^2 - H\, \theta^0 \wedge \theta^2\,, \\
\Omega^1{}_3 &= \tilde{F}\, \theta^1 \wedge \theta^3 - H\, \theta^0 \wedge \theta^3\,, \\
\Omega^2{}_3 &= F\, \theta^2 \wedge \theta^3\,,
\end{aligned} \tag{3.208}$$

where

$$
\begin{aligned}
E &:= \mathrm{e}^{-2a}(\dot{b}^2 - \dot{a}\,\dot{b} + \ddot{b}) - \mathrm{e}^{-2b}(a'^2 - a'\,b' + a'')\,,\\
\tilde{E} &:= \frac{1}{R}\mathrm{e}^{-2a}(\ddot{R} - \dot{a}\dot{R}) - \frac{1}{R}\mathrm{e}^{-2b}a'R'\,,\\
H &:= \frac{1}{R}\mathrm{e}^{-a-b}(\dot{R}' - a'\dot{R} - \dot{b}R')\,,\\
F &:= \frac{1}{R^2}(1 - R'^2\,\mathrm{e}^{-2b} + \dot{R}^2\mathrm{e}^{-2a})\,,\\
\tilde{F} &:= \frac{1}{R}\mathrm{e}^{-2a}\,\dot{b}\dot{R} + \frac{1}{R}\mathrm{e}^{-2b}(b'R' - R'')\,.
\end{aligned} \tag{3.209}
$$

The Ricci tensor is immediately obtained from (3.208):

$$
\begin{aligned}
&R_{00} = -E - 2\tilde{E}\,, \quad R_{01} = -2H\,, \quad R_{02} = R_{03} = 0\,,\\
&R_{11} = E + 2\tilde{F}\,, \quad R_{12} = R_{13} = 0\,, \quad R_{23} = 0\,,\\
&R_{22} = R_{33} = \tilde{E} + \tilde{F} + F\,.
\end{aligned} \tag{3.210}
$$

The Riemann curvature scalar is then found to be

$$
R = -2(E + F) - 4(\tilde{E} + \tilde{F})\,, \tag{3.211}
$$

and hence the Einstein tensor is given by

$$
\begin{aligned}
&G^0{}_0 = F + 2\tilde{F}\,, \quad G^0{}_1 = -2H\,, \quad G^0{}_2 = G^0{}_3 = 0\,,\\
&G^1{}_1 = 2\tilde{E} + F\,, \quad G^1{}_2 = G^1{}_3 = 0\,, \quad G^2{}_3 = 0\\
&G^2{}_2 = G^3{}_3 = \tilde{E} + \tilde{F} + E.
\end{aligned} \tag{3.212}
$$

The Generalized Birkhoff Theorem

We now consider spherically symmetric solutions of the vacuum field equations, and prove the following

Theorem 3.1 (Birkhoff's Theorem). *Every (C^2-)solution of the Einstein vacuum equations which is spherically symmetric in an open subset U is locally isometric to a domain of the Schwarzschild-Kruskal solution.*

Proof. The local solution depends on the nature of the surfaces $\{R(t,r) = \text{const}\}$. We first consider the case

a) The surfaces $\{R(t,r) = \text{const}\}$ are timelike in U and $dR \neq 0$. In this case, we may choose $R(t,r) = r$. If we now use $\dot{R} = 0$ and $R' = 1$ in $G^0{}_1 = 0$, we find from (3.209) and (3.212) that $\dot{b} = 0$. This information, together with $G_{00} + G_{11} = 0$, i.e. $\tilde{E} - \tilde{F} = 0$, implies $a' + b' = 0$. Hence $a(t,r) = -b(r) + f(t)$. Thus, by choosing a new time coordinate, one can arrange things such that a is also independent of time. The equation $G_{00} = 0$ then gives $(r\mathrm{e}^{-2b})' = 1$, or $\mathrm{e}^{-2b} = 1 - 2m/r$. Thus, for case a) we obtain the Schwarzschild solution for $r > 2m$:

$$
g = -\left(1 - \frac{2m}{r}\right)dt^2 + \frac{dr^2}{1 - 2m/r} - r^2(d\vartheta^2 + \sin^2\vartheta d\varphi^2)\,. \tag{3.213}
$$

b) The surface $\{R(t,r) = \text{const}\}$ are spacelike in U and $dR \neq 0$. In this case, we may choose $R(t,r) = t$. Now the equation $G^0_{\ 1} = 0$ gives the condition $a' = 0$, so that a depends only on t. We now obtain $\dot{a} + \dot{b} = 0$ from $G_{00} + G_{11} = 0$, and we can introduce a new r-coordinate such that b also depends only on t. Finally we obtain from $G_{00} = 0$ the equation $(t\,\mathrm{e}^{-2a})^{\cdot} = -1$ or

$$\mathrm{e}^{-2a} = \mathrm{e}^{2b} = \frac{2m}{t} - 1\,, \quad R = t\,, \tag{3.214}$$

where $t < 2m$. This is the Schwarzschild solution for $r < 2m$, for if we interchange the variables r and t, we obtain (3.213) with $r < 2m$.

c) Suppose, that $\langle dR, dR\rangle = 0$ in U. Consider first the special case, that R is constant in U. Then $G_{00} = 0$ implies $R = \infty$. We may therefore assume that dR is lightlike and not equal to zero. The coordinates (t,r) can be chosen such that $R(t,r) = t - r$. If dR is lightlike, then we must have $a = b$. The equation $G^0_{\ 1}$ is then automatically satisfied and from $G_{00} + G_{11} = 0$ we obtain the condition $a' = -\dot{a}$. Hence a is a function of $t - r$. The equation $G_{00} = 0$ then implies the ridiculous condition $t^{-1} - r = 0$. Case c) is therefore incompatible with the field equations.

d) Suppose finally, that $\{R(t,r) = \text{const}\}$ is spacelike in some part of U and timelike in another part. In this case one obtains the solutions a), respectively b). As in Sect. 3.7.2, one can join these smoothly along the surfaces $\langle dR, dR\rangle = 0$, thereby obtaining a domain of the Schwarzschild-Kruskal spacetime.

□

Spherically Symmetric Metrics for Fluids

We now consider the spherically symmetric field of a (not necessarily static) fluid, whose four-velocity U is also assumed to be invariant with respect to the isotropy group $SO(3)$. Hence U is perpendicular to the orbits and its integral curves lie in the surfaces N. We may choose t such that U is proportional to $\partial/\partial t$. Since $\langle U, U\rangle = -1$, we have

$$U = \mathrm{e}^{-a}\frac{\partial}{\partial t}\,. \tag{3.215}$$

We could have chosen the time coordinate such that $U = \partial/\partial t$, but then the (t,r)-part of the metric would not necessarily be diagonal, but would have the form

$$\tilde{g} = -dt^2 + 2\,g_{rt}\,dt\,dr + g_{rr}\,dr^2\,. \tag{3.216}$$

For an ideal fluid with $p = 0$ (incoherent dust), the field equations imply $\nabla_U U = 0$. The r-component of this equation implies $\Gamma^r{}_{tt} = 0$ or, in other words, $\partial g_{tr}/\partial t = 0$. Let us therefore choose the new coordinates

$$t' = t + \int^r g_{tr}(r)\,dr\,, \qquad r' = r\,.$$

Then the metric takes the form (suppressing the primes)

$$g = -dt^2 + \left(e^{2b}dr^2 + R^2(t,r)d\Omega^2\right) \tag{3.217}$$

and

$$U = \frac{\partial}{\partial t}\,. \tag{3.218}$$

However, this is only possible when $p = 0$.

Of course, for a *static* situation, the four-velocity U (in (3.217)) should be proportional to a timelike Killing field, which can choose to be $\partial/\partial t$ (by a suitable choice of coordinates). In this case, R, a and b are independent of t. In the basis (3.205), the energy-momentum tensor has the diagonal form

$$T^{\mu\nu} = (p+\rho)U^\mu U^\nu + p\,g^{\mu\nu} = \mathrm{diag}(\rho, p, p, p)\,. \tag{3.219}$$

If we would have $R' = 0$, then, because of $G_{00} - G_{11} = 8\pi G(\rho + p)$, this would imply $\rho + p = 0$. This is excluded for normal fluids. We may therefore choose $R(r) = r$. For a static situation, we can thus always introduce coordinates such that the metric has the form

$$g = -e^{2a(r)}dt^2 + \left(e^{2b(r)}dr^2 + r^2(d\vartheta^2 + \sin^2\vartheta\,d\varphi^2)\right)\,. \tag{3.220}$$

This will be important in our study of relativistic stars (e.g., neutron stars) in Chap. 6 (see Sect. 6.10).

* * *

Exercises.

Exercise 3.11. Let (M, g) be a spherically symmetric, stationary Lorentz manifold. Show that a timelike Killing field K exists locally, which is perpendicular to the orbits $\Omega(q)$ of $SO(3)$. In this case, the t-coordinate in the surfaces N can be chosen such that $K = \partial/\partial t$. Then all the functions in (3.204) are independent of t. If we exclude the degenerate case $R' = 0$, the metric can again be written in the form (3.220).

Exercise 3.12. Derive (3.210) and (3.212) from the general formulas for warped products, obtained in the Appendix B of DG.

4 Weak Gravitational Fields

Most gravitational fields encountered in the physical universe are weak. Exceptions are the strong fields near compact objects (black holes and neutron stars) or in the very early universe. It is remarkable that Einstein investigated weak gravitational fields quite exhaustibly only one month after his first systematic exposition of GR (see [58]). Because of a computational error in his derivation of the so-called quadrupole formula for the power, emitted by a material source in the form of gravitational waves[1], he took the subject up again somewhat later and added some important considerations (see [59]).

4.1 The Linearized Theory of Gravity

In this chapter, we consider systems for which the metric field (at least in a certain region of spacetime) is nearly flat. Thus a coordinate system exists for which

$$g_{\mu\nu} = \eta_{\mu\nu} + h_{\mu\nu}\,, \quad |h_{\mu\nu}| \ll 1\,. \tag{4.1}$$

For example, in the solar system, we have $|h_{\mu\nu}| \sim |\phi|/c^2 \lesssim G M_\odot/c^2 R_\odot \sim 10^{-6}$. However, the field can vary rapidly with time, as is the case for weak gravitational waves.

For such fields, we expand the field equations in powers of $h_{\mu\nu}$ and keep only the linear terms. The Ricci tensor is given, up to quadratic terms, by

$$R_{\mu\nu} = \partial_\lambda \Gamma^\lambda{}_{\nu\mu} - \partial_\nu \Gamma^\lambda{}_{\lambda\mu}\,. \tag{4.2}$$

Here, we can use the linearized approximation

$$\Gamma^\alpha{}_{\mu\nu} = \frac{1}{2}\eta^{\alpha\beta}\left(h_{\mu\beta,\nu} + h_{\beta\nu,\mu} - h_{\mu\nu,\beta}\right) = \frac{1}{2}\left(h^\alpha{}_{\mu,\nu} + h^\alpha{}_{\nu,\mu} - h_{\mu\nu}{}^{,\alpha}\right). \tag{4.3}$$

We adopt the convention that indices are raised or lowered with $\eta^{\mu\nu}$ or $\eta_{\mu\nu}$, respectively. Then the Ricci curvature can be written as

---

[1] The term 'gravitational waves' (onde gravifique) first appeared in 1905 in a paper by Poincaré, in which he proposed the first Lorentz-invariant equation for the gravitational field.

$$R_{\mu\nu} = \frac{1}{2}\left(h^{\lambda}{}_{\mu,\lambda\nu} - \Box h_{\mu\nu} - h^{\lambda}{}_{\lambda,\mu\nu} + h^{\lambda}{}_{\nu,\lambda\mu}\right), \tag{4.4}$$

and for the scalar Riemann curvature we obtain

$$R = h^{\lambda\sigma}{}_{,\lambda\sigma} - \Box h\,, \tag{4.5}$$

where

$$h := h^{\lambda}_{\lambda} = \eta^{\alpha\beta} h_{\alpha\beta}\,. \tag{4.6}$$

In the linear approximation, the Einstein tensor is given by

$$2G_{\mu\nu} = -\Box h_{\mu\nu} - h_{,\mu\nu} + h^{\lambda}{}_{\mu,\lambda\nu} + h^{\lambda}{}_{\nu,\lambda\mu} + \eta_{\mu\nu}\Box h - \eta_{\mu\nu} h^{\lambda\sigma}{}_{,\lambda\sigma}\,. \tag{4.7}$$

The contracted Bianchi identity reduces to

$$G^{\mu\nu}{}_{,\nu} = 0\,, \tag{4.8}$$

as one also sees by direct computation. As a consequence of (4.8) and the **linearized field equations**

$$\Box h_{\mu\nu} + h_{,\mu\nu} - h^{\lambda}{}_{\mu,\lambda\nu} - h^{\lambda}{}_{\nu,\lambda\mu} - \eta_{\mu\nu}\Box h + \eta_{\mu\nu} h^{\lambda\sigma}{}_{,\lambda\sigma} = -16\pi G\, T_{\mu\nu}\,, \tag{4.9}$$

we obtain

$$T^{\mu\nu}{}_{,\nu} = 0\,. \tag{4.10}$$

Thus in this approximation the gravitational field generated by $T_{\mu\nu}$ does not react back on the source. For example, if we consider incoherent dust, i.e., $T_{\mu\nu} = \rho\, u^{\mu} u^{\nu}$ and $(\rho u^{\mu})_{,\mu} = 0$, Equation (4.10) implies $u^{\nu} u^{\mu}{}_{,\nu} = 0$ or $du^{\mu}/ds = 0$, so that the integral curves of u^{μ} are straight lines.

As a first step of an iterative procedure for weak fields, one can determine $h_{\mu\nu}$ from the linearized field equations (4.9) with the "flat" $T_{\mu\nu}$ and compute the reaction on physical systems by setting $g_{\mu\nu} = \eta_{\mu\nu} + h_{\mu\nu}$ in the basic equations of Sect. 1.4. This procedure makes sense as long as the reaction is small. We shall discuss some specific examples further on.

In the following, it will be useful to work with the quantity

$$\gamma_{\mu\nu} := h_{\mu\nu} - \frac{1}{2}\eta_{\mu\nu} h\,, \tag{4.11}$$

in terms of which $h_{\mu\nu}$ is given by

$$h_{\mu\nu} = \gamma_{\mu\nu} - \frac{1}{2}\eta_{\mu\nu}\gamma\,, \tag{4.12}$$

where $\gamma := \gamma^{\lambda}_{\lambda}$. The linearized field equations become

$$-\Box\gamma_{\mu\nu} - \eta_{\mu\nu}\gamma_{\alpha\beta}{}^{,\alpha\beta} + \gamma_{\mu\alpha,\nu}{}^{,\alpha} + \gamma_{\nu\alpha,\mu}{}^{,\alpha} = 16\pi G\, T_{\mu\nu}\,. \tag{4.13}$$

We may regard the linearized theory as a Lorentz covariant field theory (for the field $h_{\mu\nu}$) in flat space. (This point of view was adopted in Sect. 1.2).

If we consider a global Lorentz transformation, we obtain from the transformation properties of $g_{\mu\nu}$ with respect to coordinate transformations, that $h_{\mu\nu}$ transforms as a tensor with respect to the Lorentz group. Indeed, if

$$x^\mu = \Lambda^\mu{}_\nu x'^\nu\,, \quad \Lambda^T \eta \Lambda = \eta\,,$$

where $\eta := \mathrm{diag}(-1,1,1,1)$, then we can write

$$\eta_{\alpha\beta} + h'_{\alpha\beta} = \frac{\partial x^\mu}{\partial x'^\alpha}\frac{\partial x^\nu}{\partial x'^\beta}\, g_{\mu\nu} = \Lambda^\mu{}_\alpha \Lambda^\nu{}_\beta (\eta_{\mu\nu} + h_{\mu\nu}) = \eta_{\alpha\beta} + \Lambda^\mu{}_\alpha \Lambda^\nu{}_\beta\, h_{\mu\nu}\,,$$

so that

$$h'_{\alpha\beta} = \Lambda^\mu{}_\alpha \Lambda^\nu{}_\beta\, h_{\mu\nu}\,. \tag{4.14}$$

As is the case in electrodynamics, there is a **gauge group**. The linearized Einstein tensor (4.7) is *invariant* with respect to gauge transformations of the form

$$h_{\mu\nu} \longrightarrow h_{\mu\nu} + \xi_{\mu,\nu} + \xi_{\nu,\mu}\,, \tag{4.15}$$

where ξ^μ is a vector field, as one can easily verify by direct computation. This transformation can be regarded as an **infinitesimal coordinate transformation**. In fact, let

$$x'^\mu = x^\mu + \xi^\mu(x)\,, \tag{4.16}$$

where $\xi^\mu(x)$ is an infinitesimal vector field (it must be infinitesimal to preserve (4.1)). This transformation induces infinitesimal changes in all fields. With the exception of the gravitational field, which is described by the infinitesimal $h_{\mu\nu}$, it is possible to neglect these changes in the linearized theory. From the transformation law

$$g_{\mu\nu}(x) = \underbrace{\frac{\partial x'^\sigma}{\partial x^\mu}}_{(\delta^\sigma_\mu + \xi^\sigma{}_{,\mu})} \cdot \underbrace{\frac{\partial x'^\rho}{\partial x^\nu}}_{(\delta^\rho_\nu + \xi^\rho{}_{,\nu})}\, g'_{\sigma\rho}(x')$$

we obtain

$$g_{\mu\nu}(x) = g'_{\mu\nu}(x') + g'_{\sigma\nu}(x')\xi^\sigma{}_{,\mu} + g'_{\mu\sigma}(x')\xi^\sigma{}_{,\nu}\,.$$

To first order in ξ^μ and $h_{\mu\nu}$ this gives

$$h_{\mu\nu}(x) = h'_{\mu\nu}(x) + \xi_{\mu,\nu} + \xi_{\nu,\mu}\,.$$

Gauge invariance can be expressed more geometrically as follows: Previously, we emphasized that the metric g is physically equivalent to $\varphi^* g$ for any diffeomorphism φ. In particular, g is equivalent to $\phi_s^* g$, where ϕ_s denotes a local flow of some vector field X. For small s we have

$$\phi_s^* g = g + s\, L_X g + O(s^2)\,.$$

If we express this in terms of $h := g - \eta$, and set $\xi = s X$, this amounts to

$$h \longrightarrow h + L_\xi g + O(s^2) = h + L_\xi \eta + L_\xi h + O(s^2)\,.$$

Thus, it is clear that for weak fields h and $h + L_\xi \eta$ are physically equivalent for an infinitesimal vector field ξ. Since

$$(L_\xi \eta)_{\mu\nu} = \xi_{\mu,\nu} + \xi_{\nu,\mu}\,,$$

this amounts to the substitution (4.15). With this interpretation of gauge transformations the gauge invariance of the linearized Riemann tensor $R^{(1)}$ can be understood conceptually as follows: The Riemann tensor $R[g]$ fulfills the covariance property $\phi_s^*(R[g]) = R[\phi_s^* g]$. Hence, we have

$$L_X(R[g]) = \left.\frac{d}{ds}\right|_{s=0} (R[g + s\,L_X g])\,. \tag{4.17}$$

If we now set $g = \eta$ on both sides, we find $0 = R^{(1)}(L_X \eta)$, and thus

$$R^{(1)}(h + L_X \eta) = R^{(1)}(h)$$

for every vector field X. This argument shows: Since the Riemann tensor vanishes to zeroth order, it is gauge invariant to first order.

* * *

Generalization

For later use we generalize this discussion to small perturbations h around some curved background metric $\bar{g}$. The gauge transformations are

$$h \longrightarrow h + L_\xi \bar{g} \quad \text{or} \quad h_{\mu\nu} \longrightarrow h_{\mu\nu} + \xi_{\mu|\nu} + \xi_{\nu|\mu}\,, \tag{4.18}$$

where an upright line denotes covariant derivative with respect to the background metric.

If $\bar{g}$ satisfies the vacuum equations $\mathrm{Ric}[\bar{g}] = 0$, then the linearized Ricci tensor is gauge invariant. This follows as before from

$$L_\xi(\mathrm{Ric}[\bar{g}]) = \left.\frac{d}{ds}\right|_{s=0} (\mathrm{Ric}[\bar{g} + s\,L_\xi \bar{g}])\,. \tag{4.19}$$

If $\mathrm{Ric}[\bar{g}]$ does not vanish, (4.19) implies for the linearized Ricci tensor $\delta R_{\mu\nu}[h]$ the transformation law

$$\delta R_{\mu\nu} \longrightarrow \delta R_{\mu\nu} + L_\xi \bar{R}_{\mu\nu}\,,$$

where $\bar{R}_{\mu\nu}$ is the Ricci tensor of the background metric. We can easily find an explicit expression for the linearized Ricci tensor $\delta R_{\mu\nu}[h]$ by using the Palatini identity (2.86)

$$\delta R_{\mu\nu} = ({\Gamma^\alpha}_{\mu\nu})_{|\alpha} - ({\Gamma^\alpha}_{\mu\alpha})_{|\nu} ,$$

and (2.93)

$$\delta {\Gamma^\mu}_{\alpha\beta} = \frac{1}{2} g^{\mu\nu} \left(h_{\nu\alpha|\beta} + h_{\nu\beta|\alpha} - h_{\alpha\beta|\nu} \right) .$$

This gives

$$\delta R_{\mu\nu}[h] = \frac{1}{2} g^{\alpha\beta} \left(h_{\beta\mu|\nu\alpha} - h_{\alpha\beta|\mu\nu} + h_{\beta\nu|\mu\alpha} - h_{\mu\nu|\beta\alpha} \right) . \tag{4.20}$$

(Compare this with (4.4).)

The linearized field equations are in obvious notation

$$\delta R_{\mu\nu}[h] = 8\pi G \left(\delta T_{\mu\nu} - \frac{1}{2} g_{\mu\nu} g^{\alpha\beta} \delta T_{\alpha\beta} + \frac{1}{2} g_{\mu\nu} h_{\alpha\beta} \bar{T}^{\alpha\beta} - \frac{1}{2} h_{\mu\nu} {\bar{T}^\lambda}_\lambda \right) \tag{4.21}$$

and are invariant under infinitesimal diffeomorphisms

$$h_{\mu\nu} \longrightarrow h_{\mu\nu} + L_\xi h_{\mu\nu} , \qquad \delta T_{\mu\nu} \longrightarrow \delta T_{\mu\nu} + L_\xi \bar{T}_{\mu\nu} \tag{4.22}$$

for any vector field ξ. These substitutions can be written as

$$\begin{aligned} h_{\mu\nu} &\longrightarrow h_{\mu\nu} + \xi_{\mu|\nu} + \xi_{\nu|\mu} , \\ \delta T_{\mu\nu} &\longrightarrow \delta T_{\mu\nu} + {\bar{T}^\lambda}_\mu \xi_{\lambda|\nu} + {\bar{T}^\lambda}_\nu \xi_{\lambda|\mu} + \bar{T}_{\mu\nu|\lambda} \xi^\lambda . \end{aligned} \tag{4.23}$$

The linearized field equations describe the propagation of small disturbances on a given spacetime background, satisfying the field equations with energy-momentum tensor $\bar{T}_{\mu\nu}$. In vacuum they reduces to

$$\bar{g}^{\alpha\beta} \left(h_{\beta\mu|\nu\alpha} - h_{\alpha\beta|\mu\nu} + h_{\beta\nu|\mu\alpha} - h_{\mu\nu|\beta\alpha} \right) = 0 . \tag{4.24}$$

* * *

Returning to weak perturbations of Minkowski spacetime, we now make use of gauge invariance to simplify the field equations. One can always find a gauge such that[2]

$$\gamma^{\alpha\beta}{}_{,\beta} = 0 . \tag{4.25}$$

This gauge is called the **Hilbert gauge**. To see that (4.25) can be achieved we note that the gauge transformation

$$h_{\mu\nu} \longrightarrow h_{\mu\nu} + \xi_{\mu,\nu} + \xi_{\nu,\mu}$$

translates to

$$\gamma_{\mu\nu} \longrightarrow \gamma_{\mu\nu} + \xi_{\mu,\nu} + \xi_{\nu,\mu} - \eta_{\mu\nu} \xi^\lambda{}_{,\lambda} . \tag{4.26}$$

---

[2] Einstein used this gauge already in his research note book in 1912, after he returned from Prag to Zürich, see [57], as well as the editorial note to this document.

Thus, we obtain

$$\gamma^{\mu\nu}{}_{,\nu} \longrightarrow \gamma^{\mu\nu}{}_{,\nu} + \Box\xi^{\mu} + \xi^{\nu,\mu}{}_{,\nu} - \xi^{\lambda}{}_{,\lambda}{}^{,\mu} = \gamma^{\mu\nu}{}_{,\nu} + \Box\xi^{\mu}\,.$$

If $\gamma^{\mu\nu}{}_{,\nu} \neq 0$, we merely need to choose ξ^{μ} to be a solution of

$$\Box\xi^{\mu} = -\gamma^{\mu\nu}{}_{,\nu}$$

in order to satisfy the Hilbert condition. Such solutions always exist (retarded potentials).

For the Hilbert gauge, (4.13) simplifies considerably:

$$\Box\gamma_{\mu\nu} = -16\pi G\, T_{\mu\nu}\,. \tag{4.27}$$

The most general solution of (4.27) satisfying the subsidiary condition (4.25), is

$$\gamma_{\mu\nu} = 16\pi G\, D_R * T_{\mu\nu} + \text{solution of the homogeneous wave equation}\,, \tag{4.28}$$

where D_R is the retarded Green's function

$$D_R(x) = \frac{1}{4\pi|\boldsymbol{x}|}\delta(x^0 - |\boldsymbol{x}|)\theta(x^0)\,.$$

Note that

$$\partial_\nu(D_R * T_{\mu\nu}) = D_R * \partial_\nu T_{\mu\nu} = 0\,,$$

so that the first term of (4.28) indeed satisfies the Hilbert condition (4.25). We shall discuss the solutions to the homogeneous equation (weak gravitational waves) in Sect. 4.3.

The retarded solution is explicitly

$$\gamma_{\mu\nu}(x) = 4G\int \frac{T_{\mu\nu}(x^0 - |\boldsymbol{x} - \boldsymbol{x}'|, \boldsymbol{x}')}{|\boldsymbol{x} - \boldsymbol{x}'|}\, d^3x'\,. \tag{4.29}$$

We interprete this as the field generated by the source, while the second term in (4.28) represents gravitational waves coming from infinity. As in electrodynamics, we conclude that gravitational effects propagate at the speed of light.

* * *

Exercises.

Exercise 4.1. Compute the Riemann tensor in the linearized theory and show by an explicit calculation that it is invariant under the gauge transformations (4.15).

Exercise 4.2. Show that the linearized vacuum field equations (4.9) can be derived from the Lagrangian density

$$\mathcal{L} = \frac{1}{4} h_{\mu\nu,\sigma} h^{\mu\nu,\sigma} - \frac{1}{2} h_{\mu\nu,\sigma} h^{\sigma\nu,\mu} - \frac{1}{4} h^{,\sigma} + \frac{1}{2} h_{,\sigma} h^{\nu\sigma}{}_{,\nu} \,. \tag{4.30}$$

For this, show that

$$\delta \int_D \mathcal{L}\, d^4x = \int_D G^{\mu\nu} \delta h_{\mu\nu}\, d^4x \tag{4.31}$$

for variations $\delta h_{\mu\nu}$ which vanish on the boundary of D.

Exercise 4.3. Show that the Lagrangian density (4.30) changes only by a divergence under gauge transformations (4.15). Thus if ξ^μ vanishes on the boundary of D, the action integral over D is invariant under gauge transformations. Using this, derive the Bianchi identity (4.8) from (4.31).

Remark. Since $\mathcal{L}$ changes if the gauge is changed, the canonical energy-momentum tensor (from SR) is *not* gauge invariant. This is a reflection of the non-localizability of the gravitational energy. Even in the linearized theory, it is not possible to find gauge invariant expressions for the energy and momentum densities.

Exercise 4.4. Show that Einstein's vacuum field equations are uniquely determined by the following three requirements:

a) General invariance;
b) the linearized approximation agrees with (4.9) for $T_{\mu\nu} = 0$ (massless spin-2 equation in Minkowski spacetime);
c) the field equation is linear in the second order derivatives.

Remark. This says, roughly speaking, that the linearized theory determines already the full theory, as a result of its general invariance. This fact is quite remarkable.

4.2 Nearly Newtonian Gravitational Fields

We now consider nearly Newtonian sources, with $T_{00} \gg |T_{0j}|, |T_{ij}|$ and such small velocities that retardation effects are negligible. In this case (4.29) becomes to leading order

$$\gamma_{00} = -4\phi\,, \quad \gamma_{0j} = \gamma_{ij} = 0\,, \tag{4.32}$$

where ϕ is the Newtonian potential

$$\phi(\boldsymbol{x}) = -G \int \frac{T_{00}(t, \boldsymbol{x}')}{|\boldsymbol{x} - \boldsymbol{x}'|}\, d^3x'\,. \tag{4.33}$$

For the metric

$$g_{\mu\nu} = \eta_{\mu\nu} + h_{\mu\nu} = \eta_{\mu\nu} + \left(\gamma_{\mu\nu} - \frac{1}{2}\eta_{\mu\nu}\gamma\right) \tag{4.34}$$

this gives

$$g_{00} = -(1+2\phi)\,, \quad g_{0i} = 0\,, \quad g_{ij} = (1-2\phi)\delta_{ij}\,, \tag{4.35}$$

or

$$g = -(1+2\phi)dt^2 + (1-2\phi)\big(dx^2+dy^2+dz^2\big)\,. \tag{4.36}$$

This **Newtonian approximation** was already obtained in Sect. 1.5 (see Exercise 1.17). At large distances from the source, the monopole contribution to (4.33) dominates and we obtain

$$g = -\left(1-\frac{2m}{r}\right)dt^2 + \left(1+\frac{2m}{r}\right)\big(dx^2+dy^2+dz^2\big)\,, \tag{4.37}$$

where, as in (3.14), $m = GM/c^2$.

The errors in the metric (4.36) are:

a) Terms of order ϕ^2 are neglected. These arise as a result of the non-linearities in GR.
b) γ_{0j} vanishes up to terms of order $v\,\phi$, where $v \sim |T_{0j}|/T_{00}$ is a typical velocity of the source. The first order terms will be treated below.
c) γ_{ij} vanishes, up to terms of order $\phi|T_{ij}|/T_{00}$.

In the solar system, all of these corrections are of order 10^{-12}, while $\phi \sim 10^{-6}$.

Remark. We have seen in Sect. 1.7 that Fermat's principle (1.88) and (4.36) imply that light rays behave in almost Newtonian fields exactly as in an optical medium with index of refraction (1.90):

$$n = 1 - 2\phi\,. \tag{4.38}$$

Later in this Chapter (see Sect. 4.8) we shall use this to develop the foundations of gravitational lensing theory.

Gravitomagnetic Field and Lense-Thirring Precession

In the next order (in $1/c$) we encounter for rotating sources the "gravitomagnetic field". If the spatial stresses T_{ij} can be neglected, the field equations (4.27) reduce to

$$\Box\gamma_{ij} = 0\,, \qquad \Box\gamma_{0\mu} = -16\pi G\, T_{0\mu}\,. \tag{4.39}$$

Thus $A_\mu := \frac{1}{4}\gamma_{0\mu}$ satisfy Maxwell type equations:

$$\Box A_\mu = -4\pi\, J_\mu\,, \tag{4.40}$$

where $J_\mu = G\,T_{0\mu}$ is proportional to the mass-energy current density. (Note that $A_0 = -\phi$.) It is natural to define "gravitational electric and magnetic fields" $\boldsymbol{E}$ and $\boldsymbol{B}$ by the same formulas in terms of A_μ as in electrodynamics.

Let us now assume that the time derivatives of $\gamma_{\mu\nu}$ can be neglected (quasi-stationary situations). Then $\Delta\gamma_{ij} = 0$ in all space, whence $\gamma_{ij} = 0$, and hence A_μ describes the gravitational field completely. These potentials are given by

$$A_0 = -\phi\,, \quad A_i(\boldsymbol{x}) = G \int \frac{T_{0i}(\boldsymbol{x}')}{|\boldsymbol{x}-\boldsymbol{x}'|}\, d^3x'\,. \tag{4.41}$$

Next, we show that the geodesic equation of motion for a neutral test particle can be expressed in this approximation entirely in terms of the **gravitoelectric** and **gravitomagnetic fields** $\boldsymbol{E}$ and $\boldsymbol{B}$:

$$\ddot{\boldsymbol{x}} = \boldsymbol{E} + 4\,\dot{\boldsymbol{x}} \wedge \boldsymbol{B}\,. \tag{4.42}$$

The factor 4 in the "magnetic term" reflects the spin-2 character of the gravitational field.

For the derivation of (4.42) we first note that the metric can be expressed in terms of A_μ as

$$g_{00} = -1 + 2A_0\,, \quad g_{0i} = 4A_i\,, \quad g_{ij} = (1+2A_0)\delta_{ij}\,. \tag{4.43}$$

The geodesic equation of motion follows from the variational principle

$$\delta \int \sqrt{-g_{\mu\nu}\frac{dx^\mu}{dt}\frac{dx^\nu}{dt}}\, dt = 0\,.$$

If $v^i = dx^i/dt$ we have, neglecting higher order terms,

$$g_{\mu\nu}\frac{dx^\mu}{dt}\frac{dx^\nu}{dt} = -1 + \boldsymbol{v}^2 - 2A^0 + 8\boldsymbol{A}\cdot\boldsymbol{v}\,.$$

Hence, the effective Lagrangian is

$$L = \frac{1}{2}\boldsymbol{v}^2 - A^0 + 4\boldsymbol{A}\cdot\boldsymbol{v}\,. \tag{4.44}$$

As in electrodynamics, the corresponding Euler-Lagrange equations are found to agree with (4.42).

With these results it is easy to determine the Lense-Thirring precession of a gyroscope, due to the gravitomagnetic field. One can simply translate the well-known spin precession formula in electrodynamics, $\dot{\boldsymbol{S}} = \boldsymbol{\mu} \wedge \boldsymbol{B}$, where $\boldsymbol{\mu} = \frac{e}{2m}\boldsymbol{S}$, by the substitution $e \to m$ and $\boldsymbol{B} \to 4\boldsymbol{B}$. This gives the precession frequency

$$\boldsymbol{\Omega}_{LT} = -2\,\boldsymbol{\nabla} \times \boldsymbol{A}\,. \tag{4.45}$$

This agrees with (1.164).

* * *

Exercises.

Exercise 4.5. Show that the metric (4.37) implies the correct deflection of light rays.

Exercise 4.6. Show that the metric (4.37) gives 4/3 times the Einstein value for the perihelion precession. This demonstrates once more that the precession of the perihelion is sensitive to non-linearities in GR.

Exercise 4.7. Consider a stationary rotating mass like the Sun or the Earth. Show that the leading order of $A_i(\boldsymbol{x})$ in a multipole expansion of (4.41) is given by

$$A_i(\boldsymbol{x}) = \frac{1}{2}\frac{1}{r^3}\varepsilon_{ijk}x^j J^k , \tag{4.46}$$

where $\boldsymbol{J}$ is the angular momentum of the body,

$$J_k = \int_{body} \varepsilon_{krs} x^r \, T^{s0} \, d^3x \, . \tag{4.47}$$

(The integrand is the density of the k-component of the angular momentum.)

Show that in this approximation the Lense-Thirring precession frequency is

$$\boldsymbol{\Omega}_{LT} = \frac{G}{r^3}\left(-\boldsymbol{J} + 3\frac{(\boldsymbol{J}\cdot\boldsymbol{x})\boldsymbol{x}}{r^2}\right) . \tag{4.48}$$

Hint. Use the zeroth order conservation law $\partial_\nu T^{\nu\mu} = 0$ to derive

$$\int x^k T^{l0} \, dx^3 = \frac{1}{2}\varepsilon^{klm} J_m \, . \tag{4.49}$$

4.3 Gravitational Waves in the Linearized Theory

We now consider the linearized theory in vacuum. In the Hilbert gauge, the field equations then read (see (4.27))

$$\Box\gamma_{\mu\nu} = 0 \, . \tag{4.50}$$

For the free field, it is possible to find, even within the Hilbert gauge class, a gauge such that we also have

$$\gamma = 0 \, . \tag{4.51}$$

This can be seen as follows. According to (4.26), a change of gauge has the effect

$$\gamma_{\mu\nu} \longrightarrow \gamma_{\mu\nu} + \xi_{\mu,\nu} + \xi_{\nu,\mu} - \eta_{\mu\nu}\xi^{\lambda}{}_{,\lambda} \, ,$$

and hence $\gamma \longrightarrow \gamma - 2\xi^{\lambda}{}_{,\lambda}$ as well as

$$\gamma^{\mu\nu}{}_{,\nu} \longrightarrow \gamma^{\mu\nu}{}_{,\nu} + \Box\xi^{\mu} \, .$$

In order to preserve the Hilbert condition $\gamma^{\mu\nu}{}_{,\nu} = 0$, we have to impose

$$\Box\xi^{\mu} = 0\,. \tag{4.52}$$

If $\gamma \neq 0$, then we must seek a ξ^{μ} which satisfies (4.52) as well as $2\xi^{\mu}{}_{,\mu} = \gamma$. Note that (4.52) requires the consistency condition $\Box\gamma = 0$, which is generally satisfied only in vacuum. However, this condition is also *sufficient*. In order to prove this we show that if ϕ is a scalar field such that $\Box\phi = 0$, then there exists a vector field ξ^{μ} with the properties $\Box\xi^{\mu} = 0$ and $\xi^{\mu}{}_{,\mu} = \phi$.

Construction of the vector field ξ^{μ}: Let η^{μ} be a solution of $\eta^{\mu}{}_{,\mu} = \phi$. Such a solution exists; one may take $\eta_{\mu} = \Lambda_{,\mu}$, with $\Box\Lambda = \phi$. Now let $\zeta^{\mu} = \Box\eta^{\mu}$. Since $\zeta^{\mu}{}_{,\mu} = 0$, there exists[3] an antisymmetric tensor field $f^{\mu\nu} = -f^{\nu\mu}$, such that $\zeta^{\mu} = f^{\mu\nu}{}_{,\nu}$. Now let $\sigma^{\mu\nu} = -\sigma^{\nu\mu}$ be a solution of $\Box\sigma^{\mu\nu} = f^{\mu\nu}$. Then

$$\xi^{\mu} := \eta^{\mu} - \sigma^{\mu\nu}{}_{,\nu}$$

satisfies the equations

$$\begin{aligned} \xi^{\mu}{}_{,\mu} &= \eta^{\mu}{}_{,\mu} = \phi\,, \\ \Box\xi^{\mu} &= \Box\eta^{\mu} - \Box\sigma^{\mu\nu}{}_{,\nu} = \zeta^{\mu} - f^{\mu\nu}{}_{,\nu} = 0\,. \end{aligned}$$

Within the gauge class determined by (4.25) and (4.51), only gauge transformations which satisfy the additional conditions

$$\Box\xi^{\mu} = 0 \quad \text{and} \quad \xi^{\mu}{}_{,\mu} = 0 \tag{4.53}$$

are allowed. For this gauge class we have $\gamma_{\mu\nu} = h_{\mu\nu}$.

Plane Waves

The most general solution of (4.50) can be represented as a superposition of plane waves

$$h_{\mu\nu} = \mathrm{Re}\left(\varepsilon_{\mu\nu}\mathrm{e}^{-i\langle k,x\rangle}\right). \tag{4.54}$$

The field equations (4.50) are satisfied provided

$$k^2 := \langle k, k\rangle = 0\,. \tag{4.55}$$

The Hilbert condition $\gamma^{\mu\nu}{}_{,\nu} = h^{\mu\nu}{}_{,\nu} = 0$ implies

$$k_{\mu}\varepsilon^{\mu}{}_{\nu} = 0\,, \tag{4.56}$$

and (4.51) gives

$$\varepsilon^{\nu}{}_{\nu} = 0\,. \tag{4.57}$$

[3] This follows from Poincaré's Lemma, written in terms of the codifferential (see [illegible], Lemma 12.2).

The matrix $\varepsilon_{\mu\nu}$ is called the **polarization tensor**. The conditions (4.56) and (4.57) imply that at most five components are independent. We shall now prove that, due to the residual gauge invariance, only two of them are independent.

Under a gauge transformation belonging to the vector field

$$\xi^\mu(x) = \mathrm{Re}\left(i\varepsilon^\mu \mathrm{e}^{i\langle k,x\rangle}\right) \tag{4.58}$$

the polarization tensor $\varepsilon_{\mu\nu}$ changes according to

$$\varepsilon_{\mu\nu} \longrightarrow \varepsilon_{\mu\nu} + k_\mu \varepsilon_\nu + k_\nu \varepsilon_\mu \,. \tag{4.59}$$

Now, consider a wave which propagates in the positive z-direction:

$$k^\mu = (k, 0, 0, k)\,. \tag{4.60}$$

From (4.56) we have

$$\varepsilon_{0\nu} = \varepsilon_{3\nu}\,, \quad \varepsilon_{01} = \varepsilon_{31}\,, \quad \varepsilon_{02} = \varepsilon_{32}\,,$$

Note that the first equation is equivalent to $\varepsilon_{00} = \varepsilon_{30} = \varepsilon_{03} = \varepsilon_{33}$. Equation (4.57) implies $\varepsilon_{00} - \varepsilon_{11} - \varepsilon_{22} - \varepsilon_{33} = 0$, so that

$$\varepsilon_{11} + \varepsilon_{22} = 0\,.$$

These relations permit us to express all components in terms of

$$\varepsilon_{00}, \varepsilon_{11}, \varepsilon_{01}, \varepsilon_{02} \text{ and } \varepsilon_{12}\,. \tag{4.61}$$

In particular, we have

$$\varepsilon_{03} = \varepsilon_{00}\,, \quad \varepsilon_{13} = \varepsilon_{01}\,, \quad \varepsilon_{22} = -\varepsilon_{11}\,, \quad \varepsilon_{23} = \varepsilon_{02}\,, \quad \varepsilon_{33} = \varepsilon_{00}\,. \tag{4.62}$$

For a gauge transformation (4.59), the components (4.61) transform as

$$\begin{aligned} &\varepsilon_{00} \longrightarrow \varepsilon_{00} + 2k\,\varepsilon_0\,, \quad \varepsilon_{11} \longrightarrow \varepsilon_{11}\,, \quad \varepsilon_{01} \longrightarrow \varepsilon_{01} + k\,\varepsilon_1\,, \\ &\varepsilon_{02} \longrightarrow \varepsilon_{02} + k\,\varepsilon_2\,, \quad \varepsilon_{12} \longrightarrow \varepsilon_{12}\,. \end{aligned} \tag{4.63}$$

If we require that (4.58) satisfies the conditions (4.53), then we have $k^\mu \varepsilon_\mu = 0$, i.e., $\varepsilon_0 = \varepsilon_3$. It then follows from (4.63) that one can choose ε^μ such that only ε_{12} and $\varepsilon_{11} = -\varepsilon_{22}$ do not vanish. As for light, one has *only two linearly independent polarization states*. Under a rotation about the z-axis, these transform according to

$$\varepsilon'_{\mu\nu} = R^\alpha{}_\mu R^\beta{}_\nu \varepsilon_{\alpha\beta}\,,$$

where

$$R(\varphi) = \left(\begin{array}{c|ccc} 1 & & 0 & \\ \hline & \cos\varphi & \sin\varphi & 0 \\ 0 & -\sin\varphi & \cos\varphi & 0 \\ & 0 & 0 & 1 \end{array}\right).$$

One easily finds

$$\begin{aligned} \varepsilon'_{11} &= \varepsilon_{11}\cos 2\varphi + \varepsilon_{12}\sin 2\varphi\,, \\ \varepsilon'_{12} &= -\varepsilon_{11}\sin 2\varphi + \varepsilon_{12}\cos 2\varphi\,, \end{aligned}$$

or for $\varepsilon_\pm := \varepsilon_{11} \mp i\,\varepsilon_{12}$,

$$\varepsilon'_\pm = \mathrm{e}^{\pm 2i\varphi}\varepsilon_\pm\,. \tag{4.64}$$

This shows that the polarization states $\varepsilon_\pm$ have **helicity** ± 2 (left, respectively right handed **circular polarization**).

Transverse and Traceless Gauge

In a gauge in which only ε_{12} and $\varepsilon_{11} = -\varepsilon_{22}$ do not vanish, we obviously have

$$h_{\mu 0} = 0\,, \quad h_{kk} = 0 \quad h_{kj,j} = 0\,, \tag{4.65}$$

where $h_{kk} := \sum_{k=1}^{3} h_{kk}$.

We now consider a *general* gravitational wave $h_{\mu\nu}$ of the linearized theory. For each plane wave of the Fourier decomposition, we choose the special gauge (4.65). One can achieve this by first choosing the special gauge $h^{\mu\nu}{}_{,\nu} = 0$, $h^\mu{}_\mu = 0$, and then carrying out a superposition of gauge transformations of the form (4.58) defined by the vector field

$$\xi^\mu(x) = \mathrm{Re}\left(\int i\,\varepsilon^\mu(k)\mathrm{e}^{i\langle k,x\rangle}\,d^4k\right),$$

Since the gauge conditions (4.65) are all linear, they will then be satisfied for the general wave under consideration. We have thus shown that the gauge conditions (4.65) can be satisfied for an arbitrary gravitational wave. If a symmetric tensor satisfies (4.65), we call it a **transverse, traceless tensor**, and the gauge (4.65) is known as the **transverse, traceless gauge (TT gauge)**. Note that for the TT gauge only the h_{ij} are non-vanishing, and hence we have *only six* wave equations

$$\Box h_{ij} = 0\,. \tag{4.66}$$

We now compute the linearized Riemann tensor in the TT gauge. For any gauge, we have

$$R^\mu{}_{\sigma\nu\rho} = \partial_\nu \varGamma^\mu{}_{\rho\sigma} - \partial_\rho \varGamma^\mu{}_{\nu\sigma} + \text{quadratic terms in } \varGamma\,.$$

Using (4.3) for the Christoffel symbols, we obtain ($R_{\mu\sigma\nu\rho} = g_{\mu\lambda}R^\lambda{}_{\sigma\nu\rho}$)

$$R_{\mu\sigma\nu\rho} = \frac{1}{2}\left(h_{\sigma\mu,\nu\rho} + h_{\rho\mu,\sigma\nu} - h_{\rho\sigma,\mu\nu}\right) - \frac{1}{2}\left(h_{\sigma\mu,\nu\rho} + h_{\nu\mu,\sigma\rho} - h_{\nu\sigma,\mu\rho}\right)$$
$$= \frac{1}{2}\left(h_{\nu\sigma,\mu\rho} + h_{\rho\mu,\sigma\nu} - h_{\nu\mu,\sigma\rho} - h_{\rho\sigma,\mu\nu}\right). \tag{4.67}$$

Below we need the components $R_{i0j0} = R_{0i0j} = -R_{i00j} = -R_{0ij0}$. In the TT gauge, indicated by the subscripts TT, we find

$$R_{i0j0} = \frac{1}{2}\left(h_{0j,0i} + h_{i0,j0} - h_{ij,00} - h_{00,ij}\right) = -\frac{1}{2}\, h^{TT}_{ij,00}\,. \tag{4.68}$$

Since the Riemann tensor is gauge invariant, one sees from (4.68) that $h_{\mu\nu}$ cannot be reduced to fewer components than we achieved in the TT gauge.

Geodetic Deviation in the Metric Field of a Gravitational Wave

For an understanding of the action of a gravitational wave on matter (a detector) it is helpful to look at the induced oscillations of the separations between neighboring freely falling test particles. As in Sect. 2.1 we consider a congruence of timelike geodesics, representing a swarm of freely falling small bodies. The separation vector satisfies the equation for geodesic deviation (2.9). Relative to an orthonormal triad $\{e_j\}$ perpendicular to the four-velocity field, which is parallel transported along an arbitrary geodesic $\gamma(\tau)$ of the congruence, we can write this in the form (2.12):

$$\frac{d^2}{d\tau^2}\boldsymbol{n} = K\,\boldsymbol{n}\,, \tag{4.69}$$

where $K_{ij} = R^i{}_{00j}$.

There exists a **TT-coordinate system**, such that $h_{\mu\nu}$ satisfy the TT conditions (4.65), for which, to leading order in the h_{ij}, we have along γ

$$\partial/\partial x^i = e_i \;\; (i = 1, 2, 3)\,, \qquad \partial/\partial t = e_0 := \dot{\gamma}\,. \tag{4.70}$$

Such a system can be constructed in two steps. One first constructs a local inertial system along γ (for which $t = \tau$), as in Sect. 1.10.5. For this (4.70) holds. Then one performs an infinitesimal coordinate transformation, such that the TT conditions (4.65) are satisfied.

According to (4.68) and (4.69), we have (since t is equal to the proper time τ along γ) to first order in the h_{ij}

$$\frac{d^2 n^i}{dt^2} = \frac{1}{2}\frac{\partial^2 h^{TT}_{ij}}{\partial t^2}\, n^j\,. \tag{4.71}$$

This equation describes the oscillations of test bodies near γ induced by weak gravitational waves. For example, if the particles were at rest relative to one another before the wave arrives ($\boldsymbol{n} = \boldsymbol{n}_{(0)}$ for $h_{ij} = 0$), then integration of (4.71) results in

$$n^i(\tau) \simeq n^i_{(0)} + \frac{1}{2} h_{ij}\big(\gamma(\tau)\big) n^j_{(0)}\,. \tag{4.72}$$

Let us consider the special case of a plane wave propagating in the z-direction. In the TT gauge the non-vanishing components are

$$h_{xx}^{TT} = -h_{yy}^{TT} = A(t-z)\,,$$
$$h_{xy}^{TT} = h_{yx}^{TT} = B(t-z)\,. \tag{4.73}$$

The particle will not oscillate if it is moving in the direction of propagation ($\boldsymbol{n}_{(0)} = (0,0,a)$), since then $h_{ij}n_{(0)}^j = 0$. Hence only *transverse oscillations are possible.* We shall now discuss these for various polarizations of the incident plane wave. The transverse components of the displacement vector $\boldsymbol{n}$ satisfy

$$\ddot{n}_\perp = K_\perp n_\perp\,, \tag{4.74}$$

where

$$K_\perp = \frac{1}{2}\begin{pmatrix} \ddot{h}_{xx} & \ddot{h}_{xy} \\ \ddot{h}_{yx} & \ddot{h}_{yy} \end{pmatrix}, \qquad K_\perp^T = K_\perp\,, \qquad \mathrm{Tr}\, K_\perp = 0\,. \tag{4.75}$$

According to (4.72), an approximate solution is

$$n_\perp \simeq n_\perp^{(0)} + \frac{1}{2}\begin{pmatrix} h_{xx} & h_{xy} \\ h_{yx} & h_{yy} \end{pmatrix} n_\perp^{(0)}\,. \tag{4.76}$$

If we now transform the matrix on the right-hand side to principal axes,

$$\frac{1}{2}\begin{pmatrix} h_{xx} & h_{xy} \\ h_{yx} & h_{yy} \end{pmatrix} = R\begin{pmatrix} \Omega & 0 \\ 0 & -\Omega \end{pmatrix} R^T\,, \tag{4.77}$$

and introduce the displacements (ξ, η) in the principle directions,

$$n_\perp =: R\begin{pmatrix} \xi \\ \eta \end{pmatrix}, \qquad n_\perp^{(0)} = R\begin{pmatrix} \xi_0 \\ \eta_0 \end{pmatrix}, \tag{4.78}$$

the motion is described by

$$\xi \simeq \xi_0 + \Omega(t)\xi_0\,,$$
$$\eta \simeq \eta_0 - \Omega(t)\eta_0\,. \tag{4.79}$$

Thus the point (ξ, η) performs a "quadrupole oscillation" about the point (ξ_0, η_0).

Consider the special case of a *periodic* plane wave propagating in the z-direction

$$h_{xx} = -h_{yy} = \mathrm{Re}\left\{A_+ \mathrm{e}^{-i\omega(t-z)}\right\},$$
$$h_{xy} = h_{yx} = \mathrm{Re}\left\{A_\times \mathrm{e}^{-i\omega(t-z)}\right\}. \tag{4.80}$$

If $A_\times = 0$, the principle axes coincide with the x- and y-axes. If $A_+ = 0$, the principle axes are rotated by 45°. For obvious reasons, we say that the wave is **linear polarized** in these two cases. If $A_\times = \pm iA_+$, the wave is said to be (right, respectively left) **circularly polarized**. For the upper sign, the principle axes rotate in the positive direction (counter clockwise for a wave approaching the reader), and in the negative direction for the lower sign. Note that the principal axes rotate at the frequency $\omega/2$.

Figure 4.1 shows the motion of a ring of test particles moving about a central particle in the transverse plane, for the two linearly polarized states.

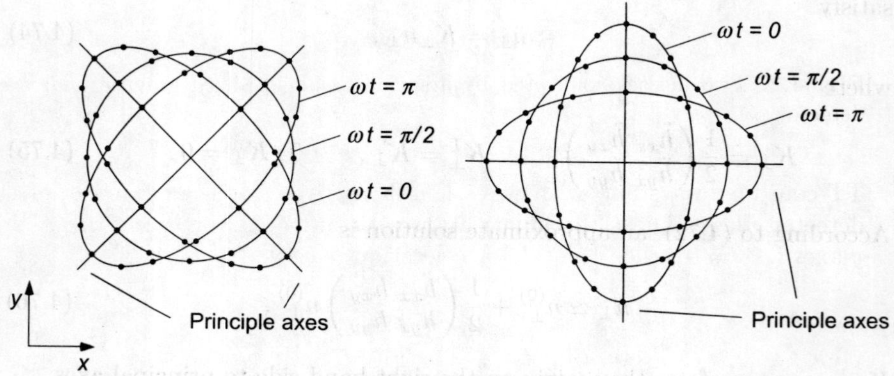

Fig. 4.1. Motion of a ring of test particles for the two linearly polarized states.

A Simple Mechanical Detector

Gravitational wave detectors are instruments that are highly sensitive to the extremely weak tidal forces induced by a gravitational wave. The best hopes for the first direct detections of gravitational radiation lies with large-scale interferometers. Several such detectors are now ready or will soon be operative. We shall say more on this later (see Sect. 4.6). At this point it may be helpful to discuss a simple idealized oscillator.

Consider two masses m connected by a massless spring of equilibrium length $2l_0$. We choose the orientation of the system as shown in Fig. 4.2. Let ω_0 be the frequency of the oscillator, and assume that the damping time τ_0 is much larger than $1/\omega_0$. If the separation of the two masses is $2(l_0 + \xi(t))$, then the relative separation $\xi(t)$ satisfies the differential equation

$$\ddot{\xi} + \frac{1}{\tau_0}\dot{\xi} + \omega_0^2 \xi = \text{tidal acceleration.} \tag{4.81}$$

The tidal acceleration can easily be obtained from our previous results. Consider an incoming periodic wave (4.80) propagating in the z-direction of a

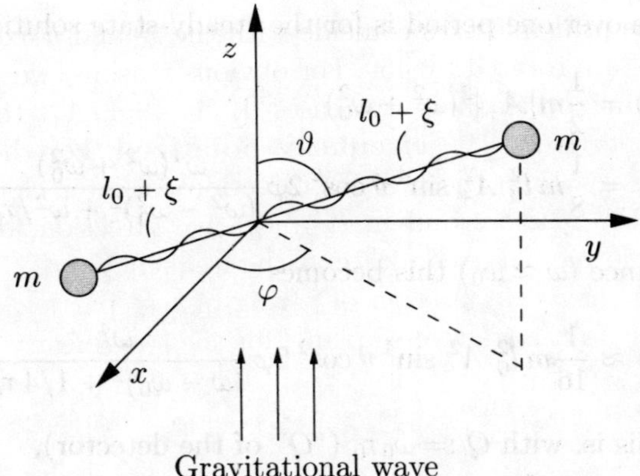

Fig. 4.2. Two masses connected by a spring as an idealized gravitational wave detector.

TT-coordinate system, with wave length much larger than the size of the detector. If we choose the polarization with principle axis along the x- and y-axes ($A_\times = 0$), then the tidal acceleration matrix $K_\perp$ in (4.75) is

$$K_\perp = -\frac{1}{2}\omega^2 \mathrm{Re}(A_+ \mathrm{e}^{-i\omega t}) \begin{pmatrix} 1 & 0 \\ 0 & -1 \end{pmatrix} .$$

The tidal acceleration on the right of (4.81) is therefore equal to

$$-\frac{1}{2}\omega^2 \mathrm{Re}(A_+ \mathrm{e}^{-i\omega t}) l_0 \left(\sin^2\vartheta \cos^2\varphi - \sin^2\vartheta \sin^2\varphi\right) ,$$

and the **driven oscillator equation** becomes

$$\ddot{\xi} + \frac{1}{\tau_0}\dot{\xi} + \omega_0^2 \xi = -\frac{1}{2}\omega^2 l_0 \mathrm{Re}(A_+ \mathrm{e}^{-i\omega t}) \sin^2\vartheta \cos 2\varphi . \tag{4.82}$$

The angular dependence $\cos 2\varphi$ of the driving force again reflects the spin-2 character. The factor $\sin^2\vartheta$ results from the transversal action of the gravitational wave.

The steady-state solution of (4.82) is

$$\xi = \mathrm{Re}(\mathcal{A}_+ \mathrm{e}^{-i\omega t}) , \tag{4.83}$$

with

$$\mathcal{A}_+ = \frac{1}{2} A_+ l_0 \sin^2\vartheta \cos 2\varphi \, \frac{\omega^2}{\omega^2 - \omega_0^2 + i\,\omega/\tau_0} . \tag{4.84}$$

The energy of the oscillator is

$$E = 2 \cdot \left(\frac{1}{2} m \, \dot{\xi}^2 + \frac{1}{2} m \omega_0^2 \, \xi^2 \right) .$$

Its time average over one period is for the steady-state solution

$$\begin{aligned}\langle E\rangle &= \frac{1}{2}m|\mathcal{A}_+|^2(\omega^2+\omega_0^2)\\ &= \frac{1}{8}m\,l_0^2\,A_+^2\sin^4\vartheta\cos^2 2\varphi\,\frac{\omega^4(\omega^2+\omega_0^2)}{(\omega^2-\omega_0^2)^2+\omega^2/\tau_0^2}\,. \end{aligned}\tag{4.85}$$

Near the resonance ($\omega \approx \omega_0$) this becomes

$$\langle E\rangle \approx \frac{1}{16}m\,l_0^2\,A_+^2\sin^4\vartheta\cos^2 2\varphi\,\frac{\omega_0^4}{(\omega-\omega_0)^2+1/4\,\tau_0^2}\,. \tag{4.86}$$

At resonance this is, with $Q := \omega_0\tau_0$ ("Q" of the detector),

$$\langle E\rangle_{res} = \frac{1}{4}m\,l_0^2\omega_0^2\,A_+^2\sin^4\vartheta\cos^2 2\varphi\,Q^2\,. \tag{4.87}$$

As always, a high quality factor Q enhances the response at resonance.

* * *

Exercise 4.8. Consider the following ansatz for a gravitational plane wave

$$g = -dt^2 + dx^2 + L^2\left(\mathrm{e}^{2\beta}\,dy^2 + \mathrm{e}^{-2\beta}\,dz^2\right), \tag{4.88}$$

where L and β are functions of $u := \frac{1}{\sqrt{2}}(x-t)$. Show that Einstein's vacuum equations are fulfilled if the following relation between $L(u)$ and $\beta(u)$ holds:

$$L'' + (\beta')^2 L = 0\,. \tag{4.89}$$

Hints. Introduce also the null coordinate $v := \frac{1}{\sqrt{2}}(x+t)$, and work relative to the tetrad

$$\theta^0 = du\,,\quad \theta^1 = dv\,,\quad \theta^2 = L\mathrm{e}^{\beta}\,dy^2\,,\quad \theta^3 = L\mathrm{e}^{-\beta}\,dz^2\,.$$

Show that the connection forms are given by

$$\omega^2{}_0 = \frac{1}{L}(L' + L\beta')\theta^2\,,\quad \omega^3{}_0 = \frac{1}{L}(L' - L\beta')\theta^3\,,$$

and all other $\omega^\mu{}_\nu = 0$.

A routine calculation then shows that only the component R_{00} of the Ricci tensor is non-vanishing, and is given by

$$R_{00} = -2\frac{1}{L}(L'' + \beta'^2 L)\,.$$

(For a detailed discussion of the class of plane wave solutions, see Sect. 35.9 of [13].)

4.4 Energy Carried by a Gravitational Wave

We have learned in Sect. 2.4 that in GR it is impossible to define energy- and momentum densities of a gravitational field. On the other hand, we have just seen that gravitational waves put energy into material systems (detectors) they pass through. Therefore, it should be possible to associate energy flux to a plane wave, at least in an *averaged* approximate sense.

A very physical way to arrive at such an expression is described in the text book [12]. We repeat here only the general idea; the details are fully given in [12].

Consider an array of idealized oscillator detectors, discussed at the end of the last section, nearly continuously distributed on a plane on which an incident plane wave is vertically impinging. After some transients the oscillators will carry out steady oscillations under the influence of the tidal forces of the wave. Thus the wave supplies energy to each oscillator, whose time average we have computed in (4.85). From this it is easy to obtain the time averaged energy put into the oscillators per unit surface and time. It is very reasonable to define this as the change $\delta\mathcal{F}$ of the net energy flux $\mathcal{F}$ of the wave.

On the other hand, each oscillator will radiate waves (see Sect. 4.5 below), which interfere with the primary wave. Exactly downstream this interference is destructive, and a calculation (familiar from electrodynamics) leads to a certain expression for the change δA of the amplitude A of the gravitational wave. The crucial result is that the ratio $\delta\mathcal{F}/\delta A$ is given by the following simple formula, in which the properties of the detector have dropped out:

$$\frac{\delta\mathcal{F}}{\delta A} = \frac{1}{16\pi G}\,\omega^2 A\,. \tag{4.90}$$

Here, ω is the circular frequency of the gravitational wave. Integration gives

$$\mathcal{F} = \frac{1}{32\pi G}\,\omega^2 A^2\,. \tag{4.91}$$

Now, A^2 is equal to the time average over one period of $h^{TT}_{\mu\nu}h^{TT\,\mu\nu}$. Denoting this average by an angle bracket, we thus obtain for the **time averaged energy flux** the simple expression

$$\mathcal{F} = \frac{1}{32\pi G}\,\omega^2\langle h^{TT}_{\mu\nu}h^{TT\,\mu\nu}\rangle = \frac{1}{32\pi G}\langle \dot{h}^{TT}_{\mu\nu}\dot{h}^{TT\,\mu\nu}\rangle\,. \tag{4.92}$$

In the remainder of this section we give a more formal justification of (4.92). We begin with a few qualitative remarks on the notion of a gravitational wave. This is an idealized approximate concept, as is the case for wave phenomena in other fields of physics. By "gravitational waves" we mean propagating ripples in curvature on scales much smaller than the characteristic scales of the background spacetime. A typical example is the time dependent field due to an aspheric core collapse of a massive star. This produces outgoing fluctuations in the gravitational field outside the star (supernova) on a background which is approximately described by the Schwarzschild metric. A good analogy to this are water waves as small ripples rolling on the ocean's surface with its large scale curvature.

Clearly, the separation into ripples and background cannot be precise, but nearly so, because of the widely different length scales involved. We proceed with a more precise discussion (see also Sect. 35 of [13]).

4.4.1 The Short Wave Approximation

For a gravitational wave propagating through a vacuum background with metric $\bar{g}$ we introduce two **length scales**: L denotes a typical curvature radius of the background, and λ a typical wavelength of the wave (see Fig. 4.3). We assume that $\lambda/L \ll 1$. For the decomposition of the full metric $g_{\mu\nu}$

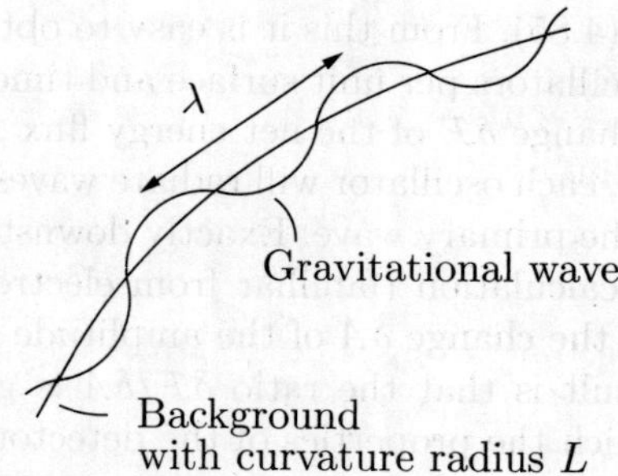

Fig. 4.3. Characteristic length scales L, λ for a gravitational wave.

into the background $\bar{g}_{\mu\nu}$ plus fluctuation $h_{\mu\nu}$,

$$g_{\mu\nu} = \bar{g}_{\mu\nu} + h_{\mu\nu}\,, \tag{4.93}$$

we assume that the magnitude ε of a typical amplitude of the ripples described by $h_{\nu\mu}$ is much smaller than 1. More precisely, we assume that there are coordinate systems (called "*steady*") such that

$$\begin{aligned} |h_{\mu\nu}| &\lesssim \varepsilon(\bar{g}_{\mu\nu}) \qquad (\varepsilon \ll 1)\,, \\ |\bar{g}_{\mu\nu,\lambda}| &\lesssim \frac{1}{L}(\bar{g}_{\mu\nu})\,, \\ |h_{\mu\nu,\lambda}| &\lesssim \frac{1}{\lambda}(h_{\mu\nu})\,, \end{aligned}$$

where the round brackets stand for "typical value of".

With this set up we expand the Ricci tensor in powers of h about the background $\bar{R}_{\mu\nu}$:

$$R_{\mu\nu} = \bar{R}_{\mu\nu} + R^{(1)}_{\mu\nu} + R^{(2)}_{\mu\nu} + \ldots\,. \tag{4.94}$$

Note that $R^{(1)}_{\mu\nu} \sim \varepsilon/\lambda^2, R^{(2)}_{\mu\nu} \sim \varepsilon^2/\lambda^2$, etc. The background contribution $\bar{R}_{\mu\nu}$ is a *non-linear* phenomenon: We expect that the energy density contained in the wave, averaged over regions much larger than λ but small in comparison to L, is of order $\frac{1}{G}(\varepsilon/\lambda)^2$. The Einstein equation then implies that the background curvature ($\sim L^{-2}$) is about G times this energy density, thus of the same order as $R^{(2)}_{\mu\nu}$. In certain situations the background may even be *entirely* due to the waves.

The linearized approximation of the field equations has already been discussed briefly in Sect. 4.1. Recall that

$$R^{(1)}_{\mu\nu} = \frac{1}{2}\left(-h_{\mu\nu|\alpha}{}^{\alpha} + h_{\alpha\mu|\nu}{}^{\alpha} + h_{\alpha\nu|\mu}{}^{\alpha} - h_{|\mu\nu}\right) . \tag{4.95}$$

In vacuum this has to vanish

$$R^{(1)}_{\mu\nu}[h] = 0 , \tag{4.96}$$

since $R^{(1)}_{\mu\nu}$ is the only linear term in (4.94). In second order ($\sim \varepsilon^2/\lambda^2$) the vacuum field equation implies

$$\bar{R}_{\mu\nu} = -R^{(2)}_{\mu\nu} . \tag{4.97}$$

At this point we introduce averages over four-dimensional spacetime regions that are much larger than λ, but small in comparison to L. Denoting these by $\langle\cdots\rangle$, we can split (4.97) into two equations, namely an averaged (smooth) part

$$\bar{R}_{\mu\nu} = -\langle R^{(2)}_{\mu\nu}\rangle \tag{4.98}$$

and a fluctuation part

$$R^{(2)}_{\mu\nu} - \langle R^{(2)}_{\mu\nu}\rangle = 0 . \tag{4.99}$$

The slowly varying part (4.98) is equivalent to the following equation for the background Einstein tensor,

$$\bar{G}_{\mu\nu} = 8\pi G\, T^{(GW)}_{\mu\nu} , \tag{4.100}$$

where

$$T^{(GW)}_{\mu\nu} = -\frac{1}{8\pi G}\left(\langle R^{(2)}_{\mu\nu}\rangle - \frac{1}{2}\bar{g}_{\mu\nu}\langle R^{(2)}\rangle\right) . \tag{4.101}$$

This equation for the background suggests that $T^{(GW)}_{\mu\nu}$ has to be regarded as the **effective energy-momentum tensor** for the gravitational wave. For high frequency waves this notion is well-defined in an averaged sense. It looses meaning as soon as λ is no more much smaller than L. (Analogous averaged quantities are used at many places in physics, for instance in the electrodynamics of media.)

4.4.2 Discussion of the Linearized Equation $R^{(1)}_{\mu\nu}[h] = 0$

We know from Sect. 4.1 that under a gauge transformation

$$h_{\mu\nu} \longrightarrow h_{\mu\nu} + \xi_{\mu|\nu} + \xi_{\nu|\mu} \tag{4.102}$$

$R^{(1)}_{\mu\nu}$ transforms as

$$R^{(1)}_{\mu\nu} \longrightarrow R^{(1)}_{\mu\nu} + L_\xi \bar{R}_{\mu\nu} \,. \tag{4.103}$$

Therefore, the linearized equation $R^{(1)}_{\mu\nu}[h] = 0$ is only gauge invariant if the background Ricci tensor $\bar{R}_{\mu\nu}$ vanishes. Because of (4.98) this is not the case, but $\bar{R}_{\mu\nu}$ is of higher order than $R^{(1)}_{\mu\nu}$. Moreover, since $L_\xi \bar{R}_{\mu\nu}$ involves partial derivatives, the violation of gauge invariance is very small. Up to such small terms which we can safely ignore, we can thus impose gauge conditions on $h_{\mu\nu}$, while keeping (4.96). In the *ideal case* when we consider the propagation equation (4.96) on a fixed background with vanishing Ricci curvature, the following discussion will be exact.

In a fist step we impose the **Hilbert gauge condition**:

$$\gamma_{\mu\nu}{}^{|\nu} = 0 \,, \tag{4.104}$$

where

$$\gamma_{\mu\nu} = h_{\mu\nu} - \frac{1}{2}\bar{g}_{\mu\nu} h \qquad (h = h^\lambda{}_\lambda) \,. \tag{4.105}$$

This can always be enforced. Indeed, $\gamma_{\mu\nu}{}^{|\nu}$ transforms under (4.102) as (use the Ricci identity)

$$\gamma_\mu{}^\nu{}_{|\nu} \longrightarrow \gamma_\mu{}^\nu{}_{|\nu} + \xi_\mu{}^{|\nu}{}_{|\nu} - \bar{R}_{\mu\nu}\xi^\nu \,. \tag{4.106}$$

Hence, if the ξ_μ are chosen such that

$$\xi_\mu{}^{|\nu}{}_{|\nu} - \bar{R}_{\mu\nu}\xi^\nu = -\gamma_\mu{}^\nu{}_{|\nu} \,,$$

then the transformed $\gamma_{\mu\nu}$ satisfies (4.104). But this hyperbolic system of partial differential equations has locally always solutions.

Now we write equation (4.96), i.e.

$$-h_{\mu\nu|\alpha}{}^{|\alpha} + h_{\alpha\mu|\nu}{}^{|\alpha} + h_{\alpha\nu|\mu}{}^{|\alpha} - h_{|\mu\nu} = 0 \,, \tag{4.107}$$

in terms of $\gamma_{\mu\nu}$, and impose (4.104). A straightforward calculation (see Exercise 4.9) leads to

$$\gamma_{\mu\nu|\alpha}{}^{|\alpha} + 2\bar{R}_{\alpha\mu\beta\nu}\gamma^{\alpha\beta} - \bar{R}_{\alpha\mu}\gamma^\alpha{}_\nu - \bar{R}_{\alpha\nu}\gamma^\alpha{}_\mu = 0 \,. \tag{4.108}$$

Here, we ignore the last two terms because they are of order ε^2/λ^2, while the leading order of (4.107) is ε/λ^2. This leads us to the **propagation equation**

$$\gamma_{\mu\nu|\alpha}{}^{|\alpha} + 2\bar{R}_{\alpha\mu\beta\nu}\gamma^{\alpha\beta} = 0 \tag{4.109}$$

within the Hilbert gauge class. Note again, that this is exact in the idealized limit mentioned earlier.

Is it possible to require also $h = 0$, as for the Minkowski background? To answer this question we first note that the Hilbert condition is preserved if ξ_μ satisfy

$$\xi_\mu{}^{|\nu}{}_{|\nu} - \bar{R}_{\mu\nu}\xi^\nu = 0\,. \tag{4.110}$$

It will be useful to write this equation in terms of differential forms. Let $\xi = \xi_\mu dx^\mu$ and let $\Box = \delta \circ d + d \circ \delta$ be the wave operator for differential forms with respect to the background metric, then (4.110) is equivalent to

$$\Box\xi = 0\,. \tag{4.111}$$

Since $\gamma := \gamma^\lambda_\lambda \longrightarrow \gamma - 2\xi_\mu{}^{|\mu}$, γ can be made equal to zero within the Hilbert gauge, if the 1-form ξ can be chosen such that it satisfies (4.111) as well as $2\delta\xi = \gamma$. For this to be possible, the following consistency condition

$$\Box\gamma = 2\Box\delta\xi = 2\delta\Box\xi = 0$$

must be satisfied. This is not strictly consistent with (4.108), which implies (taking the trace)

$$\Box\gamma + 2\bar{R}_{\alpha\beta}\gamma^{\alpha\beta} = 0\,. \tag{4.112}$$

However, the consistency condition is violated by very small terms. (For the idealized case there is again no inconsistency.)

Adopting the argument in Sect. 4.3 for weak fields, we now show conversely, that the condition $\Box\gamma = 0$ is also sufficient. In other words, we prove the

Lemma 4.1. *Given a function ϕ with $\Box\phi = 0$, then there exists locally a 1-form ξ with*

$$\Box\xi = 0\,, \qquad \delta\xi = \phi\,. \tag{4.113}$$

Proof. Choose first a 1-form η with $\delta\eta = \phi$. This is always possible: Take $\eta = d\Lambda$ of a function Λ which satisfies $\Box\Lambda = \phi$. Next, let $\zeta = \Box\eta$ and note that $\delta\zeta = \Box\delta\eta = \Box\phi = 0$. Hence, by Poincaré's lemma there exists locally a 2-form φ with $\zeta = \delta\varphi$. Now, choose a 2-form σ such that $\Box\sigma = \varphi$. We claim that $\xi := \eta - \delta\sigma$ satisfies the equations (4.113). Indeed, $\delta\xi = \delta\eta = \phi$, and $\Box\xi = \Box\eta - \delta\Box\sigma = \zeta - \delta\varphi = \zeta - \zeta = 0$. □

Summarizing, to sufficient accuracy we can use (4.94) for the Hilbert gauge class, and impose – up to higher order inconsistencies – in addition the gauge condition $\gamma = 0$.

4.4.3 Averaged Energy-Momentum Tensor for Gravitational Waves

A direct calculation of (4.101) is very involved (see [121]). It is much simpler to compute the right-hand side $\langle G^{(2)}_{\mu\nu}\rangle$ in (4.101) with the help of the identity (2.223) which relates the Einstein tensor to the Landau-Lifshitz pseudotensor. We take the average of the quadratic part in $h_{\mu\nu}$ of this equation:

$$\left\langle \{(-g)(G^{\mu\nu} + 8\pi G\, t^{\mu\nu}_{LL})\}^{(2)} \right\rangle = \langle \text{divergence} \rangle . \tag{4.114}$$

In evaluating averages we can make use of the following rules. First, we can ignore averages of spacetime derivatives, because these can be transformed to surface integrals, whereby the order with respect to $1/\lambda$ is increased by 1. For this reason we can "integrate by parts": $\langle AB_{,\alpha}\rangle = -\langle A_{,\alpha}B\rangle$. Furthermore, we can leave out terms which (possibly after partial integration) vanish as a result of the linearized equation (4.96).

In a first step we obtain from (4.114)

$$\left\langle \{(-g)G^{\mu\nu}\}^{(2)} \right\rangle = -8\pi G \left\langle \{(-g)t^{\mu\nu}_{LL}\} \right\rangle . \tag{4.115}$$

Using the first order equation $G^{\mu\nu(1)} = 0$, the left-hand side is equal to $(-\bar{g})\langle G^{\mu\nu(2)}\rangle$. Turning to the right-hand side, we recall that $t^{\mu\nu}_{LL}$ is quadratic in the first derivatives of $\tilde{g}^{\mu\nu}$ (see (2.215)). In order to determine the quadratic part in $h_{\mu\nu}$ we need

$$g^{\mu\nu} = \bar{g}^{\mu\nu} - h^{\mu\nu} + O(h^2) , \quad \tilde{g}^{\mu\nu} = \sqrt{-\bar{g}}\,\bar{g}^{\mu\nu} - \sqrt{-\bar{g}}\left(h^{\mu\nu} - \frac{1}{2}\bar{g}^{\mu\nu} h\right) + O(h^2) , \tag{4.116}$$

thus the first order deviation of $\tilde{g}^{\mu\nu}$ from the background is equal to $-\sqrt{-\bar{g}}\,\gamma^{\mu\nu}$. The leading contribution ($\sim \varepsilon^2/\lambda^2$) inside the average on the right of (4.115) is obtained by replacing in (2.215) $\tilde{g}^{\alpha\beta}{}_{,\mu}$ by $-\sqrt{-\bar{g}}\,\gamma^{\alpha\beta}{}_{,\mu}$, and the other metric factors (without derivatives) by the background metric. This gives

$$\left\langle G^{\mu\nu(2)} \right\rangle = -8\pi G \left\langle t^{\mu\nu(2)} \right\rangle ,$$

and thus, as expected,

$$T^{\alpha\beta}_{(GW)} = \left\langle t^{\alpha\beta(2)}_{LL} \right\rangle . \tag{4.117}$$

Moreover, to leading order,

$$\begin{aligned} 16\pi G\, t^{\alpha\beta(2)} = {} & \gamma^{\alpha\beta}{}_{,\lambda}\gamma^{\lambda\mu}{}_{,\mu} - \gamma^{\alpha\lambda}{}_{,\lambda}\gamma^{\beta\mu}{}_{,\mu} + \frac{1}{2}\bar{g}^{\alpha\beta}\bar{g}_{\lambda\mu}\gamma^{\lambda\nu}{}_{,\rho}\gamma^{\rho\mu}{}_{,\nu} \\ & - \left(\bar{g}^{\alpha\lambda}\bar{g}_{\mu\nu}\gamma^{\beta\nu}{}_{,\rho}\gamma^{\mu\rho}{}_{,\lambda} + \bar{g}^{\beta\lambda}\bar{g}_{\mu\nu}\gamma^{\alpha\nu}{}_{,\rho}\gamma^{\mu\rho}{}_{,\lambda}\right) \\ & + \bar{g}_{\lambda\mu}\bar{g}^{\nu\rho}\gamma^{\alpha\lambda}{}_{,\nu}\gamma^{\beta\mu}{}_{,\rho} + \frac{1}{8}\left(2\bar{g}^{\alpha\lambda}\bar{g}^{\beta\mu} - \bar{g}^{\alpha\beta}\bar{g}^{\lambda\mu}\right) \\ & \times \left(2\bar{g}_{\nu\rho}\bar{g}_{\sigma\tau} - \bar{g}_{\rho\sigma}\bar{g}_{\nu\tau}\right)\gamma^{\nu\tau}{}_{,\lambda}\gamma^{\rho\sigma}{}_{,\mu} . \end{aligned} \tag{4.118}$$

We can replace everywhere the partial derivatives $\gamma^{\mu\nu}{}_{,\rho}$ by covariant $\gamma^{\mu\nu}{}_{|\rho}$ with respect to the background metric, because the relative difference is of order $\lambda/L \ll 1$. So, again to leading order,

$$\begin{aligned}16\pi G\, t_{LL}^{\alpha\beta(2)} &= \gamma^{\alpha\beta}{}_{|\lambda}\gamma^{\lambda\mu}{}_{|\mu} - \gamma^{\alpha\lambda}{}_{|\lambda}\gamma^{\beta\mu}{}_{|\mu} + \frac{1}{2}\bar{g}^{\alpha\beta}\gamma_{\mu}{}^{\nu}{}_{|\rho}\gamma^{\rho\mu}{}_{|\nu} \\ &-\gamma^{\beta\nu}{}_{|\rho}\gamma_{\nu}{}^{\rho|\alpha} + \gamma^{\alpha\nu}{}_{|\rho}\gamma_{\nu}{}^{\rho|\beta} + \gamma^{\alpha}{}_{\mu|\nu}\gamma^{\beta\mu|\nu} + \frac{1}{2}\gamma_{\rho\sigma}{}^{|\alpha}\gamma^{\rho\sigma|\beta} \\ &-\frac{1}{4}\gamma^{|\alpha}\gamma^{|\beta} - \frac{1}{4}\bar{g}^{\alpha\beta}\left(\gamma^{\mu\nu|\rho}\gamma_{\mu\nu|\rho} - \frac{1}{2}\gamma^{|\rho}\gamma_{|\rho}\right). \end{aligned} \tag{4.119}$$

(Indices of $\gamma^{\alpha\beta}$ and $\gamma^{\alpha\beta}{}_{|\mu}$ are always raised and lowered with the background metric.)

Further simplifications can be made for the average of (4.119) by using the first order equation (4.96), after some partial integrations. Note that partial integrations under $\langle\cdots\rangle$ can also be done with respect to covariant derivatives. Since we kept in (4.119) only the leading contributions of order ε^2/λ^2, we can replace (4.96) by any equation that differs by terms which are a positive power of λ/L times smaller. We can thus use (4.112), if the Hilbert condition is imposed. We may even leave out the curvature term in (4.112), because this induces only a fractional error of order $(\lambda/L)^2$. On the right-hand side of (4.119) we can also interchange covariant derivatives, making again a fractional error of order $(\lambda/L)^2$. With this and the leading order of the wave equation

$$\gamma_{\mu\nu|\alpha}{}^{|\alpha} = 0, \tag{4.120}$$

we have, for example, to leading order

$$\left\langle\gamma^{\alpha}{}_{\mu|\nu}\gamma^{\beta\mu|\nu}\right\rangle = -\left\langle\gamma^{\alpha}{}_{\mu}\gamma^{\beta\mu|\nu}{}_{|\nu}\right\rangle = 0\,.$$

Similarly, using the Hilbert condition, we get

$$\left\langle\bar{g}^{\alpha\beta}\gamma_{\mu}{}^{\nu}{}_{|\rho}\gamma^{\rho\mu}{}_{|\nu}\right\rangle = -\left\langle\bar{g}^{\alpha\beta}\gamma_{\mu}{}^{\nu}\gamma^{\rho\mu}{}_{|\nu\rho}\right\rangle = -\left\langle\bar{g}^{\alpha\beta}\gamma_{\mu}{}^{\nu}\gamma^{\rho\mu}{}_{|\rho\nu}\right\rangle = 0\,.$$

In this way we can considerably simplify the average of (4.119). For the Hilbert gauge class we can immediately read off that

$$\left\langle t_{LL}^{\alpha\beta(2)}\right\rangle = \frac{1}{32\pi G}\left\langle\gamma_{\mu\nu}{}^{|\alpha}\gamma^{\mu\nu|\beta} - \frac{1}{2}\gamma^{|\alpha}\gamma^{|\beta}\right\rangle. \tag{4.121}$$

According to (4.117) the effective energy-momentum tensor for gravitational waves is thus

$$T_{\alpha\beta}^{(GW)} = \frac{1}{32\pi G}\left\langle\gamma_{\mu\nu|\alpha}\gamma^{\mu\nu}{}_{|\beta} - \frac{1}{2}\gamma_{|\alpha}\gamma_{|\beta}\right\rangle. \tag{4.122}$$

Note that in leading order $T_{\alpha\beta}^{(GW)}$ is trace-free.

Beside the Hilbert condition, we can impose (in leading order) the additional gauge condition $\gamma = 0$. Then $\gamma_{\mu\nu} = h_{\mu\nu}$, so

$$T^{(GW)}_{\alpha\beta} = \frac{1}{32\pi G}\left\langle h_{\mu\nu|\alpha} h^{\mu\nu}{}_{|\beta}\right\rangle \qquad \text{for } h^{\mu\nu}{}_{|\nu} = 0\,,\ \ h = 0\,. \tag{4.123}$$

For a weak plane wave this gives the result (4.92) that was obtained in a completely different manner. We shall apply (4.122) or (4.123) only for weak gravitational waves on a flat background with $\varepsilon \lesssim 10^{-21}$, say.

Effective Energy-Momentum Tensor for a Plane Wave

As an important application of (4.122) we consider a weak plane wave $h_{\mu\nu}(x^1 - t)$ propagating in the x^1-direction. We impose the Hilbert condition $\gamma_\mu{}^\nu{}_{,\nu} = 0$, which is equivalent to

$$(\gamma_\mu{}^0)' = (\gamma_\mu{}^1)'\,. \tag{4.124}$$

According to (4.122) the only non-vanishing components of $T^{(GW)}_{\alpha\beta}$ are

$$\begin{aligned} T^{(GW)}_{00} &= T^{(GW)}_{11} = -T^{(GW)}_{01} = -T^{(GW)}_{10} \\ &= \frac{1}{32\pi G}\left\langle (\gamma_{\mu\nu})'(\gamma_{\mu\nu})' - \frac{1}{2}(\gamma')^2\right\rangle . \end{aligned}$$

By making use of (4.124) one readily finds

$$\begin{aligned} T^{(GW)}_{00} &= -T^{(GW)}_{01} = \frac{1}{32\pi G}\left\langle 2(\gamma'_{23})^2 + \frac{1}{2}(\gamma'_{22} - \gamma'_{33})^2\right\rangle \\ &= \frac{1}{32\pi G}\left\langle 2(h'_{23})^2 + \frac{1}{2}(h'_{22} - h'_{33})^2\right\rangle . \end{aligned} \tag{4.125}$$

In the solved Exercise 4.11 we shall obtain this result more directly. (Recall that (4.122) is based on the expression (2.215) for the Landau-Lifshitz pseudotensor, which was the result of rather heavy calculations.) For a harmonic wave (4.80) this gives

$$T^{00}_{(GW)} = T^{01}_{(GW)} = \frac{1}{32\pi G}\,\omega^2\left(|A_+|^2 + |A_\times|^2\right) . \tag{4.126}$$

It is not difficult to generalize (4.122) to an arbitrary gauge. Instead of (4.120) we then use

$$\gamma_{\mu\nu|\alpha}{}^{|\alpha} + \bar{g}_{\mu\nu}\gamma^{\alpha\beta}{}_{|\alpha\beta} - \gamma_{\alpha\mu}{}^{|\alpha}{}_{|\nu} - \gamma_{\alpha\nu}{}^{|\alpha}{}_{|\mu} = 0\,. \tag{4.127}$$

This is obtained from (4.107) by freely commuting covariant derivatives; compare with (4.13). Proceeding as before, one finds (see Exercise 4.10)

$$T^{(GW)}_{\alpha\beta} = \frac{1}{32\pi G}\left\langle \gamma_{\mu\nu|\alpha}\gamma^{\mu\nu}{}_{|\beta} - \frac{1}{2}\gamma_{|\alpha}\gamma_{|\beta} - \gamma^{\mu\nu}{}_{|\nu}(\gamma_{\mu\alpha|\beta} + \gamma_{\mu\beta|\alpha})\right\rangle . \quad (4.128)$$

* * *

Exercises.

Exercise 4.9. Derive (4.108) from (4.107) and the Hilbert condition (4.104).

Exercise 4.10. Derive (4.128), valid for any gauge.

Hints. For the simplification of (4.119) we use in leading order (4.127), and we are free to interchange the order of covariant derivatives. This equation implies

$$\gamma_{|\rho}{}^{|\rho} + 2\gamma^{\rho\sigma}{}_{|\rho\sigma} = 0 . \quad (4.129)$$

Consider, for example, the three terms in (4.119) proportional to $\bar{g}^{\alpha\beta}$:

$$\frac{1}{2}\gamma_{\mu}{}^{\nu}{}_{|\rho}\gamma^{\rho\mu}{}_{|\nu} - \frac{1}{4}\left(\gamma^{\mu\nu|\rho}\gamma_{\mu\nu|\rho} - \frac{1}{2}\gamma^{|\rho}\gamma_{|\rho}\right) .$$

Under the average this sum can be replaced , with partial integrations and use of (4.124), by

$$-\frac{1}{8}\gamma\gamma_{|\rho}{}^{|\rho} + \frac{1}{4}\gamma^{\mu\nu}\left(\gamma_{\mu\nu|\rho}{}^{|\rho} - \gamma_{\mu\rho|\nu}{}^{|\rho} - \gamma_{\nu\rho|\mu}{}^{|\rho}\right)$$
$$= -\frac{1}{8}\gamma\gamma_{|\rho}{}^{|\rho} + \frac{1}{4}\gamma^{\mu\nu}\bar{g}_{\mu\nu}\gamma^{\rho\sigma}{}_{|\rho\sigma} = -\frac{1}{8}\gamma\gamma_{|\rho}{}^{|\rho} - \frac{1}{4}\gamma\gamma^{\rho\sigma}{}_{|\rho\sigma} = 0 ,$$

according to (4.127). Similar manipulations lead to (4.125).

Exercise 4.11 (Energy and momentum of a plane wave). Consider in linearized theory a not necessarily harmonic wave $h_{\mu\nu}(x^1 - t)$, propagating in the x^1-direction. Show that it is possible to find a gauge within the Hilbert gauge class such that at most h_{23} and $h_{22} = -h_{33}$ do not vanish (TT gauge). Compute the energy-momentum complex $t^{\mu\nu}_{LL}$ in this particular gauge by using formula (2.197) for $*t^{\alpha}_{LL}$.

Solution. The Hilbert condition $\gamma^{\nu}{}_{\mu,\nu} = 0$ implies $(\gamma^0{}_\mu)' = (\gamma^1{}_\mu)'$. Apart from an irrelevant integration constant (we are only interested in time-dependent fields), this equation is also valid for the $\gamma_\mu{}^\nu$. Hence

$$\gamma^0{}_0 = \gamma^1{}_0 , \gamma^0{}_1 = \gamma^1{}_1 , \gamma^0{}_2 = \gamma^1{}_2 , \gamma^0{}_3 = \gamma^1{}_3$$

and thus $\gamma^0{}_0 = -\gamma^1{}_1$. The independent components are therefore γ_{00} , γ_{02} , γ_{03}, γ_{22} , γ_{23} and γ_{33}.

By introducing a gauge transformation with a vector field of the form $\xi^\mu(x^1 - t)$, so that $\Box\xi^\mu = 0$ and $\xi^\lambda{}_{,\lambda} = \xi_0' + \xi_1'$, the independent components transform as

$$\gamma_{00} \longrightarrow \gamma_{00} - \xi_0' + \xi_1' , \quad \gamma_{02} \longrightarrow \gamma_{02} - \xi_2' , \quad \gamma_{03} \longrightarrow \gamma_{03} - \xi_3' ,$$
$$\gamma_{22} \longrightarrow \gamma_{22} - \xi_0' - \xi_1' , \quad \gamma_{23} \longrightarrow \gamma_{23} , \quad \gamma_{33} \longrightarrow \gamma_{33} - \xi_0' - \xi_1' . \tag{4.130}$$

Among these, γ_{23} and the combination $\gamma_{22} - \gamma_{33}$ remain invariant. Obviously, it is possible to make γ_{00}, γ_{02}, γ_{03}, and $\gamma_{22} + \gamma_{33}$ vanish by a suitable choice of ξ_μ. Then $\gamma = 0$ and thus $\gamma_{\alpha\beta} = h_{\alpha\beta}$. Moreover,

$$(h_{\alpha\beta}) = \left(\begin{array}{cc|cc} 0 & 0 & 0 & 0 \\ 0 & 0 & 0 & 0 \\ \hline 0 & 0 & h_{22} & h_{23} \\ 0 & 0 & h_{23} & -h_{22} \end{array}\right) .$$

We now compute the energy-momentum 3-forms (2.197) in this gauge. We shall perform the calculation with respect to an orthonormal basis. In the linearized theory, such a basis has the form

$$\theta^\alpha = dx^\alpha + \varphi^\alpha{}_\beta \, dx^\beta , \tag{4.131}$$

where

$$\varphi_{\alpha\beta} + \varphi_{\beta\alpha} = h_{\alpha\beta} . \tag{4.132}$$

Equation (4.132) determines only the symmetric part of $\varphi_{\alpha\beta}$. Under infinitesimal local Lorentz transformations, $\varphi_{\alpha\beta}$ is changed by an antisymmetric contribution. We may thus choose $\varphi_{\alpha\beta}$ to be symmetric, i.e., $\varphi_{\alpha\beta} = \frac{1}{2} h_{\alpha\beta}$. Since

$$d\theta^\alpha = \varphi^\alpha{}_{\beta,\gamma} \, dx^\gamma \wedge dx^\beta \simeq \varphi^\alpha{}_{\beta,\gamma} \, \theta^\gamma \wedge \theta^\beta$$

one finds, to first order in $\varphi_{\alpha\beta}$, for the connection forms

$$\omega^\alpha{}_\beta = (\varphi^\alpha{}_{\gamma,\beta} - \varphi_{\beta\gamma}{}^{,\alpha}) \, dx^\gamma . \tag{4.133}$$

For our special plane wave, the $\varphi_{\alpha\beta}$ are functions only of $u = x^1 - t$, and hence

$$\varphi_{\alpha\beta,\gamma} = \varphi'_{\alpha\beta} n_\gamma , \qquad (n_\gamma) := (-1, 1, 0, 0) .$$

Therefore,

$$\omega_{\alpha\beta} = \phi_\alpha n_\beta - \phi_\beta n_\alpha , \quad \phi_\alpha := \varphi'_{\alpha\gamma} \, dx^\gamma . \tag{4.134}$$

Since only φ_{23} and $\varphi_{22} = -\varphi_{33}$ do not vanish, we note that

$$n^\alpha \phi_\alpha = 0 , \qquad n^\alpha n_\alpha = 0 . \tag{4.135}$$

From this it follows that $\omega_{\alpha\beta} \wedge \omega^\alpha{}_\gamma = 0$ and the first term of the Landau-Lifshitz forms (2.197),

$$*t^\alpha_{LL} = -\frac{1}{16\pi G} \eta^{\alpha\beta\gamma\delta} \left(\omega_{\alpha\beta} \wedge \omega^\sigma{}_\gamma \wedge \theta_\delta - \omega_{\beta\gamma} \wedge \omega_{\sigma\delta} \wedge \theta^\sigma \right) ,$$

vanishes. Using this and (4.134), we obtain

$$*t^{\alpha}_{LL} = \frac{1}{16\pi G}\, 2\,\eta^{\alpha\beta\gamma\delta}\phi_\beta n_\gamma \wedge (\phi_\sigma n_\delta - \phi_\delta n_\sigma)\wedge\theta^\sigma$$
$$= -\frac{1}{8\pi G}\,\eta^{\alpha\beta\gamma\delta}\phi_\beta\wedge\phi_\delta\wedge n_\gamma n_\sigma\theta^\sigma\,,$$

so that

$$*t^{\alpha}_{LL} = -\frac{1}{8\pi G}\, n_\gamma\eta^{\alpha\beta\gamma\delta}\phi_\beta\wedge\phi_\delta\wedge du\,. \tag{4.136}$$

In particular, we have for $\alpha = 0$:

$$*t^0_{LL} = \frac{1\cdot 2}{8\pi G}\, n_1\eta^{0213}\,\phi_2\wedge\phi_3\wedge du = +\frac{1}{4\pi G}\,\phi_2\wedge\phi_3\wedge du$$
$$= -\frac{1}{4\pi G}(\varphi'_{22}\,dx^2 + \varphi'_{23}\,dx^3)\wedge(\varphi'_{32}\,dx^2 + \varphi'_{33}\,dx^3)\wedge du$$
$$= -\frac{1}{4\pi G}\left(\varphi'_{22}\varphi'_{33} - (\varphi'_{23})^2\right)dx^2\wedge dx^3\wedge du\,.$$

This gives, with $\theta^\mu\wedge *t^\alpha = t^{\alpha\mu}\eta$,

$$t^{00}_{LL} = t^{01}_{LL} = \frac{1}{16\pi G}\left((h'_{23})^2 + \frac{1}{4}(h'_{22} - h'_{33})^2\right)\,. \tag{4.137}$$

Here we used $h'_{22} = -h'_{33}$. These expressions are invariant under a gauge transformations of the type (4.130). One obtains the same result relative to the coordinate basis dx^α (show this). After taking averages, this agrees with (4.125).

4.5 Emission of Gravitational Radiation

In this section we investigate, in the framework of the linearized theory, the energy radiated in the form of gravitational waves by time-dependent sources. For this we need the retarded solution (4.29) in the Hilbert gauge

$$\gamma_{\mu\nu}(t,\boldsymbol{x}) = 4G\int\frac{T_{\mu\nu}(t - |\boldsymbol{x}-\boldsymbol{x}'|,\boldsymbol{x}')}{|\boldsymbol{x}-\boldsymbol{x}'|}\,d^3x' \tag{4.138}$$

at large distances from the localized sources.

Far away from the matter sources we can replace the denominator of (4.138) by $|\boldsymbol{x}| \equiv r$, while the retarded time can be approximated by

$$t - |\boldsymbol{x}-\boldsymbol{x}'| \simeq t - r + \boldsymbol{x}'\cdot\hat{\boldsymbol{x}}\,, \quad \hat{\boldsymbol{x}} := \boldsymbol{x}/r\,.$$

The asymptotic field is thus

$$\gamma^{\mu\nu}(t,\boldsymbol{x}) = \frac{4G}{r}\int T^{\mu\nu}(t - r + \hat{\boldsymbol{x}}\cdot\boldsymbol{x}',\boldsymbol{x}')\,d^3x'\,. \tag{4.139}$$

4.5.1 Slow Motion Approximation

Later we shall calculate the energy loss for arbitrary time variations of $T^{\mu\nu}$, but let us first study the case when the time variation of $T^{\mu\nu}$ is sufficiently slow that we can approximate the retarded time in (4.139) by

$$\gamma^{\mu\nu}(t,\boldsymbol{x}) = \frac{4G}{r}\int T^{\mu\nu}(t-r,\boldsymbol{x}')\,d^3x' \,. \tag{4.140}$$

For the calculation of the energy flux we only need (4.140) for space-space indices. Then the integral can be conveniently transformed by making use of $T^{\mu\nu}{}_{,\nu} = 0$ (in linearized approximation). We claim that this zeroth order conservation law implies the relation

$$\int T^{kl} d^3x = \frac{1}{2}\frac{\partial^2}{\partial t^2}\int T^{00}x^k x^l\, d^3x \,. \tag{4.141}$$

To show this we start from

$$0 = \int x^k \partial_\nu T^\nu{}_\mu\, d^3x = \frac{\partial}{\partial t}\int x^k T^0{}_\mu\, d^3x + \int x^k \partial_l T^l{}_\mu\, d^3x \,.$$

Integrating the last term by parts gives

$$\int T^k{}_\mu\, d^3x = \frac{\partial}{\partial t}\int T^0{}_\mu x^k\, d^3x \,. \tag{4.142}$$

Making use of Gauss' theorem we also have

$$\begin{aligned}\frac{\partial}{\partial t}\int T^{00}x^k x^l\, d^3x &= \int \partial_\nu (T^{\nu 0}x^k x^l)\, d^3x = \int T^{\nu 0}\partial_\nu (x^k x^l) d^3x \\ &= \int (T^{k0}x^l + T^{l0}x^k)\, d^3x \end{aligned} \tag{4.143}$$

Combining (4.142) and (4.143) gives (4.141):

$$\frac{1}{2}\frac{\partial^2}{\partial t^2}\int T^{00}x^k x^l\, d^3x = \frac{1}{2}\frac{\partial}{\partial t}\int (T^{k0}x^l + T^{l0}x^k)\, d^3x = \int T^{kl}\, d^3x \,.$$

We consider only nearly Newtonian sources for which the proper energy density is dominated by the matter density ρ: $T^{00} \simeq \rho$. Then we obtain from (4.140) and (4.141)

$$\gamma^{kl}(t,\boldsymbol{x}) \simeq \frac{2G}{r}\frac{\partial^2}{\partial t^2}\int \rho(t-r,\boldsymbol{x}')x'^k x'^l\, d^3x' \,. \tag{4.144}$$

We can write this in terms of the trace-free **quadrupole tensor**[4]

[4] Note that some authors use a different normalization (e.g., 1/3 of (4.145)).

$$Q_{kl}(t) = \int (3x^k x^l - r^2\delta_{kl})\rho(t, \boldsymbol{x})\, d^3x \tag{4.145}$$

as

$$\gamma_{kl}(t, \boldsymbol{x}) = \frac{2G}{3}\frac{1}{r}\left(\frac{\partial^2}{\partial t^2}Q_{kl}(t-r) + \delta_{kl}\frac{\partial^2}{\partial t^2}\int r'^2\rho(t-r, \boldsymbol{x}')\, d^3x'\right). \tag{4.146}$$

This will now be used to compute the average energy flux. At large distances from the source, the wave (4.146) can be considered 'locally' to be a plane wave. Consider first the x^1-direction. According to (4.125) the energy flux in this direction is

$$T^{01}_{(GW)} = \frac{1}{32\pi G}\left\langle 2(\dot\gamma_{23})^2 + \frac{1}{2}(\dot\gamma_{22} - \dot\gamma_{33})^2\right\rangle. \tag{4.147}$$

Here the second term in (4.146) does not contribute to the radiation, while the first term gives

$$T^{01}_{(GW)} = \frac{G}{72\pi}\frac{1}{r^2}\left\langle 2(\dddot{Q}_{23})^2 + \frac{1}{2}(\dddot{Q}_{22} - \dddot{Q}_{33})^2\right\rangle. \tag{4.148}$$

The angle brackets always denote the average over several characteristic periods of the source.

The quantity inside the square bracket can be interpreted invariantly as follows. Consider in a first step the part of Q_{ij} transversal to the direction $\boldsymbol{n} = (1, 0, 0)$,

$$Q^T = \begin{pmatrix} Q_{22} & Q_{23} \\ Q_{32} & Q_{33} \end{pmatrix}$$

and then form its trace-free part

$$Q^{TT} = Q^T - \frac{1}{2}(Q_{22} + Q_{33})\mathbf{1} = \begin{pmatrix} \frac{1}{2}(Q_{22} - Q_{33}) & Q_{23} \\ Q_{23} & -\frac{1}{2}(Q_{22} - Q_{33}) \end{pmatrix}.$$

Then

$$\mathrm{Tr}(Q^{TT}Q^{TT}) = 2Q_{23}^2 + \frac{1}{2}(Q_{22} - Q_{33})^2. \tag{4.149}$$

For an arbitrary direction $\boldsymbol{n}$ and any symmetric Euclidean tensor Q_{ij} this construction can be written in terms of the projection operator $P_{ij} = \delta_{ij} - n_i n_j$ in matrix notation as follows:

$$Q^{TT} = PQP - \frac{1}{2}P\,\mathrm{Tr}(PQ). \tag{4.150}$$

Hence, we have for an arbitrary direction

$$T^{0s}_{(GW)}n^s = \frac{G}{72\pi}\frac{1}{r^2}\left\langle \mathrm{Tr}\left(\dddot{Q}^{TT}\right)^2\right\rangle. \tag{4.151}$$

Using that Q_{ij} is trace-free, one can easily show that

$$\mathrm{Tr}\left(\dddot{Q}^{TT}\right)^2 = \dddot{Q}_{kl}\dddot{Q}_{kl} - 2\dddot{Q}_{kl}\dddot{Q}_{km}\, n^l n^m + \frac{1}{2}\left(\dddot{Q}_{kl}\, n^k n^l\right)^2 . \tag{4.152}$$

(Note that it suffices to verify this for $\boldsymbol{n} = (1, 0, 0)$).

The averaged power per unit solid angle emitted in the direction $\boldsymbol{n}$, denoted by $dL_{GW}/d\Omega$, is equal to (4.151) times r^2:

$$\frac{dL_{GW}}{d\Omega} = \frac{G}{72\pi}\left\langle \mathrm{Tr}\left(\dddot{Q}^{TT}\right)^2\right\rangle . \tag{4.153}$$

For the total power (**gravitational luminosity**) we use (4.152) and the following averages, familiar from electrodynamics,

$$\begin{aligned} \frac{1}{4\pi}\int_{S^2} n^l n^m \, d\Omega &= \frac{1}{3}\delta_{lm} , \\ \frac{1}{4\pi}\int_{S^2} n^k n^l n^m n^r \, d\Omega &= \frac{1}{15}(\delta_{kl}\delta_{mr} + \delta_{km}\delta_{lr} + \delta_{kr}\delta_{lm}) . \end{aligned} \tag{4.154}$$

This leads to the famous **quadrupole formula**, first derived by Einstein (see [58] and [59]):

$$L_{GW} = \frac{G}{45}\langle \dddot{Q}_{kl}\dddot{Q}_{kl}\rangle . \tag{4.155}$$

In contrast to electrodynamics, the lowest multipole is quadrupole radiation. This, of course, is due to the spin-2 nature of the gravitational field. For the same reason the prefactor in (4.155) differs from the one for the quadrupole radiation in electrodynamics.

Let us express the total energy E_{GW} radiated by the source in terms of the Fourier transform

$$\hat{Q}_{ij}(\omega) = \frac{1}{2\pi}\int Q_{ij}(t) \mathrm{e}^{i\omega t} dt \tag{4.156}$$

of the quadrupole tensor:

$$E_{GW} = \frac{G}{45}\int_{-\infty}^{+\infty} dt\, \dddot{Q}_{kl}\dddot{Q}_{kl} = \frac{4\pi G}{45}\int_0^\infty \omega^6 \hat{Q}^*_{kl}(\omega)\hat{Q}_{kl}(\omega) .$$

(Beside the Parseval equation we have used the reality condition $\hat{Q}_{ij}(-\omega) = \hat{Q}^*_{ij}(\omega)$.) The spectral distribution of the quadrupole radiation is thus

$$\frac{dE_{GW}}{d\omega} = \frac{4\pi G}{45}\hat{Q}^*_{kl}(\omega)\hat{Q}_{kl}(\omega) . \tag{4.157}$$

For later use, we note that (4.146) gives in the TT gauge

$$h^{TT}_{kl}(t, \boldsymbol{x}) = \frac{2G}{3}\frac{1}{r}\ddot{Q}^{TT}_{kl}(t - |\boldsymbol{x}|) . \tag{4.158}$$

This quantity is measurable via the equation for geodesic deviation (see Sect. 4.3).

4.5.2 Rapidly Varying Sources

Let us generalize these results, making only the assumption that the fields are weak, but may be rapidly varying in time. Then we have to use the expression (4.139) for the asymptotic field.

It is useful to replace $T^{\mu\nu}$ in (4.139) by its Fourier transform with respect to time

$$T^{\mu\nu}(t,\boldsymbol{x}) = \int_{\mathbb{R}} \hat{T}^{\mu\nu}(\omega,\boldsymbol{x})\mathrm{e}^{-i\omega t}\,d\omega\,. \tag{4.159}$$

Then

$$\gamma^{\mu\nu}(t,\boldsymbol{x}) = \int d\omega\,\mathrm{e}^{-i\omega(t-r)} \int d^3x'\,\hat{T}^{\mu\nu}(\omega,\boldsymbol{x}')\mathrm{e}^{-i\boldsymbol{k}\cdot\boldsymbol{x}'},$$

where $\boldsymbol{k} = \omega\hat{\boldsymbol{x}}$ is the wave vector in the direction $\hat{\boldsymbol{x}}$. In terms of the space-time Fourier transform

$$\tilde{T}^{\mu\nu}(\omega,\boldsymbol{k}) = \int_{\mathbb{R}^3} \hat{T}^{\mu\nu}(\omega,\boldsymbol{x}')\mathrm{e}^{-i\boldsymbol{k}\cdot\boldsymbol{x}'}\,d^3x' \tag{4.160}$$

we get

$$\gamma^{\mu\nu}(t,\boldsymbol{x}) = \frac{4G}{r}\int d\omega\,\tilde{T}^{\mu\nu}(\omega,\boldsymbol{k})\mathrm{e}^{i(\boldsymbol{k}\cdot\boldsymbol{x}-\omega t)}\,. \tag{4.161}$$

The conservation law ${T^{\mu\nu}}_{,\nu} = 0$ translates to

$$k_\nu \tilde{T}^{\mu\nu}(k) = 0\,, \quad k = (\omega,\boldsymbol{k})\,. \tag{4.162}$$

From this we obtain

$$\tilde{T}_{0i}(\omega,\boldsymbol{k}) = -\hat{k}^j\,\tilde{T}_{ji}(\omega,\boldsymbol{k})\,, \tag{4.163a}$$

$$\tilde{T}_{00}(\omega,\boldsymbol{k}) = \hat{k}^i\hat{k}^j\,\tilde{T}_{ij}(\omega,\boldsymbol{k})\,. \tag{4.163b}$$

We want to compute the *total* energy emitted per unit solid angle

$$\frac{dE_{GW}}{d\Omega} = \int_{-\infty}^{+\infty} L_{GW}(t)\,dt = \int_{-\infty}^{+\infty} r^2\hat{x}^i T^{0i}_{(GW)}\,dt\,.$$

Inserting (4.161) into (4.122), and carrying out the time integration gives

$$\frac{dE_{GW}}{d\Omega} = G\int_{-\infty}^{+\infty} d\omega\,\omega^2\left(\tilde{T}^*_{\mu\nu}(\omega,\boldsymbol{k})\tilde{T}^{\mu\nu}(\omega,\boldsymbol{k}) - \frac{1}{2}|T^\lambda_{\ \lambda}(\omega,\boldsymbol{k})|^2\right).$$

Since $T^{\mu\nu}$ is real this is equal to twice the integral over all positive frequencies:

$$\frac{dE_{GW}}{d\Omega} = \int_0^\infty \frac{d^2E_{GW}}{d\Omega d\omega}\,d\omega\,, \tag{4.164}$$

with

$$\frac{d^2 E_{GW}}{d\Omega d\omega} = 2G\omega^2 \left(\tilde{T}^*_{\mu\nu}(\omega, \boldsymbol{k}) \tilde{T}^{\mu\nu}(\omega, \boldsymbol{k}) - \frac{1}{2} |T^\lambda_\lambda(\omega, \boldsymbol{k})|^2 \right) . \tag{4.165}$$

With the help of (4.163a) and (4.163b) we can express this in terms of the space-space components $\tilde{T}^{ij}(\omega, \boldsymbol{k})$. Not surprisingly, the result can be written in terms of the TT-components of $\tilde{T}^{ij}(\omega, \boldsymbol{k})$ with respect to the direction $\hat{k}^i$:

$$\frac{d^2 E_{GW}}{d\Omega d\omega} = 2G\omega^2 \tilde{T}^{TT*}_{ij}(\omega, \boldsymbol{k}) \tilde{T}^{TT}_{ij}(\omega, \boldsymbol{k}) . \tag{4.166}$$

From this general result one can derive the earlier formulas in the slow motion approximation, i.e. for

$$\tilde{T}_{ij}(\omega, \boldsymbol{k}) \simeq \int \hat{T}_{ij}(\omega, \boldsymbol{x}) \, d^3 x . \tag{4.167}$$

4.5.3 Radiation Reaction (Preliminary Remarks)

Within the approximations assumed in this section, the concept as well as the formulas for gravitational wave energy loss far away from the sources will presumably satisfy most readers. A much more difficult question is, however, whether and exactly how the gravitational luminosity L_{GW} is related to the time derivative of some energy E of the sources, given as an explicit functional of the *instantaneous* state of the sources.

For Newtonian sources (like our planet system) it is certainly very reasonable to expect that the Newtonian energy $E(t)$ suffers a secular change given by

$$\frac{dE}{dt} = -L_{GW} , \tag{4.168}$$

as we are used to in electrodynamics. Below we shall explore the consequences of this for a binary star system. The justification of (4.168), and its extension beyond the Newtonian dynamics, is a difficult problem. For this it is necessary to derive the full equations of motion of a binary system, say, to sufficiently high orders at least until time-odd terms show up. Since L_{GW} is in leading order proportional to c^{-5} (like the quadrupole radiation in electrodynamics) a post-Newtonian expansion for the motion of the sources at least to order c^{-5} is required.

In recent recent years, after many previous attempts, this demanding work has been carried out by several groups, using different methods. More about this will be said in Chap. 5. This kind of analytic work on post-Newtonian celestial mechanics is a field of active research. Unavoidably, the details are exceedingly involved, and there are still loose ends. These investigations are important for the interpretation of future observations of binary neutron star or black hole coalescences.

4.5.4 Simple Examples and Rough Estimates

Let us now apply our basic formulas, in particular (4.155) and (4.158), to some concrete systems. This will allow us to estimate the gravitational radiation of interesting astrophysical sources.

Rigidly Rotating Body

Consider a body, rigidly rotating about the 3-axis with frequency ω. An important example is a binary star system in circular orbits about their common center of mass. If $\rho(\boldsymbol{x}')$ denotes the mass density of the body relative to Euclidean coordinates $\boldsymbol{x}'$ fixed in the body, then the time dependent mass density in the inertial (laboratory) frame is given by $\rho(\boldsymbol{x},t) = \rho(\boldsymbol{x}')$, where

$$\begin{aligned} x_1 &= x_1' \cos\omega t - x_2' \sin\omega t\,, \\ x_2 &= x_1' \sin\omega t + x_2' \cos\omega t\,, \\ x_3 &= x_3'\,. \end{aligned} \tag{4.169}$$

To simplify things we assume that the rotation is around one of the principle axis of the ellipsoid of inertia. We can then also choose the other two body fixed axis along principle directions. Then the moment-of-inertia tensor in body fixed coordinates,

$$I'_{ij} = \int \rho(\boldsymbol{x}') x'_i x'_j \, d^3x'\,, \tag{4.170}$$

is diagonal. Relative to the inertial system, the components of this tensor are

$$\begin{aligned} I_{11} &= \frac{1}{2}(I'_{11} + I'_{22}) + \frac{1}{2}(I'_{11} - I'_{22})\cos 2\omega t\,, \\ I_{12} &= \frac{1}{2}(I'_{11} - I'_{22})\sin 2\omega t\,, \\ I_{22} &= \frac{1}{2}(I'_{11} + I'_{22}) - \frac{1}{2}(I'_{11} - I'_{22})\cos 2\omega t\,, \\ I_{13} &= I_{23} = 0\,, \\ I_{33} &= I'_{33}\,. \end{aligned} \tag{4.171}$$

Note that its time dependence goes with *twice* the rotation frequency. The trace-free quadrupole tensor is given by

$$Q_{ij} = 3I_{ij} - \delta_{ij} I_{kk}\,. \tag{4.172}$$

This determines h_{ij}^{TT} by (4.158), and L_{GW} according to (4.155).

Before proceeding with this example, we recall that according to (4.76) the relative amplitude $\Delta l/l$ of spatial oscillations between test particles, induced by a linearly polarized harmonic plane wave (4.80), is equal to $\frac{1}{2}A_+$ or $\frac{1}{2}A_\times$, where A_+ and $A_\times$ are the amplitudes of $h^{TT}_{\mu\nu}$ for the two linear polarizations. It is natural to introduce as a measure the **total strain**

$$h := \sqrt{A_+^2 + A_\times^2}\,. \tag{4.173}$$

Note also, that for a plane wave (4.80) the average gravitational energy flux is according to (4.126) given by

$$\mathcal{F}_{GW} = \frac{1}{32\pi G}\omega^2\left(|A_+|^2 + |A_\times|^2\right) = \frac{1}{32\pi G}\omega^2 h^2\,. \tag{4.174}$$

Returning to our example, we work out (4.158), and first consider h^{TT}_{ij} in the 1-direction at a distance D:

$$\begin{aligned} h^{TT}_{22} &= -h^{TT}_{33} = \frac{1}{2}(h_{22} - h_{33}) = \frac{1}{2}G\theta\, e\,(2\omega)^2\frac{1}{D}\cos 2\omega t\,, \\ h^{TT}_{23} &= h_{23} = 0\,, \end{aligned} \tag{4.175}$$

where

$$\theta = I'_{11} + I'_{22}\,, \quad e = \frac{I'_{11} - I'_{22}}{\theta}\,. \tag{4.176}$$

This is a linearly polarized wave with

$$A_+ = h = \frac{2}{D}G\,\theta\, e\cdot\omega^2\,. \tag{4.177}$$

Along the 3-axis the wave is circularly polarized, with $A_\times = iA_+$,

$$A_+ = \frac{1}{D}G\,\theta\, e(2\omega)^2\,. \tag{4.178}$$

The total gravitational wave strain is thus

$$h = 4\sqrt{2}\frac{1}{D}G\,\theta\, e\cdot\omega^2\,. \tag{4.179}$$

One day measurements of the polarization of waves from the orbital motion of a binary system may reveal the inclination of the orbit to the line of sight, a crucial unknown in the modeling of such systems.

The quadrupole formula (4.155) gives for the total power, after a routine calculation,

$$L_{GW}(2\omega) = \frac{32G}{5c^5}\theta^2\, e^2\,\omega^6\,. \tag{4.180}$$

Consider, as a laboratory example, a homogeneous rotating rod of length l and mass M. Then $\theta = \frac{1}{12}Ml^2$, $e = 1$, thus

$$L_{GW}(2\omega) = \frac{2}{45}\frac{G}{c^5}M^2 l^4\omega^6\,. \tag{4.181}$$

Take for instance, an iron rod with $l = 100\,m$, $M = 1000$ tons, and a frequency of $3Hz$. Then $L_{GW} \simeq 10^{-26} W$, a very small power indeed.

More interesting is a binary star system of masses m_1 and m_2 in circular orbits about their common center of mass. If r is the separation of the two orbits, we have $e = 1$, $\theta = \mu r^2$, where μ is the reduced mass. If we also use Kepler's third law

$$\omega^2 = \frac{GM}{r^3}\,, \qquad M = m_1 + m_2\,,$$

then (4.180) gives

$$L_{GW} = \frac{32}{5}\frac{G^4}{c^5 r^5} M^3 \mu^2\,. \tag{4.182}$$

We write this expression in units of the power

$$L_0 := \frac{c^5}{G} = 3.63 \times 10^{59} erg/sec = 2.03 \times 10^5 M_\odot c^2/sec\,, \tag{4.183}$$

determined by G and c:

$$L_{GW} = \frac{32}{5}\left(\frac{GM}{c^2 r}\right)^5 \frac{\mu^2}{M^2} L_0\,. \tag{4.184}$$

Note that $GM^2/c^2 r$ is, up to a factor 2, the ratio of the Schwarzschild radius belonging to the total mass to the separation of the system.

When, for instance, an inspiraling binary neutron star system, caused by the emission of gravitational radiation, becomes very narrow, L_{GW} becomes huge. Take, for example, two neutron stars with masses $\simeq 1.4 M_\odot$ at a separation $r = 100km$ (orbital period $\simeq 10^{-2} sec$), then (4.184) gives $L_{GW} \simeq 10^{52} erg/sec$.

This example suggests the following rule of thumb: An astronomical orbiting system of mass M and size R will have a luminosity of gravitational radiation

$$L_{GW} \sim L_0 \left(\frac{GM}{c^2 R}\right)^5 . \tag{4.185}$$

The 'compactness parameter' $GM/c^2 R$, entering with the fifth power, is usually very small, but will not be much smaller than 1 when two black holes coalesce at the end of an inspiraling phase. There are high hopes to observe such events with laser interferometers. Gravitational waves provide the only way to make *direct* observations of black holes.

Let us compute the total strain h for a binary star system. Using in (4.179) Kepler's third law we arrive at

$$h = 4\sqrt{2}\frac{1}{D}\frac{G^2 M\mu}{r} = 4\sqrt{2}\,\frac{G\,m_1/c^2}{D}\frac{G\,m_2/c^2}{r}\,. \tag{4.186}$$

For the above example of two neutron stars we find for $D \simeq 30'000\,ly$ (distance to the center of the Milky Way) the total strain $h \approx 10^{-18}$. Laser interferometers will soon reach in the kHz region a sensitivity of $\sim 10^{-21}$, and in some years 10^{-22} should be achieved.

Let us again abstract from this example a rule of thumb. If the motion inside a astrophysical source is highly non-spherical, then a typical component of $\ddot{Q}_{kl}$ will have magnitude Mv_{ns}^2, where v_{ns} is the non-spherical part of the velocity inside the source. So a typical component of h_{kl}, and thus of the total strain h, is according to (4.158)

$$h \sim \frac{1}{D} 2GM\, v_{ns}^2 . \tag{4.187}$$

By the virial theorem for self-gravitating bodies we have the estimate

$$2v_{ns}^2 \lesssim \phi_{int} ,$$

where ϕ_{int} is a characteristic value of the Newtonian gravitational potential inside the system. So

$$h \lesssim \phi_{int}\phi_{ext} , \tag{4.188}$$

where ϕ_{ext} is the gravitational potential at the distance D. This agrees with (4.186). For a coalescing binary neutron star system, we have $\phi_{int} \sim 0.2$. If we observe such an event in the Virgo cluster, say, we can expect $h \sim 5 \times 10^{-22}$. This is about the Bohr radius measured in astronomical units, and illustrates the goal of detector development and the heroic efforts that have to be invested.

Estimates of the strain induced by various astrophysical sources in ground based and space based detectors as a function of frequency are shown in Fig. 4.4.

Radiation From Binary Star Systems in Elliptic Orbits

We now compute the radiation loss from a double star system, and the induced secular change of the orbital period. The implications for binary pulsars will be discussed in the next chapter (see Sect. 5.8). We shall perform the calculation using units with $G = c = 1$. Let the masses be m_1 and m_2. The semi-major axis a and the eccentricity e of the relative motion are related to the total energy E ($E < 0$) and angular momentum L by

$$a = -\frac{m_1\, m_2}{2E} , \tag{4.189a}$$

$$e^2 = 1 + \frac{2\, E\, L^2 (m_1 + m_2)}{m_1^3\, m_2^3} . \tag{4.189b}$$

Any (secular) change of energy and angular momentum leads to time dependences of the orbital parameters a, e:

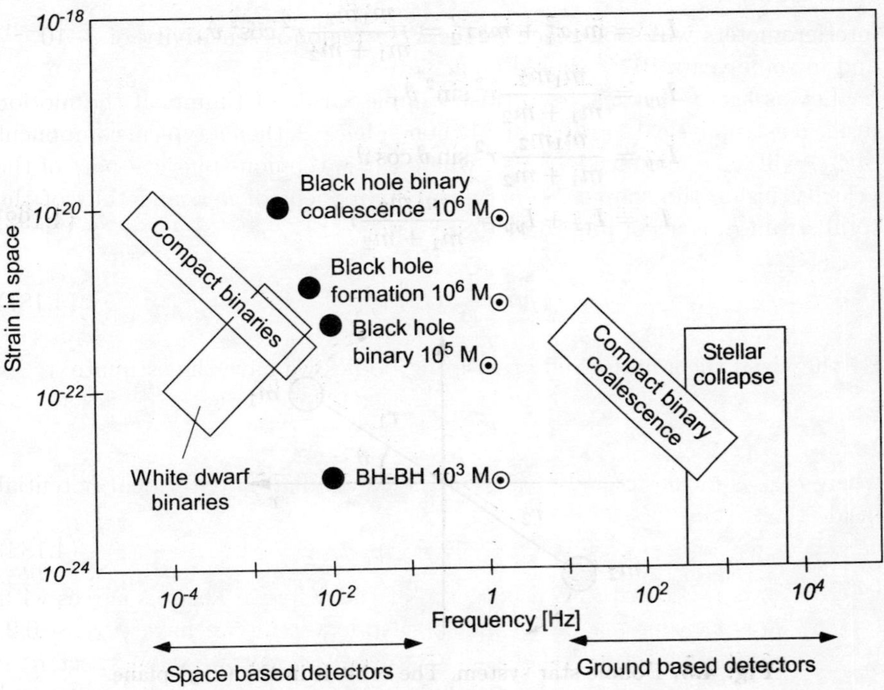

Fig. 4.4. Strain as a function of frequency in detectors for some interesting astrophysical sources.

$$\frac{da}{dt} = \frac{m_1 m_2}{2E^2}\frac{dE}{dt}\,, \tag{4.190a}$$

$$e\,\frac{de}{dt} = \frac{m_1 + m_2}{m_1^3\, m_2^3}\left(L^2\frac{dE}{dt} + 2\,E\,L\frac{dL}{dt}\right). \tag{4.190b}$$

Let r be the distance between the two masses. Then (see Fig. 4.5)

$$r = \frac{a(1-e^2)}{1 + e\cos\vartheta} \tag{4.191}$$

and

$$r_1 = \frac{m_2}{m_1 + m_2}r\,, \qquad r_2 = \frac{m_1}{m_1 + m_2}r\,.$$

The components of the moment-of-inertia tensor

$$I_{kl} = \int \rho(\boldsymbol{x})x^k x^l\, d^3x$$

are given by

$$
\begin{aligned}
I_{xx} &= m_1 x_1^2 + m_2 x_2^2 = \frac{m_1 m_2}{m_1 + m_2} r^2 \cos^2 \vartheta \,, \\
I_{yy} &= \frac{m_1 m_2}{m_1 + m_2} r^2 \sin^2 \vartheta \,, \\
I_{xy} &= \frac{m_1 m_2}{m_1 + m_2} r^2 \sin \vartheta \cos \vartheta \,, \\
I :&= I_{xx} + I_{yy} = \frac{m_1 m_2}{m_1 + m_2} r^2 \,.
\end{aligned} \tag{4.192}
$$

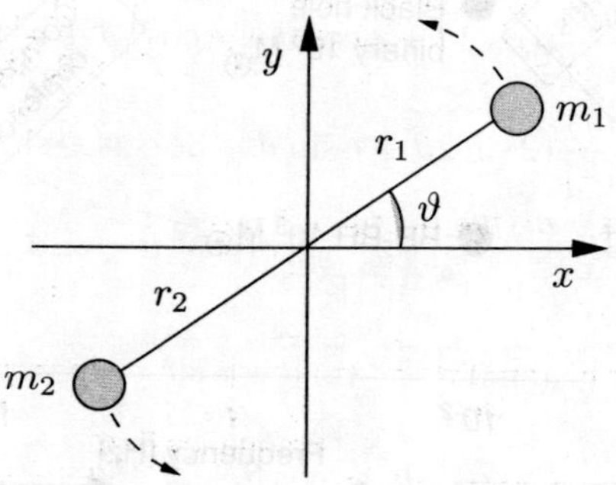

Fig. 4.5. Double star system. The orbit is in the (x, y)-plane.

The angular momentum is

$$L = \frac{m_1 m_2}{m_1 + m_2} r^2 \dot{\vartheta} \,,$$

and hence we have from (4.189a) and (4.189b)

$$\dot{\vartheta} = \frac{\sqrt{(m_1 + m_2) a (1 - e^2)}}{r^2} \,. \tag{4.193}$$

From this and (4.191) it follows that

$$\dot{r} = e \sin \vartheta \left(\frac{m_1 + m_2}{a(1 - e^2)} \right)^{1/2} \,. \tag{4.194}$$

We now compute the time derivatives of the I_{kl}, simplifying the resulting expressions using (4.191), (4.193) and (4.194). The result is

$$\dot{I}_{xx} = \frac{-2m_1m_2}{\left((m_1+m_2)a(1-e^2)\right)^{1/2}} r\cos\vartheta\sin\vartheta\,, \tag{4.195a}$$

$$\ddot{I}_{xx} = \frac{-2m_1m_2}{a(1-e^2)}(\cos 2\vartheta + e\cos^3\vartheta)\,, \tag{4.195b}$$

$$\dddot{I}_{xx} = \frac{2m_1m_2}{a(1-e^2)}(2\sin 2\vartheta + 3e\cos^2\vartheta\sin\vartheta)\,\dot{\vartheta}\,, \tag{4.195c}$$

$$\dot{I}_{yy} = \frac{2m_1m_2}{\left((m_1+m_2)a(1-e^2)\right)^{1/2}} r(\sin\vartheta\cos\vartheta + e\sin\vartheta)\,, \tag{4.195d}$$

$$\ddot{I}_{yy} = \frac{2m_1m_2}{a(1-e^2)}(\cos 2\vartheta + e\cos\vartheta + e\cos^3\vartheta + e^2)\,, \tag{4.195e}$$

$$\dddot{I}_{yy} = \frac{-2m_1m_2}{a(1-e^2)}(2\sin 2\vartheta + e\sin\vartheta + 3e\cos^2\vartheta\sin\vartheta)\,\dot{\vartheta}\,, \tag{4.195f}$$

$$\dot{I}_{xy} = \frac{m_1m_2}{\left((m_1+m_2)a(1-e^2)\right)^{1/2}} r(\cos^2\vartheta - \sin^2\vartheta + e\cos\vartheta)\,, \tag{4.195g}$$

$$\ddot{I}_{xy} = \frac{-2m_1m_2}{a(1-e^2)}(\sin 2\vartheta + e\sin\vartheta + e\sin\vartheta\cos^2\vartheta)\,, \tag{4.195h}$$

$$\dddot{I}_{xy} = \frac{-2m_1m_2}{a(1-e^2)}(2\cos 2\vartheta - e\cos\vartheta + 3e\cos^3\vartheta)\,\dot{\vartheta}\,, \tag{4.195i}$$

$$\dddot{I} = \dddot{I}_{xx} + \dddot{I}_{yy} = \frac{-2m_1m_2}{a(1-e^2)} e\sin\vartheta\,\dot{\vartheta}\,. \tag{4.195j}$$

Equation (4.155), expressed in terms of I_{kl}, can be written in the form

$$L_{GW} = \frac{1}{5}\left\langle \dddot{I}_{kl}\dddot{I}_{kl} - \frac{1}{3}\dddot{I}^2 \right\rangle = \frac{1}{5}\left\langle \dddot{I}^2_{xx} + 2\dddot{I}^2_{xy} + \dddot{I}^2_{yy} - \frac{1}{3}\dddot{I}^2 \right\rangle .$$

Inserting the explicit expressions (4.195a)–(4.195j) gives

$$L_{GW} = \frac{8\,m_1^2m_2^2}{15\,a^2(1-e^2)^2}\left\langle \left(12(1+e\cos\vartheta)^2 + e^2\sin^2\vartheta\right)\dot{\vartheta}^2 \right\rangle .$$

We take the time average over one period T:

$$\langle\ldots\rangle = \frac{1}{T}\int_0^T \ldots dt = \frac{1}{T}\int_0^{2\pi} \ldots \frac{d\vartheta}{\dot{\vartheta}}\,.$$

According to Kepler's third law, the orbital period is

$$T = \frac{2\pi a^{3/2}}{(m_1+m_2)^{1/2}}\,. \tag{4.196}$$

After some calculations, making also use of (4.165), one finds

$$L_{GW} = \frac{32}{5}\frac{m_1^2m_2^2(m_1+m_2)}{a^5(1-e^2)^{7/2}}\left(1 + \frac{73}{24}e^2 + \frac{37}{96}e^4\right). \tag{4.197}$$

From this and (4.190a) we obtain, under the *assumption* (4.168),

$$\left\langle \frac{da}{dt} \right\rangle = \frac{2a^2}{m_1 m_2} \left\langle \frac{dE}{dt} \right\rangle = -\frac{64}{5} \frac{m_1 m_2 (m_1 + m_2)}{a^3 (1-e^2)^{7/2}} \left(1 + \frac{73}{24} e^2 + \frac{37}{96} e^4 \right). \tag{4.198}$$

Together with (4.196) we then find for the secular change of the orbital period

$$\left\langle \frac{\dot{T}}{T} \right\rangle = \frac{3}{2} \left\langle \frac{\dot{a}}{a} \right\rangle = -\frac{96}{5} \frac{1}{a^4} m_1 m_2 (m_1 + m_2) f(e), \tag{4.199}$$

where

$$f(e) = \left(1 + \frac{73}{24} e^2 + \frac{37}{96} e^4 \right) (1 - e^2)^{-7/2}. \tag{4.200}$$

If we replace a by T in (4.199), we arrive at

$$\left\langle \frac{\dot{T}}{T} \right\rangle = -\frac{96}{5} \frac{m_1 m_2}{\left(\frac{T}{2\pi} \right)^{8/3} (m_1 + m_2)^{1/3}} f(e). \tag{4.201}$$

Note that the dependence on the two masses is proportional to $\mathcal{M}^{5/3}$, where $\mathcal{M}$ is the so-called **chirp mass** given by

$$\mathcal{M} = \eta^{5/3} m,$$

where $m = m_1 + m_2$ and $\eta = m_1 m_2 / (m_1 + m_2)^2$.

In an effort over more than twenty years, Hulse and Taylor succeeded to verify this expression by observations of a remarkable binary system, which consists of a pulsar (PSR 1913+16) and a compact partner, which is also a neutron star. A detailed discussion of this system will be given in Sect. 5.8.

We shall see that the very precise measurements of the pulse arrival times over years made it possible to determine not only the Keplerian parameters, but also certain post-Keplerian parameters, such as the precession of the periastron. These then fix the masses m_1 and m_2 of the two neutron stars to an amazing precision:

$$m_1 = 1.4411 \pm 0.0007 \; M_\odot, \qquad m_2 = 1.3873 \pm 0.0007 \; M_\odot.$$

Using also the 'measured' values

$$T(\text{day}) = 0.322997462736(7), \qquad e = 0.6171308(4)$$

in (4.201), the predicted secular orbital period change is the tiny number

$$\dot{T}_{theor} = (-2.40243 \pm 0.00005) \times 10^{-12}.$$

For the precise comparison with the observed value, it is necessary to take into account a small correction due to the relative acceleration between the binary pulsar system and the solar system caused by the differential rotation of the galaxy. Then theory and observation agree within the errors of about 0.5%. This is a great triumph of GR, which leaves little room for doubt in the validity of the quadrupole formula for other systems.

* * *

Exercises.

Exercise 4.12. Compute the radiated energy (4.197) for the binary pulsar system PSR 1913+16 in erg/s.

Exercise 4.13. Under the assumption that (4.199) also holds for $a \downarrow 0$, compute the time τ until the two stars coalesce if $e = 0$. Work this out numerically for two neutron stars with masses $1.4M_\odot$ and an initial separation $a_0 = R_\odot$.

Exercise 4.14. Compare the quadrupole formula (4.155) with the quadrupole contribution to electromagnetic radiation. Use this to estimate, in a naive linearized quantum theory of gravity, the graviton emissing rate for the $3d \to 1s$ transition in hydrogen. Compare the corresponding lifetime with the age of the universe.

The transition rate is ($\hbar = c = 1$):

$$\Gamma_{\text{grav}}(3d \to 1s) = \frac{1}{2^3 \cdot 3^2 \cdot 5}\alpha^6 G\, m_e^3,$$

corresponding to a lifetime $\tau \simeq 0.5 \times 10^{32} yr$.

4.6 Laser Interferometers

Laser interferometry is based on the Michelson interferometer. Laser light is used to measure changes in the difference between the lengths of two (nearly) perpendicular arms. Beside many sources of noise, such differences are induced by the strain of a gravitational wave passing through the instrument. A Michelson interferometer operating between freely suspended mirrors is ideally suited to measure these.

Figure 4.6 is a schematic drawing of an equal-arm Michelson interferometer. For LIGO, the largest operating detector, the two arms have lengths of $4\,km$. Light sent from the laser light source to the beam splitter is divided evenly between the two arms. Having traversed the arms, the light is reflected back to the splitter by mirrors at their far ends. On the return journeys to the photodetector (photodiode), the roles of reflection and transmission in the splitter are interchanged for the two beams and, furthermore, the phase of the reflected beam is inverted by 180°. Therefore the recombined beams heading towards the photodetector interfere destructively, while the beams

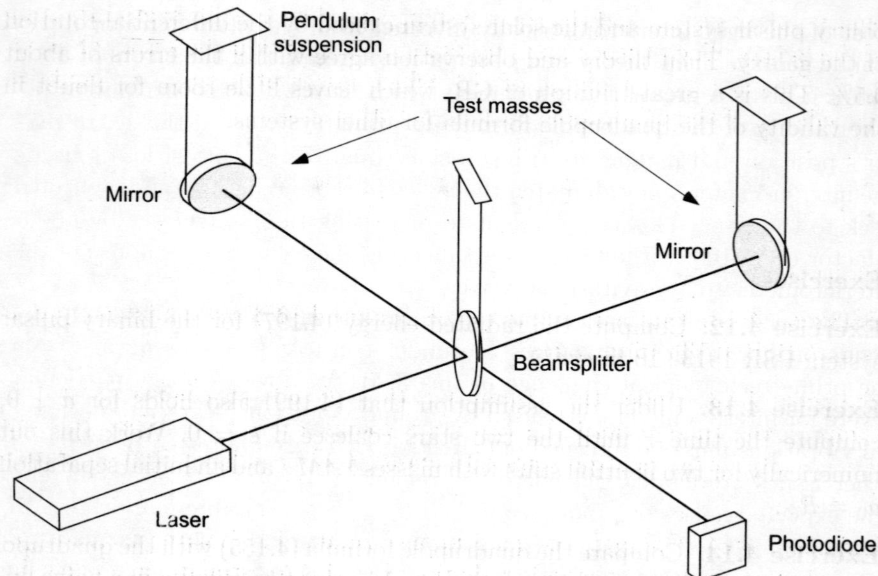

Fig. 4.6. Schematic of gravitational wave detector using laser interferometry.

heading back to the light source interfere constructively. The interference arms are adjusted such that the photodetector ideally sees no light, all of it having been diverted, by perfect interference, back to the source. One would get this kind of perfect interference if the beam geometry provides a single phase over the propagating wavefront. An idealized uni-phase plane wave has this property, as does the Gaussian wavefront in the lowest order spatial mode of a laser.

If, in the absence of any disturbances, the interferometer is carefully balanced so that no light appears at the photodetector, a sufficiently strong gravitational wave passing through the interferometer can disturb this balance and cause light to fall on the detector. That is the principle how laser interferometers will sense gravitational waves. To obtain the required sensitivity, various refinements have to be added. We indicate only two, namely **multi-pass arrangement** and **power recycling** (for details, see the review article [111]).

It is beneficial to arrange for the light to remain in an arm longer than the time to go up and down, since the intensity change at the photodetector due to a gravitational wave depends on the interaction time of the wave in the arms. A multi-pass arrangement is used in the arms of the interferometer. This can be achieved with a Fabry-Perot resonant cavity. In practice this trick results in a light-storage time of about $1\,msec$. That is about 50 times longer than a single straight transit through a $3\,km$ arm.

A second trick is called power recycling. By making the entire interferometer a resonant optical storage cavity the interfering light intensity is increased.

Most of the light interferometrically diverted from the photodetector direction (when the arms are unstrained) returns toward the light source. That makes it possible to achieve a significant gain by placing another mirror between the laser and the beam splitter. By properly choosing this extra mirror's position and making its transmission equal to the optical losses inside the interferometer, one can match the losses so that no light at all is reflected back to the laser. This is equivalent to increasing the laser power by about a factor of 50-100, without adversely affecting the frequency response of the interferometer to a gravitational wave.

The success of an interferometer (or any other detector) ultimately depends on how well noise sources are under control. That has always been the prime technological challenge in this field. An obvious noise, particular severe for interferometers, is caused by ground vibrations. External mechanical vibrations must be screened out. Readers, who are interested in such and other issues are again referred to the review [111], and references therein. The expected noise contributions of the LIGO II interferometer are shown in Fig. 4.7.

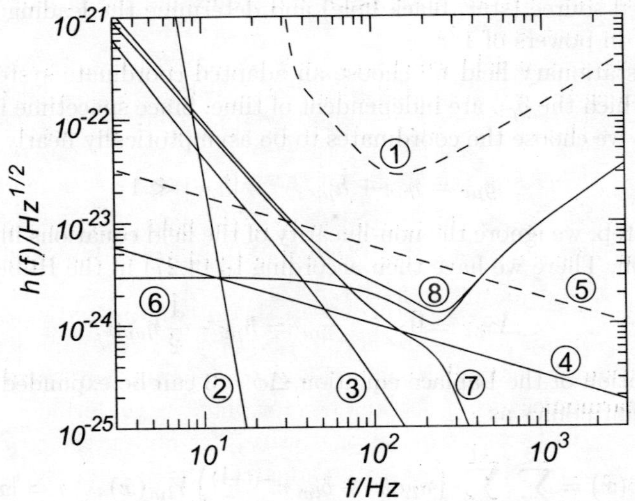

Fig. 4.7. Noise contributions to the expected sensitivity of the LIGO II interferometer. Curve 1 gives the total noise for LIGO I and 8 shows the expected improvement for LIGO II. (The total strain $h(f)$ was defined in (4.173). For the individual contributions 2-7, see [111].)

Presumably, the first generation of large gravitational detectors (GEO 600, LIGO I, VIRGO, TAMA 300) will not yet have the sensitivity for serious astronomical observations, and perhaps not even for a first detection. The observations are planned to go through several upgrades with ever increasing sensitivity. Efforts are on the way with the goal of a future pan-European project called EURO.

A detector in space would not notice the Earth's noisy environment. The LISA project, studied by both ESA and NASA would open up the frequency window between $0.1\, mH$ and $0.1\, Hz$ for the first time (see [111] and [112]). LISA, with arms of $5 \times 10^6\, km$ long would see, for example, the coalescence of gigantic black holes in the centers of galaxies with extraordinary sensitivity (see, in this context Fig. 4.4). There is also the exciting prospect that LISA may discover gravitational waves that have been produced in the very early universe.

As useful supplements of our brief treatment of gravitational waves, their sources and detection, we refer to the review articles [113] and [114], which also contain lots of references.

4.7 Gravitational Field at Large Distances from a Stationary Source

We now consider a *stationary* gravitational field far away from a stationary isolated source (star, black hole) and determine the leading terms of its expansion in powers of $1/r$.

For a stationary field we choose an adapted coordinate system, with respect to which the $g_{\mu\nu}$ are independent of time. Since spacetime is asymptotically flat, we choose the coordinates to be asymptotically nearly Lorentzian:

$$g_{\mu\nu} = \eta_{\mu\nu} + h_{\mu\nu}\,, \qquad |h_{\mu\nu}| \ll 1\,.$$

In a first step, we ignore the non-linearity of the field equations in the asymptotic region. There we have then according to (4.27) in the Hilbert gauge,

$$\Delta\gamma_{\mu\nu} = 0\,, \qquad \gamma_{\mu\nu} := h_{\mu\nu} - \frac{1}{2}\,\eta_{\mu\nu} h\,. \tag{4.202}$$

Every solution of the Laplace equation $\Delta\phi = 0$ can be expanded in terms of spherical harmonics as

$$\phi(\boldsymbol{x}) = \sum_{l=0}^{\infty} \sum_{m=-l}^{+l} \left(a_{lm}\, r^l + b_{lm}\, r^{-(l+1)} \right) Y_{lm}(\hat{\boldsymbol{x}})\,, \quad r = |\boldsymbol{x}|\,.$$

For our problem, the a_{lm} vanish and we may take the $\gamma_{\mu\nu}$ to have the form (writing explicitly only the terms with $l = 0, 1$)

$$\begin{aligned}
\gamma_{00} &= \frac{A^0}{r} + \frac{B^j\, n^j}{r^2} + O\left(\frac{1}{r^3}\right) \quad (n^j = x^j/r)\,, \\
\gamma_{0j} &= \frac{A^j}{r} + \frac{B^{jk}\, n^k}{r^2} + O\left(\frac{1}{r^3}\right), \\
\gamma_{jk} &= \frac{A^{jk}}{r} + \frac{B^{jkl}\, n^l}{r^2} + O\left(\frac{1}{r^3}\right),
\end{aligned} \tag{4.203}$$

where $A^{jk} = A^{kj}$ and $B^{jkl} = B^{kjl}$.

The Hilbert condition $\gamma^{\mu\nu}{}_{,\nu} = 0$ gives the restrictions

$$\begin{aligned} A^j &= 0\,, \qquad B^{jk}(\delta^{jk} - 3n^j n^k) = 0\,, \\ A^{jk} &= 0\,, \qquad B^{jkl}(\delta^{kl} - 3n^k n^l) = 0\,. \end{aligned} \tag{4.204}$$

We now decompose B^{jk} and B^{jkl} into irreducible pieces with respect to the rotation group. For B^{jk} we have

$$B^{jk} = \delta^{jk} B + S^{jk} + A^{jk}\,, \tag{4.205}$$

where B is the trace of B^{jk}, S^{jk} is symmetric and traceless, $S^{jj} = 0$, and $A^{jk} = \varepsilon^{jkl} F^l$ is antisymmetric. The gauge conditions (4.204) imply

$$(\delta^{jk} B + S^{jk} + A^{jk})(\delta^{jk} - 3n^j n^k) = -3\, S^{jk} n^j n^k = 0\,,$$

which means that

$$S^{jk} = 0\,.$$

For B^{jkl} we use the result of Exercise 4.15, which shows that, after imposing the Hilbert condition,

$$B^{jkl} = \delta^{jk} A^l - (A^j \delta^{kl} - A^k \delta^{jl})\,.$$

If one now uses these results in (4.203), one obtains

$$\begin{aligned} \gamma_{00} &= \frac{A^0}{r} + \frac{B^j n^j}{r^2} + O\left(\frac{1}{r^3}\right), \\ \gamma_{0j} &= \frac{\varepsilon^{jkl} n^k F^l}{r^2} + \frac{B\, n^j}{r^2} + O\left(\frac{1}{r^3}\right), \\ \gamma_{jk} &= \frac{\delta^{jk} A^l n^l - A^j n^k - A^k n^j}{r^2} + O\left(\frac{1}{r^3}\right). \end{aligned} \tag{4.206}$$

Within the Hilbert gauge, one can choose B in γ_{0j} and A^j in γ_{jk} to vanish. This is achieved with the gauge transformation defined by the vector field with covariant components $\xi_j = -A^j/r$ and $\xi_0 = B/r$. Then (4.206) becomes

$$\begin{aligned} \gamma_{00} &= \frac{A^0}{r} + \frac{(B^j + A^j)\, n^j}{r^2} + O\left(\frac{1}{r^3}\right), \\ \gamma_{0j} &= \frac{\varepsilon^{jkl} n^k F^l}{r^2} + O\left(\frac{1}{r^3}\right), \\ \gamma_{jk} &= O\left(\frac{1}{r^3}\right). \end{aligned} \tag{4.207}$$

Finally, we shift the origin of the coordinate system such that

$$x^j_{new} = x^j_{old} - \frac{(B^j + A^j)}{A^0}.$$

The second term in γ_{00} then vanishes. Rewriting everything in terms of $h_{\mu\nu}$, we end up with

$$\begin{aligned} h_{00} &= \frac{2m}{r} + O\left(\frac{1}{r^3}\right), \\ h_{0i} &= -2G\,\varepsilon_{ijk}\frac{S^j x^k}{r^3} + O\left(\frac{1}{r^3}\right), \\ h_{ij} &= \frac{2m}{r}\delta_{ij} + O\left(\frac{1}{r^3}\right). \end{aligned} \tag{4.208}$$

The constants m and S^j will be interpreted further on.

Up to now we have considered only the linear approximation in the asymptotic region. The dominant non-linear terms must be proportional to the square, $(m/r)^2$, of the dominant linear terms. In order to calculate them, it is simplest to consider the Schwarzschild solution in appropriate coordinates and expend it in powers of m/r.

First of all, we note that the Hilbert gauge condition is the linearized form of the **harmonic gauge condition** $\Box x^\mu = 0$. According to (2.239) this can be written as

$$\left(\sqrt{-g}\,g^{\mu\nu}\right)_{,\nu} = 0\,. \tag{4.209}$$

Now $\sqrt{-g} \simeq 1 + \frac{1}{2}h$ and $g^{\mu\nu} \simeq \eta^{\mu\nu} - h^{\mu\nu}$, and hence

$$\sqrt{-g}\,g^{\mu\nu} \simeq \eta^{\mu\nu} - \gamma^{\mu\nu}. \tag{4.210}$$

Thus, in the linear approximation, (4.209) does in fact reduce to the Hilbert gauge condition $\gamma^{\mu\nu}{}_{,\nu} = 0$.

For this reason, we transform the Schwarzschild solution

$$ds^2 = \left(1 - \frac{2m}{r}\right)dt^2 - \frac{dr^2}{1 - 2m/r} - r^2(d\vartheta^2 + \sin^2\vartheta\, d\varphi^2)$$

to harmonic coordinates. These are constructed using the results of Exercise 4.16. The transformed Schwarzschild metric then reads, if now $r = (x^i x^i)^{1/2}$,

$$\begin{aligned} g_{00} &= -\frac{1 - m/r}{1 + m/r}, \\ g_{i0} &= 0, \\ g_{ij} &= \left(1 + \frac{m}{r}\right)^2 \delta_{ij} + \left(\frac{m}{r}\right)^2 \left(\frac{1 + m/r}{1 - m/r}\right)\frac{x^i x^j}{r^2}. \end{aligned} \tag{4.211}$$

The quadratic term in g_{00} is ${}^{(2)}g_{00} = 2(m/r)^2$. We thus obtain the expansion

$$g_{00} = -\left(1 - \frac{2m}{r} + 2\frac{m^2}{r^2}\right) + O\left(\frac{1}{r^3}\right),$$

$$g_{i0} = -2G\,\varepsilon_{ijk}\frac{S^j x^k}{r^3} + O\left(\frac{1}{r^3}\right),$$

$$g_{ij} = \left(1 + \frac{2m}{r}\right)\delta_{ij} + O\left(\frac{1}{r^2}\right). \tag{4.212}$$

We shall obtain this result more systematically in the post-Newtonian approximation (see Sect. 5.3).

Now, we compute the energy and angular momentum of the system, using the asymptotic expressions (4.212). We use the expressions (2.207) and (2.212) for these quantities in terms of flux integrals. In our coordinates, the connection forms are given by (4.3)

$$\omega^\alpha{}_\beta = \Gamma^\alpha{}_{\gamma\beta}\,dx^\gamma = \frac{1}{2}\left(h^\alpha{}_{\gamma,\beta} + h^\alpha{}_{\beta,\gamma} - h_{\gamma\beta}{}^{,\alpha}\right)dx^\gamma.$$

According to (2.207) the total momentum is given by

$$\begin{aligned} P^\rho &= -\frac{1}{16\pi G}\int_{S^2_\infty}\sqrt{-g}\,\omega_{\alpha\beta}\wedge\eta^{\alpha\beta\rho} \\ &= -\frac{1}{16\pi G}\,\varepsilon^{\rho\alpha\beta}{}_\sigma\int_{S^2_\infty}\frac{1}{2}\left(h_{\alpha\gamma,\beta} - h_{\gamma\beta,\alpha}\right)dx^\gamma\wedge dx^\sigma \\ &= \frac{1}{16\pi G}\,\varepsilon^{\rho\alpha\beta}{}_\sigma\int_{S^2_\infty}h_{\beta\gamma,\alpha}\,dx^\gamma\wedge dx^\sigma. \end{aligned}$$

Only P^0 does not vanish:

$$P^0 = \frac{1}{16\pi G}\,\varepsilon^{0ij}{}_l\int_{S^2_\infty}h_{jk,i}\,dx^k\wedge dx^l.$$

According to (4.212), we have

$$h_{jk,i} = -\frac{2m}{r^3}x^i\delta_{jk}.$$

If we now integrate over a large spherical surface, we then obtain

$$P^0 = \frac{M}{8\pi}\int\frac{1}{r^3}\underbrace{x^i\varepsilon_{ijk}\,dx^j\wedge dx^k}_{2r^3\,d\Omega} = M,$$

where $M = m/G$. Thus, as expected, we have

$$P^0 = \frac{m}{G}. \tag{4.213}$$

According to (2.212), the angular momentum is given by the following flux integral:

$$J^{\rho\alpha} = \frac{1}{16\pi G}\int_{S^2_\infty} \sqrt{-g}\Big((x^\rho\eta^{\alpha\beta\gamma} - x^\alpha\eta^{\rho\beta\gamma})\wedge\omega_{\beta\gamma} + \eta^{\rho\alpha}\Big). \qquad (4.214)$$

For a "surface at infinity", the first term is

$$\begin{aligned}
J^{\rho\alpha}_{(1)} &= \frac{1}{16\pi G}\int_{S^2_\infty} \sqrt{-g}\,(x^\rho\eta^{\alpha\beta\gamma} - x^\alpha\eta^{\rho\beta\gamma})\wedge\omega_{\beta\gamma} \\
&= \frac{1}{2}\frac{1}{16\pi G}\int_{S^2_\infty} (x^\rho\eta^{\alpha\beta\gamma} - x^\alpha\eta^{\rho\beta\gamma})\wedge(h_{\beta\sigma,\gamma} - h_{\sigma\gamma,\beta})\,dx^\sigma \\
&= \frac{1}{16\pi G}\int_{S^2_\infty} \left(x^\rho\varepsilon^{\alpha\beta\gamma}{}_\mu - (\rho\longleftrightarrow\alpha)\right) h_{\beta\sigma,\gamma}\,dx^\mu\wedge dx^\sigma .
\end{aligned}$$

The spatial components are

$$\begin{aligned}
J^{ri}_{(1)} &= \frac{1}{16\pi G}\,\varepsilon^{i0k}{}_l\int_{S^2_\infty} x^r\,h_{0j,k}\,dx^l\wedge dx^j - (r\longleftrightarrow i) \\
&= \frac{1}{16\pi G}\,\varepsilon_{ikl}\int_{S^2_\infty}\left(x^r h_{0j,k} - (i\longleftrightarrow r)\right)dx^l\wedge dx^j .
\end{aligned}$$

The components of the angular momentum vector are given by

$$J_s = \frac{1}{2}\,\varepsilon_{ris}\,J^{ri} .$$

For these, we obtain

$$\begin{aligned}
J^{(1)}_s &= \frac{1}{16\pi G}\,\underbrace{\varepsilon_{ris}\varepsilon_{ikl}}_{-(\delta_{rk}\delta_{sl} - \delta_{rl}\delta_{ks})}\int_{S^2_\infty} x^r\,h_{0l,k}\,dx^l\wedge dx^j \\
&= \frac{1}{16\pi G}\int_{S^2_\infty}\Big(-x^k\,h_{0j,k}\,dx^s\wedge dx^j + h_{0j,s}\,\overbrace{x^l\,dx^l}^{d(r^2)/2}\wedge dx^j\Big). \qquad (4.215)
\end{aligned}$$

The second term in (4.215) does not contribute to the surface integral over a large sphere, and hence

$$\begin{aligned}
J^{(1)}_s &= -\frac{1}{16\pi G}\int_{S^2_\infty} x^k\,h_{0j,k}\,dx^s\wedge dx^j \\
&= -\frac{1}{16\pi G}\int_{S^2_\infty} 2G\,\varepsilon_{jmn}\,S^m\left(\frac{x^n}{r^3}\right)_{,k} x^k\,dx^s\wedge dx^j \\
&= -\frac{1}{16\pi G}\int_{S^2_\infty} 2G\,\varepsilon_{jmn}\,S^m\left(\frac{x^n}{r^3} - 3\,\frac{x^n x^k}{r^5}x^k\right)dx^s\wedge dx^j \\
&= -\frac{1}{4\pi}\int_{S^2_\infty}\varepsilon_{jmn}\,S^m\,\frac{x^n}{r^3}\,dx^s\wedge dx^j .
\end{aligned}$$

After an integration by parts, we have

$$\begin{aligned} J_s^{(1)} &= \frac{1}{4\pi}\int_{S^2_\infty} \frac{x^s}{r^3}\,\varepsilon_{jmn}\,S^m\,dx^n\wedge dx^j = \frac{1}{4\pi}2\int_{S^2_\infty} S^m\,\frac{x^s}{r^3}r^2\hat{x}^m\,d\Omega \\ &= \frac{1}{4\pi}2S^m\int_{S^2_\infty}\hat{x}^s\hat{x}^m\,d\Omega = \frac{2}{3}S^s \end{aligned}$$

or

$$J_k^{(1)} = \frac{2}{3}S^k. \tag{4.216}$$

The second term of (4.214) contributes

$$\begin{aligned} J^{ik}_{(2)} &= \frac{1}{16\pi G}\int_{S^2_\infty}\frac{1}{2!}\,\varepsilon^{ik}{}_{\alpha\beta}\,dx^\alpha\wedge dx^\beta \\ &= \frac{1}{16\pi G}\int_{S^2_\infty} 2\frac{1}{2!}\,\varepsilon^{ikol}\,g_{or}\,g_{ls}\,dx^r\wedge dx^s \\ &= -\frac{1}{16\pi G}\int_{S^2_\infty}\varepsilon_{ikl}\delta_{ls}\,g_{or}\,dx^r\wedge dx^s \end{aligned}$$

or

$$\begin{aligned} J_l^{(2)} &= -\frac{1}{16\pi G}\int_{S^2_\infty} g_{or}\,dx^r\wedge dx^s = \frac{1}{8\pi}S^m\int_{S^2_\infty}\varepsilon_{rmn}\,\frac{x^n}{r^3}\,dx^r\wedge dx^s \\ &= \frac{1}{8\pi}S^m\int_{S^2_\infty}\frac{x^l}{r^3}\,\varepsilon_{rmn}\,dx^n\wedge dx^r = \frac{1}{8\pi}S^m\int_{S^2_\infty}\frac{x^l}{r}\,2\hat{x}^m\,d\Omega = \frac{1}{3}\,S^l\,. \end{aligned}$$

Hence, we obtain

$$J_k^{(2)} = \frac{1}{3}\,S^k\,. \tag{4.217}$$

The total angular momentum is the sum of (4.216) and (4.217):

$$J_k = S^k\,. \tag{4.218}$$

We have thus shown that the parameters S^j in (4.212) are the components of the system's angular momentum.

On the other hand, we have representations of P^0 and J_k in terms of volume integrals (see (2.205) and (2.206)). For the particular case that the field is weak *everywhere*, these reduce to

$$P^0 \simeq \int_{\{x^0=\text{const}\}} *T_0 \simeq \int_{\{x^0=\text{const}\}} T^{00}\,d^3x\,, \tag{4.219}$$

and

$$J^{ij} \simeq \int_{\{x^0=\text{const}\}}\sqrt{-g}\,(x^i * T^j - x^j * T^i) \simeq \int_{\{x^0=\text{const}\}}(x^iT^{j0}-x^jT^{i0})\,d^3x\,. \tag{4.220}$$

The angular momentum S^k of a system can be measured with a gyroscope. Suppose this is far away from the source and at rest in the coordinate system in which (4.212) is valid. The gyroscope then rotates relative to the basis $\partial/\partial x^i$ with the angular velocity (see (1.164))

$$\boldsymbol{\Omega} \simeq -\frac{1}{2}\boldsymbol{\nabla} \times \boldsymbol{g}\,,$$

where $\boldsymbol{g} = (g_{01}, g_{02}, g_{03})$. If we now use (4.212) for the g_{i0}, we obtain

$$\boldsymbol{\Omega} = -\frac{G}{r^3}\left(\boldsymbol{S} - 3\,\frac{(\boldsymbol{S}\cdot\boldsymbol{x})\boldsymbol{x}}{r^2}\right) \tag{4.221}$$

for the precession frequency of a gyroscope relative to an asymptotically Lorentzian system (the "fixed stars").

The Komar Formula

For a stationary asymptotically flat metric of an isolated system there is a nice intrinsic formula for the total mass M in terms of the stationary Killing field, known as the **Komar formula**: If k denotes the 1-form belonging to the Killing field, then

$$M = -\frac{1}{8\pi G}\int_{S^2_\infty} *dk\,, \tag{4.222}$$

where the integral is taken over a spacelike 2-surface at spacelike infinity.

Let us prove this formula. Far away we get from (4.212)

$$\begin{aligned}
k &= (\partial_t)^\flat = g_{0\mu}\,dx^\mu = -\left(1 - \frac{2m}{r}\right)dt + \ldots\ ,\\
dk &= -\frac{2m}{r^3}\,x^i\,dx^i \wedge dt + \ldots\ ,\\
*dk &= -\frac{m}{r^3}\,x^i\varepsilon_{ijk}\,dx^j \wedge dx^k + \ldots
\end{aligned}$$

Therefore, we obtain

$$-\frac{1}{8\pi G}\int_{S^2_\infty} *dk = \lim_{r\to\infty}\frac{1}{8\pi G}\frac{m}{r^3}\int \varepsilon_{ijk}\,dx^j \wedge dx^k = M.$$

With Stoke's theorem we can convert the integral (4.222) into a volume integral if there are no horizons present. For this we need the identity

$$d * dk = 2 * R(k)\,, \tag{4.223}$$

where $R(k)$ is the 1-form with components $R_{\mu\nu}k^\nu$. To prove this we contract in the Ricci identity

$$k_{\sigma;\rho\mu} - k_{\sigma;\mu\rho} = R^\lambda{}_{\sigma\rho\mu}\,k_\lambda$$

the indices σ and ρ and use the consequence $k^{\sigma}{}_{;\sigma} = 0$ of the Killing equation $k_{\sigma;\rho} + k_{\rho;\sigma} = 0$. This gives

$$-k_{\sigma;\mu}{}^{;\sigma} = k_{\mu;\sigma}{}^{;\sigma} = -R_{\mu\sigma}k^{\sigma} . \tag{4.224}$$

This implies (see DG, (13.102))

$$\delta\, dk = -2\, R(k) , \tag{4.225}$$

where δ is the codifferential. But this last equation is equivalent to (4.223).

From (4.222) and (4.223) we obtain

$$M = -\frac{1}{4\pi G}\int_{\Sigma} *R(k) , \tag{4.226}$$

where Σ is a spacelike hypersurface extending to infinity. Adopting the argument which led to DG, (12.51) and using DG, (12.55), we obtain from the last equation

$$M = \frac{1}{4\pi G}\int_{\Sigma} R_{\mu\nu} n^{\mu} k^{\nu}\, \mathrm{Vol}_{\Sigma} , \tag{4.227}$$

where n^{μ} is the future directed unit normal on Σ. On the right we can use the field equations and obtain the **Tolman formula**:

$$M = 2\int_{\Sigma}\left(T_{\mu\nu} - \frac{1}{2}g_{\mu\nu}T^{\lambda}{}_{\lambda}\right) n^{\nu} k^{\nu}\mathrm{Vol}_{\Sigma} . \tag{4.228}$$

With the help of this formula we shall derive later (see Sect. 6.4) the general relativistic virial theorem for spherically symmetric stars.

* * *

Exercises.

Exercise 4.15. Show that the B^{jkl} in (4.203) can be written after decomposition in irreducible pieces as follows (round brackets indicate symmetrization in the enclosed indices):

$$B^{jkl} = \delta^{jk}A^{l} + C^{(j}\delta^{k)l} + \varepsilon^{ml(j}E^{k)m} + S^{jkl} ,$$

where E^{km} is symmetric and traceless and S^{jkl} is completely antisymmetric and traceless in all pairs of indices. From (4.204) we then have

$$C^{j} = -2\, A^{j} , \qquad E^{km} = S^{jkl} = 0 .$$

Exercise 4.16. Insert the ansatz

$$\begin{aligned} x^1 &= R(r)\sin\vartheta\cos\varphi , \\ x^2 &= R(r)\sin\vartheta\sin\varphi , \\ x^3 &= R(r)\cos\varphi , \\ x^0 &= t \end{aligned}$$

into the gauge condition $\Box x^\mu = 0$ and obtain the following differential equation for $R(r)$:

$$\frac{d}{dr}\left[r^2\left(1-\frac{2m}{r}\right)\frac{dR}{dr}\right] - 2R = 0\,.$$

A useful solution is $R = r - m$.

4.8 Gravitational Lensing

Gravitational Lensing has become an important field in astronomy and astrophysics. In recent years the rate and quality of the data have rapidly improved, thanks to considerable improvements of the observational capabilities (wide-field cameras, new telescopes, etc.).

Gravitational lensing produces magnifications and distortions of light beams from distant sources. It has the distinguishing feature of being independent of the nature and the physical state of the deflecting mass distributions. Therefore, it is perfectly suited to study dark matter on all scales. Moreover, the theoretical basis of lensing is very simple. For all practical purposes we can use the ray approximation for the description of light propagation. In this limit the rays correspond to null geodesics in a given gravitational field.

In this section we present an introduction to gravitational lensing that should help the reader to absorb the current literature.

4.8.1 Three Derivations of the Effective Refraction Index

In Chap. 1 we already showed in different ways that for almost Newtonian, asymptotically flat situations gravitational lensing theory is just standard ray optics with the **effective refraction index**

$$n(\boldsymbol{x}) = 1 - 2\,\phi(\boldsymbol{x})\,, \tag{4.229}$$

where $\phi(\boldsymbol{x})$ is the Newtonian potential. Let us recall how this result was obtained.

In Sect. 1.7 we saw that for a static metric the spatial path of a light ray is a geodesic of the Fermat metric $g^F_{ik} = g_{ik}/(-g_{00})$. For an almost Newtonian situation it was shown in Sect. 4.2 that

$$g = -(1 + 2\phi)\,dt^2 + (1 - 2\phi)\,d\boldsymbol{x}^2\,. \tag{4.230}$$

Therefore, the variational principle for geodesics of g^F_{ik} reduces to Fermat's principle of *ordinary* optics

$$\delta \int n(\boldsymbol{x}(\lambda))|\dot{\boldsymbol{x}}(\lambda)|\,d\lambda = 0\,, \tag{4.231}$$

with refraction index (4.229).

The result (4.229) was also obtained in Sect. 1.8 in our study of the geometrical optics limit. If we insert the eikonal ansatz for the Maxwell field

$$F_{\mu\nu} = \mathrm{Re}\left(f_{\mu\nu}\,\mathrm{e}^{i\psi}\right), \tag{4.232}$$

with a slowly varying amplitude $f_{\mu\nu}$ and a real rapidly varying ψ, we obtain in leading order the general relativistic eikonal equation (1.102):

$$g^{\mu\nu}\partial_\mu\psi\partial_\nu\psi = 0\,. \tag{4.233}$$

For an almost Newtonian situation we can set $\psi(x) = S(\boldsymbol{x}) - \omega t$, and we obtain the standard form (1.109) of the eikonal equation in ray optics with the refraction index (4.229).

We derive (4.233) once more slightly differently (working only with the field strengths). Write (4.232) in terms of differential forms as

$$F = f\,\mathrm{e}^{i\psi}\,.$$

(We omit to indicate that the real part on the right has to be taken.) From $dF = 0$ we get

$$df + i\,f \wedge d\psi = 0\,,$$

and $d * F = 0$ implies

$$d * f + i(*f) \wedge d\psi = 0\,.$$

In leading order we can neglect in these equations the differentials of f and $*f$,

$$f \wedge d\psi = 0\,, \qquad (*f) \wedge d\psi = 0\,. \tag{4.234}$$

Taking the interior product $i_{\nabla\psi}$ of the first equation we obtain

$$0 = i_{\nabla\psi}(f \wedge d\psi) = (i_{\nabla\psi} f) \wedge d\psi + f(\nabla\psi)^2\,.$$

Since the second equation in (4.234) is equivalent to $i_{\nabla\psi} f = 0$, we see that (4.233) holds.

We also repeat the Hamilton-Jacobi method for light propagation. The light rays $\gamma(\lambda)$ are the integral curves of the vector field $\nabla\psi$,

$$\dot{\gamma}(\lambda) = \nabla\psi(\gamma(\lambda))\,, \tag{4.235}$$

and are thus orthogonal to the wavefronts $\{\psi = \mathrm{const}\}$. The integral curves are null geodesics with affine parameter λ. Indeed, $\dot{\gamma}$ is a null vector because of (4.233):

$$g(\dot{\gamma}, \dot{\gamma}) = g(\nabla\psi, \nabla\psi) = 0\,.$$

Moreover,

$$\nabla_{\dot{\gamma}}\dot{\gamma} = \nabla_{\nabla\psi}(\nabla\psi)|_{\gamma(\lambda)}$$

vanishes, because the right-hand side is

$$(\nabla^\nu\psi)\nabla_\nu(\nabla^\mu\psi) = \frac{1}{2}\nabla^\mu(\nabla^\nu\psi\nabla_\nu\psi) = 0\,.$$

A third way of obtaining (4.229) makes use of Exercise 1.24, in which Maxwell's equations were written for a general static metric in terms of differential forms E, B, D and H. If the constitutive relations (1.185), $H = *B$, $D = *E$ are written for the almost Newtonian 3-metric $g_{ik} \simeq (1 - 2\phi)\delta_{ik}$, we obtain Maxwell's equations in *flat* spacetime for a **medium** with

$$\varepsilon = \mu \simeq 1 - 2\phi . \tag{4.236}$$

Equation (4.229) is thus found again, using Maxwell's relation:

$$n = \sqrt{\varepsilon\mu} \simeq 1 - 2\phi .$$

Finally, we recall how the ray equation in optics follows from Fermat's principle (4.231). The latter can be regarded as a Hamiltonian variational principle, with Lagrangian

$$L(\boldsymbol{x}, \dot{\boldsymbol{x}}) = n(\boldsymbol{x})\sqrt{\dot{\boldsymbol{x}}^2} . \tag{4.237}$$

When s is the Euclidean path length parameter, we have:

$$\frac{\partial L}{\partial \boldsymbol{x}} = \boldsymbol{\nabla} n , \qquad (\dot{\boldsymbol{x}}^2 = 1) ;$$
$$\frac{\partial L}{\partial \dot{\boldsymbol{x}}} = n \dot{\boldsymbol{x}} .$$

This gives for the Euler-Lagrange equation

$$\frac{d}{ds}\frac{\partial L}{\partial \dot{\boldsymbol{x}}} - \frac{\partial L}{\partial \boldsymbol{x}} = 0$$

the well known **ray equation**

$$\frac{d}{ds}\left(n \frac{d\boldsymbol{x}}{ds}\right) = \boldsymbol{\nabla} n . \tag{4.238}$$

At this point the qualitative picture sketched in Fig. 4.8 is useful for a first orientation. It shows the typical structure of wave fronts in the presence of a cluster perturbation, as well as the rays orthogonal to them.

For sufficiently strong lenses the wave fronts develop edges and self-interactions. An observer behind such folded fronts obviously sees more than one image. From this figure one also understands how the time delay of pairs of images arises: This is just the time elapsed between crossings of different sheets of the same wave front. Since this time difference is, as any other cosmological scale, depending on the Hubble parameter, gravitational lensing provides potentially a very interesting tool to measure the Hubble parameter. So far, the uncertainties of this method are still quite large, but the numbers fall into a reasonable range.

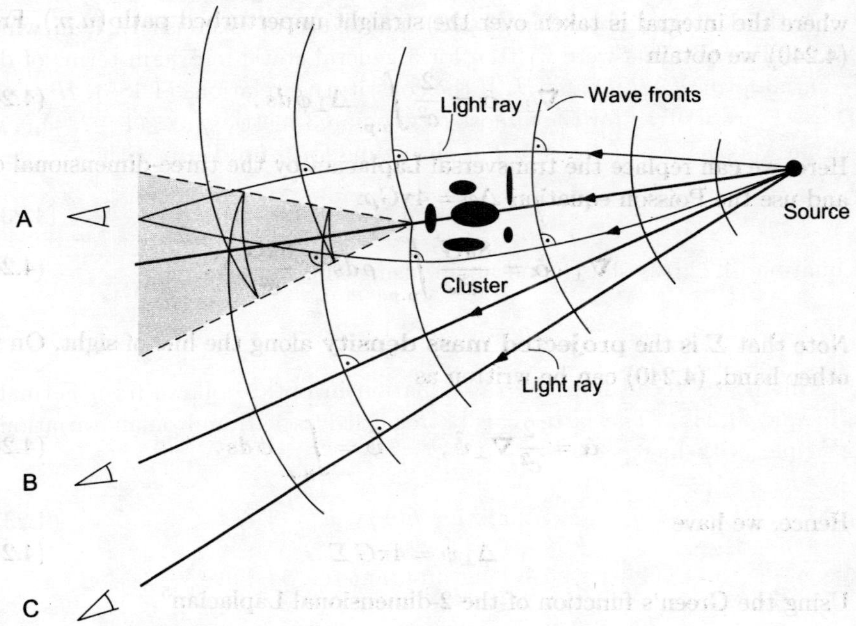

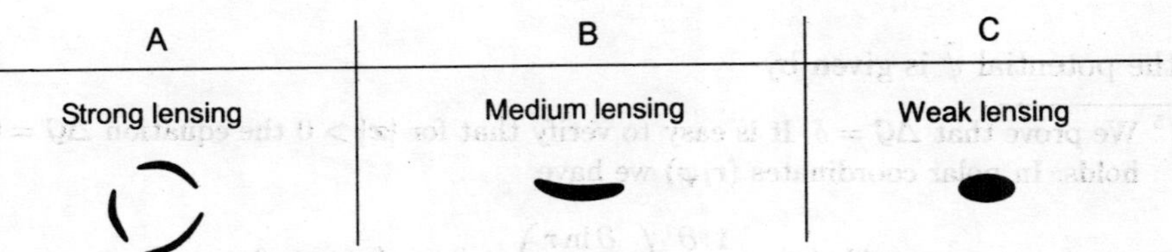

Fig. 4.8. Wave fronts and light rays in the presence of a cluster perturbation.

4.8.2 Deflection by an Arbitrary Mass Concentration

In terms of the unit tangent vector $\boldsymbol{e} = d\boldsymbol{x}/ds$, (4.238) can be approximated as (in this subsection we not set $c = 1$)

$$\frac{d}{ds}\boldsymbol{e} = -\frac{2}{c^2}\boldsymbol{\nabla}_\perp \phi\,, \tag{4.239}$$

where $\boldsymbol{\nabla}_\perp$ denotes the transverse derivative, $\boldsymbol{\nabla}_\perp = \boldsymbol{\nabla} - \boldsymbol{e}(\boldsymbol{e}\cdot\boldsymbol{\nabla})$. This gives for the **deflection angle**

$$\hat{\boldsymbol{\alpha}} = \boldsymbol{e}_{in} - \boldsymbol{e}_{fin}\,,$$

with initial and final directions $\boldsymbol{e}_{in}$ and $\boldsymbol{e}_{fin}$, respectively,

$$\hat{\boldsymbol{\alpha}} = \frac{2}{c^2}\int_{u.p.} \boldsymbol{\nabla}_\perp \phi\, ds\,, \tag{4.240}$$

where the integral is taken over the straight unperturbed path (*u.p.*). From (4.240) we obtain

$$\boldsymbol{\nabla}_\perp \cdot \hat{\boldsymbol{\alpha}} = \frac{2}{c^2}\int_{u.p.} \Delta_\perp \phi \, ds \,. \tag{4.241}$$

Here, we can replace the transversal Laplacian by the three-dimensional one and use the Poisson equation $\Delta\phi = 4\pi G\rho$:

$$\boldsymbol{\nabla}_\perp \cdot \hat{\boldsymbol{\alpha}} = \frac{8\pi G}{c^2}\int_{u.p.} \rho \, ds = \frac{8\pi G}{c^2}\Sigma \,. \tag{4.242}$$

Note that Σ is the **projected mass density** along the line of sight. On the other hand, (4.240) can be written as

$$\hat{\boldsymbol{\alpha}} = \frac{2}{c^2}\boldsymbol{\nabla}_\perp \hat{\psi} \,, \qquad \hat{\psi} = \int_{u.p.} \phi \, ds \,. \tag{4.243}$$

Hence, we have

$$\Delta_\perp \hat{\psi} = 4\pi G \, \Sigma \,. \tag{4.244}$$

Using the Green's function of the 2-dimensional Laplacian[5],

$$\mathcal{G}(\boldsymbol{\xi}) = \frac{1}{2\pi}\ln|\boldsymbol{\xi}| \,, \tag{4.245}$$

the potential $\hat{\psi}$ is given by

[5] We prove that $\Delta\mathcal{G} = \delta$. It is easy to verify that for $|\boldsymbol{x}| > 0$ the equation $\Delta\mathcal{G} = 0$ holds: In polar coordinates (r, φ) we have

$$\Delta \ln|\boldsymbol{x}| = \frac{1}{r}\frac{\partial}{\partial r}\left(r\frac{\partial \ln r}{\partial r}\right) = 0 \qquad \text{for } r > 0 \,.$$

Let f be a test function with $supp\, f \subset D_R$ (disk of radius R). Then

$$\langle \Delta \ln|\boldsymbol{x}|, f\rangle = \langle \ln|\boldsymbol{x}|, \Delta f\rangle = \lim_{\varepsilon\to 0}\int_{\varepsilon<|\boldsymbol{x}|<R} \ln|\boldsymbol{x}|\Delta f \, d^2\boldsymbol{x} \,.$$

For the last integral the second Green's formula gives

$$\int_{\varepsilon<|\boldsymbol{x}|<R} \Delta\ln|\boldsymbol{x}| \, f \, d^2x + \left(\int_{|\boldsymbol{x}|=\varepsilon} + \int_{|\boldsymbol{x}|=R}\right)\left(\ln|\boldsymbol{x}|\frac{\partial f}{\partial n} - f\frac{\partial \ln|\boldsymbol{x}|}{\partial n}\right) ds$$
$$= \int_{|\boldsymbol{x}|=\varepsilon}\left(-\ln|\boldsymbol{x}|\frac{\partial f}{\partial r} + f\frac{1}{r}\right) ds \,.$$

Here, only the second term survives in the limit $\varepsilon \longrightarrow 0$ and becomes equal to $2\pi\, f(0)$. This shows that

$$\langle \Delta \ln|\boldsymbol{x}|, f\rangle = 2\pi\langle \delta, f\rangle$$

for all test functions f, which proves our claim.

$$\hat{\psi}(\boldsymbol{\xi}) = 2G \int_{\mathbb{R}^2} \ln|\boldsymbol{\xi} - \boldsymbol{\xi}'|\Sigma(\boldsymbol{\xi}')\, d^2\boldsymbol{\xi}' . \tag{4.246}$$

From this we obtain for the deflection angle

$$\hat{\boldsymbol{\alpha}} = \frac{4G}{c^2} \int_{\mathbb{R}^2} \frac{\boldsymbol{\xi} - \boldsymbol{\xi}'}{|\boldsymbol{\xi} - \boldsymbol{\xi}'|^2}\, \Sigma(\boldsymbol{\xi}')\, d^2\boldsymbol{\xi}' . \tag{4.247}$$

For a point mass M located at the origin of the transversal plane, $\Sigma(\boldsymbol{\xi})$ is equal to $M\,\delta^2(\boldsymbol{\xi})$, and thus (4.247) gives Einstein's famous result for the deflection angle:

$$\hat{\alpha}(\boldsymbol{\xi}) = \frac{4G}{c^2} \frac{1}{|\boldsymbol{\xi}|} . \tag{4.248}$$

Consider, more generally, an axially symmetric lens with mass $M(\xi)$ located inside a cylinder defined by the impact parameter ξ, we expect from (4.248) that

$$\hat{\alpha}(\boldsymbol{\xi}) = \frac{4G}{c^2} \frac{M(\xi)}{\xi} . \tag{4.249}$$

This can easily be obtained by integrating (4.242) over a disk with radius ξ. Using the 2-dimensional version of Gauss' theorem we obtain for an axially symmetric lens

$$\int \boldsymbol{\nabla} \cdot \hat{\boldsymbol{\alpha}}\, d^2\xi = \int \hat{\boldsymbol{\alpha}} \cdot \boldsymbol{n}\, ds = 2\pi\xi\, \hat{\alpha}(\xi) = \frac{8\pi G}{c^2} M(\xi) ,$$

where $\boldsymbol{n}$ is the unit outward normal to the disk. The last equality just gives (4.249). (One also can obtain this result by introducing polar coordinates in (4.247) and working out the angular integration, but the above derivation is much simpler.)

We shall study axisymmetric lens models in more detail later, but proceed now with general lenses.

4.8.3 The General Lens Map

Figure 4.9 summarizes some of the notation we are using. (We will follow as much as possible the beautiful monograph [19].) Simple geometry shows that

$$\boldsymbol{\eta} = \frac{D_s}{D_d}\boldsymbol{\xi} - D_{ds}\hat{\boldsymbol{\alpha}}(\boldsymbol{\xi}) , \tag{4.250}$$

where $\boldsymbol{\eta}$ is the source position and $\boldsymbol{\xi}$ is the lens position. This defines a map $\boldsymbol{\xi} \longrightarrow \boldsymbol{\eta}$ from the lens plane to the source plane. Figure 4.9 also shows that

$$\boldsymbol{\xi} = D_d\boldsymbol{\theta} , \qquad \boldsymbol{\eta} = D_s\boldsymbol{\beta} , \tag{4.251}$$

hence (4.250) can be written as

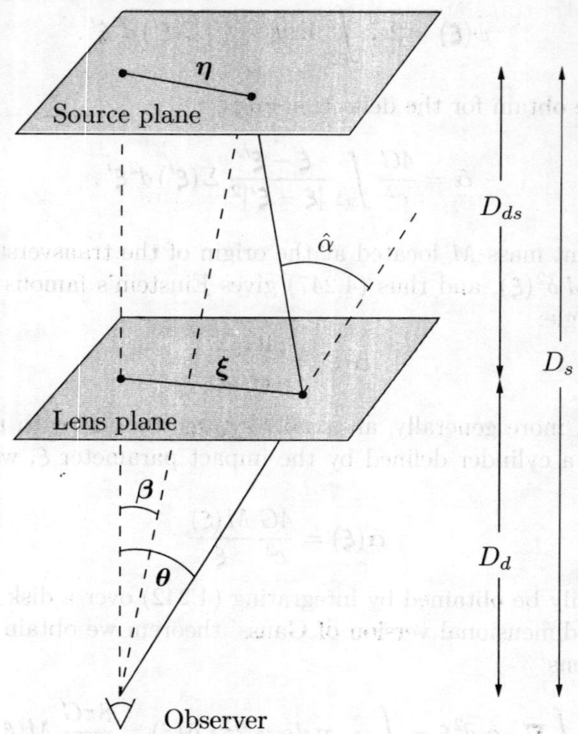

Fig. 4.9. Notation adopted for the description of the lens geometry.

$$\boldsymbol{\beta} = \boldsymbol{\theta} - \frac{D_{ds}}{D_s}\hat{\boldsymbol{\alpha}}. \tag{4.252}$$

This or (4.250) is what is called the **lens equation**. It turns out, not unexpectedly, that (4.252) holds also in cosmology.

It is convenient to write (4.250) in dimensionless form. Let ξ_0 be a length parameter in the lens plane (whose choice will depend on the specific problem), and let η_0 be the corresponding scaled length in the source plane, $\eta_0 = (D_s/D_d)\xi_0$. We set $\boldsymbol{x} = \boldsymbol{\xi}/\xi_0$, $\boldsymbol{y} = \boldsymbol{\eta}/\eta_0$, and

$$\kappa(\boldsymbol{x}) = \frac{\Sigma(\xi_0\boldsymbol{x})}{\Sigma_{crit}}, \qquad \boldsymbol{\alpha}(\boldsymbol{x}) = \frac{D_d D_{ds}}{\xi_0 D_s}\hat{\boldsymbol{\alpha}}(\xi_0\boldsymbol{x}), \tag{4.253}$$

with the **"critical surface density"**

$$\Sigma_{crit} = \frac{c^2}{4\pi G}\,\frac{D_s}{D_d D_{ds}} = 0.35 g\,cm^{-2}\left(\frac{1\,Gpc}{D_d D_{ds}/D_s}\right). \tag{4.254}$$

Then (4.250) reads as follows

$$\boldsymbol{y} = \boldsymbol{x} - \boldsymbol{\alpha}(x), \tag{4.255}$$

whereby (4.247) translates to

$$\boldsymbol{\alpha}(\boldsymbol{x}) = \frac{1}{\pi}\int_{\mathbb{R}^2} \frac{\boldsymbol{x}-\boldsymbol{x}'}{|\boldsymbol{x}-\boldsymbol{x}'|^2}\,\kappa(\boldsymbol{x}')\,d^2x' \,. \tag{4.256}$$

Note that for $\xi_0 = D_d$ we have $\boldsymbol{x} = \boldsymbol{\theta}$ and $\boldsymbol{y} = \boldsymbol{\beta}$.

As in (4.243) we can write $\boldsymbol{\alpha}$ as a gradient

$$\boldsymbol{\alpha} = \boldsymbol{\nabla}\psi \,, \tag{4.257}$$

where according to (4.244) it follows that ψ satisfies the 2-dimensional Poisson equation

$$\Delta\psi = 2\,\kappa \,. \tag{4.258}$$

The **lens map** $\varphi : \boldsymbol{x} \longmapsto \boldsymbol{y}$ defined by (4.250) is thus a gradient map

$$\varphi(\boldsymbol{x}) = \boldsymbol{\nabla}\left(\frac{1}{2}\boldsymbol{x}^2 - \psi(\boldsymbol{x})\right) . \tag{4.259}$$

From (4.258) and (4.245) we obtain

$$\psi(\boldsymbol{x}) = \frac{1}{\pi}\int_{\mathbb{R}^2} \ln(|\boldsymbol{x}-\boldsymbol{x}'|)\,\kappa(\boldsymbol{x}')\,d^2x' \,. \tag{4.260}$$

The differential $D\varphi$ will often be used. A standard parameterization is

$$D\varphi = \begin{pmatrix} 1-\kappa-\gamma_1 & -\gamma_2 \\ -\gamma_2 & 1-\kappa+\gamma_1 \end{pmatrix} , \tag{4.261}$$

where $\gamma_1 = \frac{1}{2}(\partial_{11}\psi - \partial_{22}\psi)$, and $\gamma_2 = \partial_{12}\psi = \partial_{21}\psi$. Note, that the matrix elements of $D\varphi$ are (see, e.g., (4.259))

$$(D\varphi)_{ij} = \delta_{ij} - \partial_i\partial_j\psi \,, \tag{4.262}$$

where $\partial_i = \partial/\partial x_i$. In particular, the trace of (4.261) is chosen correctly (see (4.258)), and $D\varphi$ is clearly symmetric. The dimensionless projected mass density κ is often called the **mean (Ricci) curvature**. Note that the **complex shear**

$$\gamma = \gamma_1 + i\,\gamma_2$$

describes the trace-free part of $D\varphi$ and does, therefore, *not* transform like a tensor, but like h_{kl}^{TT} of a weak gravitational wave in the TT gauge (see Sect. 4.3).

Without proof we remark that $\kappa > 1$ ($\Sigma > \Sigma_{crit}$) somewhere implies that for certain source positions there are multiple images. (For a proof see Sect. 5.4.3 of [19].) This condition is, however, not necessary.

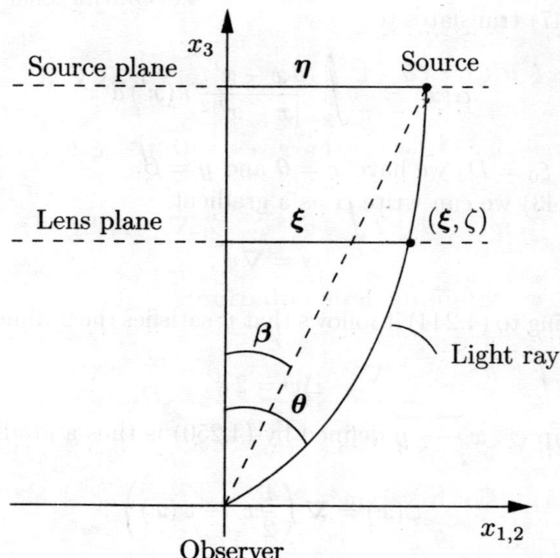

Fig. 4.10. Notation adopted for light rays.

Alternative Derivation of the Lens Equation

We give here another derivation of the basic equations (4.247) and (4.252) that can be generalized to a cosmological situation (see [115]). By making use of the notation in the Fig. 4.10, the differential equation for a light ray $\boldsymbol{\xi} = \boldsymbol{\xi}(\boldsymbol{\theta}, \zeta)$ ($\boldsymbol{\theta}$ denotes the observing angles) is, to first order in the Newtonian potential:

$$\frac{d^2\boldsymbol{\xi}}{d\zeta^2} = -2\boldsymbol{\nabla}_\perp \phi(\boldsymbol{\xi}(\boldsymbol{\theta}, \zeta), \zeta) . \tag{4.263}$$

(Note that $d\zeta/ds = 1$, up to quadratic terms in the small ξ_i.)

With the help of the Green's function $\mathcal{G}(\zeta, \zeta')$ of the operator $d^2/d\zeta^2$, given by

$$\mathcal{G}(\zeta, \zeta') = (\zeta - \zeta')\, \theta(\zeta - \zeta') , \tag{4.264}$$

we transform (4.263) into an integral equation,

$$\boldsymbol{\xi}(\boldsymbol{\theta}, \zeta) = \zeta \boldsymbol{\theta} - 2 \int_0^\zeta d\zeta' \, (\zeta - \zeta') \boldsymbol{\nabla}_\perp \phi(\boldsymbol{\xi}(\boldsymbol{\theta}, \zeta'), \zeta') . \tag{4.265}$$

Here, the first term is a solution of the homogeneous equation, which would describe the light ray without deflection. Clearly,

$$\left. \frac{d}{d\zeta} \right|_{\zeta=0} \boldsymbol{\xi}(\boldsymbol{\theta}, \zeta) = \boldsymbol{\theta} . \tag{4.266}$$

To first order in ϕ, we can replace the argument in $\boldsymbol{\nabla}_\perp \phi$ on the right-hand side of (4.265) by the unperturbed ray and obtain the explicit solution

$$\boldsymbol{\xi}(\boldsymbol{\theta},\zeta) = \zeta\boldsymbol{\theta} - 2\int_0^\zeta d\zeta'\,(\zeta-\zeta')\boldsymbol{\nabla}_\perp\phi(\zeta'\boldsymbol{\theta},\zeta'). \tag{4.267}$$

At the source plane ($\zeta = \zeta_s$) this gives, with $\boldsymbol{\eta} \equiv \boldsymbol{\xi}(\boldsymbol{\theta},\zeta_s)$,

$$\boldsymbol{\eta} = \zeta_s\boldsymbol{\theta} - 2\int_0^{\zeta_s} d\zeta\,(\zeta_s-\zeta)\boldsymbol{\nabla}_\perp\phi(\zeta\boldsymbol{\theta},\zeta). \tag{4.268}$$

Since $\boldsymbol{\eta} = \zeta_s\boldsymbol{\beta}$, we obtain the **lens equation**

$$\boldsymbol{\beta} = \boldsymbol{\theta} - 2\int_0^{\zeta_s} \underbrace{d\zeta\,\frac{\zeta_s-\zeta}{\zeta_s}}_{(1)}\boldsymbol{\nabla}_\perp\underbrace{\phi\,(u.p.)}_{(2)} \simeq \boldsymbol{\theta} - \frac{D_{ds}}{D_s}\,2\boldsymbol{\nabla}_\perp\hat{\psi}(D_d\boldsymbol{\theta})\,; \tag{4.269}$$

we used that (1) is slowly varying $\simeq \frac{\zeta_s-\zeta_d}{\zeta_s} = \frac{D_{ds}}{D_s}$, and that the unperturbed path (2) is given by $(\zeta_d\boldsymbol{\theta},\zeta) = (D_d\boldsymbol{\theta},\zeta)$. The potential $\hat{\psi}$ is defined by

$$\hat{\psi}(D_d\boldsymbol{\theta}) = \int d\zeta\,\phi(D_d\boldsymbol{\theta},\zeta)\,. \tag{4.270}$$

Thus

$$\boldsymbol{\beta} = \boldsymbol{\theta} - \frac{D_{ds}}{D_s}\hat{\boldsymbol{\alpha}}\,, \qquad \hat{\boldsymbol{\alpha}} = 2\boldsymbol{\nabla}_\perp\hat{\psi}\,. \tag{4.271}$$

From here, we again obtain

$$\boldsymbol{\nabla}_\perp\cdot\hat{\boldsymbol{\alpha}} = 2\int \Delta_\perp U\,d\zeta = 8\pi G\,\Sigma\,, \tag{4.272a}$$

$$\Delta_\perp\hat{\psi} = 4\pi G\,\Sigma\,. \tag{4.272b}$$

These results can be expressed in terms of the *rescaled quantities*

$$\boldsymbol{\alpha}(\boldsymbol{\theta}) = \frac{D_{ds}}{D_s}\,\hat{\boldsymbol{\alpha}}(D_d\boldsymbol{\theta})\,, \tag{4.273a}$$

$$\psi(\boldsymbol{\theta}) = \frac{2\,D_{ds}}{D_s\,D_d}\,\hat{\psi}(D_d\,\boldsymbol{\theta}) \tag{4.273b}$$

as before (for $\xi_0 = D_d$):

$$\boldsymbol{\beta} = \boldsymbol{\theta} - \boldsymbol{\alpha}(\boldsymbol{\theta})\,, \tag{4.274a}$$

$$\boldsymbol{\alpha} = \boldsymbol{\nabla}\psi\,, \tag{4.274b}$$

$$\Delta\psi = 2\,\kappa\,. \tag{4.274c}$$

4.8.4 Magnification, Critical Curves and Caustics

Next, we show that the **magnification** μ, that is, the ratio of the flux of an image to the flux of the unlensed source, is given by

$$\mu = \frac{1}{|\det D\varphi|}\,. \tag{4.275}$$

In order to derive this result, we recall a simple but important fact from ray optics.

Consider a ray L and any two points along the ray and construct areas dA_1 and dA_2 normal to the ray at these points. Let $dE_1 = dE_2$ be the energy of all rays passing through both dA_1 and dA_2 during the time dt. Since

$$dE_1 = I_{\nu_1}\, dA_1 dt\, d\Omega_1\, d\nu_1\,, \qquad dE_2 = I_{\nu_2}\, dA_2 dt\, d\Omega_2\, d\nu_2\,,$$

where $d\Omega_{1,2}$ is the solid angle subtended by $dA_{2,1}$ at $dA_{1,2}$, and because $d\Omega_1 = dA_2/R^2$, $d\Omega_2 = dA_1/R^2$ (R denotes the distance between dA_1 and dA_2), we obtain

$$I_{\nu_1}\, d\nu_1 = I_{\nu_2}\, d\nu_2\,. \tag{4.276}$$

If there is no frequency shift, the specific intensity is thus constant along a ray

$$I_\nu = \text{const}\,. \tag{4.277}$$

This holds also for gravitational light deflection by localized, nearly static lenses, because this does not introduce an additional frequency shift between source and observer, beside the cosmological one.[6] Now, the flux of an image of an infinitesimal source is the product of its surface brightness I and the solid angle $d\Omega$ it subtends on the sky. From (4.277) we conclude that the magnification μ is the ratio of $d\Omega$ and the solid angle $d\Omega_0$ for the undeflected situation. On the other hand $d\Omega_0/d\Omega$ is equal to the area distortion of the lens map φ, and thus equal to the Jacobian $|\det D\varphi|$. This proves our claim (4.275). (For a more sophisticated derivation, which applies also in cosmology, see [19], Sects. 3.4–3.6, (3.81), (3.82) in particular.)

The lens map φ becomes singular along **critical curves** in the lens plane. These are characterized by

$$\det(D\varphi) = 0 \tag{4.278}$$

or, using (4.261), by

$$(1-\kappa)^2 - |\gamma|^2 = 0\,. \tag{4.279}$$

The **caustics** are the images of these critical curves[7].

---

[6] More general argument: Without absorption and scattering processes the distribution function f for photons satisfies the Liouville equation (for details see, e.g., Chap. 6 of [79])

$$L_{X_g} f = 0\,,$$

where X_g is the vector field of the geodesic spray. Since the intensity $I(\omega)$ is proportional to $\omega^3 f$, we conclude that $I(\omega)/\omega^3$ remains *constant along null-geodesics.*

[7] A famous theorem of Sard tells us that these critical values of φ form a set of measure zero.

In the vicinity of these source points the magnification becomes very large. On caustics it diverges formally, but this is of no physical significance, because the magnification remains finite for any extended source (see [19], Sect. 6.4). For a point–like source, the ray approximation breaks down and we would have to use wave optics. (One finds a discussion of this in Chap. 7 of [19].)

4.8.5 Simple Lens Models

It is now high time to study some simple, but important examples of specific types of lenses. Although they are simple, they turn out to be very useful to better understand the lensing phenomenon.

Axially Symmetric Lenses: Generalities

If the lens is axially symmetric, our general lensing equations simplify considerably. For the deflection angle, this was already shown in Sect. 4.8.2. According to (4.249)

$$\hat{\alpha}(\xi) = \frac{4G}{c^2}\frac{M(\xi)}{\xi} . \tag{4.280}$$

Note that only the modulus of the angle counts. For the rescaled angle $\alpha(x)$ in (4.253) this translates to

$$\alpha(x) = \frac{m(x)}{x} , \qquad \boldsymbol{x} = (x, 0) , x > 0 , \tag{4.281}$$

where

$$m(x) = 2 \int_0^x \kappa(x')x' \, dx' . \tag{4.282}$$

The lens equation (4.255) can be written in *scalar* form

$$y = x - \alpha(x) = x - \frac{m(x)}{x} , \tag{4.283}$$

where now $x \in \mathbb{R}$ and $m(\boldsymbol{x}) = m(|x|)$. From (4.257) we obtain

$$\alpha = \frac{d\psi}{dx} . \tag{4.284}$$

The Poisson equation (4.258) for ψ becomes

$$\frac{1}{x}\frac{d}{dx}\left(x \frac{d\psi}{dx}\right) = 2\,\kappa . \tag{4.285}$$

Inserting here (4.284) and (4.281) leads to

$$\frac{dm}{dx} = 2x\,\kappa(x) , \tag{4.286}$$

which of course also follows from (4.282). From (4.284), (4.281) and (4.282) we have

$$\frac{d\psi}{dx} = \frac{2}{x}\int_0^x \kappa(x')x'\,dx' \,. \tag{4.287}$$

In this equation, the right-hand side is also equal to

$$\frac{d}{dx}\,2\int_0^x \kappa(x')x' \ln\left(\frac{x}{x'}\right) dx' \,.$$

Thus, provided that $\kappa(x)$ decreases faster than x^{-1}, we find

$$\psi(x) = 2\int_0^x \kappa(x')x' \ln\left(\frac{x}{x'}\right) dx' \,. \tag{4.288}$$

Let us also look at the differential $D\varphi$ of the lens map. According to (4.283), φ is given by

$$\boldsymbol{y} = \boldsymbol{x} - \frac{m(x)}{x^2}\boldsymbol{x} \,, \qquad x = |\boldsymbol{x}| \,. \tag{4.289}$$

Hence,

$$D\varphi = 1 - \frac{m(x)}{x^4}\begin{pmatrix} x_2^2 - x_1^2 & -2x_1x_2 \\ -2x_1x_2 & x_1^2 - x_2^2 \end{pmatrix} - \frac{m'(x)}{x^3}\begin{pmatrix} x_1^2 & x_1x_2 \\ x_1x_2 & x_2^2 \end{pmatrix} . \tag{4.290}$$

Because of (4.286), the trace is correct (see (4.261)), and the components of the complex shear are

$$\gamma_1 = \frac{1}{2}\left(x_2^2 - x_1^2\right)\left(\frac{2m}{x^4} - \frac{m'}{x^3}\right), \qquad \gamma_2 = x_1\,x_2\left(\frac{m'}{x^3} - \frac{2m}{x^4}\right) . \tag{4.291}$$

This gives

$$|\gamma|^2 = \left(\frac{m}{x^2} - \kappa\right)^2$$

and

$$\begin{aligned} \det D\varphi = (1-\kappa)^2 - |\gamma|^2 &= \left(1 - \frac{m}{x^2}\right)\left(1 + \frac{m}{x^2} - 2\kappa\right) \\ &= \left(1 - \frac{1}{x}\frac{d\psi}{dx}\right)\left(1 - \frac{d^2\psi}{dx^2}\right) . \end{aligned} \tag{4.292}$$

The last two factors are the eigenvalues of $D\varphi$.

This implies that there are two types of critical curves in the lense plane

$$\begin{aligned} \frac{m(x)}{x^2} &= 1 \qquad \textbf{tangential critical curve}\,, \\ \frac{d}{dx}\left(\frac{m(x)}{x}\right) &= 1 \qquad \textbf{radial critical curve}\,. \end{aligned} \tag{4.293}$$

(The reasons for this terminology will soon become clear.) The image of a tangential critical curve degenerates according to (4.283) into the point $y = 0$ in the source plane.

We can look at the critical points on the x_1-axis with $\boldsymbol{x} = (x, 0)$, $x > 0$. Then

$$D\varphi = 1 - \frac{m(x)}{x^2}\begin{pmatrix} -1 & 0 \\ 0 & +1 \end{pmatrix} - \frac{m'}{x}\begin{pmatrix} 1 & 0 \\ 0 & 0 \end{pmatrix} \tag{4.294}$$

and this matrix must have an eigenvector $\boldsymbol{X}$ with eigenvalue zero. For symmetry reasons, the vector must be either tangential, $\boldsymbol{X} = (0, 1)$, or normal, $\boldsymbol{X} = (1, 0)$, to the critical curve (which must be a circle). We see readily that the first case occurs for a tangential critical curve, and the second for a radial critical curve. It turns out that the radial critical curve consists of folds.

For a tangential critical curve $\{|\boldsymbol{x}| = x_t\}$, we have by (4.293) and (4.282)

$$m(x_t) = \int_0^{x_t} 2x\,\kappa(x)\,dx = x_t^2\,. \tag{4.295}$$

With (4.253) this translates to

$$\int_0^{\xi_t} 2\xi\,\Sigma(\xi)\,d\xi = \xi_t^2\,\Sigma_{crit}\,. \tag{4.296}$$

The total mass $M(\xi_t)$ inside the critical curve is thus

$$M(\xi_t) = \pi\,\xi_t^2\,\Sigma_{crit}\,. \tag{4.297}$$

This shows that the average density $\langle\Sigma\rangle_t$ inside the tangential critical curve is equal to the critical density,

$$\langle\Sigma\rangle_t = \Sigma_{crit}\,. \tag{4.298}$$

(Correspondingly, $\langle\kappa\rangle_t = 1$). This is one of the reasons for introducing Σ_{crit}.

This can be used to estimate the mass of a deflector if the lens is sufficiently strong and the geometry is such that almost complete Einstein rings[8] are formed. If θ_{arc} is the angular distance of the arc, we obtain numerically

$$\begin{aligned} M(\langle\theta_{arc}\rangle) &= \pi\,(D_d\theta_{arc})^2\,\Sigma_{crit} \\ &\approx (1.1\times 10^{14} M_\odot)\left(\frac{\theta_{arc}}{30''}\right)^2\left(\frac{D}{1Gpc}\right), \end{aligned} \tag{4.299}$$

where D is defined in (4.310) below.

---

[8] If the source is on the symmetry axis behind an axially symmetric lens it is seen as a ring. There are beautiful examples where this ideal condition is nearly satisfied. Ring-shaped images are today called 'Einstein rings', but this possibility was already described by O. Chwolson in 1924.

The Schwarzschild Lens: Microlensing

This is the simplest lens and is most relevant for the search of massive galactic halo objects (MACHOs).

We shall soon see that a convenient length scale ξ_0 is provided by the **Einstein radius**

$$R_E = \sqrt{\frac{4GM}{c^2}\frac{D_d D_{ds}}{D_s}} = 610\ R_\odot \left(\frac{M}{M_\odot}\frac{D_s}{kpc}\frac{D_d}{D_s}\left(1 - \frac{D_d}{D_s}\right)\right)^{1/2}. \tag{4.300}$$

Since $\Sigma(\boldsymbol{\xi}) = M\,\delta^2(\boldsymbol{\xi})$, we then have $\kappa(\boldsymbol{x}) = \pi\delta^2(\boldsymbol{x})$, according to (4.253), and thus $m(x) \equiv 1$. The latter equation follows also from (4.280), (4.281) and (4.253). Thus, since (4.284) implies $\psi(x) = \ln x$, we have

$$\alpha(x) = \frac{1}{x}, \tag{4.301}$$

and the lens map is given by

$$y = x - \frac{1}{x}\,. \tag{4.302}$$

If the source is on the symmetry axis ($y = 0$), then $x = \pm 1$ (Einstein ring). For a given source position y, (4.302) has two solutions

$$x_{1,2} = \frac{1}{2}\left(y \pm \sqrt{y^2+4}\right). \tag{4.303}$$

The magnification $\mu = |\det D\varphi|^{-1}$ follows immediately from (4.292)

$$\mu^{-1} = \left|1 - \frac{1}{x^4}\right|, \tag{4.304}$$

which gives for the two images

$$\mu_{1,2} = \frac{1}{4}\left|\frac{y}{\sqrt{y^2+4}} + \frac{\sqrt{y^2+4}}{y} \pm 2\right|. \tag{4.305}$$

The total magnification $\mu_p = \mu_1 + \mu_2$ is found to be

$$\mu_p = \frac{y^2+2}{y\sqrt{y^2+4}}\,. \tag{4.306}$$

It is this function which one observes for MACHOs. Note that

$$y = \frac{\eta}{\eta_0} = \frac{\eta}{(D_s/D_d)\xi_0} = \frac{\eta/D_s}{\xi_0/D_d} = \beta\,\theta_E^{-1}, \tag{4.307}$$

where $\theta_E = R_E/D_d$, i.e.,

$$\theta_E = \left(\frac{4GM}{c^2} \frac{D_{ds}}{D_d D_s} \right)^{1/2} \tag{4.308}$$

is the angular separation corresponding to the Einstein radius, and β (see Fig. 4.9) is the angular separation of the source from the optical axis. Numerically, (4.308) reads

$$\begin{aligned} \theta_E &= (0.9'' \times 10^{-3}) \left(\frac{M}{M_\odot} \right)^{1/2} \left(\frac{D}{10kpc} \right)^{-1/2} \\ &= (0.9'') \left(\frac{M}{10^{12} M_\odot} \right)^{1/2} \left(\frac{D}{1Gpc} \right)^{-1/2} , \end{aligned} \tag{4.309}$$

where

$$D := \frac{D_d D_s}{D_{ds}} \tag{4.310}$$

is the **effective distance.**

Even when it is not possible to see multiple images, the magnification can still be detected if the lens and source move relative to each other giving rise to lensing-induced time variability in the magnification. This kind of variability is called **microlensing**. Microlensing was first observed in the multiple-imaged QSO2237+0305 (see [116]). In 1986, Paczyński proposed to monitor millions of stars in the LMC to look for such magnifications in a fraction of the sources. In the meantime, this has been successfully implemented. The time scale for microlensing-induced variations is obviously given by $t_0 = D_d \theta_E / v$, where v is a typical virial velocity of the galactic halo. Numerically

$$t_0 = 0.214yr \left(\frac{M}{M_\odot} \right)^{1/2} \left(\frac{D_d}{10kpc} \right)^{1/2} \left(\frac{D_{ds}}{D_s} \right)^{1/2} \left(\frac{200km\, s^{-1}}{v} \right) . \tag{4.311}$$

Note that the ratio D_{ds}/D_s is close to unity.

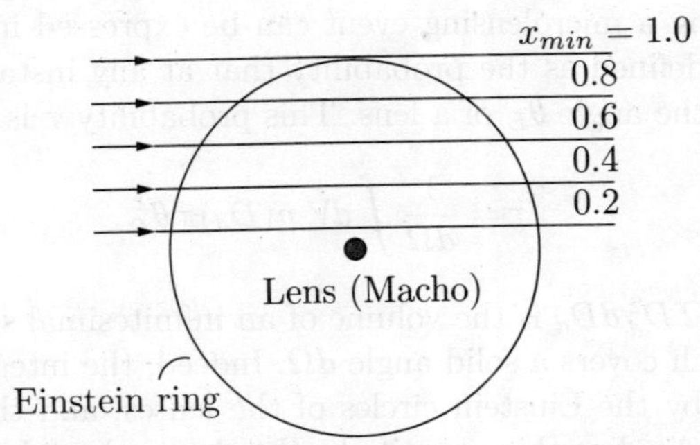

Fig. 4.11. Einstein ring and five possible relative orbits of a background star with projected minimal distances $x_{\min} = \xi_{\min}/R_E = 0.2, 0.4, \ldots, 1.0$.

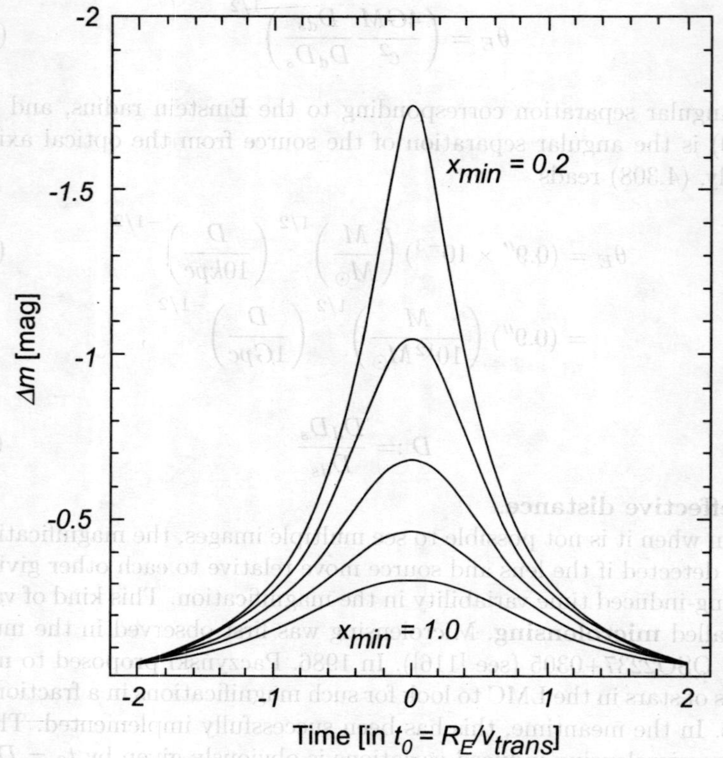

Fig. 4.12. Light curves for the five cases in Fig. 4.11. The maximal magnification is $\mu = 1.34$ or $\Delta m = -0.32 mag$, if the star just touches the Einstein radius ($x_{min} = 1.0$). For smaller values of x_{min} the maximal magnification is larger.

Typical light variation curves corresponding to (4.306) are shown in Fig. 4.11 and Fig. 4.12. Note that t_0 does not directly give the mass. The chance of seeing a microlensing event can be expressed in terms of the **optical depth**, defined as the probability that at any instant of time a given star is within the angle θ_E of a lens. This probability τ is given by

$$\tau = \frac{1}{d\Omega} \int dV \, n(D_d) \pi \, \theta_E^2 \,, \tag{4.312}$$

where $dV = d\Omega \, D_d^2 \, dD_d$ is the volume of an infinitesimal spherical shell with radius D_d which covers a solid angle $d\Omega$. Indeed, the integral gives the solid angle covered by the Einstein circles of the lenses, and the probability τ is obtained upon dividing this quantity by the observed solid angle $d\Omega$. Inserting the expression (4.308) for θ_E gives

$$\tau = \int_0^{D_s} \frac{4\pi G \, \rho}{c^2} \frac{D_d D_{ds}}{D_s} \, dD_d = \frac{4\pi G}{c^2} D_s^2 \int_0^1 \rho(x) x (1-x) \, dx \,, \tag{4.313}$$

where $x = D_d D_s^{-1}$ and ρ is the mass density of the MACHOs. It is this density that determines τ. Using a realistic density profile one finds $\tau_{LMC} \simeq 5\times 10^{-7}$. The results of several teams searching for the dark matter around our Galaxy in the form of MACHOs are puzzling. Twenty candidates of $\approx 0.5 M_\odot$ were so far detected, indicating that at most 20% of the mass of the halo can be incorporated in objects in the mass range 10^{-7} to 10 solar masses. For a recent report we refer to [117].

Singular Isothermal Sphere

The so-called singular isothermal sphere is often used as a simple model for the mass distribution in elliptical galaxies. One arrives at this model by assuming an ideal isothermal gas law $p = (\rho/m)k_B T$ for the equation of state, where ρ is the mass density of stars and m the (average) mass of a star. The equation of hydrostatic equilibrium then gives

$$\frac{k_B T}{m}\frac{d\rho}{dr} = -\rho \frac{GM(r)}{r^2} . \tag{4.314}$$

If we multiply this by $r^2(m/k_B T)$ and then differentiate with respect to r, we obtain, using

$$\frac{dM(r)}{dr} = 4\pi r^2 \rho , \tag{4.315}$$

the differential equation

$$\frac{d}{dr}\left(r^2 \frac{d}{dr}\ln\rho\right) = -\frac{Gm}{k_B T} 4\pi r^2 \rho . \tag{4.316}$$

One arrives at this equation also in the kinetic theory as follows. Start from the Jeans equation, which one obtains by taking the first moment of the collisionless Boltzmann equation. In the stationary, spherically symmetric case this reads (see [20])

$$\frac{d}{dr}\left(n\sigma_r^2\right) + \frac{2n}{r}\left(\sigma_r^2 - \sigma_t^2\right) = -n\frac{d\phi}{dr} , \tag{4.317}$$

where n is the density of particles, and σ_r, σ_t are, respectively, the radial and transversal, velocity dispersions.

For the special case $\sigma_r^2 = \sigma_t^2 \equiv \sigma^2 = \text{const}$, (4.317) reduces to

$$\sigma^2 \frac{dn}{dr} = -n\frac{GM(r)}{r^2} , \tag{4.318}$$

which is identical to (4.314) for $\rho = n\,m$, if we make the identification

$$k_B T = m\sigma^2 . \tag{4.319}$$

(Note, that σ^2 is the 1-dimensional velocity dispersion.) One solution, with a power dependence for $\rho(r)$, is easily found

$$\rho(r) = \frac{\sigma^2}{2\pi G r^2} \,. \tag{4.320}$$

Because the density is singular at the origin, it is called the **singular isothermal sphere** (regular solutions are only known numerically, see [20]).

The projected mass density for (4.320) is easily found to be

$$\Sigma(\xi) = \frac{\sigma^2}{2G}\frac{1}{\xi} \,. \tag{4.321}$$

For the length scale ξ_0 we choose

$$\xi_0 = 4\pi\left(\frac{\sigma}{c}\right)^2 \frac{D_d D_{ds}}{D_s} \tag{4.322}$$

and obtain from (4.321) and (4.253)

$$\kappa(x) = \frac{1}{2|x|} \,, \tag{4.323}$$

thus

$$m(x) = |x| \,, \qquad \alpha(x) = \frac{x}{|x|} \,, \tag{4.324}$$

and

$$y = x - \frac{x}{|x|} \,. \tag{4.325}$$

Note that the second equation in (4.324) implies that $\hat{\alpha} = 4\pi\sigma^2/c^2 = \text{const.}$ The Einstein ring is given by $|x| = 1$, $|\boldsymbol{\xi}| = \xi_0$. The corresponding angle is thus

$$\theta_E = 4\pi\left(\frac{\sigma}{c}\right)^2 \frac{D_{ds}}{D_s} \simeq (29'')\left(\frac{\sigma}{10^3 km\ s^{-1}}\right)^2 \frac{D_{ds}}{D_s} \,. \tag{4.326}$$

This is often used for clusters of galaxies.

We also note the mass $M(<\xi_0)$ inside the Einstein radius

$$\begin{aligned} M(<\xi_0) &= \int_0^1 \kappa(x)\Sigma_{crit}\, 2\pi x\, dx = \pi\,\xi_0^2\,\Sigma_{crit} = \pi(D_d\theta_E)^2\Sigma_{crit} \\ &\simeq 1.1\times 10^{14} M_\odot \left(\frac{\theta_E}{30''}\right)^2 \left(\frac{D}{1Gpc}\right) \,, \end{aligned} \tag{4.327}$$

where D is again given by (4.310).

The magnification for an image at x is easily found from (4.292):

$$\mu = \left|\frac{|x|}{|x|-1}\right| \,. \tag{4.328}$$

For $|y| > 1$ one sees from (4.324) that there is only one image at $x = y + 1$ (take $y > 0$). When $|y| < 1$, there are two images, at $x = y + 1$ and $x = y - 1$. Using this in (4.328) we find for the total magnification of a point source

$$\mu_p = \begin{cases} 2/y & \text{for } y \le 1\,, \\ (1+y)/y & \text{for } y \ge 1\,. \end{cases} \tag{4.329}$$

Note that the inner image becomes very faint for $y \longrightarrow 1$.

Isothermal Sphere with Finite Core Radius

Since no analytic solutions of (4.316) without a central singularity are known, the surface mass density (4.321) is often modified parametrically by introducing a finite core radius ξ_c

$$\Sigma(\xi) = \frac{\sigma^2}{2G} \frac{1}{\sqrt{\xi^2 + \xi_c^2}}\,. \tag{4.330}$$

Using the same scale length ξ_0 as before, the corresponding dimensionless surface mass density is

$$\kappa(x) = \frac{1}{2\sqrt{x^2 + x_c^2}}\,, \qquad x_c := \xi_c/\xi_0\,. \tag{4.331}$$

With the help of the formulas above, one can easily work out everything one is interested in. When does the lens produce multiple images? Consider, in particular, an *extended* source near $y = 0$ and discuss the form of the images (arcs and counter arcs, etc.). In Chapter 8 of [19] other examples are studied.

4.8.6 Relation Between Shear and Observable Distortions

Consider a situation as sketched in Fig. 4.8. Typically, a population of distant galaxies behind the gravitational field of a deflector (galaxy, cluster of galaxies) is observed. Equation (4.262) tells us how the Hessian of the gravitational potential ψ of the lens (the tidal field) distorts the *shapes* of galaxy images. However, since the source galaxies are not intrinsically circular, it is impossible to extract useful information from individual images (except for giant arcs).

This problem can be circumvented by observing a sufficiently dense sample of galaxies, because it is reasonable to assume that the ensemble average of the source ellipticity (precisely defined below) vanishes. Intrinsic alignments, caused for example by tidal fields, are not plausible. (There are also ways to test this hypothesis.) Under this assumption the (reduced) complex shear can be determined, as we show next.

Consider an individual galaxy with surface brightness $I(\boldsymbol{\theta})$ of its image at position $\boldsymbol{\theta}$. With this distribution we define the tensor Q_{ij} of second brightness moments. A useful quantity, characterizing the shape, is the **complex ellipticity**

$$\varepsilon = \frac{Q_{11} - Q_{22} + 2i\,Q_{12}}{\mathrm{Tr}\,Q + 2(\det Q)^{1/2}} \,. \tag{4.332}$$

The linearized lens map $D\varphi$ provides a relation between the observed ε and the source ellipticity $\varepsilon^{(s)}$ (defined in terms of the intrinsic surface brightness). This relation is completely determined by the **reduced shear**

$$g(\boldsymbol{\theta}) := \frac{\gamma(\boldsymbol{\theta})}{1 - \kappa(\boldsymbol{\theta})} \,, \tag{4.333}$$

and not by $\gamma(\boldsymbol{\theta})$ alone. This becomes obvious by writing the linearized lens map (4.261) as

$$D\varphi = (1-\kappa)\begin{pmatrix} 1-g_1 & -g_2 \\ -g_2 & 1+g_1 \end{pmatrix} . \tag{4.334}$$

(The pre-factor $(1-\kappa)$ does not affect the shape.) For $|g| \leq 1$ the relation between $\varepsilon^{(s)}$ and ε is given by

$$\varepsilon^{(s)} = \frac{\varepsilon - g}{1 - g^* \varepsilon} \,. \tag{4.335}$$

(See Exercise 4.17.)

There are good reasons to assume that the local ensemble average $\langle \varepsilon^{(s)} \rangle$ vanishes. Let us here consider only the case of **weak lensing**:

$$\kappa \ll 1 \quad \text{and} \quad g \approx \gamma \,.$$

Then (4.335) becomes $\varepsilon \approx \varepsilon^{(s)} + \gamma$ and thus $\gamma = \langle \varepsilon \rangle$. This shows that the shear γ can be measured statistically for weak lensing. On the basis of (4.335) this consideration can be generalized.

* * *

Exercise 4.17. Derive the relation (4.335).

Hints. Regard $Q = (Q_{ij})$ as a linear map $\mathbb{R}^2 \longrightarrow \mathbb{R}^2$, and translate this to a linear transformation of $\mathbb{C}$. Similarly, regard the linearized lens map as the linear transformation

$$z \longmapsto w = (1-\kappa)z - \gamma z^* \,, \qquad z = \theta_1 + i\,\theta_2 \,.$$

The amount of algebra to obtain (4.335) then becomes modest. It is actually simplest to first derive the transformation law for the following alternative ellipticity

$$\chi = \frac{Q_{11} - Q_{22} + 2i\,Q_{12}}{\mathrm{Tr}\,Q} \,, \tag{4.336}$$

which is also in use.

4.8.7 Mass Reconstruction from Weak Lensing

There is a population of distant blue galaxies in the universe whose spatial density reaches 50–100 galaxies per square arc minute at faint magnitudes. The images of these distant galaxies are *coherently distorted* by any foreground cluster of galaxies. Since they cover the sky so densely, the *distortions can be determined statistically*, as was shown in the last subsection. Typical separations between arclets are $\sim (5 \div 10)''$ and this is much smaller than the scale over which the gravitational cluster potential changes appreciably.

Initiated by an influential paper of N. Kaiser and G. Squires [118], a considerable amount of theoretical work on various parameter-free reconstruction methods has been carried out. The main problem consists in the task to make optimal use of limited noisy data, without modeling the lens. For a review, see [119].

The reduced shear g, is in principle observable over a large region. What we are really interested in, however, is the mean curvature κ, which is related to the surface mass density by (4.253).

We first look for relations between the shear γ and κ. Recall that

$$\kappa = \frac{1}{2}\Delta\psi\,, \qquad \gamma_1 = \frac{1}{2}(\psi_{,11} - \psi_{,22}) \equiv D_1\psi\,, \qquad \gamma_2 = \psi_{,12} \equiv D_2\psi\,, \quad (4.337)$$

where

$$D_1 := \frac{1}{2}\left(\partial_1^2 - \partial_2^2\right)\,, \qquad D_2 := \partial_1\partial_2\,. \quad (4.338)$$

Note the identity

$$D_1^2 + D_2^2 = \frac{1}{4}\Delta^2\,. \quad (4.339)$$

Hence

$$\Delta\kappa = 2\sum_{i=1,2} D_i\gamma_i\,. \quad (4.340)$$

In the limit of weak lensing this is all we need. If the difference between γ and g cannot be neglected, we can substitute the reduced shear, given by (4.333), on the right in (4.340) for γ. This gives the important equation

$$\Delta\kappa = -2\sum_i D_i\left(g_i(1-\kappa)\right)\,. \quad (4.341)$$

For a given (measured) g this equation does not determine κ uniquely, reflecting a famous *mass-sheet degeneracy* (a homogeneous mass sheet does not produce any shear). For a given g, (4.341) remains invariant under the substitution

$$\kappa \longrightarrow \lambda\kappa + (1-\lambda)\,, \quad (4.342)$$

where λ is a real constant. This degeneracy can be lifted with magnification information.

Equation (4.341) can be turned into an integral equation, by making use of the fundamental solution

$$\mathcal{G} = \frac{1}{2\pi} \ln |\boldsymbol{x}| , \tag{4.343}$$

introduced in (4.245). One solution of (4.340) is

$$\kappa = 2\mathcal{G} * \left(\sum_i D_i \gamma_i \right) + \kappa_0 , \tag{4.344}$$

with a real constant κ_0. The most general solution is obtained if κ_0 is replaced by any harmonic function. For physical reasons, this function must, however, be bounded, and is thus a constant. Replacing again γ by the reduced shear, we obtain an integral equation for κ. We write this in a different form by noting that

$$D_1 \ln |\boldsymbol{x}| = \frac{x_2^2 - x_1^2}{|\boldsymbol{x}|^4} \equiv \mathcal{D}_1 , \qquad D_2 \ln |\boldsymbol{x}| = -\frac{2x_1 x_2}{|\boldsymbol{x}|^4} \equiv \mathcal{D}_2 . \tag{4.345}$$

Since

$$\mathcal{G} * (D_i \gamma_i) = (D_i \mathcal{G}) * \gamma_i = \frac{1}{2\pi} \mathcal{D}_i * \gamma_i ,$$

we obtain from (4.344)

$$\kappa = \kappa_0 + \frac{1}{\pi} \big(\mathcal{D}_1 * \gamma_1 + \mathcal{D}_2 * \gamma_2 \big) , \tag{4.346}$$

and thus the integral equation becomes

$$\kappa = \kappa_0 - \frac{1}{\pi} \mathcal{D}_1 * \big(g_1 (1 - \kappa) \big) - \frac{1}{\pi} \mathcal{D}_2 * \big(g_2 (1 - \kappa) \big) . \tag{4.347}$$

Equation (4.346) appears the first time in [118]. The integral equation (4.347) has been used, for instance, in [122] for non-linear cluster inversions. Note also

$$\boldsymbol{\nabla} \kappa = \begin{pmatrix} \kappa_{,1} \\ \kappa_{,2} \end{pmatrix} = \begin{pmatrix} \frac{1}{2} (\psi_{,111} + \psi_{,221}) \\ \frac{1}{2} (\psi_{,112} + \psi_{,222}) \end{pmatrix} = \begin{pmatrix} \gamma_{1,1} + \gamma_{2,2} \\ \gamma_{2,1} - \gamma_{1,2} \end{pmatrix} . \tag{4.348}$$

This expression for the gradient $\boldsymbol{\nabla} \kappa$ in terms of the shear has been translated by N. Kaiser, [120] into a relation involving the reduced shear.

We proceed as follows. Let $K := \ln(1 - \kappa)$, then

$$\boldsymbol{\nabla} K = -(1 - \kappa)^{-1} \boldsymbol{\nabla} \kappa . \tag{4.349}$$

In addition, we have

$$\partial_j \gamma_i = -\partial_j \big((1 - \kappa) g_i \big) = -(1 - \kappa) \partial_j g_i + g_i \partial_j \kappa ,$$

and thus, with (4.348),

$$\partial_1\kappa = \partial_1\gamma_1 + \partial_2\gamma_2 = -(1-\kappa)\big(\partial_1 g_1 + \partial_2 g_2\big) + g_1\partial_1\kappa + g_2\partial_2\kappa\,.$$

Hence, we have

$$(1-g_1)\frac{\partial_1\kappa}{1-\kappa} - g_2\frac{\partial_2\kappa}{1-\kappa} = -\partial_1 g_1 - \partial_2 g_2\,.$$

Similarly,

$$-g_2\frac{\partial_1\kappa}{1-\kappa} + (1+g_1)\frac{\partial_2\kappa}{1-\kappa} = -\partial_1 g_2 + \partial_2 g_1\,.$$

The left-hand side of this linear system for $\boldsymbol{\nabla}\kappa/(1-\kappa)$ is given by the matrix

$$M = \begin{pmatrix} 1-g_1 & -g_2 \\ -g_2 & 1+g_1 \end{pmatrix}. \tag{4.350}$$

with the inverse

$$M^{-1} = \frac{1}{1-g_1^2-g_2^2}\begin{pmatrix} 1+g_1 & g_2 \\ g_2 & 1-g_1 \end{pmatrix}. \tag{4.351}$$

This, together with (4.349), gives

$$\boldsymbol{\nabla} K = \boldsymbol{u}\,, \tag{4.352}$$

where

$$\boldsymbol{u} = \frac{1}{1-g_1^2-g_2^2}\begin{pmatrix} 1+g_1 & g_2 \\ g_2 & 1-g_1 \end{pmatrix}\begin{pmatrix} \partial_1 g_1 + \partial_2 g_2 \\ \partial_1 g_2 - \partial_2 g_1 \end{pmatrix}. \tag{4.353}$$

In principle, the gradient of $K = \ln(1-\kappa)$ is thus observable.

In an ideal world (without measuring errors) Eq. (4.352) (with $\boldsymbol{u}$ given by (4.353) in terms of the reduced shear) can be solved in various ways. For instance, by taking the divergence, we get

$$\Delta K = \boldsymbol{\nabla}\cdot\boldsymbol{u}\,, \tag{4.354}$$

inside a domain Ω covered by observations. Taking the scalar product of (4.352) with the outward unit normal $\boldsymbol{n}$ on $\partial\Omega$, we find for the normal derivative

$$\frac{\partial K}{\partial n} = \boldsymbol{n}\cdot\boldsymbol{u}\,. \qquad (\text{on } \partial\Omega) \tag{4.355}$$

Equations (4.352) and (4.355) constitute a Neumann boundary problem for K, which determines K up to a constant. There are efficient and fast methods for a numerical solution of the Neumann problem.

This is not the place to discuss practical difficulties. For this, and further discussion, we refer again to the review [119]. Weak lensing is the most reliable method to determine the mass distribution of clusters of galaxies. In contrast to other methods, no assumptions on the physical state and symmetries of the matter distributions have to be made. In particular gravitational lensing allows us to directly investigate the (projected) dark-matter distribution.

5 The Post-Newtonian Approximation

"Il y a seulement des problèmes plus ou moins résolus."
—H. Poincaré (1908)

5.1 Motion and Gravitational Radiation (Generalities)

The study of the motion of multiple isolated bodies under their mutual gravitational interaction and of the accompanying emission of gravitational radiation is obviously of great astrophysical interest. The first investigations on this subject by Einstein, Droste, de Sitter and others were made very soon after the completion of GR. Since then a vast amount of work has been invested to tackle the problems of motion and gravitational radiation of isolated systems (for an instructive review see [123]). For testing GR, and applying it to interesting astrophysical systems, such as binary systems of neutron stars and black holes, we need analytical approximations and reliable numerical methods to construct global, asymptotically flat solutions of the coupled field and matter equations, satisfying some intuitively obvious properties: The matter distributions should describe a system of celestical bodies (planets, white dwarfs, neutron stars, ...), and there should be no relevant incoming gravitational radiation. It goes without saying that this is a formidable task.

In this chapter we shall only discuss approximation methods that make use of some small parameters for sufficiently separated bodies. But before starting with this, we want to address some conceptual issues, in particular the notion of asymptotic flatness. We shall afterwards treat in detail only the first post-Newtonian approximation, but an outline of the general strategies of approximation methods will also be given. The detailed implementations by various groups become – with increasing order – rapidly very complicated. For interested readers some important recent papers and reviews will be cited. In the final section we shall apply the developed tools to analyze very precise binary pulsar data.

Asymptotic Flatness

For the description of an isolated system, like an oscillating star, it is very sensible to ignore the rest of the universe. Then spacetime should resemble far away from the matter source more and more Minkowski spacetime. What that exactly means was finally formulated by R. Penrose in an elegant geometrical manner (see [124]). His definition of asymptotic flatness emphasizes the conformal structure of spacetime. Guided by the examples discussed in Sect. 3.8, the following concepts emerged:

Definition. *A spacetime (M, g) is* ***asymptotically simple****, provided all null geodesics are complete and (M, g) can conformally be embedded into a Lorentz manifold $(\tilde{M}, \tilde{g})$ such that the following holds:*

(i) M is an open submanifold of $\tilde{M}$ with smooth boundary $\partial M =: \mathscr{I}$.
(ii) The function Ω in the conformal relation $\tilde{g} = \Omega^2 g$ on M can be extended smoothly to $\tilde{M}$ such that $\Omega = 0$, $d\Omega \neq 0$ on $\mathscr{I}$.
(iii) The boundary $\mathscr{I}$ is the disjoint union of two sets $\mathscr{I}^+$ and $\mathscr{I}^-$ with the property that each null geodesic in M has an endpoint in the past on $\mathscr{I}^-$ and an endpoint in the future on $\mathscr{I}^+$.

An asymptotically simple spacetime is called ***asymptotically flat****, if in addition the Ricci tensor for g vanishes in a neighborhood of $\mathscr{I}$.*

Remarks.
1. The **unphysical spacetime** $(\tilde{M}, \tilde{g})$ is to some extend arbitrary, however, not the boundary of M in $\tilde{M}$. This arbitrariness has no consequence for the physics, because the extension is causally disconnected from the **physical spacetime** (M, g).
2. The second condition in the definition fixes the behavior of the conformal factor as $\mathscr{I}$ is approached from M.
3. The third property should be regarded as a completeness condition that is, however, not always satisfied. We know, for example, that for the Schwarzschild solution there are null geodesics whose spatial orbits are circles outside the horizon. For this reason the notion of **weakly asymptotically simple** spacetimes has been introduced, by requiring that such a spacetime is isometric to an asymptotically simple spacetime in a neighborhood of $\mathscr{I}$.
4. The concepts introduced above make no use of Einstein's field equations. It appears plausible that there is a large class of solutions of the field equations which satisfy Penrose's condition of asymptotic flatness. To prove this is exceedingly difficult, but it is now known that this is the case for Einstein's vacuum equations in a neighborhood of null infinity (see [125]).

Apart from such existence questions, there is also a more practical aspect that should be mentioned. The translation of the field equations by

H. Friedrich, [127] to what he called the **conformal field equations** for the unphysical spacetime may turn out to be important for numerical studies of gravitational radiation. One of the reasons is that the asymptotic structure can be calculated since $\mathscr{I}$ is within a bounded domain of the unphysical spacetime (see Fig. 5.1). For an excellent introduction and review of the potentially powerful numerical method based on the conformal field equations see [126]. A lot of work remains to be done to develop the numerical algorithms.

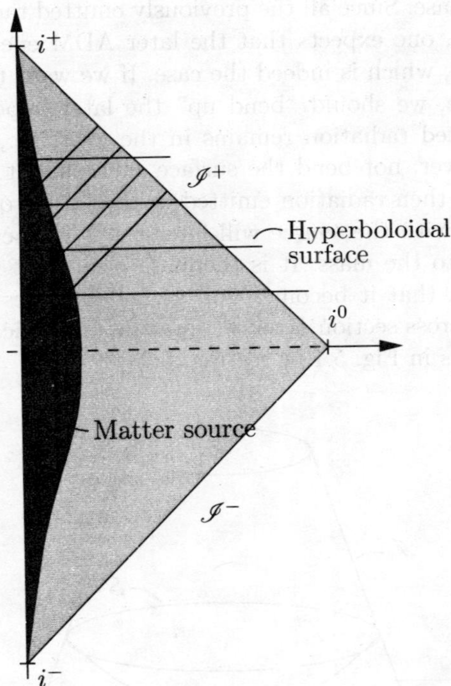

Fig. 5.1. Unphysical spacetime of the conformal approach. The "horizontal" lines correspond to a foliation by hypersurfaces that have asymptotically constant negative curvature (hyperboloidal surfaces).

Bondi-Sachs Energy and Momentum

If a spacetime is asymptotically flat one can define beside the ADM four-momentum (introduced in Sect. 2.7) also a second four-momentum which represents the energy and momentum remaining in the system after radiation has been emitted. Unlike the ADM four-momentum P^{μ}_{ADM}, which remains constant, this so-called **Bondi-Sachs four-momentum** P^{μ}_{BS} is a function of retarded time. This is a function (like u in Sect. 3.8.1) whose level surfaces are null hypersurfaces opening up towards the future. P^{μ}_{BS} is defined in terms of an asymptotic (topological) 2-sphere integral at large *null* separations.

To illustrate the difference of the two concepts, imagine an isolated initially stationary source which suddenly emits a burst of gravitational radiation. Consider first a spacelike hypersurface Σ which intersects the source during its quiescent state and the corresponding ADM four-momentum as defined in (2.207), in terms of an integral of a 2-form over a sphere $S^2_\infty \subset \Sigma$ at spatial infinity. (This can be regarded as a vector in the tangent space at i^0.) Suppose a later spacelike hypersurface intersects the source during or after its active phase. Since all the previously emitted radiation will intersect this hypersurface, one expects that the later ADM energy and momentum have not changed, which is indeed the case. If we want to measure the mass loss of the source, we should "bend up" the later hypersurfaces such that the already emitted radiation remains in the past, i.e., never intersects it. We should, however, not bend the surface such that it eventually becomes timelike, because then radiation emitted in the future of the intersection of the hypersurface with the source will intersect it further out, and so would contribute twice to the mass. It is therefore clear that we should bend the hypersurface such that it becomes null at far distances. The hypersurfaces then approach a cross section $\mathcal{S}$ on $\mathscr{I}^+$ (on the unphysical spacetime). Draw such hypersurfaces in Fig. 5.1 or see Fig. 5.2.

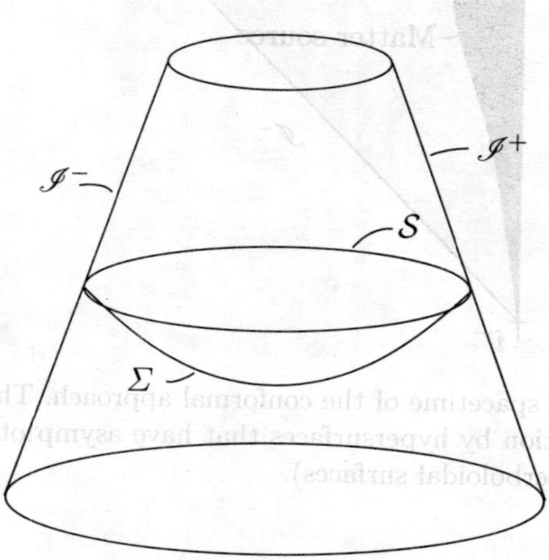

Fig. 5.2. Cross section $\mathcal{S}$ of $\mathscr{I}^+$ bounding an asymptotically null hypersurface Σ used to define the Bondi-Sachs four-momentum.

There are various 2-forms for the 2-sphere integrand which lead to the same energy and momentum. For P^{μ}_{BS} an integrand involving the connection coefficients and certain components of the Weyl tensor was given in a classical paper by Bondi, Van der Burg and Metzner, [128]. It is, however, much more convenient to use an expression in terms of spinors since this has become the bases of a relatively simple proof of the positive energy theorem for P^μ_{BS}. The same is true for P^μ_{ADM}, and one can even use the same two-form for both quantities. This will be described in detail later in Chap. 8, but we indicate the construction already here.

First, we consider the ADM four-momentum. Let χ be a two-component left-handed Weyl spinor relative to an orthonormal coframe $\{\theta^\mu\}$, which is on the hypersurface Σ asymptotically constant:

$$\chi = {}^{(0)}\chi + O(1/r)\,,$$

${}^{(0)}\chi$ being a constant spinor. Consider the **Nester two-form**

$$\omega = i\,\chi^\dagger \sigma \wedge \nabla\chi\,, \qquad \sigma := \sigma_\mu\,\theta^\mu\,,$$

with $\sigma_\mu = (\mathbf{1}, \boldsymbol{\sigma})$, $\boldsymbol{\sigma}$ being the Pauli matrices, and perform the integral of ω over the 2-sphere S^2_∞ bounding Σ at spacelike infinity. We shall see that this is proportional to $P^\mu_{ADM} \cdot {}^{(0)}l_\mu$, where ${}^{(0)}l_\mu$ is the null covector ${}^{(0)}\chi^\dagger\sigma_\mu\,{}^{(0)}\chi$. Precisely, we have (see Sect. 8.1)

$${}^{(0)}l_\mu P^\mu_{ADM} = \frac{1}{4\pi G}\int_{S^2_\infty} \omega\,.$$

It is natural to use the same formula for P^μ_{BS}, except that we integrate over a 2-sphere at large distances along an asymptotic null cone of constant retarded time. (There is actually a slight difference; the notion of an "asymptotically constant" has to be adapted to the different setting.) It turns out (see [129]) that one indeed obtains in this way the original Bondi-Sachs momentum. Moreover, the following version of the positive energy theorem can be proven with the same technique (references are given in Chap. 8).

Positive Energy Theorem. *Let Σ be a non-singular spacelike hypersurface which asymptotically approaches a null cone of fixed retarded time. If the dominant energy condition is satisfied, then P^μ_{BS} is a future directed timelike or null vector. Furthermore, $P^\mu_{BS} = 0$ if and only if the spacetime is flat in a neighborhood of Σ.*

The close similarity of the formulas for P^μ_{ADM} and P^μ_{BS} facilitates a derivation of the connection between the two momenta (see [130]), and the **Bondi-Sachs mass-loss-formula** (see [128]) is recovered. This formula shows that the energy radiated away is always positive. Thus ${}^{(0)}l_\mu P^\mu_{BS}$ is a decreasing function of retarded time, but stays non-negative because of the positive energy theorem. (Note that ${}^{(0)}l^\mu$ is past-directed.)

We shall give a detailed proof of the positive energy theorem in Chap. 8. The required spinor analysis in GR will also be developed there (see Appendix 8.5).

The Effacement Property

Analytical approximations for describing systems such as radiation-reaction driven inspiral of compact binaries (neutron stars or black holes) are based on the smallness of the ratio v/c, where v is a typical relative velocity. This **slow motion approximation** allows in the near zone an expansion in powers of $1/c$ of the field equations. On the other hand, in the radiation zone (far zone) spacetime is approximately flat and the field equations can be expanded in powers of Newton's constant G. In some approaches it is also assumed that the field is everywhere weak, so that – using the virial theorem – $(v/c)^2 \sim GM/R \sim p/\rho \ll 1$, where R is a typical size of the system of sources. A priori this excludes compact objects with strong field regions. However, in GR – as in Newton's theory – there is a drastic reduction of the influence of the internal structure of the bodies on the external motion from a priori estimates. It is this "effacement" property[1] which also allows to reduce the problem of motion of the centers of mass of N bodies to the problem of motion of N point-masses, subject to ordinary differential equations of celestical mechanics that involve effective masses (and spins) of the bodies.

We repeat here an intuitive argument given in [123] for the "effacement property". Suppose that each body of the system is spherically symmetric when isolated. Gravitational interactions between the bodies will induce tidal distortions with ellipticity (see Exercise 1.2)

$$\varepsilon \sim \frac{\text{tidal gravity}}{\text{self gravity}} \simeq \frac{GM\,L/R^3}{GM/L^2} = \left(\frac{L}{R}\right)^3 ,$$

where L is characteristic linear dimension of the bodies and R a typical separation between them. The corresponding tidally induced quadrupole moments $\sim \varepsilon\, ML^2$ lead to structure dependent interbody forces $\sim G\,\varepsilon\, M^2L^2/R^4$ (see (3.54)). Compared to the main Newtonian forces $\sim GM^2/R^2$ these are by a factor $\sim \varepsilon L^2/R^2 \sim (L/R)^5$ times smaller. This is very small for compact bodies for which $L \sim GM/c^2$, so that $L/R \sim GM/c^2R$. We conclude that structure dependent forces are of order $(GM/c^2R)^5 = O(c^{-10})$ relative to the dominant Newtonian forces.

We shall come back to the "point-mass limit" in connection with the first post-Newtonian approximation.

One of the key issues is to combine the two expansions in the near plus induction zone and the far field region to arrive at a global (approximate) solution that links the asymptotic behavior with the structure and motion of the sources. There are several strategies to achieve this. A few remarks about these will be made in Sect. 5.7. But let us now turn to the simplest problem, namely to the first post-Newtonian approximation, which is the first step in all approximation schemes. (Einstein was the first who used this approximation when he calculated the perihelion motion of Mercury in November 1915.)

---

[1] This term is due to Brillouin and Levi-Civita.

5.2 Field Equations in Post-Newtonian Approximation

We assume that the field is everywhere weak and make the slow motion approximation, so that the parameter (not to be confused with ε in the previous section)

$$\varepsilon \sim \left(\frac{GM}{c^2R}\right)^{1/2} \sim \frac{v}{c} \sim \frac{p}{\rho} \tag{5.1}$$

is much smaller than 1. From the results of Sect. 4.2 we expect that in the first post-Newtonian approximation we should keep in the expansions of the metric components in powers of ε

$$\begin{aligned} g_{00} &= -1 + {}^{(2)}g_{00} + {}^{(4)}g_{00}\Big| + {}^{(6)}g_{00} + \dots\,, \\ g_{ij} &= \delta_{ij} + {}^{(2)}g_{ij}\Big| + {}^{(4)}g_{ij} + {}^{(6)}g_{ij} + \dots\,, \\ g_{0i} &= {}^{(3)}g_{0i}\Big| + {}^{(5)}g_{0i} + \dots\,, \end{aligned} \tag{5.2}$$

the terms left to the vertical lines. The last line of (5.2) contains only odd powers of ε because g_{0i} must change sign under the time-reversal transformation $t \longmapsto -t$. At the moment, this should be considered as a reasonable ansatz, which has to be justified by showing that is leads to consistent solutions of the field equations. We shall see that a consistent post-Newtonian limit requires determination of g_{00} correct to $O(\varepsilon^4)$, g_{0i} to $O(\varepsilon^3)$ and g_{ij} to $O(\varepsilon^2)$. From $g^{\mu\lambda}g_{\lambda\nu} = \delta^\mu{}_\nu$ we easily find

$$\begin{aligned} g^{00} &= -1 + {}^{(2)}g^{00} + {}^{(4)}g^{00} + {}^{(6)}g^{00} + \dots\,, \\ g^{ij} &= \delta^{ij} + {}^{(2)}g^{ij} + {}^{(4)}g^{ij} + {}^{(6)}g^{ij} + \dots\,, \\ g^{0i} &= {}^{(3)}g^{0i} + {}^{(5)}g^{0i} + \dots\,, \end{aligned} \tag{5.3}$$

with

$${}^{(2)}g^{00} = -\,{}^{(2)}g_{00}\,, \qquad {}^{(2)}g^{ij} = -\,{}^{(2)}g_{ij}\,, \qquad {}^{(3)}g^{0i} = {}^{(3)}g_{0i}\,, \quad \text{etc.} \tag{5.4}$$

In computing the Christoffel symbols

$$\Gamma^\mu{}_{\nu\lambda} = \frac{1}{2}\, g^{\mu\rho}\left(g_{\rho\nu,\lambda} + g_{\rho\lambda,\nu} - g_{\nu\lambda,\rho}\right)$$

one should regard space and time derivatives as being of order

$$\frac{\partial}{\partial x^i} \sim \frac{1}{R}\,, \qquad \frac{\partial}{\partial x^0} \sim \frac{v}{R} \sim \frac{\varepsilon}{R}\,.$$

Inserting the expansion for $g_{\mu\nu}$ and $g^{\mu\nu}$ leads to

$$\Gamma^\mu{}_{\nu\lambda} = {}^{(2)}\Gamma^\mu{}_{\nu\lambda} + {}^{(4)}\Gamma^\mu{}_{\nu\lambda} + \dots\,, \qquad \text{for } \Gamma^i{}_{00}\,, \Gamma^i{}_{jk}\,, \Gamma^0{}_{0i}\,; \tag{5.5a}$$

$$\Gamma^\mu{}_{\nu\lambda} = {}^{(3)}\Gamma^\mu{}_{\nu\lambda} + {}^{(5)}\Gamma^\mu{}_{\nu\lambda} + \dots\,, \qquad \text{for } \Gamma^i{}_{0j}\,, \Gamma^0{}_{00}\,, \Gamma^0{}_{ij}\,, \tag{5.5b}$$

where ${}^{(n)}\Gamma^{\mu}_{\ \nu\lambda}$ denotes the term in $\Gamma^{\mu}_{\ \nu\lambda}$ of order ε^n/R. The explicit expressions which will be needed are

$$\begin{aligned}
{}^{(2)}\Gamma^{i}_{\ 00} &= -\frac{1}{2}\,{}^{(2)}g_{00,i}\,,\\
{}^{(4)}\Gamma^{i}_{\ 00} &= -\frac{1}{2}\,{}^{(4)}g_{00,i} + {}^{(3)}g_{0i,0} + \frac{1}{2}\,{}^{(2)}g_{ij}\,{}^{(2)}g_{00,j}\,,\\
{}^{(3)}\Gamma^{i}_{\ 0j} &= \frac{1}{2}\left({}^{(3)}g_{i0,j} + {}^{(2)}g_{ij,0} - {}^{(3)}g_{0j,i}\right),\\
{}^{(2)}\Gamma^{i}_{\ jk} &= \frac{1}{2}\left({}^{(2)}g_{ij,k} + {}^{(2)}g_{ik,j} - {}^{(2)}g_{jk,i}\right),\\
{}^{(3)}\Gamma^{0}_{\ 00} &= -\frac{1}{2}\,{}^{(2)}g_{00,0}\,,\\
{}^{(2)}\Gamma^{0}_{\ 0i} &= -\frac{1}{2}\,{}^{(2)}g_{00,i}\,.
\end{aligned} \tag{5.6}$$

Next, we calculate the Ricci tensor

$$R_{\mu\nu} = \Gamma^{\alpha}_{\ \mu\nu,\alpha} - \Gamma^{\alpha}_{\ \mu\alpha,\nu} + \Gamma^{\rho}_{\ \mu\nu}\Gamma^{\alpha}_{\ \rho\alpha} - \Gamma^{\rho}_{\ \mu\alpha}\Gamma^{\alpha}_{\ \rho\nu}\,.$$

From (5.5a) and (5.5b) we get the expansions

$$\begin{aligned}
R_{00} &= {}^{(2)}R_{00} + {}^{(4)}R_{00} + \dots\,,\\
R_{0i} &= {}^{(3)}R_{0i} + {}^{(5)}R_{0i} + \dots\,,\\
R_{ij} &= {}^{(2)}R_{ij} + {}^{(4)}R_{ij} + \dots\,,
\end{aligned} \tag{5.7}$$

with the following expressions of ${}^{(n)}R_{\mu\nu}$ (term of order ε^n/R^2 of $R_{\mu\nu}$) in terms of the Christoffel symbols

$$\begin{aligned}
{}^{(2)}R_{00} &= {}^{(2)}\Gamma^{i}_{\ 00,i}\,,\\
{}^{(4)}R_{00} &= {}^{(4)}\Gamma^{i}_{\ 00,i} - {}^{(3)}\Gamma^{i}_{\ 0i,0} + {}^{(2)}\Gamma^{i}_{\ 00}\,{}^{(2)}\Gamma^{j}_{\ ij} - {}^{(2)}\Gamma^{0}_{\ 0i}\,{}^{(2)}\Gamma^{i}_{\ 00}\,,\\
{}^{(3)}R_{0i} &= {}^{(2)}\Gamma^{0}_{\ 0i,0} + {}^{(3)}\Gamma^{j}_{\ 0i,j} - {}^{(3)}\Gamma^{0}_{\ 00,i} - {}^{(3)}\Gamma^{j}_{\ 0j,i}\,,\\
{}^{(2)}R_{ij} &= {}^{(2)}\Gamma^{k}_{\ ij,k} - {}^{(2)}\Gamma^{0}_{\ i0,j} - {}^{(2)}\Gamma^{k}_{\ ik,j}\,.
\end{aligned} \tag{5.8}$$

Inserting the expressions (5.6) for the Christoffel symbols gives

$${}^{(2)}R_{00} = -\frac{1}{2}\Delta\,{}^{(2)}g_{00}\,, \tag{5.9a}$$

$$\begin{aligned}
{}^{(4)}R_{00} = &-\frac{1}{2}\Delta\,{}^{(4)}g_{00} + {}^{(3)}g_{0i,0i} - \frac{1}{2}\,{}^{(2)}g_{ii,00} + \frac{1}{2}\,{}^{(2)}g_{ij}\,{}^{(2)}g_{00,ij}\\
&+\frac{1}{2}\,{}^{(2)}g_{ij,j}\,{}^{(2)}g_{00,i} - \frac{1}{4}\,{}^{(2)}g_{00,i}\,{}^{(2)}g_{jj,i} - \frac{1}{4}\,{}^{(2)}g_{00,i}\,{}^{(2)}g_{00,i}\,,
\end{aligned} \tag{5.9b}$$

$${}^{(3)}R_{0i} = -\frac{1}{2}\,{}^{(2)}g_{jj,0i} + \frac{1}{2}\,{}^{(3)}g_{j0,ij} + \frac{1}{2}\,{}^{(2)}g_{ij,0j} - \frac{1}{2}\Delta\,{}^{(3)}g_{0i}\,, \tag{5.9c}$$

$${}^{(2)}R_{ij} = \frac{1}{2}\,{}^{(2)}g_{00,ij} - \frac{1}{2}\,{}^{(2)}g_{kk,ij} + \frac{1}{2}\,{}^{(2)}g_{ik,kj} + \frac{1}{2}\,{}^{(2)}g_{kj,ki} - \frac{1}{2}\Delta\,{}^{(2)}g_{ij}\,. \tag{5.9d}$$

At this point we impose four gauge conditions. It is convenient to work with the **standard post-Newtonian gauge**:

$$g_{0j,j} - \frac{1}{2}\, g_{jj,0} = O(c^{-5})\,, \tag{5.10a}$$

$$g_{ij,j} - \frac{1}{2}\,(g_{jj} - g_{00})_{,i} = O(c^{-4})\,. \tag{5.10b}$$

This means

$${}^{(3)}g_{0k,k} - \frac{1}{2}\,{}^{(2)}g_{kk,0} = 0\,, \tag{5.11a}$$

$$\frac{1}{2}\,{}^{(2)}g_{00,i} + {}^{(2)}g_{ij,j} - \frac{1}{2}\,{}^{(2)}g_{jj,i} = 0\,. \tag{5.11b}$$

(For the origin of these gauge conditions see [131]. They should be seen as analogous of the Coulomb gauge in electrodynamics.) We differentiate (5.11b) with respect to x^k

$$\frac{1}{2}\,{}^{(2)}g_{00,ik} + {}^{(2)}g_{ij,jk} - \frac{1}{2}\,{}^{(2)}g_{jj,ik} = 0\,.$$

Interchanging i and k and adding the two equations gives

$${}^{(2)}g_{00,ik} + {}^{(2)}g_{ij,jk} + {}^{(2)}g_{kj,ji} - {}^{(2)}g_{jj,ik} = 0\,. \tag{5.12}$$

Hence ${}^{(2)}R_{ij}$ in (5.9d) simplifies to

$${}^{(2)}R_{ij} = -\frac{1}{2}\,\Delta\,{}^{(2)}g_{ij}\,. \tag{5.13}$$

Next, we use (5.11a) to simplify ${}^{(4)}R_{00}$. Equation (5.11a) implies

$${}^{(3)}g_{0i,i0} + \frac{1}{2}\,{}^{(2)}g_{ii,00} = 0$$

and thus

$$\begin{aligned} {}^{(4)}R_{00} = & -\frac{1}{2}\,\Delta\,{}^{(4)}g_{00} + \frac{1}{2}\,{}^{(2)}g_{ij}\,{}^{(2)}g_{00,ij} + \frac{1}{2}\,{}^{(2)}g_{ij,j}\,{}^{(2)}g_{00,i} \\ & -\frac{1}{4}\,{}^{(2)}g_{00,i}\,{}^{(2)}g_{00,i} - \frac{1}{4}\,{}^{(2)}g_{00,i}\,{}^{(2)}g_{jj,i}\,. \end{aligned} \tag{5.14}$$

If we also use the gauge condition (5.11a) in (5.9c) we obtain

$${}^{(3)}R_{0i} = -\frac{1}{4}\,{}^{(2)}g_{jj,0i} + \frac{1}{2}\,{}^{(2)}g_{ij,0j} - \frac{1}{2}\,\Delta\,{}^{(3)}g_{0i}\,. \tag{5.15}$$

The energy-momentum tensor has the expansion

$$\begin{aligned} T^{00} &= {}^{(0)}T^{00} + {}^{(2)}T^{00} + \dots\,, \\ T^{i0} &= {}^{(1)}T^{i0} + {}^{(3)}T^{i0} + \dots\,, \\ T^{ij} &= {}^{(2)}T^{ij} + {}^{(4)}T^{ij} + \dots\,, \end{aligned} \tag{5.16}$$

where ${}^{(n)}T^{\mu\nu}$ denotes the term in $T^{\mu\nu}$ of order $\varepsilon^n M/R$. For the field equations we need

$$S_{\mu\nu} := T_{\mu\nu} - \frac{1}{2} g_{\mu\nu} T^{\lambda}{}_{\lambda} . \tag{5.17}$$

Using (5.2) one finds from (5.16) the following expansions

$$\begin{aligned} S_{00} &= {}^{(0)}S_{00} + {}^{(2)}S_{00} + \dots , \\ S_{i0} &= {}^{(1)}S_{i0} + {}^{(3)}S_{i0} + \dots , \\ S_{ij} &= {}^{(0)}S_{ij} + {}^{(2)}S_{ij} + \dots , \end{aligned} \tag{5.18}$$

with

$$\begin{aligned} {}^{(0)}S_{00} &= \frac{1}{2} {}^{(0)}T^{00} , \\ {}^{(2)}S_{00} &= \frac{1}{2}\left({}^{(2)}T^{00} - 2\, {}^{(2)}g_{00}\, {}^{(0)}T^{00} + {}^{(2)}T^{ii} \right) , \\ {}^{(1)}S_{0i} &= -\, {}^{(1)}T^{0i} , \\ {}^{(0)}S_{ij} &= \frac{1}{2} \delta_{ij}\, {}^{(0)}T^{00} . \end{aligned} \tag{5.19}$$

With our expressions for ${}^{(n)}R_{\mu\nu}$ and ${}^{(n)}S_{\mu\nu}$ the field equations $R_{\mu\nu} = 8\pi G\, S_{\mu\nu}$ read (note $GM/R \sim \varepsilon^2$)

$$\Delta\, {}^{(2)}g_{00} = -8\pi G\, {}^{(0)}T^{00} , \tag{5.20a}$$

$$\begin{aligned} \Delta\, {}^{(4)}g_{00} = {}&{}^{(2)}g_{ij}\, {}^{(2)}g_{00,ij} + {}^{(2)}g_{ij,j}\, {}^{(2)}g_{00,i} - \frac{1}{4} {}^{(2)}g_{00,i}\, {}^{(2)}g_{00,i} - \frac{1}{4} {}^{(2)}g_{00,i}\, {}^{(2)}g_{jj,i} \\ &- 8\pi G \left({}^{(2)}T^{00} - 2\, {}^{(2)}g_{00}\, {}^{(0)}T^{00} + {}^{(2)}T^{ii} \right) , \end{aligned} \tag{5.20b}$$

$$\Delta\, {}^{(3)}g_{0i} = -\frac{1}{2} {}^{(2)}g_{jj,0i} + {}^{(2)}g_{ij,0j} + 16\pi G\, {}^{(1)}T^{i0} , \tag{5.20c}$$

$$\Delta\, {}^{(2)}g_{ij} = -8\pi G\, \delta_{ij}\, {}^{(0)}T^{00} . \tag{5.20d}$$

The second order equations (5.20a) and (5.20d) give

$${}^{(2)}g_{00} = -2\phi , \qquad {}^{(2)}g_{ij} = -2\delta_{ij}\, \phi , \tag{5.21}$$

where

$$\phi = -G \int \frac{{}^{(0)}T^{00}(t, \boldsymbol{x}')}{|\boldsymbol{x} - \boldsymbol{x}'|}\, d^3x' . \tag{5.22}$$

is the Newtonian potential. According to (5.14), (5.15) and (5.21) the field equations in third and fourth order are

$$\Delta\, {}^{(3)}g_{0i} = 16\pi G\, {}^{(1)}T^{i0} + \phi_{,0i} , \tag{5.23a}$$

$$\Delta\, {}^{(4)}g_{00} = -8\pi G \left({}^{(2)}T^{00} + 4\,\phi\, {}^{(0)}T^{00} + {}^{(2)}T^{ii} \right) + 4\phi\, \Delta\phi - 4(\boldsymbol{\nabla}\phi)^2 . \tag{5.23b}$$

We use Poisson's equation $\Delta\phi = 4\pi G\,{}^{(0)}T^{00}$ and $(\boldsymbol{\nabla}\phi)^2 = \frac{1}{2}\Delta(\phi^2) - \phi\,\Delta\phi$ to rewrite (5.23b) in the following form

$$\Delta\left({}^{(4)}g_{00} + 2\phi^2\right) = -8\pi G\left({}^{(2)}T^{00} + {}^{(2)}T^{ii}\right). \tag{5.24}$$

We define the potential ψ by

$${}^{(4)}g_{00} = -2\,\phi^2 - 2\,\psi\,. \tag{5.25}$$

Using (5.24) the potential ψ satisfies the equation

$$\Delta\psi = 4\pi G\left({}^{(2)}T^{00} + {}^{(2)}T^{ii}\right). \tag{5.26}$$

Since ${}^{(4)}g_{00}$ must vanish at infinity, the solution of (5.26) is

$$\psi = -G\int\frac{d^3x'}{|\boldsymbol{x}-\boldsymbol{x}'|}\left({}^{(2)}T^{00}(\boldsymbol{x}',t) + {}^{(2)}T^{ii}(\boldsymbol{x}',t)\right). \tag{5.27}$$

We also need the potentials ξ_i and χ defined by

$$\xi_i(\boldsymbol{x},t) = -4G\int\frac{d^3x'}{|\boldsymbol{x}-\boldsymbol{x}'|}\,{}^{(1)}T^{i0}(\boldsymbol{x}',t)\,, \tag{5.28a}$$

$$\chi(\boldsymbol{x},t) = -\frac{G}{2}\int|\boldsymbol{x}-\boldsymbol{x}'|\,{}^{(0)}T^{00}(\boldsymbol{x}',t)\,d^3x'\,. \tag{5.28b}$$

They satisfy

$$\Delta\xi_i = 16\pi G\,{}^{(1)}T^{i0}, \tag{5.29a}$$

$$\Delta\chi = \phi\,. \tag{5.29b}$$

From (5.23a) we obtain

$${}^{(3)}g_{i0} = \xi_i + \chi_{,i0}\,. \tag{5.30}$$

For later use we summarize the results obtained so far. In the first post-Newtonian order of the expansion (5.2) the components of the metric can be expressed in terms of the potentials ϕ, ξ_i, χ and ψ as follows:

$$\begin{aligned} {}^{(2)}g_{00} &= -2\phi\,, & {}^{(4)}g_{00} &= -2(\phi^2+\psi)\,,\\ {}^{(2)}g_{ij} &= -2\delta_{ij}\,\phi\,, & {}^{(3)}g_{i0} &= \xi_i + \chi_{,i0}\,. \end{aligned} \tag{5.31}$$

The potentials are given by (5.22), (5.27), (5.28a) and (5.28b). The second order terms have been obtained previously in the Newtonian approximation (see 4.35).

Note that g_{00} can be expressed very simply:

$$g_{00} = \mathrm{e}^{-2U} + O(\varepsilon^6)\,, \tag{5.32}$$

where $U = -(\phi + \psi)$. The potential U satisfies

$$\Delta U = -4\pi G\,\sigma + O(\varepsilon^6)\,, \tag{5.33}$$

with

$$\sigma := T^{00} + T^{ii}\,. \tag{5.34}$$

This metric, expressed in terms of the source $({}^{(0)}T^{00},\ {}^{(1)}T^{i0},\ {}^{(2)}T^{00},\ {}^{(2)}T^{ii})$, was obtained as a solution of the "reduced" field equation that is obtained after imposing the gauge conditions (5.11a) and (5.11b). It must be emphasized that the reduced equation does *not* imply $T^{\alpha\beta}{}_{;\beta} = 0$. This matter equation has to be imposed as an additional equation. Below we show that it implies the gauge condition (5.11a). (The gauge condition (5.11b) is satisfied by the solution (5.31).)

The gauge condition (5.11a) reads (using (5.21), (5.29a) and (5.29b))

$$\nabla\cdot\boldsymbol{\xi} + 4\,\frac{\partial\phi}{\partial t} = 0\,. \tag{5.35}$$

We want to show that this condition follows from $T^{\mu\nu}{}_{;\nu} = 0$.

We need the Christoffel symbols (5.6) in terms of ϕ, ξ_i, χ and ψ:

$$\begin{aligned}
{}^{(2)}\Gamma^{i}{}_{00} &= \frac{\partial\phi}{\partial x^i}\,,\\
{}^{(4)}\Gamma^{i}{}_{00} &= \frac{\partial}{\partial x^i}\,(2\phi^2 + \psi) + \frac{\partial\xi_i}{\partial t} + \frac{\partial^3\chi}{\partial t^2\partial x^i}\,,\\
{}^{(3)}\Gamma^{i}{}_{0j} &= -\delta_{ij}\,\frac{\partial\phi}{\partial t} + \frac{1}{2}\left(\frac{\partial\xi_i}{\partial x^j} - \frac{\partial\xi_j}{\partial x^i}\right),\\
{}^{(3)}\Gamma^{0}{}_{00} &= \frac{\partial\phi}{\partial t}\,,\\
{}^{(2)}\Gamma^{0}{}_{0i} &= \frac{\partial\phi}{\partial x^i}\,,\\
{}^{(2)}\Gamma^{i}{}_{jk} &= -\delta_{ij}\,\frac{\partial\phi}{\partial x^k} - \delta_{ik}\,\frac{\partial\phi}{\partial x^j} + \delta_{jk}\,\frac{\partial\phi}{\partial x^i}\,.
\end{aligned} \tag{5.36}$$

In order to show that $T^{\mu\nu}{}_{;\nu} = 0$ implies the gauge condition (5.35), we write this matter equation more explicitly as

$$T^{\mu\nu}{}_{,\nu} = -\Gamma^{\mu}{}_{\nu\lambda}T^{\lambda\nu} - \Gamma^{\nu}{}_{\nu\lambda}T^{\mu\lambda}\,. \tag{5.37}$$

Since all Christoffel symbols are at least of order ε^2/R, we obtain in lowest order

$$\frac{\partial\,{}^{(0)}T^{00}}{\partial t} + \frac{\partial\,{}^{(1)}T^{i0}}{\partial x^i} = 0\,. \tag{5.38}$$

Using the Poisson equations for ϕ and ξ_i this gives

$$\Delta\left(\nabla\cdot\boldsymbol{\xi} + 4\,\frac{\partial\phi}{\partial t}\right) = 0\,.$$

Since ϕ and ξ_i vanish at infinity we conclude that (5.35) indeed follows.

Note that in the next approximation we obtain from (5.37) and (5.36) a momentum conservation in post-Newtonian order

$$\frac{\partial\, {}^{(1)}T^{i0}}{\partial t} + \frac{\partial\, {}^{(2)}T^{ij}}{\partial x^j} = -\frac{\partial \phi}{\partial x^i}\, {}^{(0)}T^{00}\,. \tag{5.39}$$

There are no other conservation laws which involve *only* terms in $T^{\mu\nu}$ needed to calculate the fields in the post-Newtonian approximation. Furthermore, (5.38) and (5.39) involve $g_{\mu\nu}$ only through the Newtonian potential.

Equations of Motion for a Test Particle

For a particle in an external post-Newtonian field $(\phi, \xi_i, \chi, \psi)$ the equation of motion follows from the variational principle

$$\delta \int \left(\frac{d\tau}{dt}\right) dt = 0\,.$$

With $v^i = dx^i/dt$ we have

$$\begin{aligned}\left(\frac{d\tau}{dt}\right)^2 &= -g_{\mu\nu}\left(\frac{dx^\mu}{dt}\right)\left(\frac{dx^\nu}{dt}\right)\\ &= 1 - \boldsymbol{v}^2 - {}^{(2)}g_{00} - {}^{(4)}g_{00} - 2\,{}^{(3)}g_{0i}\, v^i - {}^{(2)}g_{ij}\, v^i v^j\,.\end{aligned}$$

This gives for $\mathcal{L} := 1 - d\tau/dt$, using the expansion of $\sqrt{1+x}$,

$$\begin{aligned}\mathcal{L} &= \frac{1}{2}\boldsymbol{v}^2 + \frac{1}{8}\left(\boldsymbol{v}^2\right)^2 + \frac{1}{2}\,{}^{(2)}g_{00} + \frac{1}{2}\,{}^{(4)}g_{00}\\ &\quad + {}^{(3)}g_{0i}\, v^i + \frac{1}{2}\,{}^{(2)}g_{ij}\, v^i v^j + \frac{1}{8}\left({}^{(2)}g_{00}\right)^2 + \frac{1}{4}\,{}^{(2)}g_{00}\, \boldsymbol{v}^2\\ &= \frac{1}{2}\boldsymbol{v}^2 - \phi + \frac{1}{8}\left(\boldsymbol{v}^2\right)^2 - \frac{1}{2}\phi^2 - \psi - \frac{3}{2}\phi\, \boldsymbol{v}^2 + v^i\left(\xi_i + \frac{\partial^2 \chi}{\partial t\, \partial x^i}\right).\end{aligned} \tag{5.40}$$

The Euler-Lagrange equations for the Lagrangian $\mathcal{L}$ are the equations of motion for a test particle. Explicitly they read in a 3-dimensional notation

$$\begin{aligned}\frac{d\boldsymbol{v}}{dt} &= -\boldsymbol{\nabla}\left(\phi + 2\phi^2 + \psi\right) - \frac{\partial \boldsymbol{\xi}}{\partial t} - \frac{\partial^2}{\partial t^2}\boldsymbol{\nabla}\chi + \boldsymbol{v} \times (\boldsymbol{\nabla} \times \boldsymbol{\xi})\\ &\quad + 3\,\boldsymbol{v}\,\frac{\partial \phi}{\partial t} + 4\,\boldsymbol{v}\,(\boldsymbol{v}\cdot\boldsymbol{\nabla})\phi - \boldsymbol{v}^2\boldsymbol{\nabla}\phi\,.\end{aligned} \tag{5.41}$$

*   *   *

Exercise 5.1. Derive (5.41) from (5.40).

Hint. Use in the post-Newtonian terms of the Euler-Lagrange equation which are proportional to the time derivative of $\boldsymbol{v}$ the Newtonian approximation for $d\boldsymbol{v}/dt$.

5.3 Stationary Asymptotic Fields in Post-Newtonian Approximation

Far away from an isolated distribution of energy and momentum we can replace

$$\frac{1}{|\boldsymbol{x}-\boldsymbol{x}'|} = \frac{1}{r} + \frac{\boldsymbol{x}\cdot\boldsymbol{x}'}{r^3} + \dots ,$$

where $r = |\boldsymbol{x}|$. We then obtain

$$\phi = -\frac{G\,{}^{(0)}M}{r} - \frac{G\boldsymbol{x}\cdot{}^{(0)}\boldsymbol{D}}{r^3} + O\left(\frac{1}{r^3}\right), \tag{5.42a}$$

$$\xi_i = -4\frac{G\,{}^{(1)}P^i}{r} - 2G\frac{x^j}{r^3}\,{}^{(1)}J_{ij} + O\left(\frac{1}{r^3}\right), \tag{5.42b}$$

$$\psi = -\frac{G\,{}^{(2)}M}{r} - \frac{G\boldsymbol{x}\cdot{}^{(2)}\boldsymbol{D}}{r^3} + O\left(\frac{1}{r^3}\right), \tag{5.42c}$$

where

$${}^{(0)}M := \int {}^{(0)}T^{00}\,d^3x\,, \tag{5.43a}$$

$${}^{(0)}\boldsymbol{D} := \int \boldsymbol{x}\,{}^{(0)}T^{00}\,d^3x\,, \tag{5.43b}$$

$${}^{(1)}\boldsymbol{P}^i := \int {}^{(1)}T^{i0}\,d^3x\,, \tag{5.43c}$$

$${}^{(1)}\boldsymbol{J}_{ij} := 2\int x^i\,{}^{(1)}T^{j0}\,d^3x\,, \tag{5.43d}$$

$${}^{(2)}M := \int \left({}^{(2)}T^{00} + {}^{(2)}T^{ii}\right)d^3x\,, \tag{5.43e}$$

$${}^{(2)}\boldsymbol{D} := \int \boldsymbol{x}\left({}^{(2)}T^{00} + {}^{(2)}T^{ii}\right)d^3x\,. \tag{5.43f}$$

From the mass-conservation equation (5.38) it follows that

$$\frac{d\,{}^{(0)}M}{dt} = 0\,, \quad \frac{d\,{}^{(0)}\boldsymbol{D}}{dt} = {}^{(1)}\boldsymbol{P}\,. \tag{5.44}$$

Let us now specialize to a *time independent* energy-momentum tensor. Then (5.38) reduces to $(\partial/\partial x^i)\,{}^{(1)}T^{i0} = 0$. With partial integrations we obtain from this equation

$$0 = \int x^i \frac{\partial}{\partial x^j}\,{}^{(1)}T^{j0}\,d^3x = -\,{}^{(1)}P^i\,,$$

$$0 = 2\int x^i x^j \frac{\partial}{\partial x^k}\,{}^{(1)}T^{k0}\,d^3x = -\,{}^{(1)}J_{ij} - {}^{(1)}J_{ji}\,.$$

This shows that ${}^{(1)}\boldsymbol{P}$ vanishes and that ${}^{(1)}J_{ij}$ is antisymmetric. Hence,

$$ {}^{(1)}J_{ij} = \varepsilon_{ijk}\, {}^{(1)}J_k \,, \tag{5.45} $$

where

$$ {}^{(1)}J_k = \frac{1}{2}\varepsilon_{ijk}\, {}^{(1)}J_{ij} = \int \varepsilon_{ijk}\, x^i\, {}^{(1)}T^{j0}\, d^3x \tag{5.46} $$

is the angular momentum vector. We conclude from these results and (5.42b) that

$$ \boldsymbol{\xi} = \frac{2G}{r^3}\,(\boldsymbol{x} \times \boldsymbol{J}) + O\left(\frac{1}{r^3}\right). \tag{5.47} $$

In order to bring the metric into a convenient form, we perform the gauge transformation

$$ t_{old} = t_{new} - \frac{\partial \chi}{\partial t}\,, \qquad \boldsymbol{x}_{old} = \boldsymbol{x}_{new}\,. \tag{5.48} $$

This leads to

$$ g_{00} \longrightarrow g_{00} - 2\,\frac{\partial^2 \chi}{\partial t^2}\,, \quad g_{0i} \longrightarrow g_{0i} - 2\,\frac{\partial^2 \chi}{\partial t\,\partial \chi^i}\,. \tag{5.49} $$

In the *new gauge* we obtain from (5.30) and (5.25)

$$ {}^{(3)}g_{0i} = \xi_i\,, \tag{5.50a} $$

$$ {}^{(4)}g_{00} = -\left(2\phi^2 + 2\psi + 2\,\frac{\partial^2 \chi}{\partial t^2}\right) \tag{5.50b} $$

and hence, up to fourth order

$$ \begin{aligned} -g_{00} &= 1 + 2\phi + 2\psi + 2\,\frac{\partial^2 \chi}{\partial t^2} + 2\phi^2 + \ldots \\ &= 1 + 2\left(\phi + \psi + \frac{\partial^2 \chi}{\partial t^2}\right) + \left(\phi + \psi + \frac{\partial^2 \chi}{\partial t^2}\right)^2 + O(\varepsilon^6)\,. \end{aligned} \tag{5.51} $$

For the physically significant field $\phi + \psi + \partial^2\chi/\partial t^2$ we have from (5.26) and (5.29b)

$$ \Delta\left(\phi + \psi + \frac{\partial^2 \chi}{\partial t^2}\right) = 4\pi G\left({}^{(0)}T^{00} + {}^{(2)}T^{00} + {}^{(2)}T^{ii} + \frac{1}{4\pi G}\,\frac{\partial^2 \phi}{\partial t^2}\right), $$

and hence asymptotically

$$ \phi + \psi + \frac{\partial^2 \chi}{\partial t^2} = -\frac{GM}{r} - \frac{G\boldsymbol{x}\cdot\boldsymbol{D}}{r^3} + O\left(\frac{1}{r^3}\right) \tag{5.52} $$

with[2]

[2] The term $\partial^2\phi/\partial t^2$ does not contribute to M because it equals $-\frac{1}{4}\boldsymbol{\nabla}\cdot(\partial\xi/\partial t)$ (see (5.35)), and hence vanishes upon integration.

$$M := \int \left({}^{(0)}T^{00} + {}^{(2)}T^{00} + {}^{(2)}T^{ii} \right) d^3x = {}^{(0)}M + {}^{(2)}M\,, \tag{5.53a}$$

$$\boldsymbol{D} := \int \boldsymbol{x} \left({}^{(0)}T^{00} + {}^{(2)}T^{00} + {}^{(2)}T^{ii} + \frac{1}{4\pi G}\frac{\partial^2 \phi}{\partial t^2} \right) d^3x\,. \tag{5.53b}$$

We can write (5.52) in the form

$$\phi + \psi + \frac{\partial^2 \chi}{\partial t^2} = -\frac{GM}{|\boldsymbol{x} - \boldsymbol{D}/M|} + O\left(\frac{1}{r^3}\right). \tag{5.54}$$

This shows that we can always choose the coordinate system such that $\boldsymbol{D} = 0$. Then the metric has the form (use (5.51), (5.54), (5.50a) and (5.47))

$$\begin{aligned} g_{00} &= -1 + \frac{2GM}{r} - 2\frac{G^2M^2}{r^2} + O\left(\frac{1}{r^3}\right), \\ g_{0i} &= -2G\,\varepsilon_{ijk}\,\frac{J^j\,x^k}{r^3} + O\left(\frac{1}{r^3}\right), \\ g_{ij} &= \left(1 + \frac{2GM}{r}\right)\delta_{ij} + O\left(\frac{1}{r^2}\right). \end{aligned} \tag{5.55}$$

This agrees with (4.212). (See also the discussion of these results in Sect. 4.7.)

5.4 Point-Particle Limit

In GR it is important to use for the matter distribution a continuum (fluid) model. Point masses are not compatible with the field equations because of their non-linearities. As is well-known, the point-particle limit is already a fundamental problem in electrodynamics. The coupled system of Maxwell and Lorentz equations for a system of point particles is mathematically ill-defined. For distributional charge and current densities the electromagnetic fields are distributions. In particular, their values at the positions of the point charges are not defined, and this causes well-known problems when back-reactions have to be taken into account (infinite self energies). For a careful discussion of this issue, that includes also recent contributions, see [18].

If the matter sources are described by a perfect fluid, the matter equation $T^{\alpha\beta}{}_{;\beta} = 0$ imply in particular the basic equations of post-Newtonian hydrodynamics. (A detailed derivation and discussion can be found, for example, in [131] and [132].) This continuum form of the equations of motion is, however, not what we need to describe post-Newtonian celestial mechanics. For such problems the sizes of the celestial bodies are much smaller that the orbital separation and this allows us to approximate the partial differential equations of the continuum description by a system of ordinary differential equations for the centers of mass which involve only a few moments (effective mass, spin, ...) of the bodies. A rigorous way to achieve this in GR is not easy. It is instructive to perform the transition from a fluid description to a "point-particle" formulation first in Newtonian theory. In [123] this is done in two different ways. We repeat here a method that has been adapted to GR by L. Blanchet, T. Damour and coworkers.

The basic equations of Newtonian theory are the Euler equations

$$\rho\left(\partial_t v^i + v^j \partial_j v^i\right) = -\partial_i p + \rho\,\partial_i \phi\,, \tag{5.56}$$

the equation of continuity

$$\partial_t \rho + \partial_i(\rho\, v^i) = 0 \tag{5.57}$$

and the Poisson equation for the Newtonian potential

$$\Delta\phi = -4\pi G\,\rho\,. \tag{5.58}$$

It will turn out to be useful to rewrite (5.56) with the help of (5.57) and (5.58) in the form of a conservation law

$$\partial_t(\rho\, v^i) + \partial_j(\rho\, v^i v^j) + \partial_j\left(T^{ij} + t^{ij}\right) = 0\,, \tag{5.59}$$

where

$$T^{ij} = p\,\delta^{ij}\,, \tag{5.60}$$

and

$$t^{ij} = \frac{1}{4\pi G}\left(\partial_i\phi\,\partial_j\phi - \frac{1}{2}\delta_{ij}(\boldsymbol{\nabla}\phi)^2\right) \tag{5.61}$$

is a gravitational stress tensor. Its divergence is given by

$$\partial_j t^{ij} = \frac{1}{4\pi G}\Delta\phi\,\partial_i\phi\,. \tag{5.62}$$

Note that this vanishes, as a result of the field equation (5.58), outside the sources.

Now we integrate (5.59) over a Volume V that contains the a-the body, however, no other body of the system. Due to Gauss' theorem the second and third terms in (5.59) give no contribution, and we obtain

$$\frac{d}{dt}\int_V \rho\, v^i\, dV = -\int_{\partial V} t^{ij}\, dS_j\,. \tag{5.63}$$

Now we introduce the center of mass position of the a-th body

$$z_a^i = \frac{1}{m_a}\int_V x^i \rho(\boldsymbol{x},t)\, d^3x\,, \tag{5.64}$$

where m_a is the mass of the body. Moreover, we make use of the following "transport theorem", that follows from the continuity equation: For any function F

$$\frac{d}{dt}\int_V F\,\rho\,dV = \int_V \left(\partial_t F + v^i\partial_i F\right)\rho\,dV\,. \tag{5.65}$$

In particular, this implies

$$m_a\,\frac{dz_a^i}{dt} = \int_V v^i\rho\,dV\,. \tag{5.66}$$

From this and (5.63) we obtain

$$m_a\,\frac{d^2 z_a^i}{dt^2} = F^i_{(e)}\,, \quad F^i_{(e)} = -\int_S t^{ij}\,dS_j\,, \tag{5.67}$$

where S is *any* surface enclosing the a-th body, and no other one. From what was said, it is clear that the external force $F^i_{(e)}$ is independent of S. Furthermore, it depends only on the gravitational field *outside* the bodies, whose potential is in obvious notation

$$\phi^{out}(\boldsymbol{x},t) = G\sum_a \int_{V_a} \frac{\rho_a(\boldsymbol{y}_a,t)}{|\boldsymbol{x}-\boldsymbol{z}_a(t)-\boldsymbol{y}_a|}\,d^3y_a\,. \tag{5.68}$$

For this we make a multipole expansion

$$\phi^{out} = \phi^{mon} + \phi^{quad} + \dots\,,$$
$$\phi^{mon}(\boldsymbol{x},t) = \sum_a \frac{G\,m_a}{|\boldsymbol{x}-\boldsymbol{z}_a(t)|}\,, \quad \text{etc.} \tag{5.69}$$

(The dipole part vanishes.) This expansion can be used in (5.67) if the arbitrary surface S is chosen to have dimension $\sim R$ (the interbody separation). Then the monopole part dominates, the quadrupole contribution being suppressed by $O((L/R)^2)$, where L again denotes the characteristic linear dimension of the bodies. So

$$F^i_{(e)} = F^i_{mon} + O\left((L/R)^2\right)\,, \tag{5.70}$$

with

$$F^i_{mon} = -\int_S t^{ij}_{mon}\,dS_j\,, \tag{5.71}$$

where

$$t^{ij}_{mon} = \frac{1}{4\pi G}\left(\phi^{mon}_{,i}\phi^{mon}_{,j} - \frac{1}{2}\delta_{ij}\phi^{mon}_{,k}\phi^{mon}_{,k}\right)\,. \tag{5.72}$$

This result shows that the equations of motion (5.67) contain, up to order $(L/R)^2$, only the centers of mass and the masses.

Now we turn to the calculation of the dominant force F^i_{mon}. In the expression (5.71) we can again choose any surface S containing the position $\boldsymbol{z}_a$, but no $\boldsymbol{z}_b$, $b \neq a$, because the integral does not depend on S if this property is satisfied. We naturally let S shrink towards $\boldsymbol{z}_a$. However, since t^{ij}_{mon}

becomes singular at $\boldsymbol{z}_a$, we have to be careful in taking this limit. Following the treatment in [123], we introduce a complex regularization ϕ_α^{mon} of ϕ^{mon} defined by

$$\phi_\alpha^{mon} = G\, m_a \frac{1}{r_a^{1-\alpha}} + \phi_{(e)a}^{mon}\,, \quad \alpha \in \mathbb{C}\,, \tag{5.73}$$

where $r_a = |\boldsymbol{x} - \boldsymbol{z}_a(t)|$ and

$$\phi_{(e)a}^{mon} = \sum_{b\neq a} \frac{G\, m_b}{|\boldsymbol{x} - \boldsymbol{z}_b(t)|}\,. \tag{5.74}$$

Replacing ϕ^{mon} in (5.72) and (5.71) by (5.73) leads to well-defined quantities even at $\boldsymbol{z}_a$ if $\operatorname{Re}\alpha > 3$. Indeed, since

$$\partial_i r^{\alpha-1} = (\alpha - 1)\left(x^i - z_a^i\right) r_a^{\alpha-3}\,, \tag{5.75}$$

the regularized $t_{mon}^{ij}(\alpha)$ behaves as

$$(x^i - z_a^i)(x_j - z_a^j) r_a^{2\alpha-6} - \frac{1}{2}\,\delta_{ij}\, r_a^{2\alpha-4}$$

and is thus smooth inside S. We can then transform the regularized surface integral (5.71) into a volume integral

$$F_{mon}^i(\alpha) = -\int_V \partial_j t_{mon}^{ij}(\alpha)\, d^3x\,. \tag{5.76}$$

Now Eq. (5.62) also holds for $t_{mon}^{ij}(\alpha)$, thus

$$\begin{aligned} \partial_j t_{mon}^{ij}(\alpha) &= \frac{1}{4\pi G} \Delta\phi_\alpha^{mon}\, \partial_i \phi_\alpha^{mon} \\ &= \frac{m_a}{4\pi} \Delta(r_a^{\alpha-1}) \left(G\, m_a \partial_i r^{\alpha-1} + \partial_i \phi_{(e)a}^{mon} \right). \end{aligned} \tag{5.77}$$

Using also

$$\Delta(r_a^{\alpha-1}) = \alpha(\alpha-1) r_a^{\alpha-3}\,, \tag{5.78}$$

we see that the first term in (5.77) gives the following contribution to $F_{mon}^i(\alpha)$:

$$-\int_V \frac{G\, m_a^2}{4\pi}\, \alpha(\alpha-1)^2 \left(x^i - z_a^i\right) r_a^{2\alpha-6}\, d^3x\,. \tag{5.79}$$

For symmetry reasons this vanishes if we choose for S a sphere of radius R_a centered on $\boldsymbol{z}_a$. Therefore,

$$F_{mon}^i(\alpha) = -\frac{m_a}{4\pi}\, \alpha(\alpha-1) \int_V r_a^{\alpha-3}\, \partial_i \phi_{(e)a}^{mon}\, dV\,. \tag{5.80}$$

Since both sides are analytic functions of α for $\operatorname{Re}\alpha > 0$ which agree for $\operatorname{Re}\alpha > 3$, they must be equal for all α with $\operatorname{Re}\alpha > 0$. At this point we can remove the regularization. In the limit $\alpha \longrightarrow 0$ we find[3]

[3] Note that since the prefactor in (5.80) approaches zero as $\alpha \longrightarrow 0$, only an infinitesimal neighborhood of $\boldsymbol{z}_a$ contributes to the integral.

$$
\begin{aligned}
F^i_{mon}(0) &= F^i_{mon} \\
&= \lim_{\alpha\to 0}\left(-\frac{m_a}{4\pi}\,\alpha(\alpha-1)\,\partial_i\phi^{mon}_{(e)a}(\boldsymbol{z}_a)\int_0^{R_a} 4\pi r_a^2\, r_a^{\alpha-3}\,dr_a\right) \\
&= m_a\,\partial_i\phi^{mon}_{(e)a}(\boldsymbol{z}_a)\,. \qquad (5.81)
\end{aligned}
$$

The result of all this is that up to $O((L/R)^2)$ we obtain the **equations of celestical mechanics**:

$$
m_a\frac{d^2 z^i_a}{dt^2} = m_a\sum_{b\neq a}\frac{\partial}{\partial z^i_a}\left(\frac{G\,m_b}{|\boldsymbol{z}_a-\boldsymbol{z}_b|}\right) + O\left((L/R)^2\right)\,. \qquad (5.82)
$$

The derivation of this result shows explicitly the internal structure effects on the equations of motion.

In [133] T. Damour has generalized this method to GR. It turned out that it can efficiently be used for higher order approximations (see [134]).

5.5 The Einstein-Infeld-Hoffmann Equations

Since the post-Newtonian field and fluid equations resemble the basic Newtonian equations (5.56)–(5.58), it should not be necessary to extend the discussion of the last section. Instead, we shall formally work with point particles. In the first Newtonian approximation there is no danger that something goes wrong. In higher orders point particles should, however, be avoided.

In SR the energy-momentum tensor of a system of point particles with positions $x^\mu_a(\tau_a)$ as a function of their proper times τ_a is given by

$$
T^{\mu\nu}(\boldsymbol{x},t) = \sum_a m_a\int\frac{dx^\mu_a}{d\tau_a}\frac{dx^\nu_a}{d\tau_a}\,\delta^4\left(x-x_a(\tau_a)\right)\,d\tau_a\,.
$$

Since $\sqrt{-g}\,d^4x$ is an invariant measure, it follows that $\delta^4(x-y)/\sqrt{-g}$ is an invariant distribution. From this we conclude that the general relativistic energy-momentum tensor for a system of point particles is

$$
\begin{aligned}
T^{\mu\nu}(\boldsymbol{x},t) &= \frac{1}{\sqrt{-g}}\sum_a m_a\int\frac{dx^\mu_a}{d\tau_a}\frac{dx^\nu_a}{d\tau_a}\,\delta^4\left(x-x_a(\tau_a)\right)\,d\tau_a \\
&= \frac{1}{\sqrt{-g}}\sum_a m_a\int\frac{dx^\mu_a}{d\tau_a}\frac{dx^\nu_a}{d\tau_a}\left(\frac{d\tau_a}{dt}\right)^{-1}\delta^3\left(\boldsymbol{x}-\boldsymbol{x}_a(t)\right)\,. \qquad (5.83)
\end{aligned}
$$

One easily finds the expansion for the determinant of the metric

$$
-g = 1 + {}^{(2)}g + {}^{(4)}g + \ldots\,, \qquad (5.84)
$$

with

$$^{(2)}g = -\,^{(2)}g_{00} + \,^{(2)}g_{ii} = -4\phi\,. \tag{5.85}$$

Using this in (5.83) and $d\tau/dt = 1 - \mathcal{L}$, where $\mathcal{L}$ is given by (5.40), gives

$$\begin{aligned}
^{(0)}T^{00} &= \sum_a m_a\, \delta^3(\boldsymbol{x} - \boldsymbol{x}_a)\,, \\
^{(2)}T^{00} &= \sum_a m_a \left(\frac{1}{2}\boldsymbol{v}_a^2 + \phi\right) \delta^3(\boldsymbol{x} - \boldsymbol{x}_a)\,, \\
^{(1)}T^{i0} &= \sum_a m_a\, v_a^i\, \delta^3(\boldsymbol{x} - \boldsymbol{x}_a)\,, \\
^{(2)}T^{ij} &= \sum_a m_a\, v_a^i v_a^j\, \delta^3(\boldsymbol{x} - \boldsymbol{x}_a)\,.
\end{aligned} \tag{5.86}$$

One verifies easily that the conservation laws (5.38) and (5.39) are satisfied, provided each particle obeys the Newtonian equations of motion

$$\frac{d\boldsymbol{v}_a}{dt} = -\boldsymbol{\nabla}\phi(\boldsymbol{x}_a)\,. \tag{5.87}$$

Clearly, we have

$$\phi(\boldsymbol{x}, t) = -G\sum_a \frac{m_a}{|\boldsymbol{x} - \boldsymbol{x}_a|}\,. \tag{5.88}$$

From (5.26) and (5.43a) we get

$$\Delta\psi = 4\pi G \sum_a m_a \left(\phi'_a + \frac{3}{2}\boldsymbol{v}_a^2\right) \delta^3(\boldsymbol{x} - \boldsymbol{x}_a)\,.$$

On the right-hand side we replaced the undefined Newtonian potential at the position $\boldsymbol{x}_a$ by

$$\phi'_a := -G\sum_{b\neq a} \frac{m_b}{|\boldsymbol{x}_b - \boldsymbol{x}_a|}\,. \tag{5.89}$$

This corresponds to the replacement of ϕ by $\phi^{mon}_{(e)a}$ in the last section. We obtain

$$\psi = -G\sum_a \frac{m_a\,\phi'_a}{|\boldsymbol{x} - \boldsymbol{x}_a|} - \frac{3G}{2}\sum_a \frac{m_a\,\boldsymbol{v}_a^2}{|\boldsymbol{x} - \boldsymbol{x}_a|}\,. \tag{5.90}$$

From (5.28a) and (5.86) we obtain

$$\xi_i = -4G\sum_a \frac{m_a\,v_a^i}{|\boldsymbol{x} - \boldsymbol{x}_a|}\,, \tag{5.91}$$

and (5.28b) gives

$$\chi = -\frac{G}{2}\sum_a m_a|\boldsymbol{x} - \boldsymbol{x}_a|\,. \tag{5.92}$$

Using (5.91) and (5.92) we find from (5.30)

$$ {}^{(3)}g_{0i} = \frac{G}{2} \sum_a \frac{m_a}{|\boldsymbol{x} - \boldsymbol{x}_a|} \left(7\, v_a^i + (\boldsymbol{v}_a \cdot \boldsymbol{n}_a) n_a^i \right) , \tag{5.93} $$

where $\boldsymbol{n}_a = \boldsymbol{x} - \boldsymbol{x}_a / |\boldsymbol{x} - \boldsymbol{x}_a|$.

The Lagrangian $\mathcal{L}_a$ of particle a in the field of the other particles is according to (5.40), (5.88), (5.90) and (5.93)

$$ \begin{aligned} \mathcal{L}_a = {} & \frac{1}{2}\boldsymbol{v}_a^2 + \frac{1}{8}\boldsymbol{v}_a^4 + G \sum_{b \neq a} \frac{m_b}{r_{ab}} \\ & - \frac{1}{2} G^2 \sum_{b,c \neq a} \frac{m_b\, m_c}{r_{ab}\, r_{ac}} - G^2 \sum_{b \neq a} \sum_{c \neq a,b} \frac{m_b\, m_c}{r_{ab}\, r_{bc}} \\ & + \frac{3}{2} \boldsymbol{v}_a^2 G \sum_{b \neq a} \frac{m_b}{r_{ab}} + \frac{3}{2} G \sum_{b \neq a} \frac{m_b\, \boldsymbol{v}_b^2}{r_{ab}} \\ & - \frac{G}{2} \sum_{b \neq a} \frac{m_b}{r_{ab}} \left(7 \boldsymbol{v}_a \cdot \boldsymbol{v}_b + (\boldsymbol{v}_a \cdot \boldsymbol{n}_{ab})(\boldsymbol{v}_b \cdot \boldsymbol{n}_{ab}) \right) , \end{aligned} \tag{5.94} $$

where $r_{ab} := |\boldsymbol{x}_a - \boldsymbol{x}_b|$ and $\boldsymbol{n}_{ab} := (\boldsymbol{x}_a - \boldsymbol{x}_b)/r_{ab}$. The total Lagrangian $\mathcal{L}$ of the N-body system must be a symmetric expression of m_a, $\boldsymbol{x}_a$, $\boldsymbol{v}_a$, for $a = 1, 2, \ldots, N$, with the property that $\lim_{m_a \to 0} \mathcal{L}/m_a = \mathcal{L}_a$. This latter condition just says that in the limit $m_a \longrightarrow 0$ the a-th particle moves on a geodesic in the field of the other particles. One easily verifies that $\mathcal{L}$ is uniquely given by the following expression

$$ \begin{aligned} \mathcal{L} = {} & \sum_a \frac{1}{2}\, m_a \boldsymbol{v}_a^2 + \sum_a \frac{1}{8}\, m_a \boldsymbol{v}_a^4 + \frac{1}{2} G \sum_{a \neq b} \frac{m_a\, m_b}{r_{ab}} \\ & + \frac{3G}{2} \sum_a m_a \boldsymbol{v}_a^2 \sum_{b \neq a} \frac{m_b}{r_{ab}} \\ & - \sum_{a \neq b} \frac{G\, m_a\, m_b}{4\, r_{ab}} \left(7 \boldsymbol{v}_a \cdot \boldsymbol{v}_b + (\boldsymbol{v}_a \cdot \boldsymbol{n}_{ab})(\boldsymbol{v}_b \cdot \boldsymbol{n}_{ab}) \right) \\ & - \frac{G^2}{2} \sum_a \sum_{b \neq a} \sum_{c \neq a} \frac{m_a\, m_b\, m_c}{r_{ab}\, r_{ac}} . \end{aligned} \tag{5.95} $$

The corresponding Euler-Lagrange equations are the **Einstein-Infeld-Hoffmann equations**. They read

$$\dot{\boldsymbol{v}}_a = -G\sum_{b\neq a} m_b \left(\frac{\boldsymbol{x}_{ab}}{r_{ab}^3}\right)\left[1 - 4G\sum_{c\neq a}\frac{m_c}{r_{ac}} + G\sum_{c\neq a,b} m_c\left(-\frac{1}{r_{bc}} + \frac{\boldsymbol{x}_{ab}\,\boldsymbol{x}_{bc}}{2\,r_{bc}^3}\right)\right.$$
$$\left. -5G\frac{m_a}{r_{ab}} + \boldsymbol{v}_a^2 - 4\,\boldsymbol{v}_a\cdot\boldsymbol{v}_b + 2\,\boldsymbol{v}_b^2 - \frac{3}{2}\left(\frac{\boldsymbol{v}_b\cdot\boldsymbol{x}_{ab}}{r_{ab}}\right)^2\right]$$
$$-\frac{7}{2}G\sum_{b\neq a}\left(\frac{m_b}{r_{ab}}\right)G\sum_{c\neq a,b}\frac{m_c\,\boldsymbol{x}_{bc}}{r_{bc}^3} + G\sum_{b\neq a} m_b\left(\frac{\boldsymbol{x}_{ab}}{r_{ab}^3}\right)\cdot(4\boldsymbol{v}_a - 3\boldsymbol{v}_b)(\boldsymbol{v}_a - \boldsymbol{v}_b)\,. \tag{5.96}$$

The Two-Body Problem in the Post-Newtonian Approximation

For two particles Eq. (5.95) reduces to

$$\mathcal{L} = \frac{m_1}{2}\boldsymbol{v}_1^2 + \frac{m_2}{2}\boldsymbol{v}_2^2 + \frac{G\,m_1 m_2}{r} + \frac{1}{8}\left(m_1\boldsymbol{v}_1^4 + m_2\boldsymbol{v}_2^4\right)$$
$$+\frac{G\,m_1 m_2}{2\,r}\left(3(\boldsymbol{v}_1^2 + \boldsymbol{v}_2^2) - 7\,\boldsymbol{v}_1\cdot\boldsymbol{v}_2 - (\boldsymbol{v}_1\cdot\boldsymbol{n})(\boldsymbol{v}_2\cdot\boldsymbol{n})\right)$$
$$-\frac{G^2}{2}\frac{m_1 m_2(m_1 + m_2)}{r^2}\,, \tag{5.97}$$

where $r := r_{12}$ and $\boldsymbol{n} := \boldsymbol{n}_{12}$. The corresponding Einstein-Infeld-Hoffmann equations imply that the center of mass

$$\boldsymbol{X} = \frac{m_1^*\,\boldsymbol{x}_1 + m_2^*\,\boldsymbol{x}_2}{m_1^* + m_2^*}\,, \tag{5.98}$$

with

$$m_a^* := m_a + \frac{1}{2}m_a\boldsymbol{v}_a^2 - \frac{1}{2}\frac{m_a m_b}{r_{ab}}\,,\quad a\neq b\,, \tag{5.99}$$

is not accelerated, i.e.,

$$\frac{d^2\boldsymbol{X}}{dt^2} = 0\,. \tag{5.100}$$

If we choose $\boldsymbol{X} = 0$, then

$$\boldsymbol{x}_1 = \left[\frac{m_2}{m} + \frac{\mu\,\delta m}{2\,m^2}\left(\boldsymbol{v}^2 - \frac{m}{r}\right)\right]\boldsymbol{x}\,,$$
$$\boldsymbol{x}_2 = \left[-\frac{m_1}{m} + \frac{\mu\,\delta m}{2\,m^2}\left(\boldsymbol{v}^2 - \frac{m}{r}\right)\right]\boldsymbol{x}\,, \tag{5.101}$$

where

$$\boldsymbol{x} := \boldsymbol{x}_1 - \boldsymbol{x}_2\,,\quad \boldsymbol{v} := \boldsymbol{v}_1 - \boldsymbol{v}_2\,,\quad m := m_1 + m_2\,,$$
$$\delta m := m_1 - m_2\,,\qquad \mu := m_1\,m_2/m\,. \tag{5.102}$$

For the relative motion we obtain from (5.97) with (5.101), after dividing by μ:

$$\mathcal{L}_{rel} = \mathcal{L}_0 + \mathcal{L}_1 \,, \tag{5.103}$$

with the Newtonian part

$$\mathcal{L}_0 = \frac{1}{2}\boldsymbol{v}^2 + \frac{G\,m}{r} \,, \tag{5.104}$$

and the post-Newtonian perturbation

$$\mathcal{L}_1 = \frac{1}{8}\left(1 - \frac{3\,\mu}{m}\right)\boldsymbol{v}^4 + \frac{G\,m}{2r}\left(3\,\boldsymbol{v}^2 + \frac{\mu}{m}\boldsymbol{v}^2 + \frac{\mu}{m}\left(\frac{\boldsymbol{v}\cdot\boldsymbol{x}}{r}\right)^2\right) - \frac{G^2 m^2}{2\,r^2} \,. \tag{5.105}$$

The corresponding Euler-Lagrange equation is

$$\begin{aligned}\dot{\boldsymbol{v}} = &-\frac{G\,m}{r^3}\boldsymbol{x}\left(1 - \frac{G\,m}{r}(4 + 2\,\mu/m) + (1 + 3\,\mu/m)\boldsymbol{v}^2 - (3\,\mu/2m)\left(\frac{\boldsymbol{v}\cdot\boldsymbol{x}}{r}\right)^2\right)\\ &+\frac{G\,m}{r^3}\boldsymbol{v}(\boldsymbol{v}\cdot\boldsymbol{x})(4 - 2\,\mu/m) \,.\end{aligned} \tag{5.106}$$

In the Newtonian limit $\mathcal{L}_1 \longrightarrow 0$ we choose the solution corresponding to a Keplerian orbit in the plane $z = 0$ with periastron on the x-axis. In standard notations we have for $G = 1$:

$$\boldsymbol{x} = r(\cos\varphi, \sin\varphi, 0) \,, \tag{5.107a}$$

$$r = \frac{p}{1 + e\cos\varphi} \,, \tag{5.107b}$$

$$r^2\frac{d\varphi}{dt} = \sqrt{m\,p} \,. \tag{5.107c}$$

The post-Newtonian solution is obtained in the following way: We write

$$r^2\,\frac{d\varphi}{dt} = |\boldsymbol{x}\times\boldsymbol{v}| = \sqrt{m\,p}\,(1 + \delta h) \,, \tag{5.108a}$$

$$\boldsymbol{v} = \frac{d\boldsymbol{x}}{dt} = \left(\frac{m}{p}\right)^{1/2}(-\sin\varphi, e + \cos\varphi, 0) + \delta\boldsymbol{v} \,. \tag{5.108b}$$

Substituting (5.108a) into the identity

$$\frac{d}{dt}\left(r^2\,\frac{d\varphi}{dt}\right) \equiv |\boldsymbol{x}\times\dot{\boldsymbol{v}}|$$

and using (5.106) for $\dot{\boldsymbol{v}}$ gives

$$\begin{aligned}\frac{d}{dt}\,\delta h &= \frac{m}{r^3}\left(4 - \frac{2\mu}{m}\right)|\boldsymbol{v}\cdot\boldsymbol{x}| = \left(4 - \frac{2\mu}{m}\right)\left(\frac{m}{p}\right)^{1/2}\frac{m}{r^2}\,e\,\sin\varphi\\ &= -\frac{m\,e}{p}(4 - 2\,\mu/m)(\cos\varphi)^{\cdot} \,,\end{aligned}$$

where we have used $e\,\sin\varphi = -(\cos\varphi)^{\cdot}\,r^2/\sqrt{m\,p}$. Hence

$$r^2\,\frac{d\varphi}{dt} = \sqrt{m\,p}\left(1 - \frac{m\,e}{p}(4 - 2\,\mu/m)\cos\varphi\right). \tag{5.109}$$

Inserting (5.108b) into (5.106), one finds by simple integration

$$\begin{aligned} \boldsymbol{v} = \sqrt{m/p}\Big\{ & -\sin\varphi\,\boldsymbol{e}_x + (e+\cos\varphi)\boldsymbol{e}_y \\ & + \frac{m}{p}\Big(\boldsymbol{e}_x\big[-3\,e\varphi + (3-\mu/m)\sin\varphi - (1+21\,\mu/8\,m)e^2\sin\varphi \\ & + \frac{1}{2}(1-2\,\mu/m)e\,\sin 2\varphi - (\mu/8\,m)e^2\sin 3\varphi\big] \\ & + \boldsymbol{e}_y\big[-(3-\mu/m)\cos\varphi - (3-31\,\mu/8\,m)e^2\cos\varphi \\ & - \frac{1}{2}(1-2\,\mu/m)e\,\cos 2\varphi + (\mu/8\,m)e^2\cos 3\varphi\big]\Big)\Big\}. \end{aligned} \tag{5.110}$$

If one substitutes (5.109) and (5.110) into the identity

$$\frac{d}{d\varphi}\frac{1}{r} \equiv -\frac{1}{r^2\,d\varphi/dt}\left(\frac{\boldsymbol{x}\cdot\boldsymbol{v}}{r}\right)$$

and integrates with respect to φ, then one finds for the orbit again formula (5.107a), but instead of (5.107b) we now have

$$\begin{aligned} \frac{p}{r} = 1 + e\,\cos\varphi + \frac{m}{p}\Big(& -(3-\mu/m) + (1+9\,\mu/4\,m)e^2 \\ & + \frac{1}{2}(7-2\,\mu/m)e\,\cos\varphi + 3\,e\varphi\,\sin\varphi - (\mu/4m)e^2\cos 2\varphi\Big). \end{aligned} \tag{5.111}$$

The next to last coefficient of m/p gives the secular periastron motion

$$\delta\varphi = \frac{6\pi\,m}{p}. \tag{5.112}$$

This is the same expression as for the Schwarzschild solution (see (3.47)), but now m is the *sum* of the two masses. In view of the non-linearities involved, this simple result is far from obvious. The results of this section are important in the analysis of binary pulsars (see Sect. 5.8).

* * *

The result (5.112) can be obtained faster with the Hamilton-Jacobi theory. The Hamiltonian corresponding to (5.97) is

$$H = H_0 + H_1\,, \tag{5.113}$$

where H_0 is the Hamiltonian function corresponding to the unperturbed problem

$$H_0 = \frac{1}{2\,m_1}\boldsymbol{p}_1^2 + \frac{1}{2\,m_2}\,\boldsymbol{p}_2^2 - \frac{G\,m_1 m_2}{r} \tag{5.114}$$

and

$$H_1 = -\mathcal{L}_1 . \tag{5.115}$$

In (5.115) the Lagrangian $\mathcal{L}_1$ has to be expressed in terms of the positions and momenta. (We can use the lowest order relation $\boldsymbol{p}_a = m_a\,\boldsymbol{v}_a$.) Equation (5.115) follows from a general result of perturbation theory[4]. Inserting (5.97) gives

$$\begin{aligned} H_1 = &-\frac{1}{8}\left(\frac{\boldsymbol{p}_1^4}{m_1^3} + \frac{\boldsymbol{p}_2^4}{m_2^3}\right) \\ &-\frac{G}{2\,r}\left(3\left(\frac{m_2}{m_1}\boldsymbol{p}_1^2 + \frac{m_1}{m_2}\boldsymbol{p}_2^2\right) - 7\,\boldsymbol{p}_1\cdot\boldsymbol{p}_2 - (\boldsymbol{p}_1\cdot\boldsymbol{n})(\boldsymbol{p}_2\cdot\boldsymbol{n})\right) \\ &+\frac{G^2}{2}\frac{m_1\,m_2(m_1+m_2)}{r^2} . \end{aligned} \tag{5.116}$$

For the relative motion we have $\boldsymbol{p}_1 = -\boldsymbol{p}_2 =: \boldsymbol{p}$ and

$$\begin{aligned} H_{rel} = &\frac{1}{2}\left(\frac{1}{m_1} + \frac{1}{m_2}\right)\boldsymbol{p}^2 - \frac{G\,m_1 m_2}{r} - \frac{\boldsymbol{p}^4}{8}\left(\frac{1}{m_1^3} + \frac{1}{m_2^3}\right) \\ &-\frac{G}{2\,r}\left(3\boldsymbol{p}^2\left(\frac{m_2}{m_1} + \frac{m_1}{m_2}\right) + 7\,\boldsymbol{p}^2 + (\boldsymbol{p}\cdot\boldsymbol{n})^2\right) + \frac{G^2\,m_1\,m_2(m_1+m_2)}{r^2} . \end{aligned} \tag{5.117}$$

We use polar angles and consider the motion in the plane $\vartheta = \frac{\pi}{2}$. Replacing $\boldsymbol{p}^2$ by $p_r^2 + p_\varphi^2/r^2$, with $p_\varphi := L = \text{const}$, the equation $H(p_r, r) = E$ becomes[5]

---

[4] Let $\mathcal{L}(q, \dot{q}, \lambda)$ be a Lagrangian which depends on a parameter λ. The canonical momenta are $p = \partial\mathcal{L}/\partial\dot{q}$ and it is assumed that these equations can uniquely be solved for $\dot{q} = \phi(q, p, \lambda)$ for every λ. Now $H(p, q, \lambda) = p\,\phi(q, p, \lambda) - \mathcal{L}(q, \phi(q, p, \lambda), \lambda)$ and hence

$$\frac{\partial H}{\partial \lambda} = p\frac{\partial \phi}{\partial \lambda} - \frac{\partial \mathcal{L}}{\partial \dot{q}}\frac{\partial \phi}{\partial \lambda} - \frac{\partial \mathcal{L}}{\partial \lambda} = -\frac{\partial \mathcal{L}}{\partial \lambda} .$$

[5] In the perturbation terms we can replace $\boldsymbol{p}^2$ by

$$\frac{2\,m_1 m_2}{m_1 + m_2}\left(E + \frac{G\,m_1 m_2}{r}\right) .$$

$$
\begin{aligned}
H_{rel}(p_r, r) = & \frac{1}{2}\left(\frac{1}{m_1} + \frac{1}{m_2}\right)\left(p_r^2 + \frac{L^2}{r^2}\right) - \frac{G\,m_1 m_2}{r} \\
& -\frac{1}{8}\left(\frac{1}{m_1^3} + \frac{1}{m_2^3}\right)\left(\frac{2\,m_1 m_2}{m_1 + m_2}\right)^2 \left(E + \frac{G\,m_1 m_2}{r}\right)^2 \\
& -\frac{G}{2\,r}\left(3\left(\frac{m_2}{m_1} + \frac{m_1}{m_2}\right) + 7\right)\frac{2\,m_1 m_2}{m_1 + m_2}\left(E + \frac{G\,m_1 m_2}{r}\right) \\
& -\frac{G}{2\,r}p_r^2 + \frac{G^2\,m_1\,m_2(m_1 + m_2)}{2\,r^2} = E\,.
\end{aligned}
\tag{5.118}
$$

The solution of this equation for p_r has the form

$$
p_r^2 = -\frac{L^2}{r^2} + \frac{L^2}{r^2}\frac{G}{2\,r}\frac{2\,m_1 m_2}{m_1 + m_2} + A + \frac{B}{r} + \frac{C}{r^2}\,. \tag{5.119}
$$

We choose a new radial variable r' such that

$$
\frac{L^2}{r^2}\left(1 - \frac{G}{r}\frac{m_1\,m_2}{m_1 + m_2}\right) = \frac{L^2}{r'^2}\,.
$$

Up to higher orders

$$
r' = r + \frac{G}{2}\frac{m_1\,m_2}{m_1 + m_2}\,. \tag{5.120}
$$

The term in the expression for p_r^2 proportional to $1/r'^2$ is equal to

$$
-L^2 + \frac{B\,G\,m_1\,m_2}{2(m_1 + m_2)} + C\,,
$$

whereby only the lowest order terms in B have to be taken into account. A simple calculation shows that

$$
p_r^2 = A + \frac{B}{r'} - (L^2 - 6\,G^2 m_1^2\,m_2^2)\,\frac{1}{r'^2}\,.
$$

with new constants A and B. The Hamilton-Jacobi function S is

$$
S = S_r + S_\varphi - E\,t = S_r + L\,\varphi - E\,t\,, \tag{5.121}
$$

with

$$
S_r := \int \sqrt{A + \frac{B}{r} - \frac{L^2 - 6\,G^2 m_1^2\,m_2^2}{r^2}}\,dr\,. \tag{5.122}
$$

The orbit equation follows from $\partial S/\partial L = \text{const}$, i.e.,

$$
\varphi + \frac{\partial S_r}{\partial L} = \text{const}\,.
$$

From the first to the second periastron we have the change

$$\Delta\varphi = -\frac{\partial}{\partial L}\Delta S_r \,. \tag{5.123}$$

Without the correction term to L^2 in (5.122) the orbit would be a Kepler ellipse and

$$-\frac{\partial}{\partial L}\Delta^{(0)}S_r = \Delta^{(0)}\varphi = 2\pi \,. \tag{5.124}$$

Since

$$\begin{aligned} \Delta S_r &= \Delta^{(0)}S_r - 6\,G^2\,m_1^2\,m_2^2\frac{\partial}{\partial(L^2)}\Delta^{(0)}S_r \\ &= \Delta^{(0)}S_r - \frac{3\,G^2 m_1^2\,m_2^2}{L}\frac{\partial}{\partial L}\Delta^{(0)}S_r \\ &= \Delta^{(0)}S_r + \frac{6\pi G^2\,m_1^2\,m_2^2}{L} \,, \end{aligned} \tag{5.125}$$

it follows from (5.123), (5.124) and (5.125)

$$\Delta\varphi = 2\pi - \frac{\partial}{\partial L}\left(\frac{6\pi G^2\,m_1^2\,m_2^2}{L}\right) = 2\pi + \left(\frac{6\pi G^2\,m_1^2\,m_2^2}{L^2}\right) .$$

Hence the periastron advance is

$$\delta\varphi = \frac{6\pi G^2\,m_1^2\,m_2^2}{L^2} = \frac{6\pi G(m_1+m_2)}{c^2\,a(1-e^2)} \,, \tag{5.126}$$

which agrees with (5.112).

* * *

For the binary pulsar PSR 1913+16 the measured periastron shift is (see Sect. 5.8)

$$\dot\omega = 4.226607 \pm 0.000007\, deg/yr \,. \tag{5.127}$$

The general relativistic prediction for $\dot\omega$ is, using the known values of the orbital elements

$$\dot\omega_{GR} = 2.11\left(\frac{m_1+m_2}{M_\odot}\right)^{2/3} \; deg/yr \,. \tag{5.128}$$

If $\dot\omega_{observed} = \dot\omega_{GR}$, then

$$m_1 + m_2 = 2.83\, M_\odot \,. \tag{5.129}$$

5.6 Precession of a Gyroscope in the Post-Newtonian Approximation

The equation of transport for the gyroscope spin S is (see (1.139))

$$\nabla_u S = \langle S, a\rangle u\,, \quad \langle S, u\rangle = 0\,, \quad a := \nabla_u u\,. \tag{5.130}$$

We are interested in the change of the components of S relative to a comoving frame. We first establish the relation between these components $\mathcal{S}^i$ and the coordinate components S^μ (in the gauge (5.11a) and (5.11b)). We calculate up to the third order. The metric is, to sufficient accuracy (see (5.31)),

$$g = -(1+2\phi)dt^2 + (1-2\phi)\delta_{ij}\,dx^i dx^j + 2h_i\,dtdx^i\,, \tag{5.131}$$

where according to (5.31)

$$h_i = \xi_i + \frac{\partial^2\chi}{\partial t\,\partial x^i}\,. \tag{5.132}$$

It is useful to introduce the orthonormal basis of 1-forms

$$\begin{aligned}\tilde\theta^0 &= (1+\phi)\,dt - h_i\,dx^i\,,\\ \tilde\theta^j &= (1-\phi)\,dx^j\,.\end{aligned} \tag{5.133}$$

The comoving frame $\hat\theta^\mu$ is obtained by a boost. We need the 3-velocity $\tilde v_j$ of the gyroscope with respect to $\tilde\theta^\mu$. If $\tilde u^\mu$ denotes the corresponding 4-velocity, we have, using (5.133),

$$\tilde v_j = \frac{\tilde u^j}{\tilde u^0} = \frac{\langle\tilde\theta^j, u\rangle}{\langle\tilde\theta^0, u\rangle} = \frac{(1-\phi)u^j}{(1+\phi)u^0 - h_i u^i} \simeq (1-2\phi)\,\frac{u^j}{u^0}\,.$$

Hence

$$\tilde v_j = (1-2\phi)\,v_j\,, \tag{5.134}$$

where v_j is the coordinate velocity.

The relation between $\tilde S^\mu := \langle\tilde\theta^\mu, S\rangle$ and $\mathcal{S}^i$ is, to a sufficient approximation,

$$\tilde S^\mu = \left(\tilde v_j \mathcal{S}^j, \mathcal{S}^i + \frac{1}{2}\tilde v_i(\tilde v_j \mathcal{S}^j)\right),$$

but $\tilde S^i = \langle\tilde\theta^i, S\rangle = (1-\phi)S^i$. Hence, to the required accuracy,

$$S^i = (1+\phi)\mathcal{S}^i + \frac{1}{2}v_i(v_k\,\mathcal{S}^k)\,.$$

The covariant components S_i are obtained by multiplication with $(1-2\phi)$:

$$S_i = (1-\phi)\mathcal{S}_i + \frac{1}{2}v_i(v_k\,\mathcal{S}_k)\,. \tag{5.135}$$

(Note that $\mathcal{S}^i = \mathcal{S}_i$.)

Next we derive an equation of motion for S_i and translate this with (5.135) into an equation for $\boldsymbol{\mathcal{S}}$. In components, (5.130) reads

$$\frac{dS_\mu}{d\tau} = \Gamma^\lambda{}_{\mu\nu} S_\lambda\, u^\nu + a^\lambda S_\lambda g_{\mu\nu}\, u^\nu\,. \tag{5.136}$$

We set $\mu = i$, multiply with $d\tau/dt$ and use

$$S_0 = -v_i\, S_i\,, \tag{5.137}$$

which follows from $\langle S, u\rangle = 0$, to obtain

$$\begin{aligned}\frac{dS_i}{dt} &= \Gamma^j{}_{i0}\, S_j - \Gamma^0{}_{i0} v_j\, S_j + \Gamma^j{}_{ik} v_k\, S_j - \Gamma^0{}_{ik} v_k v_j\, S_j \\ &\quad + (g_{0i} + g_{ik} v_k)\left(-a^0 v_j\, S_j + a^j\, S_j\right).\end{aligned} \tag{5.138}$$

Up to third order we obtain (with $a_0 = -v_i a_i$)

$$\frac{dS_i}{dt} = \left({}^{(3)}\Gamma^j{}_{i0} - {}^{(2)}\Gamma^0{}_{i0} v_j + {}^{(2)}\Gamma^j{}_{ik} v_k\right) S_j + v_i a^j\, S_j\,. \tag{5.139}$$

The Christoffel symbols are given in (5.36). We find for $\boldsymbol{S} = (S_1, S_2, S_3)$

$$\begin{aligned}\frac{d}{dt}\boldsymbol{S} &= \frac{1}{2}\boldsymbol{S} \times (\boldsymbol{\nabla} \times \boldsymbol{\xi}) - \boldsymbol{S}\,\frac{\partial\phi}{\partial t} - 2(\boldsymbol{v}\cdot\boldsymbol{S})\boldsymbol{\nabla}\phi - \boldsymbol{S}(\boldsymbol{v}\cdot\boldsymbol{\nabla}\phi) \\ &\quad + \boldsymbol{v}(\boldsymbol{S}\cdot\boldsymbol{\nabla}\phi) + \boldsymbol{v}(\boldsymbol{a}\cdot\boldsymbol{S})\,,\end{aligned} \tag{5.140}$$

where $\boldsymbol{a} = (a^1, a^2, a^3)$.

To the required order, we obtain from (5.135)

$$\boldsymbol{\mathcal{S}} = (1+\phi)\boldsymbol{S} - \frac{1}{2}\boldsymbol{v}(\boldsymbol{v}\cdot\boldsymbol{S})\,. \tag{5.141}$$

We know that $\boldsymbol{\mathcal{S}}^2 = \text{const}$ (see (1.153)). The rate of change of $\boldsymbol{\mathcal{S}}$ is given to third order by

$$\dot{\boldsymbol{\mathcal{S}}} = \dot{\boldsymbol{S}} + \boldsymbol{S}\left(\frac{\partial\phi}{\partial t} + \boldsymbol{v}\cdot\boldsymbol{\nabla}\phi\right) - \frac{1}{2}\dot{\boldsymbol{v}}(\boldsymbol{v}\cdot\boldsymbol{S}) - \frac{1}{2}\boldsymbol{v}(\dot{\boldsymbol{v}}\cdot\boldsymbol{S})\,.$$

Here we can use $\dot{\boldsymbol{v}} \simeq -\boldsymbol{\nabla}\phi + \boldsymbol{a}$ and obtain

$$\begin{aligned}\dot{\boldsymbol{\mathcal{S}}} &= \dot{\boldsymbol{S}} + \boldsymbol{S}\left(\frac{\partial\phi}{\partial t} + \boldsymbol{v}\cdot\boldsymbol{\nabla}\phi\right) + \frac{1}{2}\boldsymbol{\nabla}\phi(\boldsymbol{v}\cdot\boldsymbol{S}) + \frac{1}{2}\boldsymbol{v}(\boldsymbol{S}\cdot\boldsymbol{\nabla}\phi) \\ &\quad - \frac{1}{2}\boldsymbol{a}(\boldsymbol{v}\cdot\boldsymbol{S}) - \frac{1}{2}\boldsymbol{v}(\boldsymbol{S}\cdot\boldsymbol{a})\,.\end{aligned} \tag{5.142}$$

Now we substitute (5.140), and find to the required order

$$\dot{\boldsymbol{\mathcal{S}}} = \boldsymbol{\Omega} \times \boldsymbol{\mathcal{S}}\,, \tag{5.143}$$

with the precession angular velocity

$$\boldsymbol{\Omega} = -\frac{1}{2}(\boldsymbol{v} \times \boldsymbol{a}) - \frac{1}{2}\boldsymbol{\nabla} \times \boldsymbol{\xi} - \frac{3}{2}\boldsymbol{v} \times \boldsymbol{\nabla}\phi . \tag{5.144}$$

The first term is just the Thomas precession. This term is not present for a geodesic motion of the gyroscope. The third term is the geodetic precession, while the second term gives the Lense-Thirring precession (see (1.164) and the end of Sect. 4.7).

Gyroscope in Orbit Around the Earth

As a first application of (5.144) we consider a gyroscope placed in a circular orbit about the Earth. We need the potential $\boldsymbol{\xi}$ of the rotating Earth. For this we note that a system which is at rest and spherically symmetric, but rotates with angular velocity $\boldsymbol{\omega}(r)$, has the momentum density

$$^{(1)}T^{i0}(\boldsymbol{x}, t) = {}^{(0)}T^{00}(\boldsymbol{\omega}(r) \times \boldsymbol{x})^i . \tag{5.145}$$

Using this in (5.28a) gives

$$\boldsymbol{\xi}(\boldsymbol{x}) = -4G \int \frac{d^3x'}{|\boldsymbol{x} - \boldsymbol{x}'|}\, \boldsymbol{\omega}(r') \times \boldsymbol{x}' \, {}^{(0)}T^{00}(r') . \tag{5.146}$$

The solid-angle integral is

$$\int d\Omega' \frac{\boldsymbol{x}'}{|\boldsymbol{x} - \boldsymbol{x}'|} = \begin{cases} \dfrac{4\pi\, r'^2}{3\, r^3}\, \boldsymbol{x} & \text{for } r' < r , \\[2ex] \dfrac{4\pi}{3\, r'}\, \boldsymbol{x} & \text{for } r' > r . \end{cases} \tag{5.147}$$

Thus the field outside the sphere is

$$\boldsymbol{\xi}(\boldsymbol{x}) = \frac{16\pi G}{3\, r^3}\, \boldsymbol{x} \times \int \boldsymbol{\omega}(r') \, {}^{(0)}T^{00}(r') r'^4 \, dr' . \tag{5.148}$$

On the other hand, the angular momentum is

$$\begin{aligned} \boldsymbol{J} &= \int \Big(\boldsymbol{x}' \times (\boldsymbol{\omega}(r') \times \boldsymbol{x}')\Big) \, {}^{(0)}T^{00}(r') \, d^3\boldsymbol{x}' \\ &= \int \Big(r'^2 \boldsymbol{\omega}(r') - \boldsymbol{x}' \, (\boldsymbol{x}' \cdot \boldsymbol{\omega}(r'))\Big) \, {}^{(0)}T^{00}(r') \, d^3\boldsymbol{x}' \\ &= \frac{8\pi}{3} \int \boldsymbol{\omega}(r') \, {}^{(0)}T^{00}(r') r'^4 \, dr' . \end{aligned} \tag{5.149}$$

Comparing this with (5.148) shows that

$$\boldsymbol{\xi}(\boldsymbol{x}) = \frac{2G}{r^3}\, (\boldsymbol{x} \times \boldsymbol{J}) . \tag{5.150}$$

This agrees with (5.47), but now the formula holds *everywhere* outside the sphere. Inserting (5.150) and $\phi = -GM/r$ into (5.144) gives

$$\boldsymbol{\Omega} = \frac{G}{r^3}\left(-\boldsymbol{J} + \frac{3(\boldsymbol{J}\cdot\boldsymbol{x})\boldsymbol{x}}{r^2}\right) + \frac{3\,GM\,\boldsymbol{x}\times\boldsymbol{v}}{2\,r^3}\,. \tag{5.151}$$

The last term, which is independent of $\boldsymbol{J}$, represents the geodetic precession of the gyroscope. The first term was already obtained in Exercise 4.7, Eq. (4.48).

If $\boldsymbol{n}$ denotes the normal to the plane of the orbit, we have

$$\boldsymbol{v} = -\left(\frac{GM}{r^3}\right)^{1/2}\boldsymbol{x}\times\boldsymbol{n}\,, \tag{5.152}$$

and the precession rate, averaged over a revolution, is

$$\langle\boldsymbol{\Omega}\rangle = \frac{G}{2\,r^3}\left(\boldsymbol{J} - \boldsymbol{n}(\boldsymbol{n}\cdot\boldsymbol{J})\right) + \frac{3(GM)^{3/2}\boldsymbol{n}}{2\,r^{5/2}}\,. \tag{5.153}$$

For the Earth and $r \simeq R_\oplus$ we get

$$\frac{\text{Lense-Thirring}}{\text{geodetic}} \simeq \frac{J_\oplus G}{3(M_\oplus G)^{3/2}R_\oplus^{1/2}} = 6.5\times 10^{-3}\,. \tag{5.154}$$

This shows that the main effect is a precession around the orbital angular momentum with an averaged angular velocity

$$|\langle\boldsymbol{\Omega}\rangle| \simeq \frac{3G\,M_\oplus^{3/2}}{2\,r^{5/2}} \simeq 8.4\left(\frac{R_\oplus}{r}\right)^{5/2} sec/yr\,. \tag{5.155}$$

This may be measurable in the coming years with the Stanford Gyroscope Experiment.

Precession of Binary Pulsars

As an interesting application of (5.144) we determine $\boldsymbol{\Omega}$ for the binary pulsar. The field of the companion (indexed by 2) is (see (5.88) and (5.91))

$$\phi(\boldsymbol{x}) = -\frac{G\,m_2}{|\boldsymbol{x}-\boldsymbol{x}_2|}\,,\quad \boldsymbol{\xi}(\boldsymbol{x},t) = -4\,G\,m_2\frac{\boldsymbol{v}_2}{|\boldsymbol{x}-\boldsymbol{x}_2|}\,. \tag{5.156}$$

Using (5.144) we find

$$\boldsymbol{\Omega} = -2G\,m_2\frac{\boldsymbol{x}\times\boldsymbol{v}_2}{r^3} + \frac{3}{2}G\,m_2\frac{\boldsymbol{x}\times\boldsymbol{v}_1}{r^3}\,,$$

where now $\boldsymbol{x} := \boldsymbol{x}_1 - \boldsymbol{x}_2$ and $r := |\boldsymbol{x}|$. Since

$$\boldsymbol{v}_1 = \frac{m_2}{m_1+m_2}\dot{\boldsymbol{x}}\,, \quad \boldsymbol{v}_2 = -\frac{m_1}{m_1+m_2}\dot{\boldsymbol{x}}\,,$$

we obtain

$$\boldsymbol{\Omega} = \left(\frac{\boldsymbol{L}}{\mu}\right)\left\{2G\,m_2\frac{m_1}{m_1+m_2} + \frac{3}{2}\,G\,m_2\,\frac{m_2}{m_1+m_2}\right\}\frac{1}{r^3}\,, \tag{5.157}$$

where $\boldsymbol{L} = \mu\,\boldsymbol{x}\times\dot{\boldsymbol{x}}$ is the angular momentum and $\mu = m_1 m_2/(m_1+m_2)$ the reduced mass. We average $\boldsymbol{\Omega}$ over a period:

$$\langle\boldsymbol{\Omega}\rangle = \hat{\boldsymbol{L}}\Big\{\dots\Big\}\frac{L}{\mu}\frac{1}{T}\int_0^{2\pi}\frac{1}{r^3}\frac{1}{\dot{\varphi}}\,d\varphi\,.$$

Using

$$L = \mu\,r^2\,\dot{\varphi}\,, \quad r = \frac{a(1-e^2)}{1+e\cos\varphi}\,,$$

we find

$$\begin{aligned}\langle\boldsymbol{\Omega}\rangle &= \hat{\boldsymbol{L}}\Big\{\dots\Big\}\frac{1}{T}\int_0^{2\pi}\frac{1}{r}\,d\varphi \\ &= \hat{\boldsymbol{L}}\Big\{\dots\Big\}\frac{1}{a(1-e^2)}\frac{1}{T}\int_0^{2\pi}(1+e\cos\varphi)\,d\varphi\,.\end{aligned}$$

Thus

$$\langle\boldsymbol{\Omega}\rangle = \hat{\boldsymbol{L}}\,\frac{3\pi G\,m_2}{T}\,\frac{1}{a(1-e^2)}\left(\frac{m_2}{m_1+m_2} + \frac{4}{3}\frac{m_1}{m_1+m_2}\right)\,, \tag{5.158}$$

where the first term in the parentheses represents the geodetic precession and the second the Lense-Thirring precession. We compare this with the periastron motion (see (5.126))

$$\dot{\omega} = \frac{6\pi G(m_1+m_2)}{T\,a(1-e^2)} \tag{5.159}$$

and obtain

$$\frac{|\langle\boldsymbol{\Omega}\rangle|}{\dot{\omega}} = \frac{1}{2}\frac{m_2\left(1+\frac{1}{3}\frac{m_1}{m_1+m_2}\right)}{m_1+m_2}\,. \tag{5.160}$$

For $m_1 \simeq m_2 \simeq (m_1+m_2)/2$ we have

$$\frac{|\langle\boldsymbol{\Omega}\rangle|}{\dot{\omega}} \simeq \frac{7}{24}\,. \tag{5.161}$$

As was mentioned in Sect. 3.6 this has recently been detected for the binary pulsar PSR 1913+16 with relatively low precision (see [104]).

5.7 General Strategies of Approximation Methods

It has been realized since quite some time that the leading candidate sources of detectable gravitational waves are radiation-reaction driven inspiral of binary systems of compact objects. For the detection and interpretation of signals from such systems theoretical template waveforms of high accuracy are needed. In particular, the evolution of the orbital frequency or phase must be known with a precision well beyond the leading-order prediction of the quadrupole formula (see (4.155)). This challenge has been taken up by several groups.

In this section we make a few remarks about the general procedure of the various ongoing attempts. It would be out of place for a textbook like this to go into any further details, because of the enormous complexity of the calculations. These are well documented, and we shall give some references to the recent literature.

We describe now schematically the main steps that are common to all approximation schemes. (Spacetime indices are dropped, and we use units with $G = c = 1$.)

1. Impose four gauge conditions which we write symbolically as

$$\Gamma[g] = 0\,. \tag{5.162}$$

(The harmonic gauge condition (2.243) is often used.) Within the gauge class (5.162) Einstein's equations

$$G[g] = 8\pi\, T[m_A; g]\,, \tag{5.163}$$

where m_A denote the matter variables, are then equivalent to the "relaxed" (or reduced) system, which no longer implies $\nabla \cdot T = 0$ and can be solved (iteratively) for arbitrary matter variables. (For the harmonic gauge conditions this reduced system played and important role in our discussion of the initial value problem in Sect. 2.8. We come back to this below.) We write the reduced equations as

$$\hat{G}[g] = 8\pi\, T[m_A; g]\,. \tag{5.164}$$

2. Solve the reduced field equations (5.164) in some iteration scheme that provides a sequence of approximate metrics $g^{(N)}[m_A]$ as functionals of matter variables, obeying the "right" boundary condition (no incoming radiation).
3. Impose *afterwards* for $g^{(N)}[m_A]$ the gauge condition

$$\Gamma\left[g^{(N)}[m_A]\right] = 0\,. \tag{5.165}$$

If the matter variables m_A and the $(N-1)$-iterated metric $g^{(N-1)}$ are used to determine $T^{(N-1)}$, then (5.165) is equivalent to

$$\nabla^{(N-1)} \cdot T^{(N-1)} = 0\,, \tag{5.166}$$

where $\nabla^{(N)}$ denotes the covariant derivative belonging to $g^{(N)}$. (For the post-Newtonian approximation we saw that in Sect. 5.2.) The matter equations (5.165) and (5.166) describe for a perfect fluid the motion of the sources. (In general we have to add further matter equations.)

4. Perform a point-mass limit. We showed in detail how this can be done for the lowest Newtonian approximation. It seems to us that in high orders there is not yet a satisfactory procedure that would also allow to determine finite-size or structure-dependent terms in an unambiguous fashion. The equations of motion are now known to order $(v/c)^4$ beyond Newtonian theory, and the gravitational radiation reaction effects have been computed at orders $(v/c)^5$ and $(v/c)^7$.

After this general outline we want to become more specific.

As already indicated, a popular choice is to impose the harmonic gauge condition (2.243). Expressed in terms of

$$h^{\alpha\beta} := \eta^{\alpha\beta} - \sqrt{-g}\, g^{\alpha\beta} \tag{5.167}$$

this reads

$$h^{\alpha\beta}{}_{,\beta} = 0\,. \tag{5.168}$$

Using the Landau-Lifshitz form (2.214) of Einstein's field equations, and imposing (5.168) we obtain the **relaxed field equations**

$$\Box h^{\alpha\beta} = -16\pi G\, \tau^{\alpha\beta}\,, \tag{5.169}$$

where $\Box$ is the flat spacetime d'Alembertian, and the source $\tau^{\alpha\beta}$ is the pseudotensor

$$\tau^{\alpha\beta} = (-g)T^{\alpha\beta} + \frac{1}{16\pi G}\Lambda^{\alpha\beta}\,, \tag{5.170}$$

with

$$\Lambda^{\alpha\beta} = 16\pi G(-g)t^{\alpha\beta}_{LL} + \left(h^{\alpha\mu}{}_{,\nu}h^{\beta\nu}{}_{,\mu} + h^{\alpha\beta}{}_{,\mu\nu}h^{\mu\nu}\right)\,. \tag{5.171}$$

According to (2.215) the Landau-Lifshitz pseudo-tensor is given by

$$(-g)t^{\alpha\beta}_{LL} = \frac{1}{16\pi G}\Big\{ g_{\lambda\mu}g^{\nu\rho}h^{\alpha\lambda}{}_{,\nu}h^{\beta\mu}{}_{,\rho} + \frac{1}{2}g_{\lambda\mu}g^{\alpha\beta}h^{\lambda\nu}{}_{,\rho}h^{\rho\mu}{}_{,\nu} \tag{5.172}$$

$$-2\, g_{\mu\nu}g^{\lambda(\alpha}h^{\beta)\nu}{}_{,\rho}h^{\rho\mu}{}_{,\lambda}$$

$$+\frac{1}{8}(2g^{\alpha\lambda}g^{\beta\mu} - g^{\alpha\beta}g^{\lambda\mu})(2g_{\nu\rho}g_{\sigma\tau} - g_{\rho\sigma}g_{\nu\tau})h^{\nu\tau}{}_{,\lambda}h^{\rho\sigma}{}_{,\mu}\Big\}\,. \tag{5.173}$$

Since we used (5.168) to arrive at (5.169) the latter equation does *not* imply $T^{\alpha\beta}{}_{;\beta} = 0$, as was already emphasized earlier. This equation follows only when the gauge condition (5.168) is imposed on solutions of (5.169). Indeed, Einstein's field equation is then also satisfied, and

$$T^{\alpha\beta}{}_{;\beta} = 0 \tag{5.174}$$

follows as a result of the Bianchi identity. Equations (5.168) and (5.169) also imply

$$\tau^{\alpha\beta}{}_{,\beta} = 0 \,. \tag{5.175}$$

If (5.168) is imposed the last equation is equivalent to

$$\Big((-g)(T^{\alpha\beta} + t^{\alpha\beta}_{LL})\Big)_{,\beta} = 0 \,. \tag{5.176}$$

At this point different strategies have been chosen by the main groups of workers. We begin with the work of C. Will and collaborators (see [135] and [136]), that is conceptually easy to understand.

The relaxed field equation (5.169) has on the right-hand side the second derivative term $-h^{\alpha\beta}{}_{,\mu\nu}h^{\mu\nu}$ that one would naturally put on the left-hand side, thus changing the principal term and thus modifying the propagation characteristic from the flat null cones. It would, of course, be desirable to work with the null cones of the curved spacetime, but for explicit calculations this is not suitable. For an iterative solution of the relaxed field equation it is useful to write this in the form of an integral equation, imposing an outgoing boundary condition:

$$h^{\alpha\beta} = 4\, G_{ret} * \tau^{\alpha\beta} \,. \tag{5.177}$$

Here G_{ret} is the retarded flat spacetime Green's function, normalized as

$$\Box G_{ret} = -4\pi\, \delta^{(4)} \,.$$

Explicitly,

$$h^{\alpha\beta}(t, \boldsymbol{x}) = 4 \int \frac{\tau^{\alpha\beta}(t - |\boldsymbol{x} - \boldsymbol{x}'|, \boldsymbol{x}')}{|\boldsymbol{x} - \boldsymbol{x}'|}\, d^3x' \,. \tag{5.178}$$

The iteration procedure naturally begins by putting $h^{\alpha\beta} = 0$ in $\tau^{\alpha\beta}$. The right-hand side of (5.178) then gives us the first iterated ${}_1h^{\alpha\beta}$ as a retarded integral of $T^{\alpha\beta}$ (familiar from the linearized theory of Chap. 4). When substituted into $\tau^{\alpha\beta}$ we obtain the second iteration ${}_2h^{\alpha\beta}$, and so on. The matter variables m_A and the $(N-1)$-iterated field ${}_{N-1}h^{\alpha\beta}$ are used to determine ${}_{N-1}T^{\alpha\beta}$ and ${}_{N-1}\Lambda^{\alpha\beta}$, thus ${}_{N-1}\tau^{\alpha\beta}({}_{N-1}T^{\alpha\beta}; {}_{N-1}h^{\alpha\beta})$. The retarded integral gives then ${}_Nh^{\alpha\beta}$ as a functional of the matter variables. If matter is modeled as a perfect fluid, the matter equations in the N-th order are obtained from ${}_N\nabla_\beta({}_NT^{\alpha\beta}) = 0$. On the other hand, the N-th iterated gravitational field as a function of spacetime is obtained by solving the matter equations ${}_{N-1}\nabla_\beta({}_{N-1}T^{\alpha\beta}) = 0$. These are equivalent to the N-th iterated gauge conditions ${}_Nh^{\alpha\beta}{}_{,\beta} = 0$. (In one direction this is obvious; see also Sect. 5.2).

The integration over the past flat spacetime null cone is handled differently, depending on whether the field point is in the far zone or the near zone. In the radiation zone one can perform an expansion in powers of $1/|\boldsymbol{x}|$, in a multipole expansion, as is familiar from electrodynamics. The leading term in $1/|\boldsymbol{x}|$ is the gravitational wave form detected with interferometers (for example). On the other hand, the field in the near and induction zone is obtained by performing in (5.178) an expansion in powers of $|\boldsymbol{x} - \boldsymbol{x}'|$ about the local time t, yielding potentials that go into the equation of motion.

Now the following circumstance has in the past often been a source of trouble. Since the source $\tau^{\alpha\beta}$ contains $h^{\alpha\beta}$ itself, it is not confined to a compact region. Because of this there is no guarantee that all the integrals in the various expansions are well-defined. Indeed, in earlier attempts at some point divergences always showed up, in spite of a continuum matter model, simply as a consequence of the non-linearity of Einstein's equations. The "Direct Integration of the Relaxed Einstein Equations", called DIRE by Will, Wiseman and Pati (see [135]), splits the integration into one over the near zone and another over the far zone, and uses different integration variables to perform the explicit integrals over the two zones. It turns out that all integrals are convergent.

The DIRE approach assumes that gravity is sufficiently weak everywhere, also inside the bodies. In order to apply the results also to systems with neutron stars or black holes, one relies on the effacement property discussed earlier. It is very reasonable to argue that if the equations of motion are expressed in terms of the masses and spins (ignoring tidal forces), one can simply extrapolate the results unchanged to the situation where fields in the neighborhood of the bodies are strong.

The mass-point limit is performed in [136] on the basis of a somewhat unphysical continuum model, but this should determine correctly all contributions to the equations of motion that are independent of the internal structure, size and shape of the bodies.

We add very few remarks about other approaches. The post-Minkowskian method of Blanchet, Damour and Iyer is roughly this: First, Einstein's vacuum equations are solved in the exterior weak field zone in an expansion in powers of Newton's constant, and asymptotic solutions in terms of a set of formal, time dependent, symmetric and tracefree multipole moments are obtained. Then, in a near zone within one characteristic wavelength of radiation, the equations including the material sources are solved in a slow motion approximation (expansion in powers of $1/c$) that yields both equations of motion for the bodies, as well as a set of source multipole moments expressed as integrals over the "effective" source, including both matter and gravitational field contributions. Finally, the solutions involving two different sets of moments are matched in an intermediate overlap zone, resulting in a connection between the two sets of moments. In the matching process a regularization method is introduced to define integrals involving contributions of gravitational stress-energy with non-compact support which are not a priori well-defined[6]. As a result one obtains the exterior multipole moments as integrals over the source, and also the radiation reaction forces acting in-

[6] The method is similar to the one used in Sect. 5.4.

side the source. The calculations confirmed the expectation that the reaction force corresponds to the energy-momentum flux carried away by the waves to infinity. (For an extensive review we refer to [134].)

A third method, championed by G. Schäfer [137], is based on the ADM Hamiltonian formalism of GR, and has some advantages. At the third post-Newtonian level, the use of distributional sources leads to badly divergent integrals which cannot unambiguously be regularized. At the time of writing it is not yet clear how to get an unambiguous point-mass limit (see [138]).

Radiation Damping

As a concrete and important result of all this, we quote the final equations of motion at the 2.5th post-Newtonian level (some higher orders are now also known).

The post-Newtonian approximation of the equation of motion (5.106) for the relative motion of a binary system of compact objects is part of

$$\begin{aligned}\ddot{\boldsymbol{x}} = & -\frac{Gm}{r^2}\frac{\boldsymbol{x}}{r} + \frac{Gm}{r^2}\left(\frac{\boldsymbol{x}}{r}\left(A_{PN} + A_{2PN}\right) + \dot{r}\boldsymbol{v}\left(B_{PN} + B_{2PN}\right)\right) \\ & + \frac{8}{5}\eta\frac{G^2 m\, m}{r^2}\frac{m}{r}\left(\dot{r}\frac{\boldsymbol{x}}{r}A_{2.5PN} - \boldsymbol{v}B_{2.5PN}\right);\end{aligned} \tag{5.179}$$

where we used the same notation as before (recall that $\eta = m_1 m_2/m^2$ is the chirp mass). The coefficients A_{PN} and B_{PN} can be read off from (5.106)

$$A_{PN} = -(1+3\eta^2)v^2 + \frac{3}{2}\eta\dot{r}^2 + 2(2+\eta)\frac{Gm}{r^2}, \tag{5.180a}$$

$$B_{PN} = 2(2-\eta). \tag{5.180b}$$

The expressions for A_{2PN} and B_{2PN} can be found, for instance, in [136] and will not be used in this book. The leading radiation-reaction contributions at 2.5th post-Newtonian order are given by

$$A_{2.5PN} = 3v^2 + \frac{17}{3}\frac{Gm}{r}, \tag{5.181a}$$

$$B_{2.5PN} = v^2 + \frac{Gm}{r}. \tag{5.181b}$$

These dissipative terms lead in particular to a secular change of the orbital period. Using the method of 'variations of constants', one finds for the 'relativistic mean anomaly' l (an angle such that $l_N = 2\pi N$ for the Nth return of each body to its periastron) the following time dependence

$$\frac{l}{2\pi} = \text{const} + \frac{t}{P_b} - \frac{1}{2}\frac{\dot{P}_b}{P_b^2}t^2 + \text{periodic terms}. \tag{5.182}$$

The secular term is proportional to $\dot{P}_b$. For this period change one finds, as expected, the previous result (4.201) (for details see, e.g., [139]). The derivation along the lines sketched above is, however, much more satisfactory (although far more involved). Secular changes of other orbital parameters (eccentricity, semi-major axes) can also be computed.

5.8 Binary Pulsars

At several occasions we have compared general relativistic effects with observational data of the binary radio pulsar PSR 1913+16[7]. In the present section this marvelous binary system of two neutron stars will be discussed in detail, and we shall also mention the results for other similar binary pulsar systems.

5.8.1 Discovery and Gross Features

About 30 years ago, J. Taylor and R. Hulse discovered, in a systematic search for new pulsars, the first binary radio pulsar (PSR B 1913+16) that appears to have been exquisitely designed as a laboratory for GR. (Beside the discovery paper [140], we refer once more to the Nobel prize lectures by the two authors [97] and [98].) The nominal period of the pulsar is 59 *msec*. This short period was observed to be periodically shifted, which proved that the pulsar is a member of a binary system with an orbital period $P_b \simeq 7.75$ hours. With Kepler's third law

$$P_b = \frac{0.116\,(a/R_\odot)^{3/2}}{[(m_1+m_2)/M_\odot]^{1/2}}\,\text{days}\,, \tag{5.183}$$

where m_1, m_2 are the two masses and a the semi-major axis, and reasonable masses (e.g., $m_1 \approx m_2 \simeq 1.4 M_\odot$), one concludes that the orbit is not much larger than the Sun's diameter.

Correspondingly, the velocity of the pulsar is about $10^{-3}c$ and it moves through a relatively strong gravitational field (about 100 times stronger than the gravitational field of the Sun near Mercury). These numbers show that several special and general relativistic effects should be observable – a goal that has been explored since 1974 with increasing accuracy. It turned out that the Hulse-Taylor pulsar is one partner of a "clean" binary system of two neutron stars, each having approximately 1.4 solar masses $M_\odot$ (see Fig. 5.3). In addition, PSR B 1913+16 behaved very well all the time, for it has not made any discontinuous jumps (glitches) nor generated noise since being discovered.

[7] The following readily identifiable nomenclature is used for pulsars: The prefix PSR (abbreviation of "pulsar") is followed by a four-digit number indicating right ascension (in 1950.0 coordinates). After that a sign and two digits indicate degrees of diclination. If the pulsar is a member of a binary system, this is sometimes indicated by the letter B after the prefix PSR.

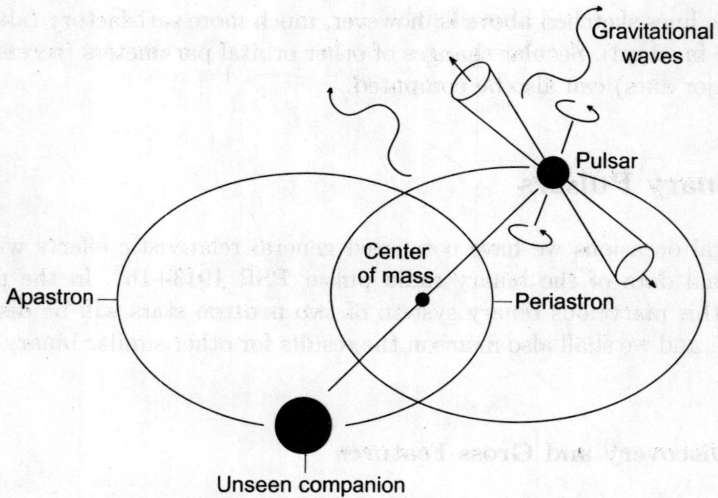

Fig. 5.3. Binary system of two neutron stars.

In September 1974 Hulse and Taylor had already obtained quite an accurate radial velocity curve from the periodically changing pulse arrival rate due to the Doppler shift, and were thus able to determine five Keplerian parameters [140]. Since this is the first order step of a much more sophisticated pulse arrival time analysis, to be discussed later, we show how this is done. (This tool will also be used in Sect. 7.6, when we shall present the current evidence for stellar-mass black holes.)

* * *

We use the notation indicated in Fig. 5.4 for the description of the pulsar orbit. The z-axis is along the line of sight, and the unit vector $\boldsymbol{N}$ is normal to the orbit plane. The angle i between these two directions is known as the **inclination angle** of the orbit. Another important line, called the **line of nodes**, is the intersection of the orbit plane with the plane perpendicular to the line of sight. There are two points on the line of nodes at which the orbit intersects the (x, y)-plane; the point at which the mass enters from below into the upper hemisphere is known as the **ascending node**. The angle Ω in Fig. 5.4 is the **longitude of the ascending node**, a time-honored name. Furthermore, the angle ω is the **argument of the periastron**, and φ is called the **true anomaly**.

The five astronomical elements i, Ω, a, e and ω play an important role in celestical mechanics. As origin of the coordinate system x, y, z we choose the center of mass of the two bodies. The corresponding triad is denoted as $\{\boldsymbol{e}_x, \boldsymbol{e}_y, \boldsymbol{e}_z\}$. Let $\boldsymbol{K}$ be a unit vector in the direction of the ascending node,

$$\boldsymbol{K} = \cos\Omega\, \boldsymbol{e}_x + \sin\Omega\, \boldsymbol{e}_y \,. \tag{5.184}$$

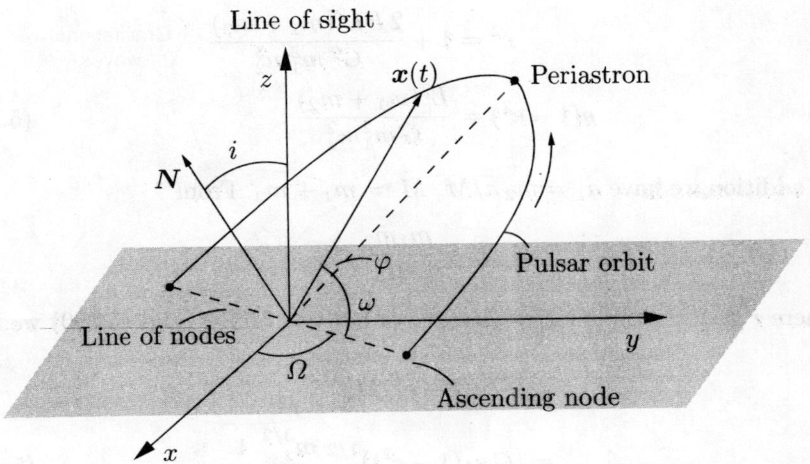

Fig. 5.4. Notation for the description of the pulsar orbit.

Then $\{\boldsymbol{K}, \boldsymbol{N}\wedge\boldsymbol{K}, \boldsymbol{N}\}$ form another orthonormal triad, whose first two vectors lie in the plane of the pulsar orbit. From Fig. 5.4 one easily sees that

$$\begin{aligned}\boldsymbol{N} &= \langle \boldsymbol{N}, \boldsymbol{e}_x\rangle \boldsymbol{e}_x + \langle \boldsymbol{N}, \boldsymbol{e}_y\rangle \boldsymbol{e}_y + \langle \boldsymbol{N}, \boldsymbol{e}_z\rangle \boldsymbol{e}_z \\ &= \sin i \sin \Omega\, \boldsymbol{e}_x - \sin i \cos \Omega\, \boldsymbol{e}_y + \cos i\, \boldsymbol{e}_z\,. \end{aligned} \tag{5.185}$$

Let $\boldsymbol{x}_1(t) = r_1\, \hat{\boldsymbol{x}}_1(t)$ be the orbit of the pulsar (an index 1 always refers to the pulsar and the index 2 to its companion). Obviously,

$$\hat{\boldsymbol{x}}_1 = \cos(\omega+\varphi)\boldsymbol{K} + \sin(\omega+\varphi)\boldsymbol{N}\wedge\boldsymbol{K}\,. \tag{5.186}$$

Inserting (5.184) and (5.185) gives, with $\theta := \omega + \varphi$,

$$\begin{aligned}\boldsymbol{x}_1 = r_1 \Big[&(\cos\theta\, \cos\Omega - \sin\theta\, \sin\Omega\, \cos i)\, \boldsymbol{e}_x \\ &+ (\cos\theta\, \sin\Omega + \sin\theta\, \cos\Omega\, \cos i)\, \boldsymbol{e}_y \\ &+ \sin\theta\, \sin i\, \boldsymbol{e}_z \Big]\,. \end{aligned} \tag{5.187}$$

In particular, the component z_1 of $\boldsymbol{x}_1$ along the line of sight is given by

$$z_1 = r_1\, \sin(\omega+\varphi)\sin i\,. \tag{5.188}$$

For the angular dependence of r_1 we have for the Kepler problem

$$r_1 = \frac{a_1(1-e^2)}{1+e\, \cos\varphi}\,, \tag{5.189}$$

where e is the eccentricity and a_1 the semi-major axis of the pulsar orbit. We recall that e and the semi-major axis a of the relative motion are related to the total energy E and the angular momentum L of the system by

$$e^2 = 1 + \frac{2\,EL^2(m_1+m_2)}{G^2\,m_1^3 m_2^3}\,,$$
$$a(1-e^2) = \frac{L^2(m_1+m_2)}{G\,m_1^2 m_2^2}\,. \tag{5.190}$$

In addition,we have $a_1 = m_2\,a/M$, $M := m_1 + m_2$. From

$$L = \frac{m_1 m_2}{M}\,r^2\dot{\varphi}\,,$$

where r denotes the relative distance of the two bodies, and (5.190) we find

$$\dot{\varphi} = \big(G\,a(1-e^2)M\big)^{1/2}\,\frac{1}{r^2}$$
$$= \big(G\,a_1(1-e^2)\big)^{1/2}\,\frac{m_2^{3/2}}{M}\,\frac{1}{r_1^2}\,. \tag{5.191}$$

If we use here Kepler's third law for the pulsar orbital period P_b,

$$\frac{P_b}{2\pi} = \frac{a^{3/2}}{G^{1/2}\,M^{1/2}}\,, \tag{5.192}$$

we obtain

$$\dot{\varphi} = \frac{2\pi}{P_b(1-e^2)^{3/2}}\,(1+e\,\cos\varphi)^2\,. \tag{5.193}$$

For $\dot{r}_1$ we find from (5.189) and (5.191)

$$\dot{r}_1 = \frac{G^{1/2} m_2^{3/2}}{M\big(a_1(1-e^2)\big)^{1/2}}\,e\,\sin\varphi\,. \tag{5.194}$$

With the help of (5.191) and (5.194) the time derivative of (5.188) is readily found to be given by

$$\dot{z}_1 = V_1\big[\cos(\omega+\varphi) + e\,\cos\omega\big]\,, \tag{5.195}$$

where (using also (5.192))

$$V_1 = \frac{2\pi}{P_b}\,\frac{a_1\,\sin i}{(1-e^2)^{1/2}}\,. \tag{5.196}$$

An early velocity curve for PSR B 1913+16, determined by Taylor and Hulse, is shown in Fig. 5.5. From this one can read off the orbital period P_b. Note that the expression in the square bracket of (5.195) oscillates in the interval $[-1+e\,\cos\omega, 1+e\,\cos\omega]$. Hence, one can immediately determine the amplitude V_1 as well as $e\,\cos\omega$, and thus the time dependence of the function $\cos(\omega+\varphi)$. The latter can be transformed to a periodic function of the **eccentric anomaly** u, defined by

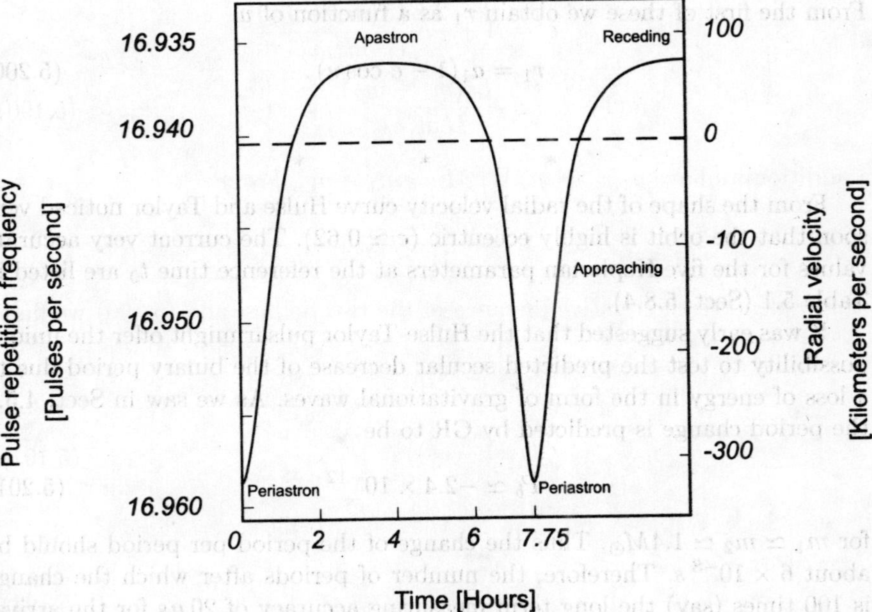

Fig. 5.5. Radial velocity curve for PSR B 1913+16.

$$\varphi = A_e(u) := 2\,\mathrm{arctg}\left[\left(\frac{1+e}{1-e}\right)^{1/2} \tan\frac{u}{2}\right]. \tag{5.197}$$

The dependence of u on the time t is determined by the famous **Kepler equation**

$$\frac{2\pi}{P_b}(t-t_0) = u - e\sin u\,, \tag{5.198}$$

where t_0 is a reference time of periastron passage. From the knowledge of $\cos(\omega + A_e(u))$ and of $e\cos\omega$, we can then also determine e, ω and t_0. So we obtain the five Keplerian parameters

$$\{P_b, a_1 \sin i, e, \omega, t_0\}.$$

The inclination angle i remains undetermined. This will change when general relativistic effects will be included. We shall see that we will then also be able to determine the individual masses, and these turn out to be close to the Chandrasekhar mass.

For future reference we add the following relations that can easily be derived:

$$\cos\varphi = \frac{\cos u - e}{1 - e\cos u}\,, \tag{5.199a}$$

$$\sin\varphi = \frac{\sqrt{1-e^2}\,\sin u}{1 - e\cos u}\,. \tag{5.199b}$$

From the first of these we obtain r_1 as a function of u:

$$r_1 = a_1(1 - e\cos u)\,. \tag{5.200}$$

* * *

From the shape of the radial velocity curve Hulse and Taylor noticed very soon that the orbit is highly eccentric ($e \simeq 0.62$). The current very accurate values for the five Keplerian parameters at the reference time t_0 are listed in Table 5.1 (Sect. 5.8.4).

It was early suggested that the Hulse-Taylor pulsar might offer the unique possibility to test the predicted secular decrease of the binary period due to a loss of energy in the form of gravitational waves. As we saw in Sect. 4.5.4 the period change is predicted by GR to be

$$\dot{P}_b \simeq -2.4 \times 10^{-12}\,, \tag{5.201}$$

for $m_1 \simeq m_2 \simeq 1.4M_\odot$. Thus the change of the period per period should be about $6 \times 10^{-8}\,s$. Therefore, the number of periods after which the change is 100 times (say) the long term measuring accuracy of $20\,\mu s$ for the arrival times is found to be about 3×10^4. This large number is reached after roughly 20 years of timing observations. That was indeed the time it took Taylor and coworkers to arrive at an accurate ($\sim 1\%$) test of the predicted value for $\dot{P}_b$.

We conclude this introduction by emphasizing an important reason why the binary Hulse-Taylor pulsar has become an ideal testing ground for GR. In this two-body system there are *regions of very strong gravitational fields.* The characteristic measure of the field strength $GM/(c^2R)$ becomes about 0.2 at the surface of neutron star (see Chap. 6), which is close to the maximum of 0.5 at the horizon of a (spherically symmetric) black hole. Pulsars in neutron star binaries provide the best known laboratories for experimental tests of GR in the radiative and strong field regimes.

5.8.2 Timing Measurements and Data Reduction

Pulsars are remarkably stable natural clocks. Mainly due to magnetic dipole radiation the period P_p of the rotating magnetized neutron star increases. For PSR 1913+16 the first time derivative $\dot{P}_p$ is very small, but accurately measured: $\dot{P}_p = 8.62713(8) \times 10^{-18}$. The pulsar is thus almost as precise as the best atomic clock.

Arrival Time

What is measured are the pulse arrival times at the telescope over a long period of time. Since PSR 1913+16 is one of the weakest pulsar ($\approx 0.7\,mJy$ at $1400\,MHz$), Taylor and coworkers used for their observations the large

305 m spherical reflector of the Arecibo Observatory in Puerto Rico. Background noise, propagation effects in the interstellar medium and the steep radio frequency spectrum limit accurate pulsar timing observations to the one-decade frequency interval $\approx 300\,MHz - 3\,GHz$.

Individual pulses are weak and have a complicated profile, but averaging over several minutes leads to a relatively regular mean-pulse profile. This is then fitted by the method of least squares to a long-term standard template, thereby yielding an effective pulse arrival time. All timing measurements since February 1981 have uncertainties not greater than 20 μs. Because the estimated uncertainties of individual times of arrival are subjected to errors, daily average values of uncertainties are taken and then scaled by the known dependence of telescope gain and system noise temperature on the zenith angle. The observatory's local time standard is corrected retroactively to the universal coordinated timescale (UTC), using data from the Global Positioning Satellites (GPS). The observational results are so accurate that an unambiguous assignment of pulse numbers is possible, despite the fact that some of the observations may be separated for months or even by several years (during the middle 1990s upgrade of the Arecibo radio telescope) from their nearest neighbors. During such observational gaps the pulsar may have rotated through as many as $10^7 - 10^{10}$ turns. For a detailed discussion of the observational procedures and instrumentation we refer to [141].

The observational arrival times of the train of pulses emitted by the orbiting pulsar contains a wealth of information because this sequence is slightly distorted by interesting effects.

An uninteresting distortion that has to be removed from the data is the dispersion delay due to the interstellar plasma. This can be determined from the measurements at two frequencies (see Sect. 3.5). The dispersion measure is rather large ($DM = 168.77\,cm^{-3}pc$), leading to differential time delays of 70 ms across a typical bandwidth of 4 MHz at 430 MHz. (The observed frequency is Doppler shifted by the Earth's motion.) This is more than the pulsar period. For this reason, the dispersion of the signals had to be compensated from the very beginning. The large dispersion measure shows that PSR 1913+16 is quite far away; the best estimate is $8.3 \pm 1.4\,kpc$.

Besides the almost periodic Doppler modulation which is influenced by general relativistic effects of the orbital motion, the pulse originating from the vicinity of the pulsar (strong field region) travels afterwards through the relatively weak field region between the compact objects, and finally through the weak field of the solar system. The arrival time τ_N^{Earth} of the Nth pulse on Earth depends therefore on many parameters p_i, which include a detailed description of the orbit of the binary system:

$$\tau_N^{Earth} = F(N; p_1, p_2, \ldots)\,. \tag{5.202}$$

In this and the next section we shall work out this timing formula to an accuracy which matches the observational precision.

Solar System Corrections

The solar system corrections are the same as for the arrival times of an isolated pulsar. So let us consider such a pulsar at a fixed distant position $\boldsymbol{x}_p$ in the weak gravitational field of the solar system

$$g = -(1+2\phi(\boldsymbol{x}))dt^2 + (1-2\phi(\boldsymbol{x}))d\boldsymbol{x}^2 . \tag{5.203}$$

Let t_e be the emission time of a radio pulse. The topocentric arrival time (measured by an atomic clock at the radio telescope) will be denoted by τ_a, and for the corresponding coordinate time of the metric (5.203) we use the symbol t_a. We are interested in the difference $\tau_a - t_e = (\tau_a - t_a) + (t_a - t_e)$. The difference $t_a - \tau_a$ is the so-called **Einstein time delay**, $\Delta_{E\odot}$, and is obtained as follows.

From (5.203) we get

$$d\tau_a^2 = (1+2\phi_\oplus)\, dt_a^2 - (1-2\phi_\oplus)\, d\boldsymbol{x}_\oplus^2 .$$

Thus to sufficient accuracy

$$\frac{d\tau_a}{dt_a} \simeq 1 + \phi_\oplus - \frac{1}{2}v_\oplus^2 ,$$

where $\phi_\oplus$ is the Newtonian potential at the position of the telescope at the time t_a, $\phi_\oplus = \phi(\boldsymbol{x}_\oplus(t_a))$, and $v_\oplus$ is the coordinate velocity of the Earth at the same time. Thus $\Delta_{E\odot}$ satisfies

$$\frac{d\Delta_{E\odot}}{dt_a} = -\phi_\oplus + \frac{1}{2}v_\oplus^2 . \tag{5.204}$$

The terms $-\phi_\oplus$ and $\frac{1}{2}v_\oplus^2$ can be interpreted as the gravitational redshift and the transverse Doppler shift, respectively. The Einstein time delay $\Delta_{E\odot}$ is obtained by integrating (5.204) from some arbitrary initial time, because additive constants to not matter. In $\phi_\oplus$ we have to sum over all relevant bodies of the solar system, except the Earth which gives just an additive constant. Because additive constants depend on some conventions, we write

$$t_a = \tau_a + \Delta_{E\odot} + \text{const.} \tag{5.205}$$

Note that the Newtonian potential is time dependent, but during the traversal time of a radio pulse through the solar system, we can neglect the motion of the bodies (adiabatic approximation).

The coordinate time difference $t_a - t_e$ is, after having corrected for dispersion effects, given by

$$t_a - t_e = \int_{\boldsymbol{x}_p}^{\boldsymbol{x}_\oplus(t_a)} (1-2\phi)|d\boldsymbol{x}| , \tag{5.206}$$

where the integral can in sufficient approximation be taken along a straight coordinate path from $\boldsymbol{x}_p$ to $\boldsymbol{x}_\oplus(t_a)$. The dominant contribution of (5.206) is equal to $|\boldsymbol{x}_p - \boldsymbol{x}_\oplus(t_a)|$. Let $\boldsymbol{x}_b$ be the position of the barycentre of the solar system. Then

$$\begin{aligned}|\boldsymbol{x}_p - \boldsymbol{x}_\oplus(t_a)| &= |(\boldsymbol{x}_p - \boldsymbol{x}_b) + (\boldsymbol{x}_b - \boldsymbol{x}_\oplus)| \\ &\simeq |\boldsymbol{x}_p - \boldsymbol{x}_b| + (\boldsymbol{x}_\oplus - \boldsymbol{x}_b)\cdot \boldsymbol{n}\,,\end{aligned}$$

where $\boldsymbol{n}$ is the unit vector in the direction from the pulsar to the barycentre. So we have

$$t_a \simeq t_e + |\boldsymbol{x}_p - \boldsymbol{x}_b| + (\boldsymbol{x}_\oplus - \boldsymbol{x}_b)\cdot \boldsymbol{n} - 2\int_{\boldsymbol{x}_p}^{\boldsymbol{x}_\oplus} \phi |d\boldsymbol{x}|\,.$$

It is convenient to introduce the (fictitious) "infinite-frequency barycentre arrival time" $\tau_a^{bary} = t_e + |\boldsymbol{x}_p - \boldsymbol{x}_b|$. This would be the (infinite-frequency) arrival time at the barycentre of the solar system if its gravitational field could be switched off. According to the last equation we have for this intermediate quantity

$$\tau_a^{bary} = t_a + \Delta_{R\odot} - \Delta_{S\odot}\,, \tag{5.207}$$

where

$$\Delta_{R\odot} = -(\boldsymbol{x}_\oplus - \boldsymbol{x}_b)\cdot \boldsymbol{n} \tag{5.208}$$

is the **Roemer time delay**, and $\Delta_{S\odot}$ is the **Shapiro time delay** given by

$$\Delta_{S\odot} = -2\int_{\boldsymbol{x}_p}^{\boldsymbol{x}_\oplus} \phi |d\boldsymbol{x}|\,. \tag{5.209}$$

Combining (5.205) and (5.207) we obtain

$$\tau_a^{bary} = \tau_a^{Earth} + \Delta_{R\odot} + \Delta_{E\odot} - \Delta_{S\odot}\,. \tag{5.210}$$

This relation holds, of course, also for binary pulsars, but then the relation between τ_a^{bary} and t_e is complicated. In the next section we turn to this relation.

* * *

Exercise 5.2. For the Shapiro time delay it suffices to keep the gravitational field of the Sun. Show that then $\Delta_{S\odot}$ is given by

$$\Delta_{S\odot} = 2\,GM_\odot \ln(1 + \cos\vartheta)\,,$$

where ϑ is the pulsar-Sun-Earth angle at the time of observation, if the eccentricity of the Earth's orbit is neglected.

5.8.3 Theoretical Analysis of the Arrival Times

After this reduction there remains the much more difficult task to relate the barycentric arrival times τ_a^{bary} of the pulses to the "proper time" T_e of the emission, measured in the pulsar comoving frame. (An aberration correction, related to the pulsar model, will be included later.) Once this is established, we obtain the timing formula (5.202) by using in addition the implicit parameterization of T_N by the first few terms of the Taylor series

$$N = N_0 + \nu_p T_N + \dot{\nu}_p \frac{T_N^2}{2} + \ddot{\nu}_p \frac{T_N^3}{6}, \tag{5.211}$$

where $\nu_p = 1/P_p$ is the pulsar frequency and $\dot{\nu}_p$, $\ddot{\nu}_p$ parameterize its secular change with time.

Einstein Time Delay

As an intermediate time variable we use the coordinate time t_e of emission of a pulse. The difference of t_e and the proper time T_e of emission (after an aberration correction taking into account that the pulsar is a rotating beacon) is obtained as (5.205). Adapting its derivation to the present situation, we have – up to higher post-Newtonian orders –

$$\frac{dT_e}{dt_e} \simeq 1 + \phi(\boldsymbol{x}_1) - \frac{1}{2}\boldsymbol{v}_1^2 + \cdots, \tag{5.212}$$

where $\phi(\boldsymbol{x})$ is the Newtonian potential of the binary system

$$\phi(\boldsymbol{x}) = -\frac{G m_1}{|\boldsymbol{x} - \boldsymbol{x}_1(t)|} - \frac{G m_2}{|\boldsymbol{x} - \boldsymbol{x}_2(t)|}. \tag{5.213}$$

One might expect that this approximation is not good enough, because of strong field effects on the surface of the pulsar. In GR these can, however, be absorbed by a renormalization of T_e which is unobservable (see [142]). In view of the "effacement property" of GR, discussed earlier, this is not so astonishing. Inserting (5.213) into (5.212), we obtain, up to an irrelevant additive constant (absorbable again in T_e),

$$\frac{dT_e}{dt_e} = 1 - \frac{G m_2}{r} - \frac{1}{2}\boldsymbol{v}_1^2 + \cdots. \tag{5.214}$$

For the right-hand side we can use the Newtonian approximation for the orbit. Energy conservation implies

$$\frac{1}{2}v^2 - \frac{G(m_1 + m_2)}{r} = \frac{E}{\mu} = -\frac{G(m_1 + m_2)}{2\,a},$$

where $\mu = m_1 m_2/(m_1 + m_2)$ is the reduced mass and v is the relative velocity. Since $v_1 = (m_2/m_1 + m_2)\,v$ we obtain

$$\frac{1}{2}v_1^2 = \frac{G\,m_2^2}{m_1+m_2}\left(\frac{1}{r}-\frac{1}{2\,a}\right). \tag{5.215}$$

Modulo an additive constant, we thus obtain

$$\frac{dT_e}{dt_e} = 1 - G\,\frac{m_2(m_1+2m_2)}{m_1+m_2}\frac{1}{r}. \tag{5.216}$$

For the time integration we use the orbit parameterization in terms of the anomalous eccentricity. Kepler's equation (5.198) reads differentially

$$dt = \frac{P_b}{2\pi}(1-\cos u)du\,. \tag{5.217}$$

Using also (5.200), i.e.,

$$r = a(1-e\,\cos u) \tag{5.218}$$

we obtain

$$dT_e = dt_e - \frac{P_b}{2\pi}\frac{G\,m_2(m_1+2m_2)}{a(m_1+m_2)}\,du$$

and thus with Kepler's equation

$$T_e = t_e - \frac{P_b}{2\pi}\frac{G\,m_2(m_1+2m_2)}{a(m_1+m_2)}\left(\frac{2\pi}{P_b}(t-t_0)+e\,\sin u\right).$$

Rescaling t_e, this can be written as

$$t_e = T_e + \varDelta_E + \text{const}\,, \tag{5.219}$$

where the Einstein time delay is given by

$$\varDelta_E = \gamma\,\sin u\,, \tag{5.220}$$

in which γ is the so-called **Einstein parameter**

$$\gamma = G\,\frac{P_b}{2\pi}\frac{m_2(m_1+2m_2)e}{a(m_1+m_2)}\,. \tag{5.221}$$

This important post-Keplerian parameter, that is going to play a crucial role, can be rewritten with Kepler's third law (5.192) as

$$\gamma = \frac{G^{3/2}\,m_2(m_1+2m_2)e}{(m_1+m_2)^{4/3}}\left(\frac{P_b}{2\pi}\right)^{1/3}. \tag{5.222}$$

This formula survives unchanged if strong field effects are introduced, when m_1 and m_2 are interpreted as the Schwarzschild masses.

Roemer and Shapiro Time Delays

In a next step we link t_e with the coordinate arrival time t_a at the barycentre of the solar system. The latter agrees with τ_a^{bary}, apart from a Doppler shift. We use, as in our treatment of the two-body problem in the post-Newtonian approximation (Sect. 5.5), a coordinate system satisfying the standard post-Newtonian gauge conditions, such that the barycentre of the binary system is at rest. The relative motion is then determined by the differential equation (5.106), which is the Euler-Lagrange equation for the Lagrangian (5.103).

In this coordinate system the barycentre of the solar system is moving with the velocity $\boldsymbol{v}_b$, hence

$$\tau_a^{bary} = (1 - v_b^2)^{1/2} t_a + \text{const} . \tag{5.223}$$

The coordinate time difference $t_a - t_e$ is, as in (5.206), given by

$$t_a - t_e = \int_{\boldsymbol{x}_1(t_e)}^{\boldsymbol{x}_b(t_a)} (1 - 2\phi)|d\boldsymbol{x}| = |\boldsymbol{x}_b(t_a) - \boldsymbol{x}_1(t_e)| + \Delta_S \, , \tag{5.224}$$

where

$$\Delta_S = -2 \int_{\boldsymbol{x}_1(t_e)}^{\boldsymbol{x}_b(t_a)} \phi |d\boldsymbol{x}| \tag{5.225}$$

is the Shapiro delay. The first term on the right in (5.224) depends on t_a through the motion of the solar system: $\boldsymbol{x}_b(t_a) = \boldsymbol{x}_b(0) + t_a \, \boldsymbol{v}_b$. Expanding it to first order in the small quantity $|t_a \, \boldsymbol{v}_b - \boldsymbol{x}_1(t_e)|/|\boldsymbol{x}_b(0)|$, we obtain with (5.223)

$$D\tau_a^{bary} = t_e + \Delta_R(t_e) + \Delta_S(t_e) + \text{const} \, , \tag{5.226}$$

where D is the "Doppler factor"

$$D = (1 - v_b^2)^{-1/2}(1 - \boldsymbol{n} \cdot \boldsymbol{v}_b) \simeq 1 - \boldsymbol{n} \cdot \boldsymbol{v}_b \tag{5.227}$$

($-\boldsymbol{n}$ is as before the unit vector along the line of sight), and Δ_R denotes the Roemer delay across the orbit

$$\Delta_R(t_e) = -\boldsymbol{n} \cdot \boldsymbol{x}_1(t_e) \, . \tag{5.228}$$

For the latter we must use the post-Newtonian approximation for the orbit.

If $\boldsymbol{v}_b$ would be constant, the Doppler parameter D would be unobservable. There is, however, some relative acceleration between the two barycenters that will be discussed later in Sect. 5.8.4.

Let us now compute Δ_S under the assumption of everywhere weak gravitational fields (see [143]). In GR strong field effects again do not change the result, up to an irrelevant additive constant. Since the first term in (5.213) simply adds a constant, we obtain from (5.225)

$$\Delta_S \simeq 2\,Gm_2 \int_{t_e}^{t_a} \frac{dt}{|\boldsymbol{y}(t) - \boldsymbol{x}_2(t_e)|} + \text{const} \, , \tag{5.229}$$

where $\boldsymbol{y}(t)$ is a straight coordinate path from $\boldsymbol{x}_1(t_e)$ to $\boldsymbol{x}_b(t_a)$:

$$\boldsymbol{y}(t) = \boldsymbol{x}_1(t_e) + \frac{t - t_e}{t_a - t_e}\left(\boldsymbol{x}_b(t_a) - \boldsymbol{x}_1(t_e)\right). \tag{5.230}$$

Let $\boldsymbol{x} \equiv \boldsymbol{x}_1(t_e) - \boldsymbol{x}_2(t_e)$ and $\theta = (t - t_e)/(t_a - t_e)$, then we obtain in a first step (apart from an unobservable additive constant)

$$\Delta_S(t_e) = (t_a - t_e)\int_0^1 \frac{d\theta}{|\boldsymbol{x} + \theta(\boldsymbol{x}_b(t_a) - \boldsymbol{x}_1(t_e))|}.$$

Since $(\boldsymbol{x}_b(t_a) - \boldsymbol{x}_1(t_e)) \simeq t_a - t_e$ and $|\boldsymbol{x}_b(t_a)| \simeq |\boldsymbol{x}_b(0)| =: r_b \gg |\boldsymbol{x}_1(t_e)|$, we have

$$\Delta_S(t_e) \simeq 2\,Gm_2\int_0^1 \frac{d\theta}{|\theta\,\boldsymbol{n} + \boldsymbol{x}/r_b|} = 2\,Gm_2\int_0^1 \frac{d\theta}{\sqrt{\theta^2 + \dfrac{2\theta\,\boldsymbol{n}\cdot\boldsymbol{x}}{r_b} + \left(\dfrac{r}{r_b}\right)^2}},$$

where $r := |\boldsymbol{x}|$. This integral is elementary, and one finds

$$\begin{aligned}\Delta_S(t_e) &\simeq 2\,Gm_2\,\ln\left(\frac{2\,r_b}{r + \boldsymbol{n}\cdot\boldsymbol{x}}\right) + \text{const}\\ &= -2\,Gm_2\,\ln\Big(\boldsymbol{n}\cdot(\boldsymbol{x}_1(t_e) - \boldsymbol{x}_2(t_e)) + |\boldsymbol{x}_1(t_e) - \boldsymbol{x}_2(t_e)|\Big) + \text{const}.\end{aligned} \tag{5.231}$$

On the right-hand side we are allowed to use the Keplerian approximation. According to (5.188) the argument of the logarithm is equal to

$$r(1 + \sin i\,\sin(\varphi + \omega)) = r(1 - \sin i(\sin\omega\,\cos\varphi + \cos\omega\,\sin\varphi)).$$

Thus, using (5.199a) and (5.199b), as well as (5.200), we obtain (modulo a constant)

$$\Delta_S(t_e) = -2\,Gm_2\,\ln\Big(1 - e\,\cos u - \sin i\big(\sin\omega(\cos u - e) + \sqrt{1 - e^2}\cos\omega\,\sin u\big)\Big). \tag{5.232}$$

This is often written in terms of the "range" and "shape" parameters r and s (the letter r now changes meaning)

$$\Delta_S(t_e) = -2\,r\,\ln\Big(1 - e\,\cos u - s\big(\sin\omega(\cos u - e) + \sqrt{1 - e^2}\cos\omega\,\sin u\big)\Big). \tag{5.233}$$

In GR we have

$$r = G\,m_2\,, \tag{5.234a}$$

$$s = \sin i = \frac{x\,M}{m_2\,a}, \quad x := a_1\sin i \tag{5.234b}$$

(projected semi-major axis of the pulsar) or using Kepler's third law

$$s = G^{-1/3} x \left(\frac{P_b}{2\pi}\right)^{-2/3} M^{2/3} m_2^{-1} . \tag{5.235}$$

Numerically, if $T_\odot = G M_\odot / c^3 = 4.925490947\,\mu s$, then

$$r = \frac{m_2}{M_\odot} T_\odot , \tag{5.236a}$$

$$s = x(s) \left(\frac{P_b}{2\pi}\right)^{-2} T_\odot^{-1/3} \left(\frac{M}{M_\odot}\right)^{2/3} \left(\frac{m_2}{M_\odot}\right)^{-1} , \tag{5.236b}$$

where $x(s)$ means the value of x in light seconds, and P_b has to be taken in seconds. For the binary pulsar PSR B 1534+12 these are given in Table 5.2 at the end of this chapter.

Explicit Expression for the Roemer Delay

For the Roemer delay we obtain from (5.228) and (5.188)

$$\Delta_R(t_e) = r_1 \sin i \, \sin(\omega + \varphi) . \tag{5.237}$$

Here, we have to use the post-Newtonian approximation for the pulsar orbit. But let use first work out the Keplerian approximation $\Delta_R^{(K)}$. This is immediately obtained from the previous derivation of (5.232):

$$\Delta_R^{(K)} = a_1 \sin i \big(\sin\omega(\cos u - e) + \sqrt{1-e^2} \cos\omega \, \sin u\big) . \tag{5.238}$$

Following T. Damour and N. Deruelle (see [144] and [145]), we shall obtain a remarkably similar expression in post-Newtonian approximation. In their first paper a "quasi-Newtonian" parametric representation of the pulsar motion was derived, starting from the Lagrangian (5.104). Since this is autonomous and invariant under spatial rotations one has the usual energy and angular momentum integrals. In terms of polar coordinates (r, θ) in the (coordinate) plane of the orbit, the equations for the relative motion can then brought into the form

$$\left(\frac{dr}{dt}\right)^2 = A + \frac{2B}{r} + \frac{C}{r^2} + \frac{D}{r^2} ,$$

$$\frac{d\theta}{dt} = \frac{H}{r^2} + \frac{I}{r^3} ,$$

where the coefficients $A, B, \ldots, I$ are polynomials in the energy and total momentum involving the masses of the binary system. In terms of an anomalous eccentricity U the motion can then be described as follows. First, we have the "Keplerian equation"

$$\frac{2\pi}{P_b}(t - T_0) = U - e_t \sin U \,, \tag{5.239}$$

with a "time eccentricity" e_t. The motion of the pulsar is given by

$$r_1 = a_1(1 - e_r \cos U), \tag{5.240a}$$
$$\theta = \omega_0 + (1 + k)A_{e_\theta}(U), \tag{5.240b}$$

involving a "radial eccentricity" e_r, as well as an "angular eccentricity" e_θ. Note that P_b is the time of return to the periastron, and k is the fractional periastron precession per orbit: $k = \dot{\omega}\, P_b/2\pi$. The function $A_e(U)$ was introduced in (5.197). According to (5.199a) and (5.199b) this function satisfies the identities

$$\cos A_e(U) = \frac{\cos U - e}{1 - e \cos U}\,, \tag{5.241a}$$

$$\sin A_e(U) = \frac{\sqrt{1 - e^2}\sin U}{1 - e \cos U}\,. \tag{5.241b}$$

The quantities P_b, T_0, e_t, a_1, e_r, ω_0, k and e_θ are all constants. We also note that the "semi-major axis" a of the relative motion (in an formula analogous to (5.240a)) is $a = (M/m_2)a_1$.

Using this parametric representation we can write (5.237) as

$$\Delta_R(t_e) = a_1 \sin i(1 - e_r \cos U)\sin\big(\omega_0 + (1 + k)A_e(U)\big)\,. \tag{5.242}$$

Damour and Deruelle show in Appendix A of [145], using some clever tricks, that this can be brought to a form similar to (5.238) by introducing a *new* eccentric anomaly u. The result is (recall $x := a_1 \sin i$)

$$\Delta_R(t_e) = x\left(\sin\omega(\cos u - e_r) + \sqrt{1 - e_\theta^2}\cos\omega\, \sin u\right)\,. \tag{5.243}$$

The new eccentric anomaly is defined by

$$\frac{2\pi}{P_b}(T - T_0) = u - e_T \sin u\,, \tag{5.244}$$

where T is, as earlier, the proper time of the pulsar. e_T is yet another eccentricity, and is related to e_r and e_θ as

$$\frac{e_r - e_T}{e_T} =: \delta_r = \frac{G}{a\,M}\left(3\,m_1^2 + 6\,m_1 m_2 + 2\,m_2^2\right)\,, \tag{5.245a}$$

$$\frac{e_\theta - e_T}{e_T} =: \delta_\theta = \frac{G}{a\,M}\left(\frac{7}{2}m_1^2 + 6\,m_1 m_2 + 2\,m_2^2\right)\,. \tag{5.245b}$$

Furthermore, the argument $\omega(T)$ of the periastron is

$$\omega = \omega_0 + k\,A_e(u)\,. \tag{5.246}$$

The periastron advance was computed earlier, with the result (5.112), i.e.,

$$k = \frac{3GM}{a(1-e^2)}. \tag{5.247}$$

In what follows, an eccentricity e without an index means that it belongs already to a small (relativistic) correction and can thus be replaced by any eccentricity introduced above, e.g., by e_T of Eq. (5.244).

Before proceeding with our theoretical discussion, we show an observational result. From (5.210), (5.219) and (5.220) is clear that the sum $\Delta_R + \Delta_E + \Delta_S$ as a function of orbital phase is measurable. The result for PSR 1913+16 obtained in 1988 is shown in Fig. 5.6. Note the high precision, illustrated by the lone data point with $50\,000\sigma$ error bars.

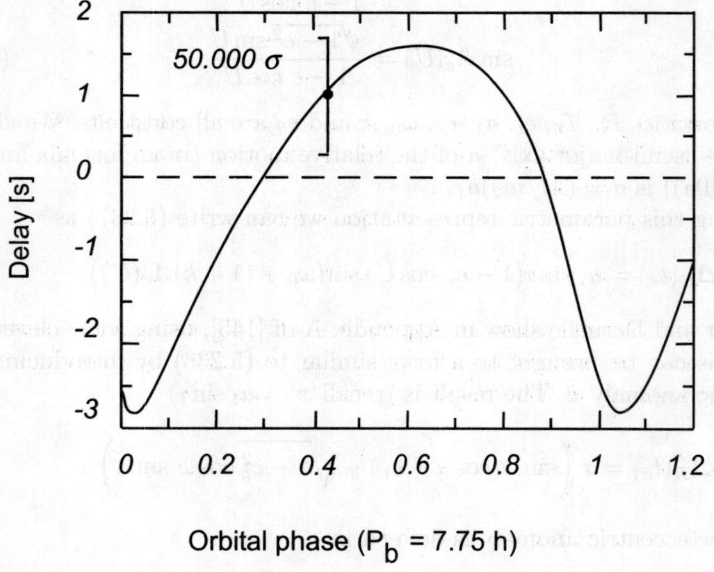

Fig. 5.6. Orbital delays ($\Delta_R + \Delta_E + \Delta_S$) observed for PSR B 1913+16 during July 1988. (From the Nobel Lecture by J.H. Taylor [98].)

Aberration Correction

Pulsars are rapidly rotating strongly magnetized neutron stars. In a manner still not completely understood very narrow beams of radio waves are emitted from the regions of the magnetic poles, precessing around the rotation axis. If one (sometimes both) of these beams happen to be aligned such that they sweep across the Earth, we see them as the characteristic pulses from a pulsar. The effect is very similar to that of a lighthouse. The **pulsar phase** is the angle Φ describing the direction in longitude of the emitted beam. As a function of the pulsar proper time T, we have according to (5.211)

$$\frac{\Phi(T)}{2\pi} = \nu_p T + \frac{1}{2}\dot{\nu}_p T^2 + \frac{1}{6}\ddot{\nu}_p T^3 \,. \tag{5.248}$$

The orbital motion of the pulsar around the center of mass of a binary system leads to an aberration effect: The direction $\boldsymbol{n}$ the observer receives light differs from the direction $\boldsymbol{n}'$ in which it was emitted in the pulsar comoving frame. To order $\boldsymbol{v}_1/c$ the relation between $\boldsymbol{n}$ and $\boldsymbol{n}'$ is given by

$$\boldsymbol{n}' - \boldsymbol{n} = -\boldsymbol{v}_1 + (\boldsymbol{v}_1 \cdot \boldsymbol{n})\boldsymbol{n} \,. \tag{5.249}$$

The unit vector $\boldsymbol{n}$ precesses around the spin axis $\boldsymbol{e}_{rot}$. The tangential component of $\boldsymbol{n}' - \boldsymbol{n}$ along this precession orbit is

$$(\boldsymbol{n}' - \boldsymbol{n}) \cdot \frac{\boldsymbol{e}_{rot} \wedge \boldsymbol{n}}{|\boldsymbol{e}_{rot} \wedge \boldsymbol{n}|} = -\boldsymbol{v}_1 \cdot \frac{\boldsymbol{e}_{rot} \wedge \boldsymbol{n}}{|\boldsymbol{e}_{rot} \wedge \boldsymbol{n}|} \,. \tag{5.250}$$

This corresponds to an "aberration angle" $\delta\Phi_A$ of the pulsar phase given by $\delta\Phi_A \cdot |\boldsymbol{e}_{rot} \wedge \boldsymbol{n}| =$(5.250), thus

$$\delta\Phi_A = \boldsymbol{v}_1 \cdot \frac{\boldsymbol{n} \wedge \boldsymbol{e}_{rot}}{(\boldsymbol{n} \wedge \boldsymbol{e}_{rot})^2} \,. \tag{5.251}$$

The corresponding **aberration delay** is (see (5.248))

$$\Delta_A = \frac{\delta\Phi_A}{2\pi\nu_p} \,. \tag{5.252}$$

The proper time T in (5.248) is related to the proper time T_e of emission by

$$T = T_e - \Delta_A \,. \tag{5.253}$$

Combining (5.219), (5.226) and (5.253), we obtain

$$D\tau_a^{bary} = T + \Delta_R + \Delta_E + \Delta_S + \Delta_A \,. \tag{5.254}$$

Together with (5.210) we arrive at our main goal

$$\tau_a^{Earth} + \Delta_{R\odot} + \Delta_{E\odot} - \Delta_{S\odot} = D^{-1}\big(T + \Delta_R + \Delta_E + \Delta_S + \Delta_A\big) \,. \tag{5.255}$$

As emphasized earlier, a constant D could be absorbed in a rescaling of the units of time, and is thus unobservable. We therefore set $D = 1$, but we shall later come back to the effect of a secular drift of D that is not always negligible. The aberration parameters A, B could be absorbed into redefinitions of T_0, $x = a_1 \sin i$, e, δ_r and δ_θ (see [145]), but this would lead to apparent secular changes of the observed values of x and e. It has also been shown in [145] that δ_r could be reabsorbed in a change of the spin phase of the pulsar, thereby introducing small changes in the observed values of the pulsar rotational parameters which need not concern us.

The Timing Formula

Before collecting the results of the previous discussion, we have to include the effect of gravitational radiation damping. This induces secular changes of the Keplerian parameters P_b, x and e. The secular change $\dot{P}_b$ modifies Eq. (5.244):

$$u - e\sin u = 2\pi\left(\frac{T-T_0}{2\pi} - \frac{1}{2}\dot{P}_b\left(\frac{T-T_0}{P_b}\right)^2\right), \tag{5.256}$$

where by definition P_b is a constant, the orbital period at epoch T_0.

Collecting our results, we arrive at the following **timing formula** of Damour and Deruelle: A pulse received on Earth at topocentric time τ^{Earth} is emitted at the time T in the comoving pulsar frame given by

$$\begin{aligned} T = \tau^{Earth} - t_0 - \frac{\mathcal{D}}{f^2} + \Delta_{R\odot} + \Delta_{E\odot} - \Delta_{S\odot} \\ - \left(\Delta_R + \Delta_E + \Delta_S + \Delta_A\right). \end{aligned} \tag{5.257}$$

Here, t_0 is a reference epoch and $\mathcal{D}/f^2$ is the dispersive delay. The orbital terms are (see (5.220), (5.233) and (5.243) and Exercise 5.3)

$$\begin{aligned} \Delta_R &= x\Big(\sin\omega\left(\cos u - e(1+\delta_r)\right) \\ &\quad + \left(1 - e^2(1+\delta_\theta)^2\right)^{1/2}\cos\omega\,\sin u\Big), \end{aligned} \tag{5.258a}$$

$$\Delta_E = \gamma\sin u, \tag{5.258b}$$

$$\begin{aligned} \Delta_S &= -2r\ln\Big(1 - e\cos u - s\big(\sin\omega(\cos u - e) \\ &\quad + \sqrt{1-e^2}\cos\omega\,\sin u\big)\Big), \end{aligned} \tag{5.258c}$$

$$\begin{aligned} \Delta_A &= A\Big(\sin(\omega + A_e(u)) + e\sin\omega\Big) \\ &\quad + B\Big(\cos(\omega + A_e(u)) + e\cos\omega\Big), \end{aligned} \tag{5.258d}$$

where

$$\omega = \omega_0 + k\,A_e(u). \tag{5.259}$$

Obviously, the splitting into the various Δ terms in (5.257) is gauge dependent, nevertheless this helps the intuition.

The five Keplerian parameters $\{P_b, T_0, e, \omega_0, x\}$ are constants (for simplicity, we neglect here the secular variations of e and x). The set of post-Keplerian parameters

$$\{\dot{\omega}, \gamma, \dot{P}_b, r, s, \delta_\theta\} \tag{5.260}$$

are in principle *separately measurable* (in this list we could include the secular changes $\dot{x}$ and $\dot{e}$), while the set

$$\{\delta_r, A, B\} \tag{5.261}$$

is not separately measurable, because they can be absorbed by suitable redefinitions of the other parameters. The proper pulsar time T is related to the pulsar phase by (5.248).

So far the timing formula holds for a broad class of gravity theories. In this so-called DD-model (for Damour and Deruelle), the six post-Keplerian parameters (5.260) are treated as free parameters in a fit to observational data.

The general relativistic expressions for (5.260) and (5.261) have all been determined, and are (in general units) given by:

$$k = \frac{3GM}{c^2 a(1-e^2)}\,, \tag{5.262a}$$

$$\dot{\omega} = \frac{2\pi}{P_b}k = 3\left(\frac{P_b}{2\pi}\right)^{-5/3}\frac{G^{2/3}}{c^2(1-e^2)}M^{2/3}\,, \tag{5.262b}$$

$$\gamma = \frac{G^{2/3}}{c^2}e\left(\frac{P_b}{2\pi}\right)^{1/3} m_2\,\frac{m_1+2\,m_2}{M^{4/3}}\,, \tag{5.262c}$$

$$\dot{P}_b = -\frac{192\pi}{5}\frac{G^{5/3}m_1 m_2}{c^5 M^{1/3}}\left(\frac{2\pi}{P_b}\right)^{5/3} f(e)\,, \tag{5.262d}$$

$$r = \frac{G\,m_2}{c^3}\,, \tag{5.262e}$$

$$s = \sin i = c\,G^{-1/3}x\left(\frac{P_b}{2\pi}\right)^{-2/3} M^{2/3}m_2^{-1}\,, \tag{5.262f}$$

$$\delta_\theta = \frac{G}{c^2\,a\,M}\left(\frac{7}{2}m_1^2 + 6\,m_1 m_2 + 2\,m_2^2\right), \tag{5.262g}$$

$$\delta_r = \frac{G}{c^2\,a\,M}\left(3\,m_1^2 + 6\,m_1 m_2 + 2\,m_2^2\right). \tag{5.262h}$$

with

$$f(e) = \left(1 + \frac{73}{24}e^2 + \frac{37}{96}e^4\right)(1-e^2)^{-7/2}\,.$$

The aberration parameters A and B are (see Ex. 5.3)

$$A = -\frac{a_1}{c\,P_b\,\nu_p\,\sin\lambda\sqrt{1-e^2}}\sin\eta\,, \tag{5.263a}$$

$$B = -\frac{a_1\cos i}{c\,P_b\,\nu_p\,\sin\lambda\sqrt{1-e^2}}\cos\eta\,. \tag{5.263b}$$

At this point the following is important: Seven parameters are needed to fully specify the dynamics of the two-body system (up to an uninteresting rotation about the line of sight). Therefore, the measurement of any two post-Keplerian parameters, beside the five Keplerian ones, allow the remaining ones to be predicted. Thus, if one is able to measure P post-Keplerian parameters then one has $P-2$ distinct tests of GR (or of alternative relativistic gravity theories). Realistically, for some systems the first five of the set (5.260) can be measured (δ_θ is difficult to be determined), in which case one has (potentially) three independent tests. In the next section we shall discuss the current situation.

* * *

Exercise 5.3. Let λ, η be the polar angles of the spin axis $\boldsymbol{e}_{rot}$ relative to the triad $\{\boldsymbol{K}, \boldsymbol{N} \wedge \boldsymbol{K}, \boldsymbol{K}\}$, introduced in Sect. 5.8.1. Show that in the Keplerian approximation for the orbit

$$\begin{aligned}\Delta_A = & A\left(\sin(\omega + A_e(u)) + e \sin\omega\right) \\ & + B\left(\cos(\omega + A_e(u)) + e \cos\omega\right) ,\end{aligned} \tag{5.264}$$

where A and B are given by (5.263a) and (5.263b).

5.8.4 Results for Keplerian and Post-Keplerian Parameters

The (phenomenological) parameters appearing in the timing formula are extracted from the set of pulse arrival times by calculating the expected times of emission $\{T_N\}$ for the pulse numbers $\{N_j\}$ from (5.256)–(5.259), and the minimizing the weighted sum of the squared residuals with respect to all parameters to be determined. Details of the procedure are described in [146] and [147].

Several thousands of pulse arrival times for PSR B 1913+16 have been recorded at Arecibo Observatory since the discovering of the pulsar in 1974. These data, especially the ones since 1981, determine the five Keplerian parameters $\{P_b, x, e, T_0, \omega_0\}$ and $\dot{\omega}$, the largest post-Keplerian parameter, very accurately, as can be seen from Table 5.1. Note, in particular, the large value of $\dot{\omega}$ (4 degrees per year), in comparison to the 43″ per century of Mercury. In hundred years the binary pulsar orbit rotates more than 360 degrees. It took some time to arrive at a precise value of the Einstein parameter γ, which is now known to better than one permille. – All quoted results are from [148].

Masses of the Two Neutron Stars

Assuming that we can ignore possible non-relativistic contributions to the periastron shift, such as tidally and rotationally induced effects caused by the companion (to be discussed later), we can use the measured values for $\dot{\omega}$ and γ to determine the masses of the two stars. Indeed, (5.262b) and (5.262c) show that these post-Newtonian parameters have different dependences on the pulsar and companion masses. This is a very interesting astrophysical application of GR, and provides the most precise values of neutron star masses. The result for the Hulse-Taylor system is

Table 5.1. Measured orbital parameters for the binary pulsar PSR B 1913+16. The reference epoch T_0 is given in modified Julian days (MJD). (From [148]).

Fitted parameter	*Value*	*Uncertainty*
$x = a_1 \sin i$ [s]	2.341774	0.000001
e	0.6171338	0.0000004
T_0 [MJD]	46443.99588317	0.00000003
P_b [d]	0.322997462727	0.000000000005
ω_0 [deg]	226.57518	0.00004
$\dot{\omega}$ [dec/yr]	4.226607	0.000007
γ [s]	0.004294	0.000001
$\dot{P}_b$ [10^{-12}s/s]	-2.4211	0.0014

$$m_1 = 1.4408 \pm 0.0003 M_\odot\,, \quad m_2 = 1.3873 \pm 0.0003 M_\odot\,. \tag{5.265}$$

It is highly significant for theories of late stellar evolution and neutron star formation in gravitational collapse that the two masses are close to the critical Chandrasekhar mass (for more on this, see Chap. 6).

Confirmation of the Gravitational Radiation Damping

We now come to the most important result of Taylor and collaborators. Knowing, beside the Keplerian parameters, also the masses of the binary system, GR predicts the secular change $\dot{P}_b$ as a result of gravitational radiation damping (see (5.262d)) to be

$$\dot{P}_b^{GR} = (-2.40247 \pm 0.00002) \times 10^{-12}\,. \tag{5.266}$$

This compares very well with the observed value of Table 5.1, but for a more precise comparison it is necessary to take into account the small effect of a relative acceleration between the binary pulsar system and the solar system, caused by the differential rotation of the galaxy.

Remember that we absorbed the Doppler factor D by a rescaling of the units of time. In particular, the ratio of the observed to the intrinsic period is given by $P_b^{obs}/P_b^{intr} = D^{-1}$. To first order we have (see (5.227))

$$\frac{P_b^{obs}}{P_b^{intr}} = 1 + \frac{v_R}{c}\,, \tag{5.267}$$

where v_R is the radial velocity of the pulsar (position $\boldsymbol{x}_1$) relative to the solar system barycentre $\boldsymbol{x}_b$:

$$v_R = \boldsymbol{n}_{1b} \cdot (\boldsymbol{v}_1 - \boldsymbol{v}_b)\,, \quad \boldsymbol{n}_{1b} := \frac{\boldsymbol{x}_1 - \boldsymbol{x}_b}{|\boldsymbol{x}_1 - \boldsymbol{x}_b|}\,. \tag{5.268}$$

Differentiating (5.267) gives

$$\left(\frac{\dot{P}_b}{P_b}\right)^{gal} = \frac{1}{c}\boldsymbol{n}_{1b}\cdot(\boldsymbol{a}_1 - \boldsymbol{a}_b) + \frac{v_T^2}{c\,d}, \tag{5.269}$$

where $\boldsymbol{a}_1$ and $\boldsymbol{a}_b$ are the accelerations of $\boldsymbol{x}_1$ and $\boldsymbol{x}_b$, v_T is the magnitude of the transversal velocity, $v_T = |\boldsymbol{n}_{1b}\wedge(\boldsymbol{v}_1 - \boldsymbol{v}_b)|$, and $d = |\boldsymbol{x}_1 - \boldsymbol{x}_b|$ the distance between the Sun and the pulsar.

Damour and Taylor estimated (5.269) using data on the location and proper motion of the pulsar, combined with the best available information on galactic rotation (see [149]). A recent reevaluation gave (see [148])

$$(\dot{P}_b)^{gal} = -(0.0125 \pm 0.0050)\times 10^{-12}. \tag{5.270}$$

Subtracting this from the observed $\dot{P}_b$ leads to the corrected

$$(\dot{P}_b)^{corr} = -(2.4086 \pm 0.0052)\times 10^{-12}, \tag{5.271}$$

to be compared with the prediction (5.266). The two numbers agree to about 0.5%:

$$\frac{(\dot{P}_b)^{GR}}{(\dot{P}_b)^{corr}} = 1.00 \pm 0.005. \tag{5.272}$$

This beautiful consistency among the measurements is shown in Fig. 5.7, in which the bands in the mass plane are shown that are allowed by the measurements of $\dot{\omega}$, γ and $\dot{P}_b$ if GR is valid. As one can see, there is a single common overlap.

Another way to show the agreement with GR is to compare the observed phase of the orbit with the theoretical prediction as a function of time. The orbit phase is given according to (5.256) by

$$\Phi_b(T) = 2\pi\left(\frac{T - T_0}{P_b} - \frac{1}{2}\dot{P}_b\left(\frac{T - T_0}{P_b}\right)^2\right).$$

Since, on the other hand, the periastron passage time T_P satisfies $\Phi_b(T_P) - T_0 = 2\pi N$, where N is an integer, this will not grow linearly with N. The commutative difference between periastron time T_P and $N\,P_b$ should vary as

$$(T_P - T_0) - N\,P_b \simeq \frac{1}{2}\dot{P}_b\,P_b^{-1}(T_P - T_0)^2.$$

Figure 5.8 shows the data and the GR prediction of this. The gap during the middle 1990s was caused by a closure of Arecibo for upgrading.

This beautiful agreement also provides strong evidence that the binary system PSR B 1913+16 is "clean". Since the companion mass is close to the Chandrasekhar limiting mass for white dwarfs, we are most probably dealing with a binary neutron star system. Such stars are so compact that we can

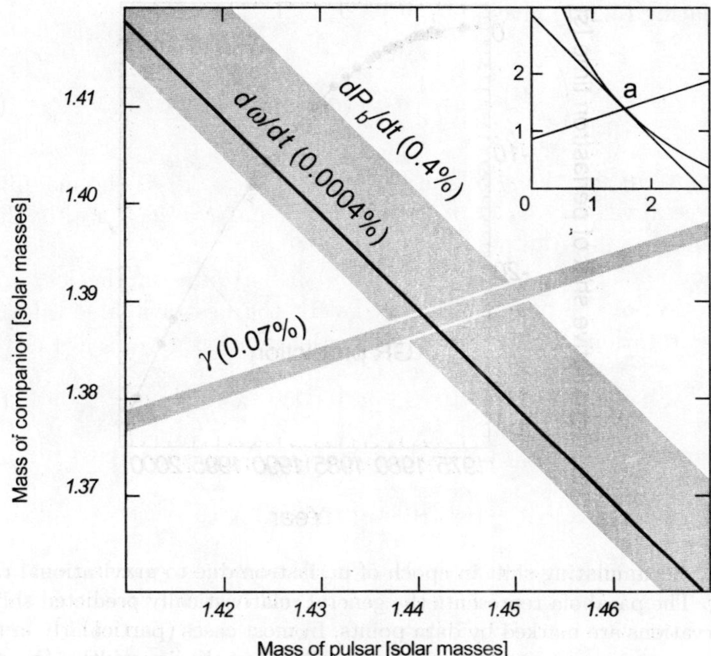

Fig. 5.7. Constraints on masses of the pulsar and its companion from data on PSR 1913+16, assuming GR to be valid. The width of each strip in the plane reflects observational accuracy, shown as a percentage. An inset shows the three constraints on the full mass plane; the intersection region a has been magnified 400 times for the full figure (adapted from [74]).

completely ignore tidally or rotationally induced Newtonian effects on the orbit. (These effects were estimated in Sect. 5.8.1). Early after the discovery of PSR 1913+16 one could not a priori exclude that the companion was a helium star or a rapidly rotating white dwarf, but this has become more than unlikely. (For an early discussion of this issue see [150] and [14].)

* * *

No relativist had even dreamed of a system like the eight-hour binary Hulse-Taylor pulsar. The high expectations held shortly after its discovery have been surpassed during the long detailed study by Taylor and his associates. GR has passed new tests with complete success.

Results for the Binary PSR B 1534+12

Since 1990 other massive binary pulsars were discovered. Of particular interest for testing GR is the binary system PSR B 1534+12 (see [151]), a bright, relatively nearby pulsar, also with a neutron star companion. Its pulses are

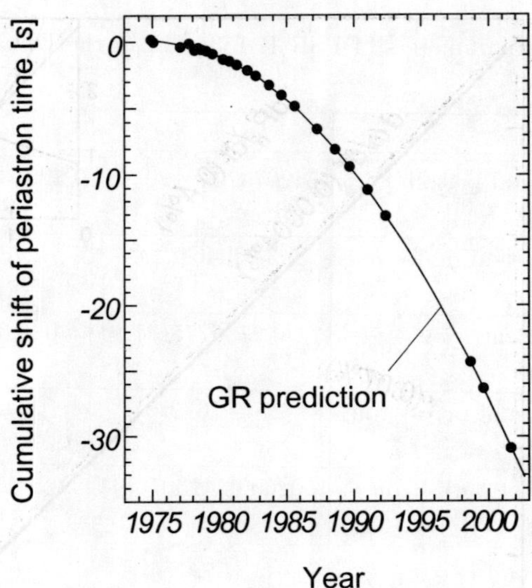

Fig. 5.8. Accumulating shift in epoch of periastron due to gravitational radiation damping. The parabola represents the general relativistically predicted shift, while the observations are marked by data points. In most cases (particularly in the later data), the measurement uncertainties are smaller than the line widths. (From [148].)

significantly stronger and narrower than those of PSR B 1913+16, allowing more precise timing measurements. Because the orbit, with $P_b \simeq 10.1$ hours, is nearly edge-on as viewed from the Earth, the Shapiro parameters r and s are more easily measurable. This led to new experimental tests.

The latest results, presented in [152], are based on an 11.5-year data set taken with the Arecibo $305\,m$ telescope. Over recent years a new data acquisition system (see [153]) was used that fully removes the dispersion effects of the interstellar medium from the signal. The accuracy of the pulse arrival times has also been improved.

The five post-Keplerian parameters $\{\dot{\omega}, \gamma, \dot{P}_b, r, s\}$ can be measured as described before and are listed in the DD column of Table 5.2, taken from [152]. If these agree within the uncertainties with the general relativistic values (5.262b)–(5.262f), we obtain bands in the mass plane shown in Fig. 5.9 (note that $\dot{\omega}$, γ are known very accurately). Leaving out for a moment $\dot{P}_b$, the remaining four bands intersect in a common region. The best mass estimates of the two neutron stars are

$$m_1 = 1.3332 \pm 0.0010 M_\odot\,, \quad m_2 = 1.3452 \pm 0.0010 M_\odot\,.$$

This provides a precise test of GR which makes only use of non-radiative parameters. The general relativistic values of the five post-Keplerian parameters (5.262b)–(5.262f) for these two masses and the Keplerian parameters

Table 5.2. Orbit parameters of PSR B 1534+12 in the DD and DDGR models. (From [152].)

	DD model	*DDGR model*
Orbital period P_b [d]	0.420737299122(10)	0.420737299123(10)
Projected semi-major axis x [s]	3.729464(2)	3.7294641(4)
Eccentricity e	0.2736775(3)	0.27367740(14)
Longitude of periastron ω [deg]	274,57679(5)	274.57680(4)
Epoch of periastron T_0 [MJD]	50260.92493075(4)	50260.92493075(4)
Advance of periastron $\dot{\omega}$ [deg yr^1]	1.755789(9)	1.7557896
Gravitational redshift γ [ms]	2.070(2)	2.069
Orbital period derivative $(\dot{P}_b)^{obs}$ 10^{12}	-0.137(3)	-0.1924
Shape of Shapiro delay s	0.975(7)	0.9751
Range of Shapiro delay r [μs]	6.7(1.0)	6.626
Derivative of x $\lvert\dot{x}\rvert$ [10^{12}]	<0.68	<0.015
Derivative of e $\lvert\dot{e}\rvert$ [10^{-15}s^{-1}]	<3	<3
Total mass $M = M_1 + m_2$ [$M_\odot$]		2.678428(18)
Companion mass m_2 [$M_\odot$]		1.3452(10)
Excess $\dot{P}_b$ [10^{-12}]		0.055(3)

are given in italic numbers in the "DDGR" column of Table 5.2. Note the excellent agreement with the DD fit parameters.

Unfortunately, the observed $\dot{P}_b$ has a galactic contribution that is dominated by the last term in (5.269), due to the transverse velocity of the pulsar binary. This involves the pulsar's distance, which can only roughly be estimated on the basis of the dispersion measure. Using a model of Taylor and

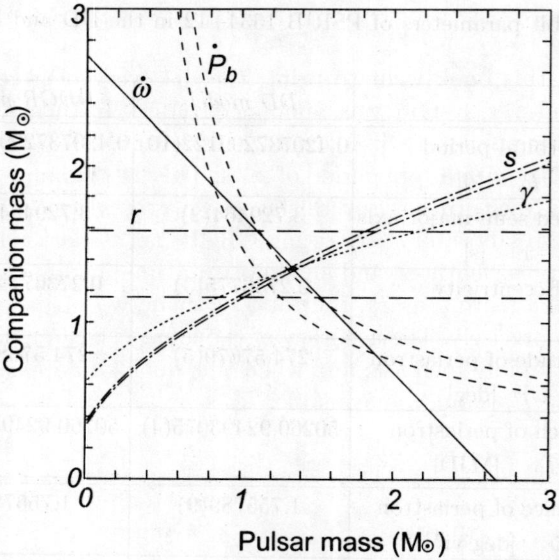

Fig. 5.9. Mass-mass diagram for the PSR B 1534+12.

Cordes (see [154]) for the free electron distribution of the Galaxy, one finds the small distance $d = 0.7 \pm 0.2 kpc$. Including also estimates of the other terms in (5.269), the following value for $(\dot{P}_b)^{gal}$ was obtained in [152]:

$$(\dot{P}_b)^{gal} = (0.037 \pm 0.011) \times 10^{-12} ,$$

where the uncertainty is dominated by the one in the distance.

The intrinsic rate of orbit decay is

$$(\dot{P}_b)^{obs} - (\dot{P}_b)^{gal} = (-0.174 \pm 0.011) \times 10^{-12} .$$

On the other hand one obtains from (5.262d)

$$(\dot{P}_b)^{GR} = -0.192 \times 10^{-12} .$$

The two values differ by some 1.7 standard deviations.

We can, however, easily bring the two quantities into agreement, by changing the pulsar distance to $d = 1.02 \pm 0.05\, kpc$ (68% confidence limit). This is valuable, but a precise test of $(\dot{P}_b)^{GR}$ would have been more interesting.

Binary pulsar data have also been analyzed in the framework of alternative theories of gravity, in particular for a broad class of scalar tensor theories (see [155]). How are close binary neutron star systems formed? Evolutionary scenarios, involving two supernova explosions, will be discussed in Sect. 6.12.5.

* * *

Double-Pulsar

Shortly before this book went to print, the discovery of a very interesting double neutron star system was announced [156]. It turned out that the companion of the previously discovered 22-ms pulsar PSR J0737-3039 [157] is also a radio pulsar with a spin period of 2.7 s. This first example of a **double-pulsar binary system** has several favorable properties which promise to make it a superb laboratory for testing GR. It has a short orbital period of 2.4 hr, narrow pulse features, and an almost maximal inclination angle. Since both neutron stars are pulsars, the ratio of their two masses could already been measured precisely. The accuracy of the four post-Keplerian parameters $\dot{\omega}$, γ, r, s, will rapidly increase, and when this book will appear a significant measurement of $\dot{P}_b$ should be available.

6 White Dwarfs and Neutron Stars

"With all reserve we suggest the view that supernovae represent the transitions from ordinary stars to neutron stars, which in their final stages consist of extremely closely packet neutrons."

—W. Baade and F. Zwicky (1934)

6.1 Introduction

The bewildering variety of normal stars covers the mass range

$$0.07\, M_\odot \lesssim M \lesssim 60 \text{ to } 100\, M_\odot\,.$$

Masses smaller than the lower limit cannot become sufficiently hot to ignite nuclear burning and stars more massive than 60 to 100 solar masses would be unstable. The many different and sometimes brilliant evolutionary paths lead to the following few end-products of stellar evolution (see Fig. 6.1).

1. Low-mass stars with masses $M \lesssim 4\, M_\odot$ evolve into **white dwarfs**. White dwarfs progenitor stars often undergo relatively gentle mass ejection (forming "planetary nebulae") at the end of their evolutionary lifetimes and thereby reduce their mass below the Chandrasekhar limit for white dwarfs of about $1.4 M_\odot$. The slowly cooling white dwarf has a size approximately equal to that of the Earth. The Sun will become a white dwarf in about 5×10^9 years. We understand the structure of these objects in great detail.
2. It is possible that some stars in the range 4 to $8 M_\odot$ ignite carbon under very degenerate conditions and are *completely* disrupted in some sort of supernova. (This is a very delicate problem for numerical simulations.)
3. For stars with masses larger than about $11 M_\odot$ nuclear fusion proceeds all the way up to iron. At this point the central core runs into an instability and collapses catastrophically. This comes about as follows: In the last stages of nuclear energy generation an "iron core" of burned out elements is formed, which has the structure of a white dwarf. Around this core of iron peak elements an onion structure is built up as a result of shell burning in various shells. The central white dwarf-like region finally becomes unstable due to electron capture and/or photo-disintegration of

the iron-peak elements into α-particles. At this point the core starts to collapse in practically free fall. There are now two possibilities.

a) For some mass range, **neutron star** residues with increasing masses of $1.4 - 2.0 M_\odot$, say, will be left behind a prompt or delayed supernova explosion.

b) For sufficiently massive stars, the core will, however, most likely accrete too much mass to be stable and will then collapse very quickly to a **black hole**. In this picture we do not expect a collapse directly to a black hole. A proto-neutron star is formed first, which accretes sufficient mass through a stalled shock until it becomes unstable and undergoes a general relativistic collapse. It is difficult to say for which mass range of stars this is going to happen, but we do expect the formation of black holes for some very massive stars. There may be a million of such stellar mass black holes in our Galaxy.

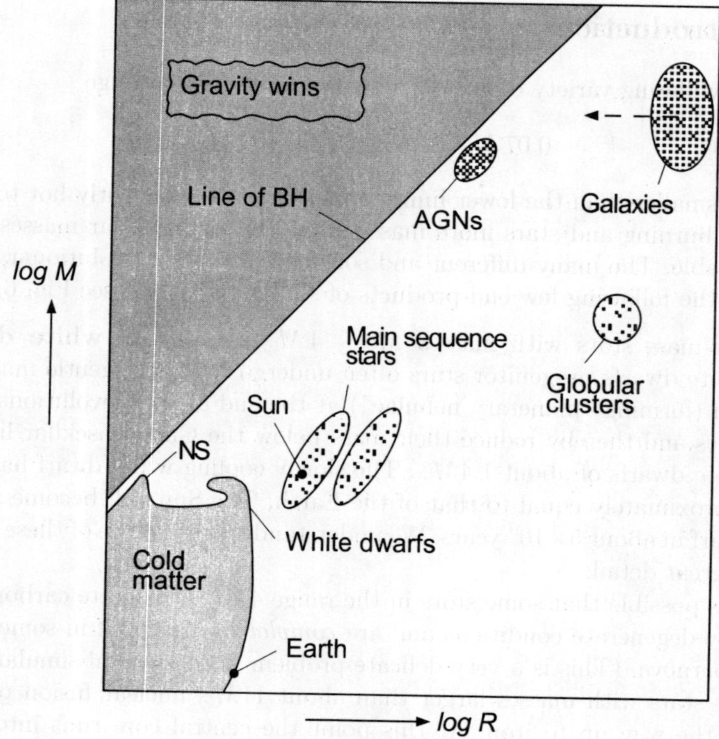

Fig. 6.1. Mass-radius diagram for some astrophysical objects.

It is natural that we understand the three final compact stationary end-products much better than the highly dynamical processes which lead to them. Supernova events belong to the most complex phenomena physicists have ever attempted to study. The physics is clearly very complicated in detail and our present models are comparably still rather primitive. It will take generations of astrophysicists to come, in order to develop a satisfactory detailed understanding. However, some important aspects of the physics of supernovas are qualitatively and even semi-quantitatively understood.

Compact objects – white dwarfs, neutron stars and black holes – are often not just dead stars. In some cases they are members of close binary semi-detached systems, collecting matter from a normal companion star. This leads to an enormously rich variety of interesting systems, showing a broad range of fascinating phenomena.

6.2 White Dwarfs

> *"The message of the companion of Sirius when it was decoded ran: 'I am composed of material 3000 times denser than anything you have ever come across; a ton of my material would be a little nugget that you could put in a matchbox'. What reply can one make to such a message? The reply which most of us made 1914 was – 'Shut up. Don't talk nonsense."*
>
> —S.A. Eddington (1927)

It is very remarkable that the quantum statistics of identical particles found its first application in astrophysics. In the same year when E. Schrödinger discovered his wave equation R.H. Fowler realized that "the black-dwarf material is best likened to a single gigantic molecule in its lowest quantum state" and he developed the non-relativistic theory of white dwarfs. It was afterwards recognized independently by several people (I. Frenkel (1928), E. Stoner (1930), S. Chandrasekhar (1931) and L.D. Landau (1932)) that relativistic kinematics weakens the quantum mechanical kinematic energy (zero-point pressure) to the extent that there is a limiting mass for white dwarfs.

The existence of a limiting mass for white dwarfs can be qualitatively understood as an immediate consequence of special relativity and the Pauli principle. It is important to understand this, before going into any quantitative investigation.

Matter in white dwarfs is completely ionized. The Coulomb forces establish local neutrality to a very high degree. For this reason the Coulomb interactions play energetically almost no role. (Coulomb corrections can be estimated and are on a few percent level.) The spatial distribution of nuclei and hence their momentum distribution is much the same as those of the electrons. Therefore, the ground state energy of a white dwarf with N electrons and N_Z nuclei with charge Ze and mass m_Z can be estimated as follows:

$$E_0(N) \approx \min\left\{ N\sqrt{p^2+m^2} - \frac{1}{2}\left(\frac{N}{Z}\right)^2 Gm_Z^2 \frac{p}{N^{1/3}\hbar} \right\}, \tag{6.1}$$

where we have made use of the Pauli principle for the contribution of the gravitational energy. (The average momentum p of a particle satisfies $p \geq N^{1/3}\hbar/R$, where R is the radius of the star, because there can be at most one electron in a de Broglie cube $(\hbar/p)^3$.) The minimum in (6.1) exists only for

$$N < N_f := \left(\frac{Gm_N}{2\hbar c}\right)^{-3/2}\left(\frac{Z}{A}\right)^3, \tag{6.2}$$

where $m_Z = A\,m_N$ (m_N denotes the nucleon mass). For the ground state the momentum and the energy are, respectively,

$$p_0 \approx mc\left(\frac{N}{N_f}\right)^{3/2}\left[1-\left(\frac{N}{N_f}\right)^{4/3}\right]^{-2}, \tag{6.3a}$$

$$E_0(N) \approx N\,mc^2\left[1-\left(\frac{N}{N_f}\right)^{4/3}\right]^{1/2}. \tag{6.3b}$$

For $N > N_f$ the expression in the curly bracket of (6.1) is not bounded from below, since in the extreme relativistic limit it becomes

$$N\left[1-\left(\frac{N}{N_f}\right)^{4/3}\right]pc\,.$$

Therefore, the system collapses for $N > N_f$. The mass M_f corresponding to N_f gives the following estimate of the **Chandrasekhar mass**

$$M_{Ch} \approx \frac{N_f}{Z}\,m_Z \approx 2.8\,\frac{M_{Pl}^3}{m_N^2}\left(\frac{Z}{A}\right)^2, \tag{6.4}$$

where M_{Pl} is the Planck mass. – In the Chandrasekhar theory the prefactor in the last expression is replaced by the number 3.1 (see below).

6.2.1 The Free Relativistic Electron Gas

Consider N non-interacting electrons (more generally, spin-1/2 particles) in a cube with volume $V = L^3$. If we impose periodic boundary conditions, the momenta of the 1-particle wave functions are quantized:

$$\boldsymbol{p}/\hbar = \frac{2\pi}{L}\boldsymbol{n}\,, \quad \boldsymbol{n} \in \mathbb{Z}^3\,, \tag{6.5}$$

and their energies are

$$\varepsilon_{\boldsymbol{p}} = \sqrt{\boldsymbol{p}^2 c^2 + m^2 c^4}\,. \tag{6.6}$$

According to the Pauli principle each momentum state is occupied at most twice, corresponding to the two spin values $s_z = \pm\hbar/2$. In the ground state, i.e., at temperature $T = 0$, all states with $|\boldsymbol{p}| \leq p_F$ (Fermi momentum) are occupied. The number of states in the Fermi sphere characterized by $|\boldsymbol{p}| \leq p_F$ is obviously $2 \cdot (4\pi/3)(p_F/\hbar^3)/(2\pi/L)^3$. Since this is also equal to the number N of electrons, we obtain for the particle number density $n = N/V$:

$$n = \frac{(p_F/\hbar)^3}{3\pi^2}\,. \tag{6.7}$$

The total kinetic energy of the ground state is

$$E_{kin} = 2 \sum_{|\boldsymbol{p}| \leq p_F} \varepsilon_{\boldsymbol{p}} = 2 \left(\frac{L}{2\pi\hbar} \right)^3 \int_{|\boldsymbol{p}| \leq p_F} \varepsilon_{\boldsymbol{p}}\, d^3p\,. \tag{6.8}$$

The maximal energy of an electron in the ground state of the gas is the Fermi energy ε_F and corresponds to the Fermi momentum p_F.

6.2.2 Thomas-Fermi Approximation for White Dwarfs

It is not so well-known that the Chandrasekhar theory of white dwarfs is just Thomas-Fermi theory, whereby the white dwarf is considered as a big "atom" with about 10^{57} electrons. Below we show that the famous Chandrasekhar equation is indeed just the relativistic Thomas-Fermi equation.

For white dwarfs the Thomas-Fermi approximation is ideally justified. The density in these stars is so high (typically one ton per cm^3) that the electrons can be treated as an ideal gas in its ground state. Thermal excitations can be neglected, as we shall soon see. Moreover, Coulomb interactions between the electrons are largely screened.

The electrons of white dwarfs are moving in a strong gravitational field which is generated by the nuclei that generate a neutralizing background. Let $\mu_e = \langle A/Z \rangle$ be the average mass number per nuclear charge (≈ 2) of these nuclei, then the matter density ρ is

$$\rho = \mu_e m_N n_e\,, \tag{6.9}$$

where n_e is the electron number density

$$n_e(\boldsymbol{x}) = \frac{1}{3\pi^2\hbar^3}\, p_F^3(\boldsymbol{x})\,. \tag{6.10}$$

This statistical description is well justified since per cm^3 there are in the average very many electrons, whose states are local plane waves in an almost constant Newtonian potential $\phi(\boldsymbol{x})$. (A Newtonian description of the white dwarf structure suffices completely.) The Fermi momentum p_F depends on the local potential ϕ that is connected with the Fermi energy ε_F, the highest energy of the occupied states in the ground state, as follows

$$\sqrt{p_F^2(\boldsymbol{x})c^2 + m_e^2c^4} + \mu_e m_N \phi(\boldsymbol{x}) = \varepsilon_F \,. \tag{6.11}$$

(Note that the potential energy per electron is $\mu_e m_N \phi$.) In equilibrium ε_F must be a *constant*. Otherwise the total energy of all electrons could be lowered by suitable displacements of electrons. (In the language of statistical mechanics, the Fermi energy is the chemical potential at $T = 0$.) The self-consistent (mean) field ϕ satisfies the Poisson equation

$$\Delta\phi = 4\pi G\,\rho \tag{6.12}$$

with ρ given by (6.9) and (6.10). Note that

$$\rho = \mu_e \frac{(m_e c)^3}{3\pi^2\hbar^3}\, m_N \left(\frac{p_F}{m_e c}\right)^3 = (0.97\times 10^6 g/cm^3)\left(\frac{p_F}{m_e c}\right)^3 . \tag{6.13}$$

Thus for $\rho \sim 10^6 g/cm^3$ the electrons become moderately relativistic ($p_F \sim mc$). Since mc^2 corresponds to a high temperature of about $0.5 \times 10^{10} K$, thermal effects are indeed small.

We solve the algebraic equation (6.11) for ϕ and insert the result in (6.12), with ρ taken from (6.9) and (6.10). With the abbreviation $f(\boldsymbol{x}) = p_F(\boldsymbol{x})/mc$ we obtain the relativistic **Thomas-Fermi equation**

$$\Delta\left(\sqrt{1+f^2}\right) = -K^2 f^3 , \tag{6.14}$$

where

$$K^2 = \frac{4\pi G(\mu_e m_N)^2 m_e^2 c}{3\pi^2\hbar^3} . \tag{6.15}$$

Instead of f it is convenient to use the function φ, defined by

$$z_c\varphi := \sqrt{1+f^2} , \tag{6.16}$$

where $z_c = \sqrt{1+(p_F/m_e c)^2}$ in the center of the star. Furthermore, we use the dimensionless radial variable ξ,

$$r =: \alpha\,\xi , \qquad \alpha := \frac{1}{K z_c} . \tag{6.17}$$

In the spherically symmetric case we obtain the famous **Chandrasekhar equation**

$$\frac{1}{\xi^2}\frac{d}{d\xi}\left(\xi^2\frac{d\varphi}{d\xi}\right) = -\left(\varphi^2 - \frac{1}{z_c^2}\right)^{3/2} , \tag{6.18}$$

that depends parametrically on the central value z_c of the Fermi energy in units of $m_e c^2$.

In astrophysics books, the central equation (6.18) is usually derived by using the equation of state for a relativistic electron gas, together with the equation for hydrostatic equilibrium. Our derivation shows that (6.18) is just the relativistic Thomas-Fermi equation.

The boundary conditions of (6.18), which is singular at $\xi = 0$, are obviously

$$\varphi(\boldsymbol{x}=0)=1\,, \qquad \frac{d\varphi}{d\xi}(\boldsymbol{x}=0)=0\,. \tag{6.19}$$

The radius R of the star corresponds to $\xi = \xi_1$, where $\varphi = 1/z_c$ (i.e., $p_F = 0$):

$$\varphi(\xi_1)=\frac{1}{z_c}\,. \tag{6.20}$$

Thus

$$R=\alpha\,\xi_1=\lambda_1(\xi_1/z_c)\,, \tag{6.21}$$

where

$$\lambda_1=\frac{1}{K}=\frac{3}{2}\left(\frac{\pi}{3}\right)^{1/2}\frac{1}{\mu_e}\frac{M_{Pl}}{m_N}\lambda\!\!\!^{-}_e=7.8\times10^8\frac{1}{\mu_e}cm\,. \tag{6.22}$$

Here, M_{Pl} is the Planck mass and $\lambda\!\!\!^{-}_e=\hbar/m_e c$ the Compton wavelength of the electron. Up to the composition factor μ_e, the length scale λ_1 for white dwarfs is thus determined by fundamental physical constants. This length scale is comparable to the radius of the Earth

$$R_{\oplus}=6.4\times10^8\,cm\,. \tag{6.23}$$

For the total mass of the white dwarf we obtain in a first step

$$M=\int_0^R \rho(r)4\pi r^2\,dr=4\pi\alpha^3\int_0^{\xi_1}\rho\xi^2\,d\xi\,. \tag{6.24}$$

With

$$\frac{\rho}{\rho_c}=\frac{f^3}{f_c^3}=\frac{z_c^3}{(z_c^2-1)^{3/2}}\left(\varphi^2-\frac{1}{z_c^2}\right)^{3/2} \tag{6.25}$$

we get

$$M=4\pi\alpha^3\rho_c\frac{z_c^3}{(z_c^2-1)^{3/2}}\int_0^{\xi_1}\xi^2\left(\varphi^2(\xi)-\frac{1}{z_c^2}\right)^{3/2}d\xi\,.$$

(Note that the index c always denotes central values.) Using the Chandrasekhar equation (6.18) this can be written as

$$M=4\pi\alpha^3\rho_c\frac{z_c^3}{(z_c^2-1)^{3/2}}\xi_1^2|\varphi'(\xi_1)|\,.$$

If we insert the definition (6.17) for α and use (6.13) in the center of the white dwarf, we finally arrive at

$$M = \frac{\sqrt{3\pi}}{2}\, \xi_1^2 |\varphi'(\xi_1)| \, \frac{1}{\mu_e^2} \frac{M_{Pl}^3}{m_N^2} . \tag{6.26}$$

The mass scale is determined by

$$\frac{M_{Pl}^3}{m_N^2} = 1.86 M_\odot . \tag{6.27}$$

The numerical solution of the Chandrasekhar equation shows that M increases monotonically with z_c. In the limit $z_c \longrightarrow \infty$ Eq. (6.18) becomes the Lane-Emden equation of index 3:

$$\frac{1}{\xi^2} \frac{d}{d\xi} \left(\xi^2 \frac{d\varphi}{d\xi} \right) + \varphi^3 = 0 , \tag{6.28}$$

and ξ_1 satisfies $\varphi(\xi_1) = 0$. Numerically one finds for this limiting case

$$\xi_1 = 6.8968 , \qquad \xi_1^2 |\varphi'(\xi_1)| = 2.01824 , \tag{6.29}$$

and hence the limiting **Chandrasekhar mass** is

$$M_{Ch} = \frac{5.84}{\mu_e^2} M_\odot . \tag{6.30}$$

(Note that in reality $\mu_e \approx 2$). This is a fundamental result.

Equations (6.26) and (6.21) give the **mass-radius relation** for white dwarfs, which is shown in Fig. 6.2. The **compactness parameter** GM/Rc^2

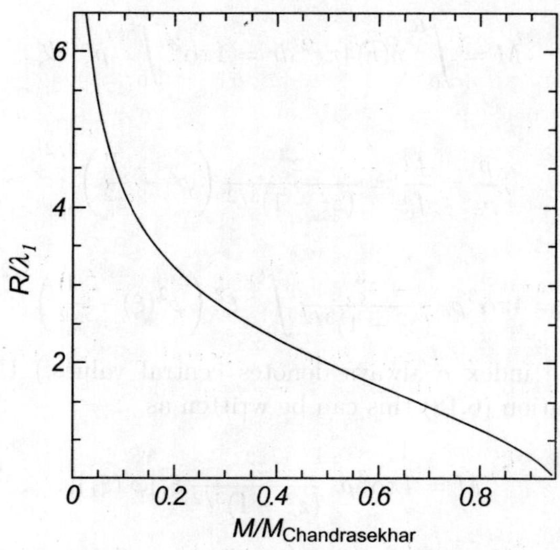

Fig. 6.2. Mass-radius relation for white dwarfs in the Chandrasekhar theory.

(half of the Schwarzschild radius divided by R), measuring the importance of GR, is for typical white dwarfs about 10^{-4}. Therefore, the Newtonian approximation is well justified. (General relativistic effects, however, play some role for stability.) – Thanks to several improvements (e.g., better estimates of distances to several white dwarfs with the Hipparcos satellite) it has become possible to test the theoretical mass-radius relation. The observational results presented in [158] show that there is good agreement between theory and observations.

The classical theory developed in this section is a good starting point for more accurate white dwarf models. Corrections due to Coulomb interactions can be computed with standard methods for dense electron gases. There are other interesting aspects, such as:

a) Effect of neutronization (inverse β-decay) on a white dwarf of homogenous composition.
b) Crystallization of the ion lattice and heat capacity of the Coulomb lattice.
c) Cooling of white dwarfs, including the effect of neutrino radiation (plasmon decay into neutrino pairs) at early times.
d) Nuclear reactions in cold dense matter (so-called pycnonuclear reactions).

For a very readable treatment of these astrophysically relevant issues we refer to [21].

6.2.3 Historical Remarks

We conclude with some historical remarks related to white dwarfs (for a detailed account we highly recommend the article [67] by W. Israel). By 1925 it was known through the work of W. Adams on the binary system of Sirius that Sirius B has the enormous average density of about $10^6 g/cm^3$. The existence of such compact stars constituted one of the major puzzles of astrophysics until the quantum statistical theory of the electron gas was worked out.

On August 26, 1926, Dirac's paper containing the Fermi-Dirac distribution was communicated by R.H. Fowler to the Royal Society. On November 3 of that year, Fowler presented his own work to the Royal Society in which he systematically worked out the quantum statistics of identical particles and in the process developed the well-known Darwin-Fowler method. Shortly thereafter, on December 10, he communicated the Royal Astronomical Society a new paper with the title "Dense Matter". In this work he showed that the electron gas in Sirius B is almost completely degenerated in the sense of the new Fermi-Dirac statistics. This paper by Fowler concludes with the following words:

> *"The black dwarf material is best likened to a single gigantic molecule in its lowest quantum state. On the Fermi-Dirac statistics, its high density can be achieved in one and only one way, in virtue*

of a correspondingly great energy content. But this energy can no more be expended in radiation than the energy of a normal atom or molecule. The only difference between black dwarf matter and a normal molecule is that the molecule can exist in a free state while the black dwarf matter can only so exist under very high external pressure."

Since Fowler treated the electrons non-relativistically, he found an equilibrium configuration for every mass (see Exercise 6.1). We have, however, seen that the momenta of the electrons in a white dwarf are comparable to $m_e c$ and the electron gas must thus be treated relativistically leading to a limiting mass. This was noticed by several authors (mentioned earlier), in particular by S. Chandrasekhar, who discovered the limiting mass for a white dwarf (see [183]). In 1934, Chandrasekhar derived the exact relation between mass and radius for completely degenerated configurations (see [184]; this can also be found in his classic text book [24]). He concluded his paper with the following statement:

"The life-history of a star of small mass must be essentially different from the life-history of a star of large mass. For a star of small mass, the natural white-dwarf stage is an initial step towards complete extinction. A star of large mass cannot pass into the white-dwarf stage and one is left speculating on other possibilities."

This conclusion was not accepted by the leading astrophysicists of the time. A comment by S.A. Eddington contains the following correct conclusion:

"Chandrasekhar shows that a star of mass greater than a certain limit remains a perfect gas and can never cool down. The star has to go on radiating and radiating and contracting and contracting, until, I suppose, it gets down to a few kilometers radius when gravity becomes strong enough to hold the radiation and the star can at last find peace."

If Eddington had stopped at this point, he would have been the first to predict the existence of black holes. However, he did not take this conclusion seriously, and continuous:

"I felt driven to the conclusion that this was almost a reductio ad absurdum of the relativistic degeneracy formula. Various accidents may intervene to save the star, but I want more protection than that. I think that there should be a law of nature to prevent the star from behaving in this absurd way."

And here is L.D. Landau (see [159]) in similar vein:

"For $M > 1.5M_\odot$ there exists in the whole quantum theory no cause preventing the system from collapsing to a point. As in reality such masses exist quietly as stars and do not show any such ridiculous

tendencies we must conclude that all stars heavier than $1.5M_\odot$ *certainly possess regions in which the laws of quantum mechanics (and therefore of quantum statistics) are violated."*

Under the impression of the quantum revolution a few years ago, some people expected already a new one in the domain of relativistic quantum theory.

* * *

Exercises.

Exercise 6.1. Consider the *non-relativistic* Thomas-Fermi theory of white dwarfs, and derive the mass-radius relation

$$MR^3 = \text{const}\,, \tag{6.31}$$

that was already obtained by R.H. Fowler in 1926 (see [185]). Compute the constant on the right of this important relation.

Exercise 6.2. Solve the Lane-Emden equation (6.28) on a computer and determine the Chandrasekhar mass. Extend this to the Chandrasekhar equation, and reproduce the graph in Fig. 6.2.

6.3 Formation of Neutron Stars

We already remarked that the central regions of sufficiently massive stars eventually become unstable. For rather massive stars this instability is induced by photodisintegration of the iron-peak elements to α-particles. The nuclear disintegration of iron to helium costs a lot of internal energy and reduces the relevant adiabatic index below a critical value. For stars in the lower of the mass range, electron capture is a further destabilizing factor (see, e.g., [21]). The result of this is that the dynamic collapse of the iron-core is unavoidable.

We do not discuss here the physics of the core collapse[1]. This is finally halted when nuclear densities are reached. At these densities the nuclei overlap and undergo a transition to a degenerate nucleon Fermi liquid with strong repulsive forces at short distances. The equation of state stiffens suddenly and within tenths of milliseconds the inner core rebounds as a unit. This generates a shock with an initial energy of roughly $5 \times 10^{51} erg$. Whether and how this shock wave is capable to produce mass ejection in the overlying mantle and envelope, is a burning question (see, e.g., [174]).

The initially formed central proto-neutron star does not resemble the final cold star. It takes tens of seconds of quasi-hydrostatic thermal, structural and

[1] The relevant physics is treated, for instance, in [160].

compositional readjustments until the residue becomes a strongly degenerate neutron star. In particular, electron capture deleptonizes the core region. This and other neutrino processes cool the proto-neutron star very rapidly. In the resulting 'cold' neutron star a β-equilibrium is established. We shall see that in this equilibrium only a few percent of protons and electrons are present. This scenario for the origin of neutron stars is strongly supported by the fact that all well determined masses in six double neutron star binaries have masses in the narrow range $1.36 \pm 0.08 M_{\odot}$. If approximately 10% binding energy is added to this one comes close to the Chandrasekhar mass for iron-cores of the progenitor stars.

The predicted cooling through neutrino diffusion over a time scale of several seconds was dramatically confirmed by the observed neutrino pulse from the famous supernova 1987A (for a review see [161]).

Neutron stars are so compact that GR has to be used for a quantitative treatment. In oder to see this, let us first estimate the structure of neutron stars in Newtonian theory using the Thomas-Fermi approximations for the neutrons[2]. Masses and radii are then obtained from our results for white dwarfs in Sect. 6.2.2 by the simple substitutions $m_e \longrightarrow m_n$ (mass of neutron) and $\mu_e \longrightarrow 1$. In particular, the characteristic size λ_1 in (6.22) now becomes approximately m_n/m_e times smaller:

$$\lambda_1 = \frac{3}{2}\left(\frac{\pi}{3}\right)^{1/2} \frac{M_{Pl}}{m_n} \lambda\!\!\!^{-} = 4.3. \times 10^5 cm\,. \qquad (6.32)$$

The expression (6.26) for the mass remains, up to the factor $1/\mu_e^2$. Especially the limiting mass becomes

$$M_{lim} \simeq 3\,\frac{M_{Pl}^3}{m_n^2}\,. \qquad (6.33)$$

So, the compactness parameter GM/c^2R is not much smaller than for black holes (realistically it is ~ 0.2).

Remark. It would be more appropriate to adopt a semi-Newtonian treatment by including in the energy density ρ in the Poisson equation (6.12) also the kinetic energy for the neutrons:

$$\rho(\boldsymbol{x}) = \frac{2}{(2\pi\hbar)^3} \int_0^{p_F(\boldsymbol{x})} \sqrt{m_n^2 c^4 + p^2 c^2}\, 4\pi p^2\, dp\,. \qquad (6.34)$$

In this treatment one finds that the mass of a neutron star as a function of the central density reaches a maximum at a finite value of ρ_c, and then decreases (see Fig. 6.4). The decreasing branch turns out to be unstable (see Sect. 6.5).

[2] Such estimates were made by L.D. Landau immediately after the neutron was discovered (see [159]).

The first model calculations for neutron stars in GR were performed by J.R. Oppenheimer and G. Volkoff in 1939 (see [162]). In this pioneering work the equation of state of a completely degenerate ideal neutron gas was used. In those early days the effects of strong interactions could not be estimated. Afterwards, theoretical interest in neutron stars dwindled, since no relevant observations existed. For two decades, F. Zwicky was one of the few who took seriously the probable role of neutron stars as final states of massive stars.

Interest in the subject we reawakened at the end of the 1950s and beginning of the 1960s. When pulsars were discovered in 1967, especially when a pulsar with a short period of 0.033 was found in the Crab Nebula, it became clear that neutron stars can be formed in type II supernova events through the collapse of the stellar core to nuclear densities.

* * *

Exercise 6.3. Derive in the non-relativistic approximation for the neutrons the following expressions for the masses and radii of neutron stars, using Newtonian gravity and the Thomas-Fermi approximation for the neutrons:

$$M = 1.47\,\frac{M_{Pl}^3}{m_n^2}\left(\frac{\rho_c}{\rho_0}\right)^{1/2}, \tag{6.35a}$$

$$R = 3.97\,\frac{M_{Pl}}{m_n}\,\frac{\hbar}{m_n c}\left(\frac{\rho_0}{\rho_c}\right)^{1/6}, \tag{6.35b}$$

where ρ_c is the central density, and

$$\rho_0 := \frac{m_n c^2}{(\hbar/m_n c)^3} = 6.11\times 10^{15} g/cm^3\,.$$

6.4 General Relativistic Stellar Structure Equations

We proceed to derive the general relativistic stellar structure equations for non-rotating, static, spherically symmetric (compact) stars. For such stars, the metric has the form (see Appendix 3.10)

$$g = -\mathrm{e}^{2a(r)}dt^2 + \left(\mathrm{e}^{2b(r)}dr^2 + r^2(d\vartheta^2 + \sin^2\vartheta\, d\varphi^2)\right)\,. \tag{6.36}$$

With respect to the orthonormal tetrad

$$\theta^0 = \mathrm{e}^a\, dt\,, \quad \theta^1 = \mathrm{e}^b\, dr\,, \quad \theta^2 = r\, d\vartheta\,, \quad \theta^3 = \sin\vartheta\, d\varphi \tag{6.37}$$

we parameterize the energy-momentum tensor as

$$T^{\mu\nu} = \mathrm{diag}\,(\rho, p, p, p)\,, \tag{6.38}$$

where ρ is the total mass-energy density and p is the pressure. Anisotropic stresses and heat conduction are thus neglected. For the overall structure of compact stars this is an excellent approximation.

The Einstein tensor corresponding to (6.36) was computed in Sect. 3.1, with the result

$$
\begin{aligned}
G^0_{\ 0} &= -\frac{1}{r^2} + \mathrm{e}^{-2b}\left(\frac{1}{r^2} - \frac{2b'}{r}\right), \\
G^1_{\ 1} &= -\frac{1}{r^2} + \mathrm{e}^{-2b}\left(\frac{1}{r^2} + \frac{2a'}{r}\right), \\
G^2_{\ 2} &= G^3_{\ 3} = \mathrm{e}^{-2b}\left(a'^2 - a'b' + a'' + \frac{a' - b'}{r}\right), \\
G_{\mu\nu} &= 0 \text{ for all other components}.
\end{aligned}
\tag{6.39}
$$

The field equations then give ($c = 1$)

$$\frac{1}{r^2} - \mathrm{e}^{-2b}\left(\frac{1}{r^2} - \frac{2b'}{r}\right) = 8\pi G\,\rho\,, \tag{6.40a}$$

$$\frac{1}{r^2} - \mathrm{e}^{-2b}\left(\frac{1}{r^2} + \frac{2a'}{r}\right) = -8\pi G\,p\,. \tag{6.40b}$$

If we use the notation $u/r := \mathrm{e}^{-2b}$, Eq. (6.40a) reads

$$u' = -8\pi G\,\rho\,r^2 + 1\,.$$

Integration of this equation gives $u = r - 2GM(r)$, where

$$M(r) = 4\pi \int_0^r \rho(r')r'^2\,dr'\,. \tag{6.41}$$

Hence

$$\mathrm{e}^{-2b} = 1 - \frac{2GM(r)}{r}\,. \tag{6.42}$$

If we subtract (6.40b) from (6.40a), we obtain

$$\mathrm{e}^{-2b}(a' + b') = 4\pi G(\rho + p)\,r \tag{6.43}$$

and thus

$$a = -b + 4\pi G \int_\infty^r \mathrm{e}^{-2b(r')} r'(\rho + p)\,dr'\,. \tag{6.44}$$

If ρ and p are known the gravitational field is thus determined.

An additional useful relation follows from the "conservation law" $D{*}T^\alpha = 0$ (see (2.167)). Equation (6.38) gives

$$*T^\alpha = q^\alpha\,\eta^\alpha\,, \tag{6.45}$$

with $q^0 = -\rho$ and $q^i = p$ (in the last equation we do not sum). Now

$$
\begin{aligned}
D * T^\alpha &= d(q^\alpha \eta^\alpha) + \sum_\beta \omega^\alpha_{\ \beta} \wedge (q^\beta \eta^\beta) \\
&= dq^\alpha \wedge \eta^\alpha + \sum_\beta \omega^\alpha_{\ \beta} \wedge \eta^\beta q^\beta + q^\alpha \, d\eta^\alpha \\
&= dq^\alpha \wedge \eta^\alpha + \sum_\beta \omega^\alpha_{\ \beta} \wedge \eta^\beta (q^\beta - q^\alpha) = 0 \, .
\end{aligned}
$$

For $\alpha = 1$, this gives, after making use of the connection forms (3.8),

$$
dp \wedge \eta^1 = \omega^1_{\ 0} \wedge \eta^0 (\rho + p)
$$

or

$$
\frac{dp}{dr} \, \mathrm{e}^{-b} \underbrace{\theta^1 \wedge \eta^1}_{\eta} = a' \mathrm{e}^{-b} \underbrace{\theta^0 \wedge \eta^0}_{-\eta} (\rho + p)
$$

so that

$$
a' = -\frac{p'}{\rho + p} \, . \tag{6.46}
$$

On the other hand, we obtain from (6.41), (6.42) and (6.43)

$$
a' = \frac{G}{1 - 2GM(r)/r} \left(\frac{M(r)}{r^2} + 4\pi \, r \, p \right). \tag{6.47}
$$

If we compare this with (6.46), we obtain the **Tolman-Oppenheimer-Volkoff equation** (TOV equation)

$$
-p' = \frac{G(\rho + p)\big(M(r) + 4\pi \, r^3 \, p\big)}{r^2 \big(1 - 2GM(r)/r\big)} \, . \tag{6.48}
$$

At the stellar radius R, the pressure vanishes. Outside the star, the metric is given by the Schwarzschild solution. The gravitational mass is

$$
M = M(R) = \int_0^R \rho \, 4\pi r^2 \, dr \, . \tag{6.49}
$$

Equation (6.48) generalizes the hydrostatic equation

$$
-p' = \frac{G \, \rho \, M(r)}{r^2} \tag{6.50}
$$

of the Newtonian theory. The pressure gradient towards the center increases for three reasons in GR:

a) Since pressure also acts as a source of a gravitational field, there is a term proportional to p in addition to $M(r)$.

b) Since gravity also acts on p, the density ρ is replaced by $(\rho + p)$.
c) The gravitational force increases faster than $1/r^2$; therefore this quantity is replaced by $r^{-2}\left[1 - 2GM(r)/r\right]^{-1}$ in GR.

These modifications will lead us to the conclusion (see Sect. 6.9) that an arbitrarily massive neutron star cannot exist, even if the equation of state becomes extremely stiff at high densities.

In order to construct a stellar model, one needs an **equation of state** $p = p(\rho)$. If this is known the density profile of the star is determined by the central density ρ_c.

We summarize the relevant equations:

$$-p' = \frac{G(\rho + p)\big(M(r) + 4\pi\, r^3\, p\big)}{r^2\big(1 - 2GM(r)/r\big)}\,, \tag{6.51a}$$

$$M' = 4\pi r^2\, \rho\,, \tag{6.51b}$$

$$a' = \frac{G}{1 - 2GM(r)/r}\left(\frac{M(r)}{r^2} + 4\pi\, r\, p\right). \tag{6.51c}$$

The initial condition

$$M(r = 0) = 0 \tag{6.52}$$

is obtained from (6.41). The other two boundary conditions are

$$p(r = 0) = p(\rho_c) \qquad \text{and} \qquad \mathrm{e}^{2a(R)} = 1 - 2GM(R)/R\,. \tag{6.53}$$

Interpretation of M

We compare the quantity M appearing in (6.49) with the baryonic mass $M_0 = N\, m_N$, where N is the total number of nucleons (baryons) in the star. If J denotes the baryon current density, then

$$N = \int_{\{t=\text{const}\}} *J \tag{6.54}$$

If $J = J_\mu \theta^\mu$, then $*J = J_\mu \eta^\mu$ and we obtain

$$N = \int_{\{t=\text{const}\}} J^0\, \eta_0 = \int_{\{t=\text{const}\}} J^0\, \theta^1 \wedge \theta^2 \wedge \theta^3$$

or, using (6.37),

$$N = \int_0^R J^0 4\pi r^2 \mathrm{e}^{b(r)}\, dr\,.$$

The component J^0 with respect to the basis (6.37) is the baryon number density

$$n := -u_\mu J^\mu = J^0\,, \tag{6.55}$$

since the four-velocity is $u^\mu = (1, 0, 0, 0)$. Hence

$$N = \int_0^R 4\pi r^2 \mathrm{e}^{b(r)}\, n(r)\, dr = \int_0^R \frac{4\pi r^2}{\sqrt{1 - 2GM(r)/r}}\, n(r)\, dr\,. \tag{6.56}$$

The proper internal material energy density is defined by

$$\varepsilon(r) = \rho(r) - m_N n(r)\,, \tag{6.57}$$

and the total internal energy is correspondingly

$$E = M - M_0 = M - m_N N\,. \tag{6.58}$$

We use (6.56) to decompose E as

$$E = T + V\,, \tag{6.59}$$

where

$$T = \int_0^R \frac{4\pi r^2}{\sqrt{1 - 2GM(r)/r}}\, \varepsilon(r)\, dr\,, \tag{6.60a}$$

$$V = \int_0^R 4\pi r^2 \left(1 - \frac{1}{\sqrt{1 - 2GM(r)/r}}\right) \rho(r)\, dr\,. \tag{6.60b}$$

In order to find the connection with Newtonian theory, we expand the square roots in (6.60a) and (6.60b), assuming that $GM(r)/r \ll 1$. This gives

$$T = \int_0^R 4\pi r^2 \left(1 + \frac{GM(r)}{r} + \dots\right) \varepsilon(r)\, dr\,, \tag{6.61a}$$

$$V = -\int_0^R 4\pi r^2 \left(\frac{GM(r)}{r} + \frac{3G^2M^2(r)}{2r^2} + \dots\right) \rho(r)\, dr\,. \tag{6.61b}$$

The leading terms in T and V are the Newtonian values for the internal and gravitational energy of the star.

General Relativistic Virial Theorem

Beside the expression (6.49) for the total mass, we now derive another formula based on the Tolman mass formula (4.228)

$$M = 2\int_\Sigma \left(T_{\mu\nu} - \frac{1}{2}\, g_{\mu\nu} T\right) n^\mu \xi^\nu\, dV_\Sigma\,. \tag{6.62}$$

For a spherically symmetric static star we have $\xi^\mu = \mathrm{e}^a\, n^\mu$, thus

$$\left(T_{\mu\nu} - \frac{1}{2}\, g_{\mu\nu} T\right) \xi^\mu n^\nu = \mathrm{e}^a \left(T_{00} - \frac{1}{2}\, T\right)$$
$$= \mathrm{e}^a \left(\rho - \frac{1}{2}(\rho - 3\,p)\right)$$
$$= \frac{1}{2} \mathrm{e}^a (\rho + 3\,p)\,.$$

This gives

$$M = \int (\rho + 3\,p) \mathrm{e}^a\, dV\,, \tag{6.63}$$

with

$$dV = \theta^1 \wedge \theta^2 \wedge \theta^3 = \mathrm{e}^b\, r^2\, d\vartheta \wedge \sin\vartheta\, d\varphi\,.$$

Equating (6.49) and (6.63) leads to the general relativistic **virial theorem**

$$\int (\rho + 3\,p) \mathrm{e}^a\, dV = \int \rho \left(1 - \frac{2GM(r)}{r}\right)^{1/2} dV\,. \tag{6.64}$$

* * *

Exercises.

Exercise 6.4. Solve the stellar structure equations for a star having uniform density $\rho = \text{const}$ and show that its mass is limited by the inequality

$$M < \frac{4}{9} (3\pi\rho)^{-1/2}\,. \tag{6.65}$$

Exercise 6.5. Show that in the non-relativistic limit the general relativistic virial theorem (6.64) reduces to the well-know virial theorem of Newtonian theory:

$$3 \int p\, dV = - \int \rho \frac{GM(r)}{r}\, dV\,.$$

Show also that (6.64) can be derived from the TOV equations.

6.5 Linear Stability

An equilibrium solution is physically relevant only if it is stable against small perturbations. The existence of instabilities can generically be established with a linear analysis, thanks to general theorems. One thereby considers time-dependent perturbations and expands all basic equations about the equilibrium solution, keeping only linear deviations. For conservative systems the time dependence of the perturbations can be separated. The frequencies in the time-dependent factor $\mathrm{e}^{-i\omega t}$ are determined by eigenvalue equations and must be symmetrically distributed under reflection in the real and imaginary axis.

In the hyperbolic case (no real frequencies) the Grobman-Hartman theorem implies that the system is also non-linearly unstable. Linear (spectral) stability is only possible when all frequencies are real (oscillatory time-dependence for each mode). This need, however, not imply stability, as can be demonstrated by simple finite-dimensional Hamiltonian systems (for instance of two non-linearly coupled harmonic oscillators).

To summarize, spectral instability (some frequencies have negative imaginary parts) generically implies instability, but a linearly stable equilibrium state may well be non-linearly unstable. It is well-known from Hamiltonian mechanics that a stability proof can be very difficult, even for low dimensional systems, as for instance for the two equilateral solutions of the restricted three body problem. In what follows we shall only investigate *linear* stability.

* * *

For spherically symmetric configurations, one considers as a first step only *radial* perturbations, which are also adiabatic. Since hydrodynamic time scales are usually much shorter than the characteristic times for energy transport, this is reasonable. In this case, one obtains an eigenvalue problem[3] of the Sturm-Liouville type for ω^2, since the equations for adiabatic perturbations are time reversal invariant (there is no dissipation). The equilibrium is linearly stable provided $\omega_0^2 > 0$ for the lowest mode frequency ω_0.

One can often find *sufficient* criteria for instability by making use of the Rayleigh-Ritz variational principle and simple trial functions.

Sometimes it is possible to decide, simply by examining the equilibrium solution, whether or not it is stable against radial, adiabatic pulsations. Here we only prove the proposition below and refer the interested reader to [174].

We consider cold matter. For a given equation of state, there is then a one-parameter family of equilibrium solutions. We choose the central density ρ_c as the most suitable parameter.

Proposition 6.1. *At each critical point of the function $M(\rho_c)$, precisely one radial adiabatic normal mode changes its stability properties; elsewhere, the stability properties do not change.*

Proof. Suppose that for a given equilibrium configuration a radial mode changes its stability property, i.e., the frequency ω of this mode passes through zero. This implies that there exist infinitesimally nearby equilibrium configurations into which the given one can be transformed, without changing the gravitational mass or energy. Hence if ω passes trough zero we have $M'(\rho_c) = N'(\rho_c) = 0$. According to the next Proposition, $N'(\rho_c) = 0$ already follows from $M'(\rho_c) = 0$. □

[3] A detailed derivation of the eigenvalue equation for radial adiabatic pulsations of relativistic stars can be found in [13].

Proposition 6.2. *For a radial adiabatic variation of a cold equilibrium solution, we have*

$$\delta M = \frac{\rho + p}{n} \, e^{a} \, \delta N . \tag{6.66}$$

In particular, if $\delta M = 0$ *then also* $\delta N = 0$.

Proof. With the notation $m(r) = GM(r)$, it follows from (6.49) and (6.55) that

$$\delta M = \int_0^\infty 4\pi r^2 \, \delta\rho(r) \, dr , \tag{6.67a}$$

$$\delta N = \int_0^\infty 4\pi r^2 \left(1 - \frac{2m(r)}{r}\right)^{-1/2} \delta n(r) \, dr + \int_0^\infty 4\pi r^2 \left(1 - \frac{2m(r)}{r}\right)^{-3/2} n(r) \delta m(r) \, dr . \tag{6.67b}$$

For an adiabatic variation we have

$$\delta(\rho/n) + p\,\delta(1/n) = T\,\delta s = 0 \tag{6.68}$$

or

$$\delta n = \frac{n}{p + \rho} \delta\rho . \tag{6.69}$$

Obviously (in this proof we now use units with $G = 1$)

$$\delta m(r) = \int_0^r 4\pi r'^2 \delta\rho(r') \, dr' . \tag{6.70}$$

We now insert these last two relations into (6.67b) and interchange in one term the order of the integrations over r and r', with the result

$$\delta N = \int_0^\infty 4\pi r^2 \left[\left(1 - \frac{2m(r)}{r}\right)^{-1/2} \frac{n(r)}{p(r) + \rho(r)} + \int_r^\infty 4\pi r' n(r') \left(1 - \frac{2m(r')}{r'}\right)^{-3/2} dr' \right] \delta\rho(r) \, dr . \tag{6.71}$$

The expression in the square bracket is independent of r. In order to see this, we differentiate with respect to r, obtaining

$$\frac{d}{dr}\Big[\ldots\Big] = \left(\frac{n'}{p+\rho} - \frac{n(p'+\rho')}{(p+\rho)^2}\right)\left(1 - \frac{2m(r)}{r}\right)^{-1/2} + \frac{n}{p+\rho}\left(4\pi r\,\rho - \frac{m(r)}{r^2}\right)\left(1 - \frac{2m(r)}{r}\right)^{-3/2} - 4\pi r\, n \left(1 - \frac{2m(r)}{r}\right)^{-3/2} .$$

From (6.69) we have for cold matter

$$n' = \frac{n}{p+\rho}\rho' . \tag{6.72}$$

Using this, we obtain

$$\frac{d}{dr}\Big[\dots\Big] = -\frac{n}{p+\rho}\left(1-\frac{2m(r)}{r}\right)^{-1/2} \times \left[\frac{p'}{p+\rho} + \left(1-\frac{2m(r)}{r}\right)^{-1}\frac{1}{r^2}(4\pi r^3 p + m)\right].$$

According to the TOV equation (6.48), the right-hand side vanishes. The constant expression in the square bracket in (6.71) is most conveniently evaluated at the stellar surface $r = R$ where $p = 0$:

$$\Big[\dots\Big] = \frac{n(R)}{\rho(R)}\left(1-\frac{2M}{R}\right)^{-1/2} = \frac{n}{\rho+p}\,\mathrm{e}^{-a}\Big|_R .$$

To prove (6.66) it remains to show that $(p+\rho)\mathrm{e}^a/n$ is independent of r, or that $\ln(p+\rho) - \ln(n) + a = \text{const}$, which is equivalent to

$$\frac{\rho'+p'}{\rho+p} - \frac{n'}{n} + a' = 0 .$$

From (6.46) and (6.72) we know, however, that this is indeed correct. □

6.6 The Interior of Neutron Stars

We do not have an accurate quantitative understanding of neutron stars, because the physics involved is very complex and uncertain. Before discussing some special aspects – especially the equation of state – we start with a brief qualitative overview of the interior of neutron stars.

6.6.1 Qualitative Overview

Figure 6.3 shows a cross section through a typical neutron star. The **outer crust** is found underneath an atmosphere of just about $1cm$ thickness. (Estimate the pressure scale height for an ideal gas.) It consists of a lattice of completely ionized nuclei and a highly degenerate relativistic electron gas. In this layer the density ρ increases to about $3\times10^{11}\,g/cm^3$; its thickness is typically a few hundred meters. The conductivity in the crust is very high which implies a very long lifetime (about 10^7 years) of the magnetic field. It should be mentioned in this connection that field decay is a very controversial

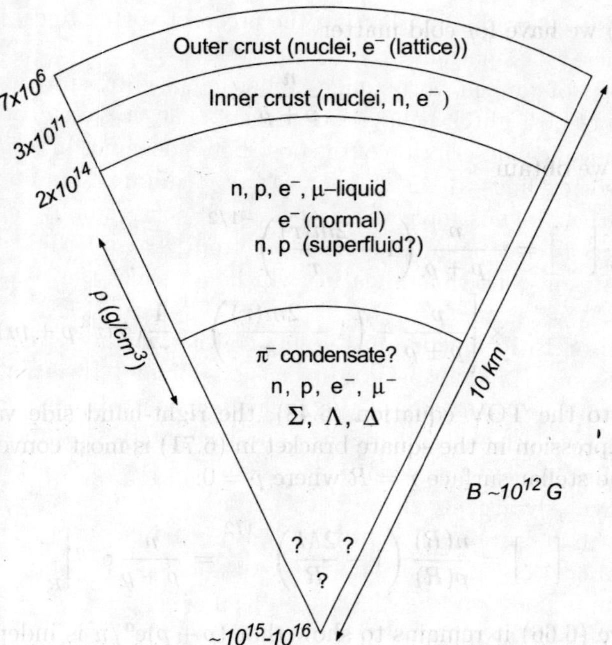

Fig. 6.3. Interior structure of neutron stars.

subject. At least in some cases there seems to be practically no field decay of neutron stars. There are, for instance, binary pulsars with white dwarf companion which are cool and hence very old (several 10^9 years). The **inner crust** starts at $\rho \sim 3 \times 10^{11} g/cm^3$ and is somewhat less than $1km$ thick, at which point the density has reached the value $\rho \sim 2 \times 10^{14} g/cm^3$. In addition to increasingly neutron rich nuclei (determined by β-equilibrium) and degenerate relativistic electrons, it contains also a degenerate neutron gas, which may be superfluid. The reason is that the interaction between neutrons at distances larger than $1fm$ is attractive. It is difficult to calculate the energy gaps reliably, because they depend exponentially on the strength of an effective interaction which is determined by the influence of the background liquid in which the neutrons move.

The crustal neutron superfluid is most probably crucial for an understanding of glitches and post glitch behavior of pulsars. The vortices of the neutron superfluid are *pinned* to the lattice of neutron-rich nuclei, because the free energy per unit length of a vortex line varies with location and brings them to preferred locations. (Note that the nuclei in the lattice have dimensions and spacing comparable to the correlation length ξ.) Glitches are probably the result of a sudden unpinning of vortex lines. The intrinsic pinning provides a natural reservoir of both angular momentum and rotational energy. Released catastrophically, and suddenly, it can give rise to a glitch in the pulsar rotation; released gradually, through the creep of vortex lines, it provides a source of energy dissipation.

The creep of pinned vortex lines following a glitch provides a natural explanation of post glitch behavior of pulsars. The **vortex creep theory** is a phenomenological description of the post-glitch behavior in terms of thermal motion of vortex lines in an inhomogeneous medium of pinning centers. As a result of the Magnus force the vortex lines move radially outward which leads to a transfer of angular momentum from the superfluid to the crust. This generates also frictional heating which can affect the thermal emission of pulsars.

The **inner fluid** or **core** in Fig. 6.3 consists mainly of neutrons, which are possibly superfluid in a certain density domain. In order to maintain β-equilibrium a few percents of the nucleons have to be protons, which are neutralized by electrons and muons. The proton fraction might be crucial for the cooling of neutron stars. In regions where the protons are superfluid, they are obviously also superconducting. With standard BCS theory one can estimate that the protons form a type II superconductor.

The **central core** is a region within a few kilometers of the center, which may contain in addition to hyperous, a pion condensate, and, for some of the more massive stars, a quark phase near the center. Indeed, the neutrons begin to touch when their density is $n_t \simeq ((4\pi/3)r_N^3)^{-1}$, where r_N is an effective nucleon radius. In terms of nuclear matter density $n_t/n_0 \simeq 1.4/r_N^3$, where $n_0 \simeq 0.17 fm^{-3}$ and r_N is measured in fm (Fermi). Thus, one might expect the transition to quark matter to occur somewhere in the range $(3-10)n_0$. – Clearly, the astrophysics of neutron stars is very rich and involves several major fields of physics.

6.6.2 Ideal Mixture of Neutrons, Protons and Electrons

This brief overview shows that the physics of the interior of a neutron star is extremely complicated. A major problem is to establish a reliable equation of state. In order to get a feeling for the orders of magnitude involved, let us first consider an *ideal* mixture of nucleons and electrons in β-equilibrium.

The energy density ρ, number density n and pressure p of an ideal Fermi gas at temperature $T = 0$ are given by $(c = 1)$:

$$\rho = \frac{8\pi}{(2\pi\hbar)^3} \int_0^{p_F} \sqrt{p^2 + m^2}\, p^2\, dp\,, \tag{6.73a}$$

$$n = \frac{8\pi}{(2\pi\hbar)^3} \int_0^{p_F} p^2\, dp\,, \tag{6.73b}$$

$$p = \frac{1}{3}\frac{8\pi}{(2\pi\hbar)^3} \int_0^{p_F} \frac{p^2}{\sqrt{p^2 + m^2}}\, p^2\, dp\,. \tag{6.73c}$$

The equilibrium reactions

$$e^- + p \longrightarrow n + \nu_e\,, \qquad n \longrightarrow p + e^- + \bar{\nu}_e$$

conserve the baryon number density $n_n + n_p$ and maintain charge neutrality $n_e = n_p$.

The condition for chemical equilibrium is

$$\mu_n = \mu_p + \mu_e\,, \tag{6.74}$$

where μ_i, for $i = n, p, e$, are the chemical potentials of the three Fermi gases. For an ideal gas at $T = 0$ we have

$$\mu = \varepsilon_F = \sqrt{p_F^2 + m^2} = \sqrt{\Lambda^2 n^{2/3} + m^2}\,, \tag{6.75}$$

where

$$\Lambda := (3\pi^2\hbar)^{1/3}\,. \tag{6.76}$$

If we now use this in the equilibrium condition (6.74), we can determine the ratio n_p/n_n as a function of n_n. A short computation results in

$$\frac{n_p}{n_n} = \frac{1}{8}\left(\frac{1 + \dfrac{2(m_n^2 - m_p^2 - m_e^2)}{\Lambda^2 n_n^{2/3}} + \dfrac{(m_n^2 - m_p^2)^2 - 2m_e^2(m_n^2 + m_p^2) + m_e^4}{\Lambda^4 n_n^{4/3}}}{1 + \dfrac{m_n^2}{\Lambda^2 n_n^{4/3}}}\right)^{3/2}. \tag{6.77}$$

Let $Q = m_n - m_p$. Since $Q, m_e \ll m_n$ we may simplify (6.77), obtaining

$$\frac{n_p}{n_n} \simeq \frac{1}{8}\left(\frac{1 + \dfrac{4Q}{m_n}\left(\dfrac{\rho_0}{m_n n_n}\right)^{2/3} + \dfrac{4(Q^2 - m_e^2)}{m_n^2}\left(\dfrac{\rho_0}{m_n n_n}\right)^{4/3}}{1 + \left(\dfrac{\rho_0}{m_n n_n}\right)^{2/3}}\right)^{3/2}, \tag{6.78}$$

where

$$\rho_0 := m_n^4/\Lambda^3 = 6.11 \times 10^{15}\,g/cm^3\,. \tag{6.79}$$

We now calculate the Fermi momentum of the electrons

$$p_{F,e}^2 = \Lambda^2 n_e^{2/3} = \Lambda^2 n_p^{2/3} = m_n^2\left(\frac{m_n n_n}{\rho_0}\right)^{2/3}\left(\frac{n_p}{n_n}\right)^{2/3}$$

$$= \frac{\dfrac{m_n^2}{4}\left(\dfrac{m_n n_n}{\rho_0}\right)^{4/3} + Q m_n\left(\dfrac{m_n n_n}{\rho_0}\right)^{2/3} + Q^2 - m_e^2}{1 + \left(\dfrac{m_n n_n}{\rho_0}\right)^{2/3}}\,. \tag{6.80}$$

We readily see from this that for $n_n > 0$, the Fermi momentum of the electrons is larger than the maximum momentum of an electron $p_{max} \simeq \sqrt{Q^2 - m_e^2} = 1.19 MeV$ in neutron β-decay. Hence the neutrons become *stable*.

The proton-neutron ratio (6.78) is large for small n_n and decreases with increasing n_n, reaching a minimum value for

$$\rho \simeq \rho_0 \left(\frac{4(Q^2 - m_e^2)}{m_n^2} \right)^{3/4} = 1.28 \times 10^{-4} \rho_0 \,, \tag{6.81}$$

with

$$\left(\frac{n_p}{n_n} \right)_{min} \simeq \left(\frac{Q + (Q^2 - m_e^2)^{1/2}}{m_n} \right)^{3/2} = 0.002 \,. \tag{6.82}$$

Afterwards n_p/n_n increases monotically and approaches the asymptotic value of 1/8.

As a typical example, we take $\rho \simeq m_n n_n = 0.107\rho_0$ and find $n_p/n_n = 0.013$ and $p_{F,e} = 105 MeV/c$. Above this density, muons are stable. – Note also that $\mu_p \ll \mu_e$ since $p_{F,e} = p_{F,p}$ and thus we have $\mu_n \simeq \mu_e$.

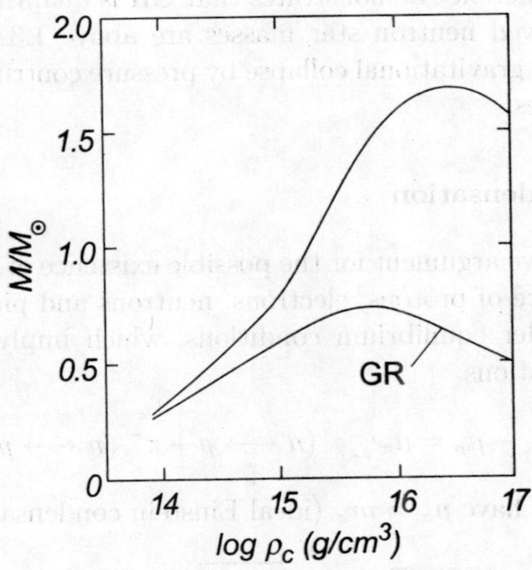

Fig. 6.4. $M(\rho_c)$ for an ideal completely degenerate neutron gas in GR and in semi-Newtonian theory.

6.6.3 Oppenheimer-Volkoff Model

The ideal equation of state $p(\rho)$, described implicitly by (6.73a) and (6.73c) was used in the first general relativistic model calculation by Oppenheimer and Volkoff in 1939 (see [162]). In Fig. 6.4 the gravitational mass of this model of neutron stars as a function of the central density ρ_c is shown. The maximum mass is $M_m = 0.71 M_\odot$ and is reached at the high density $\rho_m = 4\times10^{15} g/cm^3$. The corresponding radius is $R_m = 9.6km$. For $\rho > \rho_m$ the star is *unstable* according to Proposition 6.1. A Newtonian stability analysis shows that for $\rho \ll \rho_m$ the star is stable since the adiabatic index is then 5/3, well above the critical value 4/3. Hence, the entire rising branch in Fig. 6.4 is stable.

To show how M_m and R_m scale with the fermion mass m_f, we write the result as follows

$$M_m = 0.36 M_0\,, \qquad R_m = 3.2 R_0\,,$$

where

$$R_0 = \frac{c}{(8\pi G\,\rho_0)^{1/2}}\,, \qquad M_0 = \frac{c^2 R_0}{G}\,,$$

with

$$\rho_0 = \frac{8\pi\, m_n^4 c^3}{3(2\pi\hbar)^3}\,.$$

So, we have $M_m, R_m \propto m_f^{-2}$ and $(2G\,M_m)/(R_m c^2) = 0.23$.

In Fig. 6.4 we also show the result of the semi-Newtonian theory. The relatively large difference demonstrates that GR is quantitatively important. Since most observed neutron star masses are above $1.3 M_\odot$, they must be supported against gravitational collapse by pressure contributions originating from nuclear forces.

6.6.4 Pion Condensation

We also give a naive argument for the possible existence of a pion condensate. Consider a mixture of protons, electrons, neutrons and pions π^- at temperature $T = 0$ under equilibrium conditions, which imply for the chemical potentials the relations

$$\mu_e = \mu_n - \mu_p = \mu_\pi \qquad (n \longleftrightarrow p + \pi^-\,, n \longleftrightarrow p + e^-)\,.$$

For ideal gases we have $\mu_\pi = m_\pi$ (ideal Einstein condensate) and thus

$$m_n\sqrt{1+x_n^2} - m_p\sqrt{1+x_p^2} = m_e\sqrt{1+x_e^2}\,,$$
$$m_e\sqrt{1+x_e^2} = m_\pi\,,$$

where $x = p_F/m$. Charge neutrality implies in addition

$$\frac{1}{3\pi^2\lambda_e^3}\,x_e^3 + n_\pi = \frac{1}{3\pi^2\lambda_p^3}\,x_p^3\,.$$

This gives for the threshold of π^- production, since $m_\pi/m_e \gg 1$,

$$x_p = \frac{m_\pi}{m_p}\,, \qquad x_n = 0.584\,,$$

i.e.,

$$\rho|_{\pi^- - threshold} = 1.36 \times 10^{15}\, g/cm^3 \simeq 5\, \rho_{nuc}\,.$$

This is not far from more realistic calculations. For a discussion of the present status and references to the literature, we refer to the review [163].

* * *

Exercise 6.6. As was discussed in Sect. 3.7.2 there is a mass concentration in the center of our galaxy of $3 \times 10^6 M_\odot$ within 17 light hours. Consider the hypothetical possibility that this could be a completely degenerated star of non-interacting fermions (e.g. neutrinos). For which range of fermion masses would this be possible?

6.7 Equation of State at High Densities

The equation of state at sub-nuclear densities is reasonable well known (for a brief review and references, see [164]). Since the densities in the central regions of neutron stars can be almost an order of magnitude *higher* than the nuclear density

$$\rho_{nuc} \simeq 2.7 \times 10^{14}\, g/cm^3 \simeq 0.16 \text{ nucleons}/fm^3\,,$$

we need a reliable equation of state at super-nuclear densities. This is a very difficult problem and large uncertainties remain. Below nuclear densities the nuclear gas is dilute and one can use perturbative methods. In the relevant region above nuclear densities we are still far from the asymptotic free region of quantum chromodynamics. Furthermore, quantum chromodynamic lattice calculations are not feasible for the time being. The most useful approaches are still based on a many-body Schrödinger description, with phenomenological potentials. Before we say more about this, we briefly discuss another popular approach.

6.7.1 Effective Nuclear Field Theories

In this approach one starts from a local relativistic field theory with baryon and meson degrees of freedom. Such models have the advantage to be relativistic but sacrifice the connection to the large body of nucleon-nucleon scattering data. The coupling constants and mass parameters of the effective Lagrangian are constraint by empirical properties of nuclear matter at saturation (binding, density, compression moduls, symmetry energy). As a starting point one chooses a mean field approximation that should be reasonably good at very high density. In a second step 1-loop vacuum fluctuations are included (relativistic Hartree approximation), and further refinements have been worked out.

This approach is probably best viewed as a reasonable way of parameterizing the equation of state. It has, however, a number of drawbacks. For textbook reviews we refer to [22] and [23].

6.7.2 Many-Body Theory of Nucleon Matter

In this traditional approach one uses two-body potentials which are fitted to nucleon-nucleon scattering data and three-body terms whose form is suggested in part by theory (two-pion exchange interaction) and purely phenomenological contributions, whose parameters are determined by the binding of few-body nuclei and saturation properties of nuclear matter. There are systematic improvements in this program.

In the early 1990s the Nijmegen group carefully examined all data on elastic nucleon-nucleon scattering below the pion production threshold. Nucleon-nucleon interaction potentials which fit the data base have been constructed by various groups. These differ mainly because some of them include non-localities suggested by boson exchange contributions. This leads to different predictions for the many-body systems, but these are much smaller than those of older models.

One of the two-nucleon potentials in the Hamiltonian

$$H = -\sum_i \frac{1}{2m}\Delta_i + \sum_{i<j} \mathcal{V}_{ij} + \sum_{i<j<k} V_{ijk} \tag{6.83}$$

is the so-called Argonne $\mathcal{V}_{18}$ (A18) potential, because it consists of sums of 18 operators like $\boldsymbol{\sigma}_i \cdot \boldsymbol{\sigma}_j$, etc. (see [165]). We shall show below results for neutron stars based on this potential.

This and other 'modern' potentials underbind the triton and other light nuclei and predict too high equilibrium density for symmetric nuclear matter. To correct this, as well as for theoretical reasons, one has to include three-nucleon interactions V_{ijk} in (6.83). Such potentials are constructed by fitting binding energies of light nuclei and the estimated equilibrium properties of nuclear matter. Some of the results shown below are based on the Urbana models U-IX (see [166]).

Beside these contributions, some relativistic corrections (so-called boost interactions) have to be included (see [167]). In the graphs shown below, these are denoted by $\delta\mathcal{V}$.

There are two main methods to solve approximately the many-body problem. In the method developed by Brueckner, Bethe and Goldstone the perturbation expansion of the nuclear matter energy is ordered according to the number of independent hole lines. The other many-body calculations are based on the variational principle, whereby the energy expectation value is evaluated using cluster expansion. The energies obtained with the two methods are in reasonable agreement at densities below $0.6 fm^{-3}$. However, at higher densities the convergence of the hole line expansion is expected to deteorate and the results are significantly below the variational bound. For more details and references we refer to [163].

The equation of state can be obtained parametrically from the energy ε per baryon as a function of the baryon number density n_B as follows

$$\rho(n_B) = n_B \varepsilon(n_B), \qquad p(n_B) = n_B^2 \frac{\partial \varepsilon}{\partial n_B}. \tag{6.84}$$

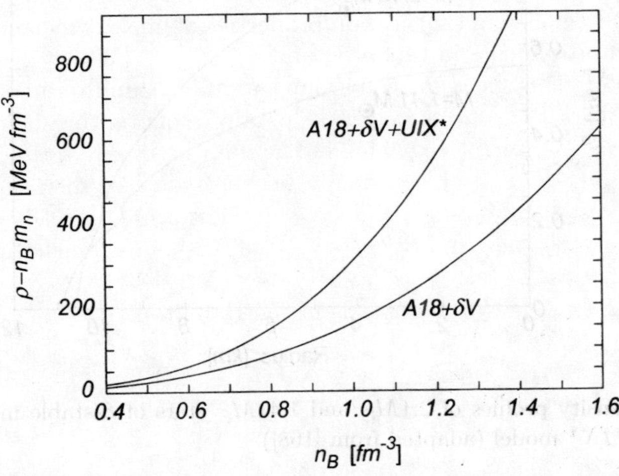

Fig. 6.5. Energy density of β-stable matter for the models A18+$\delta\mathcal{V}$ and A18+$\delta\mathcal{V}$+ UIX^* as a function of the baryon number density (adapted from [168]).

In Fig. 6.5 we show the result for $\varepsilon(n_B)$ based on one of the modern realistic models for nuclear forces (A18+$\delta\mathcal{V} + UIX^*$). The increasing difference between the two curves at large densities shows the importance of three-body interactions (the star on UIX indicates its modification when the boost interaction $\delta\mathcal{V}$ is included). The calculations use "variational chain summation methods".

6.8 Gross Structure of Neutron Stars

Using the result for cold, catalyzed β-stable matter the gross structure of neutron stars can be calculated. For A18+$\delta\mathcal{V} + UIX^*$ the maximum mass M_m turns out to be $M_m = 2.20 M_{\odot}$. The corresponding radius is

$R(M_m) = 10.1km$ and the central baryon density is $n_c(R_m) = 7.2fm^{-3}$. Figure 6.6 shows the density profiles of 2.1 $M_\odot$ and 1.41 $M_\odot$ stars. Of great observational interest is the mass-radius relation. This is plotted in Fig. 6.7. The importance of the boost and three-body interactions is also shown. For further information we refer the reader to the reviews [163] and [169].

The mass-radius relation of neutron stars is difficult to test. While a few masses are known very accurately, observational methods for estimating neu-

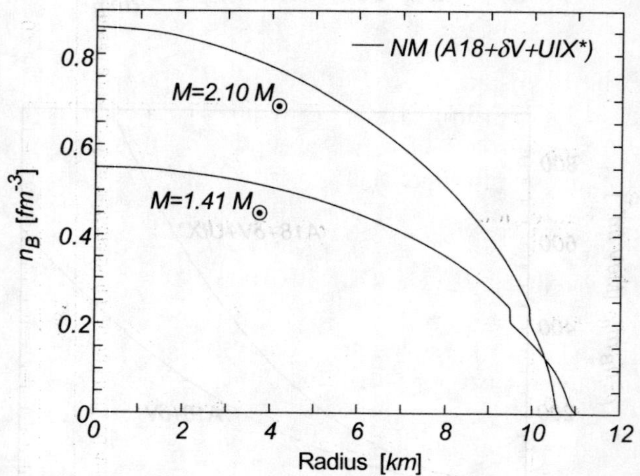

Fig. 6.6. Density profiles of $2.1M_\odot$ and $1.41M_\odot$ stars of β-stable matter for the A18+$\delta\mathcal{V} + UIX^*$ model (adapted from [168]).

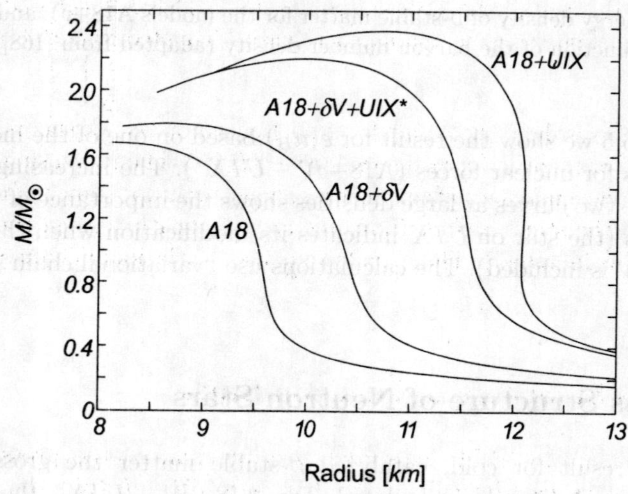

Fig. 6.7. Mass-radius relation for the A18+$\delta\mathcal{V}+UIX^*$ model (adapted from [168]).

tron star radii are so far not sufficiently accurate to differentiate between realistic equations of state. Some information comes from X-ray bursts (see Sect. 6.12) or from the thermal emission of quiescent neutron stars in binaries. An interesting case is the nearest known neutron star RX J1856.5-3754, which is non-pulsating and almost thermally radiating. It has been studied with the Hubble space telescope (see [170]). The distance D has been estimated from parallax measurements with conflicting results: $D \sim 61\,pc$ or approximately twice as large. From the measured flux F and the Stephan-Boltzmann law

$$F = \sigma_{SB} \frac{T^4 R^2}{D^2}$$

one obtains the radius $R_\infty := R/(1 - 2GM/R)^{1/2}$ of $\simeq 7\,km$, respectively $12\,km$ for $M = 1.4\,M_\odot$. In June 2003 the neutron star will pass within approximately 0.3 arcseconds of a background star and gravitational lensing might lead to a mass determination.

6.9 Bounds for the Mass of Non-rotating Neutron Stars

The value of the largest possible mass M_{max} of a non-rotating neutron star is interesting for at least two reasons. First of all, it is important for the theory of stellar evolution. If M_{max} were fairly small ($\simeq 1.5 M_\odot$) then black holes could be formed with high probability, as mentioned at the beginning of this chapter. Its value also plays a decisive role in the observational identification of black holes. If one can find a very compact object which can be shown convincingly not to rotate very fast and to have a mass which is definitely larger than M_{max}, then one has a serious candidate for a black hole. Precisely such arguments have been used to identify some binary X-ray sources as black hole candidates (see Sect. 7.6).

In view of the large uncertainties of the equation of state for high densities, it is important to find reliable limits for M_{max}. This is what we are going to do next (see also [106]). We shall assume that for densities less than some given density ρ_0 (which may depend on our current knowledge), the equation of state is reasonably well known; for densities higher than ρ_0, the equation of state will be required to satisfy only quite general conditions.

6.9.1 Basic Assumptions

We shall make the following minimal assumptions about the matter in the interior of a non-rotating neutron star:

(i) The energy-momentum tensor is described in terms of the mass-energy density and an isotropic pressure p, which satisfies an equation of state

$p(\rho)$. Stresses, which may have arisen as a result of the slowing down of the star's rotation, are neglected.

(ii) The mass-energy density is positive, i.e., $\rho \geq 0$.

(iii) The matter is microscopically stable, i.e., $dp/d\rho \geq 0$. Since the pressure is certainly positive at low density, we then have $p \geq 0$.

(iv) The equation of state is known for $\rho \leq \rho_0$. We may choose for ρ_0 a value which is not significantly below nuclear densities.

It is reasonable to assume that the velocity of sound $(dp/d\rho)^{1/2}$ is smaller than the velocity of light. A priori one may question this assumption since the sound velocity is a *phase* velocity and sound waves propagating through neutron stars are subject to dispersion and absorption. However, if one accepts a fluid description of matter in a neutron star, then the sound velocity provides the characteristics of the hydrodynamic equations for acceptable relativistic formulations. Hence, causality implies $dp/d\rho \leq 1$.

We repeat the basic equations from Sect. 6.4, since these form the basis for the following derivations (again $G = c = 1$). The metric is

$$g = -\mathrm{e}^{2a(r)}dt^2 + \frac{dr^2}{1-2m(r)/r} + r^2\left(d\vartheta^2 + \sin^2\vartheta\, d\varphi^2\right), \tag{6.85}$$

and the structure equations are

$$-\frac{dp}{dr} = (\rho + p)\frac{m + 4\pi r^3 p}{r^2(1-2m/r)}, \tag{6.86a}$$

$$\frac{dM}{dr} = 4\pi r^2 \rho, \tag{6.86b}$$

$$\frac{da}{dr} = \frac{m + 4\pi r^3 p}{r^2(1-2m/r)}. \tag{6.86c}$$

The boundary conditions are, if ρ_c is the central density,

$$p(r=0) = p(\rho_c) =: p_c, \tag{6.87a}$$

$$m(0) = 0, \tag{6.87b}$$

$$\mathrm{e}^{2a(R)} = 1 - \frac{2m(R)}{R}, \tag{6.87c}$$

where R is the stellar radius, i.e., the point at which the pressure vanishes, and $M = m(R)$ is the gravitational mass of the star.

Examination of (6.86a) leads one to expect that

$$2\,m(r)/r < 1. \tag{6.88}$$

The justification of the conclusion is as follows. Let r_* be the first point at which, starting from the center of the star, $2m(r_*) = r_*$. In a neighborhood of r_*, the Tolman-Oppenheimer-Volkoff equation (6.86a) becomes

$$\frac{1}{\rho+p}\frac{dp}{dr}=\frac{1}{2(r_*-r)}\frac{1+8\pi r_*^2\,p(r_*)}{1-8\pi r_*^2\,\rho(r_*)}+\mathcal{O}(1)\,. \tag{6.89}$$

The left-hand side of this equation is the logarithmic derivative of the relativistic enthalpy η, defined by

$$\eta:=\frac{\rho+p}{n}\,, \tag{6.90}$$

where n is the baryon number density. This is an immediate consequence of the first law of thermodynamics

$$\frac{d\rho}{dn}=\frac{\rho+p}{n}\,. \tag{6.91}$$

Thus, if we integrate (6.89) and use the fact that the right-hand side is negative for $r<r_*$, we obtain $\eta(r_*)=0$. Examination of (6.90) shows that this is not possible for a realistic equation of state.

We can also conclude from our assumptions that the density does not increase with increasing radius. From (6.86a) we have, in fact

$$\frac{d\rho}{dr}=\frac{d\rho}{dp}\frac{dp}{dr}=-\left(\frac{dp}{d\rho}\right)^{-1}(\rho+p)\frac{m+4\pi r^3\,p}{r^2(1-2m/r)}\,.$$

According to (ii) and (iii), the quantities ρ, p and $dp/d\rho$ are positive. The positive sign of the last factor is a consequence of (6.88) and

$$m(r)=\int_0^r 4\pi r^2\,\rho\,dr\,. \tag{6.92}$$

We have thus shown that $d\rho/dr\le 0$. – This allows us to divide the star into two regions.

The **envelope** is the part having $\rho<\rho_0$ and $r>r_0$, for which the equation of state is known. The interior region with $\rho>\rho_0$ and $r<r_0$ is called the **core**. In this region we assume only the general requirements (i)-(iv).

The mass M_0 of the core is given by

$$M_0=\int_0^{r_0} 4\pi r^2\,\rho\,dr\,. \tag{6.93}$$

If M_0 is given, the structure equations (6.86a) and (6.86b) can be integrated outward from r_0. The corresponding boundary conditions are $p(r_0)=p(\rho_0)=:p_0$ and $m(r_0)=M_0$. The total mass M of the star is

$$M=M_0+M_{env}(r_0,M_0)\,. \tag{6.94}$$

We can regard the mass of the envelope M_{env} as a known function of r_0 and M_0, since the equation of state is known there. We may now obtain bounds for M by the following strategy:

1. Determine the range of values (M_0, r_0) which is allowed by the assumptions (i)-(iv). This range of possible core values will be called the *allowed region* in the (r_0, M_0)-plane.
2. Look for the maximum of the function M defined by (6.94), in the allowed region of the variables r_0 and M_0.

6.9.2 Simple Bounds for Allowed Cores

One obtains a simple, but not optimal, bound for the allowed region as follows: For $r = r_0$, we find from (6.88)

$$M_0 < \frac{1}{2} r_0 . \tag{6.95}$$

If the density does not increase with r, then

$$M_0 \geq \frac{4\pi}{3} r_0^3 \rho_0 . \tag{6.96}$$

These two inequalities imply the following bounds

$$M_0 < \frac{1}{2}\left(\frac{3}{8\pi\rho_0}\right)^{1/2} , \qquad r_0 < \left(\frac{3}{8\pi\rho_0}\right)^{1/2} . \tag{6.97}$$

A simple numerical example shows that these are not completely uninteresting. If $\rho_0 = 5 \times 10^{14} g/cm^3$, then Eq. (6.97) gives $M_0 \leq 6 M_\odot$ and $r_0 < 18 km$.

6.9.3 Allowed Core Region

In order to determine precisely the allowed region for the cores, we investigate the quantity

$$\zeta(r) = \mathrm{e}^{a(r)} , \tag{6.98}$$

which is finite and positive everywhere. The quantity $\zeta(r)$ can vanish only at the center, in the limit in which the pressure approaches infinity. This is a consequence of (6.86c), which can be integrated inward from the surface, subject to the boundary condition (6.87c), without encountering a singularity.

One can derive a differential equation relating $\zeta(r)$ and $m(r)$ from the structure equations (6.86a)–(6.86c):

$$\left(1 - \frac{2m}{r}\right)^{1/2} \frac{1}{r}\frac{d}{dr}\left[\left(1 - \frac{2m}{r}\right)^{1/2} \frac{1}{r}\frac{d\zeta}{dr}\right] = \frac{\zeta}{r}\frac{d}{dr}\left(\frac{m}{r^3}\right) . \tag{6.99}$$

(It is easy to verify that this equation is correct.) Since ρ does not increase with increasing r, the mean density does not either. Therefore, the right-hand side of (6.99) is not positive. If one introduces the new independent variable

$$\xi = \int_0^r r\left(1 - \frac{2m}{r}\right)^{-1/2} dr \tag{6.100}$$

the resulting inequality can be written in the simple form

$$\frac{d^2\zeta}{d\xi^2} \leq 0 . \tag{6.101}$$

Using the mean value theorem, we conclude

$$\frac{d\zeta}{d\xi} \leq \frac{\zeta(\xi) - \zeta(0)}{\xi} . \tag{6.102}$$

This inequality is optimal, since equality holds for a star having constant density (the proof is an exercise). Since $\zeta(0) \geq 0$, we have

$$\frac{1}{\zeta}\frac{d\zeta}{d\xi} \leq \frac{1}{\xi} . \tag{6.103}$$

Equality holds for a star having constant density when the pressure at the center diverges.

When rewritten in terms of r and $a(r)$, Eq. (6.103) reads

$$\left(1 - \frac{2m}{r}\right)^{1/2} \frac{1}{r}\frac{da}{dr} \leq \left[\int_0^r r\left(1 - \frac{2m}{r}\right)^{-1/2} dr\right]^{-1} . \tag{6.104}$$

We now estimate the right-hand side in an optimal fashion. Since m/r^3 does not increase outwards, we have

$$\frac{m(r')}{r'} \geq \frac{m(r)}{r}\left(\frac{r'}{r}\right)^2$$

for all $r' \leq r$ and hence

$$\int_0^r r'\left(1 - \frac{2m(r')}{r'}\right)^{-1/2} dr' \geq \int_0^r r'\left(1 - \frac{2m(r)}{r^3} r'^2\right)^{-1/2} \tag{6.105}$$

$$= \frac{r^3}{2m(r)}\left[1 - \left(1 - \frac{2m(r)}{r}\right)^{1/2}\right] . \tag{6.106}$$

Again, equality holds for a star having uniform density. If we now use the structure equation (6.86c) for the left-hand side of (6.104), we obtain

$$\frac{m + 4\pi r^3 p}{r^3(1 - 2m/r)^{1/2}} \leq \frac{2m(r)}{r^3}\left[1 - \left(1 - \frac{2m(r)}{r}\right)^{1/2}\right]^{-1}$$

and this gives the following bound on $m(r)/r$:

$$\frac{m(r)}{r} \leq \frac{2}{9}\left[1 - 6\pi r^2\, p(r) + \left(1 + 6\pi r^2\, p(r)\right)^{1/2}\right]. \qquad (6.107)$$

Equality holds for a uniformly dense star with infinite central pressure.

The first interesting consequence of (6.107) is obtained for $r = R$, where $p(R) = 0$:

$$\frac{2M}{R} \leq \frac{8}{9}. \qquad (6.108)$$

Hence the red shift at the surface satisfies the interesting inequality

$$z_{surf} \leq 2\,. \qquad (6.109)$$

If (6.107) is evaluated at the core boundary, one obtains

$$M_0 \leq \frac{2}{9}\, r_0 \left[1 - 6\pi r_0^2\, p_0 + \left(1 + 6\pi r_0^2\, p_0\right)^{1/2}\right]. \qquad (6.110)$$

This represents the optimal improvement of (6.95). Together with (6.96), which is already optimal, Eq. (6.110) determines the allowed region in the (r_0, M_0)-plane for cores which satisfy the assumptions (i)-(iv). This region is showed in Fig. 6.8.

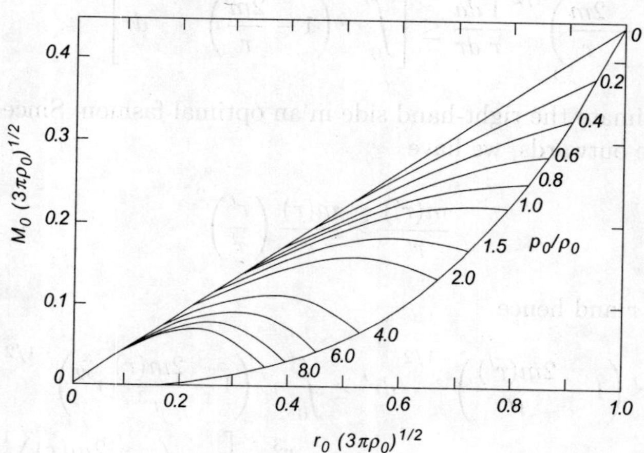

Fig. 6.8. Allowed region for cores in the (r_0, M_0)-plane. The lower curve is determined by (6.96) and the upper curve by (6.110) for $p_0 = 0$. The upper boundary (6.110) is shown for various values of the ratio p_0/ρ_0. (From [106].) curve

One finds the largest core mass for the stiffest equation of state, namely, for incompressible matter with constant density ρ. Even then the mass of the core inside a radius r_0 is limited; otherwise the pressure becomes infinite at the center and equilibrium is lost.

As long as ρ_0 is not significantly larger than nuclear matter density, then $p_0 \ll \rho_0$ (see Sect. 6.7) and we have with high accuracy the bounds (to be compared with the cruder bound (6.97))

$$M_0 < \frac{4}{9}\left(\frac{1}{3\pi\rho_0}\right)^{1/2}, \qquad r_0 < \left(\frac{1}{3\pi\rho_0}\right)^{1/2}. \tag{6.111}$$

6.9.4 Upper Limit for the Total Gravitational Mass

As an illustration, let us choose $\rho_0 = 5.1 \times 10^{14}\ g/cm^3$ and use an equation of state due to G. Baym, H.A. Bethe, C. Pethick and P.G. Sutherland given in [176]. We then have $p_0 = 7.4 \times 10^{33}\ dyn/cm^2$ and thus $p_0/\rho_0 = 0.016 \ll 1$. Recall that for nuclear matter the density is given by $\rho_{nuc} = 2.8 \times 10^{14}\ g/cm^3$. The allowed region corresponding to these values of ρ_0 and p_0 is shown in Fig. 6.9. Superimposed are contours of constant total mass, which is by (6.94) a function of M_0 and r_0. The optimum upper bound is $5M_\odot$. If one requires in addition $dp/d\rho < 1$, then one would obtain $3M_\odot$, instead, which explains why one obtains an upper limit of approximately $3M_\odot$ for "realistic" equations of state.

The calculation shows that the contribution of the envelope to the limiting mass is less than 1%. The limit is reached for the maximum value of the core mass. One can show (see [106]) that this is true as long as $p_0 \ll \rho_0$.

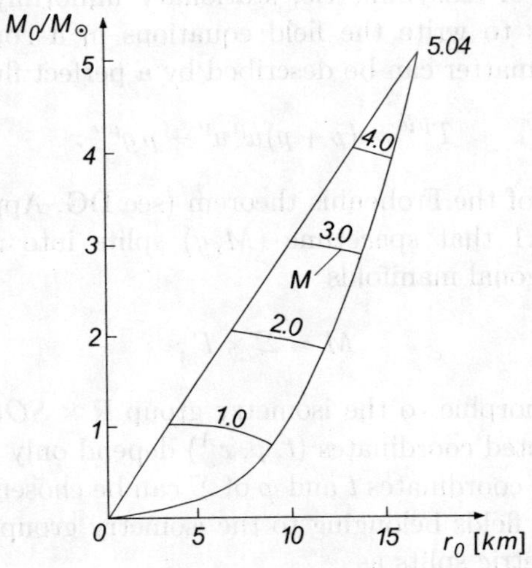

Fig. 6.9. The function $M(r_0, \rho_0)$ for the Baym, Bethe, Pethick and Sutherland equation of state and $\rho_0 = 5.1 \times 10^{14} g/cm^3$. The allowed region is also shown. The maximum mass is $5\,M_\odot$. (From [106].)

We thus obtain, to a good approximation

$$M \leq \frac{4}{9}\left(\frac{1}{3\pi\rho_0}\right)^{1/2} \tag{6.112}$$

or

$$M \leq 6.8\left(\frac{\rho_{nuc}}{\rho_0}\right)^{1/2} M_\odot . \tag{6.113}$$

The additional causality assumption $dp/d\rho \leq 1$ would have the effect of replacing the factor 6.8 by 4.0.

6.10 Rotating Neutron Stars

Most periods of pulsars are in the range $0.25 - 1\,sec$, but there is also a subclass of *millisecond* pulsars. The rotation of even the fastest millisecond pulsar PSR 1937+214, with a rotation period $P = 2\pi/\Omega = 1.558\,msec$ is not fast enough to change the internal structure of the neutron star very much. This should not be surprising in view of the fact that the Keplerian angular velocity Ω_K of a test body around the equator of a typical $1.4M_\odot$ neutron star is $\Omega_K = (GM/R^3)^{1/2} \simeq 0.9 \times 10^4 sec^{-1}$. If pulsars with $\Omega \gtrsim 0.8 \times 10^4 sec$ should one day be found this would have stringent implications for the equation of state for neutron star matter.

For the study of axisymmetric, stationary uniformly rotating neutron stars one first has to write the field equations in a convenient form. We shall assume that matter can be described by a perfect fluid

$$T^{\mu\nu} = (\rho + p)u^\mu u^\nu + p\,g^{\mu\nu} . \tag{6.114}$$

As an application of the Frobenius theorem (see DG, Appendix C) we shall show in Sect. 7.2.1 that spacetime (M, g) splits into a product of two-dimensional orthogonal manifolds

$$M = \Sigma \times \Gamma ,$$

where Σ is diffeomorphic to the isometry group $\mathbb{R} \times SO(2)$ and the metric coefficients in adapted coordinates (t, φ, x^A) depend only on the two coordinates x^A of Γ. The coordinates t and φ of Σ can be chosen such that the two commuting Killing fields belonging to the isometry group are $k = \partial/\partial t$ and $m = \partial/\partial\varphi$. The metric splits as

$$g = \left(g_{tt}\,dt^2 + 2g_{t\varphi}\,dt\,d\varphi + g_{\varphi\varphi}\,d\varphi^2\right) + g_{AB}\,dx^A dx^B . \tag{6.115}$$

Note that the three metric coefficients g_{tt}, $g_{t\varphi}$ and $g_{\varphi\varphi}$ have an invariant meaning as scalar products of the Killing fields:

$$g_{tt} = \langle k, k\rangle , \qquad g_{t\varphi} = \langle k, m\rangle , \qquad g_{\varphi\varphi} = \langle m, m\rangle . \tag{6.116}$$

A convenient parameterization is

$$g_{tt} = -\mathrm{e}^{2U} , \qquad g_{t\varphi} = -\mathrm{e}^{2U} a , \qquad g_{\varphi\varphi} = \mathrm{e}^{-2U} W^2 - \mathrm{e}^{2U} a^2 , \qquad (6.117)$$

so that

$$a = g_{t\varphi}/g_{tt} , \qquad W^2 = -g_{tt}g_{\varphi\varphi} + g_{t\varphi}^2 . \qquad (6.118)$$

The coordinates x^A of Γ are chosen such that the induced metric is conformally flat

$$g_{AB}\, dx^A dx^B = \mathrm{e}^{-2U} \mathrm{e}^{2k} (d\rho^2 + d\zeta^2) . \qquad (6.119)$$

We are thus led to the **Lewis-Papapetrou parameterization**

$$g = -\mathrm{e}^{2U} (dt + a\, d\varphi)^2 + \mathrm{e}^{-2U} W^2\, d\varphi^2 + \mathrm{e}^{-2U} \mathrm{e}^{2k} (d\rho^2 + d\zeta^2) , \qquad (6.120)$$

with the four potentials U, a, W and k that are functions of ρ and ζ. (For the moment we ignore possible problems connected with 'ergospheres', i.e., regions, where k becomes spacelike.)

The four-velocity u of the fluid is a linear combination of the Killing fields

$$u = \mathrm{e}^{-V} (k + \Omega\, m) , \qquad (6.121)$$

where Ω is the angular velocity of the fluid, assumed to be constant (rigid rotation). From $\langle u, u \rangle = -1$ one finds that

$$\mathrm{e}^{2V} = \mathrm{e}^{2U} \left[(1 + \Omega a)^2 - \Omega^2 W^2 \mathrm{e}^{-4U} \right] . \qquad (6.122)$$

The rigid rotation has restrictive consequences, as we now show. For this we look at the Euler equations for the ideal fluid that follow from the field equations (see 1.39):

$$(\rho + p) \nabla_u u = -\nabla p - (\nabla_u p)\, u .$$

Here, the last term vanishes for rigid rotation

$$L_u p = \mathrm{e}^{-V} \big(L_k p + \Omega\, L_m p \big) = 0 .$$

Next we note that for the 1-form $\boldsymbol{u}$ belonging to u the following holds

$$(L_u \boldsymbol{u})_\mu = u^\nu u_{\mu,\nu} + u_\nu u^\nu{}_{,\mu} = u^\nu u_{\mu;\nu} + u_\nu u^\nu{}_{;\mu} = (\nabla_u \boldsymbol{u})_\mu .$$

Therefore, the Euler equation reduces to

$$(\rho + p) L_u \boldsymbol{u} = -dp .$$

Now we have, using the identity $L_{fX}\omega = f\, L_X\omega + df \wedge i_X\omega$ for a vector field X and a differential form ω,

$$\begin{aligned} L_u \boldsymbol{u} &= d(\mathrm{e}^{-V}) \wedge i_{k+\Omega\, m} \boldsymbol{u} + \mathrm{e}^{-V} L_{k+\Omega\, m} \boldsymbol{u} \\ &= -dV \wedge i_u \boldsymbol{u} = dV . \end{aligned}$$

Therefore, Euler's equation becomes simply

$$(\rho + p)dV = -dp\,. \tag{6.123}$$

But this implies that ρ and p must be functions of V alone. For a given equation of state $\rho = \rho(p)$ the function $p(V)$ is determined by the differential equation

$$\rho(p) + p = -\frac{dp}{dV}\,. \tag{6.124}$$

The boundary of the star satisfies $p(V_0) = 0$. Hence,

$$\int_0^p \frac{dp'}{\rho(p') + p'} = -(V - V_0)\,,$$

or

$$\mathrm{e}^V \exp\left(\int_0^p \frac{dp'}{\rho(p') + p'}\right) = \mathrm{e}^{V_0} = \text{const.} \tag{6.125}$$

In Chap. 7 we shall derive in detail the field equations for the potentials in the vacuum region (a particular solution will be the Kerr metric). In the interior of the body it is natural to pass to comoving coordinates

$$\varphi \longrightarrow \varphi' = \varphi - \Omega\, t$$

and all other coordinates unchanged. With respect to these the Killing field $k' := k + \Omega\, m$, proportional to u, is just $\partial/\partial t$ and is always timelike. This simple coordinate transformation does not change the Lewis-Papapetrou form (6.120). The role of k in (6.116) is now taken by k', hence

$$\mathrm{e}^{2U'} := -g'_{tt} = -\langle k', k'\rangle = \mathrm{e}^{2V}\,,$$

thus $U' = V$. It is a simple exercise to derive also the other of the following relations:

$$\begin{aligned}
\mathrm{e}^{2U'} &= \mathrm{e}^{2U}\left[(1+\Omega\, a)^2 - \Omega^2 W^2 \mathrm{e}^{-4U}\right]\,,\\
(1-\Omega\, a')\mathrm{e}^{2U'} &= (1+\Omega\, a)\,\mathrm{e}^{2U}\,,\\
k' - U' &= k - U\,,\\
W' &= W\,.
\end{aligned} \tag{6.126}$$

These are used to match the interior and the exterior vacuum solutions.

Much of the work that will be done in Sect. 7.2 for rewriting the field equations in terms of the potentials also applies to the interior of the body. In comoving coordinates the basic equations are given in [171].

The Jena group has developed a code for rotating relativistic stars, based on a "multi-domain spectral method", that is of unprecedented accuracy (see [172]). Applications for rapidly rotating neutron stars with realistic equations of state are underway. Some examples of highly flattened bodies are presented in [172]. A crucial question will be to find out how much the maximal mass for *stable* configurations increases, compared to the non-rotating case.

* * *

Exercise 6.7. Derive the rest of (6.126).

6.11 Cooling of Neutron Stars

When a neutron star is formed in the collapse of a stellar core it is very rapidly cooled by neutrino radiation. The interior temperature drops to less than $1MeV$ within minutes and to about $10^9 K$ within one year. Neutrino emission dominates the further cooling of the neutron star, until the interior temperature falls to about $10^8 K$, with a corresponding surface temperature of about $10^6 K$. Only then does photoemission also begin to play an important role.

The cooling curves, i.e., observed temperature as a function of time, depend on a number of interesting aspects of the physics of neutron stars. It turns out that the equation of state as well as the mass of neutron stars do not influence the cooling in a sensitive manner, but the possible existence of a superfluid state of the nucleons plays some role and the existence of a pion condensation or quark phase in the central region would have dramatic effects. For conventual cooling scenarios neutrino emission dominates the cooling for about 10^5 years. – We briefly discuss the dominant neutrino processes.

The so-called standard model of neutron star cooling is based upon neutrino emission from the interior that is dominated by the **modified URCA process**:

$$(n,p) + p + e^- \longrightarrow (n,p) + n + \nu_e \,, \tag{6.127a}$$
$$(n,p) + n \longrightarrow (n,p) + p + e^- + \bar{\nu}_e \,. \tag{6.127b}$$

The **direct URCA process**

$$n \longrightarrow p + e^- + \bar{\nu}_e \,, \qquad p + e^- \longrightarrow n + \nu_e \,, \tag{6.128}$$

is strongly suppressed in degenerate matter by energy and momentum conservation. The fermions n, p and e^- participating in the process have energies within kT of the Fermi surfaces. By energy conservation the neutrino and anti-neutrino energies are then also approximately kT. But the Fermi momenta of the electrons and protons are small compared with the neutron Fermi momentum and thus the processes (6.128) are strongly suppressed by momentum conservation.

This is no longer the case when an additional nucleon, which can absorb energy and momentum, takes part in the process, as in (6.127a) and (6.127b). A pion condensate would have the same effect as such a spectator nucleon.

In order to determine the cooling curves for neutron stars, we need to know both the internal energy and the luminosity as functions of the internal temperature. We must also know the relation between interior temperature and surface temperature. A vast amount of work has been invested over the years in these issues. For an introduction we refer to [21] and [14]. (For an extensive recent review on most theoretical aspects see [173]). Beside typical cooling curves (see Fig. 6.10) with various assumptions concerning the role of superfluidity, the estimated surface temperatures of a hand-full of thermally emitting isolated neutron stars are shown. Although the temperature measurements are not yet accurate, the data support the theoretical prediction that the nucleons in the core are in a superfluid state.

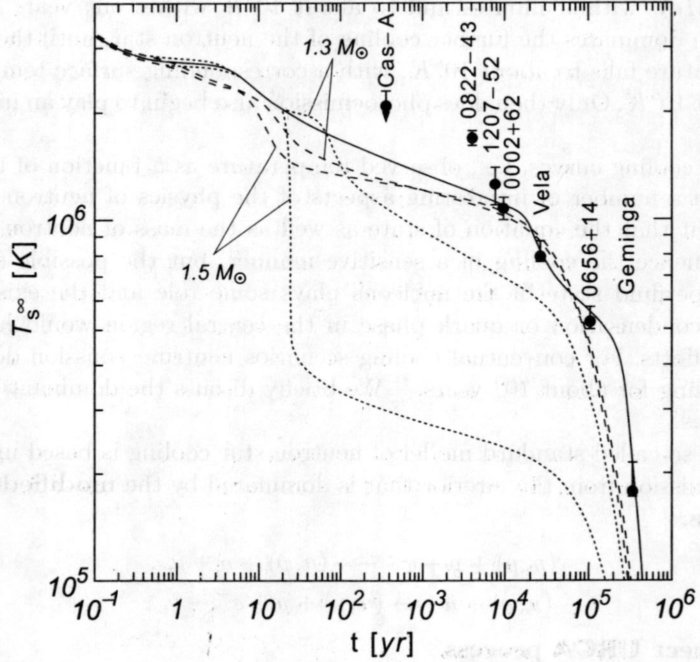

Fig. 6.10. Cooling curves of the superfluid models with the moderate equation of state. Dots and dashes are for the cases of weak and strong superfluidity, respectively, for the 1.3 and 1.5 $M_\odot$ models. Solid line is for the 1.4 $M_\odot$ model with weak neutron superfluidity in the crust and constant critical temperatures of the baryon superfluidity in the core, $T_{cn} = 2 \times 10^8\ K$, $T_{cp} = 1.3 \times 10^8\ K$. Data points with error bars show the estimates of the surface temperature of several isolated neutron stars (Fig. 36 of [173]).

6.12 Neutron Stars in Binaries

Almost half of all stars about us are members of binary systems. Some of these double stars are so close that in certain evolutionary phases they can exchange matter, and evolve as a result in strange ways. The more massive star of a binary system evolves faster and thus becomes a red giant while his partner is still on the main sequence. In a close binary it spills matter onto its companion and with reduced mass then evolves into a neutron star or a black hole. Later the companion star in turn swells up as a red giant, returning matter to the collapsed star. In this accretion process the compact star turns into an X-ray source. The initially less massive star evolves further and may, if sufficiently massive, finally collapse into either a neutron star or a black hole. In this way binary pulsars like the Hulse-Taylor pulsar presumably have been formed.

Accretion on neutron stars leads to a broad range of fascinating phenomena. For the benefit of those readers who have not yet been exposed to this subject we give here a first introduction. For further information on this vast field of modern astrophysics we recommend beside [21] the copiously illustrated book [25]. More technical literature will be quoted later.

6.12.1 Some Mechanics in Binary Systems

For a basic understanding of mass exchange in close binaries, we recall some mechanics. In particular, we introduce the so-called "Roche lobe" of a star in a binary system.

Suppose, for simplicity, that the two components of the binary system move about each other in circular orbits. The orbital period P_b and the angular frequency Ω are given by Kepler's third law:

$$\Omega^2 = \left(\frac{2\pi}{P_b}\right)^2 = G\,\frac{M_1 + M_2}{a^3}\,, \tag{6.129}$$

where M_1 and M_2 are the masses of the two stars and a is the distance between their centers of mass. We now consider the motion of a test particle in the gravitational field of the two masses (restricted three-body problem). Its motion is most conveniently described in the co-rotating system:

$$\dot{\boldsymbol{v}} = \boldsymbol{\nabla}\psi - 2\boldsymbol{\Omega} \times \boldsymbol{v}\,. \tag{6.130}$$

Here, $\boldsymbol{v}$ is the velocity and ψ is the sum of the Newtonian and centrifugal potentials:

$$\psi(\boldsymbol{x}) = -\frac{GM_1}{r_1} - \frac{GM_2}{r_2} - \frac{1}{2}(\boldsymbol{\Omega} \times \boldsymbol{x})^2\,. \tag{6.131}$$

The equilibrium positions in the comoving frame are thus the critical points of ψ. There are five of these, of which three are collinear with M_1

and M_2 (Euler 1767) and two are equilateral (Lagrange 1773). Moreover, the three collinear equilibrium points are *unstable* (Plummer 1901). The equilateral equilibrium points are stable for sufficiently small mass ratios (see Exercise 6.8).

Taking the inner product of (6.129) with $\boldsymbol{v}$ leads to the *Jacobi integral*:

$$\frac{1}{2}\boldsymbol{v}^2 + \psi = \text{const}. \tag{6.132}$$

Obviously, the potential ψ cannot increase above the value of this integral as the particle moves. For this reason, the structure of the equipotential surfaces of ψ is important. These are shown in Fig. 6.11, along with the five equilibrium positions. The equipotential surface passing through the "inner Lagrange point" L_1 is particularly important. Inside this **critical Roche surface** the equipotential surfaces, which enclose the two centers of mass, are disjoint; outside this is no more the case. The Roche limit thus determines the maximum volume, the so-called **Roche lobe**, of the star. If this is exceeded, a portion of the outer layers of the star will flow over to the companion.

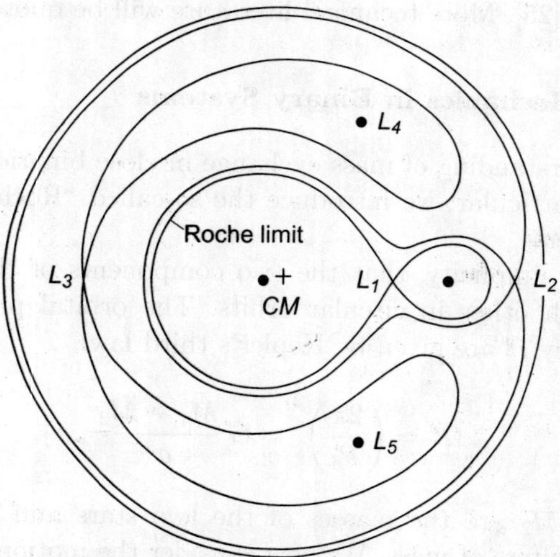

Fig. 6.11. Equipotential curves in the orbital plane of a binary star system, equilibrium positions and Roche limit for the restricted three body problem. L_1, L_2 and L_3 are unstable equilibrium positions.

For a fluid the Navier-Stokes equation becomes in the co-rotating system

$$D_t\boldsymbol{v} = -\boldsymbol{\nabla}\psi - 2\boldsymbol{\Omega}\times\boldsymbol{v} - \frac{1}{\rho}\boldsymbol{\nabla}p + \text{friction terms}, \tag{6.133}$$

where $D_t\boldsymbol{v}$ denotes the hydrodynamic derivative of the velocity field. For a stationary situation ($\boldsymbol{v} = 0$ in the co-rotating system) one obtains, as expected,

$$\nabla p = -\rho \nabla \psi . \tag{6.134}$$

Hence, the surfaces of constant pressure, and in particular the surface of the star, are equipotential surfaces of ψ.

We now imagine a situation in which one of the two stars enters a red giant phase during which it fills up its Roche lobe. If this happens, matter flows through the inner saddle point into the Roche volume of its partner. Part of it may escape and the rest will be accreted by the secondary. This mechanism will play an important role throughout this section.

* * *

Exercise 6.8.

a) Show that (6.130) is the Euler-Lagrange equation of the Lagrange function

$$\mathcal{L} = \frac{1}{2}\dot{\boldsymbol{x}}^2 + (\boldsymbol{\Omega} \times \boldsymbol{x}) \cdot \boldsymbol{x} - \psi . \tag{6.135}$$

Derive the the Hamiltonian of the corresponding autonomous system and show that it is just the Jacobi integral.

b) Show that the equilibrium positions of the corresponding Hamiltonian vector field are given by $\dot{\boldsymbol{x}} = 0$, $\nabla \psi = 0$. Determine the five equilibrium points and prove that the collinear ones are unstable. Demonstrate finally that the equilateral points are elliptic, provided that

$$\frac{M_1}{M_1 + M_2} < \frac{1}{2} - \frac{\sqrt{69}}{18} = 0.03852\ldots . \tag{6.136}$$

(We emphasize that this does not prove stability, which is a much more difficult problem that has been solved by Debrit and Debrit in 1967, using the modern KAM theory of dynamical systems.)

Solution. For a solution of this exercise, see e.g. [33], Sect. 10.2.

6.12.2 Some History of X-Ray Astronomy

One of the scientific motivations to start X-ray astronomy was to detect thermal radiation from cooling neutron stars in the X-ray energy band. The hope was to detect youngish neutron stars with surface temperatures around a few million degrees. The first X-ray source outside the solar system, Sco X-1, was discovered back in 1962 by Giacconi et al. during a short, five minute rocket exposure of a detector consisting of three Geiger counters ($\sim 20cm^2$ area). The newly discovered X-ray source did not fit the cooling scheme; the temperature was about a factor ten higher – hence the luminosity about 10^4 higher – than expected. The source Sco X-1 later turned out to be an accreting binary neutron star. This interpretation was suggested already in 1967 by Shkolovskii, shortly before pulsars were discovered.

The "X-ray window" was really opened when the famous X-ray observatory UHURU was launched from the coast of Kenya on December 12, 1970. Uhuru is the Swahili word for " freedom" and was chosen because the launch took place on Kenya's independence day. Within less than two years, about one hundred galactic and fifty extragalactic X-ray sources were discovered with this orbiting X-ray observatory in the spectral range $2 - 20\,keV$. Some of these sources show very regular short X-ray pulsations (X-ray pulsars). One of the best known examples is Hercules X-1 with a period of $1.24\,s$. This period is, however, not strictly constant, but varies with a period of 1.70017 days, showing that Her X-1 is a member of a close binary system. We now know of many such systems, for which the optical partner has also been unambiguously identified.

The completely irregular, rapidly fluctuating source Cyg X-1, which presumably contains a black hole, was also discovered with UHURU.

UHURU observations also showed that the space between galaxies in clusters of galaxies contains hot gas with a temperature of $(10 - 100) \times 10^6\,K$.

In 1975 and 1976 astronomers discovered a new class of X-ray sources, the so-called **bursters**, using the satellites ANS, SAS-3, OSO 7,8, and others. One observed brief outbursts of X-rays from sources near the galactic center of our galaxy or in globular clusters. They repeat themselves at irregular intervals which lie between a few hours and several days. Typically, the maximum intensity is reached in a few seconds and then falls back to its original value after about a minute. In this brief period some $10^{39}\,erg$ of X-ray energy are emitted. (This is comparable to the energy radiated by the Sun in about two weeks.)

The detectors which were carried by these satellites did not have a high sensitivity and thus only the strongest sources could be observed. An X-ray observatory having a sensitivity comparable to that of optical and radio telescopes at the time became available with the launch of the "Einstein Observatory" in November 1978. In addition, the telescope's resolution of four seconds of arc was 1000 times higher than the resolution of the X-ray detectors used previously. Thus, within just fifteen years, a development took place which is comparable to the progress achieved in optical astronomy from Galileo's telescope of 1610 to the five meter Hale reflector on Mount Palomar.

Since then other successful missions (EXOSAT, ROSAT, ...) have been launched. At the time of writing two powerful X-ray observatories (CHANDRA and XMM-NEWTON) are in orbit; future ones are in the planning stage.

6.12.3 X-Ray Pulsars

We now know a large number of compact galactic X-ray sources with an X-ray luminosity $L_X > 10^{34} erg/s$, with star-like optical counterparts. A considerable fraction of these are X-ray pulsars and have been identified unambiguously as binary star systems. For almost all of these, the companion is a bright O or B star having a mass of about ten to twenty solar masses (*high-mass* X-ray binaries). Her X-1 is an exception to this rule, since the mass of its companion is only about $2M_\odot$.

We believe we have a definite qualitative interpretation of the X-ray pulsars. The companion of the normal star is a neutron star which sucks up gas that is lost from its partner. The accreted[4] matter falls eventually onto the surface of the neutron star, releasing about 10 percent of its rest energy. Neutron stars, as a rule, have very strong magnetic fields, often of order $10^{12} Gauss$, since the magnetic flux is conserved when highly conductive stellar material collapses to a neutron star; this gives rise to an amplification factor of the magnetic field strength of order 10^{10}. Such strong fields cause the plasma to fall on the neutron star in the regions of the magnetic poles, giving rise to two "hot spots" of intense radiation (see Fig. 6.12). The magnetic field axis does not in general coincide with the axis of rotation, and hence the hot spots rotate with the star. The direction of the emitted X-rays also rotates, similar to the beam emitted by a lighthouse. If the rotating beam sweeps by the Earth, the star appears as an X-ray pulsar. In many cases, the X-ray emission is periodically eclipsed by the optical companion, with the orbital period of the binary system.

The observed X-ray luminosities are in the range $L_X \simeq 10^{36} - 10^{38} erg/s$. Thermal radiation of this magnitude from a source with the small surface area A of the radiating polar cap requires a temperature

$$k_B T = k_B \left(\frac{L_X}{\sigma A}\right)^{1/4} = 10 \left(\frac{L_X}{10^{38} erg/s^{-1}}\right)^{1/4} \left(\frac{A}{1km^2}\right)^{-1/4} keV\,, \quad (6.137)$$

where σ is the Stephan-Boltzmann constant. We now show that $A \sim 1km^2$, whence the X-rays must have energies of a few keV, in agreement with the observations.

To estimate the surface area A, we need the extension of the magnetosphere. The "magnetospheric radius" r_A, also called **Alfvén radius**, is roughly the distance from the neutron star where the magnetic pressure $B^2/8\pi$ is balanced by the ram pressure ρv_r^2 of the accreting gas. Approximating the radial velocity v_r by the free-fall velocity $v_f = (2GM_{NS}/r)^{1/2}$, and using the continuity equation for spherical accretion, $4\pi r^2 \rho\, v_f = \dot{M}$, where $\dot{M}$ is the mass accretion rate, one obtains for typical values $\dot{M} \simeq 10^{17} g/s$, $M_{NS} \simeq 1.4 M_\odot$, and a magnetic field strength at the pole of $B = 5 \times 10^{12} G$ an Alfvén radius $r_A \simeq 100 R_{NS} \simeq 10^8 cm$. (See also Exercise 6.9.)

Since the integral curves of a magnetic dipole field satisfy in polar coordinates $\sin^2\vartheta/r = $ const, we expect for the surface area of the radiating polar cap: $A \simeq \pi(R_{NS}\sin\vartheta_0)^2$ with $r_A \sin^2\vartheta_0 = R_{NS}$, thus $A \simeq \pi R_{NS}^2 (R_{NS}/r_A)^2 \simeq 1km^2$.

---
[4] accrecere: increase by accumulation

The mass-flow rates required to produce the observed X-ray luminosities can be sustained in close high-mass binary systems quite readily. Possible mechanisms are:

1. *Stellar winds:* The winds from O or B super-giants have typically mass-loss rates of $10^{-7} - 10^{-6} M_\odot/yr$, and this can easily lead to an accretion rate of $10^{-9} M_\odot/yr$ onto the compact companion. Indeed, the orbiting collapsed star is an obstacle in the supersonic stellar wind, and hence a bow-shaped shock front is formed around it. Some of the material flowing through this shock is decelerated enough to be captured by the compact star. Since about 10 percent of the rest energy can be converted to radiation, this is sufficient to account for the observed luminosity.
2. *Roche lobe overflow:* The normal companion can expand to the Roche limit if it is a blue or red supergiant. In this case a considerable mass transfer of $10^{-8} - 10^{-3} M_\odot/yr$ can occur. However, since column densities larger than about $1 g/cm^2$ are opaque to keV X-rays, it is possible that this Roche overflow extinguishes the X-ray source. This type of mass transfer probably produces an X-ray source only if the normal companion has a mass not larger than a few solar masses. It certainly plays a crucial role in burst sources (see Sect. 6.12.4). Fig. 6.12 shows schematically these two canonical pictures of X-ray sources.
3. *X-ray heating:* A transfer of mass can also be induced if the X-rays heat up the outer layers of the optical companion. Such a self-sustaining mechanism may be operating in Her X-1.

The details of how accretion takes place, the conversion of potential energy to radiation and the transport of the radiation through the hot plasma that is above the poles are extremely complicated problems. Much of the relevant physics is treated in detail in the book [26]. For the study of accretion disks we recommend [27].

The Eddington Limit

There are no steady X-ray sources brighter than about $5 \times 10^{38} erg\ s^{-1}$. This can roughly be understood as follows.

Let L be the luminosity assumed to be generated uniformly over a sphere of radius R. When is L large enough such that the radiation pressure stops the accretion on a compact object? To answer this we note that the radiation pressure acting on an electron produces a force of magnitude equal to $\sigma_T \times flux/c$, i.e.,

$$\frac{\text{force}}{\text{electron}} = \frac{\sigma_T (L/4\pi R^2)}{c},$$

where σ_T is the Thomson cross section. For a hydrogen plasma this balances the gravitational attraction for the **Eddington luminosity** L_E determined by (m_p is the proton mass)

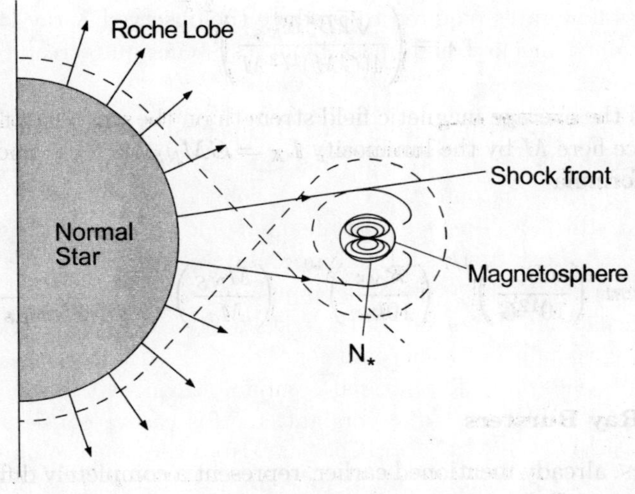

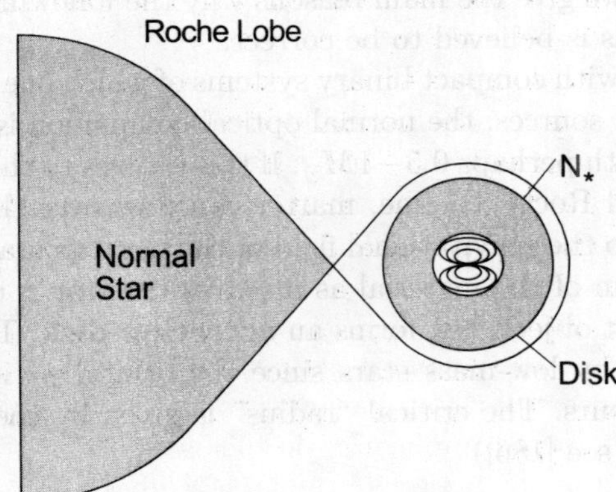

Fig. 6.12. Schematic picture of wind and disk accretion.

$$\frac{\sigma_T L_E}{4\pi R^2 c} = \frac{GMm_p}{R^2}.$$

Thus,

$$L_E = \frac{4\pi GMm_p c}{\sigma_T} = 1.3 \times 10^{38} (M/M_\odot) erg/s\,. \qquad (6.138)$$

We conclude that a neutron star of $M = 1.4M_\odot$ cannot produce a steady luminosity greater than about $2\times10^{38} erg/s$. Because of the assumed spherical symmetry, this argument is, of course, an over-simplification, but it gives the right order of magnitude.

* * *

Exercise 6.9. Use the considerations and notation of this section to show that the Alfvén radius is given by

$$r_A = \left(\frac{\sqrt{2} B_*^2 R_{NS}^6}{4(GM)^{1/2} \dot{M}} \right)^{2/7} , \quad (6.139)$$

where B_* is the average magnetic field strength on the surface of the neutron star. Replace here $\dot{M}$ by the luminosity $L_X = GM_{NS}\dot{M}/R_{NS}$, and verify the numerical formula

$$r_A = (2.6 \times 10^8 cm) \left(\frac{B_*}{10^{12} G} \right)^{4/7} \left(\frac{R_{NS}}{10 km} \right)^{10/7} \left(\frac{M_{NS}}{M_\odot} \right)^{1/2} \left(\frac{L_X}{10^{37} erg\ s^{-1}} \right)^{-2/7} . \quad (6.140)$$

6.12.4 X-Ray Bursters

The bursters, already mentioned earlier, represent a completely different class of X-ray sources. Below we give the main reasons why the following interpretation of the observations is believed to be correct.

We are again dealing with compact binary systems of which one member is a neutron star. For burst sources, the normal optical companion is, however, a rather low mass star with perhaps $0.5 - 1 M_\odot$. If this evolves to the red giant stage and fills its critical Roche volume, matter can flow over through the inner Lagrange point into the gravitational field of the neutron star. Because of the angular momentum of the material as it leaves the star it cannot fall directly onto the compact object, but forms an **accretion disk**. The critical volume can be attained by low-mass stars since the orbital periods are in most cases only a few hours. The critical "radius" is given by the following approximate expression (see [186])

$$R_{cr} = 0.46 \left(\frac{M_1}{M_1 + M_2} \right)^{1/3} a \qquad \text{for } 0 < \frac{M_1}{M_2} < 0.8 . \quad (6.141)$$

Estimate R_{cr} for periods of a few hours.

Evidence for accretion disks for *low-mass* X-ray binaries has been established for a subclass, the so-called **X-ray transients**. For these one can observe also absorption lines in the optical spectrum during quiescent periods, which are characteristic for a low mass star. This indicates that the dominant portion of the optical spectrum is emitted from the hot accretion disk, producing the emission lines, and that the luminosity of the optical member is low. In this connection it is interesting that the X-ray sources which do not pulsate and which do not show eclipses are almost always found close to the center of the Galaxy (galactic bulge sources) or in globular clusters, i.e., in rather old stellar populations. For this reason, one may expect that the neutron star's magnetic field has partially decayed[5] or that the dipole

[5] In the magnetohydrodynamic approximation, the characteristic diffusion time of the magnetic field is $\tau \simeq 4\pi\sigma R^2/c^2$. Inserting typical values for the conductivity

axis is aligned with the axis of rotation. This would explain the absence of pulsations. For a rather low field, matter falling onto the neutron star is distributed more or less uniformly over the surface of the neutron star and the gravitational energy released is emitted in the form of an approximately constant X-ray radiation.

The accreted matter consists mainly of H and He and thus fusion reactions can take place. In fact, it is now established that the "normal" bursts are due to thermonuclear explosions. Underneath a layer of hydrogen about $1m$ thick, the density has risen to about $10^{4-6} g/cm^3$ (verify this by a simple estimate). If the neutron star is still relatively hot, the hydrogen can ignite at a depth of approximately one meter. As detailed calculations show, an approximately equally thick layer of helium is produced below this. Still deeper, helium becomes unstable with respect to fusion into carbon. Thus at least two thin shells are burning, as indicated in Fig. 6.13. As K. Schwarzschild and R. Härm discovered in a different context, these shells are unstable against thermonuclear reactions[6]. The existence and strength of this instability are consequences of the very strong temperature dependence of the thermonuclear reaction rates. For the situation under discussion, it is augmented by the partial degeneracy of the burning matter, implying that the pressure does less strongly respond to the rising temperature.

The $p - p$ chains are not sufficiently temperature sensitive to induce a runaway in the hydrogen-burning layer. For the CNO cycle the temperature sensitivity is high, but this is saturated at high reaction rates by the long ($\sim 100s$) lifetimes of the β-unstable nuclei ^{13}N, ^{14}O, ^{15}O and ^{17}F, which participate in the cycle.

On the other hand, the helium burning layer is unstable for a wide range of conditions, as has been shown by detailed calculations (see [178]). After about $10^{21} g$ of matter have accumulated on the surface, a helium flash is produced in which almost all of the combustible matter is used up. Most of the energy ($\leq 10^{39} erg$) is transported to the photosphere and radiated away mostly as X-rays. The calculated properties of such X-ray bursts (rise time, maximum luminosity, decay time, etc.) are quite similar to those observed (see [178]). For a typical accretion rate of $10^{17} g/s$, the interval between bursts is a few hours. In the fusion process of helium into carbon, the released nuclear energy is only about $1 MeV$ per nucleon. The time average of the energy in the burst should, therefore, be roughly one percent of the energy released in the steady flow of X-rays. This is indeed observed (with exceptions).

This thermonuclear flash model has been supported by further observations. During the cooling off phase of a burst, the spectrum is nearly that of a black body. The emitting surface can then be estimated from the luminosity and the approximate distance of the X-ray source. In all cases one obtains

---

σ (see [177]), one obtains cosmologically long times. Nevertheless, the field of very old neutron stars may be relatively small, but this is a controversial subject.

[6] An analytical discussion of this type of instability has been given in [179].

radii of about 10 km after fifteen seconds. (During the initial phase the radius is about 100 km.) In addition, it has been shown that an outburst in the optical region occurs nearly simultaneously with the X-ray burst. Its maximum is, however, delayed by a few seconds. The interpretation is obvious: The X-ray burst heats up the accretion disk and a delayed optical "echo" results. The data also show that the extension of the accretion disk is not far from $10^6\,km$.

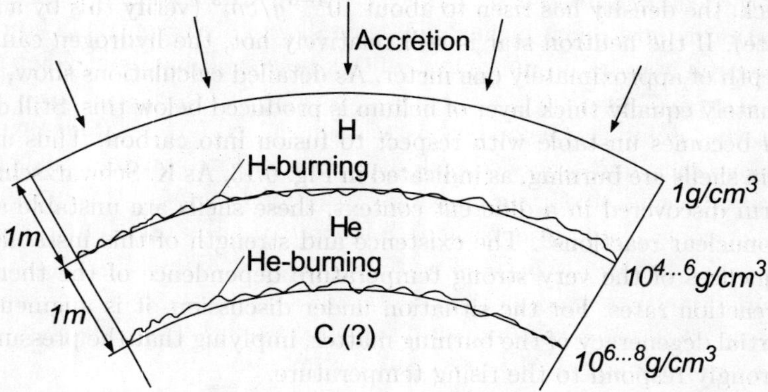

Fig. 6.13. Outer layers of an accreting weakly magnetized neutron star.

Finally, a unique source, the so-called *Rapid Burster* MXB 1730-335, should be mentioned. During its active periods, which are separated by about six months, it displays every 3-4 hours X-ray bursts of the type already described. In addition, it emits in rapid succession another kind of X-ray bursts, often a few thousands in one day. This machine-gun fire is probably due to some sort of instability in the accretion flow.

6.12.5 Formation and Evolution of Binary Systems

In this section we give a sketch of evolutionary scenarios that lead to the formation of binary X-ray sources. We begin with the formation of *massive* X-ray binaries, and describe the result of a particular computation (see Fig. 6.14).

The simulation [180] starts at time 0 with a massive pair $(14.4 + 8.0M_\odot)$ of main sequence stars in a 100 day binary, assumed to have been formed together. The more massive star evolves rapidly through hydrogen burning, generates a helium core and then within 13.3 million years expands to fill its Roche lobe. The outer layers of the bloated red giant spill over onto the companion star. In less than 100′000 years $8.5M_\odot$ are transferred onto the originally less massive star. Only a $3.5M_\odot$ helium star remains of the originally more massive component. The originally less massive star has now become a very massive main sequence star with a mass of $16.5M_\odot$. As a result of angular momentum conservation, the orbital period has increased to 416 days. Furthermore, an equatorial ring or disk around the newly massive star is formed because it is rotating rapidly. The further evolution of the helium star proceeds with an accelerating rate. (In more narrow binaries it will once more fill its Roche lobe, thus initiating a second mass transfer.) After 1.7 million years, the core becomes unstable and collapses. A portion of the outer shell may be driven off in a supernova explosion. The system will probably not be disrupted because the secondary has now a much larger mass, but the separation is now much bigger and generally highly eccentric. The remnant is a neutron star (radio pulsar) or a black hole orbiting about what is called a **Be star**.

At this point the secondary is still unevolved. At a later stage it will also leave the main sequence and after about 10 million years it becomes a supergiant with a strong stellar wind. Accretion of some of this matter by the compact companion turns the latter into a strong X-ray source (high mass X-ray binary). This stage lasts, however, less than 100 000 years, after which the supergiant has filled its Roche lobe and extinguishes the X-ray source by excessive accretion. The compact companion can accept only a small fraction of the out-flowing gas. The rest will be lost from the system. At the end of this common envelope and mass loss stage, there remains a close binary system of P_b =2.6 hours, consisting of a neutron star and a $4M_\odot$ helium star. The latter evolves rapidly. Roche lobe overflow transfers angular momentum to the neutron star turns it into a rapidly rotating (recycled) pulsar. Finally, there will be a second supernova explosion and a system like the Hulse-Taylor pulsar may come into being.

The origin of low-mass X-ray binaries is less clear. One problem is that a supernova explosion in a system with a low-mass partner would in most cases be disrupted. Perhaps the more massive of the two stars first becomes a white dwarf, and only in a later accretion process it may reach the critical mass limit, collapsing to a neutron star without disrupting the binary system. However, other scenarios are possible. Some low mass X-ray binaries are presumably the progenitors of millisecond pulsars to which we now turn.

6.12.6 Millisecond Pulsars

The first millisecond pulsar (1937+21) was discovered in 1982 with a rotational period of 1.558 *ms*. This pulsar was very mysterious, because it has no companion which could have spun the neutron star. Its estimated magnetic field (from the decrease of the pulsar period due to magnetic dipole radiation) is 10 000 times smaller than that of typical pulsars before this discovery.

Later, the discovery of the eclipsing millisecond pulsar 1957+20 suggested a plausible mechanism for the formation of a single millisecond pulsar like 1937+21. The pulsar has a companion of only $\sim 10^{-2}M_\odot$ which is most probably evaporating through the action of a strong pulsar energy flux.

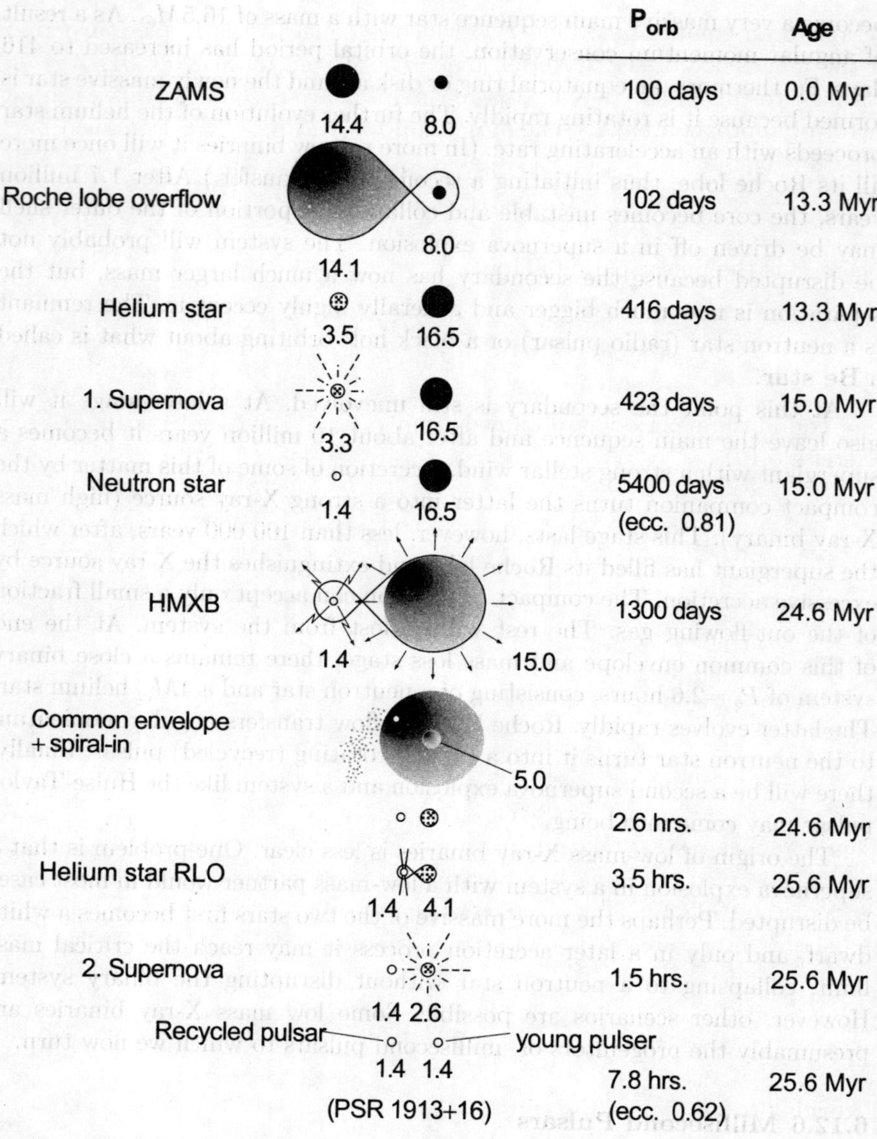

Fig. 6.14. Main evolutionary stages in the formation of a high-mass X-ray binary (HMXB) and finally of a double neutron star system (binary pulsar). (From [180].)

As mentioned before, a neutron star in a binary can be spun-up during an accretion phase since the in-falling matter has angular momentum. The population of millisecond pulsars are such recycled pulsars. At the time of writing one knows more than 40 binary millisecond pulsars, and this number will rapidly increase. An accreting neutron star will adjust its rotational period to an equilibrium value P_{eq}, which is equal to the Keplerian period of the inner edge of the accretion disk. The Alfvén radius (6.139) gives an estimate to this inner edge. (A more careful analysis confirms this within a factor of approximately 2.) The **synchronous radius** r_{synch} is defined by

$$\Omega_{Kepler}(r_{synch}) = \frac{2\pi}{P} ,$$

where P is the pulsar period. Thus,

$$r_{synch} = (1.5 \times 10^8 cm)\left(\frac{M}{M_\odot}\right)^{1/3}\left(\frac{P}{1s}\right)^{2/3} .$$

The equilibrium period is obtained from $r_{synch} = r_A$ and one finds (see Exercise 6.10)

$$P_{eq} \simeq 1.5ms\left(\frac{B}{10^9 G}\right)^{6/7}\left(\frac{R_{NS}}{10^6 cm}\right)^{15/7}\left(\frac{M_{NS}}{1.4 M_\odot}\right)^{-5/7}\left(\frac{\dot{M}}{\dot{M}_{Edd}}\right)^{-3/7} , \tag{6.142}$$

where $\dot{M}_{Edd}$ is the accretion rate corresponding to the Eddington luminosity.

The shortest spin period that can be reached is obtained for $\dot{M} = \dot{M}_{Edd}$ (the largest possible accretion rate). For canonical values of M_{NS} and R_{NS} we get

$$P_{min} \simeq 1.5ms\left(\frac{B}{10^9 G}\right)^{6/7} . \tag{6.143}$$

This defines the so-called **spin-up line** in the B versus P diagram for pulsars.

The observed phenomena in binary systems are exceedingly rich and were most of the time not foreseen. For more information we refer the reader to the quoted literature. Useful additional references are [182] and [181].

* * *

Exercise 6.10. Derive Eq. (6.142).

7 Black Holes

"In my entire scientific life (...) the most shattering experience has been the realization that an exact solution of Einstein's equations of general relativity, discovered by the New Zeeland mathematician Roy Kerr, provides the absolutely exact representation of untold numbers of massive black holes that populate the Universe."

—S.Chandrasekhar (1975)

Introduction

All stars rotate more or less rapidly. When a horizon is formed during gravitational collapse, a Schwarzschild black hole is thus never produced. One expects, however, that the horizon will quickly settle down to a stationary state as a result of the emission of gravitational waves. The geometry of the stationary black hole is of course no longer spherically symmetric.

It is remarkable that we know all stationary black hole solutions of Einstein's vacuum equations. Surprisingly, they are fully characterized by just two parameters, namely the mass and angular momentum of the hole. These quantities can be determined, in principle, by distant observers.

Thus when matter disappears behind a horizon, an exterior observer sees almost nothing of its individual properties. One can no longer say for example how many baryons formed the black hole. A huge amount of information is thus lost. The mass and angular momentum completely determine the external field, which is known analytically (Kerr solution). This led J.A. Wheeler to say "A black hole has no hair", and the previous statement is now known as the the **no-hair-theorem**.

The proof of this fact is an outstanding contribution of mathematical physics, and was completed only in the course of a number of years by various authors (W. Israel, B. Carter, S. Hawking and Robinson). A first decisive step was made by W. Israel (see [188]) who was able to show that a *static* black hole solution of Einstein's vacuum equation has to be *spherically symmetric* and, therefore, agree with the Schwarzschild solution. We shall give Israel's proof in Sect. 7.1. In a second paper Israel extended this result to black hole solutions of the coupled Einstein-Maxwell system. The Reissner-Nordstrøm 2-parameter family (see Sect. 3.9) turned out to exhaust all static electrovac black holes. It was then conjuctered by Israel, Penrose and Wheeler that in the stationary case the electrovac black holes should all be given by the 3-parameter Kerr-Newman family, discussed later in this chapter. After a number of steps by various authors this conjecture could finally be proven. For a very readable text book presentation of the proof we refer to [189].

With this black hole uniqueness theorem it was natural to conjecture further generalizations for more complicated matter models, for instance for the Einstein-Yang-Mills system. Surprisingly, it turned out that generalized no-hair conjectures do not always hold (for reviews, see [190] and [191]). While this development led to a whole sequence of surprises and remarkable insights, it is hardly of any astrophysical significance. For this reason we leave it with that.

7.1 Proof of Israel's Theorem

We give here Israel's original demonstration of the statement that a static black hole solution of Einstein's vacuum equation has to be spherically symmetric, although there are more recent proofs that need less 'technical assumptions'. The main reason is that Israel's argument is both ingenious and entirely elementary, while more modern proofs make use of the positive energy theorem (to be proven in the next chapter).

Since the spacetime $(M,\ {}^{(4)}g)$ is assumed to be static outside the horizon, it has (locally) the form (see Sect. 1.9)

$$M = \mathbb{R} \times \Sigma\,, \qquad {}^{(4)}g = -S^2 dt^2 + g\,, \tag{7.1}$$

where g is a Riemannian metric on Σ and S a smooth (lapse) function on Σ. We assume that (7.1) holds globally in the exterior spacetime region of the black hole. The Killing field k with respect to which $(M,\ {}^{(4)}g)$ is static is $k = \partial_t$. Note also that $S^2 = -\langle k, k\rangle$.

We recall some of the formulas derived in the solution of Exercise 2.4 in Sect. 2.1. Relative to an adapted orthonormal tetrad $\{\theta^\mu\}$, with $\theta^0 = S\,dt$ and $\{\theta^i\}$ an orthonormal triad of (Σ, g), we have (in the present notation)

$$^{(4)}R^i{}_{jkl} = R^i{}_{jkl}\,, \quad ^{(4)}R^0{}_{ij0} = \frac{1}{S}\,S_{|ij}\,, \quad ^{(4)}R^0{}_{ijk} = 0\,, \tag{7.2a}$$

$$^{(4)}R_{00} = \frac{1}{S}\,\Delta S\,, \qquad ^{(4)}R_{0i} = 0\,, \tag{7.2b}$$

$$^{(4)}R_{ij} = -\frac{1}{S}\,S_{|ij} + R_{ij}\,, \tag{7.2c}$$

$$^{(4)}R = -\frac{2}{S^2}\,\Delta S + R\,, \tag{7.2d}$$

$$^{(4)}G_{ij} = G_{ij} - \frac{1}{S}\left(S_{|ij} - g_{ij}\Delta S\right)\,, \tag{7.2e}$$

$$^{(4)}G_{i0} = 0\,, \tag{7.2f}$$

$$^{(4)}G_{00} = \frac{1}{2}\,R\,. \tag{7.2g}$$

(A stroke denotes the covariant derivative of (Σ, g), and Δ the corresponding Laplace-Beltrami operator.) Note that (7.2g) is a special case of Gauss' equation (see DG, (A.21)).

Foliation of Σ, Ricci Tensor, etc.

As one of the 'technical assumptions' we require that the lapse function S has no critical points: $dS \neq 0$ on Σ. Then

$$\rho := \langle dS, dS\rangle^{-1/2} > 0\,, \tag{7.3}$$

and Σ is foliated by the leaves $\{S = \text{const}\}$.

We introduce in Σ adapted coordinates. Consider for every point $p \in \Sigma$ the 1-dimensional subspace of $T_p\Sigma$ perpendicular to the tangent space of the leave $S = $ const through p. This 1-dimensional distribution is, of course, involutive in the sense of DG, Definition C.3. As a special case of the Frobenius theorem (see DG, Proposition C.4) we can introduce coordinates $\{x^i\}$ in Σ, such that x^A $(A = 2, 3)$ are constant along the integral curves of the distribution. Moreover, for x^1 we can choose S and obtain the parameterization

$$g = \rho^2 dS^2 + \tilde{g}\,, \qquad \tilde{g} = \tilde{g}_{AB}\,dx^A dx^B \quad (A, B = 2, 3)\,, \tag{7.4}$$

where ρ and $\tilde{g}_{AB}$ depend in general on all three coordinates $x^1 = S$ and x^A.

Next, we express the quantities appearing on the right of (7.2a)–(7.2g) by objects which belong to the foliation of Σ introduced above. One of these is the second fundamental form K_{AB} of the leaves, given by the Weingarten equation (see DG, (A.11)):

$$K_{AB} = \langle N, \nabla_{\partial_A}\partial_B\rangle\,. \tag{7.5}$$

Here, N is the normalized normal vector $N = \frac{1}{\rho}\,\partial_1$. (In this subsection we work relative to the coordinate basis $\{\partial_i\}$.) From (7.5) we get

$$K_{AB} = \frac{1}{\rho}\langle \partial_1, \Gamma^i{}_{AB}\partial_i\rangle = \frac{1}{\rho}\Gamma^1{}_{AB}\langle \partial_1, \partial_1\rangle ,$$

i.e.,

$$K_{AB} = \rho\, \Gamma^1{}_{AB} . \tag{7.6}$$

From

$$\Gamma^i{}_{ab} = \frac{1}{2} g^{ij}\left(g_{aj,b} + g_{bj,a} - g_{ab,j}\right)$$

we obtain, in particular

$$\Gamma^1{}_{AB} = -\frac{1}{2\rho^2}\,\tilde{g}_{AB,1} , \tag{7.7}$$

so that

$$K_{AB} = -\frac{1}{2\rho}\,\tilde{g}_{AB,1} . \tag{7.8}$$

Next we write the equations of Gauss and Codazzi-Mainardi[1] in the form DG, (A.19) and (A.20):

$$\rho^{-2}G_{11} = -\frac{1}{2}\tilde{R} + \frac{1}{2}(K^2 - K_{AB}K^{AB}) , \tag{7.9a}$$

$$\rho^{-1}G_{1A} = \rho^{-1}R_{1A} = K_{,A} - \tilde{\nabla}_B K^B_A , \tag{7.9b}$$

where $K = K^A_A$ and $\tilde{\nabla}$ denotes the covariant derivative belonging to $\tilde{g}$. – We could now proceed by adapting the results of the 3+1 formalism in Sect. 2.9 to the present situation. Instead, we proceed with explicit calculations.

A short computation gives for the Christoffel symbols

$$\Gamma^1{}_{AB} = \frac{1}{\rho}K_{AB} , \quad \Gamma^A{}_{BC} = \tilde{\Gamma}^A{}_{BC} ,$$

$$\Gamma^A{}_{11} = -\tilde{g}^{AB}\rho\,\rho_{,B} , \quad \Gamma^1{}_{11} = \frac{1}{\rho}\rho_{,1} ,$$

$$\Gamma^1{}_{1A} = \frac{1}{\rho}\rho_{,A} , \qquad \Gamma^A{}_{B1} = -\rho K^A_B . \tag{7.10}$$

The calculation of the Ricci tensor

$$R_{ij} = \partial_l\Gamma^l{}_{ij} - \partial_j\Gamma^l{}_{li} + \Gamma^s{}_{ij}\Gamma^l{}_{ls} - \Gamma^s{}_{lj}\Gamma^l{}_{is} \tag{7.11}$$

is now straightforward. For R_{AB} we have

[1] In DG, Appendix A these were derived for Lorentz manifolds. Check that the signs in (7.9a) and (7.9b) for the Riemannian case (with the sign convention for K_{AB} adopted in (7.5)).

$$\begin{aligned}R_{AB} &= \tilde{R}_{AB} + \partial_1 \Gamma^1{}_{AB} - \partial_B \Gamma^1{}_{1A} + \Gamma^1{}_{AB}\Gamma^l{}_{l1} + \Gamma^C{}_{AB}\Gamma^1{}_{1C} \\ &\quad - \Gamma^C{}_{1B}\Gamma^1{}_{AC} - \Gamma^1{}_{lB}\Gamma^l{}_{A1} \\ &= \tilde{R}_{AB} + (\rho^{-1}K_{AB})_{,1} - (\rho^{-1}\rho_{,A})_{,B} + \rho^{-1}K_{AB}(\rho^{-1}\rho_{,1} + \rho K) \\ &\quad + \tilde{\Gamma}^C_{AB}\,\rho^{-1}\,\rho_{,C} + \rho K^C_B\,\rho^{-1}K_{AC} - \rho^{-1}\rho_{,B}\,\rho^{-1}\rho_{,B} \\ &\quad + \rho^{-1}K_{BC}\,\rho K^C_A\,.\end{aligned}$$

(Note that $\Gamma^l{}_{l1} = \Gamma^1{}_{11} + \Gamma^C{}_{C1}$.) Several terms compensate and we find

$$R_{AB} = \tilde{R}_{AB} - \frac{1}{\rho}\,\rho_{;AB} + \frac{1}{\rho}\,K_{AB,1} - K\,K_{AB} + 2\,K_{AC}K^C_B\,. \tag{7.12}$$

This can be written somewhat differently. Consider for this

$$-\frac{1}{\rho}\,\tilde{g}_{AC}K^C_{B,1} = -\left(\frac{1}{\rho}\,K_{AB}\right)_{,1} + \left(\frac{1}{\rho}\,\tilde{g}_{AC}\right)_{,1} K^C_B\,,$$

and use in the last term

$$\left(\frac{1}{\rho}\,\tilde{g}_{AC}\right)_{,1} = \frac{1}{\rho}\,\tilde{g}_{AC,1} - \frac{1}{\rho^2}\,\rho_{,1}\,\tilde{g}_{AC} = -2\,K_{AC} - \frac{1}{\rho^2}\,\rho_{,1}\,\tilde{g}_{AC}$$

to get

$$2\,K_{AC}K^C_B + \frac{1}{\rho}\,K_{AB,1} = \frac{1}{\rho}\,\tilde{g}_{AC}K^C_{B,1}\,. \tag{7.13}$$

If this is inserted in (7.12) we obtain

$$R_{AB} = \tilde{R}_{AB} - \frac{1}{\rho}\,\tilde{\nabla}_A\tilde{\nabla}_B\,\rho - K\,K_{AB} + \frac{1}{\rho}\,\tilde{g}_{AC}K^C_{B,1}\,. \tag{7.14}$$

(This is the translation of (2.301) to the present situation.)

With these results it is easy to compute R_{11}. From

$$G_{11} = R_{11} - \frac{1}{2}\,g_{11}(R^1_1 + R^A_A) = \frac{1}{2}R_{11} - \frac{1}{2}\,\rho^2 R^A_A$$

we get, using the Gauss equation (7.9a) and (7.14),

$$R_{11} = -\rho(\tilde{\Delta}\rho - K_{,1} + \rho\,K_{AB}K^{AB})\,. \tag{7.15}$$

(This is the translation of (2.298)). From (7.14) and (7.15) we obtain for the scalar curvature

$$R = R^1_1 + R^A_A = \tilde{R} - (K^2 + K_{AB}K^{AB}) - \frac{2}{\rho}(\tilde{\Delta}\rho - K_{,1})\,. \tag{7.16}$$

In some of the equations (7.2a)–(7.2g) we also need ΔS and $S_{|ij}$. Specializing the general formula

$$S_{|ij} = S_{,ij} - \Gamma^l{}_{ij} S_{,l}$$

to $S = x^1$ ($S_{,1} = \delta^1_i$, $S_{,ij} = 0$), we get $S_{|ij} = -\Gamma^1{}_{ij}$ and hence from (7.10)

$$S_{|11} = -\frac{1}{\rho}\,\rho_{,1}\,, \quad S_{|1A} = -\frac{1}{\rho}\,\rho_{,A}\,, \quad S_{|AB} = -\frac{1}{\rho}\,K_{AB}\,. \tag{7.17}$$

This gives

$$\Delta S = -\frac{1}{\rho}\left(K + \frac{\rho_{,1}}{\rho^2}\right). \tag{7.18}$$

In (7.2e) we need also G_{AB}. (Note that G_{11} is given by (7.9a).) This is readily obtained from (7.14) and (7.16):

$$\begin{aligned} G_{AB} &= R_{AB} - \frac{1}{2}\,\tilde{g}_{AB}\,R \\ &= \tilde{G}_{AB} + \frac{1}{\rho}\left(\tilde{g}_{AB}\tilde{\Delta}\rho - \tilde{\nabla}_A\tilde{\nabla}_B\,\rho\right) - \frac{1}{\rho}\left(\tilde{g}_{AB}K_{,1} - \tilde{g}_{AC}K^C{}_{B,1}\right) \\ &\quad - K\,K_{AB} + \frac{1}{2}\tilde{g}_{AB}\left(K^2 + K_{AB}K^{AB}\right). \end{aligned} \tag{7.19}$$

Now we have all ingredients to express the quantities appearing on the right of (7.2a)–(7.2g) by objects which belong to the foliation of Σ. In particular we can write (7.2b) with the help of (7.18) as

$${}^{(4)}R_{00} = -\frac{1}{\rho\,S}\left(K + \frac{\rho_{,1}}{\rho^2}\right). \tag{7.20}$$

The Invariant ${}^{(4)}R_{\alpha\beta\gamma\delta}\,{}^{(4)}R^{\alpha\beta\gamma\delta}$

This invariant will play a certain role in the subsequent proof of Israel's theorem. From (7.2a) we find for this

$$\begin{aligned} \frac{1}{8}\,{}^{(4)}R_{\alpha\beta\gamma\delta}\,{}^{(4)}R^{\alpha\beta\gamma\delta} &= \frac{1}{8}\,R_{ijkl}R^{ijkl} + \frac{1}{2}\,{}^{(4)}R_{0i0j}\,{}^{(4)}R^{0i0j} \\ &= \frac{1}{8}\,R_{ijkl}R^{ijkl} + \frac{1}{2}\,S^{-2}S_{|ij}S^{|ij}\,. \end{aligned}$$

Now, a short calculation shows that the following identity holds in three dimensions (exercise)

$$\frac{1}{8}\,R_{ijkl}R^{ijkl} = \frac{1}{2}\,G_{ij}G^{ij}\,.$$

Thus we obtain

$$J := \frac{1}{8}\,{}^{(4)}R_{\alpha\beta\gamma\delta}\,{}^{(4)}R^{\alpha\beta\gamma\delta} = \frac{1}{2}\,S^{-2}S_{|ij}S^{|ij} + \frac{1}{2}\,G_{ij}G^{ij}\,. \tag{7.21}$$

We could insert here the expressions for G_{ik}. Instead, we want to use the vacuum equations ${}^{(4)}R_{\mu\nu} = 0$. According to (7.2a) these imply that S is harmonic, $\Delta S = 0$, and thus (7.2e) becomes $G_{ij} = \frac{1}{S} S_{|ij}$, whence

$$J = \frac{1}{S^2} S_{|ij} S^{|ij} . \tag{7.22}$$

Here we use the expressions (7.17). Since for a harmonic S Eq. (7.18) reduces to $\rho_{,1} = -\rho^2 K$, we find

$$S_{|ij} S^{|ij} = \left(\frac{K}{\rho}\right)^2 + 2\,\frac{1}{\rho^4}\,\rho_{,A}\,\rho^{,A} + \frac{1}{\rho^2}\,K_{AB}K^{AB},$$

and thus

$$\frac{1}{8}\,{}^{(4)}R_{\alpha\beta\gamma\delta}\;{}^{(4)}R^{\alpha\beta\gamma\delta} = \frac{1}{\rho^2 S^2}\left(K^2 + K_{AB}K^{AB} + 2\,\frac{\langle\tilde\nabla\rho, \tilde\nabla\rho\rangle}{\rho^2}\right). \tag{7.23}$$

We emphasize that this holds only if the vacuum equations are imposed. We shall have to require that this invariant remains finite at the horizon. Possible singularities should be behind the horizon.

The Proof (W. Israel, 1967)

The hypersurfaces $\{S^2 = \text{const}\}$ of M are invariant under the isometric flow belonging to the Killing field k (show that the Lie derivative of S^2 with respect to k vanishes). Hence, k is tangent to these hypersurfaces. For positive constants these are thus timelike (a timelike vector cannot be orthogonal to a timelike or lightlike vector). Obviously, such hypersurfaces can be crossed in both directions. Therefore, the assumed horizon must coincide with the hypersurface $\{S = 0\}$ and this must be a null hypersurface (see DG, Definition A.2). Only then it is tangent to the light cone at each of its points, thus allowing crossing in one direction only. The null vector field k is orthogonal to this null hypersurface (see Exercise 7.5).

We also note the following about this null hypersurface. From DG, Proposition A.1 it follows that the integral curves of k, when suitable parameterized, are null geodesics. These are called **null geodesic generators** of the horizon.

With Israel we assume that $\{S^2 = -\langle k, k\rangle = \text{const}\}$ are submanifolds of Σ which are topological 2-spheres. Asymptotically $S \longrightarrow 1$, because Σ is assumed to be asymptotically flat in the sense of the definition given in Sect. 2.7 (see p. 117). In what follows the Riemannian volume form of $\{S = \text{const}\}$, belonging to $\tilde g$, is denoted by $\tilde\eta$.

First we want to indicate the main idea of Israel's proof. Using the previously developed results, we shall derive as a consequence of the vacuum field equations two differential inequalities with the following structure

$$d(f\,\tilde{\eta}) \leq dS \wedge g\,\tilde{\eta}\,, \tag{7.24}$$

where f and g are functions on Σ. (Note that, as 3-forms, both sides are proportional to the volume form η on Σ, and the inequality sign means that the proportionality factors obey the inequality.) Furthermore, the equality sign in (7.24) holds if and only if, the trace-free part of the second fundamental form K_{AB} and $\tilde{\nabla}\rho$ both vanish. We shall then integrate the two inequalities over Σ, and replace with Stokes theorem the left-hand sides by boundary integrals over the horizon and a 2-sphere S^2_∞ 'at infinity'. We will be able to work out the integrals on both sides, and conclude from the results that in both inequalities of the form (7.24) the equality sign must hold. According to what has been said, this implies

$$K_{AB} - \frac{1}{2}\,\tilde{g}_{AB}K = 0 \quad \text{and} \quad \tilde{\nabla}\rho = 0\,. \tag{7.25}$$

With this information it will be quite easy to deduce the spherical symmetry.

As the main point we now derive the announced differential inequalities. For this, let us first collect some of the previous results. By (7.1) and (7.4) the metric of the static spacetime is parameterized as

$$^{(4)}g = -S^2dt^2 + \rho^2 dS^2 + \tilde{g}_{AB}\,dx^A dx^B\,. \tag{7.26}$$

The orthonormal tetrads for which (7.2a)–(7.2g) are valid are further specified, by adapting them to (7.26): $\theta^0 = S\,dt$ (as above), $\theta^1 = \rho\,dS$ and $\{\theta^A\}$ an orthonormal 2-bein for $\tilde{g}$. In order to avoid confusions we put hats on indices which are connected to the orthonormal tetrad $\{\theta^\mu\}$, while for the indices connected to the coordinate basis we use standard notation. From (7.2c) and (7.16) we get, without using the field equations,

$$^{(4)}G_{\hat{0}\hat{0}} = \frac{1}{2}\left(\tilde{R} - (K^2 + K_{AB}K^{AB}) - \frac{2}{\rho}(\tilde{\Delta}\rho - K_{,1})\right), \tag{7.27}$$

and (7.2e) gives

$$^{(4)}G_{\hat{1}\hat{1}} = G_{\hat{1}\hat{1}} + \frac{1}{S}(\Delta S - S_{|\hat{1}\hat{1}})\,.$$

For the second term on the right we use (7.17) and (7.18) to get

$$\Delta S - S_{|\hat{1}\hat{1}} = \Delta S - \rho^{-2}S_{|11} = -\rho^{-1}K\,.$$

Inserting also the expressions (7.9a) for $G_{\hat{1}\hat{1}}$ we obtain

$$^{(4)}G_{\hat{1}\hat{1}} = -\frac{1}{2}\tilde{R} + \frac{1}{2}(K^2 - K_{AB}K^{AB}) - \frac{1}{\rho\,S}K\,. \tag{7.28}$$

Instead of (7.27) and (7.28) we consider two linear combinations. The first is

$$^{(4)}G_{\hat{0}\hat{0}} + {}^{(4)}G_{\hat{1}\hat{1}} = -K_{AB}K^{AB} + \frac{1}{\rho}\left(-\frac{K}{S} - \tilde{\Delta}\rho + K_{,S}\right).$$

It is useful to introduce at this point the trace-free part of K_{AB}:

$$\bar{K}_{AB} = K_{AB} - \frac{1}{2}\,\tilde{g}_{AB}K\,. \tag{7.29}$$

Beside $K_{AB}K^{AB} = \bar{K}_{AB}\bar{K}^{AB} + \frac{1}{2}K^2$ we also use the decomposition

$$\frac{1}{\rho}\tilde{\Delta}\rho = 2\,\frac{\tilde{\Delta}(\sqrt{\rho})}{\sqrt{\rho}} + \frac{1}{2}\frac{\langle\tilde{\nabla}\rho, \tilde{\nabla}\rho\rangle}{\rho^2},$$

and obtain the first decisive identity

$$\begin{aligned} ^{(4)}G_{\hat{0}\hat{0}} + {}^{(4)}G_{\hat{1}\hat{1}} &= \frac{1}{\rho}\left(-\frac{K}{S} + K_{,S} - \frac{1}{2}\,\rho\,K^2\right) - \frac{2}{\sqrt{\rho}}\tilde{\Delta}(\sqrt{\rho}) \\ &\quad - \left(\frac{\langle\tilde{\nabla}\rho, \tilde{\nabla}\rho\rangle}{\rho^2} + \bar{K}_{AB}\bar{K}^{AB}\right). \end{aligned} \tag{7.30}$$

It is important that the expression in the last parenthesis is ≥ 0, and vanishes if and only if the two equations in (7.25) hold.

The second useful combination is

$$^{(4)}G_{\hat{0}\hat{0}} + 3\,{}^{(4)}G_{\hat{1}\hat{1}} = \left({}^{(4)}G_{\hat{0}\hat{0}} + {}^{(4)}G_{\hat{1}\hat{1}}\right) + 2\,{}^{(4)}G_{\hat{1}\hat{1}}\,.$$

Inserting (7.30) and (7.28) we arrive at

$$\begin{aligned} ^{(4)}G_{\hat{0}\hat{0}} + {}^{(4)}G_{\hat{1}\hat{1}} &= \frac{1}{\rho}\left(-\frac{3\,K}{S} + K_{,S} - \frac{2}{\sqrt{\rho}}\tilde{\Delta}(\sqrt{\rho})\right) - \tilde{R} \\ &\quad - \left(\frac{\langle\tilde{\nabla}\rho, \tilde{\nabla}\rho\rangle}{2\,\rho^2} + 2\,\bar{K}_{AB}\bar{K}^{AB}\right) \end{aligned}$$

or

$$\begin{aligned} ^{(4)}G_{\hat{0}\hat{0}} + 3\,{}^{(4)}G_{\hat{1}\hat{1}} &= \frac{1}{\rho}\left(-\frac{3\,K}{S} + K_{,S}\right) - \tilde{R} - \tilde{\Delta}\ln\rho \\ &\quad - \left(\frac{\langle\tilde{\nabla}\rho, \tilde{\nabla}\rho\rangle}{\rho^2} + 2\,\bar{K}_{AB}\bar{K}^{AB}\right). \end{aligned} \tag{7.31}$$

Again, the expression in the last parenthesis is ≥ 0, and equality holds under the same conditions as for (7.30).

Before proceeding, we recall (7.20) in the form

$$\rho_{,S} = -\rho^2\left(K + S\,{}^{(4)}R_{\hat{0}\hat{0}}\right). \tag{7.32}$$

We shall also need

$$\frac{1}{\sqrt{\tilde{g}}}\,\partial_S\left(\sqrt{\tilde{g}}\right)=\frac{1}{2}\tilde{g}^{AB}\partial_S\tilde{g}_{AB}\overset{(7.8)}{=}-\tilde{g}^{AB}\rho\,K_{AB}=-\rho\,K\,. \tag{7.33}$$

Only at this point we now use the vacuum field equations. Then, as already emphasized, $\Delta S=0$ and (7.18) reduces to

$$\rho_{,S}=-\rho^2\,K\,. \tag{7.34}$$

Together with (7.33) we then have

$$\partial_S\left(\frac{\sqrt{\tilde{g}}}{\sqrt{\rho}}\right)=\frac{1}{2}\sqrt{\tilde{g}}\,\rho\,K\,, \tag{7.35}$$

and

$$\partial_S\left(\frac{\sqrt{\tilde{g}}}{\rho}\right)=0\,. \tag{7.36}$$

With this we can rewrite the first parenthesis in (7.30), since

$$-S\,\frac{\sqrt{\rho}}{\sqrt{\tilde{g}}}\left(\frac{\sqrt{\tilde{g}}\,K}{\sqrt{\rho}\;S}\right)_{,S}=\left(\frac{K}{S}-K_{,S}-\frac{\rho}{2}K^2\right), \tag{7.37}$$

and (7.30) is equivalent to

$$\frac{S}{\sqrt{\rho}}\frac{1}{\sqrt{\tilde{g}}}\left(\frac{\sqrt{\tilde{g}}\,K}{\sqrt{\rho}\;S}\right)_{,S}=\frac{2}{\sqrt{\rho}}\tilde{\Delta}(\sqrt{\rho})+\left(\frac{\langle\tilde{\nabla}\rho,\tilde{\nabla}\rho\rangle}{2\,\rho^2}+\bar{K}_{AB}\bar{K}^{AB}\right). \tag{7.38}$$

This implies

$$-\left(\frac{\sqrt{\tilde{g}}\,K}{\sqrt{\rho}\;S}\right)_{,S}\leq\frac{2\sqrt{\tilde{g}}}{S}\tilde{\Delta}(\sqrt{\rho})\,, \tag{7.39}$$

whereby the equality sign holds if and only if the conditions in (7.25) holds.

We also try to deduce from (7.31) a similar inequality. For the first parenthesis on the right use

$$\begin{aligned}\left[\frac{\sqrt{\tilde{g}}}{\rho}\left(-K\,S+\frac{4}{\rho}\right)\right]_{,S}&=\frac{\sqrt{\tilde{g}}}{S}\left(-S\,K_{,S}-K-\frac{4}{\rho^2}\,\rho_{,S}\right)\\&=\frac{S\sqrt{\tilde{g}}}{\rho}\left(-K_{,S}+\frac{3\,K}{S}\right).\end{aligned} \tag{7.40}$$

(Note that $\rho_{,S}=-\rho^2K$.) Then we deduce from (7.29)

$$\left[\frac{\sqrt{\tilde{g}}}{\rho}\left(-K\,S+\frac{4}{\rho}\right)\right]_{,S}\leq-S\sqrt{\tilde{g}}\left(\tilde{\Delta}(\ln\rho)+\tilde{R}\right), \tag{7.41}$$

with the same conditions for the equality sign as in (7.39).

Now we bring the inequalities (7.39) and (7.41) into the desired form (7.24). For this we multiply (7.39) with $dS \wedge dx^2 \wedge dx^3$ to obtain

$$-d\left(\frac{K}{\sqrt{\rho}S}\,\tilde{\eta}\right) \le -\frac{2}{S}\tilde{\Delta}(\sqrt{\rho})\,dS \wedge \tilde{\eta}\,. \tag{7.42}$$

In the same manner we deduce from (7.41)

$$d\left(\frac{-K+4/\rho}{\rho}\,\tilde{\eta}\right) \le -S\,dS \wedge \left(\tilde{\Delta}(\ln\rho) + \tilde{R}\right)\tilde{\eta}\,. \tag{7.43}$$

Integration of (7.24) over the manifold Σ, which is bounded by the horizon $H = \{S = 0\}$ and the 2-sphere S^2_∞, gives

$$\int_{S^2_\infty} f\,\tilde{\eta} - \int_H f\,\tilde{\eta} \le \int_0^1 dS \int_{\{S=\mathrm{const}\}} g\,\tilde{\eta}\,. \tag{7.44}$$

Let us first work out the right-hand side for (7.42) and (7.43). For (7.42) Gauss' theorem implies

$$-\int_0^1 \frac{2}{S}\,dS \int_{\{S=\mathrm{const}\}} \tilde{\Delta}(\sqrt{\rho})\,\tilde{\eta} = 0\,.$$

For (7.43) we can use in addition the Gauss-Bonnet theorem (see Sect. 2.6)

$$\int_{\{S=\mathrm{const}\}} \tilde{R}\,\tilde{\eta} = 8\,\pi\,,$$

whence

$$-\int_0^1 S\,dS \int_{\{S=\mathrm{const}\}} \left(\tilde{\Delta}(\ln\rho) + \tilde{R}\right)\tilde{\eta} = -4\,\pi\,.$$

In summary, we arrive at

$$\left[\int f\,\tilde{\eta}\right]_{S=0}^{S=1} \le \begin{cases} 0 \\ -4\,\pi \end{cases} \tag{7.45}$$

for

$$f = \begin{cases} -\dfrac{K}{\sqrt{\rho}S}\,, \\[2ex] \dfrac{-K\,S + 4/\rho}{\rho}\,. \end{cases} \tag{7.46}$$

We need in (7.45) information about f at the horizon and at infinity. The general behavior of a stationary field far away has been discussed in Sect. 5.3. For almost Euclidean coordinates on Σ we obtain from (5.55)

$$S^2 = 1 - \frac{2M}{r} + O\left(\frac{1}{r^2}\right),$$

$$g_{ij} = \left(1 + \frac{2M}{r}\right)\delta_{ij} + O\left(\frac{1}{r^2}\right), \tag{7.47}$$

where $r^2 = \sum_{i=1}^{3}(x^i)^2$, and M is the total mass. From the definition of ρ, $\rho^{-2} = \langle dS, dS\rangle$, and (7.34) then follows

$$\rho^{-1} \simeq \frac{M}{r^2}, \qquad K \simeq -\frac{2}{r} \quad \text{for } r \longrightarrow \infty. \tag{7.48}$$

For the behavior at the horizon H we conclude from (7.23) that

$$K_{AB} = 0 \qquad \text{and} \qquad \tilde{\nabla}\rho = 0 \quad \text{on } H. \tag{7.49}$$

The second condition implies that $\rho_H := \rho|_H = \text{const.}$ (We shall see later in Sect. 7.5.1 that ρ_H^{-1} is the so-called surface gravity.) We use (7.49) in (7.28) and get

$$-\frac{K}{S} = \frac{1}{2}\rho\tilde{R} \qquad \text{on } H. \tag{7.50}$$

For the left-hand side of (7.45) we obtain with this information for the first case, using once more the Gauss-Bonnet theorem,

$$\begin{aligned}\int_{\{S=1\}} -\frac{K}{\sqrt{\rho}}\tilde{\eta} - \int_H \frac{1}{2}\sqrt{\rho_H}\,\tilde{R}\tilde{\eta} &= \lim_{r\to\infty}\int_{S^2}\frac{2}{r}\frac{\sqrt{M}}{r}r^2\,d\Omega - \frac{1}{2}\sqrt{\rho_H}\int_H \tilde{R}\tilde{\eta} \\ &= 8\pi\sqrt{M} - 4\pi\sqrt{\rho_H}.\end{aligned}$$

Hence the first inequality (7.45) reduces to the very simple statement

$$M \leq \frac{1}{4}\rho_H. \tag{7.51}$$

For the second inequality the contribution at infinity vanishes, and the left-hand side is (remember $\rho_H = \text{const}$) equal to $-(4/\rho_H^2)\mathcal{A}$, where $\mathcal{A}$ is the surface of the horizon

$$\mathcal{A} = \int_H \tilde{\eta}. \tag{7.52}$$

Hence we obtain

$$\frac{1}{4\pi}\rho_H^{-1}\mathcal{A} \geq \frac{1}{4}\rho_H. \tag{7.53}$$

Further below we show that the left-hand side of this equation is equal to M, whence the two inequalities (7.51) and (7.53) are *only compatible for the equality sign in both of them.* Therefore, as already announced, the equations (7.25) must be satisfied on Σ. In particular, ρ is only a function of S.

As a final step we show that these facts imply the spherical symmetry. For this we write (7.28) for ${}^{(4)}G_{\hat{1}\hat{1}} = 0$ in the form

$$\tilde{R} = \frac{1}{2}K^2 - \frac{2\,K}{\rho\,S} \tag{7.54}$$

(we made use of $\bar{K}_{AB} = 0$.) According to (7.34) the second fundamental form K is also only a function of S, whence the last equation implies that this also holds for $\tilde{R}$. Therefore, the surfaces $\{S = \text{const}\}$ have constant curvature and are thus metrical 2-spheres.

We still have to show that

$$M = \frac{1}{4\pi\,\rho_H}\mathcal{A}\,. \tag{7.55}$$

We know from (4.222) that M is given by the Komar integral

$$M = -\frac{1}{8\pi}\int_{S^2_\infty} *dk\,. \tag{7.56}$$

Now Stokes' theorem implies, using the vacuum equations,

$$\int_{S^2_\infty} *dk - \int_H *dk = \int_\Sigma d*dk = 0\,, \tag{7.57}$$

because of (4.225),

$$d*dk = 2*R(k)\,, \tag{7.58}$$

where $R(k)$ is the **Ricci form** with components $R_{\mu\nu}k^\nu$. From (7.56) and (7.57) we conclude that

$$M = -\frac{1}{8\pi}\int_H *dk\,. \tag{7.59}$$

In our case we have $k = -S^2\,dt$, $dk = -2S\,dS\wedge dt = \frac{2}{\rho}\,\theta^0\wedge\theta^1$, where $\theta^0 = S\,dt$ and $\theta^1 = \rho\,dS$. Hence

$$*dk = -\frac{2}{\rho}\,\tilde{\eta}\,,$$

and therefore

$$M = \frac{1}{4\pi}\int_H \rho_H^{-1}\,\tilde{\eta} = \frac{1}{4\pi}\rho_H^{-1}\int_H \tilde{\eta} = \frac{1}{4\pi}\rho_H^{-1}\mathcal{A}\,.$$

This completes the proof of Israel's theorem.

With this we are done. For completeness we want to show how the derived formulas imply the Schwarzschild solution. Denoting derivatives with respect to S by a dash, Eq. (7.34) reads

$$\rho' = -K\,\rho^2\,, \tag{7.60}$$

and from (7.8) we get $\tilde{g}'_{AB} = -2\rho\,K_{AB} = -\rho\,K\tilde{g}_{AB}$. We write this in terms of the Schwarzschild radial coordinate r, defined such that $\tilde{g} = r^2\,d\Omega$ with $d\Omega$ the standard metric of S^2: $(r^2)' = -\rho\,r^2 K$, i.e.,

$$K = -\frac{2\,r'}{\rho\, r}\,. \tag{7.61}$$

The last two equations imply $(r^2\,\rho^{-1})' = 0$, thus

$$\rho = \frac{r^2}{M}\,. \tag{7.62}$$

Inserting (7.61) and (7.62), as well as $\tilde{R} = 2/r^2$ into (7.54) gives the differential equation

$$\left(\frac{r'}{r}\right)^2 = \frac{r^2}{M} - \frac{2}{S}\frac{r'}{r}\,. \tag{7.63}$$

The solution of this is

$$r(S) = \frac{2\,M}{1 - S^2}\,. \tag{7.64}$$

Therefore we have $S^{-2}dr^2 = \rho^2 dS^2$, hence we obtain

$$\begin{aligned} {}^{(4)}g &= -S^2 dt^2 + S^{-2} dr^2 + r^2 d\Omega\,, \\ S^2 &= 1 - \frac{2\,M}{r}\,, \end{aligned} \tag{7.65}$$

and this is the Schwarzschild solution for $r > 2\,M$ (see (3.12)).

7.2 Derivation of the Kerr Solution

The Kerr solution, which describes stationary rotating black holes, ranks among the most important solutions of Einstein's (vacuum) equations. Originally, it was found in 1963 more or less by accident (see [192]) and its physical meaning was not recognized until later. E.T. Newman and coworkers showed in 1965 that the Kerr solution can be guessed formally from the Schwarzschild solution by means of a complex coordinate transformation (see [193]). (In [194] this was generalized to electrically charged black holes.) The Kerr (-Newman) solution was also obtained in a systematic study of algebraically degenerated solutions of the Einstein (-Maxwell) equations (see [195]). Unfortunately, there is still no physically natural route leading to the Kerr (-Newman) solution.

In this section we shall find the Kerr solution in what we regard as the most natural way. In the first six subsections we shall see that the problem of finding *stationary* and *axisymmetric* solutions of Einstein's vacuum equations can be completely reduced to a non-linear partial differential equation for a complex potential, defined in a certain region of the plane. This so-called **Ernst equation** for the complex potential $\varepsilon(x, y)$ reads as follows:

$$\left[(x^2-1)\varepsilon_{,x}\right]_{,x} + \left[(1-y^2)\varepsilon_{,y}\right]_{,y} = -\frac{2\,\bar{\varepsilon}}{1-\varepsilon\,\bar{\varepsilon}}\left[(x^2-1)\varepsilon_{,x}^2 + (1-y^2)\varepsilon_{,y}^2\right]\,. \tag{7.66}$$

This is the Euler-Lagrange equation for the Lagrange density

$$\mathcal{L} = \frac{(x^2-1)\varepsilon_{,x}\bar{\varepsilon}_{,x} + (1-y^2)\varepsilon_{,y}\bar{\varepsilon}_{,y}}{(1-\varepsilon\bar{\varepsilon})^2}. \tag{7.67}$$

So far the things are quite straightforward. What is, however, really miraculous is that the simplest non-trivial solution of the Ernst equation, namely

$$\varepsilon(x,y) = p\,x + q\,y\,, \tag{7.68}$$

where $p, q \in \mathbb{R}$ with $p^2+q^2=1$, corresponds to the Kerr black hole. Exactly this 2-parameter family of stationary axisymmetric vacuum solutions has event horizons, as is established by a uniqueness theorem. A priori there is, however, no reason why this should be so. We learn this only with hindsight through a rather involved argument. A basic reason for why there is, presumably, no more direct way to arrive at this result is connected with the fact that an event horizon is a *global* concept. It is, therefore, impossible to impose the existence of a horizon in any local manner.

The details of what follows are somewhat lengthly. We recommend to first take a glance at the line of reasoning, before going into the details. Section 7.2.1 is of more general significance. There we shall arrive at a very convenient parameterization of stationary axisymmetric metrics which is, for instance, also the starting point for the study of rotating neutron star models (see Sect. 6.10).

7.2.1 Axisymmetric Stationary Spacetimes

A spacetime (M,g) is **axisymmetric** if it admits the group $SO(2)$ as an isometry group such that the group orbits are closed spacelike curves. In what follows we consider only asymptotically flat spacetimes. The spacetime (M,g) is **stationary** and **axisymmetric**, if $\mathbb{R}\times SO(2)$ acts isometrically, such that (M,g) is axisymmetric with respect to the subgroup $SO(2)$, and the Killing field belonging to $\mathbb{R}$ (time translations) is at least asymptotically timelike. We allow, however, "ergospheres", where this Killing field becomes spacelike.

The two Killing fields belonging to $\mathbb{R}$ and $SO(2)$ will be denoted by k, respectively m. These are, of course, commuting

$$[k,m] = 0\,. \tag{7.69}$$

The orbits of the $\mathbb{R}\times SO(2)$ action are 2-dimensional submanifolds, whose tangent spaces are spanned by k and m. The collection of these tangent space define in the terminology of DG, Appendix C an involutive (integrable) 2-distribution E. Beside E we consider the orthogonal distribution $E^\perp$ and make the generic assumption that $E\cap E^\perp = \{0\}$. Then E and $E^\perp$ are complementary: $T_pM = E_p \oplus E_p^\perp$, $p\in M$. The distribution $E^\perp$ is annihilated by the 1-forms $k^\flat$ and $m^\flat$. The Frobenius theorem (proven in DG, Appendix C) tells us that $E^\perp$ is also involutive if and only if the ideal generated by $k^\flat$ and $m^\flat$ is differential. This in turn is, by the same theorem, equivalent to the **Frobenius conditions**

$$k \wedge m \wedge dk = 0\,, \qquad k \wedge m \wedge dm = 0\,. \tag{7.70}$$

(From now on we drop the symbol $\flat$.) We shall soon see that Einstein' vacuum equations imply the Frobenius conditions (7.70), so $E^\perp$ is integrable. This is also true for certain matter models, as we shall see.

If the conditions (7.70) are satisfied we say that (M, g) is **circular**. Then there exist, as shown in DG, Appendix C, adapted coordinates x^a $(a = 0, 1)$, x^A $(A = 2, 3)$, such that

$$k = \partial_t\,, \quad m = \partial_\varphi \qquad (x^0 = t\,, x^1 = \varphi)\,, \tag{7.71}$$

and

$$g = g_{ab}(x^C)dx^a dx^b + g_{AB}(x^C)dx^A dx^B\,. \tag{7.72}$$

(The $g_{\mu\nu}$ depend only on x^A $(A = 2, 3)$.)

Ricci Circularity

We now show that the Frobenius conditions are equivalent to the **Ricci circularity** properties for the Ricci forms introduced in (7.58):

$$k \wedge m \wedge R(k) = 0\,, \quad k \wedge m \wedge R(m) = 0\,. \tag{7.73}$$

To show this, we first rewrite (7.70) in terms of the **twist forms** belonging to k and m:

$$\omega_k = \frac{1}{2} * (k \wedge dk)\,, \qquad \omega_k = \frac{1}{2} * (m \wedge dm)\,. \tag{7.74}$$

For example, the first 4-form in (7.70) is proportional to $m \wedge *\omega_k = \langle m, \omega_k \rangle\, \eta$ with η the volume form of (M, g). Thus the Frobenius conditions (7.70) are equivalent to

$$\langle m, \omega_k \rangle = 0\,, \qquad \langle k, \omega_m \rangle = 0\,. \tag{7.75}$$

Using the Cartan identity $L_m = d \circ i_m + i_m \circ d$, we find for the differential of $\langle m, \omega_k \rangle$

$$d\langle m, \omega_k \rangle = d\, i_m\, \omega_k = -i_m\, d\omega_k\,,$$

where we used that m is a Killing field. In a footnote at the end of this subsection, we shall derive the important identity

$$d\omega_k = *(k \wedge R(k))\,. \tag{7.76}$$

Inserting this gives

$$d\langle m, \omega_k \rangle = -i_m * \left(k \wedge R(k)\right), \tag{7.77}$$

and, interchanging the roles of k and m,

$$d\langle k, \omega_m \rangle = -i_k * \left(m \wedge R(m)\right), \tag{7.78}$$

This shows that the Frobenius conditions imply the Ricci circularity properties (7.73), because for any p-form α and vector field X we have the useful algebraic identity (see DG, (12.32))

$$i_X * \alpha = *(\alpha \wedge X^\flat). \tag{7.79}$$

Conversely, the Ricci circularity properties (7.73) imply by (7.77) and (7.78) that $d\langle k, \omega_m \rangle = d\langle m, \omega_k \rangle = 0$. The scalar products in (7.75) are thus constant. Since spacetime is assumed to be asymptotically flat, there are fixpoints under the $SO(2)$ action in which m vanishes. Thus the Frobenius conditions hold. These are therefore, in particular, implied by the vacuum field equations. But there are other interesting cases where this is true.

Consider, for example, the interior of a rigidly rotating neutron star whose energy-momentum tensor is described by an ideal fluid:

$$T^{\mu\nu} = (\rho + p)u^\mu u^\nu + p\, g^{\mu\nu}. \tag{7.80}$$

The four-velocity field has the form (see (6.121))

$$u = \mathrm{e}^{-V}(k + \Omega\, m), \quad \Omega = \text{const}, \tag{7.81}$$

where the proportionality factor e^{-V} is determined by the normalization condition $\langle u, u \rangle = -1$. It is obvious that the field equations in the form

$$R_{\mu\nu} = 8\pi G \left(T_{\mu\nu} - \frac{1}{2}\, g_{\mu\nu} T^\lambda{}_\lambda \right)$$

imply the Ricci circularity (7.73).

In summary, a stationary and axisymmetric, asymptotically flat vacuum spacetime $(M, {}^{(4)}g)$ can be described locally as follows:

$$M = \Sigma \times \Gamma, \qquad {}^{(4)}g = \sigma + g. \tag{7.82}$$

Here Σ is diffeomorphic to $\mathbb{R} \times SO(2)$ and the metric coefficients in the adapted coordinates $x^0 = t$, $x^1 = \varphi$ depend only on the coordinates of Γ. (Σ, σ) is a two-dimensional Lorentz manifold, while (Γ, g) is a two-dimensional Riemannian manifold orthogonal to Σ. The two Killing fields $k = \partial_t$, $m = \partial_\varphi$ are tangential to Σ and orthogonal to Γ. Denoting (t, φ) by x^a, for $a = 0, 1$ (with indices always taken from the beginning of the Latin alphabet), and using the indices $i, j, k, l, \ldots$ for the coordinates x^i ($x^i = 2, 3$) chosen on Γ, the metrics σ and g have the form (see (7.72))

$$\sigma = \sigma_{ab}(x^i)dx^a dx^b\,, \qquad g = g_{ij}(x^k)dx^i dx^k\,. \tag{7.83}$$

Note that the σ_{ab} are scalar products of the two Killing fields, and thus have an invariant meaning.

* * *

Footnote: Derivation of Two Identities

In this chapter $d^\dagger$ will denote the *negative* of the codifferential δ, because $d^\dagger$ is the adjoint of Cartan's differential d in the sense of DG, (13.99).

The following identity for a Killing field K and a p-form α is often useful

$$d^\dagger(K \wedge \alpha) = -K \wedge d^\dagger\alpha - L_K\alpha\,. \tag{7.84}$$

(As above we use the same letters for K and its associated 1-form $K^\flat$.) This can be verified as follows. We have, using (7.79),

$$\begin{aligned} d^\dagger(K \wedge \alpha) &= (-1)^{p+1} *^{-1} d * (K \wedge \alpha) = - *^{-1} d * (\alpha \wedge K) \\ &= - *^{-1} d\, i_K * \alpha = - *^{-1} \left(L_K * \alpha - i_K\, d * \alpha\right) \\ &= -L_K\alpha + *^{-1} i_K\, d * \alpha\,. \end{aligned}$$

In the last term we use

$$\begin{aligned} i_K\, d * \alpha = i_K * *^{-1} d * \alpha &= i_K * \left((-1)^p d^\dagger\alpha)\right) \\ &= (-1)^p * (d^\dagger\alpha \wedge K) = - * (K \wedge d^\dagger\alpha)\,. \end{aligned} \tag{7.85}$$

If this is inserted the identity (7.84) follows.

Next, we prove the identity (7.76) for the twist form $\omega = \frac{1}{2} * (K \wedge dK)$. In

$$d\omega = \frac{1}{2}\, d * (K \wedge dK) = -\frac{1}{2} * d^\dagger(K \wedge dK)$$

we apply (7.84) for $\alpha = dK$. Since $L_K dK = d\, L_K K = 0$, we get, using also (7.58),

$$d\omega = \frac{1}{2} * \left(K \wedge d^\dagger dK\right) = *\big(K \wedge R(K)\big)\,. \tag{7.86}$$

This identity is valid for any Killing field K.

* * *

7.2.2 The Ernst Equation

For the metric components σ_{ab} we also use the notation

$$V = -\langle k, k\rangle\,, \quad W = \langle k, m\rangle\,, \quad X = \langle m, m\rangle\,. \tag{7.87}$$

Furthermore, let

$$\rho = \sqrt{-\sigma} = \sqrt{VX + W^2}\,. \tag{7.88}$$

Since among ω_k, ω_m the latter will play a more distinct role we abbreviate it by ω: $\omega \equiv \omega_m$. The following complex **Ernst form** on the two-dimensional Riemannian manifold (Γ, g)

$$\mathcal{E} = -dX + 2i\,\omega \tag{7.89}$$

will be particularly relevant. Below we shall show in a footnote, without using the field equations, that its codifferential $d^{\dagger(g)}\mathcal{E}$ for the metric g is given by

$$\frac{1}{\rho}\, d^{\dagger(g)}(\rho\,\mathcal{E}) = \frac{\langle \mathcal{E}, \mathcal{E}\rangle}{X} - 2\,R(m,m)\,. \tag{7.90}$$

For vacuum solutions, i.e., ${}^{(4)}\mathrm{Ric} = 0$, the last term is absent.

The exterior differential d of $\mathcal{E}$ is obtained with the help of (7.86):

$$d\mathcal{E} = 2i\,d\omega = 2i\, *\big(m \wedge R(m)\big)\,. \tag{7.91}$$

For vacuum solutions the Ernst form $\mathcal{E}$ is therefore closed. If the domain of $\mathcal{E}$ is simply connected, $\mathcal{E}$ is by the Poincaré's Lemma exact,

$$\mathcal{E} = dE\,. \tag{7.92}$$

We choose the free additive constant for the **Ernst potential** E such that (see (7.89))

$$\mathrm{Re}\, E = -X\,. \tag{7.93}$$

The Ernst potential satisfies the **Ernst equation**

$$\frac{1}{\rho}\, d^{\dagger(g)}(\rho\, dE) + 2\,\frac{\langle dE, dE\rangle}{E + \bar{E}} = 0\,. \tag{7.94}$$

It will turn out that this equation *decouples* from the other equations.

At this point we establish an interesting representation of the twist form ω belonging to m, the imaginary part of $\mathcal{E}$. Let

$$A = W/X\,, \tag{7.95}$$

which may be regarded as the only non-vanishing component of the shift in a 3+1 splitting (see Sect. 2.9). We claim that

$$\omega = -\frac{X^2}{2\rho}\, *^{(g)}\, dA\,. \tag{7.96}$$

This comes about as follows. From

$$\begin{aligned} m &= m_\mu\, dx^\mu = {}^{(4)}g_{\mu\nu} m^\nu\, dx^\mu = {}^{(4)}g_{\mu\varphi}\, dx^\mu \\ &= W\,dt + X\,d\varphi = X(d\varphi + A\,dt)\,, \end{aligned}$$

and

$$dm = \frac{1}{X} dX \wedge m + X dA \wedge dt$$

we get

$$\omega = \frac{1}{2} * (m \wedge dm) = \frac{1}{2} X^2 * (dA \wedge dt \wedge d\varphi) = \frac{1}{2} \frac{X^2}{\rho} * \left(dA \wedge \eta^{(\sigma)} \right),$$

where $\eta^{(\sigma)}$ is the volume form of the two-dimensional Lorentz manifold (Σ, σ). In the last expression we use the general formula

$$* \left(\alpha \wedge \eta^{(\sigma)} \right) = - *^{(g)} \alpha \,, \tag{7.97}$$

for a 1-form α on Γ, with which the claim (7.96) follows.

Equation (7.97) is best verified by multiplying both sides from the left with $\eta^{(\sigma)} \wedge \beta$, where β is an arbitrary 1-form on Γ. Then the right-hand side becomes (use DG, (12.26) and other tools developed in DG, Sect. 12.6.2)

$$-\eta^{(\sigma)} \wedge \left(\beta \wedge *^{(g)} \alpha \right) = -\eta^{(\sigma)} \wedge \langle \beta, \alpha \rangle \, \eta^{(g)} = -\langle \beta, \alpha \rangle \, \eta \,,$$

while the left-hand side becomes

$$\begin{aligned} \beta \wedge \eta^{(\sigma)} \wedge * \left(\alpha \wedge \eta^{(\sigma)} \right) &= \left\langle \beta \wedge \eta^{(\sigma)}, \alpha \wedge \eta^{(\sigma)} \right\rangle \eta \\ &= \langle \beta, \alpha \rangle \left\langle \eta^{(\sigma)}, \eta^{(\sigma)} \right\rangle \eta = -\langle \beta, \alpha \rangle \, \eta \,. \end{aligned}$$

This implies (7.97).

The inverse of (7.96) reads (check the signs)

$$dA = 2\rho *^{(g)} \left(\frac{\omega}{X^2} \right) . \tag{7.98}$$

* * *

Footnote: Derivation of Equation (7.90)

This geometrical relation rests on some general identities for Killing fields. Let K be a Killing field with norm $N = \langle K, K \rangle$ and twist form ω. We want to show that

$$d \left(\frac{K}{N} \right) = -2 * \left(K \wedge \frac{\omega}{N^2} \right) . \tag{7.99}$$

For this, consider

$$\begin{aligned} N \, dK + K \wedge dN &= (i_K K) dK + K \wedge d i_K K \\ &= (i_K K) dK + K \wedge (-i_K \, dK) \\ &= i_K (K \wedge dK) = 2 i_K * \omega \\ &= -2 * (K \wedge \omega) \,. \end{aligned} \tag{7.100}$$

The last equality sign follows from (7.79). This equation is equivalent to (7.99).

Beside the differential of ω, given in (7.86), we are also interested in the codifferential $d^\dagger\omega$. For this we apply the identity (7.84) to $\alpha = \omega/N^2$:

$$d^\dagger\left(K \wedge \frac{\omega}{N^2}\right) + K\, d^\dagger\left(\frac{\omega}{N^2}\right) = -L_K(\omega/N^2) = 0\,.$$

Here the first term vanishes because of (7.99), thus

$$d^\dagger\left(\frac{\omega}{N^2}\right) = 0 \qquad \text{or} \qquad d^\dagger\omega = -2\,\frac{\langle\omega, dN\rangle}{N}\,. \tag{7.101}$$

With this we can compute the codifferential of the Ernst form $\mathcal{E} = -dN + 2i\,\omega$,

$$d^\dagger\mathcal{E} = -d^\dagger\, dN - 4i\,\frac{\langle\omega, dN\rangle}{N}\,.$$

Hence, we have

$$d^\dagger\mathcal{E} - \frac{\langle\mathcal{E},\mathcal{E}\rangle}{N} = \Delta N - \frac{1}{N}\langle dN, dN\rangle + \frac{4}{N}\langle\omega,\omega\rangle\,. \tag{7.102}$$

Now we look for a formula for ΔN. In a first step we use an expression for dN we got in the derivation of (7.99):

$$\begin{aligned} -\Delta N = d^\dagger\, dN &= -d^\dagger i_K dK \overset{(7.79)}{=} d^\dagger * (K \wedge *dK) \\ &= *d(K \wedge *dK) = \left(\langle dK, dK\rangle + \langle K, \Delta K\rangle\right) * \eta\,. \end{aligned}$$

(Since K is a Killing field, we have $d^\dagger K = 0$ thus $\Delta K = -d^\dagger\, dK$.) With $*\eta = -1$ and (7.58) we arrive at

$$-\Delta N = 2\,R(K,K) - \langle dK, dK\rangle\,. \tag{7.103}$$

For the last scalar product we take dK from (7.99)

$$dK = -\frac{1}{N}\left(2 * (K \wedge \omega) + K \wedge dN\right). \tag{7.104}$$

The cross terms do not contribute in the scalar product (use DG, (12.26)). For the remaining terms we have

$$\langle K \wedge dN, K \wedge dN\rangle = \langle K, K\rangle\langle dN, dN\rangle - \langle K, dN\rangle^2 = N\langle dN, dN\rangle$$

(we used $\langle K, dN\rangle = i_K dN = L_K N = 0$), and

$$\langle *(K \wedge \omega), *(K \wedge \omega)\rangle = -\langle K \wedge \omega, K \wedge \omega\rangle = -N\,\langle\omega,\omega\rangle\,,$$

since $\langle K, \omega\rangle = 0$. As a result we find

$$N\langle dK, dK\rangle = \langle dN, dN\rangle - 4\langle \omega, \omega\rangle . \tag{7.105}$$

Inserting this in (7.103) and using the resulting expression for ΔN in (7.102) gives

$$d^\dagger \mathcal{E} - \frac{\langle \mathcal{E}, \mathcal{E}\rangle}{N} = -2\, R(K,K) . \tag{7.106}$$

For our situation, i.e., $K = m$ and $N = X$, this becomes

$$d^\dagger \mathcal{E} - \frac{\langle \mathcal{E}, \mathcal{E}\rangle}{X} = -2\, R(m,m) . \tag{7.107}$$

Now, the Ernst form $\mathcal{E}$ is a 1-form on Γ. But for any 1-form α on Γ the following holds[2]

$$d^\dagger \alpha = \frac{1}{\rho}\, d^{\dagger(g)}(\rho\, \alpha) . \tag{7.108}$$

If this is used in (7.107) we finally obtain the important equation (7.90).

* * *

7.2.3 Ricci Curvature

For setting up the field equations, we need the Ricci tensor for the metric of a stationary and axisymmetric, asymptotically flat vacuum spacetime:

$$^{(4)}g = \sigma_{ab}\, \theta^a \otimes \theta^b + g_{ij}\, \theta^i \otimes \theta^j , \tag{7.109}$$

where $\theta^a = dx^a$ and $\theta^i = dx^i$. Recall that all metric coefficients σ_{ab} and g_{ij} depend only on the x^i $(i = 2, 3)$. The structure of (7.109) can be regarded as a generalization of a warped product.

We need the connection forms $\omega^\nu{}_\mu$ relative to the basis $\{\theta^\mu\} = \{\theta^a, \theta^i\}$. These satisfy, beside $\omega_{\mu\nu} + \omega_{\nu\mu} = dg_{\mu\nu}$, the first structure equations. Thus

$$\begin{aligned} \omega_{ai} + \omega_{ia} &= 0 , \\ \omega_{ab} + \omega_{ba} &= d\sigma_{ab} = \sigma_{ab,i}\, \theta^i , \\ \omega_{ij} + \omega_{ji} &= dg_{ij} . \end{aligned} \tag{7.110}$$

and

---

[2] This can be seen as follows:

$$\begin{aligned} d^\dagger \alpha &= -\frac{1}{\sqrt{-{}^{(4)}g}}\partial_\mu \left(\sqrt{-{}^{(4)}g}\, \alpha^\mu\right) = -\frac{1}{\rho\sqrt{g}}\partial_\mu \left(\rho \sqrt{g}\, \alpha^\mu\right) \\ &= -\frac{1}{\rho}\frac{1}{\sqrt{g}}\partial_i \left(\sqrt{g}\, \rho\, \alpha^i\right) = \frac{1}{\rho}\, d^{\dagger(g)}(\rho\, \alpha) . \end{aligned}$$

Alternatively, $d^\dagger \alpha = d^{\dagger(g)}\alpha - \langle d\ln\rho, \alpha\rangle$.

$$d\theta^i + \omega^i{}_j \wedge \theta^j + \omega^i{}_a \wedge \theta^a = 0\,,$$
$$d\theta^a + \omega^a{}_b \wedge \theta^b + \omega^a{}_i \wedge \theta^i = 0\,. \tag{7.111}$$

It is natural to guess that

$$\omega^i{}_j = \text{connection forms of } (\Gamma, g)\,, \tag{7.112a}$$
$$\omega_{ab} = \frac{1}{2}\, d\sigma_{ab} \Longrightarrow \omega^a{}_b = \frac{1}{2}\,\sigma^{ac} d\sigma_{cb}\,, \tag{7.112b}$$
$$\omega_{ia} = -\omega_{ai} = -\frac{1}{2}\,\sigma_{ab,i}\,\theta^b\,. \tag{7.112c}$$

One readily sees that this ansatz indeed satisfies all equations in (7.110) and (7.111). Because of (7.112a) we denote the connection forms of (Γ, g) also by $\omega^i{}_j$.

It is now easy to compute the curvature forms. We concentrate on those terms which contribute to the Ricci tensor

$${}^{(4)}R_{\mu\nu} = \Omega^\alpha{}_\mu(e_\alpha, e_\nu)\,, \tag{7.113}$$

where $e_\alpha = \partial_\alpha$. Let us begin with ${}^{(4)}R_{ij} = {}^{(4)}R^a{}_{iaj} + {}^{(4)}R^k{}_{ikj}$. From

$$\Omega^i{}_j = \underbrace{d\omega^i{}_j + \omega^i{}_k \wedge \omega^k{}_j}_{{}^{(g)}\Omega^i{}_j} + \underbrace{\omega^i{}_c \wedge \omega^c{}_j}_{\propto\ \theta^a \wedge \theta^b}$$

it follows that

$${}^{(4)}R^i{}_{jkl} = {}^{(g)}R^i{}_{jkl}\,, \tag{7.114}$$

in particular

$${}^{(4)}R^k{}_{ikj} = {}^{(g)}R_{ij}\,. \tag{7.115}$$

Furthermore, we have

$$\begin{aligned}
\Omega^a{}_i &= d\omega^a{}_i + \omega^a{}_k \wedge \omega^k{}_i + \omega^a{}_c \wedge \omega^c_i \\
&= d\left(\frac{1}{2}\,\sigma^{ac}\sigma_{cb,i}\,\theta^b\right) + \frac{1}{2}\,\sigma^{ac}\sigma_{cb,k}\,\theta^b \wedge \omega^k{}_i + \omega^a{}_c \wedge \omega^c{}_i \\
&= \left(\frac{1}{2}\,\sigma^{ac}\sigma_{cb,i}\right)_{,j} \theta^j \wedge \theta^b + \frac{1}{2}\,\sigma^{ac}\sigma_{cb,k}\,\theta^b \wedge \omega^k{}_i + \omega^a{}_c \wedge \omega^c{}_i
\end{aligned}$$

or

$$\Omega^a{}_i = {}^{(g)}\nabla_j \left(\frac{1}{2}\sigma^{ac}\sigma_{cb,i}\right) \theta^j \wedge \theta^b + \omega^a{}_c \wedge \omega^c{}_i\,. \tag{7.116}$$

From this we find, denoting the covariant derivative ${}^{(g)}\nabla$ in (Γ, g) by a stroke,

$$\begin{aligned}\Omega^a{}_i(e_a, e_j) &= -\frac{1}{2}\left(\sigma^{ac}\sigma_{ac,i}\right)_{|j} - \omega^a{}_b(e_j)\omega^b{}_i(e_a) \\ &= -\left(\frac{1}{\rho}\,\rho_{,i}\right)_{|j} - \frac{1}{4}\,\sigma^{ac}\sigma_{cb,j}\sigma^{bd}\sigma_{da,i} \\ &= -\left(\frac{1}{\rho}\,\rho_{,i}\right)_{|j} + \frac{1}{4}\,\sigma^{ad}{}_{,j}\sigma_{da,i} \\ &= -\frac{1}{\rho}\,\rho_{|ij} + \frac{1}{\rho^2}\,\rho_{,i}\rho_{,j} + \frac{1}{4}\,\sigma^{ab}{}_{,i}\sigma_{ab,j}\,. \end{aligned} \tag{7.117}$$

Taken together we obtain

$$^{(4)}R_{ij} = {}^{(g)}R_{ij} - \frac{1}{\rho}\,\rho_{|ij} + \frac{1}{\rho^2}\,\rho_{,i}\rho_{,j} + \frac{1}{4}\,\sigma^{ab}{}_{,i}\sigma_{ab,j}\,. \tag{7.118}$$

For later use we express the last term by X, W and V. From

$$\sigma^{00} = -\frac{1}{\rho^2}\,\sigma_{11}\,, \qquad \sigma^{01} = \frac{1}{\rho^2}\,\sigma_{01}\,, \qquad \sigma^{11} = -\frac{1}{\rho^2}\,\sigma_{00}\,,$$

we get

$$\begin{aligned}\frac{1}{4}\,\sigma^{ab}{}_{,i}\sigma_{ab,j} &= \frac{1}{4}\left[V_{,i}\left(\frac{1}{\rho^2}\,X\right)_{,j} + X_{,i}\left(\frac{1}{\rho^2}\,V\right)_{,j} + 2\,W_{,i}\left(\frac{1}{\rho^2}\,W\right)_{,j}\right] \\ &= \frac{1}{4\rho^2}\left[V_{,i}X_{,j} + X_{,i}V_{,j} + 2\,W_{,i}W_{,j}\right] \\ &\quad + \frac{1}{4}\left(\frac{1}{\rho^2}\right)_{,j}\left[V_{,i}X + X_{,i}V + 2\,W_{,i}W_{,j}\right]. \end{aligned}$$

The last square bracket is equal to $2\rho\,\rho_{,i}$ (see (7.88)). So (7.118) can be written as

$$^{(4)}R_{ij} = {}^{(g)}R_{ij} - \frac{1}{\rho}\,\rho_{|ij} + \frac{1}{4\rho^2}\left[X_{,i}V_{,j} + V_{,i}X_{,j} + 2\,W_{,i}W_{,j}\right]. \tag{7.119}$$

Next we determine the components $R_{ab} = \Omega^c{}_a(e_c, e_b) + \Omega^i{}_a(e_i, e_b)$. According to (7.116) and the connection forms (7.112b), (7.112c) we have

$$\Omega_{ai} = \frac{1}{2}\,\sigma_{ab|ij}\,\theta^j \wedge \theta^b - \frac{1}{4}\,\sigma_{ac,j}\sigma^{cd}\,\sigma_{bd,i}\,\theta^j \wedge \theta^b\,. \tag{7.120}$$

Hence

$$^{(4)}R_{aibj} = {}^{(4)}R_{iajb} = -\frac{1}{2}\,\sigma_{ab|ij} + \frac{1}{4}\,\sigma^{cd}\,\sigma_{ac,j}\sigma_{bd,i}\,. \tag{7.121}$$

In particular we obtain

$$^{(4)}R^i{}_{aib} = -\frac{1}{2}\,\sigma_{ab}{}^{,i}{}_{|i} + \frac{1}{4}\,\sigma^{cd}\,\sigma_{ac}{}^{,i}\sigma_{bd,i}\,. \tag{7.122}$$

We still need

$$\begin{aligned}\Omega^c{}_a(e_c, e_b) &= d\omega^c{}_a(e_c, e_b) + (\omega^c{}_i \wedge \omega^i{}_a)(e_c, e_b) \\ &= (\omega^c{}_i \wedge \omega^i{}_a)(e_c, e_b) \\ &= -\frac{1}{4}\,\sigma^{cd}\,\sigma_{cd,i}\sigma_{ab}{}^{,i} + \frac{1}{4}\,\sigma^{cd}\,\sigma_{db,i}\sigma_{ac}{}^{,i}\,.\end{aligned}$$

The first term in the last line is equal to $-(1/2\rho)\,\rho_{,i}\sigma^{ab,i}$. Since the last term is manifestly symmetric in a and b, it is equal to

$$-\frac{1}{4}\,\sigma^{cd}{}_{,i}\sigma_{bd}\sigma_{ac}{}^{,i} = -\frac{1}{4}\,\sigma_{ad}\sigma^{cd}{}_{,i}\sigma_{bc}{}^{,i}\,.$$

As a result we obtain

$$^{(4)}R^c{}_{acb} = -\frac{1}{2\rho}\,\rho_{,i}\sigma_{ab}{}^{,i} - \frac{1}{4}\,\sigma_{ad}\sigma^{cd}{}_{,i}\sigma_{bc}{}^{,i}\,. \tag{7.123}$$

Combining (7.122) and (7.123) gives

$$^{(4)}R_{ab} = -\frac{1}{2}\,\sigma_{ad}\frac{1}{\rho}\left(\rho\,\sigma^{cd}\sigma_{bc}{}^{,i}\right)_{|i}\,. \tag{7.124}$$

In particular, the partial trace $R^a{}_a$ is given by

$$^{(4)}R^a{}_a = -\frac{1}{\rho}\left(\rho\,\frac{1}{2}\,\sigma^{ab}\sigma_{ab,i}\right)^{|i} = -\frac{1}{\rho}\left(\rho\,\frac{1}{\rho}\,\rho_{,i}\right)^{|i}$$

or

$$^{(4)}R^a{}_a = -\frac{1}{\rho}\,{}^{(g)}\Delta\rho\,. \tag{7.125}$$

Thus for a vacuum solution $\rho = \sqrt{-\sigma}$ is *harmonic*. The mixed components $^{(4)}R_{ai}$ vanish (see Exercise 7.2).

* * *

It will turn out to be useful to work with the two-dimensional Riemannian metric

$$\gamma_{ij} = X\,g_{ij} \tag{7.126}$$

on Γ. Using the results of Exercise 7.3 we obtain

$$^{(g)}R_{ij} = {}^{(\gamma)}R_{ij} + \frac{1}{2}\gamma_{ij}\,{}^{(\gamma)}\Delta \ln X\,, \tag{7.127}$$

and for a function f on Γ

$$\begin{aligned}{}^{(g)}\nabla_i\,{}^{(g)}\nabla_j\,f &= {}^{(\gamma)}\nabla_i\,{}^{(\gamma)}\nabla_j\,f \\ &\quad + \frac{1}{2X}\left(f_{,i}X_{,j} + f_{,j}X_{,i} - \gamma_{ij}\langle df, df\rangle^{(\gamma)}\right)\,.\end{aligned} \tag{7.128}$$

Especially

$$^{(g)}\Delta f = X\, ^{(\gamma)}\Delta f\,. \tag{7.129}$$

If we use this in (7.119), a short computation gives in a first step

$$^{(4)}R_{ij} = {}^{(\gamma)}R_{ij} - \frac{1}{\rho}\, {}^{(\gamma)}\nabla_i\, {}^{(\gamma)}\nabla_j \rho - \frac{1}{2X^2} X_{,i} X_{,j} + \frac{X^2}{2\rho^2} A_{,i} A_{,j}$$
$$+ \frac{1}{2} g_{ij} \left({}^{(g)}\Delta \ln X + \langle d\ln\rho, d\ln X\rangle^{(g)} \right),$$

where the term in the parenthesis is equal to $^{(4)}\Delta \ln X$ (see the footnote on p. 440). Now, from (7.103) and (7.105) we conclude that

$$^{(4)}\Delta \ln X = -4\, \frac{\langle\omega,\omega\rangle}{X^2} - 2\, \frac{R(m,m)}{X}\,. \tag{7.130}$$

Furthermore, from (7.98), i.e.

$$A_{,i} = \frac{2\rho}{X^2}\, {}^{(g)}\eta_{ij}\, \omega^j\,,$$

we find

$$A_{,i} A_{,j} = \left(\frac{2\rho}{X^2}\right)^2 \left(\langle\omega,\omega\rangle^{(g)} g_{ij} - \omega_i\, \omega_j \right). \tag{7.131}$$

All together gives the important formula

$$^{(\gamma)}R_{ij} - \frac{1}{\rho}\, {}^{(\gamma)}\nabla_i\, {}^{(\gamma)}\nabla_j \rho = {}^{(4)}R_{ij} + \gamma_{ij}\, \frac{R(m,m)}{X^2} + \frac{X_{,i} X_{,j} + 4\,\omega_i\,\omega_j}{2X^2}\,. \tag{7.132}$$

Here, the last term is also equal to $(2X)^{-2}(\mathcal{E}_i \bar{\mathcal{E}}_j + \mathcal{E}_j \bar{\mathcal{E}}_i)$. We also note, that in two dimensions

$$^{(\gamma)}\, R_{ij} = {}^{(\gamma)}\mathcal{K}\, \gamma_{ij}\,, \tag{7.133}$$

where $^{(\gamma)}\mathcal{K}$ is the Gaussian curvature of (Γ, γ) (see the end of Sect. 2.6).

Intermediate Summary

For further use let us collect the results obtained so far. The metric $^{(4)}g$ on $M = \Sigma \times \Gamma$ can be parameterized as follows

$$^{(4)}g = -\frac{\rho^2}{X}\, dt^2 + X(d\varphi + A\, dt)^2 + \frac{1}{X}\, \gamma\,. \tag{7.134}$$

The metric functions ρ, A and X satisfy the identities

$$\frac{1}{\rho}\, d^\dagger(\rho\mathcal{E}) = \frac{1}{X}\Big(\langle\mathcal{E},\mathcal{E}\rangle^{(\gamma)} - 2\,R(m,m)\Big)\,, \qquad (7.135a)$$

$$\frac{1}{\rho}\,{}^{(\gamma)}\Delta\rho = -\frac{1}{X}\,{}^{(4)}R^a{}_a\,, \qquad (7.135b)$$

$$\frac{1}{\rho}\, dA = 2 *^{(\gamma)}\left(\frac{\omega}{X^2}\right), \qquad (7.135c)$$

$${}^{(\gamma)}\mathcal{K}\gamma_{ij} - \frac{1}{\rho}\,{}^{(\gamma)}\nabla_i{}^{(\gamma)}\nabla_j\rho = {}^{(4)}R_{ij} + \gamma_{ij}\,\frac{R(m,m)}{X^2} + \frac{\mathcal{E}_i\bar{\mathcal{E}}_j + \mathcal{E}_j\bar{\mathcal{E}}_i}{4X^2}\,. \qquad (7.135d)$$

Here $\mathcal{E}$ is the Ernst form

$$\mathcal{E} = -dX + 2i\,\omega\,, \qquad (7.136)$$

where ω is the twist form belonging to m. Furthermore,

$$d\omega = *(m \wedge R(m))\,. \qquad (7.137)$$

In this summary no use of the field equations has been made. All these geometrical relations can, for instance, also be used for rigidly rotating neutron star models. – From *now* on we assume the vacuum equations to hold.

7.2.4 Weyl Coordinates

According to (7.135b), $\rho = \sqrt{-\sigma}$ is harmonic for a vacuum solution (this is also true for certain matter models, e.g., electrodynamics). It is natural to use ρ as a coordinate. For this we have to check, however, that ρ has no critical points. We make the assumption that the outer domain of communication of the black hole is simply connected. This is then also the case for the two-dimensional Riemannian manifold (Γ,γ) on which ρ is harmonic.

For the proof that ρ has no critical points, we consider the domain $\{0 < \rho < \rho_0\}$ of Γ, which is topologically a strip bounded by the curves $\{\rho = 0\}$ and $\{\rho = \rho_0\}$. According to the Riemannian mapping theorem there is a holomorphic diffeomorphism from this domain into the strip $\{0 < \mathrm{Re}\,\zeta < \rho_0\}$ of the complex ζ-plane, whereby the boundary curves are continuously mapped into each other[3].

As a function of ζ, ρ is again harmonic and hence also

$$h := \rho - \mathrm{Re}\,\zeta\,. \qquad (7.138)$$

[3] The Riemannian mapping theorem guarantees always the existence of a holomorphic diffeomorphism (conformal transformation) between two simply connected domains D_1 and $D_2 \neq \mathbb{C}$. In general the mapping $f : D_1 \longrightarrow D_2$ does not extend to a homeomorphism of $\bar{D}_1$ onto $\bar{D}_2$. If the boundary of D_1 is a finite union of smooth curves, one can show that f extends to a homeomorphism $f : \bar{D}_1 \longrightarrow \bar{D}_2$ (see, e.g., [48]), Proposition 6.2.

Moreover, this function vanishes on the boundary of the strip in the ζ-plane. For Im $\zeta \longrightarrow \pm\infty$ the function h vanishes also, because of the asymptotic flatness. Indeed, ρ becomes asymptotically the radial cylindrical coordinate of Minkowski spacetime, and thus the mapping into the complex ζ-plane becomes asymptotically the identity, i.e., $\rho - \text{Re}\,\zeta$ approaches zero. With this information, the maximum principle of harmonic functions implies that $h \equiv 0$. Thus $\rho = \text{Re}\,\zeta$, whence ρ has no critical points.

The possibility to use ρ as a coordinate simplifies things very much. We can choose a second coordinate z such that the metric γ becomes conformally flat[4]

$$\gamma = e^{2h}(d\rho^2 + dz^2)\,. \tag{7.139}$$

At this point the four-metric ${}^{(4)}g$ contains only the three functions X, A and h, depending on the **Weyl coordinates** ρ and z:

$${}^{(4)}g = -\frac{\rho^2}{X}\,dt^2 + X(d\varphi + A\,dt)^2 + \frac{1}{X}\,e^{2h}(d\rho^2 + dz^2)\,. \tag{7.140}$$

In the next step we have to specialize the basic equations (7.135a)–(7.137) to this situation. – For convenience, let us collect them here. The Ernst form is exact, i.e., $\mathcal{E} = dE$, with (see (7.93))

$$E = -X + i\,Y\,, \qquad dY = 2\omega \overset{(7.135c)}{=} -\frac{X^2}{\rho} *^{(\gamma)} dA\,. \tag{7.141}$$

The central Ernst equation (7.94) now decouples

$$\frac{1}{\rho}\,d^{\dagger(\gamma)}(\rho\,dE) = \frac{1}{X}\langle dE, dE\rangle^{(\gamma)}\,. \tag{7.142}$$

Note that the dependence on h is spurious; see (7.149) below. Using the vacuum equations (7.135d) becomes

$${}^{(\gamma)}\mathcal{K}\gamma_{ij} - \frac{1}{\rho}\,{}^{(\gamma)}\nabla_i\,{}^{(\gamma)}\nabla_j\rho = \frac{1}{4X^2}(E_{,i}\bar{E}_{,j} + E_{,j}\bar{E}_{,i})\,. \tag{7.143}$$

The relation in (7.141) between dY and dA becomes for the conformally flat metric (7.139) explicitly

$$Y_{,\rho} = -\frac{X^2}{\rho}\,A_{,z}\,, \qquad Y_{,z} = \frac{X^2}{\rho}\,A_{,\rho}\,. \tag{7.144}$$

We also write (7.142) and (7.143) explicitly. For this we need various quantities. Using Exercise 7.1 (with $\bar{g}$ the flat metric), we get

---

[4] In a first step we can choose z such that it is orthogonal to ρ: $\gamma = e^{2h}(d\rho^2 + f\,dz^2)$. Since ρ is harmonic, one readily concludes that f is independent of ρ. By a simple redefinition of z we can, therefore, choose $f \equiv 1$.

$$0 = {}^{(0)}R_{ij} = {}^{(\gamma)}R_{ij} + \gamma_{ij}\,{}^{(\gamma)}\Delta h = {}^{(\gamma)}R_{ij} + \gamma_{ij}\mathrm{e}^{-2h}\,{}^{(0)}\Delta h\,,$$

where the index (0) always refers to the flat metric $d\rho^2 + dz^2$. Together with (7.133), we obtain for the Gaussian curvature

$$ {}^{(\gamma)}\mathcal{K} = -\mathrm{e}^{-2h}\,{}^{(0)}\Delta h\,. \tag{7.145}$$

For the covariant derivatives of ρ in (7.143) we need the Christoffel symbols for the metric γ. These are readily found to be

$$\begin{aligned} \Gamma^{\rho}{}_{\rho\rho} &= \Gamma^{z}{}_{\rho z} = -\Gamma^{\rho}{}_{zz} = h_{,\rho}\,, \\ \Gamma^{z}{}_{zz} &= \Gamma^{\rho}{}_{\rho z} = -\Gamma^{z}{}_{\rho\rho} = h_{,z}\,. \end{aligned} \tag{7.146}$$

Since for the coordinate ρ, we have ${}^{(\gamma)}\nabla_i\,{}^{(\gamma)}\nabla_j\rho = -\Gamma^{\rho}{}_{ij}$, for $i,j=\rho,z$, we get

$$\begin{aligned} {}^{(\gamma)}\nabla_z\,{}^{(\gamma)}\nabla_z\rho &= -{}^{(\gamma)}\nabla_\rho\,{}^{(\gamma)}\nabla_\rho\rho = h_{,\rho}\,, \\ {}^{(\gamma)}\nabla_\rho\,{}^{(\gamma)}\nabla_z\rho &= {}^{(\gamma)}\nabla_z\,{}^{(\gamma)}\nabla_\rho\rho = -h_{,z}\,. \end{aligned} \tag{7.147}$$

Inserting (7.147) and (7.145) into (7.143), we obtain the following three equations

$$\frac{1}{\rho}\,h_{,\rho} = \frac{1}{4X^2}(E_{,\rho}\bar{E}_{,\rho} - E_{,z}\bar{E}_{,z})\,, \tag{7.148a}$$

$$\frac{1}{\rho}\,h_{,z} = \frac{1}{4X^2}(E_{,\rho}\bar{E}_{,z} + E_{,z}\bar{E}_{,\rho})\,, \tag{7.148b}$$

$$ {}^{(0)}\Delta h = -\frac{1}{4X^2}(E_{,\rho}\bar{E}_{,\rho} + E_{,z}\bar{E}_{,z})\,. \tag{7.148c}$$

Finally, the Ernst equation (7.142) can be written as

$$\frac{1}{\rho}\,d^{\dagger(0)}(\rho\,dE) = \frac{1}{X}\left\langle {}^{(0)}\nabla E, {}^{(0)}\nabla E\right\rangle^{(0)}$$

or

$$\frac{1}{\rho}\,{}^{(0)}\nabla\cdot\left(\rho\,{}^{(0)}\nabla E\right) = \frac{\left\langle {}^{(0)}\nabla E, {}^{(0)}\nabla E\right\rangle^{(0)}}{\mathrm{Re}\,E}\,. \tag{7.149}$$

Explicitly

$$E_{,\rho\rho} + E_{,zz} + \frac{1}{\rho}\,E_{,\rho} = \frac{(E_{,\rho})^2 + (E_{,z})^2}{\mathrm{Re}\,E}\,. \tag{7.150}$$

This equation also guarantees that the integrability conditions for the system (7.148a)–(7.148c) are consistent. Consider, for example the integrability condition

$$
\begin{aligned}
0 = \partial_z\partial_\rho h - \partial_\rho\partial_z h &= -\rho\,\frac{X_{,z}}{2X^3}(E_{,\rho}\bar{E}_{,\rho} - E_{,z}\bar{E}_{,z}) \\
&\quad -\frac{X - 2\rho\,X_{,\rho}}{4X^3}(E_{,\rho}\bar{E}_{,z} + E_{,z}\bar{E}_{,\rho}) \\
&\quad +\frac{\rho}{4X^2}\left((E_{,\rho}\bar{E}_{,\rho} - E_{,z}\bar{E}_{,z})_{,z} - (E_{,\rho}\bar{E}_{,z} + E_{,z}\bar{E}_{,\rho})_{,\rho}\right) \\
&= -E_{,z}\frac{\rho}{4X^2}\left(E_{,\rho\rho} + E_{,zz} + \frac{1}{\rho}E_{,\rho} + \frac{E_{,\rho}^2 + E_{,z}^2}{X}\right) + \text{c.c.}
\end{aligned}
$$

The last expression vanishes thanks to (7.150).

Once we have a solution of the Ernst equation, we can determine A from (7.144), and h from (7.148a) and (7.148b).

7.2.5 Conjugate Solutions

Below we shall find an extremely simple solution of the Ernst equation which is, however, at first sight useless. It will turn out that the real part $-X = \langle m, m\rangle$ of the Ernst potential E (see (7.141)) does not vanish on the rotation axis (critical points of m). Moreover, $X \not\geq 0$ and has the wrong asymptotic behavior. By a simple reinterpretation, we will nevertheless by able to arrive at the Kerr solution. For this the following observation is crucial.

Suppose we have a Ricci flat metric in the **Papapetrou form** (7.140). Perform the substitution $t \longmapsto \varphi$, $\varphi \longmapsto -t$, then the metric assumes the form

$$
{}^{(4)}g = -\frac{\rho^2}{X}\,d\varphi^2 + X(-dt + A\,d\varphi)^2 + \frac{1}{X}\,e^{2h}(d\rho^2 + dz^2)\,.
$$

This can again be brought into the form (7.140) by a suitable substitution $X \longmapsto \hat{X}$, $A \longmapsto \hat{A}$ and $h \longmapsto \hat{h}$. By identification one finds that the coefficients of $d\varphi^2$ are connected as

$$
\hat{X} = X\left(A^2 - \frac{\rho^2}{X^2}\right).
$$

In order that the coefficients of $dt\,d\varphi$ agree, we must require $\hat{X}\hat{A} = -XA$, thus

$$
\hat{A} = -A\left(A^2 - \frac{\rho^2}{X^2}\right)^{-1}.
$$

With this the coefficients of dt^2 also agree, if we set

$$
e^{2\hat{h}} = e^{2h}\frac{\hat{X}}{X} = e^{2h}\left(A^2 - \frac{\rho^2}{X^2}\right).
$$

In summary: If ${}^{(4)}g$ is a Ricci flat metric of the form (7.140), then also the **conjugate metric** that is obtained from ${}^{(4)}g$ by performing the substitutions

$$X \longmapsto \hat{X} = X\left(A^2 - \frac{\rho^2}{X^2}\right),$$
$$A \longmapsto \hat{A} = -A\left(A^2 - \frac{\rho^2}{X^2}\right)^{-1},$$
$$e^{2h} \longmapsto e^{2\hat{h}} = e^{2h}\left(A^2 - \frac{\rho^2}{X^2}\right). \tag{7.151}$$

7.2.6 Basic Equations in Elliptic Coordinates

We shall later see that outside the horizon of the black hole the 2-form $k\wedge m$ is timelike, $\rho^2 = -\langle k\wedge m, k\wedge m\rangle > 0$, except on the rotation axis where m and thus ρ vanish. Also on the horizon $\rho = 0$. In this situation it is helpful to introduce "elliptical coordinates" x and y. These are defined by

$$\rho^2 = \mu^2(x^2-1)(1-y^2) \geq 0, \qquad z = \mu\, xy, \tag{7.152}$$

with $x \geq 1$ and $|y| \leq 1$, where μ is a positive constant. The horizon corresponds to the set $\{x = 1, -1 \leq y \leq 1\}$, and the northern part of the rotational axis to the part of the boundary $\{x \geq 1, y = 1\}$. Correspondingly the southern part of the rotational axis is parameterized by $\{x \geq 1, y = -1\}$ (see Fig. 7.1). A short calculation shows that

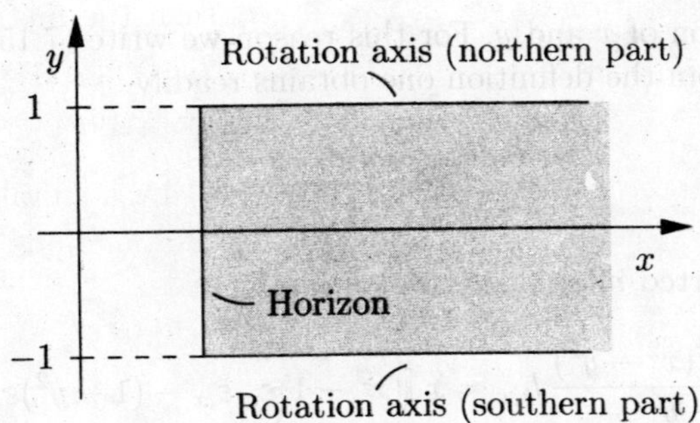

Fig. 7.1. Parameterization in terms of elliptical coordinates.

$$d\rho^2 + dz^2 = \mu^2(x^2-y^2)\left(\frac{dx^2}{x^2-1} + \frac{dy^2}{1-y^2}\right). \tag{7.153}$$

Up to the factor e^{2h} this is the metric γ in (7.126).

Now we have to rewrite the basic equations in terms of the new coordinates. The Ernst equation is best obtained from the invariant form (7.142), which can be written as

$$\left(\rho\sqrt{\gamma}\,\gamma^{ij}E_{,j}\right)_{,i} + \frac{1}{X}\left(\rho\sqrt{\gamma}\,\gamma^{ij}\right)E_{,i}E_{,j} = 0. \tag{7.154}$$

We shall write this explicitly a bit later. The transformation of (7.144) to elliptic coordinates is also obtained from the invariant form (7.142). There we can replace the Hodge dual $*^{(\gamma)}$ by the Hodge dual of the metric (7.153), and readily find

$$A_{,x} = \mu(1-y^2)\frac{Y_{,y}}{X^2}\,, \quad A_{,y} = -\mu(x^2-1)\frac{Y_{,x}}{X^2}\,. \tag{7.155}$$

The transformation of (7.148a) and (7.148b) is a bit tedious but entirely mechanical. One finds

$$\begin{aligned} h_{,x} &= \frac{1-y^2}{4X^2(x^2-y^2)}\Big[-y(x^2-1)\left(E_{,x}\bar{E}_{,y}+E_{,y}\bar{E}_{,x}\right) \\ &\quad +x(x^2-1)E_{,x}\bar{E}_{,x} - x(1-y^2)E_{,y}\bar{E}_{,y}\Big]\,, \\ h_{,y} &= \frac{x^2-1}{4X^2(x^2-y^2)}\Big[x(1-y^2)\left(E_{,x}\bar{E}_{,y}+E_{,y}\bar{E}_{,x}\right) \\ &\quad +y(x^2-1)E_{,x}\bar{E}_{,x} - y(1-y^2)E_{,y}\bar{E}_{,y}\Big]\,. \end{aligned} \tag{7.156}$$

It will turn out that the Kerr solution corresponds to a solution of the Ernst equation, for which

$$\varepsilon := \frac{1+E}{1-E} \tag{7.157}$$

is a *linear* function of x and y. For this reason we write (7.154) and (7.156) in terms of ε. From the definition one obtains readily

$$\frac{E_{,i}\bar{E}_{,j}}{4X^2} = \frac{\varepsilon_{,i}\bar{\varepsilon}_{,j}}{(1-\varepsilon\bar{\varepsilon})^2}\,. \tag{7.158}$$

When this is inserted into (7.156) we obtain

$$\begin{aligned} \frac{(1-\varepsilon\bar{\varepsilon})^2(x^2-y^2)}{1-y^2}\,h_{,x} &= x\left[(x^2-1)\varepsilon_{,x}\bar{\varepsilon}_{,x} - (1-y^2)\varepsilon_{,y}\bar{\varepsilon}_{,y}\right] \\ &\quad -y(x^2-1)(\varepsilon_{,x}\bar{\varepsilon}_{,y}+\varepsilon_{,y}\bar{\varepsilon}_{,x})\,, \end{aligned} \tag{7.159a}$$

$$\begin{aligned} \frac{(1-\varepsilon\bar{\varepsilon})^2(x^2-y^2)}{x^2-1}\,h_{,y} &= y\left[(x^2-1)\varepsilon_{,x}\bar{\varepsilon}_{,x} - (1-y^2)\varepsilon_{,y}\bar{\varepsilon}_{,y}\right] \\ &\quad +x(1-y^2)(\varepsilon_{,x}\bar{\varepsilon}_{,y}+\varepsilon_{,y}\bar{\varepsilon}_{,x})\,. \end{aligned} \tag{7.159b}$$

The transformed Ernst equation can be obtained from (7.154). Using the fact that $\sqrt{\gamma}\,\gamma^{ij}$ is conformally invariant, one obtains from (7.153)

$$\sqrt{\gamma}\,\gamma^{ij} = \sqrt{\tilde{\gamma}}\,\tilde{\gamma}^{ij}\,, \qquad \text{with } \tilde{\gamma} = \frac{dx^2}{x^2-1} + \frac{dy^2}{1-y^2}\,,$$

and thus (see (7.152))

$$\sqrt{\tilde{\gamma}} = \frac{1}{\sqrt{(x^2-1)(1-y^2)}}\,, \qquad \rho\sqrt{\tilde{\gamma}} = \mu\,. \tag{7.160}$$

With this we can write down (7.154) in terms of the elliptic coordinates x and y. It is, however, easier (and also useful) to use a variational principle for the Ernst equation (7.149). This is obviously the Euler-Lagrange equation for the Lagrangian density

$$\mathcal{L} = \rho\,\frac{\left\langle {}^{(0)}\nabla E, {}^{(0)}\nabla \bar{E}\right\rangle^{(0)}}{(\mathrm{Re}\,E)^2}\,. \tag{7.161}$$

This has a geometrical significance. According to (7.145) and (7.148c)

$$\sqrt{\gamma}\,{}^{(\gamma)}\mathcal{K} = -\,{}^{(0)}\Delta h = \frac{1}{4X^2}\left\langle {}^{(0)}\nabla E, {}^{(0)}\nabla \bar{E}\right\rangle, \tag{7.162}$$

so that the action S is

$$S := \int_\Gamma \mathcal{L}\,d\rho dz = 4\int_\Gamma \left({}^{(\gamma)}\mathcal{K}\,\rho\right){}^{(\gamma)}\eta\,. \tag{7.163}$$

It is, of course, crucial that the Gaussian curvature is multiplied by ρ. Otherwise the action S would be a topological invariant, according to the Gauss-Bonnet theorem.

The Lagrangian (7.161) can be rewritten as follows, using (7.158)

$$\mathcal{L} = \frac{4\rho\sqrt{\tilde{\gamma}}\,\tilde{\gamma}^{ij}E_{,i}\bar{E}_{,j}}{4\,X^2} \overset{(7.160)}{=} 4\mu\,\frac{(x^2-1)E_{,x}\bar{E}_{,x} + (1-y^2)E_{,y}\bar{E}_{,y}}{4X^2}$$
$$\overset{(7.158)}{=} 4\mu\,\frac{(x^2-1)\varepsilon_{,x}\bar{\varepsilon}_{,x} + (1-y^2)\varepsilon_{,y}\bar{\varepsilon}_{,y}}{(1-\varepsilon\bar{\varepsilon})^2}\,.$$

Now, the Euler-Lagrange equation to this Lagrangian density

$$\mathcal{L} = 4\mu\frac{(x^2-1)|\varepsilon_{,x}|^2 + (1-y^2)|\varepsilon_{,y}|^2}{(1-\varepsilon\bar{\varepsilon})^2} \tag{7.164}$$

reads as follows

$$\left[(x^2-1)\varepsilon_{,x}\right]_{,x} + \left[(1-y^2)\varepsilon_{,y}\right]_{,y} = -\frac{2\,\bar{\varepsilon}}{1-\varepsilon\,\bar{\varepsilon}}\left[(x^2-1)\varepsilon_{,x}^2 + (1-y^2)\varepsilon_{,y}^2\right]. \tag{7.165}$$

The last two equations have been announced in the introduction to Sect. 7.2.

Now we have everything in the desired form. The metric in elliptic coordinates is

$${}^{(4)}g = -\frac{\rho^2}{X}\,dt^2 + X(d\varphi + A\,dt)^2 + \frac{e^{2h}}{X}\,\mu^2(x^2-y^2)\left(\frac{dx^2}{x^2-1} + \frac{dy^2}{1-y^2}\right), \tag{7.166}$$

with

$$\rho^2 = \mu^2(x^2-1)(1-y^2)\,, \tag{7.167}$$

and

$$X = \frac{1-\varepsilon\bar{\varepsilon}}{|1+\varepsilon|^2}\,, \qquad Y = \frac{i(\bar{\varepsilon}-\varepsilon)}{|1+\varepsilon|^2}\,. \tag{7.168}$$

The two other potentials A and h are determined by (7.155) and (7.159a), (7.159b) in terms of the Ernst potential, which satisfies the non-linear partial differential equation (7.165) representing the Euler-Lagrange equation of (7.164). – This is a remarkable reduction for finding stationary axisymmetric solutions of the vacuum equations.

7.2.7 The Kerr Solution

The Ernst equation (7.165) has a simple linear solution

$$\varepsilon = p\,x + iq\,y\,, \qquad p^2+q^2 = 1\,, \qquad p,q \in \mathbb{R}\,. \tag{7.169}$$

We determine the corresponding potentials X, A and h. As already mentioned in the introduction, it will turn out that this provides the Kerr solution, a truly miraculous fact.

Equation (7.168) gives

$$X = \frac{1-p^2x^2-q^2y^2}{(1+px)^2+q^2y^2}\,, \qquad Y = \frac{2q\,y}{(1+px)^2+q^2y^2}\,. \tag{7.170}$$

According to (7.155) and (7.159a), (7.159b) we know the partial derivatives of A and h:

$$A_{,x} = 2q\,\mu(1-y^2)\frac{(1+px)^2-(qy)^2}{\left[1-(px)^2-(qy)^2\right]^2}\,, \tag{7.171a}$$

$$A_{,y} = 2p\,\mu(x^2-1)\frac{(1+px)2q\,y}{\left[1-(px)^2-(qy)^2\right]^2}\,, \tag{7.171b}$$

$$h_{,x} = -\frac{x(1-y^2)}{(x^2-y^2)\left[1-(px)^2-(qy)^2\right]}\,, \tag{7.171c}$$

$$h_{,y} = -\frac{y(x^2-1)}{(x^2-y^2)\left[1-(px)^2-(qy)^2\right]}\,. \tag{7.171d}$$

Integration gives (up to irrelevant integration constants)

$$A = 2\mu\frac{q}{p}\frac{(1-y^2)(1+px)}{1-(px)^2-(qy)^2}\,, \tag{7.172a}$$

$$\mathrm{e}^{2h} = \frac{1-(px)^2-(qy)^2}{p^2(x^2-y^2)}\,. \tag{7.172b}$$

Unfortunately, this is unacceptable. According to (7.170) the norm of the Killing field m does not vanish on the rotation axis $\{y=\pm1, x>1\}$.

Furthermore, X is not definite in the strip $\{x > 1, |y| \leq 1\}$, corresponding to the region outside the horizon. Fortunately, we obtain a solution with all desired properties by passing to the conjugate solution (7.151). For this we obtain after a few manipulations

$$\hat{X} = \frac{\mu^2}{p^2}(1-y^2)\frac{\left[(1+px)^2+q^2\right]^2 - p^2q^2(x^2-1)(1-y^2)}{(1+px)^2+(qy)^2}, \tag{7.173a}$$

$$\hat{A} = -2\,\frac{pq}{\mu}\,\frac{1+p\,x}{\left[(1+px)^2+q^2\right]^2 - p^2q^2(x^2-1)(1-y^2)}, \tag{7.173b}$$

$$e^{2\hat{h}} = \hat{X}\,\frac{(1+px)^2+(qy)^2}{p^2(x^2-y^2)}. \tag{7.173c}$$

The **vorticity potential** $\hat{Y}$ (defined by $d\hat{Y} = 2\hat{\omega}$) is obtained by integrating (7.155) for the conjugate quantities, with the result

$$\hat{Y} = 2\mu^2\frac{q}{p^2}\,y\left(3 - y^2 + \frac{q^2(1-y^2)^2}{(1+px)^2+(qy)^2}\right). \tag{7.174}$$

Finally, we also give the expressions for the metric functions $\hat{W}$ and $\hat{V}$. From $\hat{W} = \hat{A}\hat{X} = -AX$, we get

$$\hat{W} = -2\mu\,\frac{q}{p}\,\frac{(1-y^2)(1+px)}{(1+px)^2+(qy)^2}. \tag{7.175}$$

Similarly, using $\hat{V}\hat{X} + \hat{W}^2 = \rho^2 = \hat{V}\hat{X} + A^2X^2$, we see that

$$\hat{V} = \frac{1}{\hat{X}}(\rho^2 - A^2X^2) \overset{(7.151)}{=} -X, \tag{7.176}$$

whence

$$\hat{V} = \frac{1-(px)^2-(qy)^2}{(1+px)^2+(qy)^2}. \tag{7.177}$$

In terms of the expressions (7.173a), (7.173c), (7.175) and (7.177) for the potentials $\hat{V}, \hat{W}, \hat{X}$ and $\hat{h}$ we get the metric of the **Kerr solution in elliptical coordinates**

$$^{(4)}g = -\hat{V}\,dt^2 + 2\hat{W}\,dtd\varphi + \hat{X}\,d\varphi^2 + \frac{e^{2\hat{h}}}{\hat{X}}\mu^2(x^2-y^2)\left(\frac{dx^2}{x^2-1} + \frac{dy^2}{1-y^2}\right). \tag{7.178}$$

Kerr Solution in Boyer-Lindquist Coordinates

We transform this result to new coordinates $(t, r, \vartheta, \varphi)$, the so-called Boyer-Lindquist coordinates, which are better adapted. They are defined by

$$r = m(1+px), \qquad \cos\vartheta = y. \tag{7.179}$$

Instead of the parameters μ, p and q we introduce

$$m = \mu/p, \qquad a = \mu q/p. \tag{7.180}$$

Note that $p^2 + q^2 = 1$ implies $m^2 - a^2 = \mu^2 \geq 0$, i.e., $|a| \leq m$. The following abbreviations are useful

$$\Delta = r^2 - 2mr + a^2, \qquad \Xi = r^2 + a^2\cos^2\vartheta. \tag{7.181}$$

Using the relations

$$x^2 - 1 = \frac{\Delta}{\mu^2}, \qquad \frac{dx^2}{x^2-1} = \frac{1}{\Delta}dr^2,$$
$$1 - y^2 = \sin^2\vartheta, \qquad \frac{dy^2}{1-y^2} = d\vartheta^2,$$

and the identities

$$\begin{aligned}
(1+px)^2 + (qy)^2 &= \frac{\Xi}{m^2}, \\
(1+px)^2 + q^2 &= \frac{r^2+a^2}{m^2}, \\
(px)^2 + (qy)^2 - 1 &= \frac{\Delta - a^2\sin^2\vartheta}{m^2}, \\
1 + px = \frac{r}{m} &= \frac{(r^2+a^2) - \Delta}{2m^2},
\end{aligned}$$

one finds

$$\begin{aligned}
\hat{X} &= \Xi^{-1}\sin^2\vartheta\left[(r^2+a^2)^2 - \Delta a^2\sin^2\vartheta\right], \\
\hat{W} &= \Xi^{-1}a\sin^2\vartheta\left[\Delta - (r^2+a^2)\right], \\
\hat{V} &= \Xi^{-1}\left[\Delta - a^2\sin^2\vartheta\right], \\
e^{2\hat{h}} &= \hat{X}\,\Xi\left[\Delta + (m^2-a^2)\sin^2\vartheta\right]^{-1}, \\
\hat{Y} &= 2am\cos\vartheta\left[(3-\cos^2\vartheta) + \Xi^{-1}a^2\sin^4\vartheta\right].
\end{aligned} \tag{7.182}$$

For the Kerr solution we finally obtain the **Kerr solution in Boyer-Lindquist coordinates**

$$\begin{aligned}
{}^{(4)}g = \frac{1}{\Xi}\Big[&-(\Delta - a^2\sin^2\vartheta)dt^2 + 2a\sin^2\vartheta(\Delta - r^2 - a^2)dt\,d\varphi \\
&+ \sin^2\vartheta\left((r^2+a^2)^2 - \Delta a^2\sin^2\vartheta\right)d\varphi^2\Big] \\
&+ \Xi\left(\frac{dr^2}{\Delta} + d\vartheta^2\right).
\end{aligned} \tag{7.183}$$

Below we discuss this two-parameter family in detail. For the proof that there are no other stationary axisymmetric vacuum black holes, we have to refer to the literature (see [189], and references therein).

Interpretation of the Parameters *a* and *m*

At this point, we interprete the two parameters m and a introduced in (7.180). For this we look at the asymptotic form of (7.183) for large 'radial coordinate' r:

$$\begin{aligned} {}^{(4)}g = &-\left[1 - \frac{2m}{r} + O\left(\frac{1}{r^2}\right)\right] dt^2 - \left[\frac{4a\,m}{r}\sin^2\vartheta + O\left(\frac{1}{r^2}\right)\right] dt\,d\varphi \\ &+ \left[1 + O\left(\frac{1}{r}\right)\right]\left(dr^2 + r^2(d\vartheta^2 + \sin^2\vartheta\,d\varphi^2)\right). \end{aligned} \tag{7.184}$$

Introducing 'Cartesian coordinates'

$$x = r\sin\vartheta\sin\varphi\,, \quad y = r\sin\vartheta\cos\varphi\,, \quad z = r\cos\varphi\,,$$

(7.184) takes the general form (4.212), which was derived more systematically in Sect. 5.3. It was shown in Sect. 4.7 that m is the *total mass* of the black hole, and that the magnitude J of the angular momentum is given by

$$J = a\,m\,, \tag{7.185}$$

We know (see (4.222)) that m is also given by the Komar formula

$$m = -\frac{1}{8\pi}\int_{S^2_\infty} *\,dk\,. \tag{7.186}$$

As expected, also J can be expressed as a Komar integral:

$$J = \frac{1}{16\pi}\int_{S^2_\infty} *\,dm\,. \tag{7.187}$$

Let us show this.

To avoid notational confusions we use a capital M for the total mass. The Killing 1-form $m = g_{\mu\varphi}dx^\mu$ is asymptotically

$$m \simeq -\frac{2a\,M}{r}\sin^2\vartheta\,dt + r^2\sin^2\vartheta\,d\varphi\,,$$

thus

$$dm \simeq \frac{2a\,M}{r^2}\sin^2\vartheta\,dr\wedge dt + 2r\sin^2\vartheta\,dr\wedge d\varphi + \ldots\,.$$

The terms that are not written out do not contribute to $*dm|_{t,r}$ in the flux integral (7.187). To determine the Hodge-dual we work with the basis

$$\theta^0 = \left(1 - \frac{2M}{r}\right)dt\,, \qquad \theta^1 = dr\,, \qquad \theta^2 = r\,d\vartheta\,,$$

$$\theta^3 = r\sin\vartheta\left(d\varphi - \frac{2a\,M}{r^3}\,dt\right),$$

which, according to (7.184), is asymptotically orthonormal. Since

$$dm \simeq -\frac{2a\,M}{r^2}\sin^2\vartheta\,\theta^0\wedge\theta^1 - \frac{4a\,M}{r^2}\sin^2\vartheta\,\theta^0\wedge\theta^1 + \dots$$
$$= -\frac{6a\,M}{r^2}\sin^2\vartheta\,\theta^0\wedge\theta^1 + \dots\,,$$

we obtain

$$*\,dm \simeq \frac{6a\,M}{r^2}\sin^2\vartheta\,\theta^2\wedge\theta^3 + \dots\,.$$

Again, only terms written out contribute to

$$\frac{1}{16\pi}\int_{S^2_\infty} *\,dm = \frac{6}{16\pi}a\,M\int_{S^2}\sin^3\vartheta\,d\vartheta\wedge d\varphi = a\,M = J\,.$$

* * *

Exercises.

Exercise 7.1. Derive (7.118) in the traditional way with the use of the Christoffel symbols.

Exercise 7.2. Show that

$$^{(4)}R_{ai} \equiv 0\,. \tag{7.188}$$

Exercise 7.3. Consider on a two-dimensional manifold two conformally related metrics g and $\bar{g}$, with $\bar{g} = \mathrm{e}^{-2U}g$. Show that for a function f

$$\begin{aligned}\bar{\nabla}_i\bar{\nabla}_j f &= \nabla_i\nabla_j f + U_{,i}f_{,j} + U_{,j}f_{,,i} - g_{ij}\langle df, dU\rangle^{(g)}\,,\\ \bar{\Delta}f &= \mathrm{e}^{2U}\,\Delta f\,,\end{aligned} \tag{7.189}$$

and derive the following relation between the Ricci tensors

$$\bar{R}_{ij} = R_{ij} + g_{ij}\,\Delta U\,, \tag{7.190}$$

thus

$$\bar{R} = \mathrm{e}^{2U}(R + 2\,\Delta U)\,. \tag{7.191}$$

(In connection with this exercise see also Exercise 2.22 and its solution.)

Exercise 7.4. Use an algebraic computer program to check that the metric (7.183) is really Ricci flat.

7.3 Discussion of the Kerr-Newman Family

The analytic expression for the Kerr metric looks rather complicated. In this section we analyze this important vacuum solution and will arrive at a good qualitative understanding. In particular, we shall see that there is an event horizon. Compared to the Schwarzschild black hole, interesting new phenomena will show up. In this section we do not assume that the reader has studied the rather lengthy derivation of the Kerr solution in Sect. 7.2.

We change the notation used in Sect. 7.2.7 slightly and extend the Kerr solution to the three-parameter Kerr-Newman solution that describes also *electrically* charged black holes. This will be given in Boyer-Lindquist coordinates $(t, r, \vartheta, \varphi)$. We introduce the following abbreviations involving the three parameters M, a and $Q (c = G = 1)$

$$\begin{aligned} \Delta &= r^2 - 2M\,r + a^2 + Q^2\,, \\ \rho^2 &= r^2 + a^2\cos^2\vartheta\,, \\ \Sigma^2 &= (r^2+a^2)^2 - a^2\Delta\sin^2\vartheta\,. \end{aligned} \tag{7.192}$$

(Instead of Ξ in (7.181) we now use the letter ρ.) In terms of these functions the metric coefficients of the **Kerr-Newman solution** are

$$g_{rr} = \frac{\rho^2}{\Delta}\,, \qquad g_{\vartheta\vartheta} = \rho^2\,, \qquad g_{\varphi\varphi} = \frac{\Sigma^2}{\rho^2}\sin^2\vartheta\,, \tag{7.193}$$

$$g_{tt} = -1 + \frac{2M\,r - Q^2}{\rho^2}\,, \qquad g_{t\varphi} = -a\frac{2M\,r}{\rho^2}\sin^2\vartheta\,. \tag{7.194}$$

(Without loss of generality we can choose $a > 0$.) As special cases we have:

a) $Q = a = 0$: Schwarzschild solution;
b) $a = 0$: Reissner-Nordstrøm solution;
c) $Q = 0$: Kerr solution.

Let us write this metric also in standard 3+1 form (see Sect. 2.9)

$$g = \left[-\alpha^2\,dt^2 + g_{\varphi\varphi}\,(d\varphi + \beta^\varphi\,dt)^2\right] + \left[g_{rr}\,dr^2 + g_{\vartheta\vartheta}\,d\vartheta^2\right]\,, \tag{7.195}$$

with only the component β^φ of the shift being different from zero. Note that

$$-\alpha^2 = \frac{1}{g_{\varphi\varphi}}(g_{tt}g_{\varphi\varphi} - g_{t\varphi}^2)\,, \qquad \beta^\varphi = \frac{g_{\varphi t}}{g_{\varphi\varphi}}\,. \tag{7.196}$$

Lapse and shift are found to be

$$\alpha^2 = \frac{\rho^2}{\Sigma^2}\Delta, \qquad \beta^\varphi = -a\,\frac{2M\,r}{\Sigma^2}\,. \tag{7.197}$$

The electromagnetic field of the Kerr-Newman solution is

$$F = Q\,\rho^{-4}(r^2 - a^2\cos^2\vartheta)\,dr\wedge(dt\wedge a\sin^2\vartheta\,d\varphi)$$
$$+2Q\,\rho^{-4}a\,r\,\cos\vartheta\sin\vartheta\,d\vartheta\wedge\left((r^2+a^2)d\varphi - a\,dt\right)\,. \qquad (7.198)$$

(For a derivation along similar lines as in Sect. 7.2 we recommend [189].) For $a = 0$ this reduces to the electromagnetic field (3.167b) of the Reissner-Nordstrøm solution with $e = Q$.

7.3.1 Gyromagnetic Factor of a Charged Black Hole

The three parameters M, a and Q are most easily interpreted by looking at the asymptotic forms of the metric g and the electromagnetic field F. The leading terms in the expansion of the metric in powers of $1/r$ are the same as for $Q = 0$, and have been given earlier in (7.184). There we showed that M is the total mass and that the total angular momentum J is given by

$$J = a\,M\,. \qquad (7.199)$$

The asymptotic expressions for the electric and magnetic field components in the r, ϑ and φ directions are (a hat indicates the components relative to an orthonormal triad):

$$E_{\hat{r}} = E_r = F_{rt} = \frac{Q}{r^2} + O\left(\frac{1}{r^3}\right)\,,$$
$$E_{\hat{\vartheta}} = \frac{E_\vartheta}{r} = \frac{1}{r}F_{\vartheta t} = O\left(\frac{1}{r^4}\right)\,,$$
$$E_{\hat{\varphi}} = \frac{E_\varphi}{r\,\sin\varphi} = \frac{F_{\varphi t}}{r\,\sin\vartheta} = 0\,; \qquad (7.200a)$$
$$B_{\hat{r}} = F_{\hat{\vartheta}\hat{\varphi}} = \frac{F_{\vartheta\varphi}}{r^2\,\sin\vartheta} = \frac{2\,Q\,a}{r^3}\cos\vartheta + O\left(\frac{1}{r^4}\right)\,,$$
$$B_{\hat{\vartheta}} = F_{\hat{\varphi}\hat{r}} = \frac{F_{\varphi r}}{r\,\sin\vartheta} = \frac{Q\,a}{r^3}\sin\vartheta + O\left(\frac{1}{r^4}\right)\,,$$
$$B_{\hat{\varphi}} = F_{\hat{r}\hat{\vartheta}} = \frac{F_{r\vartheta}}{r} = 0\,. \qquad (7.200b)$$

Asymptotically the electric field is a Coulomb field with charge Q.

In general the electric charge is given by the following Gaussian flux integral

$$Q = \frac{1}{4\pi}\int_{S^2_\infty} *F\,. \qquad (7.201)$$

This formula comes about as follows. Consider a spacelike hypersurface Σ which is non-singular, and some domain $\mathcal{D}\subset\Sigma$. The electric charge $Q(\mathcal{D})$ inside $\mathcal{D}$ can be expressed in various ways:

$$Q = \int_{\mathcal{D}} *J = \frac{1}{4\pi}\int_{\mathcal{D}} d*F = \frac{1}{4\pi}\int_{\partial\mathcal{D}} *F\,. \qquad (7.202)$$

When Σ extends to spatial infinity and $\mathcal{D}$ becomes all of Σ, we obtain (7.201). It is natural to adopt this equation as a definition of the total electric charge for more general situations, in particular for black holes with singularities inside the horizon. If the asymptotic form for F in (7.198) is inserted in (7.201), one easily finds that the flux integral on the right agrees with the parameter Q in the Kerr-Newman solution. So Q is the **electric charge** of the black hole.

Equation (7.200b) shows that the magnetic field is asymptotically dipolar with magnetic dipole moment

$$\mu = Q\,a = \frac{Q}{M} J =: g\,\frac{Q}{2M} J\,. \tag{7.203}$$

Thus, we obtain the amazing result for the **gyromagnetic factor** g:

$$g = 2\,. \tag{7.204}$$

Symmetries of the Metric

The metric coefficients (7.193) are all independent of t and φ. Hence

$$k = \frac{\partial}{\partial t} \quad \text{and} \quad m = \frac{\partial}{\partial \varphi} \tag{7.205}$$

are Killing fields, i.e., $L_k g = L_m g = 0$. (In Sect. 7.2.1 this was actually the starting point for the derivation of the Kerr metric.) Therefore, the metric coefficients g_{tt}, $g_{t\varphi}$ and $g_{\varphi\varphi}$ are scalar products of these two commuting Killing fields and have thus an intrinsic meaning. We note

$$g_{tt} = \langle k, k\rangle\,, \quad g_{t\varphi} = \langle k, m\rangle\,, \quad g_{\varphi\varphi} = \langle m, m\rangle\,. \tag{7.206}$$

7.3.2 Static Limit and Stationary Observers

An observer moving along a world line with constant r, ϑ and uniform angular velocity sees an unchanging spacetime geometry, and is thus a *stationary* observer. His angular velocity, measured at infinity, is

$$\omega = \frac{d\varphi}{dt} = \frac{\dot\varphi}{\dot t} = \frac{u^\varphi}{u^t}\,, \tag{7.207}$$

where u is the four-velocity of the observer. This four-velocity is proportional to a timelike Killing field

$$u = u^t\left(\frac{\partial}{\partial t} + \omega\frac{\partial}{\partial \varphi}\right) = \frac{k + \omega\, m}{|k+\omega m|}\,, \tag{7.208}$$

where $|k+\omega m| = \left(-\langle k+\omega\, m, k+\omega\, m\rangle\right)^{1/2}$. Since $k+\omega\, m$ is timelike, we have

$$g_{tt} + 2\omega\, g_{t\varphi} + \omega^2\, g_{\varphi\varphi} < 0\,. \tag{7.209}$$

The left-hand side vanishes, i.e., $k + \omega\, m$ becomes lightlike, for

$$\omega = \frac{-g_{t\varphi} \pm \sqrt{g_{t\varphi}^2 - g_{tt} g_{\varphi\varphi}}}{g_{\varphi\varphi}}.$$

Let $\Omega = -g_{t\varphi}/g_{\varphi\varphi} = -\langle k, m\rangle / \langle m, m\rangle$, then

$$\omega_{min} = \Omega - \sqrt{\Omega^2 - g_{tt}/g_{\varphi\varphi}}\,, \tag{7.210a}$$

$$\omega_{max} = \Omega + \sqrt{\Omega^2 - g_{tt}/g_{\varphi\varphi}}\,. \tag{7.210b}$$

Note that

$$\Omega = -\beta^{\varphi} = a\,\frac{2M\,r - Q^2}{\Sigma^2}\,. \tag{7.211}$$

For an interpretation of Ω consider stationary observers who are non-rotating with respect to local freely falling test particles that have been dropped in radially from infinity. Since the angular momentum of such test particles vanishes, we have for these special stationary observers, so-called **Bardeen observers**, $\langle u, m\rangle = 0$, hence $\langle k + \omega\, m, m\rangle = 0$, i.e., $\omega = \Omega$. Obviously $\omega_{min} = 0$ if and only if $g_{tt} = 0$, that is for $\langle k, k\rangle = 0$:

$$r = r_0(\vartheta) := M + \sqrt{M^2 - Q^2 - a^2\cos^2\vartheta}\,. \tag{7.212}$$

We always assume that $M^2 > a^2 + Q^2$; otherwise there exists a "naked singularity" (see below).

An observer is said to be *static* (relative to the "fixed stars") if $\omega = 0$, so that u is proportional to k (see (7.208)). Static observers can exist only outside the **static limit** $\{r = r_0(\vartheta)\}$. At the static limit, k becomes lightlike. An observer would have to move at the speed of light in order to remain at rest with respect to the fixed stars. The redshift which an asymptotic observer measures for light emitted from a source "at rest" (u proportional to k) outside the static limit is (see (1.124))

$$\frac{\nu_e}{\nu_0} = \frac{\sqrt{-\langle k, k\rangle_0}}{\sqrt{-\langle k, k\rangle_e}} \simeq \frac{1}{\sqrt{-\langle k, k\rangle_e}}\,. \tag{7.213}$$

Note that this expression goes to infinity at the static limit.

7.3.3 Killing Horizon and Ergosphere

For $\Omega^2 = g_{tt}/g_{\varphi\varphi}$, ω_{min} and ω_{max} coincide and are equal to Ω. Since by definition $\Omega^2 = g_{t\varphi}^2/g_{\varphi\varphi}^2$ this is equivalent to

$$g_{tt}g_{\varphi\varphi} - g_{t\varphi}^2 = 0\,. \tag{7.214}$$

Alternatively, this critical surface can be described as a result of (7.196) and (7.197) as

$$\alpha^2 = 0 \quad \text{or} \quad \Delta = 0 . \tag{7.215}$$

Sometimes it is useful to formulate this as follows: Let

$$\xi = k + \Omega\, m , \tag{7.216}$$

then (7.214) is equivalent to ξ being lightlike, $\langle \xi, \xi \rangle = 0$. We shall see further below that this hypersurface H is an **event horizon** (for $M^2 > a^2 + Q^2$). Explicitly H is given by (see 7.192)

$$r = r_H := M + \sqrt{M^2 - a^2 - Q^2} . \tag{7.217}$$

On the hypersurface H the expression Ω becomes by (7.211)

$$\Omega|_H =: \Omega_H = \frac{a(2M\, r_H - Q)}{(r_H^2 + a^2)^2}$$

or, using (7.217),

$$\Omega_H = \frac{a}{r_H^2 + a^2} \qquad (= \text{const}) . \tag{7.218}$$

This is a very remarkable result: The black hole rotates *like a rigid body.* We have obtained this fact from the explicit formulas. In Appendix 7.7.1 we shall give a general argument why Ω has to be constant on the horizon. Because of this,

$$l := k + \Omega_H\, m \tag{7.219}$$

is a Killing field, coinciding on H with ξ, and is thus null on H. The flow of the vector field l leaves the horizon invariant, because

$$L_l \langle l, l \rangle = 0 . \tag{7.220}$$

Indeed, since l is a Killing field we have

$$L_l \langle l, l \rangle = (L_l g)(l, l) + 2 \langle [l, l], l \rangle = 0 .$$

So any hypersurface $\{\langle l, l \rangle = \text{const}\}$ is left invariant. In particular, the Killing field l is a tangent null vector of H.

Next, we show that H is a null hypersurface, that is, normal vectors are lightlike. This can again be seen from the explicit formulas for the metric. Since ξ and l coincide on H, it suffices to show that the gradient of $\langle \xi, \xi \rangle$ becomes lightlike on H. Now,

$$\langle \xi, \xi \rangle = -\alpha^2 = -\frac{\rho^2}{\Sigma^2} \Delta ,$$

and because g^{rr} is proportional to Δ, we see that $\langle \nabla \alpha^2, \nabla \alpha^2 \rangle$ indeed vanishes on H.

We now state a theorem which demonstrates what is behind these results. The formulation makes use of concepts introduced in Sect. 7.2.1, and the proof of the theorem will be given in Appendix 7.7.1 of this chapter.

Theorem 7.1 (Weak Rigidity Theorem). *Let (M, g) be a circular spacetime with commuting Killing fields k, m and let $\xi = k + \Omega m$ with $\Omega = -\langle k, m\rangle / \langle m, m\rangle$. Then Ω is constant on the hypersurface $H = \{\langle \xi, \xi\rangle = 0\}$. Moreover, H is a null hypersurface.*

The hypersurface H is a Killing horizon in the sense of the following

Definition 7.1. *Let K be a Killing field and H the set where K is null. A connected component of this set which is a null hypersurface, and any union of such null surfaces is a* ***Killing horizon****.*

We emphasize that this concept is – in contrast to the one of the event horizon – of a *local* nature. It is natural to expect that the event horizon of a stationary black hole is a Killing horizon. That this is indeed true has been proven only relatively recently as part of a corrected version of the so-called "strong rigidity theorem"[5].

Since a Killing horizon is a null hypersurface, the tangent space at each point is orthogonal to a null vector, and therefore does not contain timelike vectors. Moreover, the set of null vectors is one-dimensional and spanned by any normal vector. (This, and other algebraic facts will be proven in Exercise 7.5.) In other words, a Killing horizon is tangent to the light cone at each point. Therefore, crossing is possible in only *one* direction.

In the situation above l is as a null vector tangent to H also proportional to a normal vector field on H, i.e.,

$$dN = -2\kappa\, l\,, \qquad N := \langle l, l\rangle\,. \tag{7.221}$$

The proportionality factor κ is the **surface gravity** of the Killing horizon, and will play an important role. Instead of (7.221) we can write

$$\nabla_l l = \kappa\, l\,, \tag{7.222}$$

because for a Killing field we have

$$dN = d\, i_l l = -i_l\, dl = 2\, i_l \nabla l = -2\, \nabla_l l\,. \tag{7.223}$$

(We have used DG, Eq. (13.101); in index notation (7.223) follows immediately from the Killing equation.)

Equation (7.222) tells us that the integral curves of l are non-affinely parameterized geodesics. (Show that by an appropriate re-parameterization these integral curves satisfy the standard geodesic equation.) These null geodesics generate the Killing horizon. For a more general discussion, see DG, Appendix A, especially Proposition A.1.

[5] For a precise formulation and a complete proof we refer to [218]. Unfortunately, so far the unjustified hypothesis of analyticity of the metric in a neighborhood of the event horizon had to be made.

The static limit is timelike, except at the poles (critical points of m), because a normal vector to the static hypersurface, being orthogonal to the null vector k, cannot be timelike. Therefore one can pass through this surface in *both* directions, in contrast to the Killing horizon. The region between the static limit and this horizon is the so-called **ergosphere** (for reasons which will be clarified later). The static limit and the horizon come together at the poles (see Fig. 7.2). Inside the ergosphere, k is spacelike. This implies that "all the king's horses and all the king's men" cannot prevent on observer from rotating about the black hole. The ergosphere disappears when $a \longrightarrow 0$.

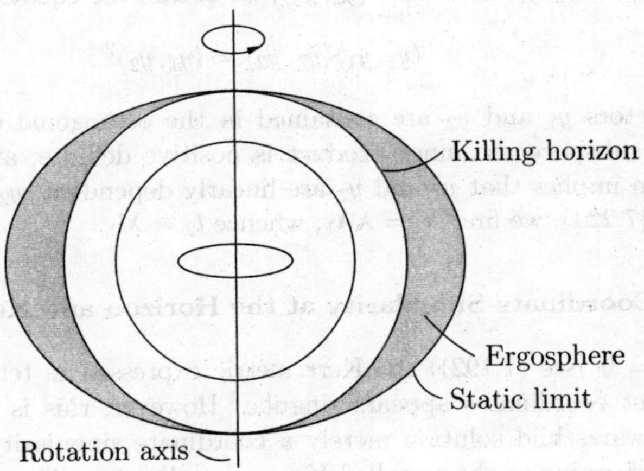

Fig. 7.2. Cross section through the axis of rotation of a Kerr-Newman solution.

* * *

Exercise 7.5. Let V be an n-dimensional Minkowski vector space with inner product $\langle\cdot,\cdot\rangle$. Prove the following facts:

a) Two timelike vectors are never orthogonal.
b) A timelike vector is never orthogonal to a null vector.
c) Two null vectors are orthogonal if and only if they are linearly dependent.
d) The orthogonal complement of a null vector is an $(n-1)$-dimensional subspace of V in which the inner product is positive semi-definite and has rank $n-2$.

Solution. Fix a timelike unit vector u. Any vector $x \in V$ can be uniquely decomposed in the form

$$x = \lambda u + y, \qquad \langle u, y\rangle = 0.$$

The scalar λ is given by

$$\lambda = -\langle x, u \rangle .$$

From this decomposition all four properties can be established. As an example we prove c). – Let l_1 and l_2 be two orthogonal null vectors. Then we have the decompositions

$$l_1 = \lambda_1 \, u + y_1 , \qquad l_2 = \lambda_2 \, u + y_2 ,$$

with

$$\langle l_1, l_2 \rangle = 0 = -\lambda_1 \lambda_2 + \langle y_1, y_2 \rangle . \tag{7.224}$$

Since $\lambda_1^2 = \langle y_1, y_1 \rangle$ and $\lambda_2^2 = \langle y_2, y_2 \rangle$, we obtain the equation

$$\langle y_1, y_1 \rangle \langle y_2, y_2 \rangle = \langle y_1, y_2 \rangle^2 .$$

The vectors y_1 and y_2 are contained in the orthogonal complement of u. In this subspace the inner product is positive definite, and hence the last equation implies that y_1 and y_2 are linearly dependent, $y_2 = \lambda \, y_1$. Inserting this in (7.224), we find $\lambda_2 = \lambda \, \lambda_1$, whence $l_2 = \lambda l_1$.

7.3.4 Coordinate Singularity at the Horizon and Kerr Coordinates

For $\Delta = 0$ (see (7.192)) the Kerr metric expressed in terms of the Boyer-Lindquist coordinates appears singular. However, this is as in the case of the Schwarzschild solution merely a coordinate singularity, as can be seen by transforming to the so-called **Kerr coordinates**. These new coordinates are generalizations of the Eddington-Finkelstein coordinates for spherically symmetric black holes and are defined by

$$\begin{aligned} d\tilde{V} &= dt + \frac{r^2 + a^2}{\Delta} dr , \\ d\tilde{\varphi} &= d\varphi + \frac{a}{\Delta} dr . \end{aligned} \tag{7.225}$$

Note that the exterior differential of the right-hand sides vanish. The metric can be written in terms of the new coordinates $(\tilde{V}, r, \vartheta, \tilde{\varphi})$ as follows

$$\begin{aligned} {}^{(4)}g = &-\left(1 - \rho^{-2}(2M\, r - Q^2)\right) d\tilde{V}^2 + 2\, dr\, d\tilde{V} + \rho^2\, d\vartheta^2 \\ &+\rho^{-2}\left((r^2 + a^2)^2 + \Delta\, a^2 sin^2 \vartheta\right) \sin^2 \vartheta\, d\tilde{\varphi}^2 - 2a \sin^2 \vartheta\, d\tilde{\varphi}\, dr \\ &-2a\, \rho^{-2}(2m\, r - Q^2) \sin^2 \vartheta\, d\tilde{\varphi}\, d\tilde{V} . \end{aligned} \tag{7.226}$$

This expression is regular at the horizon. In place of $\tilde{V}$ on often uses $\tilde{t} = \tilde{V} - r$. In terms of the Kerr coordinates, the Killing fields k and m are given by

$$k = \left(\frac{\partial}{\partial \tilde{t}} \right)_{r, \vartheta, \tilde{\varphi}} , \qquad m = \left(\frac{\partial}{\partial \tilde{\varphi}} \right)_{\tilde{t}, r, \vartheta} . \tag{7.227}$$

Singularities of the Kerr-Newman Metric

The Kerr-Newman metric has a true singularity which lies inside the horizon for $M^2 > a^2 + Q^2$. It has a rather complicated structure, which his described in detail in [8]. The maximal analytic continuation of the Kerr solution is also discussed there. The global structure turns out to be very similar to that of the Reissner-Nordstrøm solution. These aspects have no relevance for astrophysics and hence will not be treated here.

7.3.5 Structure of the Light Cones and Event Horizon

Visualizing the spacetime geometry is made easier by considering the structure of the light cones. We examine this more closely in the equatorial plane, as indicated in Fig. 7.3. Each point in this plane represents an integral curve of the Killing field k. The wave fronts of light signals which have formed shortly after being emitted from the marked points are also shown in Fig. 7.3. We note the following facts:

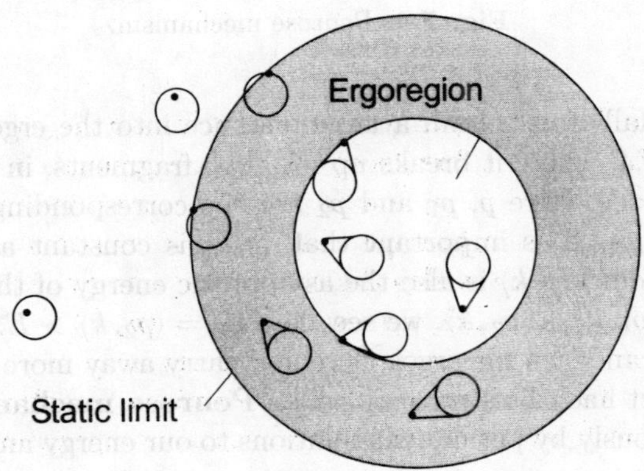

Fig. 7.3. Structure of light cones in the equatorial plane of a Kerr-Newman black hole.

a) Since k is *timelike* outside the static limit, the points of emission are inside the wave fronts.
b) At the static limit k becomes *lightlike*, and the point of emission lies thus on the wave front.
c) Inside the ergosphere k is *spacelike* and hence the emitting points are outside the wave fronts.
d) For $r = r_H$, the Killing field k is still *spacelike*, but the wave fronts arising from a point of emission on this surface lie entirely inside the surface, except for touching points, because $r = r_H$ (i.e. $\Delta = 0$) is a null surface. This demonstrates that this surface is indeed an **event horizon**. This

lightlike hypersurface is invariant with respect to k and m (prove this or see Appendix 7.7.1).

7.3.6 Penrose Mechanism

Since k is spacelike inside the ergosphere, it is possible in principle to extract energy from a black hole, thereby reducing its angular velocity and thus also the size of the ergosphere. Imagine, as a gedankenexperiment, a piece of

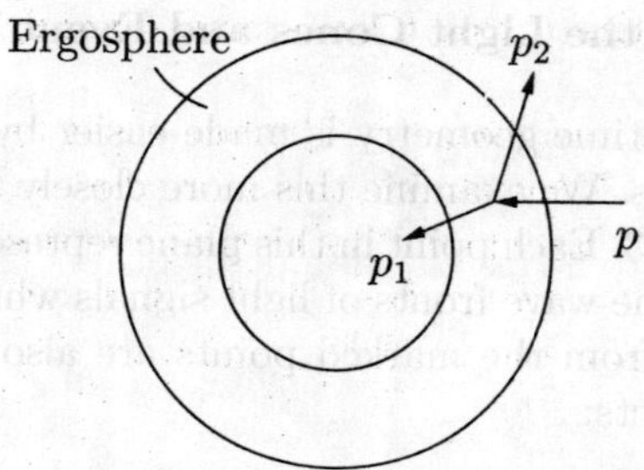

Fig. 7.4. Penrose mechanism.

matter which falls freely from a large distance into the ergosphere, as indicated in Fig. 7.4, where it breaks up into two fragments, in such a way that $E_1 := \langle p_1, k\rangle < 0$. Here p, p_1 and p_2 are the corresponding four-momenta, and $p = p_1 + p_2$. It is important that $\langle p, k\rangle$ is constant along a geodesic. The scalar product $\langle p, k\rangle$ is also the asymptotic energy of the particle. Since $E = \langle p, k\rangle = \langle p_1, k\rangle + \langle p_2, k\rangle$, we see that $E_2 = \langle p_2, k\rangle > E$. Hence, the second fragment can leave the ergosphere and carry away more energy than the incoming object had. This process, called **Penrose mechanism**, could provide simultaneously by permanent solutions to our energy and waste disposal problems.

An astrophysically interesting possibility to extract the rotational energy from a black hole has been suggested in [196]. This so-called **Blandford-Znajek process** can work if rotation and turbulence in an accretion disk around a black hole generates sufficiently large magnetic fields. A detailed treatment of this magnetic energy extraction is given in [197] and [198]. The possibility that gamma-ray bursts are powered by the Blandford-Znajek process is reviewed in [199].

7.3.7 Geodesics of a Kerr Black Hole

The description of the geodesics for the Kerr solution is far more complicated than for the Schwarzschild metric. The problem is, however, completely integrable. Surprisingly, it turns out that beside the three obvious first integrals of motion there is a fourth one. This was discovered by B. Carter (see [200]) with the help of the Hamilton-Jacobi method. Since we know of no other way to discover this hidden symmetry, we briefly remind the reader how this works.

* * *

The Hamilton-Jacobi Method

We know that the geodesic equation can be regarded as the Euler-Lagrange equation for the Lagrangian

$$\mathcal{L}(x^\mu, \dot{x}^\nu) = \frac{1}{2} g_{\mu\nu} \dot{x}^\mu \dot{x}^\nu . \tag{7.228}$$

From this we can pass to a Hamiltonian description. The canonical momenta are

$$p_\mu = \frac{\partial \mathcal{L}}{\partial \dot{x}^\mu} = g_{\mu\nu} \dot{x}^\nu , \tag{7.229}$$

and hence the Hamiltonian function is given by

$$H(x^\mu, p_\nu) = \frac{1}{2} g^{\mu\nu} p_\mu p_\nu . \tag{7.230}$$

The symplectic form on the phase space is the standard one:

$$\omega = dq^\mu \wedge dp_\mu .$$

Hamilton's canonical equations

$$\dot{x}^\mu = \frac{\partial H}{\partial p_\mu}, \qquad \dot{p}_\mu = -\frac{\partial H}{\partial x^\mu} \tag{7.231}$$

are equivalent to the geodesic equations. (There is a very geometrical way to describe all this; see, e.g., [33].)

The Hamilton-Jacobi method of solving the system of canonical equations (7.231) consists of the following (see, e.g., [28]): Construct a *complete* solution[6] $S(x^\mu, \lambda; \alpha_\nu)$ of the **Hamilton-Jacobi equation**

$$H\left(x^\mu, \frac{\partial S}{\partial x^\nu}\right) + \frac{\partial S}{\partial \lambda} = 0 , \tag{7.232}$$

depending on four parameters α_ν (equal to the number of degrees of freedom). Then the orbits $x^\mu(\lambda)$, $p_\nu(\lambda)$ are determined by

$$p_\mu = \frac{\partial S}{\partial x^\mu} , \tag{7.233}$$

[6] By this one means that $\det(\partial^2 S/\partial q^\mu \partial \alpha_\nu) \neq 0$. Such solutions always exist locally.

and

$$\beta^\mu = \frac{\partial S}{\partial \alpha_\mu}, \tag{7.234}$$

where α_μ and β^ν are constants. (S is the generating function of a canonical transformation $(x^\mu, p_\nu) \longmapsto (\beta^\mu, \alpha_\nu)$, with the property that the transformed Hamilton equations become trivial: $\dot\beta^\mu = 0$, $\dot\alpha_\nu = 0$.)

For autonomous systems, as our geodesics problem, one can separate the "time" dependence of S:

$$S = \frac{1}{2}\mu^2\lambda + W(x^\mu), \tag{7.235}$$

where μ is a constant. Then W satisfies the *reduced* Hamilton-Jacobi equation

$$H\left(x^\mu, \frac{\partial W}{\partial x^\nu}\right) = \frac{1}{2}\mu^2. \tag{7.236}$$

For our problem this becomes

$$g^{\mu\nu}\frac{\partial W}{\partial x^\mu}\frac{\partial W}{\partial x^\nu} + \mu^2 = 0. \tag{7.237}$$

Since, by (7.233), $p_\mu = \partial W/\partial x^\mu$ it is clear that μ is the rest mass of the particle,

$$g^{\mu\nu}p_\mu p_\nu + \mu^2 = 0. \tag{7.238}$$

The Fourth Integral of Motion

Now, we study (7.237) for the Kerr metric. The inverse metric $g^{\mu\nu}$ is easily found. For Boyer-Lindquist coordinates one obtains

$$\begin{aligned} g^{\mu\nu}p_\mu p_\nu &= \frac{\Delta}{\rho^2}p_r^2 + \frac{1}{\rho^2}p_\vartheta^2 + \frac{\Delta - a^2\sin^2\vartheta}{\rho^2\Delta\sin^2\vartheta}p_\varphi^2 \\ &- \frac{4Mar}{\rho^2\Delta}p_\varphi p_t - \frac{\Sigma}{\rho^2\Delta}p_t^2. \end{aligned} \tag{7.239}$$

So (7.237) becomes, replacing p_μ by $\partial W/\partial x^\mu$,

$$\begin{aligned} &\frac{\Delta}{\rho^2}(W_{,r})^2 + \frac{1}{\rho^2}(W_{,\vartheta})^2 + \frac{\Delta - a^2\sin^2\vartheta}{\rho^2\Delta\sin^2\vartheta}(W_{,\varphi})^2 \\ &- \frac{4Mar}{\rho^2\Delta}W_{,\varphi}W_{,t} - \frac{\Sigma}{\rho^2\Delta}(W_{,t})^2 + \mu^2 = 0. \end{aligned} \tag{7.240}$$

The coefficients of this partial differential equation are independent of t and φ. To find a complete solution we therefore try the separation ansatz

$$W = -E\,t + L_z\,\varphi + S_r(r) + S_\vartheta(\vartheta), \tag{7.241}$$

where E and L_z are constants. For their interpretation we note that (7.233) gives for t and φ

$$p_t = -E\,, \qquad p_\varphi = L_z\,. \tag{7.242}$$

These two conservation laws are just

$$\langle p, k\rangle = -E\,, \qquad \langle p, m\rangle = L_z \tag{7.243}$$

for the two Killing fields k and m (recall (3.35)).

Inserting (7.241) into (7.240) gives

$$\begin{aligned} &\frac{\Delta}{\rho^2}\left(\frac{dS_r}{dr}\right)^2 + \frac{1}{\rho^2}\left(\frac{dS_\vartheta}{d\vartheta}\right)^2 + \frac{\Delta - a^2\sin^2\vartheta}{\rho^2\Delta\sin^2\vartheta}\,p_\varphi^2 \\ &+ \frac{4Ma\,r}{\rho^2\Delta}\,L_z E - \frac{\Sigma}{\rho^2\Delta}\,E^2 + \mu^2 = 0\,. \end{aligned} \tag{7.244}$$

Now we follow the standard procedure. Multiply this equation with ρ^2 and collect terms that contain a ϑ-dependence. The resulting expression must be equal to a separation constant $\mathcal{C}$. This will be our fourth parameter of the complete solution. In this way we find two equations of the type

$$\begin{aligned} \left(\frac{dS_\vartheta}{d\vartheta}\right)^2 &= f(\vartheta; E, L_z, \mu^2, \mathcal{C})\,, \\ \left(\frac{dS_r}{dr}\right)^2 &= g(r; E, L_z, \mu^2, \mathcal{C})\,. \end{aligned} \tag{7.245}$$

Equation (7.233) gives for ϑ

$$p_\vartheta^2 = f(\vartheta; E, L_z, \mu^2, \mathcal{C})\,.$$

Solving this for $\mathcal{C}$ one finds

$$p_\vartheta^2 + \cos^2\vartheta\left(a^2(\mu^2 - E^2) + \frac{L_z^2}{\sin^2\vartheta}\right) = \mathcal{C}\,. \tag{7.246}$$

This is our fourth integral of motion.

The complete solution of the Hamilton-Jacobi equation is (see (7.235))

$$S = \frac{1}{2}\,\mu^2\lambda - E\,t + L_z\,\varphi + \int \sqrt{f}\,d\vartheta + \int \sqrt{g}\,dr\,. \tag{7.247}$$

We could now proceed by writing down the Eqs. (7.234), taking as parameters α^μ: $(\mathcal{C} + L_z - a\,E, \mu^2, E, L_z)$. Differentiating the result with respect to λ, we get rid of the integrals and arrive at a system of first order equations $\dot{x}^\mu = \varphi^\mu(x^\nu; E, L_z, \mu^2, \zeta)$. – It is, however, easier to obtain these by using the four integrals of motion:

$$p_t = -E, \qquad p_\varphi = L_z\,, \tag{7.248a}$$

$$p_\vartheta^2 + \cos^2\vartheta\left(a^2(\mu^2 - E^2) + \frac{L_z^2}{\sin^2\vartheta}\right) = \mathcal{C}\,, \tag{7.248b}$$

and $g^{\mu\nu} p_\mu p_\nu + \mu^2 = 0$, which becomes using (7.239) and (7.248a)

$$\begin{aligned}\frac{\Delta}{\rho^2}\, p_r^2 + \frac{1}{\rho^2}\, p_\vartheta^2 - \frac{\Delta - a^2\sin^2\vartheta}{\rho^2\Delta\sin^2\vartheta}\, L_z^2 \\ + \frac{4Ma\,r}{\rho^2\Delta}\, L_z E - \frac{\Sigma}{\rho^2\Delta} E^2 = -\mu^2\,. \end{aligned} \tag{7.249}$$

In these we use $p_r = \mu\, g_{rr}\, \dot r$, etc., and find after simple algebraic manipulations the basic system of first order equations

$$\begin{aligned}\rho^2\frac{d\vartheta}{d\lambda} &= \sqrt{\Theta}\,, \\ \rho^2\frac{dr}{d\lambda} &= \sqrt{R}\,, \\ \rho^2\frac{d\varphi}{d\lambda} &= -\left(a\,E - \frac{L_z}{\sin^2\vartheta}\right) + \frac{a}{\Delta} P\,, \\ \rho^2\frac{dt}{d\lambda} &= -a(a\,E\sin^2\vartheta - L_z) + \frac{r^2 + a^2}{\Delta}\, P\,, \end{aligned} \tag{7.250}$$

where

$$\begin{aligned}\Theta &= \mathcal{C} - \cos^2\vartheta\left(a^2(\mu^2 - E^2) + \frac{L_z^2}{\sin^2\vartheta}\right), \\ R &= P^2 - \Delta\left(\mu^2 r^2 + (L_z - a\,E)^2 + \mathcal{C}\right), \end{aligned} \tag{7.251}$$

with

$$P = E(r^2 + a^2) - L_z a\,. \tag{7.252}$$

Note that the first two equations determine $r(\lambda)$ and $\vartheta(\lambda)$ by quadrature. With these functions at hand, the remaining equations determine also $\varphi(\lambda)$ and $t(\lambda)$ through quadratures.

While this solves the problem theoretically, it is difficult to understand the orbit structure analytically. This is an example where computer animations can help. We discuss here only a very special class of orbits which is, however, of considerable astrophysical interest.

Equatorial Circular Geodesics

Let us now specialize the basic equations (7.250) to test particle motion in the equatorial plane. We are especially interested in the radial motion, determined by

$$r^2 \frac{dr}{d\lambda} = \sqrt{R}.$$

For $\vartheta = \pi/2$ the fourth integral of motion in (7.248b) vanishes and the function R becomes

$$R = P^2 - \Delta\left(\mu^2 r^2 + (L_z - a\,E)^2\right).$$

Collecting terms gives[7]

$$r^3 \left(\frac{dr}{d\lambda}\right)^2 = E^2(r^3 + a^2\,r + 2M\,a^2) - 4aM\,E\,L_z - (r - 2M)L_z^2 - \mu^2 r\,\Delta\,. \tag{7.253}$$

At this point we specialize even further and consider only *circular* motion. For these the right-hand side and its derivative in (7.252) must vanish. These equations can be solved for $\gamma := E/m$ and $l := L_z/m$ with the result (see [201]):

$$\gamma = \frac{r^2 - 2M\,r \pm a\sqrt{M\,r}}{r\left(r^2 - 3M\,r \pm 2a\sqrt{M\,r}\right)^{1/2}}, \tag{7.254a}$$

$$l = \pm \frac{\sqrt{M\,r}\left(r^2 \mp 2a\sqrt{M\,r} + a^2\right)}{r\left(r^2 - 3M\,r \pm 2a\sqrt{M\,r}\right)^{1/2}}. \tag{7.254b}$$

The upper sign refers to co-rotating test particles and the lower sign has to be taken for counter-rotating ones.

Kepler's third law for the motion of the test particle is best obtained by going back to the Lagrangian for test particles in the equatorial plane

$$\begin{aligned} 2\mathcal{L} = &-\left(1 - \frac{2M}{r}\right)\dot{t}^2 - \frac{4aM}{r}\,\dot{t}\,\dot{\varphi} \\ &+ \frac{r^2}{\Delta}\dot{r}^2 + \left(r^2 + a^2 + \frac{2Ma^2}{r}\right)\dot{\varphi}^2\,. \end{aligned} \tag{7.255}$$

The radial Euler-Lagrange equation becomes for constant r: $\partial\mathcal{L}/\partial r = 0$, i.e.,

$$-\frac{M}{r^2}\,\dot{t}^2 + \frac{2aM}{r^2}\,\dot{t}\,\dot{\varphi} + \left(r - \frac{Ma^2}{r^2}\right)\dot{\varphi}^2 = 0\,.$$

Hence the frequency $\omega = d\varphi/dt = \dot{\varphi}/\dot{t}$ satisfies

$$-\frac{M}{r^2} + \frac{2aM}{r^2}\omega + \left(r - \frac{Ma^2}{r^2}\right)\omega^2 = 0\,,$$

[7] This can, of course, be obtained directly from the three obvious conservation laws (7.248a) and (7.249).

whence Kepler's third law becomes

$$\omega = \pm \frac{M^{1/2}}{r^{3/2} \pm a\, M^{1/2}} \,. \tag{7.256}$$

For the further discussion we note that the denominator for the energy constant γ in (7.254a) develops in the co-rotating case a zero when

$$\frac{M}{2}\sqrt{\frac{r}{M}}\left(3 - \frac{r}{M}\right) = a \,. \tag{7.257}$$

For $a < M$ the function $\gamma(r)$ becomes at this point infinite. Qualitatively, $\gamma(r)$ has the shape shown in Fig. 7.5, with a minimum at some radius larger than the one determined by (7.257). We claim that this critical radius corresponds to a *marginally stable* circular orbit. To prove this, let us write (7.253) in the form

$$\dot{r}^2 + V(r; \gamma, l) = 0 \,,$$

with an effective potential V. The circular orbits correspond to critical points of this one-dimensional problem with 'energy' equal to zero: $V = 0$ and $V' \equiv \partial V / \partial r = 0$. The expressions (7.254a) and (7.254b) are the solutions of these two equations. Since the solutions are unique the determinant of

$$\begin{pmatrix} \dfrac{\partial V}{\partial \gamma} & \dfrac{\partial V}{\partial l} \\[2ex] \dfrac{\partial V'}{\partial \gamma} & \dfrac{\partial V'}{\partial l} \end{pmatrix}$$

is different from zero. The derivatives $d\gamma/dr$ and dl/dr are obtained from

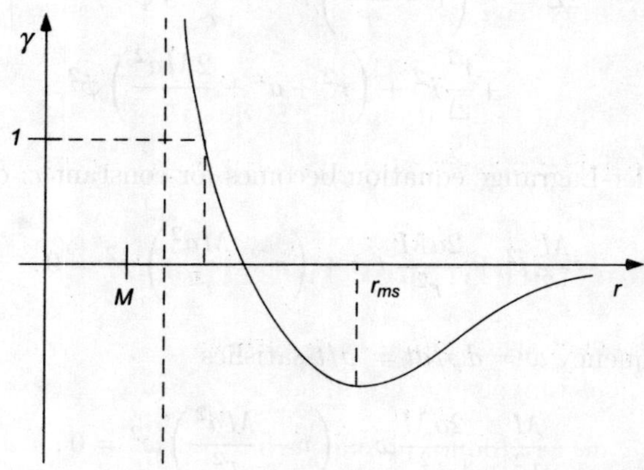

Fig. 7.5. Energy constant γ for circular orbits as a function of r for fixed a.

$$0 = \frac{dV}{dr} = \frac{\partial V}{\partial \gamma}\frac{d\gamma}{dr} + \frac{\partial V}{\partial l}\frac{dl}{dr} + V',$$
$$0 = \frac{dV'}{dr} = \frac{\partial V'}{\partial \gamma}\frac{d\gamma}{dr} + \frac{\partial V'}{\partial l}\frac{dl}{dr} + V''.$$

Stability of the circular orbit with radius r requires $V'' \geq 0$. The marginally stable orbit corresponds to $V'' = 0$. For this the last system of equations implies that $d\gamma/dr = 0$ and $dl/dr = 0$. Physically it is clear that these extrema are minima.

Marginally bound circular orbits correspond to $\gamma = 1$. Their radial coordinate is

$$r_{mb} = 2M \mp a + 2M^{1/2}(M \mp a)^{1/2}. \tag{7.258}$$

The analytical expression for the radius r_{ms} of the marginally stable circular orbits is rather complicated (see [201]). We consider here only the two limiting cases $a = 0$ and $a \longrightarrow M$. For the Schwarzschild solution $r_{ms} = 6M$, and for $a \longrightarrow M$ we have $r_{ms} = M$ (co-rotating) or $9M$ (retrograde). Of great interest is the binding energy of the marginally stable circular orbit. This is determined by

$$\frac{a}{M} = \mp\frac{4\sqrt{2}(1-\gamma^2)^{1/2} - 2\gamma}{3\sqrt{3}(1-\gamma^2)}. \tag{7.259}$$

From this one finds that γ decreases from $\sqrt{8/9}$ for $a = 0$ to $1/\sqrt{3}$ in the extreme limit for co-rotating orbits. In the other case the value increases from $\sqrt{8/9}$ to $\sqrt{25/27}$. For the first case, if a particle can spiral down in an accretion disk from far away to the innermost circular stable orbit, the fraction $1 - 1/\sqrt{3}$ of the rest energy is set free. Thus, a rotating black hole allows a gravitational energy conversion up to $\simeq 42.3\%$. This enormous efficiency is the main reason why the existence of super-massive black holes in the centers of quasars and other active galactic nuclei[8] has been suggested already quite early.

Remark. For a recent application of (7.256) to the super-massive black hole in our Galaxy, see [222].

7.4 Accretion Tori Around Kerr Black Holes

It is now generally believed that the energy source of quasars and other active galactic nuclei involves accretion of matter onto a super-massive black

[8] Active galactic nuclei produce prodigious luminosities in tiny volumes (in some cases as much as 10^4 times the luminosity of a typical galaxy in much less than $1pc^3$). This radiation can emerge over an extraordinary broad range of frequencies. For a text book presentation of active galactic nuclei, see e.g. [219].

hole, i.e., a black hole of $10^8 - 10^9$ solar masses, because this is an efficient way to convert rest mass into radiation in a relatively small region of space (approximatively the dimension of the solar system). Although we do not have a detailed picture, the formation of such holes in the cores of some galaxies is quite plausible. A super-massive black hole can easily be fed by sufficient material from its surroundings (a few solar masses per year) to generate in a small volume the enormous luminosities of about $10^{47} erg/sec$ observed from quasars. A useful number to compare with is the Eddington luminosity,

$$L_{Edd} = 1.3 \times 10^{46} (M_H/10^8 M_\odot) erg/s \,, \tag{7.260}$$

which corresponds to the mass M_H of the black hole. If the radiation efficiency of the infalling gas is 0.15, then the accretion rate $\dot{M}$ is given by $0.15\dot{M}\, c^2 = L_{Edd}$ or $\dot{M} \simeq 1.3 (M_H/10^8 M_\odot) M_\odot yr^{-1}$. When the mass accretion rate is close to L_{Edd}, the inner accretion disk around the black hole will thicken (see for instance [14]). It turns out that in that part the radiation pressure dominates and the opacity is determined mainly by electron scattering. Under these circumstances a structure may form close to the horizon which looks like a torus. – In this section we will study these accretion tori in some detail. We shall see that the relativistic tori have new, qualitatively important features. The study of such thick accretion disks is probably also important for an understanding of the formation of well collimated pairs of energetic jets.

7.4.1 Newtonian Approximation

It is instructive to first study the problem in the Newtonian approximation. – In cylindrical coordinates (s, φ, z) we assume that the velocity field of the infalling matter has the form

$$v_s = 0 \,, \qquad v_\varphi = s\,\Omega(s,z) \,, \qquad v_z = 0 \,. \tag{7.261}$$

Then the radial and the z-component of the Euler equation become

$$\frac{1}{\rho}\,\partial_s p = -\partial_s \phi + s\,\Omega^2 \,, \qquad \frac{1}{\rho}\,\partial_z p = -\partial_z \phi \,. \tag{7.262}$$

Together they can be written as

$$\frac{1}{\rho}\, dp = -d\phi + \Omega^2 s\, ds \,. \tag{7.263}$$

If we take the differential of this we find

$$\begin{aligned} d\left(\frac{1}{\rho}\right) \wedge dp &= 2\Omega s\, d\Omega \wedge ds = 2\Omega\, s\, \partial_z \Omega\, dz \wedge ds \\ &= 2\,\partial_z \Omega * v \,, \end{aligned} \tag{7.264}$$

where $*v$ is the dual of the velocity 1-form

$$v = \Omega s^2 \, d\varphi \,. \tag{7.265}$$

Now we make the model assumption that we have a *barotropic* situation: $p = p(\rho)$. It should be noted that this is not necessarily the equation of state. For instance, if $\partial_z \Omega = 0$ such a relation has to hold as a consequence of (7.264), whatever the equation of state is. For a barotrope the left-hand side of (7.264) vanishes and therefore the function Ω must be independent of z: $\Omega = \Omega(s)$, i.e., Ω is constant on cylinders. This is a so-called **von Zeipel relation**. For this situation it is convenient to introduce the rotational potential V by $dV = -\Omega^2 s \, ds$, i.e.,

$$V = -\int_{s_{in}}^{s} \Omega^2(s) s \, ds \,, \tag{7.266}$$

where s_{in} corresponds to the inner boundary of the disk. Then (7.263) becomes

$$\frac{1}{\rho} \, dp = -d\psi \,, \qquad \psi := \phi + V \,. \tag{7.267}$$

Thus, isobaric surfaces and surfaces of constant density are just equipotential surfaces of ψ. This analysis leads to the following procedure:

1. Choose a "von Zeipel relation"

$$\Omega = \Omega(s) \tag{7.268}$$

   and determine V.
2. With the Newtonian potential ϕ from the central object,

$$\phi = -\frac{GM}{r} \,, \qquad r = \sqrt{s^2 + z^2} \tag{7.269}$$

   we know ψ and can determine the equipotential surfaces.

As example, we choose

$$l := s^2 \Omega(s) = \text{const} \,, \tag{7.270}$$

which gives $V = l^2 / 2\, s^2$, and

$$\psi = -\frac{GM}{\sqrt{s^2 + z^2}} + \frac{l^2}{2s^2} \,. \tag{7.271}$$

The corresponding equipotential surfaces are shown in Fig. 7.6. One sees that the thick disk possesses a narrow funnel along the rotation axis of the central object along which large radiative pressure gradients can, perhaps, accelerate matter in collimated jets. This is the standard picture for the formation of well collimated jets in opposite directions. One should note in this connection that the binding energy of material near the surface of the funnel is relatively low and thus radiation pressure may easily propel matter to escape to infinity. Other acceleration mechanisms may, however, be more important.

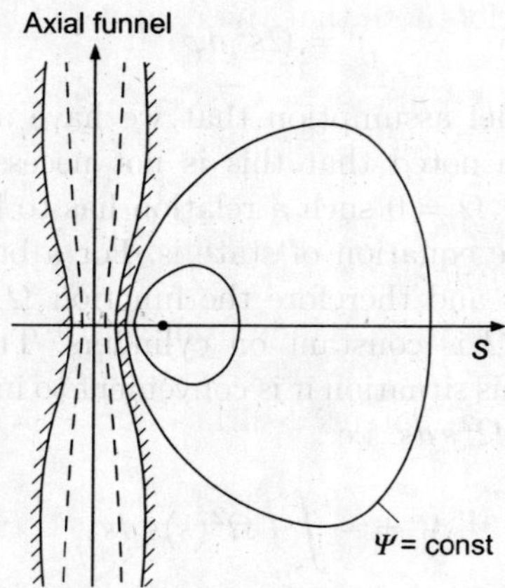

Fig. 7.6. Meridional cross-section through equipotentials of the toroidal disk.

7.4.2 General Relativistic Treatment

We consider now a stationary and axisymmetric flow of an ideal fluid in the gravitational field of a Kerr black hole. Lie derivatives of physical quantities, like the pressure p and the four-velocity u of the ideal fluid, with respect to the Killing fields $k = \partial_t$ and $m = \partial_\varphi$ have to vanish.

The four-velocity u has the form

$$u = u^t(k + \Omega\, m)\,. \tag{7.272}$$

We introduce the following invariant quantities of the ideal fluid:

$$e = -\langle u, k\rangle = -u_t\,, \qquad \text{specific energy} \tag{7.273a}$$

$$j = \langle u, m\rangle = u_\varphi\,, \qquad \text{mechanical angular momentum} \tag{7.273b}$$

$$l = \frac{j}{e} = -\frac{u_\varphi}{u_t}\,. \qquad \text{fluid angular momentum} \tag{7.273c}$$

From the definitions one derives the following relations

$$e^{-2} = \frac{g_{\varphi\varphi} + 2l\, g_{t\varphi} + l^2\, g_{tt}}{g_{t\varphi}^2 - g_{tt}\, g_{\varphi\varphi}}\,, \tag{7.274a}$$

$$\Omega = -\frac{g_{t\varphi} + l\, g_{tt}+}{g_{\varphi\varphi} + l\, g_{t\varphi}}\,, \tag{7.274b}$$

$$l = -\frac{g_{t\varphi} + \Omega\, g_{\varphi\varphi}}{g_{tt} + \Omega\, g_{t\varphi}}\,, \tag{7.274c}$$

$$u^t u_t = \frac{1}{1 - \Omega\, l}\,, \tag{7.274d}$$

where $g_{tt} = \langle k, k\rangle$, $g_{t\varphi} = \langle k, m\rangle$ and $g_{\varphi\varphi} = \langle m, m\rangle$ are metric coefficients in adapted coordinates, which have an invariant meaning.

We show below in a footnote that the relativistic Euler equation

$$(\rho+p)\nabla_u u = -\nabla p - (\nabla_u p)u \tag{7.275}$$

implies

$$\frac{dp}{\rho+p} = -d\ln e + \frac{\Omega}{1-\Omega l}\,dl\,. \tag{7.276}$$

For a barotronic situation we conclude from this, by taking the differential, that $d\Omega\wedge dl = 0$, i.e., $\Omega = \Omega(l)$ (von Zeipel relation). Therefore, the isobaric surfaces agree with the equipotential surfaces of

$$W := \ln e - \int \frac{\Omega(l)}{1-\Omega(l)l}\,dl\,. \tag{7.277}$$

For a given von Zeipel relation everything is determined.

As an example we set l = const. Then $W = \ln e$ and the equipotential surfaces are given by e = const. According to (7.274a) they are determined by

$$\frac{g_{\varphi\varphi} + 2l\,g_{t\varphi} + l^2\,g_{tt}}{g_{t\varphi}^2 - g_{tt}\,g_{\varphi\varphi}} = \text{const}\,. \tag{7.278}$$

For a Schwarzschild black hole we have

$$g_{tt} = -1 + \frac{2m}{r}\,, \qquad g_{\varphi\varphi} = r^2\sin^2\vartheta\,, \tag{7.279}$$

with $m = GM/c^2$, and (7.278) gives

$$\frac{1}{1-2GM/r} - \frac{l^2}{r^2\sin^2\vartheta} = \text{const}\,. \tag{7.280}$$

This generalizes the Newtonian result (7.271).

In Fig. 7.7 the surfaces (7.280) are shown for $l = 3.77M$. Matter can fill each of the closed equipotential surfaces (for which $W < 0$). In comparison to the Newtonian result the most important new phenomenon is the appearance of a *sharp cusp* on the inner edge of the accretion torus, which is near the marginally stable circular orbit. This acts like the inner Lagrange point in a close binary system; the gas falls from the disk over the cusp toward the black hole with no dissipation of the angular momentum. Therefore, the accretion is driven by pressure gradient forces and viscosity is not necessary. – Note that, qualitatively, one finds the same result for Kerr black holes. An example is shown in Fig. 7.8. For a more detailed discussion we refer to [215].

* * *

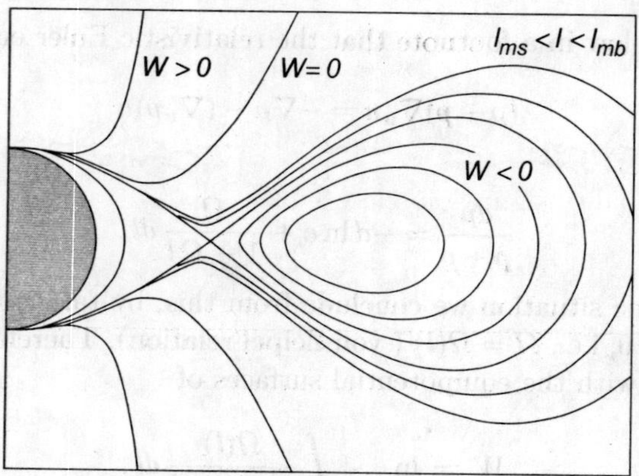

Fig. 7.7. Equipotential surfaces (meridional section) for $l = 3.77M$ around a Schwarzschild black hole. The interior of the hole is dashed. (From [215].)

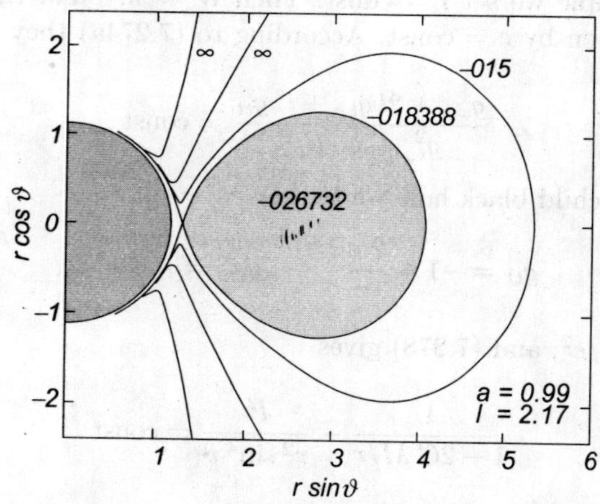

Fig. 7.8. Meridional section of the accretion torus with $l = 2.17M$ orbiting a Kerr black hole with $a = 0.99$. The numbers give the values of the potential W. The fluid can fill up the equipotential surface with the cusp. (From [215].)

Footnote: Derivation of Equation (7.276)

First we note that the last term in (7.275) vanishes because $L_u p = 0$. Let $\boldsymbol{u}$ be the 1-form belonging to u. Then we have

$$(L_u \boldsymbol{u})_\mu = u^\nu u_{\mu,\nu} + u_\nu u^\nu{}_{,\mu} = u^\nu u_{\mu;\nu} + u_\nu u^\nu{}_{;\mu} = (\nabla_u \boldsymbol{u})_\mu . \tag{7.281}$$

Next, we compute $L_{\phi u}\boldsymbol{u}$, with $\phi := 1/u^t$, in two different ways:

a) $L_{\phi u}\boldsymbol{u} = \phi\, L_u \boldsymbol{u} + d\phi \wedge i_u \boldsymbol{u} = \phi \nabla_u \boldsymbol{u} - d\phi;$
b) $L_{\phi u}\boldsymbol{u} = L_{k+\Omega m}\boldsymbol{u} = d\Omega \wedge i_m \boldsymbol{u} = j\, d\Omega\,.$

By comparison we find

$$\phi\,\nabla_u \boldsymbol{u} = j\,d\Omega + d\phi = de - \Omega\,dj\,. \tag{7.282}$$

In the last equality sign we made use of

$$\phi = -i_{\phi u}\boldsymbol{u} = -i_{k+\Omega m}\boldsymbol{u} = -e + j\,\Omega\,.$$

With the help of (7.273a) and (7.274d) we conclude from (7.282), using also $j = l\,e$,

$$\nabla_u \boldsymbol{u} = d\ln e - \frac{\Omega}{1-\Omega\,l}\,dl\,. \tag{7.283}$$

By inserting this into the Euler equation (7.275), we obtain indeed the relation (7.276).

7.5 The Four Laws of Black Hole Dynamics

In principle it is possible to extract an enormous amount of energy from a rotating black hole, thereby reducing its angular velocity and thus also the size of the ergosphere. The fraction of rotational energy that can ideally be extracted is limited by what is called the **second law of black hole dynamics**, which implies that the surface of a black hole cannot decrease. The general form of this **area law**, formulated later in this section, is in close formal analogy with the second law of thermodynamics. Not only the second law, but also the other main laws of thermodynamics have close analogies in the physics of black holes.

While the formal analogy between these two sets of laws has been established very early ($\sim$ 1973), the question of whether there is deep physical reason behind this, is still a matter of research. A first important step in this direction was made by S. Hawking's discovery of thermal radiation by black holes in a semi-classical approach (quantized matter fields on a classical black hole background).

In this section we give only an introduction to the classical aspects of the subject. For more information the reader is referred to the cited literature.

General Definition of Black Holes

Until now we never gave a technically precise definition of a black hole and its event horizon, since physically it is clear what is meant by this concepts. With the help of the conformal description of asymptotic flatness, discussed in Sect. 5.1, a concise mathematical definition can be given.

For this we consider the **causal past** of future null infinity, denoted by $J^-(\mathscr{I}^+)$. This is the set of all events $p \in M$ for which there is a future pointing **causal curve** (non-spacelike tangent vectors), starting at p and ending on $\mathscr{I}^+$. We say that the physical spacetime M contains a black hole if $J^-(\mathscr{I}^+)$ does not contain all of M. The complement

$$B := M \setminus J^-(\mathscr{I}^+)$$

is the **black hole region**, and

$$H := \partial B$$

is its (future) **event horizon**. (The time reflected notion of past event horizons is not of astrophysical interest, as was already remarked in connection with Schwarzschild black holes.) Being a causal boundary, one expects that H is a null hypersurface. We shall see later in Appendix 7.7.4 that this is indeed the case (ignoring for the moment possible breakdown of differentiability).

7.5.1 The Zeroth Law of Black Hole Dynamics

In the formal relationship, alluded to in the introductory remarks, the analogue of the temperature is the surface gravity, that was introduced in Sect. 7.3.3 for Killing horizons (see (7.221) and (7.222)). The analogue of the zeroth law of thermodynamics – the constancy of the (absolute) temperature T throughout a body in thermal equilibrium – is the constancy of the surface gravity κ over the horizon of a stationary black hole. We shall prove this constancy for Killing horizons of static and circular spacetimes. It holds, however, also under other assumptions.

The details of the general argument will be given in Appendix 7.7.2 of this chapter. Here, we compute κ for the Kerr-Newman family. Before doing this we want to interprete κ, and derive alternative formulas for this quantity.

Surface Gravity

For the interpretation of the surface gravity on an arbitrary Killing horizon H belonging to a Killing field K (see Definition 7.1), we consider integral curves of K in the region where K is timelike (outer domain of communication). Their four-velocities are

$$u = K/(-N)^{1/2}\,, \qquad N = \langle K, K\rangle\,. \tag{7.284}$$

The acceleration of a stationary observer along such an integral curve is

$$a = \nabla_u u = \frac{1}{(-N)^{1/2}}\nabla_K\left(K/(-N)^{1/2}\right) = \frac{1}{(-N)}\nabla_K K\,, \tag{7.285}$$

since $L_K N = 0$. The magnitude $\langle a, a\rangle$ diverges at the horizon, but by a suitable renormalization (measuring the time in terms of the asymptotic time) we obtain a finite result. We claim that

$$\kappa = \lim(-N)^{1/2}\langle a, a\rangle^{1/2}\,, \tag{7.286}$$

as the horizon is approached. (Recall that $(-N)^{1/2}$ is the redshift factor, as in (7.213).)

To show that (7.286) holds, we first note that according to (7.285)

$$\lim(-N)\langle a, a\rangle = \lim\left(-\frac{1}{N}\right)\langle\nabla_K K, \nabla_K K\rangle\,.$$

As in (7.223) we have

$$dN = -2\,\nabla_K K\,, \tag{7.287}$$

thus

$$\langle\nabla_K K, \nabla_K K\rangle = \frac{1}{4}\langle dN, dN\rangle\,, \tag{7.288}$$

and so

$$\lim(-N)\langle a, a\rangle = -\frac{1}{4}\lim\frac{1}{N}\langle dN, dN\rangle\,. \tag{7.289}$$

Here, we use the general identity (7.105),

$$N\langle dK, dK\rangle = \langle dN, dN\rangle - 4\,\langle\omega, \omega\rangle\,, \tag{7.290}$$

where ω is the twist form of K. On the horizon this twist *vanishes*. This follows immediately from the fact that on H the null vector K must be proportional to the gradient of N. (Introduce local coordinates such that N is one of them and the remaining ones parameterize locally the hypersurface H. Then $K \wedge dK = 0$ on H becomes obvious.) With this we obtain

$$\lim(-N)\langle a, a\rangle = -\frac{1}{4}\langle dK, dK\rangle|_H\,. \tag{7.291}$$

Finally, we show that the right-hand side of (7.291) is equal to κ^2, proving our claim (7.286). For the derivation of

$$\kappa^2 = -\frac{1}{4}\langle dK, dK\rangle|_H\,, \tag{7.292}$$

we start from the evident identity

$$i_K(dK \wedge *dK) = -dN \wedge *dK + 2\,dK \wedge \omega\,.$$

Here, the left-hand side is equal to $\langle dK, dK\rangle * K$. (To see this use (7.79) for $\alpha = 1$ and $*1 = \eta$.) For the right-hand side we obtain on the horizon, using the original definition (7.221) of κ, i.e., $dN = -2\kappa\,K$ on H,

$$\begin{aligned}-dN \wedge *dK &= 2\kappa\,K \wedge *dK = -2\kappa * i_K dK \\ &= 2\kappa * dN = -4\kappa^2 * K\,.\end{aligned}$$

This proves (7.292).

For circular spacetimes a closely related formula is easier to work out in the case of the Kerr-Newman family. Consider for a circular spacetime the vector field $\xi = k + \Omega\, m$ used earlier (see (7.216)). We know from the weak rigidity theorem that $H = \{\langle \xi, \xi \rangle = 0\}$ is a Killing horizon, and that Ω_H is constant on H. Therefore $l = k + \Omega_H m$ is a Killing field that plays the role of K in the considerations above. From $\xi = l + (\Omega - \Omega_H)m$ we get

$$d\xi = dl + (\Omega - \Omega_H)dm + d\Omega \wedge m\,.$$

But $d\Omega$ is normal to H, thus proportional to l, therefore the last term in

$$\langle d\xi, d\xi \rangle|_H = \langle dl, dl \rangle|_H + \langle d\Omega \wedge m, d\Omega \wedge m \rangle|_H$$

vanishes, being proportional to $\langle l \wedge m, l \wedge m \rangle|_H = \langle l, l \rangle \langle m, m \rangle - \langle l, m \rangle^2$, and $\langle l, m \rangle|_H = 0$. Thus we obtain the formula

$$\kappa^2 = -\frac{1}{4} \langle d\xi, d\xi \rangle|_H \tag{7.293}$$

for the surface gravity.

In adapted coordinates with $k = \partial_t$, $m = \partial_\varphi$ we have

$$\xi = -\alpha^2\, dt\,, \qquad d\xi = -2\alpha\, d\alpha \wedge dt\,,$$

and

$$\langle d\xi, d\xi \rangle = 4\alpha^2 \langle d\alpha, d\alpha \rangle \langle dt, dt \rangle = -4 \langle d\alpha, d\alpha \rangle\,.$$

Hence, we have

$$\kappa^2 = \langle d\alpha, d\alpha \rangle|_H\,. \tag{7.294}$$

We apply this to the Kerr-Newman family. For this the lapse function α is according to (7.197) given by $\alpha = \rho\sqrt{\Delta}/\Sigma$, thus (using $\langle dr, dr \rangle = \Delta/\rho^2$)

$$\kappa^2 = \lim_{r \to r_H} \left(\frac{\rho}{\Sigma} \frac{\Delta'}{2\sqrt{\Delta}} \right)^2 \langle dr, dr \rangle = \frac{1}{4\Sigma^2} (\Delta')^2|_H\,.$$

As a result we obtain

$$\kappa = \frac{r_H - M}{r_H^2 + a^2} = \frac{r_H - M}{2M\, r_H - Q^2}\,. \tag{7.295}$$

In particular we see that the surface gravity is *constant* what is called the **zeroth law** of black hole dynamics.

This explicit calculation does not shed much light on why this is so. In Appendix 7.7.2 we show that the zeroth law is a purely geometrical property of Killing horizons for static and for circular spacetimes.

7.5.2 The First Law for the Kerr-Newman Family

In thermodynamics the differential of the internal energy U of a system in thermodynamic equilibrium is given by

$$dU = T\,dS + \alpha\,, \tag{7.296}$$

where α is the differential 1-form of the reversible work. A typical example for α is, in standard notation,

$$\alpha = -p\,dV + \mu\,dN + \dots\,. \tag{7.297}$$

The combination of (7.296) and (7.297) is often called the first law, although an important part of the second law of thermodynamics is also integrated (otherwise we cannot speak of entropy).

In black hole physics, the analogue of the thermodynamic potential $U(S, V, N, \dots)$ is the total mass as a function of the surface area $\mathcal{A}$ of the horizon, the angular momentum J, the electric charge Q and perhaps other quantities for certain matter models. That $\mathcal{A}$ is the "right" choice will become clear later.

* * *

Surface Area of Kerr-Newman Horizon

In Boyer-Lindquist coordinates the induced metric $\bar{g}$ of the horizon for fixed t is according to (7.193) and (7.195) given by

$$\bar{g} = g_{\vartheta\vartheta}\,d\vartheta^2 + g_{\varphi\varphi}\,d\varphi^2 = \rho^2 d\vartheta^2 + \frac{(r^2+a^2)^2}{\rho^2}\sin^2\vartheta\,d\varphi^2\,, \tag{7.298}$$

where $r = r_H$,

$$r_H = M + \sqrt{M^2 - a^2 - Q^2}\,. \tag{7.299}$$

The corresponding volume form $\bar{\eta}$ is

$$\bar{\eta} = (r_H^2 + a^2)\,d\vartheta \wedge \sin\vartheta\,d\varphi\,,$$

so the surface area of the horizon is

$$\mathcal{A} = 4\pi(r_H^2 + a^2) = 4\pi(2M\,r_H - Q^2)\,. \tag{7.300}$$

* * *

This equation can now be solved for the total mass M of the black hole. Using the abbreviation $M_0^2 = \mathcal{A}/16\pi$, the result is

$$M(\mathcal{A}, J, Q) = \left[\left(M_0 + \frac{Q^2}{4M_0}\right)^2 + \frac{J^2}{4M_0^2}\right]^{1/2}. \tag{7.301}$$

From this one finds

$$\frac{\partial M}{\partial \mathcal{A}} = \frac{1}{8\pi}\kappa, \quad \frac{\partial M}{\partial J} = \Omega_H. \tag{7.302}$$

Before writing down the result for $\partial M/\partial Q$, we introduce the electric potential on the horizon. The Faraday form is $F = dA$, with (see (7.198))

$$A = \frac{Q}{\rho^2} r(dt - a \sin^2 \vartheta \, d\varphi). \tag{7.303}$$

The **electric potential** (relative to the Killing field l) is defined by

$$\phi = i_l A = \frac{Q}{\rho^2} r(1 - \Omega_H \, a \sin^2 \vartheta). \tag{7.304}$$

Using (7.218) we see that

$$\phi_H := \phi|_H = \frac{Q \, r_H}{r_H^2 + a^2} = \frac{Q \, r_H}{2M \, r_H - Q^2} \tag{7.305}$$

is *constant* on the horizon. This turns out to be the partial derivative $\partial M/\partial Q$,

$$\frac{\partial M}{\partial Q} = \phi_H. \tag{7.306}$$

Taken together, we find the remarkable formula for the differential of M, the so-called **first law**:

$$dM = \frac{1}{8\pi}\kappa \, d\mathcal{A} + \Omega_H \, dJ + \phi_H \, dQ. \tag{7.307}$$

The analogy with

$$dU = T \, dS - p \, dV + \mu \, dN + \dots$$

is striking. This suggests the correspondence $T \longleftrightarrow \kappa$ and $S \longleftrightarrow \mathcal{A}$. Within the classical framework proportionality factors are undetermined. The Hawking radiation suggests that one should associate a temperature and an entropy to a black hole given by

$$T_H = \frac{\hbar \kappa}{8\pi k}, \quad S = \frac{k\mathcal{A}}{4\hbar} \tag{7.308}$$

where k is Boltzmann's constant. The quantity S is the so-called **Bekenstein-Hawking entropy** and T_H the **Hawking temperature**. For this subject we refer to [204], and the literature cited there.

Again, one would like to see a more general derivation of the first law (7.307), that does not depend on the explicit solution. A textbook treatment of this is given in [189], where also the relevant original papers are quoted, and various generalizations (e.g., [202] and [203]) are mentioned.

7.5.3 The Second Law of Black Hole Dynamics

The most general form of the second law is due to S. Hawking. It states the following:

Theorem 7.2. *In any (classical) interaction of matter and radiation with black holes, the total surface area of the boundaries of these holes (as formed by their horizons) can never decrease.*

The key idea of the proof is to use the following fact which was found previously by R. Penrose: Light rays generating a horizon never intersect. If one would assume that null geodesic generators of a horizon begin to converge and the energy density is always non-negative, then the focusing effect of the gravitational field would lead to the intersection of light rays. One must therefore conclude that the light rays generating the horizon cannot converge. This, in turn, implies that the surface area of a horizon cannot decrease. We shall elaborate on this in Appendix 7.7.4. Some essential tools concerning congruences of null geodesics will be developed in Appendix 7.7.3.

The limitation to classical interactions means that we do not consider changes in the quantum theory of matter due to the presence of the strong external gravitational fields of black holes. For macroscopic black holes, this is completely justified. If "mini-holes" were to exist, quantum effects, such as spontaneous (Hawking) radiation, would become important.

As a special example, the second law implies that if two black holes collide and coalesce to a single black hole, then the surface area of the resulting black hole is larger than the sum of the surface of the event horizons of the two original black holes (see Fig. 7.9).

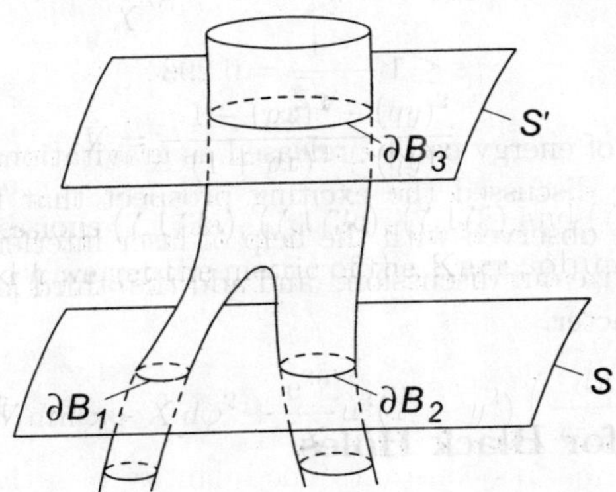

Fig. 7.9. Coalescence of two black holes as an illustration of the second law. For the surfaces of the black holes we have $\mathcal{A}(\partial B_1) + \mathcal{A}(\partial B_2) \leq \mathcal{A}(\partial B_3)$.

Applications

For a Kerr-Newman black hole we found for the surface area (see (7.300))

$$\mathcal{A} = 4\pi(r_H^2 + a^2) = 4\pi\left[\left(M + \sqrt{M^2 - a^2 - Q^2}\right)^2 + a^2\right]. \tag{7.309}$$

According to the second law, the "irreducible mass" M_0 cannot decrease when an isolated black hole interacts with matter and radiation. The maximum energy which can be extracted when $Q = 0$ is according to (7.309)

$$\Delta M = M - M_0\,, \qquad \frac{\Delta M}{M} = 1 - \frac{1}{\sqrt{2}}\left(1 + \sqrt{1 - \frac{a^2}{M^2}}\right)^{1/2}. \tag{7.310}$$

Since $a^2 \leq M^2$, this means that

$$\frac{\Delta M}{M} \leq 1 - \frac{1}{\sqrt{2}}. \tag{7.311}$$

We now consider two Kerr black holes which collide and coalesce to a single one, as shown in Fig. 7.9. The final hole is assumed to be stationary. The second law implies the inequality

$$M\left(M + \sqrt{M^2 - a^2}\right) \geq M_1\left(M_1 + \sqrt{M_1^2 - a_1^2}\right) + M_2\left(M_2 + \sqrt{M_2^2 - a_2^2}\right).$$

Hence, we have for the efficiency

$$\varepsilon := \frac{M_1 + M_2 - M}{M_1 + M_2} < \frac{1}{2}. \tag{7.312}$$

As a special case, take $M_1 = M_2 = \frac{1}{2}\mathcal{M}$, and $a_1 = a_2 = a = 0$. Then $M^2 \geq 2\left(\frac{1}{2}\mathcal{M}\right)^2$, or $M > \mathcal{M}/\sqrt{2}$ and hence

$$\varepsilon \leq 1 - \frac{1}{\sqrt{2}} = 0.293. \tag{7.313}$$

In principle, a lot of energy can be released as gravitational radiation.

In Chap. 4 we discussed the exciting prospect that the coalescence of black holes will be observed with the help of laser interferometers. – In Table 7.1 we summarize our discussion, and add the third law, which is of less fundamental character.

7.6 Evidence for Black Holes

The evidence for black holes in some X-ray binary systems and for supermassive black holes in galactic centers is still indirect, but has become overwhelming during the past few years. There is so far, however, little evidence that these collapsed objects are described by the Kerr metric. – Before we have a closer look at the observational situation, let us address the question how black holes are formed.

Table 7.1. Comparison of the laws of thermodynamics and black hole dynamics. In the right column the following notation is used: M is the mass of the black hole, $\mathcal{A}$ the surface of the horizon, J the angular momentum of the black hole, Ω_H the angular velocity of the horizon, $T_H = (8\pi k)^{-1}\hbar\kappa$ the Hawking temperature and $S = (4\hbar)^{-1}k\mathcal{A}$ is the Beckenstein-Hawking entropy.

	Thermodynamics	*Black hole*
Zeroth law	$T =$ const throughout the body in thermal equilibrium	Surface gravity $\kappa =$ const over horizon of stationary black hole
First law	$dE = T\,dS +$ reversible work	$dM = (1/8)\kappa\,d\mathcal{A} + \Omega_H\,dJ$ $= T_H\,dS + \Omega_H\,dJ$
Second law	$\delta S \geq 0$ in any process	$\delta\mathcal{A} \geq$ in any process
Third law	Impossible to reach $T = 0$ by any physical process	Impossible to reach $\kappa = 0$ by any physical process

7.6.1 Black Hole Formation

Stellar-mass black holes are inevitable end-products of the evolution of sufficiently massive stars. This is simply because 'cold' self-gravitating matter could only exist in a small mass range below a few solar masses (see Sect. 6.9) There is, on the other hand, absolutely no reason for all the many massive stars (or associations of stars) to get rid of sufficient mass in the course of their evolution in some violent processes (supernovas), in order to be able to settle down in a cold final state (white dwarf, neutron stars, quark star, ...).

Realistic collapse studies are still at an early stage. Fully relativistic computer simulations for the non-spherical case are in progress, but it will take some time before these give a reliable picture and permit us to answer questions such as:

a) When and how are black holes formed in stellar core collapse?
b) How much mass is ejected from the outer layers of a centrally collapsing star?
c) How much gravitation radiation and neutrino radiation is emitted? What are their spectra?
d) In which manner do the answer to these questions depend on the mass and the angular momentum of the progenitor star?
e) What are the effects of magnetic fields (a 'smart' question one can always ask)?

The following *qualitative* remarks about the realistic collapse can be made:

1. The formation of trapped surfaces is a stable phenomenon. A **trapped surface** is a spacelike two-surface such that both outgoing and incoming families of future-directed null geodesics orthogonal to it are converging, i.e., the expansion rate θ for both families is negative (see Appendix

7.7.3). This implies that if such a two-surface emitted an instantaneous flash of light, both the outgoing and ingoing wave fronts would be decreasing in area (see Eq. (7.377)). According to the Proposition 7.1, the outgoing wave front would shrink to zero in a finite affine distance, trapping within it all the material, because this cannot travel outwards faster than light. (Illustrate this concept for the Schwarzschild-Kruskal spacetime.) Since trapped surfaces are formed in spherical collapse to a black hole, the stability Cauchy development (see Sect. 2.8) implies that the formation of trapped surfaces is generic.

A general theorem by Hawking and Penrose then implies that a 'singularity' must occur in the sense that spacetime is *geodesically incomplete* (see [8]). According to the "cosmic censorship hypothesis", this will be hidden behind an event horizon[9]. If this is the case, then not only the development of singularities, but also the occurrence of horizons is a generic phenomenon.

2. Trapped surfaces, and thus also singularities and horizons are presumably formed whenever an isolated mass distribution becomes enclosed within a two-dimensional surface S of circumference $d \sim 2\pi GM/c^2$, i.e., if any two points on S can be joined by a curve in S whose length is smaller than d. The trapped surface then lies inside the horizon. (There are some exact results in this direction (see [187])).
3. Calculations based on perturbation theory indicate that a black hole quickly (in $10^{-3}(M/M_\odot)s$) approaches a stationary state after it is formed. The emission and partial reabsorption of gravitational waves is primarily responsible for this. The black hole is then a member of the Kerr-Newman family and thus characterized by its mass, angular momentum and charge (the latter is likely to be negligible in all realistic cases.) These are the only properties of its previous history which are retained. Reliable calculations of the gravitational radiation emitted during this relaxation process are in progress.

In summary, we have acquired a crude picture of the realistic collapse by induction from the spherically symmetric case, a mixture of rigorous mathematical results, physical arguments, extrapolation from perturbation calculations and faith. It remains for the future to refine this picture by numerical simulations.

If a black hole is surrounded by matter, originating for example from a normal star which forms a close binary system with the black hole, then the matter will be sucked in by the black hole and heated up so strongly that it becomes a source of intense X-ray radiation. This possibility leads to the next section, in which the evidence for the existence of black holes in some X-ray binaries will be discussed.

---

[9] There are various 'physical' and mathematical formulations of the cosmic censor conjecture (see, e.g., [7]). Since nothing has been proven about the validity of the conjecture, we do not discuss this issue.

How super-massive black holes are formed is not yet known, although a great deal has been written on the subject. Possible evolutionary scenarios have to take into account that there are luminous quasars even at redshift $z > 6$ (corresponding to an age of less than one billion years after the big bang), and that these show already high metal abundances. An extreme example is the recently discovered quasar SDSS J114816.64+525150.3 at a redshift $z = 6.41$. There is strong evidence that this quasar contains a supermassive black hole of about $3 \times 10^9 M_\odot$ (see [220]).

So there was some time interval between "seed" black holes, originating from an early generation of metal poor massive stars and the ignition of quasar activity. The problem is, however, that there are many e-foldings of mass that lie between such black holes and a final mass of about $10^8 M_\odot$. It is plausible that the seed black holes grow hierarchically by mergers and gas accretion, but how this happen is largely unknown. Since the formation processes of super-massive black holes are not known, we do not discuss here proposed scenarios.

7.6.2 Black Hole Candidates in X-Ray Binaries

So far all the evidence that stellar mass black holes exist comes from dynamical studies of X-ray binaries. – In 1964, long before the X-ray window was really opened by the famous satellite UHURU, Y.B. Zel'dovich first suggested the possibility that black holes may manifest themselves as compact X-ray sources in close binary systems. Cygnus X-1, for a long time the solitary example of a black hole candidate with firm supporting evidence, was located by UHURU within a few minutes of arc. By 1971 Cygnus X-1 was linked with a variable radio source which was then more accurately pin-pointed by radio astronomers in Green Bank and Holland. It was finally identified with the one-line optical binary HDE 226868 by L. Webster, P. Murdin and C.T. Bolton. Using a large body of observational data, D.R. Gies and Bolton derived a lower bound for the mass of the invisible massive compact companion of $7M_\odot$, with a preferred value of $16M_\odot$ (see [207]).

In December 1982, evidence of a massive invisible component in the extragalactic LMC X-3 was found (see [208]). In 1985, a third black hole candidate, A0620-00, was identified (see [209]).

To rule out the possibility that the compact companion of mass M_X in a binary system is a neutron star, it is necessary to show that M_X is clearly above about $3M_\odot$, the upper limit of neutron star masses (see Sect. 6.9). For this it is necessary to establish a large but save lower limit for M_X. A very useful tool for this is the **optical mass function** of the binary system, consisting of the compact object and a visible star of mass M_{opt}:

$$f(M_X, M_{opt}, i) = \frac{(M_X \sin i)^3}{(M_X + M_{opt})^2}, \tag{7.314}$$

where i is the inclination angle of the orbit (see Fig. 5.4 in Sect. 5.8.1). Using Kepler's third law (5.192) we obtain alternatively

$$f(M_X, M_{opt}, i) = \frac{4\pi^2}{G P_b^2}(a_{opt} \sin i)^3 . \tag{7.315}$$

Here $a_{opt} \sin i$ is the projected semi-major axis of the visible companion. In Sect. 5.8.1 we saw that the radial velocity along the line of sight varies periodically with an amplitude (often called K)

$$V_{opt} = \frac{2\pi\, a_{opt} \sin i}{P_b \sqrt{1 - e^2}} , \tag{7.316}$$

where e is the eccentricity of the orbit. Inserting this in (7.315) gives

$$f(M_X, M_{opt}, i) = \frac{P_b}{2\pi G}(1 - e^2)^{3/2} V_{opt}^3 . \tag{7.317}$$

Except for the eccentricity, the right-hand side contains only directly measurable quantities. We also saw in Sect. 5.8.1 that the radial velocity curve, described by (5.195), allows us also to determine e. So $f(M_X, M_{opt}, i)$ is observable, and from (7.314) we see that this provides a *lower limit* for the mass M_X:

$$M_X \geq f(M_X, M_{opt}, i) . \tag{7.318}$$

The main observational uncertainty comes from the radial velocity amplitude V_{opt}, which can be affected by X-ray heating, tidal distortion, non-synchronous rotation of the companion, and spectral contamination due to the surrounding gas and disk.

If the lower limit (7.317) is not good enough it may be possible to infer a (conservative) lower limit from additional information, for instance when no prominent X-ray eclipses are observed. However, this introduces as a rule considerable uncertainties.

Among the many (over 200) X-ray binaries known in the Galaxy, about 30 are transient sources, known as "X-ray nova" or "soft X-ray transients". These sources are characterized by episodic outbursts at X-ray, optical and radio frequencies, separated by long intervals (years to decades) of quiescence. In a quiescent state of an X-ray nova, the absorption line velocities of the secondary star can be measured precisely because the non-stellar light from the accretion flow is modest compared to the light of the secondary. For such systems the mass function can therefore be determined accurately.

At the time of writing more than a dozen X-ray nova with reliable measurements of the mass function, have been found with $f \geq 3\, M_\odot$ (see [210]).

7.6.3 The X-Ray Nova XTE J118+480

Several soft X-ray transients are now known with a mass function larger than $6 M_\odot$, greatly exceeding the maximum mass of (not very rapidly rotating) neutron stars. As an example we discuss briefly the X-ray transient XTE J1118+480, discovered in 2000 (for details see [211] and [212]). Later in that year the X-ray nova faded to near its quiescent level, and spectroscopical as well as photometric data in quiescence were soon obtained. In particular, an accurate radial velocity curve of the secondary star was obtained (see Fig. 7.10). The data are well fitted for a circular orbit. In this case Eqs. (5.193) and (5.195) give for the radial velocity

$$v_r = V_{opt} \sin\left(\frac{2\pi}{P_b}(t - T_0)\right) , \tag{7.319}$$

where T_0 corresponds to inferior conjunction of the secondary star. A sine fit to the data give the result (see [211])

$$P_b = 0.169930 \pm 0.000004\, d\,, \qquad V_{opt} = 701 \pm 10\, km\, s^{-1}\,. \tag{7.320}$$

The corresponding mass function is

$$f(M_X, M_{opt}, i) = 6.1 \pm 0.3 M_\odot\,. \tag{7.321}$$

This lower limit for M_X is an important result. The spectral type of the secondary is estimated to be $K7V$, and a plausible mass $M_{opt} = 0.09 - 0.5 M_\odot$ was obtained.

In conclusion, the spectroscopic, photometric and dynamical results give strong evidence that XTE J1118+480 is a black hole X-ray system. This is the first one found in the galactic halo (distance equal to $1.9 \pm 0.4\, kpc$ and $1.7 \pm 0.4\, kpc$ above the Galactic plane). – For a list of other transient X-ray sources that presumably contain black holes we refer to [210].

7.6.4 Super-Massive Black Holes

Until a few years ago the best one could say about the evidence of super-massive black holes in the centers of some galaxies, was that it was compelling if dynamical studies and observations of active galactic nuclei were taken together. In the meantime the situation has improved radically.

In Sect. 3.7 we already mentioned the beautiful work [109] which established a dark mass concentration of about $3 \times 10^6 M_\odot$ near the center of the Milky Way, whose extension is less than 17 light hours (see also [214]). If this would be a cluster of low mass stars or neutron stars, its central density would exceed $10^{17} M_\odot/pc^3$ and would not survive for more than a few 10^5 years. The least exotic interpretation of this enormous dark mass concentration is that it is a black hole. That this is not the only possibility was illustrated in Exercise 6.6, where we showed that a ball of fermions (e.g., neutrinos) for a certain range of fermion masses could in principle account for this mass concentration. Alternatively, one can also cook up bosonic star models. While such ways out are highly implausible, they illustrate that dynamical studies can not give an incontrovertible proof of the existence of

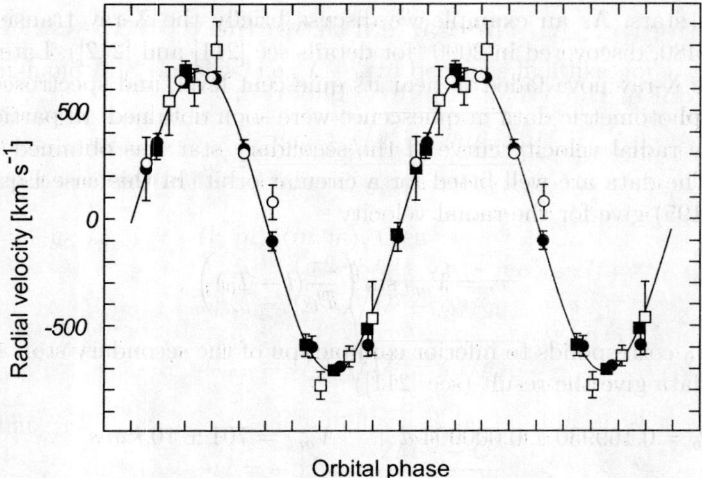

Fig. 7.10. Radial velocity of the X-ray nova XTE J118+480 and the best fitting sinusoid.

black holes. Ideally, one would have to show that some black hole candidate has an *event horizon.* There have been various attempts in this direction, but presumably only gravitational wave astronomy will reveal the essential properties of black holes.

Most astrophysicists do not worry about possible remaining doubts. The evidence for (super-massive) black holes is now so overwhelming that we can pass the burden of proof against them to the hard-core skeptics.

The concentration of dark matter in the center of the Milky Way is now the best case for a super-massive black hole. Another very good case is provided by the galaxy NGC 4258. The following observations show that the central mass of the galaxy must be a black hole (or something even more exotic). – The peculiar spiral galaxy NGC 4258 at a distance of about $6.5 Mpc$ has in its core a disk extending from about 0.2 to $0.13 pc$ in radius. This was discovered with a very precise mapping of gas motions via the $1.3\,cm$ maser-emission of H_2O with Very Long Baseline Interferometry (VLBI). The angular resolution of the array was better than 0.5 milliarc seconds. (This is 100 times better than the resolution of the Hubble Space Telescope.) The rotational velocity distribution in the disk follows an exact Keplerian law around a compact dark mass, and the velocity of the inner edge is $1080\,km/s$. From this one infers a dark mass concentration of $3.6 \times 10^7\, M_\odot$. As in the case of the Milky Way, there are no long-lived star clusters with these extreme properties. (For more details see [213].)

7.7 Appendix: Mathematical Appendix on Black Holes

7.7.1 Proof of the Weak Rigidity Theorem

In this appendix we prove the weak rigidity theorem quoted in Sect. 7.3.3, making use of the notation and tools developed in Sect. 7.2.1. For convenience, we repeat its formulation

Theorem 7.3 (Weak rigidity theorem). *Let (M, g) be a circular space-time with commuting Killing fields k, m and let $\Omega = -\langle k, m\rangle/\langle m, m\rangle$, $\xi = k + \Omega\, m$. Then the hypersurface $H = \{\langle \xi, \xi\rangle = 0\}$ is null, and Ω stays constant on H.*

Proof. First, we establish that H is a null hypersurface. For this we introduce

$$\sigma = -\langle k \wedge m, k \wedge m\rangle = VX + W^2\,, \tag{7.322}$$

where

$$V = -\langle k, k\rangle\,, \qquad W = \langle k, m\rangle\,, \qquad X = \langle m, m\rangle\,. \tag{7.323}$$

Note that H is given by $\sigma = 0$. Conceptually it is clear that the hypersurfaces $\{\sigma = \text{const}\}$ are invariant under the isometry group $\mathbb{R} \times SO(2)$, so that the vector fields k and m are tangent to these surfaces. Analytically, this follows from

$$L_k W = \big(L_k g\big)(k, m) - \langle [k, k], m\rangle - \langle k, [k, m]\rangle = 0\,,$$

and similar equations. Therefore the gradient $\nabla\sigma$ is perpendicular to k and m, i.e., $\nabla\sigma \in E^\perp$, where E is the integrable distribution spanned by the vector fields k and m. Below we establish that on H the gradient $\nabla\sigma$ is also in E. Thus $\nabla\sigma|_H \in E \cap E^\perp$. This intersection is one-dimensional on H, and generated by the null field ξ. Hence the normal $\nabla\sigma$ of H is null, proving that H is lightlike.

Let us now calculate the gradient of σ. We have

$$d\sigma = 2W\, dW + X\, dV + V\, dX\,.$$

Here we use

$$dW = d\, i_k m = -i_k dm\,, \qquad dW = -i_m dk\,,$$
$$dX = -i_m dm\,, \qquad dV = i_k dk$$

to get

$$d\sigma = -W(i_k dm + i_m dk) + X\, i_k dk - V\, i_m dm\,.$$

Now, we multiply this equation from the right with $(k \wedge m)$, and use in the resulting expressions the Frobenius conditions (see (7.70)). These and the anti-derivation rule for the interior product imply, for example

$$0 = i_k(dm \wedge k \wedge m) = (i_k dm) \wedge k \wedge m - V\, dm \wedge m - W\, dm \wedge k\,.$$

Collecting terms gives

$$d\sigma \wedge (k \wedge m) = \sigma\, d(k \wedge m)\,. \tag{7.324}$$

Especially on H we get $d\sigma \wedge (k \wedge m) = 0$. At this point we make the generic assumption that the group action is such that k and m are linearly independent on H. Then $d\sigma$ must be a linear combination of k and m. – In [205] it is shown that the non-degeneracy assumption just mentioned is not necessary, but this requires more work.

The second part of the theorem, $\Omega|_H =: \Omega_H = \text{const}$ follows with a similar argument. Let us look at the gradient of Ω. As before $\nabla\Omega \in E^\perp$. On the other hand we find in a first step

$$d\Omega = -\frac{1}{X}(dW + \Omega\, dX) = -\frac{1}{X}\, i_\xi dm\,.$$

The same manipulations which led to (7.324) give now, using $i_\xi k = \langle \xi, \xi \rangle$ and $i_\xi m = \langle \xi, m \rangle = 0$,

$$k \wedge m \wedge d\Omega = \frac{1}{X}\langle \xi, \xi \rangle m \wedge dm\,. \tag{7.325}$$

On H the right-hand side vanishes, and we again reach the conclusion that $\nabla\Omega \in E$, thus $\nabla\Omega$ is proportional to ξ. In other words $\nabla\Omega$ is normal to H, hence H is a surface of constant Ω. This completes the proof. □

7.7.2 The Zeroth Law for Circular Spacetimes

We consider the same situation as in Appendix 7.7.1. The strategy for proving that κ is constant on a Killing horizon will be similar to the one we used in proving the constancy of $\Omega|_H$.

We extend the definition of the surface gravity κ outside the horizon in a natural manner by (see (7.292))

$$\kappa^2 = -\frac{1}{4}\langle dl, dl \rangle\,. \tag{7.326}$$

Since k and m are commuting Killing fields, we have

$$L_k\langle dl, dl \rangle = 2\langle L_k dl, dl \rangle = 2\langle dL_k l, dl \rangle = 0\,,$$

and $L_m\langle dl, dl \rangle = 0$. Therefore the gradient $\nabla\kappa$ is in $E^\perp$. Below we show that it is on H also in E, so $\nabla\kappa$ has to be proportional to l. This implies that κ on H is a constant.

Once we have established that

$$m \wedge l \wedge d\kappa|_H = 0\,, \tag{7.327}$$

we are done. To show this we start from the following consequence of the two Frobenius conditions (7.70)

$$m \wedge l \wedge dl = 0\,, \tag{7.328}$$

which holds everywhere. Taking the interior product with l gives

$$0 = \langle l, m\rangle l \wedge dl - m \wedge i_l(l \wedge dl)\,.$$

On this we apply $d \circ i_Z$ for any vector field Z and obtain

$$\begin{aligned} 0 &= d\big(\langle l,m\rangle\, i_Z(l\wedge dl)\big) - d\big(\langle Z,m\rangle\, i_l(l\wedge dl)\big) \\ &\quad - dm \wedge i_Z\, i_l(l\wedge dl) + m \wedge d\, i_Z\, i_l(l \wedge dl)\,. \end{aligned}$$

In the next step we show that when this equation is restricted to H the first three terms vanish. For the first term this is clear, since both factors in the parenthesis vanish on H. Obviously the third term is also zero on H. To show this for the second term we have to consider ($N := \langle l, l\rangle$)

$$d\, i_l(l \wedge dl) = d(N\, dl + l \wedge dN) = 2\, dN \wedge dl\,,$$

which vanishes on H since dN is proportional to l and $l \wedge dl|_H = 0$. So we arrive at

$$m \wedge d\, i_Z\, i_l(l \wedge dl) = 0 \quad \text{on } H\,. \tag{7.329}$$

As above, we have

$$\begin{aligned} i_Z\, i_l(l\wedge dl) &= i_Z(N\, dl + l \wedge dN) \\ &= N\, i_Z dl + \langle Z, l\rangle\, dN - l \wedge i_Z dN\,. \end{aligned}$$

So the factor of m in (7.329) is equal to

$$\begin{aligned} &dN \wedge i_Z dl + N\, d\, i_Z\, dl + (d\, i_Z l) \wedge dN - dl \wedge i_Z dN + l \wedge d\, i_Z dN \\ &= N\, d\, i_Z dl - i_Z(dN \wedge dl) + d\langle Z, l\rangle \wedge dN + l \wedge d(i_Z dN)\,. \end{aligned}$$

In (7.329) we need the restriction of this to H. The first two terms in the last expression obviously vanish there, and in the third term we can replace dN by $-2\kappa\, l$. So (7.329) becomes

$$m \wedge \big[d\langle Z, l\rangle \wedge (-2\kappa\, l) + l \wedge d(i_Z dN)\big] = 0 \quad \text{on } H\,. \tag{7.330}$$

Is it allowed to replace dN in the last expression also by $-2\kappa\, l$, in spite of the exterior differentiation? The answer is yes, as the following simple argument shows. After multiplication with (-2κ) the last term in the square bracket of (7.330) has the form $dN \wedge df$ for a function f. If we again use adapted coordinates, with N one of them, and $\{x^i\}$ parameterizing H, we have $df = \bar{d}f + \partial_N f\, dN$, where $\bar{d}f = \partial_i f\, dx^i$ is the differential in H. So

$dN \wedge df = dN \wedge d\bar{f}$ on H, showing that on H $dN \wedge df = dN \wedge dg$ for two functions f and g that agree on H. Therefore, on H

$$l \wedge d(i_Z dN) = l \wedge d(-2\kappa\langle Z, l\rangle) \,,$$

and (7.330) becomes

$$\langle Z, l\rangle \, m \wedge l \wedge d\kappa = 0 \,.$$

Since the vector field Z is arbitrary, we arrive at (7.327).

Let us summarize: *On a Killing horizon, whose outer domain of communication is circular, the surface gravity is constant.* This is the **zeroth law** for circular spacetimes.

Note that this is a purely geometrical fact. We have never used the field equations and no assumptions on the matter content had to be made. The proof of the zeroth law for *static* spacetimes is left as an exercise.

* * *

Exercise 7.6. Prove the zeroth law for a Killing horizon, whose outer domain of communication is static.

Hints. For a static spacetime with Killing field K, we have $K \wedge dK = 0$. Apply on this $d \circ i_Z \circ i_K$ and restrict the result to the Killing horizon to conclude that K and $\nabla\kappa$ are proportional.

7.7.3 Geodesic Null Congruences

Congruences of timelike geodesics of a Lorentz manifold have been studied in Sect. 2.1. In this appendix we investigate the behavior of (infinitesimal) light beams. We shall introduce the "optical scalars" and derive their propagation equations. This material is indispensable for proving the area law. Some of it will also be used in the proof of the positive energy theorem for black holes (see Chap. 8). Null congruences also play an important role in the proofs of singularity theorems.

Consider a light beam with wave vector field k. Let $\gamma(\lambda)$ be a central null geodesic, and consider a one-parameter family of neighboring null geodesics of the beam, i.e., a variation of γ in the sense of DG, Sect. 14.4, with deviation vector field X along γ. As shown in DG, Sect. 14.4, X satisfies the Jacobi equation

$$\ddot{X} + R(X, k)k = 0 \,, \tag{7.331}$$

where a dot denotes the covariant derivation ∇_k in the direction k. Note that

$$\langle k, X\rangle^{\cdot\cdot} = \langle k, \nabla_k^2 X\rangle \overset{(7.331)}{=} 0 \,,$$

because of the symmetry properties of the Riemann tensor. Note also that along $\gamma(\lambda)$ fields of the form $(a\lambda + b)\dot{\gamma}(\lambda)$, with a, b constants, are Jacobi fields. Subtracting such a tangent field, we can always assume that

$$\langle k, X\rangle = 0 \,. \tag{7.332}$$

Optical Scalars

We choose a four-velocity field u along the central ray γ, which is parallel along γ and satisfies $\langle u, u\rangle = -1$. Furthermore, we normalize the affine parameter λ such that $\langle k, u\rangle = -1$.

Let us now introduce the following basis along γ: E_A, for $A = 1, 2$, is an orthonormal basis of the plane orthogonal to the span of k and u, which is parallel along γ. We also use the complex vector $\varepsilon = E_1 + i\, E_2$ and its conjugate complex $\bar{\varepsilon} = E_1 - i\, E_2$. The following scalar products are obvious,

$$\langle k, k\rangle = \langle k, E_A\rangle = \langle u, E_A\rangle = \langle k, \varepsilon\rangle = \langle u, \varepsilon\rangle = \langle E_1, E_2\rangle = \langle \varepsilon, \varepsilon\rangle = 0\,,$$
$$-\langle k, k\rangle = -\langle u, u\rangle = \langle E_A, E_A\rangle = \frac{1}{2}\langle \bar{\varepsilon}, \varepsilon\rangle = 1\,. \tag{7.333}$$

Let also

$$E_0 = u\,, \qquad E_3 = k - u\,. \tag{7.334}$$

Then $\{E_0, \ldots, E_3\}$ form an orthonormal vierbein along γ.

Next, we introduce the **optical scalars** θ, σ and ω defined by

$$\sigma = \frac{1}{2}\langle \varepsilon, \nabla_\varepsilon k\rangle\,, \qquad \theta = \operatorname{Re}\rho\,, \qquad \omega = \operatorname{Im}\rho\,, \tag{7.335}$$

where

$$\bar{\rho} = \frac{1}{2}\langle \varepsilon, \nabla_{\bar{\varepsilon}} k\rangle\,. \tag{7.336}$$

Thus we have

$$\rho = \theta + i\,\omega\,. \tag{7.337}$$

Transport Equation

The vector field X can be decomposed as

$$X = \xi_1\, E_1 + \xi_2\, E_2 + \tilde{\xi}\, k\,. \tag{7.338}$$

(A priori we can write $X = (\xi_1\, E_1 + \xi_2\, E_2) + (\xi_0\, E_0 + \xi_3\, E_3)$, but $\langle X, k\rangle = 0$ implies $\xi_0 = \xi_3$, hence the last term is proportional to k.) Also let

$$\xi = \xi_1 + i\,\xi_2\,, \tag{7.339}$$

thus

$$X = \frac{1}{2}(\xi\,\bar{\varepsilon} + \bar{\xi}\,\varepsilon) + \text{term proportional to } k\,. \tag{7.340}$$

Note that by (7.333)

$$\xi = \langle \varepsilon, X\rangle\,, \tag{7.341}$$

hence, recalling that k and X commute,

$$\dot{\xi} = \langle \varepsilon, \nabla_k X\rangle = \bar{\rho}\,\xi + \sigma\,\bar{\xi}\,,$$

where we used that $\nabla_k X = \nabla_X k = \frac{1}{2}\xi \nabla_{\bar{\varepsilon}} k + \frac{1}{2}\bar{\xi} \nabla_{\varepsilon} k + 0$. The resulting **transport equation**

$$\dot{\xi} = \bar{\rho}\xi + \sigma\bar{\xi} \tag{7.342}$$

is important in what follows. The real form of this equation is easily derived[10]. Let

$$\boldsymbol{\xi} = \begin{pmatrix} \xi_1 \\ \xi_2 \end{pmatrix}, \tag{7.343}$$

then

$$\dot{\boldsymbol{\xi}} = S\,\boldsymbol{\xi}, \tag{7.344}$$

where

$$S = \begin{pmatrix} \theta + \operatorname{Re}\sigma & \operatorname{Im}\sigma + \omega \\ \operatorname{Im}\sigma - \omega & \theta - \operatorname{Re}\sigma \end{pmatrix}. \tag{7.345}$$

From (7.345) the geometrical meaning of the optical scalars in obvious :

$$\begin{aligned} \theta &: \quad \textbf{isotropic expansion}, \\ \sigma &: \quad \textbf{rate of shear}, \\ \omega &: \quad \textbf{rotation rate}, \end{aligned} \tag{7.346}$$

because the **optical deformation matrix** S splits as

$$S = \begin{pmatrix} \theta & 0 \\ 0 & \theta \end{pmatrix} + \begin{pmatrix} \operatorname{Re}\sigma & \operatorname{Im}\sigma \\ \operatorname{Im}\sigma & -\operatorname{Re}\sigma \end{pmatrix} + \begin{pmatrix} 0 & \omega \\ -\omega & 0 \end{pmatrix}.$$

(In the literature θ is often defined to be the trace of S.)

We remark that if k is a gradient field, then $k_{\alpha;\beta} = k_{\beta;\alpha}$, hence

---

[10] Consider a linear map $J : \mathbb{R}^2 \longrightarrow \mathbb{R}^2$, $\boldsymbol{y} = J\,\boldsymbol{x}$. In $\mathbb{C}$ this reads

$$w = \bar{a}\,z + b\,\bar{z},$$

where $z = x_1 + i\,x_2$, $w = y_1 + i\,y_2$ and

$$\begin{aligned} \bar{a} &= \frac{1}{2}(J_{11} + J_{22}) - \frac{i}{2}(J_{12} - J_{21}), \\ b &= \frac{1}{2}(J_{11} - J_{22}) + \frac{i}{2}(J_{12} + J_{21}). \end{aligned}$$

We see that

$$\begin{aligned} J &= \begin{pmatrix} \operatorname{Re}a + \operatorname{Re}b & \operatorname{Im}b + \operatorname{Im}a \\ \operatorname{Im}b - \operatorname{Im}a & \operatorname{Re}a - \operatorname{Re}b \end{pmatrix} \\ &= \operatorname{Re}(a)\,\mathbf{1}_2 + \operatorname{Im}(b)\,\sigma_1 + i\operatorname{Im}(a)\,\sigma_2 + \operatorname{Re}(b)\,\sigma_3. \end{aligned}$$

Note that

$$\det J = |a|^2 - |b|^2, \qquad \operatorname{Tr} J = 2\operatorname{Re}a.$$

$$\langle E_1, \nabla_{E_2} k \rangle = E_1^\alpha E_2^\beta \, k_{\alpha;\beta} = \langle E_2, \nabla_{E_1} k \rangle ,$$

which is equivalent to ρ being real, that is $\omega = 0$. (See also Eq. (7.351) below).

Next we show that

$$\theta = \frac{1}{2} \, k^\alpha_{\ ;\alpha} \, . \tag{7.347}$$

From the definition of θ we have

$$\begin{aligned} \theta &= \frac{1}{4} \left(\langle \varepsilon, \nabla_{\bar{\varepsilon}} k \rangle + \langle \bar{\varepsilon}, \nabla_{\varepsilon} k \rangle \right) \\ &= \frac{1}{4} \varepsilon^\alpha \bar{\varepsilon}^\beta (k_{\alpha;\beta} + k_{\beta;\alpha}) \\ &= \frac{1}{4} \, k_{\alpha;\beta} (\varepsilon^\alpha \bar{\varepsilon}^\beta + \bar{\varepsilon}^\alpha \varepsilon^\beta) . \end{aligned} \tag{7.348}$$

Now, the metric can be decomposed as follows

$$\begin{aligned} g &= -E_0 \otimes E_0 + \sum_j E_j \otimes E_j \\ &= -u \otimes u + E_1 \otimes E_1 + E_2 \otimes E_2 + (k-u) \otimes (k-u) \end{aligned}$$

or

$$g = \frac{1}{2} (\varepsilon \otimes \bar{\varepsilon} + \bar{\varepsilon} \otimes \epsilon) + k \otimes k - k \otimes u - u \otimes k \, , \tag{7.349}$$

where the first term is the induced metric on span$\{E_1, E_2\}$. Thus

$$\frac{1}{2} (\varepsilon^\alpha \bar{\varepsilon}_\beta + \bar{\varepsilon}^\alpha \varepsilon_\beta) = \delta^\alpha_\beta - k^\alpha k_\beta + k^\alpha u_\beta + u^\alpha k_\beta \, . \tag{7.350}$$

If this is inserted in (7.348) and use is made of

$$\begin{aligned} k^\alpha k^\beta k_{\alpha;\beta} &= \langle k, \nabla_k k \rangle = 0 \, , \\ k^\alpha u^\beta k_{\alpha;\beta} &= \langle k, \nabla_u k \rangle = \frac{1}{2} \nabla_u \langle k, k \rangle = 0 \, , \\ u^\alpha k^\beta k_{\alpha;\beta} &= \langle u, \nabla_k k \rangle = 0 \, , \end{aligned}$$

we obtain indeed

$$\theta = \frac{1}{4} \, k_{\alpha;\beta} \, 2 g^{\alpha\beta} = \frac{1}{2} \, k^\alpha_{\ ;\alpha} \, .$$

We leave it as an exercise to show similarly that

$$\omega^2 = \frac{1}{2} \langle dk, dk \rangle \, . \tag{7.351}$$

Hence, the rotation rate ω vanishes if $dk = 0$. (For a more general condition, see the applications below.) Derive also an invariant expression for $|\sigma|^2$ in terms of covariant derivatives of k (consider for this $k_{(\alpha;\beta)} k^{(\alpha;\beta)}$).

The Sachs Equations

We now give a very simple (possibly novel) derivation of the propagation equations for the optical scalars. We start by differentiating the transport equation (7.342):

$$\begin{aligned}\ddot{\xi} &= \dot{\rho}^*\xi + \rho^*\dot{\xi} + \dot{\sigma}\xi^* + \sigma\dot{\xi}^* \\ &= \dot{\rho}^*\xi + \rho^*(\rho^*\xi + \sigma\xi^*) + \dot{\sigma}\xi^* + \sigma(\rho\xi^* + \sigma^*\xi) \\ &= \xi(\dot{\rho}^* + \bar{\rho}^2 + |\sigma|^2) + \xi^*(\dot{\sigma} + \rho^*\sigma + \rho\sigma)\,.\end{aligned}$$

On the other hand, the Jacobi equation (7.331) gives

$$\begin{aligned}\ddot{\xi} &= \langle \varepsilon, \ddot{X}\rangle = -\langle \varepsilon, R(X,k)k\rangle \\ &\overset{(7.340)}{=} -\left\langle \varepsilon, R\left(\frac{1}{2}\xi\,\bar{\varepsilon} + \frac{1}{2}\bar{\xi}\,\varepsilon, k\right)k\right\rangle \\ &= -\xi\,\frac{1}{2}\langle \varepsilon, R(\bar{\varepsilon},k)k\rangle - \bar{\xi}\,\frac{1}{2}\langle \varepsilon, R(\varepsilon,k)k\rangle \\ &\equiv -\mathcal{R}\,\xi - \mathcal{F}\,\bar{\xi}\,.\end{aligned} \tag{7.352}$$

Identifying, we obtain ($\mathcal{R}$ is real, see below)

$$\dot{\rho} + \rho^2 + |\sigma|^2 = -\mathcal{R}\,, \tag{7.353a}$$

$$\dot{\sigma} + 2\,\theta\,\sigma = -\mathcal{F}\,. \tag{7.353b}$$

We rewrite $\mathcal{R}$ and $\mathcal{F}$:

$$\begin{aligned}\mathcal{R} &= \frac{1}{2}\langle \varepsilon, R(\bar{\varepsilon},k)k\rangle = \frac{1}{2}\,\bar{\varepsilon}^{\alpha}\varepsilon^{\beta}\,R_{\alpha\mu\beta\nu}k^{\mu}k^{\nu} \\ &= \frac{1}{4}(\bar{\varepsilon}^{\alpha}\varepsilon^{\beta} + \varepsilon^{\alpha}\bar{\varepsilon}^{\beta})R_{\alpha\mu\beta\nu}k^{\mu}k^{\nu} \overset{(7.350)}{=} \frac{1}{2}g^{\alpha\beta}R_{\alpha\mu\beta\nu}k^{\mu}k^{\nu}\,,\end{aligned}$$

giving

$$\mathcal{R} = \frac{1}{2}\,R_{\mu\nu}\,k^{\mu}k^{\nu}\,. \tag{7.354}$$

Similarly, we find

$$\mathcal{F} = \frac{1}{2}\langle \varepsilon, R(\varepsilon,k)k\rangle = \frac{1}{2}\varepsilon^{\alpha}\varepsilon^{\beta}\,R_{\alpha\mu\beta\nu}k^{\mu}k^{\nu}\,. \tag{7.355}$$

Here, because of (7.333) only the Weyl tensor $C_{\alpha\mu\beta\nu}$ contributes (see Sect. 13.10 for the definition of the Weyl tensor),

$$\mathcal{F} = \frac{1}{2}\varepsilon^{\alpha}\varepsilon^{\beta}\,C_{\alpha\mu\beta\nu}k^{\mu}k^{\nu}\,. \tag{7.356}$$

Let us split (7.353a) into two equations for the real and imaginary parts. Together with (7.353b) we get the basic **Sachs equations** for the propagation of the optical scalars:

$$\dot{\theta} + \theta^2 - \omega^2 + |\sigma|^2 = -\mathcal{R}, \tag{7.357a}$$
$$\dot{\sigma} + 2\,\theta\sigma = \mathcal{F}, \tag{7.357b}$$
$$\dot{\omega} + 2\,\theta\omega = 0. \tag{7.357c}$$

Of particular importance is (7.357a), also known as the **Raychauduri equation**.

Let us write (7.352), i.e.,

$$\ddot{\xi} = -\mathcal{R}\xi - \mathcal{F}\xi^*, \tag{7.358}$$

in real form:

$$\ddot{\boldsymbol{\xi}} + T\boldsymbol{\xi} = 0, \tag{7.359}$$

where the **optical tidal matrix** T is given by

$$T = \begin{pmatrix} \mathcal{R} + \operatorname{Re}\mathcal{F} & \operatorname{Im}\mathcal{F} \\ \operatorname{Im}\mathcal{F} & \mathcal{R} - \operatorname{Re}\mathcal{F} \end{pmatrix}. \tag{7.360}$$

Conceptually, one should regard $\boldsymbol{\xi}$ as a representative of the "equivalence class mod k" of X (see (7.338)).

Applications

Let us consider some special families of null geodesics that play an important role in GR, for instance in proofs of singularity theorems.

First, we consider a spacelike two-dimensional surface S in the spacetime (M, g) and a congruence of null geodesics orthogonal[11] to S for some neighborhood $U \subset S$. Let k denote the tangent vector field (or the corresponding one-form), and let $\{E_i, i = 1, 2\}$ be an orthoncrmal basis of vector fields of S in U. We choose a point $p \in U$ and the null geodesics γ with $\gamma(0) = p$ (see Fig. 7.11). – We claim that for this situation the rotation rate ω vanishes at p, for any $p \in U$.

The proof of this is simple. From the definition one sees that[12]

$$\omega = \frac{1}{2}\left(\langle E_1, \nabla_{E_2} k\rangle - \langle E_2, \nabla_{E_1} k\rangle\right). \tag{7.362}$$

On the initial surface S, where $\langle E_1, k\rangle = \langle E_2, k\rangle = 0$, the Ricci identity gives

[11] Note that the orthogonal complement of T_pS in T_pM is a two-dimensional Minkowski vector space. Hence it contains two different null directions, an "ingoing" and an "outgoing" one.

[12] Similarly, the real part θ of ρ is given by

$$\theta = \frac{1}{2}\left(\langle E_1, \nabla_{E_1} k\rangle + \langle E_2, \nabla_{E_2} k\rangle\right). \tag{7.361}$$

In differential geometry $-\theta$ is called the **convergence** of S; see, e.g., [39] and (7.376) below.

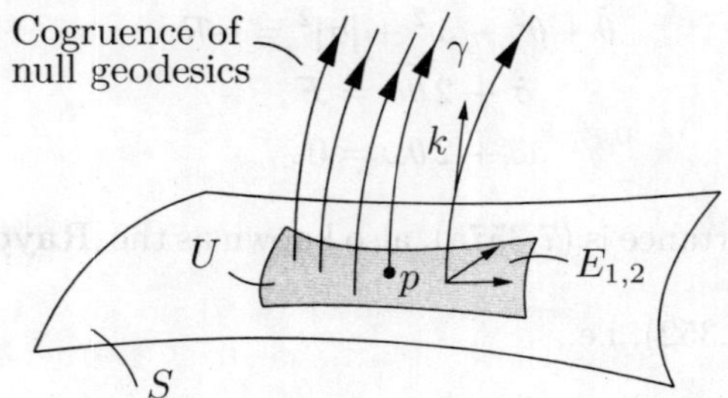

Fig. 7.11. Two-parameter family of null geodesics normal to a two-dimensional spacelike surface S.

$$2\omega = \langle \nabla_{E_1} E_2, k \rangle - \langle \nabla_{E_2} E_1, k \rangle = \langle [E_1, E_2], k \rangle = 0,$$

since $[E_1, E_2] \in TS$ by the Frobenius theorem (see DG, Appendix C).

If the initial value of ω vanishes, the propagation equation (7.357c) implies that $\omega = 0$ everywhere. So the two-parameter family sketched in Fig. 7.11 has vanishing rotation.

Consider next a congruence of null geodesics which is orthogonal to a family of hypersurfaces, the **wave fronts**, as in the eikonal approximation of wave optics. According to the Frobenius theorem, this is locally equivalent to (see DG, (C.24))

$$k \wedge dk = 0 . \tag{7.363}$$

In the notation introduced above, we evaluate $k \wedge dk$ on the triple $(\varepsilon, \bar{\varepsilon}, u)$. Using the scalar products in (7.333), and the definition of ρ, we get

$$\begin{aligned} 0 = (k \wedge dk)(\varepsilon, \bar{\varepsilon}, u) &= \langle k, u \rangle \, dk(\varepsilon, \bar{\varepsilon}) = -dk(\varepsilon, \bar{\varepsilon}) \\ &= -\langle \nabla_{\varepsilon} k, \bar{\varepsilon} \rangle + \langle \nabla_{\bar{\varepsilon}} k, \varepsilon \rangle = 2(\rho - \bar{\rho}) . \end{aligned}$$

This shows that ρ is real, hence $\omega = 0$.

For $\omega = 0$ the Raychauduri equation (7.357a) for the expansion θ becomes particular interesting, because the right-hand side of

$$\dot{\theta} + \theta^2 = -(\mathcal{R} + |\sigma|^2) \tag{7.364}$$

is non-positive if some reasonable energy condition is assumed. Indeed, Einstein' field equation implies

$$\mathcal{R} = \frac{1}{2} R_{\mu\nu} k^\mu k^\nu = \frac{1}{2} T_{\mu\nu} k^\mu k^\nu , \tag{7.365}$$

and this is non-negative if the following **weak energy condition**

$$T_{\mu\nu} \, \xi^\mu \xi^\nu \geq 0 \tag{7.366}$$

for all timelike ξ^μ is satisfied. In this case (7.364) implies

$$(1/\theta)^{\cdot} \geq 1 \tag{7.367}$$

or, integrating between λ_0 and λ,

$$\theta(\lambda) \leq \left(\theta(\lambda_0)^{-1} + (\lambda - \lambda_0)\right)^{-1} . \tag{7.368}$$

So if $\theta(\lambda_0)$ is negative, $\theta(\lambda)$ will become unboundedly negative in the interval $[\lambda_0, \lambda_0 - \theta(\lambda_0)^{-1}]$. This means that the area element between neighboring curves of a two-parameter variation as in Fig. 7.11 goes to zero somewhere in this interval (see Eq. (7.377) below). The point on the null geodesic where this happens is a **focal** or **conjugate point** of S along the null geodesic.

Let us summarize this important conclusion.

Proposition 7.1. *Let θ be the expansion rate along a null geodesic whose initial rotation rate ω vanishes. If the weak (or strong) energy condition for the energy-momentum tensor holds, then $\theta(\lambda_0) < 0$ implies that θ becomes $-\infty$ at a finite value of the affine parameter λ in the interval $[\lambda_0, \lambda_0 - \theta(\lambda_0)^{-1}]$, provided the null geodesic is defined on this interval. This holds, in particular, for any geodesic of a hypersurface orthogonal congruence of null geodesics, and for a two-parameter family of null geodesics normal to a two-dimensional spacelike surface.*

Change of Area

We come back to the situation sketched in Fig. 7.11. The two-parameter family of null geodesics defines the one-parameter family of immersions

$$\begin{aligned} \phi_\lambda &: U \longrightarrow M , \\ U \ni p &\longmapsto \exp\big(\lambda\, k(p)\big) \end{aligned}$$

for some interval of λ. Let η_λ be the induced volume forms on $\phi_\lambda(U)$. We want to know how these change with increasing λ. As expected, the rate for this is twice the expansion rate θ:

$$\dot{\eta}_\lambda := \left.\frac{d}{d\varepsilon}\right|_{\varepsilon=0} \phi_\varepsilon^* \, \eta_{\lambda+\varepsilon} = 2\,\theta\,\eta_\lambda . \tag{7.369}$$

For deriving this we need the fact that k is orthogonal to the family of immersed submanifolds $\phi_\lambda(U)$. This one may call "Gauss' lemma for null geodesics" (see Sect. 2.9). To prove it, let $\sigma(t)$ be a curve in U, and consider the one-parameter subfamily of null geodesics starting on $\sigma(t)$:

$$\sigma(t)\lambda \longmapsto F(\lambda, t) = \phi_\lambda(\sigma(t)) .$$

The tangential fields along F belonging to $\partial/\partial\lambda$ and $\partial/\partial t$ are

$$k = TF \circ \frac{\partial}{\partial \lambda}, \qquad V = TF \circ \frac{\partial}{\partial t} .$$

Obviously, V is tangential to the submanifolds $\phi_\lambda(U)$. To show that $\langle k, V\rangle = 0$ we consider

$$k\langle k, V\rangle = \langle k, \nabla_k V\rangle = \langle k, \nabla_V k\rangle = \frac{1}{2} V\langle k, k\rangle = 0\,.$$

So $\frac{\partial}{\partial\lambda}\langle k, V\rangle = 0$. Since for $\lambda = 0$, k and V are orthogonal, k is orthogonal to all surfaces $\phi_\lambda(U)$.

It will be useful to introduce an orthonormal frame $\{E_\alpha\}$ which is adapted to the foliation: E_i $(i = 1, 2)$ are tangential to the submanifolds $\phi_\lambda(U)$, and E_A $(A = 0, 3)$ are normal to $\phi_\lambda(U)$. The dual basis of 1-forms is denoted by $\{\theta^\alpha\}$. Obviously,

$$\eta_\lambda = \theta^1 \wedge \theta^2 \qquad \text{on } \phi_\lambda(U)\,. \tag{7.370}$$

We want to compute

$$\lim_{\varepsilon\to 0} \frac{1}{\varepsilon}\Big(\phi_\varepsilon^* \eta_\lambda - \eta_\lambda\Big) = L_k(\theta^1 \wedge \theta^2) \qquad \text{on } \phi_\lambda(U)\,.$$

Let $\Theta = \theta^1 \wedge \theta^2$, then

$$\dot\eta_\lambda = L_k\Theta = i_k d\Theta + d\, i_k\Theta\,. \tag{7.371}$$

The last term vanishes, because k is perpendicular to E_1 and E_2:

$$i_k\Theta = (i_k\theta^1)\theta^2 - \theta^1(i_k\theta^2) = 0\,.$$

For the first term on the right of (7.371) we use the first structure equation, giving ($\omega^\alpha_{\ \beta}$ denote the connection forms)

$$\begin{aligned} d\Theta &= d\theta^1 \wedge \theta^2 - \theta^1 \wedge d\theta^2 = -\omega^1_{\ \alpha} \wedge \theta^\alpha \wedge \theta^2 + \theta^1 \wedge \omega^2_{\ \alpha} \wedge \theta^\alpha \\ &= \theta^A \wedge (\omega^1_{\ A} \wedge \theta^2 + \theta^1 \wedge \omega^2_{\ A})\,. \end{aligned}$$

Hence,

$$\big(i_k\, d\Theta\big)(E_1, E_2) = \theta^A(k)\, \omega^j_{\ A}(E_j)\,. \tag{7.372}$$

On the other hand we have, using the notation $\varepsilon_\alpha \equiv \langle E_\alpha, E_\alpha\rangle = \pm 1$,

$$\nabla_{E_j} E_j = \omega^\alpha_{\ j}(E_j) E_\alpha = -\varepsilon_\alpha\, \omega^j_{\ \alpha}(E_j) E_\alpha\,.$$

Therefore, the **mean curvature normal**

$$H = \frac{1}{2}\left(\sum_{j=1}^{2} \varepsilon_j \nabla_{E_j} E_j\right)_\perp\,, \tag{7.373}$$

where $\perp$ denotes the normal component, can be expressed as $(\varepsilon_j = 1)$

$$2\,H = -\varepsilon_A\, \omega^j_{\ A}(E_j) E_A\,. \tag{7.374}$$

Hence

$$-\langle k, 2\,H\rangle = \theta^A(k)\,\omega^j{}_A(E_j)\,.$$

Comparison with (7.372) gives

$$i_k\,d\Theta = -\langle k, 2\,H\rangle\Theta\,,$$

so

$$L_k\Theta = -\langle k, 2\,H\rangle\Theta\,. \tag{7.375}$$

This proves (7.369), since by (7.361) and (7.373)

$$\theta = -\langle k, H\rangle\,. \tag{7.376}$$

Here, the scalar product on the right-hand side is the so-called **convergence**.

A consequence of (7.369) is this: Let $V \subset U$, then

$$\frac{d}{d\lambda}\int_{\phi_\lambda(V)} \eta_\lambda = \int_V \frac{d}{d\lambda}\phi^*_\lambda\,\eta_\lambda = \int_V \phi^*_\lambda(2\,\theta\,\eta_\lambda) = \int_{\phi_\lambda(V)} 2\,\theta\,\eta_\lambda\,. \tag{7.377}$$

In particular, if $\theta \geq 0$ the area of $\phi_\lambda(V)$ does *not decrease*. On the other hand, if θ would be negative somewhere along one of the null geodesics, Proposition 7.1 implies that one could find a neighborhood V such that the area of $\phi_\lambda(V)$ vanishes at some finite value of λ (provided that the null geodesic is defined on a sufficiently large interval). This will play an important role in the next section.

7.7.4 Area Law for Black Holes

If the spacetime (M, g) satisfies certain general conditions (global hyperbolicity and asymptotic predictability[13]) the horizon H is a closed embedded three-dimensional C^{1-}-submanifold[14] (see Proposition 6.3.1 in [8]), and is generated by null geodesics *without future endpoints*. (Null geodesic generators of null hypersurfaces are introduced in DG, Appendix A. Note that null geodesic through a point of H is unique, up to rescaling of the affine parameter.) Null geodesic generators may leave H in the past, but have to go into $J^-(\mathscr{I}^+)$. Such leaving events are called **caustics** of H. At a caustic H is not C^1 and has a cusp-like structure. But once a null geodesic has joined into H it will never encounter a caustic again, *never leave H and not intersect any other generator*. All these properties, established by R. Penrose in [221], are (partially) proved in a pedagogical manner in [13], Box 34.1. There is no point to repeat this here.

[13] In [7] the spacetime (M, g) is said to be **strongly asymptotically predictable**, if in the unphysical spacetime $(\tilde{M}, \tilde{g})$ there is an open region $\tilde{V}$ containing the closure of $M \cap J^-(\mathscr{I}^+)$ in $\tilde{M}$, such that $(\tilde{V}, \tilde{g})$ is globally hyperbolic.

[14] This means that H is a topological manifold with an atlas whose transition functions are locally Lipschitz.

Based on these properties we now "prove" the area law (the second law) for black holes. *If* the horizon would be a sufficiently smooth submanifold, the area law would follow easily. Using the differential geometric tools developed in Appendix 7.7.3, we could argue as follows.

Consider two Cauchy hypersurfaces Σ and Σ', with Σ' in the future of Σ, and let $\mathcal{H} = H \cap \Sigma$ and $\mathcal{H}' = H \cap \Sigma'$ be the horizons at time Σ, respectively Σ'. We want to compare the surface areas

$$\mathcal{A}(\mathcal{H}) = \int_{\mathcal{H}} \eta_{\mathcal{H}} \qquad \text{and} \qquad \mathcal{A}(\mathcal{H}') = \int_{\mathcal{H}'} \eta_{\mathcal{H}'} ,$$

where $\eta_{\mathcal{H}}$, $\eta_{\mathcal{H}'}$ denote the induced volume forms (measures) on $\mathcal{H}$ and $\mathcal{H}'$. At each point $p \in \mathcal{H}$ there is a unique future- and outward-pointing null direction perpendicular to $\mathcal{H}$, and we can choose a vector field l of future directed null vectors. (Show that this is possible in a smooth fashion.) The null geodesics with these initial conditions, i.e., the family

$$\lambda \longmapsto \gamma_p(\lambda) = \exp_p\big(\lambda\, l(p)\big)$$

are generators of H without future endpoints. Therefore, they intersect Σ' in unique points on $\mathcal{H}'$, with unique affine parameter $\tau(p)$ for each $p \in \mathcal{H}$. If we rescale the null field l as $k(p) := \tau(p)\, l(p)$, then we obtain a injective map

$$\begin{aligned} \Phi : \mathcal{H} &\longrightarrow \mathcal{H}' , \\ p &\longmapsto \exp_p k(p) . \end{aligned}$$

(The map Φ is in general not surjective because new generators may join the horizon.) As in Appendix 7.7.3 we introduce the one-parameter family of immersions

$$\begin{aligned} \phi : \mathcal{H} &\longrightarrow M , \\ p &\longmapsto \exp_p\big(\lambda\, k(p)\big) \end{aligned}$$

for $\lambda \in [0, 1]$. Clearly, $\Phi = \phi_{\lambda=1}$.

From Proposition 7.1 we conclude that the expansion rate θ (or the negative of the convergence) of the generating family of null geodesics, $\lambda \longmapsto \phi_\lambda(p)$, $p \in \mathcal{H}$, cannot be negative, if the weak energy condition of matter holds. (Otherwise focal points would be formed in the future, violating Penrose's theorem on the global structure of horizons (quoted above)). Formula (7.377) then implies that

$$\mathcal{A}(\mathcal{H}') \geq \mathcal{A}(\mathcal{H}) .$$

Unfortunately, this beautiful argument is based on a smoothness assumption for the horizon that is generally not satisfied, because of the formation of caustics where the manifold has a cuspy structure. As was pointed out by several authors, the general validity of the area law is for this reason still an open problem. For discussions of this delicate issue, that appears somewhat strange from a physical point of view, we refer to [216] and [217].

8 The Positive Mass Theorem

We begin by repeating some material treated at various places in earlier chapters. – In Newtonian gravity theory the total energy of a system can be indefinitely negative. (For the critical masses of the white dwarfs, discussed in Sect. 6.2, the total energy becomes $-\infty$.) This is not the case in GR. It is impossible to construct an object out of ordinary matter, i.e., matter with positive local energy density, whose total energy (including gravitational contributions) is negative. This is roughly the content of the positive energy theorem.

If objects with negative mass could exist in GR, they could *repel* rather than attract nearby matter. The positive energy theorem implies, for example, that there cannot exist a regular interior solution of the "anti-Schwarzschild" solution with negative mass. All this is, of course, only true for the correct sign on the right-hand side of the Einstein equation. Changing the sign would lead to repulsion. But then the positive energy theorem would no more hold, and we could presumably extract an unlimited amount of energy from a system with negative energy.

With the correct sign (i.e, attraction) this bizarre situation cannot occur in GR. Physically, the reason is that as a system is compressed to take advantage of the negative gravitational binding energy, a black hole is eventually formed which has positive total energy.

In this chapter we will give essentially E. Witten's proof of the positive energy theorem [223] which makes crucial use of spinor fields. To be complete, we develop the necessary tools on spinors in GR in an appendix to this chapter. It is very remarkable that spinors have turned out to be so useful in simplifying the proof of an entirely classical property of GR. The first complete proof of the positive energy theorem was found by R. Schoen and S.T. Yau [84] using minimal surface techniques. Their proof was in detail rather complicated and also quite indirect. Witten's proof, on the other hand, is not only relatively simple, but provides also an explicit positive integral expression for the energy.

Beside its basic importance for GR, the positive energy theorem provides also a useful tool for proving certain non-existence theorems for solitons and "no-hair-theorems" for black holes.

8.1 Total Energy and Momentum for Isolated Systems

As we discussed earlier, there is no general conservation law for energy and momentum in GR. For isolated systems it is, however, still possible to define the *total* energy and the *total* momentum. There are actually two different concepts of total energy-momentum which can be extracted from an asymptotically flat spacetime. The asymptotic region at null infinity yields the Bondi four-momentum (see Sect. 5.1), and the other, at spatial infinity, the Arnowitt-Deser-Misner (ADM) four-momentum (see Sect. 2.7). The corresponding, masses are the **Bondi mass** and the **ADM mass**. We prove here only the non-negativity of the ADM mass. The corresponding proof of the Bondi mass is very similar; see, e.g., [224].

There are a number of different expressions for the ADM 4-momentum, some of which will be discussed below. For the proof of the positive energy theorem it will be crucial to have a flux integral representation in terms of a (test) spinor field. But first, we recall a sufficiently precise formulation of what we mean by asymptotic flatness at spatial infinity.

Consider a Lorentz manifold (M, g) and a complete oriented 3-dimensional spacelike hypersurface Σ. We say that Σ is **asymptotically flat** if the following is true:

a) There is a compact set $\mathcal{C} \subset \Sigma$ such that $\Sigma \setminus \mathcal{C}$ is diffeomorphic to the complement of a contractible compact set in $\mathbb{R}^3$. (More generally, $\Sigma \setminus \mathcal{C}$ could be the disjoint union of a finite number of such sets.)
b) Under this diffeomorphism the metric on $\Sigma \setminus \mathcal{C}$ should be of the form

$$g_{ij} = \delta_{ij} + h_{ij}\,, \qquad (i,j = 1,2,3) \tag{8.1}$$

in the standard coordinates of $\mathbb{R}^3$, where $h_{ij} = O(1/r)$, $\partial_k h_{ij} = O(1/r^2)$ and $\partial_l \partial_k h_{ij} = O(1/r^3)$. Furthermore, the second fundamental form K_{ij} of $\Sigma \subset M$ should satisfy $K_{ij} = O(1/r^2)$ and $\partial_k K_{ij} = O(1/r^3)$.

As a starting point we use the following asymptotic flux integral representation for the ADM 4-momentum belonging to Σ, which we justified in Sect. 2.7, Eq. (2.207):

$$P^\mu = -\frac{1}{16\pi G}\int_{S^2_\infty} \sqrt{-g}\,\omega_{\alpha\beta} \wedge \eta^{\alpha\beta\mu}\,. \tag{8.2}$$

Here $\omega_{\alpha\beta}$ are the Levi-Civita connection forms relative to a tetrad $\{\theta^\mu\}$ and $\eta^{\alpha\beta\mu}$ is the Hodge-dual of $\theta^\alpha \wedge \theta^\beta \wedge \theta^\mu$:

$$\eta^{\alpha\beta\mu} = *(\theta^\alpha \wedge \theta^\beta \wedge \theta^\mu)\,. \tag{8.3}$$

The integration is extended over a 2-sphere in $\Sigma \setminus \mathcal{C}$ "at infinity" and we should use the tetrad $\{\theta^\alpha\}$ such that asymptotically $\theta^i = dx^i$ on Σ for the coordinates introduced above. Far away we have for nearly Lorentzian coordinates

$$\omega_{\alpha\beta} \simeq \frac{1}{2}\left(g_{\alpha\gamma,\beta} + g_{\alpha\beta,\gamma} - g_{\beta\gamma,\alpha}\right) dx^\gamma$$

and this gives for the energy in (8.2)

$$P^0 = -\frac{1}{16\pi G}\varepsilon_{ijl}\int_{S^2_\infty} g_{jk,i}\, dx^k \wedge dx^l \tag{8.4}$$

or

$$P^0 = \frac{1}{16\pi G}\int_{S^2_\infty}(\partial_j g_{ij} - \partial_i g_{jj})N^i\, dA\,, \tag{8.5}$$

where N^i is the unit outward normal to the sphere and dA denotes its surface element. Equation (8.5) agrees with the expression of ADM. Similarly, if one uses the following general formula for the second fundamental form

$$\omega_{0i} = K_{ij}\,\theta^j \qquad (\text{on } \Sigma) \tag{8.6}$$

for a tetrad which is adapted to Σ (see DG, (A.8a)), one finds for the momentum in (8.2) immediately

$$P_i = \frac{1}{8\pi G}\int_{S^2_\infty}(K_{ij} - \delta_{ij}K_{ll})\,N^j dA\,, \tag{8.7}$$

which again agrees with the expression given by ADM. (The ADM formulas (8.5) and (8.7) have already been derived in Sect. 2.7.)

Note that the integrals (8.5) and (8.7) involve only data on Σ. The disadvantage of these formulas is that they are not manifestly covariant. J. Nester [225] has given instead a simple covariant expression in terms of spinors which we discuss next. (We prefer, however, to use 2-component Weyl spinors instead of 4-component Majorana spinors.) Readers who are not familiar with spinors in GR should first read the Appendix 8.5 to this chapter, where we introduce all that is needed in what follows. The others should have a look at our conventions.

The covariant derivative of a fundamental Weyl spinor field ϕ, transforming as $D^{(\frac{1}{2},0)}$, is given by

$$\nabla\phi = d\phi + \omega_{\alpha\beta}\,\sigma^{\alpha\beta}\phi\,, \quad \sigma^{\alpha\beta} = -\frac{1}{8}\left(\sigma^\alpha\hat{\sigma}^\beta - \sigma^\beta\hat{\sigma}^\alpha\right), \tag{8.8}$$

where $\omega^\alpha{}_\beta$ are the connection forms relative to an *orthonormal* tetrad $\{\theta^\alpha\}$. More generally, if ψ denotes a spinor-valued p-form corresponding to the representation ρ of $SL(2,\mathbb{C})$, we have

$$\nabla\psi = d\psi + \omega_{\alpha\beta}\wedge\rho_*\left(\sigma^{\alpha\beta}\right)\psi\,, \tag{8.9}$$

where ρ_* denotes the induced representation of the Lie algebra $sl(2,\mathbb{C})$.

We shall use repeatedly the following simple fact which can easily be verified: If ϕ is a spinor-valued p-form belonging to the representation $D^{(\frac{1}{2},0)}$

and $\hat{\sigma} := \hat{\sigma}_\mu \, \theta^\mu$, then $\hat{\sigma} \wedge \phi$ is a spinor-valued $(p+1)$-form transforming as $D^{(0,\frac{1}{2})}$, and we have (exercise)

$$\nabla(\hat{\sigma} \wedge \phi) = \hat{\sigma} \wedge \nabla\phi \,. \tag{8.10}$$

For a fundamental Weyl spinor ϕ of type $D^{(\frac{1}{2},0)}$ we can introduce the 2-form

$$F = -i \, \phi^\dagger \, \hat{\sigma} \wedge \nabla\phi \,, \tag{8.11}$$

in terms of which we shall be able to express the ADM 4-momentum. For this we consider spinors ϕ which become asymptotically constant on Σ: On $\Sigma \setminus C$ (identified as above with the complement of a compact contractible set of $\mathbb{R}^3$ with its standard coordinates), we assume that asymptotically

$$\phi = \phi^{(0)} + O(1/r) \,, \tag{8.12}$$

where $\phi^{(0)}$ is a constant spinor. Now we claim that the contraction of P_μ with the null vector $l^{(0)\mu} = \phi^{(0)\dagger} \, \hat{\sigma}^\mu \phi^{(0)}$ is given by $(G = 1)$:

$$l^{(0)\mu} \, P_\mu = \frac{1}{4\pi} \int_{S^2_\infty} F \,. \tag{8.13}$$

In order to verify this, we note first that (use Eq.(8.10))

$$i \left(\phi^\dagger \hat{\sigma} \wedge \nabla\phi - (\nabla\phi)^\dagger \wedge \hat{\sigma}\phi \right) = -i d \left(\phi^\dagger \, \hat{\sigma}\phi \right)$$

and we thus have

$$\int_{S^2_\infty} F = -\frac{i}{2} \int_{S^2_\infty} \left(\phi^\dagger \hat{\sigma} \wedge \nabla\phi + (\nabla\phi)^\dagger \wedge \hat{\sigma}\phi \right) \,, \tag{8.14}$$

which is clearly real. Let us consider the contribution of $\phi^{(0)}$ to the integrand on the right-hand side:

$$\begin{aligned} \phi^{(0)\dagger} \hat{\sigma} \wedge \omega_{\alpha\beta} \, \sigma^{\alpha\beta} \phi^{(0)} &- \phi^{(0)\dagger} \omega_{\alpha\beta} \, \hat{\sigma}^{\alpha\beta} \wedge \hat{\sigma}\phi^{(0)} \\ &= -\frac{1}{4} \phi^{(0)\dagger} (\hat{\sigma}_\mu \sigma_\alpha \hat{\sigma}_\beta - \hat{\sigma}_\beta \sigma_\alpha \hat{\sigma}_\mu) \phi^{(0)} \, \theta^\mu \wedge \omega^{\alpha\beta} \,. \end{aligned} \tag{8.15}$$

Here, we use the following identity for the σ-matrices (exercise):

$$\hat{\sigma}_\mu \sigma_\alpha \hat{\sigma}_\beta - \hat{\sigma}_\beta \sigma_\alpha \hat{\sigma}_\mu = 2i \, \eta_{\mu\alpha\beta\lambda} \, \hat{\sigma}^\lambda \tag{8.16}$$

and also

$$\eta_{\alpha\beta\lambda} = -\eta_{\mu\alpha\beta\lambda} \, \theta^\mu \,, \tag{8.17}$$

where $\eta_{\mu\alpha\beta\lambda}$ is the Levi-Civita tensor (see (8.3)). The $\phi^{(0)}$-contribution to the integrand in (8.13) is thus equal to

$$\frac{1}{4} \eta_{\alpha\beta\lambda} \wedge \omega^{\alpha\beta} \, l^{(0)\lambda} \,.$$

It is now essential that the $O(1/r)$ term $\phi^{(1)}$ in (8.12) gives no contribution to the flux integral at infinity. Indeed, such a piece would come from the following additional term of the integrand in (8.14):

$$\phi^{(0)}\hat{\sigma}\wedge d\phi^{(1)} + \left(d\phi^{(1)}\right)^{\dagger}\wedge\hat{\sigma}\phi = d\left(\phi^{(0)\dagger}\hat{\sigma}\phi^{(1)} + \phi^{(1)\dagger}\hat{\sigma}\phi^{(0)}\right) + O(1/r^3)\,.$$

Here we have used

$$d\hat{\sigma} = \hat{\sigma}_\mu\, d\theta^\mu = -\hat{\sigma}_\mu\, \omega^\mu{}_\nu \wedge \theta^\nu = O(1/r^2)\,. \tag{8.18}$$

Hence we obtain

$$\frac{1}{4}\int_{S^2_\infty} F = -l^{(0)\lambda}\frac{1}{16\pi}\int_{S^2_\infty} \omega^{\alpha\beta}\wedge\eta_{\lambda\alpha\beta}\,. \tag{8.19}$$

Recalling (8.2), we see that the fully covariant formula (8.15) for the ADM 4-momentum belonging to Σ is indeed correct.

8.2 Witten's Proof of the Positive Energy Theorem

We are now ready to formulate and to prove the

Theorem 8.1 (Positive Mass Theorem). *Assume that the spacelike hypersurface Σ is asymptotically flat in the sense defined above and that the dominant energy condition holds on Σ. Then the total energy-momentum P^μ is a future directed timelike or null vector. Furthermore, $P^\mu = 0$ if and only if spacetime is flat in a neighborhood of Σ.*

Proof. We prove this theorem first for the case when there are no horizons present. In a first step we use Stoke's theorem (see DG, Sect. 12.7) to write

$$l^{(0)}_\mu P^\mu = \frac{1}{4\pi}\int_\Sigma dF \tag{8.20}$$

and work out the differential of F. Use of (8.10) gives

$$dF = -i(\nabla\phi)^\dagger\wedge\hat{\sigma}\wedge\nabla\phi + i\,\phi^\dagger\,\hat{\sigma}\wedge\nabla^2\phi\,. \tag{8.21}$$

The second covariant derivative of ϕ can be expressed in terms of the curvature forms $\Omega^\alpha{}_\beta$:

$$\nabla^2\phi = \Omega_{\alpha\beta}\sigma^{\alpha\beta}\phi\,. \tag{8.22}$$

Therefore, the second term in (8.21) involves

$$\hat{\sigma}\wedge\Omega_{\alpha\beta}\,\sigma^{\alpha\beta} = -\theta^\mu\wedge\Omega^{\alpha\beta}\,\frac{1}{4}\hat{\sigma}_\mu\sigma_\alpha\hat{\sigma}_\beta\,.$$

Here we use the relation (generalizing (8.16))

$$\hat{\sigma}_\mu \sigma_\alpha \hat{\sigma}_\beta = i\,\eta_{\mu\alpha\beta\lambda}\,\hat{\sigma}^\lambda - (g_{\mu\beta}\hat{\sigma}_\alpha - g_{\alpha\beta}\hat{\sigma}_\mu - g_{\mu\alpha}\hat{\sigma}_\beta) \tag{8.23}$$

and the first Bianchi identity $\Omega^\alpha_{\ \beta} \wedge \theta^\beta = 0$, giving

$$i\,\hat{\sigma} \wedge \Omega_{\alpha\beta}\,\sigma^{\alpha\beta} = -\frac{1}{4}\,\eta_{\alpha\beta\lambda} \wedge \Omega^{\alpha\beta}\hat{\sigma}^\lambda\,.$$

The second term in (8.21) is thus

$$i\,\phi^\dagger \hat{\sigma} \wedge \nabla^2\phi = -\frac{1}{4}\eta_{\alpha\beta\lambda} \wedge \Omega^{\alpha\beta}\, l^\lambda\,, \tag{8.24}$$

where l^λ is the null vector

$$l^\lambda = \phi^\dagger \hat{\sigma}^\lambda \phi\,. \tag{8.25}$$

Now Einstein's field equations read in the tetrad formalism (see Sect. 2.6, Eq. (2.162)):

$$-\frac{1}{4}\,\eta_{\alpha\beta\lambda} \wedge \Omega^{\alpha\beta} = 4\pi * T_\lambda \tag{8.26}$$

with

$$T_\lambda = T_{\lambda\sigma}\,\theta^\sigma\,. \tag{8.27}$$

For a tetrad which is adapted to Σ we thus have

$$i\,\phi^\dagger\hat{\sigma} \wedge \nabla^2\phi\big|_\Sigma = -4\pi\, T_{0\lambda}\, l^\lambda\, \mathrm{Vol}_\Sigma\,, \tag{8.28}$$

where $\mathrm{Vol}_\Sigma = - * \theta^0$ is the volume form of Σ. The dominant energy condition (see p.118) implies that the right-hand side of (8.28) is non-negative. (Note that l^λ is past-directed null vector.)

We consider now the first term in (8.21). Setting $\nabla\phi = (\nabla_\alpha\phi)\,\theta^\alpha$ we obtain

$$-i(\nabla\phi)^\dagger \wedge \hat{\sigma} \wedge \nabla\phi = -i(\nabla_\alpha\phi)^\dagger \hat{\sigma}_\lambda \nabla_\beta\phi\; \theta^\alpha \wedge \theta^\lambda \wedge \theta^\beta\,.$$

The restriction of this expression to Σ gives only non-vanishing contributions for terms with all indices α, λ, β spatial:

$$-i(\nabla\phi)^\dagger \wedge \hat{\sigma} \wedge \nabla\phi\big|_\Sigma = -(\nabla_a\phi)^\dagger i\,\varepsilon_{abc}\,\sigma_c \nabla_b\phi\; \mathrm{Vol}_\Sigma\,.$$

Here we use $i\,\varepsilon_{abc}\,\sigma_c = \sigma_a\sigma_b - \delta_{ab}\mathbf{1}$, giving

$$-i(\nabla\phi)^\dagger \wedge \hat{\sigma} \wedge \nabla\phi\big|_\Sigma = \left((\nabla_a\phi)^\dagger(\nabla_a\phi) - (\sigma_a\nabla_a\phi)^\dagger(\sigma_b\nabla_b\phi)\right)\, \mathrm{Vol}_\Sigma\,. \tag{8.29}$$

Inserting (8.21), (8.28) and (8.29) into (8.20) leads to the following decisive result

$$l^{(0)\mu}P_\mu = \int_\Sigma \left[-T_{0\lambda}\, l^\lambda + \frac{1}{4\pi}\left((\nabla_a\phi)^\dagger(\nabla_a\phi) - (\sigma_a\nabla_a\phi)^\dagger(\sigma_b\nabla_b\phi)\right)\right] \mathrm{Vol}_\Sigma\,. \tag{8.30}$$

At first sight this looks bad since the last term has the "wrong" sign. We have, however, some freedom in choosing the spinor field on Σ. In particular, we can try to get rid of the negative term by imposing the **Witten equation**:

$$\sigma_a \nabla_a \phi = 0 \qquad (\text{on } \Sigma)\,. \tag{8.31}$$

Suppose we could prove that this equation has a solution for any constant asymptotic value $\phi^{(0)}$, then we would have the following manifestly non-negative expression for $l^{(0)\mu} P_\mu$:

$$l^{(0)\mu} P_\mu = \int_\Sigma \left(-T_{\mu\nu}\, n^\mu l^\nu + \frac{1}{4\pi} (\nabla_a \phi)^\dagger (\nabla_a \phi) \right) \mathrm{Vol}_\Sigma\,, \tag{8.32}$$

where n^μ denotes the future directed unit normal to Σ. The second term is manifestly non-negative and for the first this is true as a consequence of the dominant energy condition.

The existence (and uniqueness) of asymptotically constant solutions of the Witten equation (8.31) has been proven in [228] and [229]. Before making some remarks about this problem, we finish the proof of the positive energy theorem. Since $\phi^{(0)}$ is arbitrary, we deduce that $l^\mu P_\mu \geq 0$ for any past directed null vector l^μ. Hence P_μ is a future directed timelike or null vector.

To prove the second part of the theorem we assume $P_\mu = 0$. Equation (8.32) implies then

$$\nabla_a \phi = 0\,. \tag{8.33}$$

Changing the asymptotic value $\phi^{(0)}$, we see that (8.33) must be satisfied for two linear independent spinor fields on Σ. Together with (8.22) this implies that $R_{\alpha\beta ij}\, \sigma^{\alpha\beta} \phi = 0$ for both fields, and therefore

$$R_{\alpha\beta ij} = 0 \qquad (i,j = 1,2,3)\,.$$

But then the dominant energy condition and Einstein's field equation give

$$8\pi\, |T_{\alpha\beta}| \leq 8\pi\, T_{00} = \frac{1}{2} R^{ij}{}_{ij} = 0\,,$$

and hence we have also $R_{\alpha\beta} = 0$ on Σ. Taken together, all curvature components must vanish and thus (M, g) is flat along Σ. □

Below we will extend the proof to the case when horizons are present. Before that we consider (8.31) a bit more closely.

Remarks on the Witten Equation

First, we rewrite the linear Witten equation in terms of the induced covariant derivative $\bar{\nabla}$ of Σ (belonging to the induced metric). Let $\{e_\mu\}$ be the orthonormal moving frame dual to the adapted basis $\{\theta^\mu\}$ of 1-forms used above. We have

$$\nabla_a \phi = \langle d\phi, e_a \rangle + \omega_{ij}(e_a)\sigma^{ij}\,\phi + 2\,\omega_{0j}(e_a)\sigma^{0j}\,\phi .$$

Using (8.6) and the Gauss formula $\omega_{ij}(e_a) = \bar{\omega}_{ij}(e_a)$, where $\bar{\omega}_{ij}$ are the connection forms of Σ belonging to $\bar{\nabla}$ (see Appendix A of DG, (A.3)), we obtain

$$\nabla_a \phi = \bar{\nabla}_a \phi - \frac{1}{2}\, K_{ab}\, \sigma^b \phi . \tag{8.34}$$

Witten's equation (8.31) is thus equivalent to

$$\sigma^a \bar{\nabla}_a \phi = \frac{1}{2}\, K\, \phi , \tag{8.35}$$

where $K = K^a_a$. Note in particular, that for a maximal hypersurface ($K = 0$) the Witten equation reduces to Weyl's equation $\sigma^a \bar{\nabla}_a \phi = 0$.

We do not give here detailed proofs for the existence and uniqueness for (8.31) or (8.35), supplemented by the requirement that ϕ is asymptotically constant, but indicate only the method.

Since the Witten equation is *elliptic* ($\sigma^a \xi_a \neq 0$ for $\xi \neq 0$), one can follow the usual strategy: First, one proves the existence of a *weak* solution in a suitable Sobolev space H^s of spinor valued functions on Σ. Well-known elliptic regularity theorems then guarantee that the weak solution is in all Sobolev spaces H^s for arbitrary large s. Finally, one proves a generalized Sobolev lemma which implies that the weak solution is automatically C^∞. The generalized Sobolev lemma rests on a generalized Sobolev inequality, whose validity has been reduced in [228] to the classical Sobolev inequality in $\mathbb{R}^d$.

Application

As an application of the positive mass theorem, we give a simple proof of the following important

Theorem 8.2 (A. Lichnerowicz). *A solution of Einstein's vacuum equations which is regular, globally stationary and asymptotically flat must be flat.*

Proof. For such a solution the ADM mass is given by the vanishing integral (4.226). Hence, spacetime is flat. □

8.3 Generalization to Black Holes

The positive mass theorem for black holes was proven in [230]. Here, we give a slightly different derivation, using again 2-component spinors.

First we rewrite $F|\Sigma$ with the help of Witten's equation, which will again be imposed. For our adapted frame we have

$$F|_\Sigma = i\,\phi^\dagger \sigma_a \nabla_b \phi\, \theta^a \wedge \theta^b = \phi^\dagger i\, \varepsilon_{abc}\, \sigma_a \nabla_b \phi\, dA^c$$
$$= \phi^\dagger \nabla_a \phi\, dA^a\,, \tag{8.36}$$

where $dA^a := \bar{*}\theta^a$ and $\bar{*}$ the Hodge-dual in Σ relative to the induced metric. With the help of (8.34) this gives

$$F|_\Sigma = \left(\phi^\dagger \bar{\nabla}_a \phi - \frac{1}{2} K_{ab}\, \phi^\dagger \sigma^b \phi\right) dA^a\,. \tag{8.37}$$

Let us restrict this to a 2-surface $S \subset \Sigma$ which later will become the "future apparent horizon" (see Fig. 8.1).

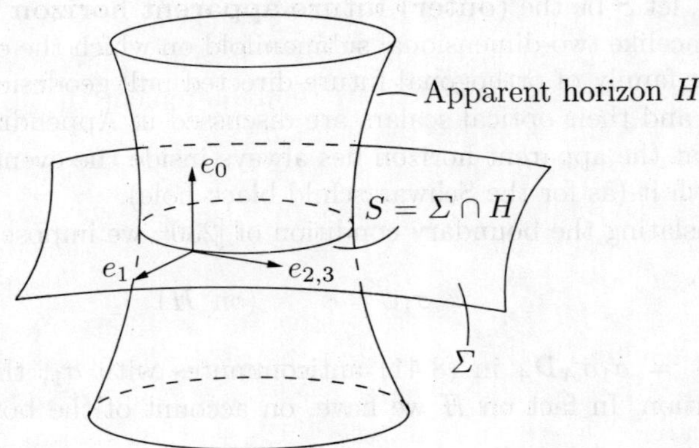

Fig. 8.1. Spacelike hypersurface Σ and apparent horizon.

We choose the dual pair $\{\theta^\mu\}$, $\{e_\mu\}$ of adapted frames such that e_1 is perpendicular to S and e_A, for $A = 2, 3$, are parallel to S. (Indices with capital letters will always run over 2 and 3.) From (8.37) we obtain

$$F|_S = \left(\phi^\dagger \bar{\nabla}_1 \phi - \frac{1}{2}\, K_{1a} \phi^\dagger \sigma^a \phi\right) \mathrm{Vol}_S\,. \tag{8.38}$$

We can find a useful expression for $\bar{\nabla}_1 \phi$ if we multiply Witten's equation in the form (8.35) by σ_1:

$$\bar{\nabla}_1 \phi = -\sigma_1 \sigma_A \bar{\nabla}_A \phi + \frac{1}{2}\, K \sigma_1 \phi\,. \tag{8.39}$$

Next we express $\bar{\nabla}_A|S$ by the induced covariant derivative $\bar{\bar{\nabla}}_A$ on S and the second fundamental form J_{AB} of S in Σ. Using the formula which corresponds to (8.6) we get

$$\bar{\nabla}_A \phi = \bar{\bar{\nabla}}_A\, \phi + 2\, \bar{\omega}_{1B}(e_A) \sigma^{1B} \phi = \bar{\bar{\nabla}}_A\, \phi + \frac{1}{2} J_{AB}\, \sigma^1 \sigma^B \phi\,. \tag{8.40}$$

In (8.39) we need

$$-\sigma_1\sigma_A\bar{\nabla}_A\phi = -\sigma_1\sigma_A \overset{=}{\nabla}_A \phi + \frac{1}{2} J \phi ,$$

where $J = J^A_A$. Taken together, (8.38) becomes

$$F|_S = \frac{1}{2}\phi^\dagger \left(J + (-K_{11} + K)\sigma_1 - 2\sigma_1\sigma_A\mathcal{D}_A\right)\phi \,\mathrm{Vol}_S , \tag{8.41}$$

where

$$\mathcal{D}_A = \overset{=}{\nabla}_A - \frac{1}{2}K_{1A}\,\sigma_1 . \tag{8.42}$$

Now, let S be the **(outer) future apparent horizon** H. This is a compact, spacelike two-dimensional submanifold on which the expansion θ of the outgoing family of orthogonal future-directed null geodesics vanishes. (Such families and their optical scalars are discussed in Appendix 7.7.3.) One can show that the apparent horizon lies always inside the event horizon or coincides with it (as for the Schwarzschild black hole).

Translating the boundary condition of [230], we impose

$$\sigma_1\phi = \phi \qquad (\text{on } H) . \tag{8.43}$$

Since $\mathcal{O} := \sigma_1\sigma_A\mathcal{D}_A$ in (8.41) anticommutes with σ_1, this term gives no contribution. In fact on H we have, on account of the boundary condition (8.43),

$$\phi^\dagger\mathcal{O}\phi = \phi^\dagger\mathcal{O}\sigma_1\phi = -\phi^\dagger\sigma_1\mathcal{O}\phi = -\phi^\dagger\mathcal{O}\phi .$$

We are thus left with

$$F|_H = \frac{1}{2}\phi^\dagger\left(K - K^1_1 + J\right)\phi . \tag{8.44}$$

We will show below that the expansion θ of the outgoing null congruence on S is given by

$$\theta = \frac{1}{\sqrt{2}}\left(-K + K^1_1 - J\right) , \tag{8.45}$$

and thus (8.44) vanishes.

This settles the positive energy theorem for black holes, because when we use Stokes theorem to transform the surface integral (8.15) at infinity into an integral over Σ with the inner boundary H, we do not pick up an additional contribution from the apparent horizon. Of course, one has to show that the Witten equation has always a solution for a given asymptotic value $\phi^{(0)}$, which satisfies also the boundary condition (8.43) on H. In [230] it is made plausible that such solutions indeed exist, but we are not aware of a rigorous proof.

Finally, we derive (8.45). Consider the two null fields

$$k = \frac{1}{\sqrt{2}}(e_0 + e_1), \qquad l = \frac{1}{\sqrt{2}}(e_0 - e_1)$$

for which $\langle k, l \rangle = 1$. The expansion of the null congruence belonging to k is $\theta = k^A_{;A}$ (see Appendix 7.7.3). We compute first

$$\sqrt{2}\,\nabla_{e_A} k = \omega^\mu{}_0(e_A)e_\mu + \omega^\mu{}_1(e_A)e_\mu ,$$

giving

$$\sqrt{2}\,k^B_{;A} = \omega^B{}_0(e_A) + \omega^B{}_1(e_A) = -K^B_A - J^B_A .$$

Thus

$$\theta = \frac{1}{\sqrt{2}}(-K^A_A - J^A_A) = \frac{1}{\sqrt{2}}(-K + K^1_1 - J) .$$

8.4 Penrose Inequality

The Penrose inequality, formulated below, can be regarded as a sharpening of the positive energy theorem for black holes.

For motivation, consider a Schwarzschild black hole. Its mass (the ADM mass for a spacelike hypersurface $\Sigma = \{t = \text{const}\}$) determines the area $\mathcal{A}(\mathcal{H})$ of the intersection of the horizon H with Σ according to $\mathcal{A}(\mathcal{H}) = 4\pi(2GM)^2$. Thus

$$M = \frac{1}{4\sqrt{\pi}G}\left(\mathcal{A}(\mathcal{H})\right)^{1/2} . \tag{8.46}$$

It is natural to guess that for more general situations the right-hand side of (8.46) will be a lower limit for the ADM mass. This was originally conjectured by R. Penrose [231], and has relatively recently been proven by G. Huisken and T. Ilmanen under conditions we now discuss.

Assume that the asymptotically flat spacelike hypersurface Σ is maximal. Gauss' equation in the form (A.21), together with Einstein's field equation and the weak energy condition (7.366) show that the Riemann scalar $\bar{R}_\Sigma$ for the induced metric on Σ is non-negative:

$$\bar{R}_\Sigma \geq 0 . \tag{8.47}$$

Let $\mathcal{H}$ be the intersection of Σ with the apparent horizon of a black hole. From (8.45) we obtain on $\mathcal{H}$ (where $\theta = 0$): $J = K^1_1$. Hence, if Σ is totally geodesic in the sense of Definition A.1, we see that $\mathcal{H}$ is a **minimal surface** of Σ (the mean curvature vanishes). These considerations lead to the assumptions of the following

Theorem 8.3 (of G. Huisken and T. Ilmanen). *Let (Σ, g) be a complete asymptotically flat Riemannian three-manifold diffeomorphic to $\mathbb{R}^3$ minus an open three-ball, with a boundary consisting of a compact minimal surface. If*

the scalar curvature is non-negative, then the ADM mass M_Σ of Σ satisfies the ***Penrose inequality***

$$M_\Sigma \geq \frac{1}{4\sqrt{\pi}G}\left(\mathcal{A}(\mathcal{H})\right)^{1/2}. \tag{8.48}$$

Remarks.

1. For a proof (and a slightly more general formulation) see [232]. Because of (8.46) the inequality (8.48) is optimal.
2. The quoted theorem ("Riemann Penrose inequality") covers only the case when Σ is totally geodesic , in other words, when the initial data on Σ are time symmetric. A general proof of the Penrose inequality is still lacking.

* * *

Exercises.

Exercise 8.1. Verify the Equation (8.10).

Exercise 8.2. Proof the identity (8.16).

8.5 Appendix: Spin Structures and Spinor Analysis in General Relativity

In the following treatment of the bare essentials we assume that the reader is familiar with spinor analysis in SR, and also with the theory of connections in principle fibre bundles (see [35]).

8.5.1 Spinor Algebra

We begin by reviewing those parts of spinor algebra that we shall need, mainly to fix various conventions and our notation. For detailed treatments see [34] and [29].

Let $(\mathbb{R}^4, \langle\cdot,\cdot\rangle)$ be the Minkowski vector space with the non-degenerate symmetric bilinear form

$$\langle x, y\rangle = \eta_{\mu\nu}\, x^\mu y^\nu = x^T \eta\, y\,, \qquad \eta = (\eta_{\mu\nu}) = \mathrm{diag}(-1,1,1,1)\,. \tag{8.49}$$

(Note that $x, y \in \mathbb{R}^4$ regarded as column matrices.) The 1-component of the homogeneous Lorentz group – the **orthochronous Lorentz group** – is denoted by $L_+^\uparrow$. Its universal covering group is $SL(2,\mathbb{C})$. The two-fold covering homomorphism $\lambda : SL(2,\mathbb{C}) \longrightarrow L_+^\uparrow$ is determined as follows:

$$\underline{\lambda(L)x} = L\underline{x}L^\dagger\,, \tag{8.50}$$

where $\underline{x}$ denotes for each $x \in \mathbb{R}^4$ the hermitian 2×2 matrix

$$\underline{x} = x^\mu \sigma_\mu \,, \qquad \sigma = (1, \sigma_k) \,. \tag{8.51}$$

(Here $\sigma_k = \sigma^k$ are the Pauli matrices, and $L^\dagger$ denotes the hermitian conjugate of L.) From

$$\underline{x} = \begin{pmatrix} x^0 + x^3 & x^1 - i\,x^2 \\ x^1 + i\,x^2 & x^0 - x^3 \end{pmatrix} \tag{8.52}$$

it follows that

$$\det \underline{x} = -\langle x, x \rangle \,. \tag{8.53}$$

Using this it is easy to see that the assignment $L \longmapsto \lambda(L)$ is a homomorphism from $SL(2,\mathbb{C})$ into $L_+^\uparrow$. One can show that the image is all of $L_+^\uparrow$ (see [34] or [29]).

The 6-dimensional Lie algebra of $SL(2,\mathbb{C})$ is

$$sl(2,\mathbb{C}) = \{ M \in gl(2,\mathbb{C}) \mid \operatorname{Tr} M = 0 \} \,. \tag{8.54}$$

A useful basis of $sl(2,\mathbb{C})$ is

$$\left\{ \frac{1}{2\,i}\sigma_j, \frac{1}{2}\sigma_j \quad (j = 1,2,3) \right\} . \tag{8.55}$$

The Lie algebra of $L_+^\uparrow$ is

$$so(1,3) = \{ M \in gl(4,\mathbb{R}) \mid \eta M + M^T \eta = 0 \} \,. \tag{8.56}$$

From (8.50) we obtain for the induced isomorphism $\lambda_* : sl(2,\mathbb{C}) \longrightarrow so(1,3)$ the formula

$$\underline{\lambda_*(M) x} = M \underline{x} + \underline{x} M^\dagger \,, \tag{8.57}$$

where $M \in sl(2,\mathbb{C})$. The reader is invited to show that the images of the basis (8.55) generate the 1-parameter subgroups of rotations about the j-th axis, respectively, the special Lorentz transformations along the j-th axis.

There are two types of fundamental representations of $SL(2,\mathbb{C})$ in terms of which all other non-trivial finite dimensional irreducible representations can be constructed (using tensor products and symmetrizations). The (fundamental) **spinor representation**, denoted by $D^{(\frac{1}{2},0)}$ is simply $L \longmapsto L$, while the **cospinor representation** $D^{(0,\frac{1}{2})}$ is defined by $L \longmapsto (L^\dagger)^{-1}$. Two-component spinors ξ, transforming with L are called **right-handed**, whereas two-component spinors η, transforming with $(L^\dagger)^{-1}$ are called **left-handed**.

Let V be the space of right-handed spinors $\xi = (\xi^A)$. We then have for $\xi \in V$

$$L : \xi \longmapsto L\xi \,, \qquad L : \xi^A \longmapsto L^A_B \, \xi^B \,. \tag{8.58}$$

Since $\det L = 1$, the following symplectic form on V

$$s(\xi, \xi') = \xi^T \varepsilon \, \xi' = \xi^A \varepsilon_{AB} \, \xi'^B \,, \tag{8.59}$$

with

$$\varepsilon = (\varepsilon_{AB}) = \begin{pmatrix} 0 & 1 \\ -1 & 0 \end{pmatrix} \tag{8.60}$$

is *invariant* under $SL(2,\mathbb{C})$. This means that $L^T \varepsilon L = \varepsilon$.

It will turn out to be useful to introduce the following operation on 2×2 complex matrices A:

$$A \longmapsto \hat{A} = \varepsilon \bar{A} \varepsilon^{-1}, \tag{8.61}$$

where $\bar{A}$ denotes the complex conjugate matrix. From the previous remark it follows

$$\hat{L} = (L^{\dagger})^{-1} \in SL(2,\mathbb{C}) \tag{8.62}$$

for $L \in SL(2,\mathbb{C})$. The cospinor representation is thus $L \longmapsto \hat{L}$.

The symplectic form (8.59) on V can be used to lower spinor indices

$$\xi_A := \varepsilon_{AB}\, \xi^B\,. \tag{8.63}$$

The inverse of this is

$$\xi^A = \varepsilon^{AB} \xi_B\,, \qquad (\varepsilon^{AB}) = \varepsilon^{-1} = -\varepsilon\,. \tag{8.64}$$

Clearly, $\xi^A \xi'_A$ is invariant under $SL(2,\mathbb{C})$ for $\xi, \xi' \in V$.

Let $\bar{V}$ be the space of 2-component left-handed spinors. Following van der Waerden, we denote the components of $\eta \in \bar{V}$ with lower dotted indices $\eta_{\dot{A}}$. Obviously, the symplectic form

$$s(\eta, \eta') = \eta_{\dot{A}}\, \varepsilon^{\dot{A}\dot{B}}\, \eta_{\dot{B}}\,, \qquad (\varepsilon^{\dot{A}\dot{B}}) = (\varepsilon^{AB})\,, \tag{8.65}$$

on $\bar{V}$ is also $SL(2,\mathbb{C})$ invariant. We can thus raise lower indices

$$\eta^{\dot{A}} = \varepsilon^{\dot{A}\dot{B}} \eta_{\dot{B}}\,. \tag{8.66}$$

The inverse is, as for right-handed spinors,

$$\eta_{\dot{A}} = \varepsilon_{\dot{A}\dot{B}} \eta^{\dot{B}}\,, \qquad (\varepsilon_{\dot{A}\dot{B}}) = (\varepsilon_{AB}) = \varepsilon\,. \tag{8.67}$$

From (8.61) it follows that $\eta^{\dot{A}}$ transforms with $\bar{L}$:

$$\eta^{\dot{A}} \longmapsto \overline{L^A_B}\, \eta^{\dot{B}}. \tag{8.68}$$

The representations $L \longmapsto \hat{L}$ and $L \longmapsto \bar{L}$ are, of course, equivalent.

It can been shown (see Chap. 2 of [29]) that all finite dimensional irreducible representations of $SL(2,\mathbb{C})$ (apart from the trivial one) are given by

$$\left(D^{(\frac{1}{2},0)}\right)^{\otimes_s m} \otimes \left(D^{(0,\frac{1}{2})}\right)^{\otimes_s n}, \tag{8.69}$$

where $\otimes_s m$ denotes the m-fold symmetric tensor product.

Dirac spinors transform according to the reducible representation

$$\rho(L) = \begin{pmatrix} L & 0 \\ 0 & \hat{L} \end{pmatrix} . \tag{8.70}$$

(If space reflections are added, the representation becomes irreducible.)

Equation (8.50) shows that the matrix elements of $\underline{x}$ transform as a spinor with one upper undotted plus one upper dotted index. Therefore, we write

$$\underline{x} = \left(x^{A\dot{B}} \right) . \tag{8.71}$$

Thus, a vector x^μ can be regarded as a spinor $x^{A\dot{B}}$. Equation (8.71) tells us that the matrix elements of σ_μ should be written as $\sigma_\mu = (\sigma_\mu^{A\dot{B}})$. The relation between x^μ and $x^{A\dot{B}}$ is then given by

$$x^{A\dot{B}} = x^\mu \sigma_\mu^{A\dot{B}} . \tag{8.72}$$

Lowering the indices in this equation or in (8.71) can be expressed as

$$(x_{A\dot{B}}) = (\hat{\underline{x}})^T . \tag{8.73}$$

We define σ^μ by $\sigma^\mu = \eta^{\mu\nu}\sigma_\nu = (-\mathbf{1}, \sigma_k)$. Note that $\hat{\sigma}^\mu = (-\mathbf{1}, -\sigma_k)$, and also the important relation

$$\sigma^\mu \hat{\sigma}^\nu + \sigma^\nu \hat{\sigma}^\mu = -2\,\eta^{\mu\nu}\mathbf{1} . \tag{8.74}$$

In particular,

$$\underline{x}\,\hat{\underline{x}} = -\langle x, x\rangle \mathbf{1} . \tag{8.75}$$

In SR one uses the differential operators

$$\underline{\partial} = \sigma^\mu \partial_\mu = -\mathbf{1}\partial_0 + \boldsymbol{\sigma}\cdot\boldsymbol{\partial} , \tag{8.76a}$$

$$\hat{\underline{\partial}} = \hat{\sigma}^\mu \partial_\mu = -(\mathbf{1}\partial_0 + \boldsymbol{\sigma}\cdot\boldsymbol{\partial}) . \tag{8.76b}$$

We have, for example, the identity

$$\partial_{A\dot{B}}\partial^{A\dot{B}} = \mathrm{Tr}(\underline{\partial}\,\hat{\underline{\partial}}) = -2\,\eta^{\mu\nu}\partial_\mu\partial_\nu ,$$

i.e.

$$\Box \equiv \eta^{\mu\nu}\partial_\mu\partial_\nu = -\frac{1}{2}\partial_{A\dot{B}}\partial^{A\dot{B}} . \tag{8.77}$$

Similarly, the divergence of a vector field u^μ can be written as

$$\partial_\mu u^\mu = -\frac{1}{2}\partial_{A\dot{B}} u^{A\dot{B}} . \tag{8.78}$$

In terms of $\partial_{A\dot{B}}$ the **Weyl equations** for fundamental spinor fields φ^A and $\chi_{\dot{B}}$ read

$$\partial_{A\dot{B}}\varphi^A = 0\,, \qquad \partial^{A\dot{B}}\chi_{\dot{B}} = 0\,. \tag{8.79}$$

A Dirac spinor field ψ is composed of two Weyl spinor fields as

$$\psi = \begin{pmatrix} \varphi^A \\ \chi_{\dot{B}} \end{pmatrix}, \tag{8.80}$$

and the **Dirac equation** is equivalent to the pair

$$\underline{\hat{\partial}}\,\varphi = i\,m\,\chi\,, \qquad \underline{\partial}\,\chi = i\,m\,\varphi\,. \tag{8.81}$$

Introducing the Dirac matrices

$$\gamma^\mu = \begin{pmatrix} 0 & \sigma^\mu \\ \hat{\sigma}^\mu & 0 \end{pmatrix}, \tag{8.82}$$

this pair can be written in the traditional form

$$(i\,\gamma^\mu\partial_\mu + m)\,\psi = 0\,. \tag{8.83}$$

8.5.2 Spinor Analysis in GR

The question is now, how to extend these elements of spinor analysis in SR to GR. In particular, we need a covariant derivative of spinor fields. To this end we need some basic concepts.

Let (M, g) be a Lorentz manifold, $P(M, L_+^\uparrow)$ the principle fibre bundle of space and time oriented Lorentz frames with the proper orthochronous Lorentz group (1-component of the homogeneous Lorentz group) as structure group and $\lambda : SL(2,\mathbb{C}) \longrightarrow L_+^\uparrow$ the universal covering homomorphism.

Definition 8.1. *A* ***spin structure*** *on M is a principle bundle $\tilde{P}(M, SL(2,\mathbb{C}))$ on M with the structure group $SL(2,\mathbb{C})$, together with a 2 : 1 bundle morphism $\pi_s : \tilde{P} \longrightarrow P$, which is compatible with λ. Thus the following diagram is commutative:*

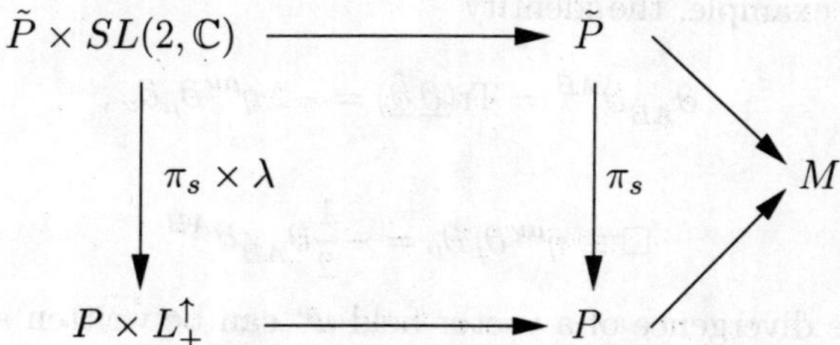

Given a representation ρ of $SL(2,\mathbb{C})$ in a vector space F we can introduce the vector bundle $E(F, SL(2,\mathbb{C}))$ which is associated to $\tilde{P}(M, SL(2,\mathbb{C}))$ and has typical fibre F.

Definition 8.2. *A* ***Weyl spinor*** *of type ρ over (M, g) is a section in the associated bundle $\pi_E : E \longrightarrow M$, i.e., a map $\phi : M \longrightarrow E$ such that $\pi_E \circ \phi = id_M$.*

We do not assume in what follows that the reader is familiar with associated bundles, because we shall use an entirely equivalent definition of Weyl spinors, which makes only use of the principle bundle $\tilde{\pi} : \tilde{P} \longrightarrow M$. A Weyl spinor can also be viewed as a map $\tilde{\phi} : \tilde{P} \longrightarrow F$ which satisfies the following covariance condition

$$\tilde{\phi}(p \cdot g) = \rho(g^{-1})\tilde{\phi}(p) \tag{8.84}$$

for each $p \in \tilde{P}$ and $g \in G$.

The Levi-Civita connection form ω on $P(M, L_+^\uparrow)$ induces a 1-form $\tilde{\omega}$ on $\tilde{P}$ with values in the Lie algebra $sl(2, \mathbb{C})$ by

$$\tilde{\omega} = \lambda_*^{-1} (\pi_s^* \omega) \ , \tag{8.85}$$

where $\lambda_* : sl(2, \mathbb{C}) \longrightarrow so(1,3)$ is the Lie algebra isomorphism belonging to λ. It is a good exercise to verify that $\tilde{\omega}$ is a connection form on the spin bundle $\tilde{P}$. With it we can now define in the usual fashion a **covariant derivative** of Weyl spinor fields by

$$\nabla\tilde{\phi} := (d\tilde{\phi})_{hor} \, ,$$

where 'hor' denotes the horizontal component belonging to $\tilde{\omega}$. Explicitely one has always

$$\nabla\tilde{\phi} = d\tilde{\phi} + \rho_*(\tilde{\omega})\tilde{\phi} \, . \tag{8.86}$$

More generally, if ϕ is a spinor-valued 'horizontal' p-form, we have

$$\nabla\phi = d\phi + \rho_*(\omega) \wedge \phi \, . \tag{8.87}$$

Relative to a local section $\tilde{\sigma} : U \longrightarrow \tilde{P}$ the pull-back of $\tilde{\omega}$ is

$$\tilde{\sigma}^*\tilde{\omega} = \lambda_*^{-1} (\tilde{\sigma}^* \circ \pi_s^* \omega) = \lambda_*^{-1} \sigma^*(\omega) , \tag{8.88}$$

where $\sigma = \pi_s \circ \tilde{\sigma} : U \longrightarrow P$ is the local section of P on U belonging to $\tilde{\sigma}$. For the local representative $\phi_U = \tilde{\sigma}^*\tilde{\phi} : U \longrightarrow P$ of $\tilde{\phi}$ we obtain from (8.86)

$$\nabla\phi_U = d\phi_U + \rho_* (\tilde{\sigma}^*\tilde{\omega}) \, \phi_U \, . \tag{8.89}$$

From now on we work with local sections and write for (8.88) and (8.89)

$$\tilde{\omega} = \lambda_*^{-1}\omega \, , \tag{8.90a}$$

$$\nabla\phi = d\phi + \rho_*(\tilde{\omega})\phi \, . \tag{8.90b}$$

The section σ defines an orthonormal basis $\{\theta^\mu\}$ of 1-forms (tetrad belonging to σ).

Since ω is $so(1,3)$-valued, we can decompose it as follows

$$\omega = \omega^{\alpha}_{\ \beta}\,\Sigma^{\beta}_{\ \alpha} = \omega_{\alpha\beta}\Sigma^{\alpha\beta}\,, \tag{8.91}$$

where $\Sigma^{\alpha\beta} = E^{\alpha\beta} - E^{\beta\alpha} = -\Sigma^{\beta\alpha}$, $E^{\alpha\beta}$ being the matrix with elements $(E^{\alpha\beta})_{\mu\nu} = \delta^{\alpha}_{\mu}\delta^{\beta}_{\nu}$. (The Σ^{β}_{α} with $\alpha < \beta$, form a basis of $so(1,3)$.) A routine calculation (exercise) gives for $\tilde{\omega}$ in (8.90a) the following decomposition

$$\tilde{\omega} = \omega_{\alpha\beta}\sigma^{\alpha\beta}\,, \quad \sigma^{\alpha\beta} = -\frac{1}{8}\left(\sigma^{\alpha}\hat{\sigma}^{\beta} - \sigma^{\beta}\hat{\sigma}^{\alpha}\right). \tag{8.92}$$

Recall that $\sigma_{\mu} = (\mathbf{1}, \sigma_k)$, where σ_k are the Pauli matrices, and $\sigma^{\mu} = \eta^{\mu\nu}\sigma_{\nu}$, $\hat{\sigma}^{\mu} = \eta^{\mu\nu}\hat{\sigma}_{\nu}$, with

$$\hat{\sigma}_{\mu} = \varepsilon\,\bar{\sigma}_{\mu}\,\varepsilon^{-1} = (\mathbf{1}, -\sigma_k)\,. \tag{8.93}$$

A bar denotes complex conjugation, and $\eta_{\mu\nu}$ is the flat metric diag$(-1,1,1,1)$.

Note that $\omega^{\alpha}_{\ \beta}$ in (8.91) and (8.92) are just the Levi-Civita connection forms relative to the tetrad $\{\theta^{\mu}\}$.

Not every Lorentz manifold has a spin structure. We quote the following famous

Theorem 8.4. *A space or time orientable pseudo-Riemannian manifold has a spin structure if and only if its second Stiefel-Whitney class vanishes.*

R.P. Geroch (see [227]) has proven the following

Theorem 8.5. *A space and time orientable open (non-compact) Lorentz manifold has a spin structure if and only if the manifold is parallelizable.*

Spinors and spinor analysis in GR are treated very extensively in [105] and [226].

* * *

Exercises.

Exercise 8.3. Compute the images of the basis (8.55) for $sl(2,\mathbb{C})$ under the Lie algebra homomorphism λ_*.

Exercise 8.4. Let $\gamma(x) = x^{\mu}\gamma_{\mu}$, $x \in \mathbb{R}^4$, and show that for the representation (8.70) for Dirac spinors the Dirac matrices satisfy

$$\gamma\big(\lambda(L)x\big) = \rho(L)\gamma(x)\rho(L)^{-1}\,,$$

where $L \in SL(2,\mathbb{C})$.

Exercise 8.5. Show that the 1-form $\tilde{\omega}$ in (8.85) has all the defining properties of a connection form on the spinor bundle $\tilde{P}$.

Exercise 8.6. Prove (8.86) by evaluating both sides for horizontal and fundamental (vertical) vector fields.

Part III

Differential Geometry

In this purely mathematical part, we develop the most important concepts and results of differential geometry which are needed for general relativity theory.

The presentation differs little from that in many contemporary mathematical text books. The language of modern differential geometry and the "intrinsic" calculus on manifolds are now frequently used by workers in the field of general relativity and are beginning to appear in textbooks on the subject. This has a number of advantages, such as:

a) It enables one to read the mathematical literature and make use of the results to attack physical problems.
b) The fundamental concepts, such as differentiable manifolds, tensor fields, affine connection, and so on, adopt a clear and intrinsic formulation.
c) Physical statements and conceptual problems are not confused by the dependence on the choice of coordinates. At the same time, the role of distinguished coordinates in the physical applications is clarified. For example, these can be adapted to symmetry properties of the system.
d) The exterior calculus of differential forms is a very powerful method for practical calculations; one often finds the results faster than with older methods.

Space does not allow us to always give complete proofs and sufficient motivation. In those cases, we give detailed references to the literature (see [30]-[48]) where these can be found. In the chapter entitled "Some Details" the reader can find some of the proofs not given in the main body of the text. Many readers will have the requisite mathematical knowledge to skip this part after familiarizing themselves with our notation (which is quite standard). This is best done by looking at the collection of important formulas in Appendix D. Readers who have no previous knowledge in differential geometry should first study all of Chaps. 9 and 10 and then go directly to Chap. 13, and absorb Sects. 13.1-13.6. These contain the most important tools for GR and form a selfcontained part. Other sections may be studied when they are needed.

9 Differentiable Manifolds

A manifold is a topological space which locally looks like the space $\mathbb{R}^n$ with the usual topology.

Definition 9.1. *An* ***n-dimensional topological manifold*** M *is a topological Hausdorff space with a countable base, which is locally homeomorphic to* $\mathbb{R}^n$. *This means that for every point* $p \in M$ *there is an open neighborhood* U *of* p *and a homeomorphism*

$$h : U \longrightarrow U'$$

which maps U *onto an open set* $U' \subset \mathbb{R}^n$.

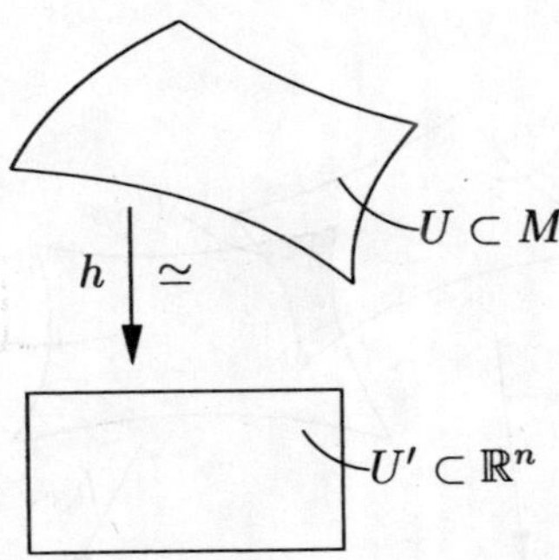

Fig. 9.1. Chart.

As an aside, we note that a topological manifold M also has the following properties:

a) M is σ-compact;
b) M is paracompact and the number of connected components is at most denumerable.

The second of these properties is particularly important for the theory of integration. For a proof, see e.g. [36], Chap. II, Sect. 15.

Definition 9.2. *If M is a topological manifold and $h : U \longrightarrow U'$ is a homeomorphism which maps an open subset $U \subset M$ onto an open subset $U' \subset \mathbb{R}^n$, then h is a* **chart** *of M and U is called the* **domain of the chart** *or* **local coordinate neighborhood***. The coordinates $(x^1, \ldots, x^n)$ of the image $h(p) \in \mathbb{R}^n$ of a point $p \in U$ are called the* **coordinates** *of p in the chart. A set of charts $\{h_\alpha | \alpha \in I\}$ with domains U_α is called an* **atlas** *of M, if $\bigcup_{\alpha \in I} U_\alpha = M$.*

If h_α and h_β are two charts, then both define homeomorphisms on the intersection of their domains $U_{\alpha\beta} := U_\alpha \cap U_\beta$; one thus obtains a homeomorphism $h_{\alpha\beta}$ between two open sets in $\mathbb{R}^n$ via the commutative diagram:

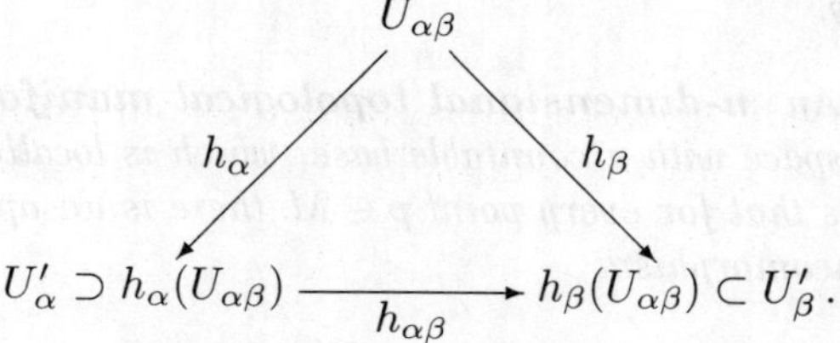

Thus $h_{\alpha\beta} = h_\beta \circ h_\alpha^{-1}$ on the domain where the mapping is defined. This mapping gives a relation between the coordinates in the two charts and is called a **change of coordinates** or **coordinate transformation** (see Fig. 9.2).

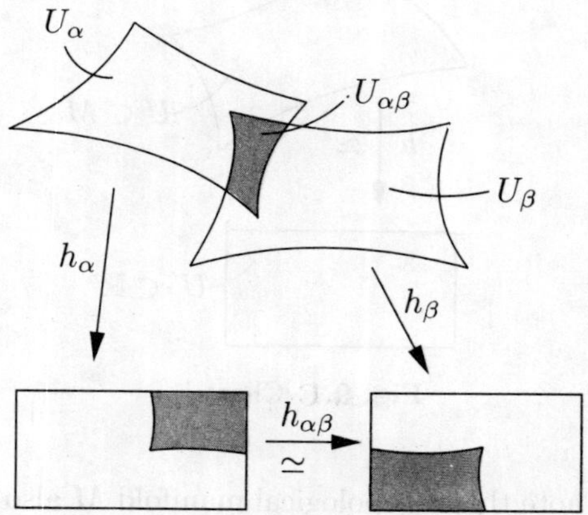

Fig. 9.2. Change of coordinates.

Sometimes, particularly in the case of charts, it is useful to include the domain of a mapping in the notation; thus, we write (h, U) for the mapping $h : U \longrightarrow U'$.

Definition 9.3. *An atlas defined on a manifold is said to be* **differentiable** *if all of its coordinate changes are differentiable mappings.*

For simplicity, unless otherwise stated, we shall always mean differentiable mappings of class C^∞ on $\mathbb{R}^n$ (the derivatives of all orders exist and are continuous). Obviously, for all coordinate transformations one has (on the domains for which the mappings are defined) $h_{\alpha\alpha} = Id$ and $h_{\beta\gamma} \circ h_{\alpha\beta} = h_{\alpha\gamma}$, so that $h_{\alpha\beta}^{-1} = h_{\beta\alpha}$, and hence the inverses of the coordinate transformations are also differentiable. They are thus diffeomorphisms.

If $\mathcal{A}$ is a differentiable atlas defined on manifold M, then the atlas $\mathcal{D}(\mathcal{A})$ contains those charts for which all coordinate changes among charts from $\mathcal{A}$ are differentiable. The atlas $\mathcal{D}(\mathcal{A})$ is then also differentiable since, locally, a coordinate change $h_{\beta\gamma}$ in $\mathcal{D}(\mathcal{A})$ can be written as a composition $h_{\beta\gamma} = h_{\alpha\gamma} \circ h_{\beta\alpha}$ of two other coordinate changes with a chart $h_\alpha \in \mathcal{A}$, and differentiability is a local property. The atlas $\mathcal{D}(\mathcal{A})$ is clearly maximal. In other words, $\mathcal{D}(\mathcal{A})$ cannot be enlarged by the addition of further charts, and is the largest atlas which contains $\mathcal{A}$. Thus, every differentiable atlas $\mathcal{A}$ determines uniquely a maximal differentiable atlas $\mathcal{D}(\mathcal{A})$ such that $\mathcal{A} \subset \mathcal{D}(\mathcal{A})$. Furthermore, $\mathcal{D}(\mathcal{A}) = \mathcal{D}(\mathcal{B})$ if and only if the atlas $\mathcal{A} \cup \mathcal{B}$ is differentiable.

Definition 9.4. *A* ***differentiable structure*** *on a topological manifold is a maximal differentiable atlas. A* ***differentiable manifold*** *is a topological manifold, together with a differentiable structure.*

In order to define a differentiable structure on a manifold, one must specify a differentiable atlas. In general, one specifies as small an atlas as possible, rather than a maximal one, which is then obtained as described above. We shall tacitly consider only charts and atlases of a manifold having a differentiable structure $\mathcal{D}$ that are contained in $\mathcal{D}$. (These are sometimes called *admissable.*)

As a shorthand notation, we write M, rather than $(M, \mathcal{D})$ to denote a differentiable manifold.

Examples.
1. $M = \mathbb{R}^n$. The atlas is formed by the single chart $(\mathbb{R}^n, Id)$.
2. Any open subset of a differentiable manifold has an obvious differentiable structure. It may have others as well.

Definition 9.5. *A continuous mapping $\phi : M \longrightarrow N$ from one differentiable manifold to another is said to be* ***differentiable at the point*** *$p \in M$ if for some (and hence for every) pair of charts $h : U \longrightarrow U'$ of M and $k : V \longrightarrow V'$ of N with $p \in U$ and $\phi(p) \in V$, the composite mapping $k \circ \phi \circ h^{-1}$ is differentiable at the point $h(p) \in U'$. Note that this mapping is defined in the neighborhood $h(\phi^{-1}(V) \cap U)$ of $h(p)$. The mapping ϕ is* ***differentiable*** *if it is differentiable at every point $p \in M$. One can regard $k \circ \phi \circ h^{-1}$ as a coordinate representation of ϕ in the charts h and k; for such a representation, the concept of differentiability is clear.*

The identity and the composites of differentiable mappings are differentiable. Hence differentiable manifolds form a category. Let $C^\infty(M, N)$ be the

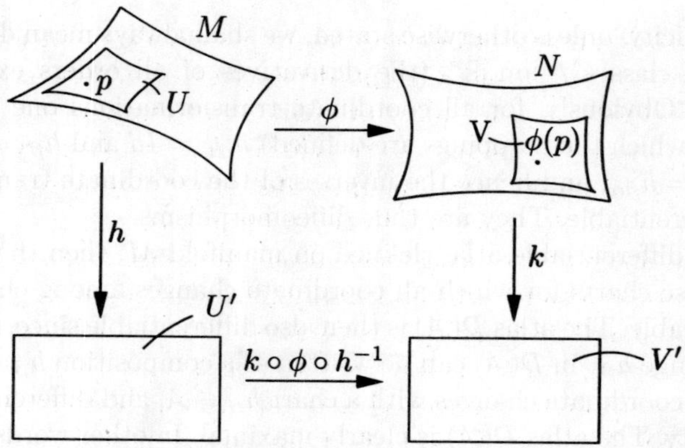

Fig. 9.3. Differentiability of a mapping.

set of differentiable mappings from M to N. We write

$$\mathcal{F}(M) = C^\infty(M) := C^\infty(M, \mathbb{R}).$$

Definition 9.6. *A bijective differentiable mapping ϕ, whose inverse ϕ^{-1} is also differentiable is called a* ***diffeomorphism***.

Every chart $h : U \longrightarrow U'$ of M is a diffeomorphism between U and U', provided U' is taken to have the standard differentiable structure as an open set in $\mathbb{R}^n$. Differential topology is the study of properties which remain invariant under diffeomorphisms.

It is a nontrivial question as to whether two different differentiable structures can be introduced on a given topological manifold in such a way that the resulting differentiable manifolds are not diffeomorphic. For example, one can show (Kervaire and Milnor, 1963), that the topological 7-sphere has exactly 15 different structures which are not mutually diffeomorphic. Recently it has been shown that even $\mathbb{R}^4$ has more than one, actually infinitely many differentiable structures.

Definition 9.7. *A differentiable mapping $\phi : M \longrightarrow N$ is called an* ***immersion*** *if (with the notation of Def. 9.5) the charts $h : U \longrightarrow U' \subset \mathbb{R}^m$ and $k : V \longrightarrow V' \subset \mathbb{R}^n$ can be chosen such that $k \circ \phi \circ h^{-1} : h(U) \longrightarrow k(V)$ is the inclusion, when we regard $\mathbb{R}^m$ as $\mathbb{R}^m \times 0 \subset \mathbb{R}^n$ (it is assumed that $m < n$). In other words, the local coordinate representation of ϕ is given by* $(x^1, \ldots, x^m) \longmapsto (x^1, \ldots, x^m, 0, \ldots, 0)$. (See also Fig. 9.4.)

Remarks.

1. An immersion is locally injective, but not necessarily globally injective.
2. If $\phi : M \longrightarrow N$ is an injective immersion, the mapping $\phi : M \longrightarrow \phi(M) \subset N$, where $\phi(M)$ has the induced topology, is not necessarily a homeomorphism. If, in addition, ϕ is in fact a homeomorphism, then ϕ is called an **embedding**.

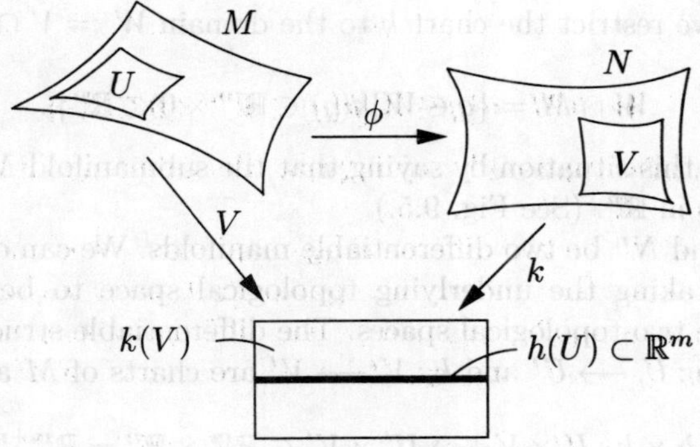

Fig. 9.4. Immersion.

Definition 9.8. *If M and N are differentiable manifolds, then M is said to be a* ***submanifold*** *of N provided that*

(i) $M \subset N$ (as sets).
(ii) The inclusion $\iota : M \hookrightarrow N$ is an embedding.

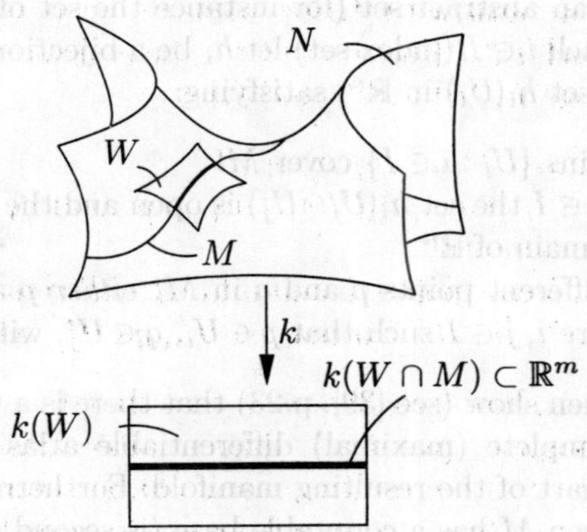

Fig. 9.5. Submanifold.

The reader is cautioned that the concepts of embedding and submanifold are not defined in a unified manner in the literature. For example, what we call a submannifold is sometimes called a proper or regular submanifold.

Since the inclusion ι of Def. 9.8 is an immersion, one can, according to Definition 9.7, choose charts $h : U \longrightarrow U' \subset \mathbb{R}^m$ and $k : V \longrightarrow V' \subset \mathbb{R}^n$ whose domains are neighborhoods of the point $p \in M \subset N$ such that locally ι has the representation $\iota : (x^1, \ldots, x^m) \longmapsto (x^1, \ldots, x^m, 0, \ldots, 0)$. Since M has the induced topology (ι is an embedding, hence also a homeomorphism), it follows that U has the form $U = \tilde{V} \cap M$, where $\tilde{V}$ is an open neighborhood of p in N. If we restrict the chart k to the domain $W := \tilde{V} \cap V$ then we see that

$$W \cap M = \{q \in W | k(q) \in \mathbb{R}^m \times 0 \subset \mathbb{R}^n\}.$$

One describes this situation by saying that the submanifold M lies locally in N as $\mathbb{R}^m$ does in $\mathbb{R}^n$. (See Fig. 9.5.)

Let M^m and N^n be two differentiable manifolds. We can define a *product manifold* by taking the underlying topological space to be the Cartesian product of the two topological spaces. The differentiable structure is defined as follows: if $h : U \longrightarrow U'$ and $k : V \longrightarrow V'$ are charts of M and N, then

$$h \times k : U \times V \longrightarrow U' \times V' \subset \mathbb{R}^m \times \mathbb{R}^n = \mathbb{R}^{m+n}$$

is a chart of $M \times N$ and the set of all charts defines the differentiable structure of $M \times N$.

Remark. The (standard) Definition 9.4 of a differentiable manifold starts with a topological space. One can alternatively begin with a set and introduce the topology with a given atlas. This approach is not only practical to construct differentiable manifolds, but is also more appropriate from a physical point of view. In detail this is done as follows.

Let M be an abstract set (for instance the set of elementary space-time events). For each $i \in I$ (index set) let h_i be a bijection from a subset $U_i \subset M$ onto an open set $h_i(U_i)$ in $\mathbb{R}^n$, satisfying:

a) The domains $\{U_i : i \in I\}$ cover M.
b) For all $i, j \in I$ the set $h_i(U_i \cap U_j)$ is open and the map $h_j \circ h_i^{-1}$ is smooth on this domain of $\mathbb{R}^n$.
c) For two different points p and q in M, *either* p and q are in a single U_i *or* there are $i, j \in I$ such that $p \in U_i$, $q \in U_j$, with U_i and U_j disjoint.

One can then show (see [39], p.23) that there is a unique Hausdorff topology and a complete (maximal) differentiable atlas on M such that each (U_i, h_i) is a chart of the resulting manifold. Furthermore, if countable many U_i cover M then M has a countable base (is second countable).

10 Tangent Vectors, Vector and Tensor Fields

At every point p of a differentiable manifold M, one can introduce a linear space, called the tangent space $T_p(M)$. A tensor field is a (smooth) map which assigns to each point $p \in M$ a tensor of a given type on $T_p(M)$.

10.1 The Tangent Space

Before defining the tangent space, let us introduce some basic concepts. Consider two differentiable manifolds M and N, and the set of differentiable mappings $\{\phi \mid \phi : U \longrightarrow N$ for a neighborhood U of $p \in M\}$. Two such mappings ϕ, ψ are called equivalent, $\phi \sim \psi$, if and only if there is a neighborhood V of p such that $\phi|V = \psi|V$. Here $\phi|V$ denotes the restriction of ϕ to the domain V. In other words, ϕ and ψ are equivalent if they coincide on some neighborhood V of p.

Definition 10.1. *An equivalence class of this relation is called a* **germ** *of smooth mappings $M \longrightarrow N$ at the point $p \in M$. A germ with representative ϕ is denoted by*

$$\bar{\phi} : (M,p) \longrightarrow N \qquad \textit{or} \qquad \bar{\phi} : (M,p) \longrightarrow (N,q) \quad \textit{if} \quad q = \phi(p) .$$

Compositions of germs are defined naturally via their representatives. A germ of a function is a germ $(M,p) \longrightarrow \mathbb{R}$. The set of all germs of functions at a point $p \in M$ is denoted by $\mathcal{F}(p)$.

The set $\mathcal{F}(p)$ has the structure of a real algebra, provided the operations are defined using representatives. A differentiable germ $\bar{\phi} : (M,p) \longrightarrow (N,q)$ defines, through composition, the following homomorphism of algebras:

$$\begin{aligned} \phi^* : \mathcal{F}(q) &\longrightarrow \mathcal{F}(p) , \\ \bar{f} &\longmapsto \bar{f} \circ \bar{\phi} . \end{aligned} \tag{10.1}$$

Obviously

$$Id^* = Id , \qquad (\psi \circ \phi)^* = \phi^* \circ \psi^* . \tag{10.2}$$

If $\bar{\phi}$ is a germ having an inverse $\bar{\phi}^{-1}$, then $\phi^* \circ (\phi^{-1})^* = Id$ and ϕ^* is thus an isomorphism.

For every point $p \in M$ of a n-dimensional differentiable manifold, a chart h having a neighborhood of p as its domain defines an invertible germ $\bar{h} : (M,p) \longrightarrow (\mathbb{R}^n, 0)$ and hence an isomorphism

$$h^* : \mathcal{F}_n \longrightarrow \mathcal{F}(p),$$

where $\mathcal{F}_n$ is the set of germs $(\mathbb{R}^n, 0) \longrightarrow \mathbb{R}$.

We now give three equivalent definitions of the tangent space at a point $p \in M$. One should be able to switch freely among these definitions. The "*algebraic definition*" is particularly handy.

Definition 10.2 (Algebraic Definition of the Tangent Space). *The* ***tangent space*** $T_p(M)$ *of a differentiable manifold* M *at a point* p *is the set of derivations of* $\mathcal{F}(p)$*. A* ***derivation*** *of* $\mathcal{F}(p)$ *is a linear mapping* $X : \mathcal{F}(p) \longrightarrow \mathbb{R}$ *which satisfies the Leibniz rule (product rule)*

$$X(\bar{f}\,\bar{g}) = X(\bar{f})\,\bar{g}(p) + \bar{f}(p)X(\bar{g}). \tag{10.3}$$

A differentiable germ $\bar{\phi} : (M,p) \longrightarrow (N,q)$ (and thus a differentiable mapping $\phi : M \longrightarrow N$) induces an algebra homomorphism $\phi^* : \mathcal{F}(q) \longrightarrow \mathcal{F}(p)$ and hence also a linear mapping

$$\begin{aligned} T_p\phi\colon T_p(M) &\longrightarrow T_q(N), \\ X &\longmapsto X \circ \phi^*. \end{aligned} \tag{10.4}$$

Definition 10.3. *The linear mapping* $T_p\phi$ *is called the* ***differential*** *(or* ***tangent map****) of* ϕ *at* p.

The set of derivations obviously forms a vector space. The Leibniz rule gives

$$X(1) = X(1 \cdot 1) = X(1) + X(1), \quad \text{i.e.} \quad X(1) = 0,$$

where 1 is the constant function having value unity. Linearity then implies that $X(c) = 0$ for every constant c.

The definition of the differential says in particular that for a given germ $\bar{f} : (N,q) \longrightarrow \mathbb{R}$

$$T_p\phi(X)(\bar{f}) = X \circ \phi^*(\bar{f}) = X(\bar{f} \circ \bar{\phi}), \qquad X \in T_p(M). \tag{10.5}$$

Hence, or from (10.2), we have for a composition $(M,p) \xrightarrow{\bar{\phi}} (N,q) \xrightarrow{\bar{\psi}} (L,r)$ the chain rule

$$T_p(\bar{\psi} \circ \bar{\phi}) = T_q\bar{\psi} \circ T_p\bar{\phi} \tag{10.6}$$

for differentials.

If $\bar{h} : (N,p) \longrightarrow (\mathbb{R}^n, 0)$ is a germ of some chart, then the induced mapping $h^* : \mathcal{F}_n \longrightarrow \mathcal{F}(p)$ is an isomorphism. This is then also true for the differential $T_p h : T_p N \longrightarrow T_0\mathbb{R}^n$. In order to describe the latter space we use the

Lemma 10.1. *If U is an open ball about the origin of $\mathbb{R}^n$, or $\mathbb{R}^n$ itself, and $f : U \longrightarrow \mathbb{R}$ is a differentiable function, then there exist differentiable functions $f_1, \ldots, f_n : U \longrightarrow \mathbb{R}$, such that*

$$f(x) = f(0) + \sum_{i=1}^{n} f_i(x) x^i .$$

Proof. We have

$$\begin{aligned} f(x) - f(0) &= \int_0^1 \frac{d}{dt} f(tx^1, \ldots, tx^n) dt \\ &= \sum_{i=1}^{n} x^i \int_0^1 D_i f(tx^1, \ldots, tx^n) dt , \end{aligned}$$

where D_i denotes the partial derivative with respect to the ith variable. It suffices to set

$$f_i(x) = \int_0^1 D_i f(tx^1, \ldots, tx^n) dt .$$

□

Particular derivations of $\mathcal{F}_n$ are the partial derivatives at the origin

$$\frac{\partial}{\partial x^i} : \mathcal{F}_n \longrightarrow \mathbb{R}, \qquad \bar{f} \longmapsto \frac{\partial}{\partial x^i} f(0) .$$

Corollary 10.1. *The $\partial / \partial x^i$, $i = 1, \ldots, n$ form a basis of $T_0\mathbb{R}^n$, the vector space of derivations of $\mathcal{F}_n$.*

Proof.

a) Linear independence: If the derivation $\sum_i a^i \partial / \partial x^i = 0$, then applying it to the germ $\bar{x}^j$ of the jth coordinate function gives

$$a^j = \sum_i a^i \frac{\partial}{\partial x^i} (\bar{x}^j) = 0$$

for all j. Hence, the $\partial / \partial x^i$ are linearly independent.

b) Let $X \in T_0\mathbb{R}^n$ and $a^i = X(\bar{x}^i)$. We now show that

$$X = \sum_i a^i \frac{\partial}{\partial x^i} .$$

For this purpose, consider the derivation $Y = X - \sum_i a^i \partial / \partial x^i$, and observe that, by construction, $Y(\bar{x}^i) = 0$ for every coordinate function x^i. If $\bar{f} \in \mathcal{F}_n$ is any germ, then, by Lemma 10.1, it is of the form

$$\bar{f} = \bar{f}(0) + \sum_i \bar{f}_i \, \bar{x}^i$$

and hence, according to the product rule

$$Y(\bar{f}) = Y\big(\bar{f}(0)\big) + \sum_i \bar{f}_i(0) Y(\bar{x}^i) = 0\,.$$

□

Remark. According to what has just been discussed, the tangent spaces of an n-dimensional differentiable manifold M also have dimension n. Thus, the dimension of M is unambiguously defined. This can also be shown to be true for topological manifolds.

The use of local coordinates $(x^1,\ldots,x^n)$ in a neighborhood of a point $p \in N^n$ enables us to write vectors in $T_p(N)$ explicitly as linear combinations of the $\partial/\partial x^i$. As a result of the isomorphism $T_p h : T_p N \longrightarrow T_0\mathbb{R}^n$, we can regard the $\partial/\partial x^i$ also as elements of $T_p(N)$. According to (10.5) we then have for $\bar{f} \in \mathcal{F}(p)$

$$\left(\frac{\partial}{\partial x^i}\right)_p \cdot \bar{f} = \frac{\partial}{\partial x^i}(f \circ h^{-1})(h(p))\,,$$

where $f \circ h^{-1}$ denotes the coordinate representation of f.

If $\bar{\phi} : (N^n, p) \longrightarrow (M^m, q)$ is a differentiable germ and if we introduce local coordinates $(y^1,\ldots,y^m)$ in a neighborhood of q, then $\bar{\phi}$ can be expressed as a germ $(\mathbb{R}^n, 0) \longrightarrow (\mathbb{R}^m, 0)$ which we denote for simplicity also by $\bar{\phi}$:

$$\begin{array}{ccc} (N^n,p) & \xrightarrow{\ \bar{\phi}\ } & (M^m,q) \\ \Big\downarrow{\scriptstyle \text{chart}} & & \Big\downarrow{\scriptstyle \text{chart}} \\ (\mathbb{R}^n,0) & \xrightarrow[\ \bar{\phi}\ ]{} & (\mathbb{R}^m,0)\,. \end{array}$$

The tangent map $T_0\bar{\phi}$ is obtained as follows: Let $\bar{f} \in \mathcal{F}_m$ and set $\phi(x^1,\ldots,x^n) = (\phi^1(x^1,\ldots,x^n),\ldots,\phi^m(x^1,\ldots,x^n))$. From (10.5) and the chain rule we can write

$$T_0\bar{\phi}\left(\frac{\partial}{\partial x^i}\right)(\bar{f}) = \frac{\partial}{\partial x^i}(\bar{f} \circ \bar{\phi}) = \frac{\partial \bar{f}}{\partial y^j}(0)\,\frac{\partial \phi^j}{\partial x^i}(0)\,.$$

Hence,

$$T_0\bar{\phi}\left(\frac{\partial}{\partial x^i}\right) = \frac{\partial \phi^j}{\partial x^i} \cdot \frac{\partial}{\partial y^j}\,. \tag{10.7}$$

The matrix

$$D\phi := \left(\frac{\partial \phi^i}{\partial x^j}\right) \tag{10.8}$$

is called the **Jacobian** of the mapping ϕ.

We can also express the result (10.7) as follows: if $v = \sum_i a^i \partial/\partial x^i$, then $T_0\bar{\phi}(v) = \sum_j b^j(\partial/\partial y^j)$, where

$$b = D\phi_0 \cdot a. \tag{10.9}$$

We summarize:

Theorem 10.1. *If one introduces local coordinates $(x^1, \ldots, x^n)$ in a neighborhood of $p \in N^n$ and $(y^1, \ldots, y^m)$ in a neighborhood of $q \in M^m$, then the derivations $\partial/\partial x^i$ and $\partial/\partial y^i$ form bases of the vector spaces T_pN and T_qM, respectively. The tangent mapping of a germ $\bar{\phi} : (N,p) \longrightarrow (M,q)$ is given by*

$$D\phi_0 : \mathbb{R}^n \longrightarrow \mathbb{R}^m$$

with respect to these bases.

We now give the "*physicist's definition*" of the tangent space which starts from the preceding result. Briefly, one often says that a contravariant vector is a real n-tuple which transforms according to the Jacobian matrix. We now wish to express this more precisely.

If $\bar{h}$ and $\bar{k} : (N,p) \longrightarrow (\mathbb{R}^n, 0)$ are germs of charts, then the coordinate transformation $\bar{g} = \bar{k} \circ \bar{h}^{-1} : (\mathbb{R}^n, 0) \longrightarrow (\mathbb{R}^n, 0)$ is an invertible differentiable germ. The set of all invertible germs of coordinate transformations forms a group G under composition. For two germs $\bar{h}$ and $\bar{k}$ there is exactly one $\bar{g} \in G$ such that $\bar{g} \circ \bar{h} = \bar{k}$. To each $\bar{g} \in G$ we assign the Jacobian at the origin Dg_0. This defines a homomorphism

$$\begin{aligned} G &\longrightarrow GL(n, \mathbb{R}), \\ \bar{g} &\longmapsto Dg_0 \end{aligned}$$

between G and the linear group of non-singular $n \times n$ matrices. This enables us to formulate the "physicist's definition" more precisely.

Definition 10.4 (Physicist's Definition of the Tangent Space). *A* ***tangent vector*** *at a point $p \in N^n$ is an assignment which associates with every germ of a chart $\bar{h} : (N,p) \longrightarrow (\mathbb{R}^n, 0)$, a vector $v = (v^1, \ldots, v^n) \in \mathbb{R}^n$ such that to the germ $\bar{g} \circ \bar{h}$ corresponds the vector $Dg_0 \cdot v$.*

Thus, if we denote by K_p the set of germs of charts $\bar{h} : (N,p) \longrightarrow (\mathbb{R}^n, 0)$, the "physicist's" tangent space $T_p(N)_{phys}$ is the set of mappings $v : K_p \longrightarrow \mathbb{R}^n$ for which $v(\bar{g} \circ \bar{h}) = Dg_0 \cdot v(\bar{h})$ for all $\bar{g} \in G$.

The set of all such mappings forms a vector space, since Dg_0 is a linear mapping. One can choose the vector $v \in \mathbb{R}^n$ arbitrarily for a given chart h, and the choice for all other charts is then fixed. The vector space $T_p(N)_{phys}$ is isomorphic to $\mathbb{R}^n$. An isomorphism is given via the choice of a local coordinate system. The canonical isomorphism

$$T_p(N) \longrightarrow T_p(N)_{phys}$$

with the algebraically defined tangent space (see Definition 10.2) assigns to the derivation $X \in T_p(N)$ the vector $(X(\bar{h}^1), \ldots, X(\bar{h}^n)) \in \mathbb{R}^n$ for a given germ $\bar{h} = (\bar{h}^1, \ldots, \bar{h}^n) : (N, p) \longrightarrow (\mathbb{R}^n, 0)$. The components of this vector are precisely the coefficients of X with respect to the basis $\partial/\partial x^i$. They transform according to the Jacobian under a change of coordinates.

The "*geometric definition*" of the tangent space is the most intuitive one. It identifies tangent vectors with "velocity vectors" of curves through the point p at that point.

Definition 10.5. *A* ***curve*** *on N passing through p is given by a differentiable mapping w which maps an open interval $I \subset \mathbb{R}$ into N with $0 \in I$ and $w(0) = p$.*

Definition 10.6 (Geometric Definition of the Tangent Space). *We introduce an equivalence relation on the set W_p of germs of paths through p. We say that $\bar{w}$ and $\bar{v}$ in W_p are equivalent ($\bar{w} \sim \bar{v}$) if and only if for every germ $\bar{f} \in \mathcal{F}(p)$*

$$\frac{d}{dt}(\bar{f} \circ \bar{w})(0) = \frac{d}{dt}(\bar{f} \circ \bar{v})(0).$$

An equivalence class $[w]$ of this relation is a ***tangent vector*** *at the point p.*

Two germs of paths define the same tangent vector if they define the same "derivative along the curve". To every equivalence class $[w]$ we can thus associate a derivation X_w of $\mathcal{F}(p)$ by

$$X_w(\bar{f}) := \frac{d}{dt}(\bar{f} \circ \bar{w})(0).$$

This relation defines the injective mapping

$$\begin{aligned} W_p/\sim =: T_p(N)_{geom} &\longrightarrow T_p(N), \\ w &\longmapsto X_w \end{aligned}$$

of the set of equivalence classes of germs of paths into the tangent space. This mapping is also surjective, since if $w(t)$ is given in local coordinates by $w(t) = (ta^1, \ldots, ta^n)$ then $X_w = \sum_i a^i \partial/\partial x^i$.

Obviously $X_w = X_v$ in precisely those cases for which $d/dt\ w^i(0) = d/dt\ v^i(0)$ for any local coordinate system. In this geometrical interpretation the tangent map is easily visualized. A germ $\bar{\phi} : (N, p) \longrightarrow (M, q)$ induces the mapping

$$\begin{aligned} T_p(N)_{geom} &\longrightarrow T_p(M)_{geom}, \\ [w] &\longmapsto [\phi \circ w]. \end{aligned}$$

The following equation shows that this definition of the tangent map is consistent with that in Definition 10.2:

$$X_{\phi \circ w}(\bar{f}) = \frac{d}{dt}(\bar{f} \circ (\phi \circ w))(0) = X_w(\bar{f} \circ \phi) \overset{(10.5)}{=} T_p\phi(X_w)(\bar{f}),$$

thus

$$X_{\phi \circ w} = T_p\phi \cdot X_w .$$

In the following, we shall regard the three different definitions of a tangent space as equivalent and use whichever of them is most convenient.

Definition 10.7. *The* ***rank*** *of a differentiable mapping $\phi : M \longrightarrow N$ at the point $p \in M$ is the number $rk_p :=$ rank of $T_p\phi$, i.e. the rank of the matrix $D\phi_p$.*

Lemma 10.2. *The rank of a mapping is lower semicontinuous: if $rk_p\phi = r$, there is a neighborhood U of p such that $rk_q\phi \geq r$ for all $q \in U$.*

Proof. We choose local coordinates and consider the Jacobian $D\phi$ corresponding to ϕ in the neighborhood of $p \in V \subset \mathbb{R}^m$. The elements of this matrix describe a differentiable mapping $V \longrightarrow \mathbb{R}^{m \cdot n}$ such that $q \longmapsto \partial\phi^i/\partial x^j(q)$. Since $rk_p\phi = r$, there exists an $r \times r$ submatrix of $D\phi_p$ (without loss of generality, we may take this to be the first r rows and r columns) which has a non-vanishing determinant at p. Therefore the mapping

$$\begin{array}{ccccccc} V & \longrightarrow \mathbb{R}^{m \cdot n} & \longrightarrow & \mathbb{R}^{r \cdot r} & \longrightarrow & \mathbb{R} \\ q & \longmapsto D_q\phi & \longmapsto & \text{submatrix} & \longmapsto & \text{determinant} \end{array}$$

does not vanish at p, and hence also not in some neighborhood of p. The rank cannot decrease there. □

Usually one define an immersion as follows:

Definition 10.8. *A differentiable mapping $\phi : M \longrightarrow N$ is called an* ***immersion*** *if its rank is equal to the dimension of M for all $p \in M$.*

One can show that this definition is equivalent to the one given in Chap. 9 (see Definition 9.7). This follows easily from the inverse function theorem (see, for example, [36], p.55).

10.2 Vector Fields

Definition 10.9. *If to every point p of a differentiable manifold M a tangent vector $X_p \in T_p(M)$ is assigned, then we call the map $X : p \longmapsto X_p$ a* ***vector field*** *on M.*

If $(x^1, \ldots, x^n)$ are local coordinates in an open set $U \subset M$, then for every point $p \in U$, X_p has a unique representation of the form

$$X_p = \xi^i(p) \left(\frac{\partial}{\partial x^i} \right)_p . \tag{10.10}$$

Definition 10.10. *The n functions ξ^i $(i = 1, \ldots, n)$ defined on U are the* ***components*** *of X with respect to the local coordinate system $(x^1, \ldots, x^n)$.*

Consider now a second local coordinate system $(\bar{x}^1, \ldots, \bar{x}^n)$ on U and let $\bar{\xi}^i$ be the components of X relative to $(\bar{x}^1, \ldots, \bar{x}^n)$, then we have, according to the results of Sect. 10.1,

$$\bar{\xi}^i(p) = \frac{\partial \bar{x}^i}{\partial x^j}(p)\, \xi^j(p)\,, \qquad p \in U\,. \tag{10.11}$$

Hence, the property that the components of X are continuous, or of class C^r at a point p does not depend on the choice of the local coordinate system. If X is continuous or of class C^r at every point of M, we say that the vector field is continuous, or of class C^r on M.

In the following, we shall consider only vector fields of class C^∞ (unless stated otherwise) and denote the set of such fields by $\mathcal{X}(M)$. $\mathcal{F}(M)$ or $C^\infty(M)$ denote the class of C^∞ functions on M, as before.

If X, $Y \in \mathcal{X}(M)$ and $f \in \mathcal{F}(M)$, the assignments $p \longmapsto f(p)X_p$ and $p \longmapsto X_p + Y_p$ define new vector fields on M. These are denoted by fX and $X + Y$, respectively. The following rules apply: If f, $g \in \mathcal{F}(M)$ and X, $Y \in \mathcal{X}(M)$, then

$$\begin{aligned} f(gX) &= (fg)X\,, \\ f(X+Y) &= fX + fY\,, \\ (f+g)X &= fX + gX\,. \end{aligned}$$

We also define the functions Xf on M by

$$(Xf)(p) = X_p f\,, \qquad p \in M\,.$$

In local coordinates, we have

$$X = \xi^i \partial/\partial x^i\,,$$

where the ξ^i are C^∞ functions. Since

$$(Xf)(p) = \xi^i(p)\, \frac{\partial f}{\partial x^i}(p)$$

it follows that Xf is also a C^∞ function. Xf is called the **derivative of f with respect to the vector field** X. The following rules hold:

$$\begin{aligned} X(f+g) &= Xf + Xg\,, \\ X(fg) &= (Xf)\,g + f\,(Xg) \qquad \text{(Leibniz rule)}\,. \end{aligned}$$

In algebraic language, $\mathcal{X}(M)$ is a module[1] over the associative algebra $\mathcal{F}(M)$.

[1] For this basic algebraic concept, and a few other ones assumed to be known, consult any book on algebra (or groups).

If we set

$$D_X f = Xf \qquad f \in \mathcal{F}(M), X \in \mathcal{X}(M), \tag{10.12}$$

then D_X is a **derivation** of the algebra $\mathcal{F}(M)$. One can prove (see [36], p.73) that conversely every derivation of $\mathcal{F}(M)$ is of the form D_X for some vector field X.

The commutator of two derivations D_1 and D_2 of an algebra $\mathcal{A}$,

$$[D_1, D_2]a := D_1(D_2 a) - D_2(D_1 a), \qquad a \in \mathcal{A},$$

is also a derivation, as one can easily show by direct computation. The commutator is antisymmetric

$$[D_1, D_2] = -[D_2, D_1]$$

and satisfies the Jacobi identity

$$[D_1, [D_2, D_3]] + [D_2, [D_3, D_1]] + [D_3, [D_1, D_2]] = 0 .$$

As a result, one is led to define the **commutator** of two vector fields X and Y according to

$$[X, Y]f = X(Yf) - Y(Xf) . \tag{10.13}$$

One can easily prove the following properties:

$$\begin{aligned} [X+Y, Z] &= [X, Z] + [Y, Z], \qquad X, Y, Z \in \mathcal{X}(M), \\ [X, Y] &= -[Y, X], \\ [fX, gY] &= fg[X, Y] + f(Xg)Y - g(Yf)X, \qquad f, g \in \mathcal{F}(M), \\ [X, [Y, Z]] &+ [Z, [X, Y]] + [Y, [Z, X]] = 0 . \end{aligned} \tag{10.14}$$

The set of vector fields $\mathcal{X}(M)$ is an $\mathbb{R}$**-Lie-algebra** with respect to the commutator product. If, in local coordinates,

$$X = \xi^i \frac{\partial}{\partial x^i}, \qquad Y = \eta^i \frac{\partial}{\partial x^i},$$

then one easily finds that

$$[X, Y] = \left(\xi^i \frac{\partial \eta^j}{\partial x^i} - \eta^i \frac{\partial \xi^j}{\partial x^i} \right) \frac{\partial}{\partial x^j} . \tag{10.15}$$

10.3 Tensor Fields

For the following, the reader is assumed to be familiar with some basic material of multilinear algebra (see [33], Sect. 1.7). In addition to the tangent space $T_p(M)$ at a point $p \in M$, we shall consider the **dual space** or **cotangent space** $T_p^*(M)$.

Definition 10.11. *Let f be a differentiable function defined on an open set $U \subset M$. If $p \in U$ and $v \in T_p(M)$ is an arbitrary vector, we set*

$$(df)_p(v) := v(f)\,. \tag{10.16}$$

*The mapping $(df)_p : T_p(M) \longrightarrow \mathbb{R}$ is obviously linear, and hence $(df)_p \in T_p^*(M)$. The linear function $(df)_p$ is the **differential** of f at the point p.*

For a local coordinate system $(x^1, \ldots, x^n)$ in a neighborhood of p, we have

$$(df)_p \left(\frac{\partial}{\partial x^i}\right)_p = \frac{\partial f}{\partial x^i}(p)\,. \tag{10.17}$$

In particular, the component function $x^i : p \longmapsto x^i(p)$ satisfies

$$(dx^i)_p \left(\frac{\partial}{\partial x^j}\right)_p = \delta^i_j\,. \tag{10.18}$$

That is , the n-tuple $\{(dx^1)_p, \ldots, (dx^n)_p\}$ is a basis of $T_p^*(M)$ which is *dual* to the basis $\{\left(\frac{\partial}{\partial x^1}\right)_p, \ldots, \left(\frac{\partial}{\partial x^n}\right)_p\}$ of $T_p(M)$.

We can write $(df)_p$ as a linear combination of the $(dx^i)_p$:

$$(df)_p = \lambda_j\,(dx^j)_p\,.$$

Now,

$$(df)_p \left(\frac{\partial}{\partial x^i}\right)_p = \lambda_j\,(dx^j)_p \left(\frac{\partial}{\partial x^i}\right)_p = \lambda_i\,.$$

From (10.17) we then have $\lambda_i = \frac{\partial f}{\partial x^i}(p)$ and thus

$$(df)_p = \left(\frac{\partial f}{\partial x^i}\right)(p)(dx^i)_p\,. \tag{10.19}$$

Definition 10.12. *Let $T_p(M)^r_s$ be the set of tensors of rank (r,s) defined on $T_p(M)$ (contravariant of rank r, covariant of rank s). If we assign to every $p \in M$ a tensor $t_p \in T_p(M)^r_s$, then the map $t : p \longmapsto t_p$ defines a **tensor field of type (r,s)**.*

Algebraic operations on tensor fields are defined pointwise; for example, the sum of two tensor fields is defined by

$$(t+s)_p = t_p + s_p\,, \qquad t, s \in T_p(M)^r_s\,.$$

Tensor products and contractions of tensor fields are defined analogously. A tensor field can also be multiplied by a function $f \in \mathcal{F}(M)$:

$$(f\,t)_p = f(p)\,t_p\,.$$

The set $\mathcal{T}^r_s(M)$ of tensor fields of type (r, s) is thus a module over $\mathcal{F}(M)$.

In a coordinate neighborhood U, having coordinates $(x^1, \ldots, x^n)$, a tensor field can be expanded in the form

$$t = t^{i_1 \ldots i_r}_{j_1 \ldots j_s} \left(\frac{\partial}{\partial x^{i_1}} \otimes \ldots \otimes \frac{\partial}{\partial x^{i_r}} \right) \otimes \left(dx^{j_1} \otimes \ldots \otimes dx^{j_s} \right) . \tag{10.20}$$

The $t^{i_1 \ldots i_r}_{j_1 \ldots j_s}$ are the **components** of t relative to the coordinate system $(x^1, \ldots, x^n)$. If the coordinates are transformed to $(\bar{x}^1, \ldots, \bar{x}^n)$, the components of t transform according to

$$\bar{t}^{i_1 \ldots i_r}_{j_1 \ldots j_s} = \frac{\partial \bar{x}^{i_1}}{\partial x^{k_1}} \cdot \ldots \cdot \frac{\partial \bar{x}^{i_r}}{\partial x^{k_r}} \cdot \frac{\partial x^{l_1}}{\partial \bar{x}^{j_1}} \cdot \ldots \cdot \frac{\partial x^{l_s}}{\partial \bar{x}^{j_s}} \cdot t^{k_1 \ldots k_r}_{l_1 \ldots l_s} . \tag{10.21}$$

A tensor field is of class C^r if all its components are of class C^r. We see that this property is independent of the choice of coordinates. If two tensor fields t and s are of class C^r, then so are $s + t$ and $s \otimes t$. In the following, we shall only consider C^∞ tensor fields.

Covariant tensors of order 1 are also called **one-forms**. The set of all one-forms will be denoted by $\mathcal{X}^*(M)$. The completely antisymmetric covariant tensors of higher order (differential forms) play an important role. We shall discuss these in detail in Chap. 12.

Let $t \in \mathcal{T}^r_s(M)$, $X_1, \ldots, X_s \in \mathcal{X}(M)$ and $\omega^1, \ldots, \omega^r \in \mathcal{X}^*(M)$. We consider, for every $p \in M$,

$$F(p) = t_p \left(\omega^1(p), \ldots, \omega^r(p), X_1(p), \ldots, X_s(p) \right) .$$

The mapping defined by $p \longmapsto F(p)$ is obviously a C^∞ function, which we denote by $t(\omega^1, \ldots, \omega^r, X_1, \ldots, X_s)$. The assignment

$$(\omega^1, \ldots, \omega^r, X_1, \ldots, X_s) \longmapsto t(\omega^1, \ldots, \omega^r, X_1, \ldots, X_s)$$

is $\mathcal{F}(M)$-multilinear. For every tensor field $t \in \mathcal{T}^r_s(M)$ there is thus an associated $\mathcal{F}(M)$-multilinear mapping:

$$\underbrace{\mathcal{X}^*(M) \otimes \ldots \otimes \mathcal{X}^*(M)}_{\text{r-times}} \otimes \underbrace{\mathcal{X}(M) \otimes \ldots \otimes \mathcal{X}(M)}_{\text{s-times}} \longrightarrow \mathcal{F}(M) .$$

One can show (see [36], p.137) that every such mapping can be obtained in this manner. In particular, every one-form can be regarded as an $\mathcal{F}(M)$-linear function defined on the vector fields.

Definition 10.13. *Let $\phi : M \longrightarrow N$ be a differentiable mapping. We define the* **pull back** *of an one-form ω on N by*

$$(\phi^* \omega)_p(X_p) := \omega_{\phi(p)}(T_p \phi \cdot X_p) , \qquad X_p \in T_p(M) . \tag{10.22}$$

Analogously, we define the pull back of a covariant tensor field $t \in \mathcal{T}_s^0(N)$ by

$$(\phi^* t)_p(v_1, \ldots, v_s) := t_{\phi(p)}(T_p\phi \cdot v_1, \ldots, T_p\phi \cdot v_s)\,, \quad v_1, \ldots, v_s \in T_p(M)\,. \tag{10.23}$$

For a function $f \in \mathcal{F}(N)$, we have

$$\phi^*(df) = d(\phi^* f)\,. \tag{10.24}$$

Indeed, from (10.5) we have for $v \in T_p(M)$,

$$\begin{aligned}(\phi^* df)_p(v) &= df_{\phi(p)}(T_p\phi \cdot v) = T_p\phi(v) f = v(f \circ \phi) \\ &= v(\phi^*(f)) = d(\phi^* f)(v)\,.\end{aligned} \tag{10.25}$$

Definition 10.14. *A* ***pseudo-Riemannian metric*** *on a differentiable manifold M is a tensor field $g \in \mathcal{T}_s^0(M)$ having the properties*

(i) $g(X,Y) = g(Y,X)$ for all X, $Y \in \mathcal{X}(M)$.

(ii) For every $p \in M$, g_p is a non-degenerate bilinear form on $T_p(M)$. This means that $g_p(X,Y) = 0$ for all $X \in T_p(M)$ if and only if $Y = 0$.

The tensor field $g \in \mathcal{T}_s^0(M)$ is a (proper) ***Riemannian metric*** *if g_p is positive definite at every point p.*

Definition 10.15. *A* ***(pseudo-)Riemannian manifold*** *is a differentiable manifold M, together with a (pseudo-)Riemannian metric g.*

If $\{\theta^i\}$, for $i = 1, \ldots, \dim M$, is a basis of one-forms on an open subset of M, we write

$$g = g_{ik}\, \theta^i \otimes \theta^k \tag{10.26}$$

or

$$ds^2 = g_{ik}\, \theta^i \theta^k\,, \qquad \text{with} \quad \theta^i\theta^k := \frac{1}{2}(\theta^i \otimes \theta^k + \theta^k \otimes \theta^i)\,. \tag{10.27}$$

Obviously,

$$g_{ik} = g(e_i, e_k)\,, \tag{10.28}$$

if $\{e_i\}$ is the basis of (local) vector fields which is dual to $\{\theta^i\}$.

11 The Lie Derivative

Before defining the Lie derivative of tensor fields, we introduce some important concepts.

11.1 Integral Curves and Flow of a Vector Field

Recall that a *curve* in a differentiable manifold M is a differentiable map $\gamma : J \longrightarrow M$ from an open interval J of $\mathbb{R}$ into M. For such a curve there is a *tangent vector* (*velocity vector*) at each $\gamma(t)$, $t \in J$, defined by

$$\dot{\gamma}(t)f = \frac{d}{dt}(f \circ \gamma)(t)\,, \qquad f \in \mathcal{F}(M)\,.$$

Equivalently, $\dot{\gamma}(t)$ can be regarded as the equivalence class $[\gamma]$ at $\gamma(t)$ in the sense of Definition 10.6.

Let X be a vector field (as usual C^∞).

Definition 11.1. *Let J be an open interval in $\mathbb{R}$ which contains 0. A differentiable curve $\gamma : J \longrightarrow M$ is called an* **integral curve** *of X, with starting point $x \in M$ (or through $x \in M$), provided*

$$\dot{\gamma}(t) := \frac{d\gamma}{dt} = X(\gamma(t))$$

for every $t \in J$ and $\gamma(0) = x$.

From the well-known existence and uniqueness theorems for ordinary differential equations one obtains

Theorem 11.1. *For every $x \in M$, there is a unique maximal (inextendible) integral curve through x of class C^∞.*

We denote this curve by $\phi_x : J_x \longrightarrow M$. Let $\mathcal{D} := \{(t,x) | t \in J_x, x \in M\} \subset \mathbb{R} \times M$. For every $t \in \mathbb{R}$, let $\mathcal{D}_t := \{x \in M | (t,x) \in \mathcal{D}\} = \{x \in M | t \in J_x\}$.

Definition 11.2. *The mapping*

$$\phi : \mathcal{D} \longrightarrow M\,,$$
$$(t,x) \longmapsto \phi_x(t)$$

is called the (maximal local) **flow** *of X, and $\mathcal{D}$ is its* **domain of definition**.

Theorem 11.2.

(i) $\mathcal{D}$ *is an open set in* $\mathbb{R} \times M$ *and* $\{0\} \times M \subset \mathcal{D}$ *(this implies that* $\mathcal{D}_t$ *is an open subset of* M*).*

(ii) The mapping $\phi : \mathcal{D} \longrightarrow M$ *is a* C^∞*-morphism.*

(iii) For every $t \in \mathbb{R}$,

$$\phi_t : \mathcal{D}_t \longrightarrow M\,,$$
$$x \longmapsto \phi(t, x)$$

is a diffeomorphism from $\mathcal{D}_t$ *to* $\mathcal{D}_{-t}$, *with inverse* ϕ_{-t}.

Corollary 11.1. *For every* $a \in M$, *there is an open interval* J, *such that* $0 \in J$ *and an open neighborhood* U *of* a *such that* $J \times U \subset \mathcal{D}$.

Definition 11.3. $\psi := \phi | J \times U : J \times U \longrightarrow M$ *(i.e. the restriction of* ϕ *to* $J \times U$*) is called a* **local flow** *of* X *at* a.

Theorem 11.3. *A local flow* $\psi : J \times U \longrightarrow M$ *of* X *at* a *has the properties:*

(i)

$$\psi_x : J \longrightarrow M\,,$$
$$t \longmapsto \psi(t, x)$$

is an integral curve of X *with starting point* x. *Thus,* $\dot{\psi}_x(t) = X(\psi_x(t))$ *for* $t \in J$ *and* $\psi_x(0) = x$.

(ii)

$$\psi_t : U \longrightarrow M\,,$$
$$x \longmapsto \psi(t, x)$$

is a diffeomorphism from U *onto* $\psi_t(U)$.

(iii) If s, t, $s + t \in J$ *and* $x \in U$, *then* $\psi_{s+t}(x) = \psi_s(\psi_t(x))$. *In other words,*

$$\psi_{s+t} = \psi_s \circ \psi_t\,.$$

For this reason, one also calls ψ *a* **local one parameter group of local diffeomorphisms***.*

Definition 11.4. *If* $\mathcal{D} = \mathbb{R} \times M$, *then* X *is said to be* **complete***.*

In this case, we have

Theorem 11.4. *If* X *is complete and* $\phi : \mathbb{R} \times M \longrightarrow M$ *is the global flow of* X, *then* $\mathcal{D}_t = M$ *for every* $t \in \mathbb{R}$ *and*

(i) $\phi_t : M \longrightarrow M$ *is a diffeomorphism.*

(ii) $\phi_s \circ \phi_t = \phi_{s+t}$ *for* $s, t \in \mathbb{R}$.

This means that the assignment $t \longmapsto \phi_t$ is a group homomorphism $\mathbb{R} \longrightarrow \mathrm{Diff}(M)$ ($\mathrm{Diff}(M)$ denotes the group of all diffeomorphisms of M); the set $(\phi_t)_{t \in \mathbb{R}}$ is a one parameter group of diffeomorphisms of M.

Remark. One can show that a vector field on a compact manifold is complete.

Complete proofs of these theorems can be found in Sect. 4.1 of [40].

11.2 Mappings and Tensor Fields

Let $\phi : M \longrightarrow N$ be a differentiable mapping. This mapping induces a mapping of covariant tensor fields (see also Sect. 10.3) by means of

Definition 11.5. *The* ***pull back*** $\phi^* : \mathcal{T}_q^0(N) \longrightarrow \mathcal{T}_q^0(M)$ *is defined by*

$$(\phi^* t)_x(u_1, \ldots, u_q) = t_{\phi(x)}(T_x\phi \cdot u_1, \ldots, T_x\phi \cdot u_q)\,,$$

for $t \in \mathcal{T}_q^0(N)$ *and for arbitrary* $x \in M$ *and* $u_1, \ldots, u_q \in T_x(M)$.

Remarks. 1. If $q = 0$, we define $\phi^* : \mathcal{F}(N) \longrightarrow \mathcal{F}(M)$ by $\phi^*(f) = f \circ \phi$, $f \in \mathcal{F}(N)$.

2. If $q = 1$, we have

$$(\phi^* \omega)(x) = (T_x\phi)^* \omega(\phi(x))$$

for $\omega \in \mathcal{T}_1^0(N)$ and $x \in M$. Here, $(T_x\phi)^*$ denotes the linear transformation dual to $T_x\phi$.

The following statement is obvious.

Theorem 11.5. *The map*

$$\phi^* : \bigoplus_{q=0}^{\infty} \mathcal{T}_q^0(N) \longrightarrow \bigoplus_{q=0}^{\infty} \mathcal{T}_q^0(M)$$

is an $\mathbb{R}$*-algebra homomorphism of the* ***covariant tensor algebras****.*

If ϕ is not a diffeomorphism we still have the following useful concept:

Definition 11.6. *The vector fields* $X \in \mathcal{X}(M)$ *and* $Y \in \mathcal{X}(N)$ *are said to be* ***ϕ-related*** *if*

$$T_x\phi \cdot X_x = Y_{\phi(x)}$$

for every $x \in M$.

The next theorem is nontrivial; we prove it in Sect. 14.1.

Theorem 11.6. *Let* $X_1, X_2 \in \mathcal{X}(M)$ *and* $Y_1, Y_2 \in \mathcal{X}(N)$. *If* X_i *and* Y_i *are* ϕ*-related (for* $i = 1, 2$*) then so are the commutators* $[X_1, X_2]$ *and* $[Y_1, Y_2]$.

In addition we obviously have

Theorem 11.7. *Let* $t \in \mathcal{T}_q^0(N)$, $X_i \in \mathcal{X}(M)$, $Y_i \in \mathcal{X}(N)$, $i = 1, \ldots, q$. *If the* X_i *and* Y_i *are* ϕ*-related then*

$$(\phi^* t)(X_1, \ldots, X_q) = \phi^*\big(t(Y_1, \ldots, Y_q)\big)\,.$$

Now let us consider in particular diffeomorphisms. First recall from linear algebra that if E and F are two vector spaces , and E^r_s, F^r_s are the vector spaces of tensors of type (r,s) on these spaces, and if the mapping $A : E \longrightarrow F$ is an isomorphism, then it induces an isomorphism $A^r_s : E^r_s \longrightarrow F^r_s$, which is defined as follows: Let $t \in E^r_s$; for arbitrary $y_1^*, \ldots, y_r^* \in F^*$ and $y_1, \ldots, y_s \in F$,

$$(A^r_s t)(y_1^*, \ldots, y_r^*, y_1, \ldots, y_s) := t(A^* y_1^*, \ldots, A^* y_r^*, A^{-1} y_1, \ldots, A^{-1} y_s) .$$

Note that $A^1_0 = A$ and $A^0_1 = (A^{-1})^*$. In addition, we set $A^0_0 = Id_{\mathbb{R}}$.

Now let $\phi : M \longrightarrow N$ be a diffeomorphism. We define two induced mappings

$$\begin{aligned} \phi_* &: \mathcal{T}^r_s(M) \longrightarrow \mathcal{T}^r_s(N) \quad \text{and} \\ \phi^* &: \mathcal{T}^r_s(N) \longrightarrow \mathcal{T}^r_s(M) , \end{aligned}$$

which are each other's inverses, by $(T^r_s\phi := (T\phi)^r_s)$

$$\begin{aligned} (\phi_* t)_{\phi(p)} &= T^r_s \phi(t_p) \qquad \text{for} \quad t \in \mathcal{T}^r_s(M) , \\ (\phi^* t)_p &= T^r_s \phi^{-1}(t_{\phi(p)}) \quad \text{for} \quad t \in \mathcal{T}^r_s(N) . \end{aligned}$$

It follows immediately that $X \in \mathcal{X}(M)$ and $\phi_* X \in \mathcal{X}(N)$ are ϕ-related.

The linear extension of ϕ_* and ϕ^* to the complete tensor algebras results in two $\mathbb{R}$-*algebra isomorphisms* which are mutually inverse and which we also denote by ϕ_*, respectively ϕ^*. Note that the definition of ϕ^* is compatible with (10.23).

Definition 11.7. *If $\phi^* t = t$ for some tensor field t, we say that t is* ***invariant*** *under the diffeomorphism ϕ. If $\phi_t^* s = s$ for every $t \in \mathbb{R}$ of a one parameter group, we say that s is* ***invariant*** *under the one parameter group $(\phi_t)_{t\in\mathbb{R}}$.*

Exercise 11.1. Derive explicit representations for ϕ_* and ϕ^* in terms of coordinates. The results are similar to the formulas for coordinate changes (as for the passive and active interpretations of similarity transformations in linear algebra).

11.3 The Lie Derivative

Let $X \in \mathcal{X}(M)$ and let ϕ_t be the flow of X.

Definition 11.8. *We set*

$$L_X T := \left.\frac{d}{dt}\right|_{t=0} \phi_t^* T = \lim_{t\to 0} t^{-1} \left(\phi_t^* T - T\right) ,$$

where $T \in \mathcal{T}(M)$ is an arbitrary element of the algebra of tensor fields on M. $L_X T$ is called the ***Lie derivative*** *of T with respect to X.*

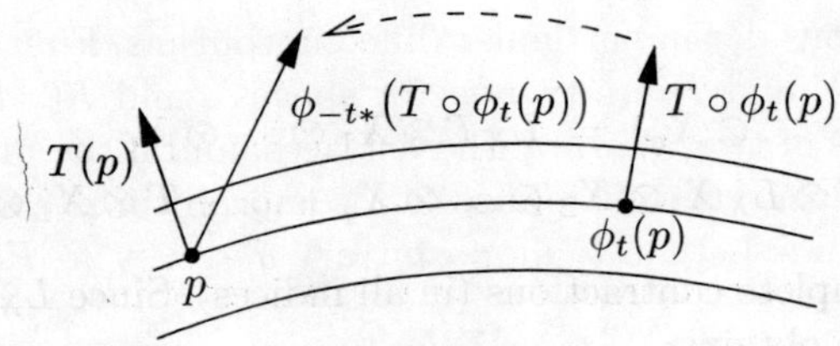

Fig. 11.1. Lie derivative.

Theorem 11.8. *The Lie derivative has the following properties:*

(i) L_X *is* $\mathbb{R}$*-linear, i.e.* $L_X(T_1 + T_2) = L_X(T_1) + L_X(T_2)$*;*
(ii) $L_X(T \otimes S) = (L_X T) \otimes S + T \otimes (L_X S)$ *(Leibniz rule);*
(iii) $L_X(\mathcal{T}^r_s(M)) \subseteq \mathcal{T}^r_s(M)$*;*
(iv) L_X *commutes with contractions;*
(v) $L_X f = Xf = df(X)$*, if* $f \in \mathcal{F}(M) = \mathcal{T}^0_0(M)$*;*
(vi) $L_X Y = [X, Y]$*, if* $Y \in \mathcal{X}(M)$*.*

The only non-obvious statement is property (vi) which we prove in Sect. 14.1.

Theorem 11.9. *If* $X, Y \in \mathcal{X}(M)$ *and* $\lambda \in \mathbb{R}$*, then*

(i) $L_{X+Y} = L_X + L_Y$, $L_{\lambda X} = \lambda L_X$*;*
(ii) $L_{[X,Y]} = [L_X, L_Y] = L_X \circ L_Y - L_Y \circ L_X$*.*

Proof. See Sect. 14.1. □

Theorem 11.10. *If* $X, Y \in \mathcal{X}(M)$ *and* ϕ*,* ψ *are the flows of* X *and* Y*, respectively, then the following statements are equivalent:*

(i) $[X, Y] = 0$*;*
(ii) $L_X \circ L_Y = L_Y \circ L_X$*;*
(iii) $\phi_s \circ \psi_t = \psi_t \circ \phi_s$ *for all* $s, t \in \mathbb{R}$ *(where both sides are defined).*

Proof. See Sect. 14.1. □

Theorem 11.11. *Let* ϕ_t *be a one-parameter transformation group and let* X *be the corresponding vector field (*"infinitesimal transformation"*). A tensor field on* M *is invariant under* ϕ_t *(i.e.* $\phi_t^* T = T$ *for all* t*) if and only if* $L_X T = 0$*.*

This follows from an important formula derived in Sect. 14.1.

Theorem 11.12. *If* $T \in \mathcal{T}^0_q(M)$ *is a covariant tensor field and* $X_1, \ldots, X_q \in \mathcal{X}(M)$*, then*

$$\begin{aligned}(L_X\, T\,)(X_1, \ldots, X_q) \\ = X\big(T(X_1, \ldots, X_q)\big) - \sum_{k=1}^{q} T\big(X_1, \ldots, [X, X_k], \ldots, X_q\big)\,. \quad (11.1)\end{aligned}$$

Proof. Consider

$$
\begin{aligned}
L_X\big(T\otimes X_1\otimes\ldots\otimes X_q\big) &= L_X T\otimes X_1\otimes\ldots\otimes X_q\\
&\quad + T\otimes L_X X_1\otimes X_2\otimes\ldots\otimes X_q+\ldots+T\otimes X_1\otimes\ldots\otimes L_X X_q
\end{aligned}
$$

and take the complete contractions (in all indices). Since L_X commutes with contractions, one obtains

$$
\begin{aligned}
L_X\big(T(X_1,\ldots,X_q)\big) &= (L_X T)(X_1,\ldots,X_q)\\
&\quad + T(\underbrace{L_X X_1}_{[X,X_1]},X_2,\ldots,X_q)+\ldots+T(X_1,\ldots,\underbrace{L_X X_q}_{[X,X_q]}),
\end{aligned}
$$

which is identical to (11.1). □

Example 11.1. Let $\omega\in\mathcal{X}^*(M)$ and $Y\in\mathcal{X}(M)$. Then, writing also $\langle\omega,Y\rangle$ for $\omega(Y)$,

$$
(L_X\omega)(Y)=X\langle\omega,Y\rangle-\omega([X,Y])\,. \tag{11.2}
$$

As an application, we show that

$$
L_X df = dL_X f \quad \text{for} \quad f\in\mathcal{F}(M). \tag{11.3}
$$

Proof. According to (11.2), we have

$$
\begin{aligned}
\langle L_X df, Y\rangle &= X\langle df,Y\rangle - df([X,Y])\\
&= XYf-[X,Y]f = YXf = YL_Xf=\langle dL_X f, Y\rangle\,.
\end{aligned}
$$

□

Local Coordinate Expressions for Lie Derivatives

We choose the coordinates $\{x^i\}$ and the dual bases $\{\partial_i:=\partial/\partial x^i\}$ and $\{dx^i\}$. First of all, we note that

$$
L_X dx^i = dL_X x^i = X^i_{\ ,j}\,dx^j \tag{11.4}
$$

and

$$
L_X\partial_i=[X,\partial_i]=-X^j_{\ ,i}\,\partial_j\,. \tag{11.5}
$$

Now let $T\in\mathcal{T}^r_s(M)$. The determination of the components of $L_X T$ is analogous to the proof of Theorem 11.12. For $\omega^k=dx^{i_k}$ and $Y_l=\partial_{i_l}$, we form

$$
L_X(T\otimes\omega^1\otimes\ldots\otimes\omega^r\otimes Y_1\otimes\ldots\otimes Y_s)\,,
$$

apply the product rule, and then contract completely. Since L_X commutes with contractions we obtain

$$L_X \left(T(dx^{i_1}, \ldots, dx^{i_r}, \partial_{j_1}, \ldots, \partial_{j_s})\right) = (L_X T)(dx^{i_1}, \ldots, \partial_{j_s}) + \\ + T(L_X dx^{i_1}, \ldots, \partial_{j_s}) + \cdots + T(dx^{i_1}, \ldots, [X, \partial_{j_s}]) .$$

The left-hand side is equal to $L_X T^{i_1 \ldots i_r}_{j_1 \ldots j_s}$. Inserting (11.4) and (11.5) yields

$$\begin{aligned} (L_X T)^{i_1 \ldots i_r}_{j_1 \ldots j_s} = X^i \, & T^{i_1 \ldots i_r}_{j_1 \ldots j_s, i} \\ & - T^{k i_2 \ldots i_r}_{j_1 \ldots j_s} \cdot X^{i_1}_{,k} - \text{all upper indices} \\ & + T^{i_1 \ldots i_r}_{k j_2 \ldots j_s} \cdot X^k_{,j_1} + \text{all lower indices} . \end{aligned} \tag{11.6}$$

In particular, if $\omega \in \mathcal{X}^*(M)$, then

$$(L_X \omega)_i = X^k \omega_{i,k} + \omega_k X^k_{,i} . \tag{11.7}$$

12 Differential Forms

Cartan's calculus of differential forms is particularly useful in general relativity (but also in other fields of physics). We begin our discussion by repeating some algebraic preliminaries.

12.1 Exterior Algebra

Let A be a commutative, associative, unitary $\mathbb{R}$-algebra and let E be an A-module. In the following we have either $A = \mathbb{R}$ and E a finite dimensional real vector space or $A = \mathcal{F}(M)$ and $E = \mathcal{X}(M)$ for some differentiable manifold M.

We consider the space $T^p(E)$ of A-valued, p-multilinear forms on E and the subspace $\bigwedge^p(E)$ of completely antisymmetric multilinear forms. Both spaces can naturally be turned into A-modules. In particular, $\bigwedge^0(E) := T^0(E) := A$ and $\bigwedge^1(E) = T^1(E) = E^*$ (dual module of E). The elements of $\bigwedge^p(E)$ are called **(exterior) forms of degree** p.

Definition 12.1. *We define the* ***alternation operator*** *on* $T^p(E)$ *by*

$$(\mathcal{A}T)(v_1, \dots, v_p) := \frac{1}{p!} \sum_{\sigma \in \mathcal{S}_p} (sgn\, \sigma) T\left(v_{\sigma(1)}, \dots, v_{\sigma(p)}\right) \quad \textit{for } T \in T^p(E)\,, \tag{12.1}$$

where the sum in (12.1) extends over the permutation group $\mathcal{S}_p$ *of* p *objects;* $sgn\,\sigma$ *denotes the signature of the permutation* $\sigma \in \mathcal{S}_p$.

The following statements obviously hold:

a) $\mathcal{A}$ is an A-linear mapping from $T^p(E)$ onto $\bigwedge^p(E)$: $\mathcal{A}(T^p(E)) = \bigwedge^p(E)$;
b) $\mathcal{A} \circ \mathcal{A} = \mathcal{A}$.

In particular, $\mathcal{A}\omega = \omega$ for any $\omega \in \bigwedge^p(E)$.

Definition 12.2. *Let* $\omega \in \bigwedge^p(E)$ *and* $\eta \in \bigwedge^q(E)$. *We define the* ***exterior product*** *by*

$$\omega \wedge \eta := \frac{(p+q)!}{p!\,q!} \mathcal{A}(\omega \otimes \eta)\,. \tag{12.2}$$

The exterior product has the following properties:

a) $\wedge$ is A-bilinear;
b) $\omega \wedge \eta = (-1)^{p \cdot q}\, \eta \wedge \omega$;
c) $\wedge$ is associative:

$$(\omega_1 \wedge \omega_2) \wedge \omega_3 = \omega_1 \wedge (\omega_2 \wedge \omega_3)\,. \tag{12.3}$$

In the finite dimensional case we have the

Theorem 12.1. *Let θ^i, for $i = 1, 2, \ldots, n < \infty$, be a basis for E^*. Then the set*

$$\theta^{i_1} \wedge \theta^{i_2} \wedge \ldots \wedge \theta^{i_p}\,, \quad 1 \leq i_1 < i_2 < \ldots < i_p \leq n \tag{12.4}$$

is a basis for the space $\bigwedge^p(E)$, $p \leq n$, which has the dimension

$$\binom{n}{p} := \frac{n!}{p!\,(n-p)!}\,, \tag{12.5}$$

if $p > n$, $\bigwedge^p(E) = \{0\}$.

The **Grassman algebra** (or **exterior algebra**) $\bigwedge(E)$ is defined as the direct sum

$$\bigwedge(E) := \bigoplus_{p=0}^{n} \bigwedge^p(E)\,,$$

where the exterior product is extended in a bilinear manner to the entire $\bigwedge(E)$. Thus $\bigwedge(E)$ is a graded, associative, unitary A-algebra.

Definition 12.3. *For every $p \in \mathbb{N}_0$ we define the mapping*

$$E \times \bigwedge^p(E) \longrightarrow \bigwedge^{p-1}(E)\,,$$
$$(v, \omega) \longmapsto i_v \omega\,,$$

where

$$(i_v\omega)(v_1, \ldots, v_{p-1}) = \omega(v, v_1, \ldots, v_{p-1})\,,$$
$$i_v\omega = 0 \quad \text{for} \quad \omega \in \bigwedge^0(E)\,. \tag{12.6}$$

The association $(v, \omega) \longmapsto i_v\omega$ is called the ***interior product*** *of v and ω.*

For every p, the interior product is an A-bilinear mapping and can thus be uniquely extended to an A-bilinear mapping

$$E \times \bigwedge(E) \longrightarrow \bigwedge(E)\,,$$
$$(v, \omega) \longmapsto i_v\omega\,.$$

For a proof of the following theorem and other unproved statements in this section see, for example, [36], Chap. III.

Theorem 12.2. *For every fixed $v \in E$, the mapping $i_v : \bigwedge(E) \longrightarrow \bigwedge(E)$ has the properties:*

(i) i_v is A-linear;
(ii) $i_v(\bigwedge^p(E)) \subseteq \bigwedge^{p-1}(E)$;
(iii) $i_v(\alpha \wedge \beta) = (i_v\alpha) \wedge \beta + (-1)^p \alpha \wedge i_v\beta$ for $\alpha \in \bigwedge^p(E)$.

12.2 Exterior Differential Forms

Let M be an n-dimensional C^∞-manifold. For every $p = 0, 1, \dots, n$ and every $x \in M$, we construct the spaces

$$\bigwedge^p(T_x(M)) \subset T_x(M)^0_p .$$

In particular,

$$\bigwedge^0(T_x(M)) = \mathbb{R},$$
$$\bigwedge^1(T_x(M)) = T^*_x(M) .$$

Furthermore,

$$\bigwedge(T_x(M)) = \bigoplus_{p=0}^{n} \bigwedge^p(T_x(M))$$

is the exterior algebra (of dimension 2^n) over $T_x(M)$.

Definition 12.4. *An **(exterior) differential form of degree** p or **differential p-form** on M is a differentiable tensor field of rank p with values in $\bigwedge^p(T_x(M))$ for every $x \in M$.*

All statements made in Sect. 10.3 about tensor fields are of course also valid for differential forms. We denote by $\bigwedge^p(M)$ the $\mathcal{F}(M)$-module of p-forms. In addition,

$$\bigwedge(M) = \bigoplus_{p=0}^{n} \bigwedge^p(M)$$

is the **exterior algebra of differential forms** on M. (Many books use instead the notation $\Omega^p(M)$, $\Omega(M)$.) In $\bigwedge(M)$ the algebraic operations, in particular the exterior product, are defined pointwise.

As in Sect. 10.3, we can assign to $\omega \in \bigwedge^p(M)$ and vector fields $X_1, \dots, X_p$ the function

$$F(x) := \omega_x\big(X_1(x), \dots, X_p(x)\big) .$$

This function will be denoted by $\omega(X_1, \dots, X_p)$. The assignment

$$(X_1, \dots, X_p) \longmapsto \omega(X_1, \dots, X_p)$$

is $\mathcal{F}(M)$-multilinear and completely antisymmetric. That is, for a given element $\omega \in \bigwedge^p(M)$, there is an element of the exterior algebra over the $\mathcal{F}(M)$-module $\mathcal{X}(M)$. We denote the latter by $\bigwedge(\mathcal{X}(M))$. One can show that this correspondence is an isomorphism which preserves all algebraic structures (see [36], p.137). In addition, we have for the interior product

$$(i_X\omega)(x) = i_{X(x)}\omega(x)\,, \qquad \text{for} \quad X \in \mathcal{X}(M), \omega \in \bigwedge(M), x \in M\,.$$

If $(x^1, \ldots, x^n)$ is a local coordinate system in $U \subset M$, then $\omega \in \bigwedge^p(M)$ can be expanded in the form

$$\begin{aligned}\omega &= \sum_{i_1<\ldots<i_p} \omega_{i_1\ldots i_p}\, dx^{i_1} \wedge dx^{i_2} \wedge \ldots \wedge dx^{i_p} \\ &= \frac{1}{p!} \sum_{i_1,\ldots,i_p} \omega_{i_1\ldots i_p}\, dx^{i_1} \wedge \ldots \wedge dx^{i_p} \end{aligned} \tag{12.7}$$

on U. Here $\{\omega_{i_1\ldots i_p}\}$ are the **components** of ω. In the second line of (12.7) they are totally antisymmetric in their indices.

Differential Forms and Mappings

Let $\phi : M \longrightarrow N$ be a morphism of manifolds. The induced mapping ϕ^* in the space of covariant tensors (see Sect. 11.2) defines a mapping

$$\phi^* : \bigwedge(N) \longrightarrow \bigwedge(M)$$

(we use the same symbol). Using Theorem 11.5, it is easy to see that

$$\phi^*(\omega \wedge \eta) = \phi^*\omega \wedge \phi^*\eta\,, \qquad \omega, \eta \in \bigwedge(N)\,.$$

Thus, ϕ^* is an $\mathbb{R}$-algebra homomorphism. If ϕ is a diffeomorphism, then ϕ^* is an $\mathbb{R}$-algebra isomorphism.

12.3 Derivations and Antiderivations

Let M be an n-dimensional differentiable manifold and

$$\bigwedge(M) = \bigoplus_{p=0}^{n} \bigwedge^p(M)$$

be the graded $\mathbb{R}$-algebra of exterior differential forms on M.

Definition 12.5. *A mapping $\theta : \bigwedge(M) \longrightarrow \bigwedge(M)$ is called* ***derivation (antiderivation) of degree*** $k \in \mathbb{Z}$, *provided*

(i) θ *is* $\mathbb{R}$*-linear;*
(ii) $\theta(\bigwedge^p(M)) \subset \bigwedge^{p+k}(M)$ *for* $p = 0, 1, \ldots, n$;
(iii)

$$\theta(\alpha \wedge \beta) = \theta\alpha \wedge \beta + \alpha \wedge \theta\beta \qquad \text{for } \alpha, \beta \in \bigwedge(M),$$

(Leibniz rule)

$$\theta(\alpha \wedge \beta) = \theta\alpha \wedge \beta + (-1)^p \alpha \wedge \theta\beta \qquad \text{for } \alpha \in \bigwedge^p(M), \beta \in \bigwedge(M).$$

("anti-Leibniz rule")

Theorem 12.3. *If* θ *and* θ' *are antiderivations of odd degree* k *and* k' *on* $\bigwedge(M)$, *then* $\theta \circ \theta' + \theta' \circ \theta$ *is a derivation of degree* $k + k'$.

Proof. Direct verification. □

Theorem 12.4. *Every derivation (antiderivation) on* $\bigwedge(M)$ *is local. This means that for every* $\alpha \in \bigwedge(M)$ *and every open submanifold* U *of* M, $\alpha|U = 0$ *implies* $\theta(\alpha)|U = 0$.

Proof. Let $x \in U$. There exists[1] a function $h \in \mathcal{F}(M)$, such that $h(x) = 1$ and $h = 0$ on $M \setminus U$. As a consequence, $h\alpha = 0$ and hence also $\theta(h\,\alpha) = 0$, so that $\theta(h) \wedge \alpha + h\,\theta(\alpha) = 0$. If we evaluate this equation at the point x, we obtain $\theta(\alpha)(x) = 0$. □

Corollary 12.1. *Let* α *and* $\alpha' \in \bigwedge(M)$. *If* $\alpha|U = \alpha'|U$, *then* $\theta(\alpha)|U = \theta(\alpha')|U$.

Thus we can define the restriction $\theta|U$ on $\bigwedge(U)$. For this, choose for $p \in U$ and $\alpha \in \bigwedge(U)$ an $\tilde{\alpha} \in \bigwedge(M)$ such that $\alpha = \tilde{\alpha}$ for some neighborhood of p, and set $(\theta|U)(\alpha)(p) = (\theta\tilde{\alpha})(p)$. According to the corollary, this definition is independent of the extension $\tilde{\alpha}$. The existence of an extension $\tilde{\alpha}$ is based on the following

Lemma 12.1. *Let* U *be an open submanifold of* M *and let* K *be a compact subset of* U. *For every* $\beta \in \bigwedge(U)$ *there is an* $\alpha \in \bigwedge(M)$ *such that* $\beta|K = \alpha|K$ *and* $\alpha|(M \setminus U) = 0$.

Proof. There exists a function $h \in \mathcal{F}(M)$ such that $h(x) = 1$ for all $x \in K$ and $h = 0$ on $M \setminus U$ (i.e., $\operatorname{supp} h \subseteq U$)[2]. With the aid of this function, define $\alpha \in \bigwedge(M)$ by

$$\alpha(x) = \begin{cases} h(x)\beta(x) & , x \in U \\ 0 & , x \notin U. \end{cases}$$

This differential form has the desired properties. □

---

[1] A proof of the existence of such a function can be found, for example, in [36], p.69.

[2] For a proof, see [36], p.92.

We thus have the following

Theorem 12.5. *Let θ be a derivation (antiderivation) on $\bigwedge(M)$ and let U be an open submanifold of M. There exists a unique derivation (antiderivation) θ_U on $\bigwedge(U)$ such that for every $\alpha \in \bigwedge(M)$,*

$$\theta_U(\alpha|U) = (\theta\alpha)|U\,.$$

θ_U is called the ***derivation (antiderivation) on $\bigwedge(U)$ induced by θ****.*

In addition to this localization theorem, we need a globalizing theorem.

Theorem 12.6. *Let $(U_i)_{i\in I}$ be an open covering of M. For every i, let θ_i be a derivation (antiderivation) on $\bigwedge(U_i)$; we denote by θ_{ij} the derivation (antiderivation) on $\bigwedge(U_i \cap U_j)$ induced by θ_i on $U_i \cap U_j$. For every $(i,j) \in I \times I$ let $\theta_{ij} = \theta_{ji}$. Then there exists precisely one derivation (antiderivation) θ on $\bigwedge(M)$ such that $\theta|U_i = \theta_i$ for every $i \in I$.*

Proof. For each $\alpha \in \bigwedge(M)$ and $i \in I$, define $(\theta\alpha)|U_i = \theta_i(\alpha|U_i)$. Since

$$((\theta\alpha)|U_i)|U_j = \theta_{ij}(\alpha|U_i \cap U_j) = \theta_{ji}(\alpha|U_i \cap U_j) = ((\theta\alpha)|U_j)|U_i\,,$$

for all $(i,j) \in I \times I$, $\theta\alpha \in \bigwedge(M)$ is well defined and $\theta : \alpha \longmapsto \theta\alpha$ is a derivation (antiderivation) which induces θ_i on every U_i by construction. □

Later on, the following theorem will also be useful:

Theorem 12.7. *Let θ be a derivation (antiderivation) of degree k on $\bigwedge(M)$. If $\theta f = 0$ and $\theta(df) = 0$ for every $f \in \mathcal{F}(M)$, then $\theta\alpha = 0$ for every $\alpha \in \bigwedge(M)$.*

Proof. Let $(h_i, U_i)_{i\in I}$ be an atlas of M and let $\theta_i = \theta|U_i$ (as in Theorem 12.5). Each θ_i satisfies the premises of the theorem, i.e., $\theta_i f_i = 0$ and $\theta_i\, df_i = 0$ for every $f_i \in \mathcal{F}(U_i)$. Let $\alpha \in \bigwedge^p(M)$ and $i \in I$. $\alpha|U_i$ has the form

$$\alpha|U_i = \sum_{i_1<\ldots<i_p} \alpha_{i_1\ldots i_p}\, dx^{i_1} \wedge \ldots \wedge dx^{i_p}\,.$$

As a consequence, since θ is a derivation,

$$(\theta\alpha)|U_i = \theta(\alpha|U_i) = \sum_{i_1<\ldots<i_p} \Big(\theta_i\alpha_{i_1\ldots i_p} \wedge dx^{i_1} \wedge \ldots \wedge dx^{i_p} + \\ + \alpha_{i_1\ldots i_p} \sum_{k=1}^{p} dx^{i_1} \wedge \ldots \wedge \theta_i\, dx^{i_k} \wedge \ldots \wedge dx^{i_p}\Big) = 0\,,$$

and similarly if θ is an antiderivation. □

Corollary 12.2. *A derivation (antiderivation) θ on $\bigwedge(M)$ is uniquely determined by its value on $\bigwedge^0(M) = \mathcal{F}(M)$ and on $\bigwedge^1(M)$. In fact, its values on $\mathcal{F}(M)$ and all differentials of $\mathcal{F}(M)$ are already sufficient.*

12.4 The Exterior Derivative

We are now ready to extend the differential of functions to a very important antiderivation of $\bigwedge(M)$.

Theorem 12.8. *There exists precisely one operator*

$$d : \bigwedge(M) \longrightarrow \bigwedge(M)$$

with the following properties:

(i) d is an antiderivation of degree 1 on $\bigwedge(M)$,
(ii) $d \circ d = 0$,
(iii) d is the differential of f for every $f \in \mathcal{F}(M)$, i.e. $\langle df, X\rangle = Xf$ for all $f \in \mathcal{F}(M)$, $X \in \mathcal{X}(M)$.

The operator d is called the ***exterior derivative***.

Proof.
a) The uniqueness of d is a consequence of Theorem 12.7.
b) Let $h : U \longrightarrow U'$ be a chart of M. According to Theorem 12.5, the exterior derivative (if it exists) induces an exterior derivative on U, which is also denoted by d. For

$$\alpha = \sum_{i_1<\ldots<i_p} \alpha_{i_1\ldots i_p}\, dx^{i_1} \wedge \ldots \wedge dx^{i_p} \in \bigwedge^p (U)$$

one has, according to (i) and (ii),

$$d\alpha = \sum_{i_1<\ldots<i_p} d\alpha_{i_1\ldots i_p} \wedge dx^{i_1} \wedge \ldots \wedge dx^{i_p} \in \bigwedge^{p+1} (U)$$

or, by (iii)

$$d\alpha = \sum_{i_0<i_1<\ldots<i_p} \sum_{k=0}^{p} (-1)^k \frac{\partial}{\partial x^{i_k}}\, \alpha_{i_0\ldots\hat{i}_k\ldots i_p}\, dx^{i_0} \wedge \ldots \wedge dx^{i_p}. \qquad (12.8)$$

In other words, we must have

$$(d\alpha)_{i_0\ldots i_p} = \sum_{k=0}^{p} (-1)^k \frac{\partial}{\partial x^{i_k}}\, \alpha_{i_0\ldots\hat{i}_k\ldots i_p} \qquad (i_0 < i_1 < \ldots < i_p). \qquad (12.9)$$

c) For every p, $0 \le p \le n$ and every $\alpha \in \bigwedge^p(U)$, define $d\alpha$ by the previous equation. By straightforward computation, one can show that $d : \bigwedge(U) \longrightarrow \bigwedge(U)$ has the three properties of the theorem (exercise). Thus d is an exterior derivative on U.

d) Let $(h_i, U_i)_{i\in I}$ be an atlas of M. For every $i \in I$, let $d_i : \bigwedge(U_i) \longrightarrow \bigwedge(U_i)$ be the exterior derivative on U_i. For $(i,j) \in I \times I$, d_i and d_j each induces an exterior derivative d_{ij} or d_{ji}, respectively, on $U_i \cap U_j$. According to *a)*, $d_{ij} = d_{ji}$. By Theorem 12.6, there exists a unique antiderivation d on $\bigwedge(M)$ which induces for every $i \in I$ the exterior derivative d_i on U_i. Obviously, $d \circ d = 0$ and df is the differential of f for every $f \in \mathcal{F}(M)$. □

Definition 12.6. *A differential form α such that $d\alpha = 0$ is called a* ***closed*** *form. A differential form α such that $\alpha = d\beta$ for some form β is called an* ***exact*** *form.*

Since $d \circ d = 0$, every exact differential form $\alpha = d\beta$ is closed, i.e., $d\alpha = d(d\beta) = 0$. The converse is valid locally:

Lemma 12.2 (Poincaré's Lemma). *Let α be a closed form on M. For every $x \in M$, there is an open neighborhood U of x such that $\alpha|U$ is exact.*

Proof. This important fact will be proven in Sect. 14.1. □

Morphisms and Exterior Derivatives

Theorem 12.9. *Let $\phi : M \longrightarrow N$ be a morphism of manifolds. The following diagram is commutative:*

$$\begin{array}{ccc} \bigwedge(M) & \xleftarrow{\ \phi^*\ } & \bigwedge(N) \\ \downarrow d & & \downarrow d \\ \bigwedge(M) & \xleftarrow[\ \phi^*\ ]{} & \bigwedge(N) \end{array}$$

so that

$$d \circ \phi^* = \phi^* \circ d\,. \tag{12.10}$$

Proof. This has already been shown for functions in Sect. 10.3 and hence we have

$$(d \circ \phi^*)df = d \circ d(\phi^* f) = 0 = \phi^* \circ d(df)\,.$$

In other words, (12.10) is also true for differentials of functions. In the special case $M = N$ both sides of (12.10) are antiderivations on $\bigwedge(M)$. By Theorem 12.7, they must be equal. For the general case we first note that it suffices to prove (12.10) locally: $d(\phi^*\omega|U) = (\phi^* d\omega)|U$ $(U \subset M, \omega \in \bigwedge(N))$, since d and ϕ^* are *natural* relative to restrictions (e.g., $d(\alpha|U) = (d\alpha)|U$). But the local statement can easily be verified by a direct calculation (exercise). □

12.5 Relations Among the Operators d, i_X and L_X

There are several important relations between the derivation L_X, the antiderivation i_X and the exterior derivative d. The most useful of these is **Cartan's equation**

$$L_X = d \circ i_X + i_X \circ d \,. \tag{12.11}$$

Proof.

a) According to Theorem 12.3 the operation

$$\theta = d \circ i_X + i_X \circ d : \bigwedge(M) \longrightarrow \bigwedge(M)$$

is a derivation of degree 0. For $f \in \mathcal{F}(M)$, one has

$$\theta f = d(i_X f) + i_X(df) = i_X df = \langle df, X \rangle = Xf$$

and

$$\theta\, df = d(i_X df) + i_X d(df) = d(i_X df) = d(Xf)\,.$$

b) The Lie derivative with respect to X, $L_X : \bigwedge(M) \longrightarrow \bigwedge(M)$ is a derivation of degree 0 and one has (see Sect. 11.3)

$$\begin{aligned} L_X f &= Xf\,, \\ (L_X df)(Y) &= L_X(df(Y)) - df(L_X Y) \\ &= L_X(Yf) - df([X,Y]) \\ &= X(Yf) - [X,Y]f \\ &= YXf = \langle d(Xf), Y \rangle\,, \end{aligned}$$

which shows that $L_X df = d(Xf)$.

c) From *a)* and *b)* we obtain, using Theorem 12.7, $L_X = \theta$. □

Corollary 12.3. *It follows from (12.11) that*

$$d \circ L_X = L_X \circ d\,. \tag{12.12}$$

Exercise 12.1. In a similar manner, prove that

$$i_{[X,Y]} = [L_X, i_Y] \quad \text{for} \quad X, Y \in \mathcal{X}(M)\,. \tag{12.13}$$

Formula for the Exterior Derivative

As an application of (12.11) we prove the following formula for the exterior derivative: if $\omega \in \bigwedge^p(M)$, then

$$\begin{aligned} d\omega(X_1,\dots,X_{p+1}) = &\sum_{1\le i\le p+1} (-1)^{i+1} X_i\, \omega\big(X_1,\dots,\hat{X}_i,\dots,X_{p+1}\big) \\ &+ \sum_{i<j} (-1)^{i+j}\, \omega\big([X_i,X_j], X_1,\dots,\hat{X}_i,\dots,\hat{X}_j,\dots,X_{p+1}\big)\,. \end{aligned} \tag{12.14}$$

Proof.

a) If $p = 0$, (12.14) states that $df(X) = Xf$, for $f \in \mathcal{F}(M)$, which is obviously correct.

b) If $\omega \in \bigwedge^1(M)$, we obtain from (12.11)

$$\begin{aligned}(L_X\omega)(Y) &= i_X d\omega(Y) + d(i_X\omega)(Y)\\ &= i_X d\omega(Y) + d\,(\omega(X))\,(Y)\\ &= d\omega(X,Y) + Y\omega(X).\end{aligned}$$

Using (11.1) for $L_X\omega$, we find

$$d\omega(X,Y) = (L_X\omega)(Y) - Y\omega(X) = X\omega(Y) - Y\omega(X) - \omega([X,Y]),$$

which agrees with the right hand site of (12.14).

b) We now complete the proof by induction. Assume that (12.14) is correct for all differential forms having degree less than or equal to $p-1$. If $\omega \in \bigwedge^p(M)$, use (12.11), the induction assumption for $i_X\omega$ and the explicit expression (11.1) for L_X. After a short calculation, one shows that (12.14) is also true for differential forms of degree p. (The details are left to the reader.) □

12.6 The $*$-Operation and the Codifferential

12.6.1 Oriented Manifolds

Definition 12.7. *An atlas $\mathcal{A}$ of a differentiable manifold M is* ***oriented*** *if for every pair of charts (h,U) and (k,V) in $\mathcal{A}$ the Jacobi determinant for the coordinate transformation $k \circ h^{-1}$ is positive.*

There are manifolds (for example, the Möbius strip) which do not have an oriented atlas.

Definition 12.8. *A manifold is* ***orientable*** *if it has an oriented atlas. Two atlases $\mathcal{A}_1$ and $\mathcal{A}_2$ have the same orientation if $\mathcal{A}_1 \cup \mathcal{A}_2$ is also an oriented atlas. This is an equivalence relation. An equivalence class of oriented atlases is called an* ***orientation of the manifold*** *M. An orientable manifold, together with a chosen orientation, is said to be an* ***oriented manifold****.*

A chart (h,U) of M is said to be positive (negative) if for every chart (k,V) belonging to some oriented atlas which defines the orientation of M, the Jacobi determinant of $k \circ h^{-1}$ is positive (negative). We state without proof

Theorem 12.10. *Let M be an n-dimensional, paracompact, orientable manifold. There exists on M an n-form of class C^∞ which does not vanish anywhere on M. This n-form is called a* ***volume form*** *of M. The converse is also true.*

Proof. See [36], p.258. □

We now consider a pseudo-Riemannian manifold (M, g). Let g_{ij} be the components of g relative to the coordinate system $(x^1, \ldots, x^n)$. We set

$$|g| := |\det(g_{ij})| . \tag{12.15}$$

Let $\bar{g}_{ij}$ be the components of g in terms of new coordinates $(y^1, \ldots, y^n)$ and let $|\bar{g}|$ be the absolute value of the corresponding determinant. From transformation law

$$\bar{g}_{ij} = \frac{\partial x^k}{\partial y^i} \frac{\partial x^l}{\partial y^j} g_{kl} ,$$

it follows that

$$|\bar{g}| = \left[\det\left(\frac{\partial x^k}{\partial y^i}\right)\right]^2 |g| .$$

Now suppose that M is oriented and that both coordinate systems are *positive*. We then have $\det(\partial x^k / \partial y^i) > 0$ and

$$\sqrt{|\bar{g}|} = \sqrt{|g|} \det\left(\frac{\partial x^k}{\partial y^i}\right) . \tag{12.16}$$

If, on the other hand, we consider an n-form ω with

$$\omega = a(x)\, dx^1 \wedge \ldots \wedge dx^n = b(y)\, dy^1 \wedge \ldots \wedge dy^n ,$$

then

$$a = b \det\left(\frac{\partial y^i}{\partial x^j}\right) . \tag{12.17}$$

Hence an n-form is defined on M by

$$\eta := \sqrt{|g|}\, dx^1 \wedge dx^2 \wedge \ldots \wedge dx^n \tag{12.18}$$

for a positive coordinate system. According to (12.16) and (12.17), this definition is independent of the coordinate system. In addition, η is a volume form. We call η the **canonical** or (**Riemannian**) **volume form** on (M, g). It plays an important role.

12.6.2 The $*$-Operation

Let (M, g) be an n-dimensional oriented pseudo-Riemannian manifold and let $\eta \in \bigwedge^n(M)$ be the canonical volume form corresponding to g. We shall now use η to associate with each form $\omega \in \bigwedge^p(M)$ another form $*\omega \in \bigwedge^{n-p}(M)$. We first do this with the help of coordinates. So consider a positive local coordinate system $(x^1, \ldots, x^n)$ and write ω in the form

$$\omega = \frac{1}{p!}\, \omega_{i_1 \ldots i_p}\, dx^{i_1} \wedge \ldots \wedge dx^{i_p} . \tag{12.19}$$

(Note that $\omega_{i_1 \dots i_p}$ is totally antisymmetric.) In the same manner, we may write

$$\eta = \frac{1}{n!}\, \eta_{i_1 \dots i_n}\, dx^{i_1} \wedge \dots \wedge dx^{i_n} . \tag{12.20}$$

From (12.18) we have

$$\eta_{i_1 \dots i_n} = \sqrt{|g|}\, \varepsilon_{i_1 \dots i_n} , \tag{12.21}$$

where

$$\varepsilon_{i_1 \dots i_n} = \begin{cases} +1 & \text{if } i_1, \dots, i_n \text{ is an even permutation of } (1, \dots, n) \\ -1 & \text{if } i_1, \dots, i_n \text{ is an odd permutation of } (1, \dots, n) \\ 0 & \text{otherwise} . \end{cases}$$

Definition 12.9. *We define the* ***Hodge star operator (*-operation)*** *by*

$$(*\omega)_{i_{p+1} \dots i_n} = \frac{1}{p!}\, \eta_{i_1 \dots i_n}\, \omega^{i_1 \dots i_p} , \tag{12.22}$$

where $\omega^{i_1 \dots i_p} = g^{i_1 j_1} \dots g^{i_p j_p}\, \omega_{j_1 \dots j_p}$.

The Definition 12.9 is obviously independent of the coordinate system. In the Exercise 12.2 below, we derive a coordinate-free expression for the $*$-operation. Equation (12.22) is equivalent to

$$*\omega = \frac{1}{p!}\, \omega^{i_1 \dots i_p}\, i(e_{i_p}) \circ \dots \circ i(e_{i_1})\, \eta , \tag{12.23}$$

where $\{e_i\}$ is a local basis of vector fields and $i(e_j)$ denotes the interior products. The correspondence $\omega \longmapsto *\omega$ defines an isomorphism $\bigwedge^p(M) \longrightarrow \bigwedge^{n-p}(M)$. A simple calculation shows that

$$*(*\omega) = (-1)^{p(n-p)}\, sgn(g)\, \omega , \tag{12.24}$$

where $sgn(g)$ is the signature of the metric (e.g., (-1) for a Lorentzian metric). The contravariant components of η are given by

$$\eta^{i_1 \dots i_n} = sgn(g)\, \frac{1}{\sqrt{|g|}}\, \varepsilon_{i_1 \dots i_n} . \tag{12.25}$$

The following exercises (with solutions) contain useful properties of the $*$-operation.

* * *

Exercises.

Exercise 12.2. Let α, $\beta \in \bigwedge^p(M)$. Show that

$$\alpha \wedge *\beta = \beta \wedge *\alpha = \langle \alpha, \beta \rangle \eta , \tag{12.26}$$

where

$$\langle \alpha, \beta \rangle = \frac{1}{p!}\, \alpha_{i_1 \dots i_p}\, \beta^{i_1 \dots i_p} \tag{12.27}$$

is the **scalar product** induced in $\bigwedge^p(M)$.

Solution. Since $i(e_j)$ is an antiderivation, we can write

$$\begin{aligned}\alpha \wedge *\beta &= \frac{1}{p!}\,\beta^{i_1\dots i_p}\alpha \wedge \left(i(e_{i_p})\circ\dots\circ i(e_{i_1})\eta\right)\\ &= \frac{1}{p!}\,\beta^{i_1\dots i_p}\left(i(e_{i_1})\alpha\right) \wedge \left(i(e_{i_p})\circ\dots\circ i(e_{i_2})\eta\right)\\ &= \frac{1}{p!}\,\beta^{i_1\dots i_p}\left(i(e_{i_p})\circ\dots\circ i(e_{i_1})\alpha\right)\eta\\ &= \frac{1}{p!}\,\beta^{i_1\dots i_p}\,\alpha_{i_1\dots i_p}\,\eta\,.\end{aligned}$$

Exercise 12.3. Let $\alpha,\,\beta \in \bigwedge^p(M)$. Show that

$$\langle *\alpha, *\beta\rangle = sgn(g)\langle\alpha,\beta\rangle\,. \tag{12.28}$$

Solution. From Exercise 12.2 and (12.24), we find

$$\begin{aligned}\langle *\alpha, *\beta\rangle\eta &= *\alpha \wedge **\,\beta = (-1)^{p(n-p)}sgn(g) * \alpha\wedge\beta\\ &= sgn(g)\beta\wedge *\alpha = sgn(g)\langle\alpha,\beta\rangle\eta\,.\end{aligned}$$

Exercise 12.4. Let $\alpha \in \bigwedge^p(M)$ and $\beta \in \bigwedge^{n-p}(M)$. Show that

$$\langle\alpha\wedge\beta,\eta\rangle = \langle *\alpha,\beta\rangle\,. \tag{12.29}$$

Solution. There is a unique p-form ϕ such that $\beta = *\phi$. The results of the previous exercises imply

$$\langle\alpha\wedge\beta,\eta\rangle = \langle\alpha\wedge *\phi,\eta\rangle = sgn(g)\langle\alpha,\phi\rangle = \langle *\alpha, *\phi\rangle = \langle *\alpha,\beta\rangle\,.$$

Exercise 12.5. If β' is an $(n-p)$-form, and if for all $\alpha \in \bigwedge^{n-p}(M)$ there is a $\beta \in \bigwedge^p(M)$ such that

$$\langle\beta',\alpha\rangle = \langle\beta\wedge\alpha,\eta\rangle\,,$$

then $\beta' = *\beta$.

Solution. According to the hypothesis and Exercise 12.4, we have $\langle *\beta - \beta',\alpha\rangle = 0$ for all $\alpha \in \bigwedge^{n-p}(M)$. Hence $\beta' = *\beta$, since the scalar product is nondegenerate.

Exercise 12.6. Let $\theta^1,\dots,\theta^n$ be an oriented basis of one-forms on an open subset $U \subset M$. Show that η can be represented in the form

$$\eta = \sqrt{|g|}\,\theta^1\wedge\dots\wedge\theta^n\,, \tag{12.30}$$

where $|g| = |\det g(e_i,e_k)|$ and $\{e_i\}$ the dual basis of $\{\theta^i\}$.

Solution. We set $\theta^i = a^i_j\,dx^j$, where $\{x^j\}$ defines a positive coordinate system. Then

$$\begin{aligned}\sqrt{|g|}\,\theta^1\wedge\ldots\wedge\theta^n &= \sqrt{|g|}\,a^1_{j_1}\ldots a^n_{j_n}\,dx^{j_1}\wedge\ldots\wedge dx^{j_n}\\ &= \sqrt{|g|}\,\varepsilon_{j_1\ldots j_n}a^1_{j_1}\ldots a^n_{j_n}\,dx^1\wedge dx^2\wedge\ldots\wedge dx^n\\ &= \sqrt{|g|}\,\det(a^i_j)\,dx^1\wedge\ldots\wedge dx^n\,;\end{aligned}$$

note that $\det(a^i_j)>0$. Now

$$\frac{\partial}{\partial x^j} = a^i_j\,e_i\,, \qquad \left\langle\frac{\partial}{\partial x^i},\frac{\partial}{\partial x^j}\right\rangle = a^k_i a^l_j\langle e_k,e_l\rangle = a^k_i a^l_j\,g_{kl}\,.$$

Therefore,

$$\left|\det\left\langle\frac{\partial}{\partial x^i},\frac{\partial}{\partial x^j}\right\rangle\right|^{1/2} = \sqrt{|g|}\,\det(a^i_j)\,.$$

If we insert this above, we obtain (12.30).

Remark. Observe also that $g^{ij}=\langle\theta^i,\theta^j\rangle$ is the matrix inverse to $g_{ij}=\langle e_i,e_j\rangle$.

Exercise 12.7. Prove that, with the notation from Exercise 12.6,

$$*(\theta^{i_1}\wedge\ldots\wedge\theta^{i_p}) = \frac{\sqrt{|g|}}{(n-p)!}\,\varepsilon_{j_1\ldots j_n}\,g^{j_1i_1}\ldots g^{j_pi_p}\,\theta^{j_{p+1}}\wedge\ldots\wedge\theta^{j_n}\,. \tag{12.31}$$

Solution. According to the result of Exercise 12.5, it suffices to show that

$$\begin{aligned}&\frac{\sqrt{|g|}}{(n-p)!}\,\varepsilon_{\,j_1\ldots j_n}\,g^{j_1i_1}\ldots g^{j_pi_p}\left\langle\theta^{j_{p+1}}\wedge\ldots\wedge\theta^{j_n},\theta^{i_{p+1}}\wedge\ldots\wedge\theta^{i_n}\right\rangle\\ &\quad=\left\langle\theta^{i_1}\wedge\ldots\wedge\theta^{i_p}\wedge\theta^{i_{p+1}}\wedge\ldots\wedge\theta^{i_n},\eta\right\rangle.\end{aligned}$$

Now, according to Exercise 12.6, the right-hand side is equal to

$$\begin{aligned}&\sqrt{|g|}\left\langle\theta^{i_1}\wedge\ldots\wedge\theta^{i_n},\theta^1\wedge\ldots\wedge\theta^n\right\rangle\\ &\quad=\frac{1}{\sqrt{|g|}}\,\varepsilon_{i_1\ldots i_n}\langle\eta,\eta\rangle = sgn(g)\,\frac{1}{\sqrt{|g|}}\,\varepsilon_{i_1\ldots i_n}\,.\end{aligned}$$

On the other hand, the left-hand side is equal to

$$\sqrt{|g|}\,\varepsilon_{j_1\ldots j_n}\,g^{j_1i_1}\ldots g^{j_ni_n} = sgn(g)\,\frac{1}{\sqrt{|g|}}\,\varepsilon_{i_1\ldots i_n}\,,$$

so that both sides agree.

Exercise 12.8. Let α be a p-form, X a vector field and $X^\flat$ its associated 1-form defined by $X^\flat(Y)=g(X,Y)$ for all $Y\in\mathcal{X}(M)$. Derive the identity

$$i_X*\alpha = *(\alpha\wedge X^\flat)\,. \tag{12.32}$$

Solution. Apply i_X to (12.23) to get $(X = X^j\, e_j)$

$$i_X * \alpha = \frac{1}{p!}\, \alpha^{i_1 \dots i_p}\, X^j\, i(e_j) \circ i(e_{i_p}) \circ \dots \circ i(e_{i_1})\, \eta$$
$$= \frac{1}{p!}\, \alpha^{[i_1 \dots i_p}\, X^{j]}\, i(e_j) \circ i(e_{i_p}) \circ \dots \circ i(e_{i_1})\, \eta \,,$$

where the square bracket denotes antisymmetrization. But

$$\frac{1}{p!}\, \alpha^{[i_1 \dots i_p}\, X^{j]} = \frac{1}{(p+1)!} \left(\alpha \wedge X^\flat\right)^{i_1 \dots i_p j}$$

(check the combinatorial coefficient on the right). Therefore using once more (12.23) we obtain

$$i_X * \alpha = \frac{1}{(p+1)!} \left(\alpha \wedge X^\flat\right)^{i_1 \dots i_p j} \quad i(e_j) \circ i(e_{i_p}) \circ \dots \circ i(e_{i_1})\, \eta$$
$$= *(\alpha \wedge X^\flat).$$

12.6.3 The Codifferential

We now introduce an important differential operator which generalizes the divergence to differential forms.

Definition 12.10. *The* ***codifferential*** $\delta : \bigwedge^p(M) \longrightarrow \bigwedge^{p-1}(M)$ *is defined by*[3]

$$\delta := sgn(g)(-1)^{np+n} * d * \,. \tag{12.33}$$

Since $d \circ d = 0$, we also have

$$\delta \circ \delta = 0\,. \tag{12.34}$$

From (12.24) we see that the equation $\delta\omega = 0$ is equivalent to $d * \omega = 0$. Therefore, by the Lemma of Poincaré 12.2, there exists locally a form ϕ such that $*\omega = d\phi$. Hence, $\omega = \pm * d\phi = \delta\psi$, where $\psi = \pm * \phi$. Thus, $\delta\omega = 0$ implies the local existence of a form ψ with $\omega = \delta\psi$.

Coordinate Expression for the Codifferential

For $\omega \in \bigwedge^p(M)$ the coordinate expression for $\delta\omega$ is given by

$$(\delta\omega)^{i_1 \dots i_{p-1}} = \frac{1}{\sqrt{|g|}} \left(\sqrt{|g|}\, \omega^{k i_1 \dots i_{p-1}}\right)_{,k} \,. \tag{12.35}$$

---

[3] Many authors use the negative of (12.33). One of the reasons for our choice is Equation (12.35).

Proof. We use

$$(*d*\omega)^{k_1\dots k_{p-1}} = \frac{1}{(q+1)!}\eta^{j_1\dots j_{q+1}k_1\dots k_{p-1}}(d*\omega)_{j_1\dots j_{q+1}}, \tag{12.36}$$

where $q = n - p$. The right-hand side of (12.36) is a consequence of (12.22). For any form $\alpha \in \bigwedge^s(M)$, we have

$$\begin{aligned}\alpha &= \frac{1}{s!}\,\alpha_{i_1\dots i_s}\,dx^{i_1}\wedge\dots\wedge dx^{i_s}\,,\\ d\alpha &= \frac{1}{s!}\,\alpha_{i_1\dots i_s,i_{s+1}}dx^{i_s+1}\wedge dx^{i_1}\wedge\dots\wedge dx^{i_s}\\ &= \frac{1}{(s+1)!}(s+1)(-1)^s\alpha_{[i_1\dots i_s,i_{s+1}]}\,dx^{i_1}\wedge\dots\wedge dx^{i_{s+1}}\,,\end{aligned}$$

so that

$$(d\alpha)_{i_1\dots i_{s+1}} = (-1)^s(s+1)\,\alpha_{[i_1\dots i_s,i_{s+1}]}\,. \tag{12.37}$$

Using this in (12.36) gives

$$(*d*\omega)^{k_1\dots k_{p-1}} = \frac{1}{(q+1)!}\,\eta^{j_1\dots j_{q+1}k_1\dots k_{p-1}}(-1)^q(q+1)\,(*\omega)_{[j_1\dots j_q,j_{q+1}]}\,. \tag{12.38}$$

We can ignore the antisymmetrization. If we once more use (12.22), we obtain

$$\begin{aligned}(*d*\omega)^{k_1\dots k_{p-1}} &= (-1)^q\frac{1}{q!}\frac{1}{p!}\left(\eta_{i_1\dots i_p j_1\dots j_q}\,\omega^{i_1\dots i_p}\right)_{,j_{q+1}}\eta^{j_1\dots j_{q+1}k_1\dots k_{p-1}}\\ &= (-1)^q\frac{1}{p!\,q!}sgn(g)\frac{1}{\sqrt{|g|}}\left(\sqrt{|g|}\omega^{i_1\dots i_p}\right)_{,j_{q+1}}\\ &\quad\cdot\underbrace{\varepsilon_{j_1\dots j_{q+1}k_1\dots k_{p-1}}}_{(-1)^{p-1}\varepsilon_{j_1\dots j_q k_1\dots k_{p-1}j_{q+1}}}\cdot\underbrace{\varepsilon_{i_1\dots i_p j_1\dots j_q}}_{(-1)^{pq}\varepsilon_{j_1\dots j_q i_1\dots i_p}}\,.\end{aligned} \tag{12.39}$$

Now, we have

$$\varepsilon_{j_1\dots j_q k_1\dots k_p}\,\varepsilon_{j_1\dots j_q i_1\dots i_p} = p!\,q!\,\delta^{k_1}_{[i_1}\delta^{k_2}_{i_2}\dots\delta^{k_p}_{i_p]}\,. \tag{12.40}$$

Using (12.40) in (12.39) gives

$$(*d*\omega)^{k_1\dots k_{p-1}} = (-1)^q sgn(g)(-1)^{p-1}(-1)^{pq}\frac{1}{\sqrt{|g|}}\left(\sqrt{|g|}\omega^{k_1\dots k_p}\right)_{,k_p}. \tag{12.41}$$

By definition,

$$\delta\omega = sgn(g)(-1)^{np+n}*d*\omega\,,$$

and hence

$$\delta\omega^{k_1\dots k_{p-1}} = (-1)^{np+n}(-1)^q(-1)^{p-1}(-1)^{pq}\frac{1}{\sqrt{|g|}}\left(\sqrt{|g|}\omega^{k_1\dots k_p}\right)_{,k_p}\,.$$

Equation (12.35) follows from this. □

* * *

Exercises.

Exercise 12.9. Show that the $*$-operation commutes with the pull-back by an orientation preserving isometry $\phi : M \longrightarrow M$, $\phi^* g = g$. Conclude that the same is true for the codifferential. What happens if the orientation is reversed?

Exercise 12.10. Discuss the invariance properties under isometries of an equation of the form

$$\delta F = J\,,$$

where F is a 2-form and J a 1-form. Specialize the discussion to Minkowski spacetime.

12.7 The Integral Theorems of Stokes and Gauss

12.7.1 Integration of Differential Forms

Let M be an oriented, n-dimensional differentiable manifold. We wish to define integrals of the type

$$\int_M \omega\,, \qquad \omega \in \bigwedge^n(M)\,.$$

We consider first the case where the support of ω is contained in the domain U of a chart.

Let $(x^1, \ldots, x^n)$ be positive coordinates in this region (in the following, we shall take all coordinates to be positive). If $\omega = f\, dx^1 \wedge \ldots \wedge dx^n$, $f \in \mathcal{F}(U)$, we define

$$\int_M \omega = \int_U f(x^1, \ldots, x^n)\, dx^1 \ldots dx^n\,, \tag{12.42}$$

where the right-hand side is an ordinary n-fold Lebesgue (Riemann) integral.

This definition makes sense, for if the support ω is contained in the domain of a second chart with coordinates $(y^1, \ldots, y^n)$ and if $\omega = g\, dy^1 \wedge \ldots \wedge dy^n$, then (since both coordinate systems are positive)

$$\int_{\phi(U)} g\, dy^1 \ldots dy^n = \int_U g \circ \phi \det\left(\frac{\partial y^i}{\partial x^k}\right) dx^1 \ldots dx^n\,,$$

where ϕ denotes the change of coordinates. But by (12.17)

$$f(x^1, \ldots, x^n) = (g \circ \phi)(x^1, \ldots, x^n) \det\left(\frac{\partial y^i}{\partial x^k}\right).$$

Hence, the integral (12.42) is independent of the choice of the coordinate system.

We now consider an arbitrary $\omega \in \bigwedge^n(M)$ with compact support. Let $(U_i, g_i)_{i\in I}$ be a C^∞ atlas of M, such that $(U_i)_{i\in I}$ is a locally finite covering of M (i.e. for every $x \in M$ there exists a neighborhood $U \subset M$ such that $U \cap U_i$ is nonempty only for a *finite* number of $i \in I$).

Definition 12.11. *Let $(h_i)_{i\in I}$ be a* ***corresponding partition of unity****. This is a family of differentiable functions $(h_i)_{i\in I}$ on M with the following properties:*

(i) $h_i(x) \geq 0$ for all $x \in M$;
(ii) $\operatorname{supp} h_i \subseteq U_i$;
(iii) $\sum_{i\in I} h_i(x) = 1$ for all $x \in M$.

Note that because of (ii) and the local finiteness of the covering, only a finite number of terms contribute to the sum in (iii).

Remark. By definition (see Chap. 9), M has a countable basis. One can show that M then has a locally finite covering and also that for every such covering, a corresponding partition of unity exists. (For proofs, see Sects. II.14, 15 of [36]).

Since the support of ω is compact and $(U_i)_{i\in I}$ is a locally finite covering, the number of U_i with non-empty $U_i \cap \operatorname{supp} \omega$ is finite. For $h_i\,\omega \neq 0$, we have $\operatorname{supp} h_i\,\omega \subset U_i$. Since also $\sum_{i\in I} h_i = 1$, we have

$$\omega = \sum_{i\in I} h_i\,\omega\,.$$

Thus ω is a *finite sum* of n-forms, each of which has its support in a single chart U_i. We then define

$$\int_M \omega = \sum_{i\in I} \int_M h_i\,\omega\,. \tag{12.43}$$

This definition is independent of the partition of unity. Indeed, let $(V_j, f_j)_{j\in J}$ be another C^∞ atlas of M, such that $(V_j)_{j\in L}$ is a locally finite covering of M and if $(k_j)_{j\in J}$ is a partition of unity corresponding to this covering, then $\{U_i \cap V_j\}_{(i,j)\in I\times J}$ is also a C^∞ atlas of M, such that $(U_i \cap V_j)$ is a locally finite covering and $(h_i \cdot k_j)$ is a corresponding partition of unity. Now

$$\sum_{i\in I} \int_M h_i\,\omega = \sum_{(i,j)\in I\times J} \int_M k_j h_i\,\omega \quad \text{and} \quad \sum_{j\in J} \int_M k_j\,\omega = \sum_{(i,j)\in I\times J} \int_M h_i k_j\,\omega\,,$$

which shows that (12.43) is independent of the partition of unity.

The integral has the following properties: Let $\bigwedge_c^n(M)$ be the set of n-forms having compact support. Then

a)
$$\int_M (a_1\omega_1 + a_2\omega_2) = a_1 \int_M \omega_1 + a_2 \int_M \omega_2 \,,$$
with $a_1, a_2 \in \mathbb{R}$ and $\omega_1, \omega_2 \in \bigwedge_c^n(M)$.

b) If $\phi : M \longrightarrow N$ is an orientation preserving diffeomorphism, then
$$\int_M \phi^*\omega = \int_N \omega \,, \qquad \omega \in \bigwedge\nolimits_c^n(M) \,.$$

c) If the orientation O of M is changed, the integral changes sign:
$$\int_{(M,-O)} \omega = -\int_{(M,O)} \omega \,.$$

12.7.2 Stokes' Theorem

Let D be a region (i.e. an open connected subset) in an n-dimensional differentiable manifold M. One says that the region D has a **smooth boundary**, provided that for every $p \in \partial D$ (point on the boundary of D) there exist an open neighborhood U of p in M and a local coordinate system $(x^1, \ldots, x^n)$ on U such that (see Fig. 12.1)
$$U \cap \bar{D} = \{q \in U \mid x^n(q) \geq x^n(p)\} \,. \tag{12.44}$$

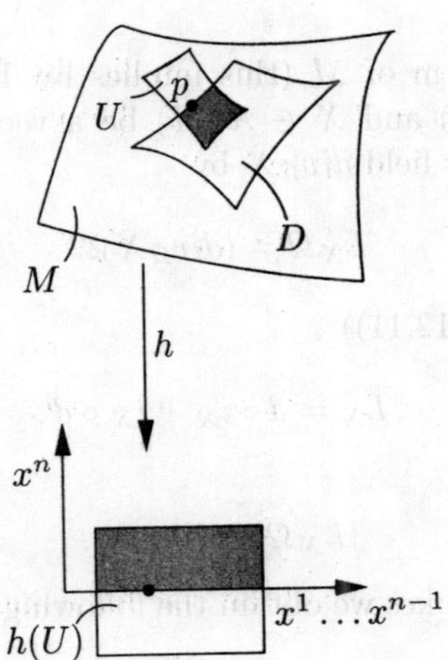

Fig. 12.1. Smooth boundary of a region.

One can show that ∂D is a closed, $(n-1)$-dimensional submanifold of M (see Sect. V.5 of [36]). If M is orientable, then ∂D is also orientable.

If $(x^1, \dots, x^n)$ is a positive coordinate system which satisfies (12.44), then we may choose an orientation of ∂D such that $(x^1, \dots, x^{n-1})$ is a positive coordinate system of ∂D. However, the following convention is more suitable: if n is even, then $(x^1, \dots, x^{n-1})$ is positive, while if n is odd, $(x^1, \dots, x^{n-1})$ is negative. Such an orientation of ∂D is called the **induced orientation** of ∂D.

If $\bar{D}$ is compact and $\omega \in \bigwedge^n(M)$, we define

$$\int_D \omega = \int_M \chi_D \, \omega \,,$$

where χ_D is the characteristic function of D. Now we are ready to formulate one of the key theorems of mathematics.

Theorem 12.11 (Stokes' Theorem). *Let M be an n-dimensional, oriented differentiable manifold (with denumerable basis) and let D be a region of M having a smooth boundary and such that $\bar{D}$ is compact. For every form $\omega \in \bigwedge^{n-1}(M)$ we have*

$$\int_D d\omega = \int_{\partial D} \omega \,. \tag{12.45}$$

More precisely, $\int_D d\omega = \int_{\partial D} \iota^*(\omega)$, where $\iota : \partial D \hookrightarrow M$ is the canonical injection. In (12.45), the induced orientation of ∂D has been chosen. The proof is standard (see, e.g., Sect. V.5 of [36]).

Application

Let Ω be a volume form of M (this implies by Theorem 12.10 that M is an orientable manifold) and $X \in \mathcal{X}(M)$ be a vector field. We define the **divergence** of a vector field $div_\Omega X$ by

$$L_X \Omega = (div_\Omega X) \Omega \,. \tag{12.46}$$

Cartan's formula (see (12.11))

$$L_X = d \circ i_X + i_X \circ d \,,$$

gives

$$L_X \Omega = d(i_X \Omega) \,. \tag{12.47}$$

From the theorem of Stokes we obtain the following form of Gauss' Theorem

$$\int_D (div_\Omega X) \Omega = \int_{\partial D} i_X \Omega \,. \tag{12.48}$$

$i_X \Omega$ defines a measure $d\mu_X$ on ∂D; $d\mu_X$ depends linearly on X. We may thus write $d\mu_X = \langle X, d\sigma \rangle$, where $d\sigma$ is a measure-valued one-form.

Theorem 12.12 (Gauss' Theorem). *With these notations, we have*

$$\int_D (div_\Omega X)\Omega = \int_{\partial D} \langle X, d\sigma\rangle\,. \qquad (12.49)$$

In particular, these equations are valid for pseudo-Riemannian manifolds with the standard volume form (in local coordinates)

$$\Omega = \eta = \sqrt{|g|}\, dx^1 \wedge \ldots \wedge dx^n\,.$$

Expression for $div_\Omega X$ in Local Coordinates

Let $\Omega = a(x)\, dx^1 \wedge \ldots \wedge dx^n$, $X = X^i\, \partial/\partial x^i$. Then

$$\begin{aligned} L_X \Omega &= (Xa)\, dx^1 \wedge \ldots \wedge dx^n + a \sum_{i=1}^n dx^1 \wedge \ldots \wedge d(Xx^i) \wedge \ldots \wedge dx^n \\ &= \sum_{i=1}^n (X^i a_{,i} + a X^i_{,i})\, dx^1 \wedge \ldots \wedge dx^n \\ &= a^{-1} \sum_{i=1}^n (aX^i)_{,i}\, \Omega\,. \end{aligned}$$

Hence, we have, locally,

$$div_\Omega X = a^{-1}(aX^i)_{,i}\,. \qquad (12.50)$$

Let g be a (positive definite) Riemannian metric and let $\eta_{\partial D}$ be the volume form on ∂D, which belongs to the induced Riemannian metric on ∂D. Then, as we show below,

$$i_X \eta | \partial D = \langle X, n\rangle\, \eta_{\partial D}\,, \qquad (12.51)$$

where n is the unit outward normal on ∂D. Hence, the theorem of Gauss 12.12 can be written in the familiar form

$$\int_D (div\, X)\eta = \int_{\partial D} \langle X, n\rangle\, \eta_{\partial D}\,. \qquad (12.52)$$

Proof (of Equation (12.51)). We choose a positive orthonormal basis $w_1, \ldots, w_{n-1}$ of $T_p(\partial D)$. Then $(n, w_1, \ldots, w_{n-1})$ is a positive orthonormal basis of $T_p(M)$. Now $X = \langle X, n\rangle n + Y$, $Y \in T_p(\partial D)$, and hence

$$\begin{aligned} i_X \eta(w_1, \ldots, w_{n-1}) &= \eta(X, w_1, \ldots, w_{n-1}) \\ &= \langle X, n\rangle\, \eta_{\partial D}(w_1, \ldots, w_{n-1})\,. \end{aligned}$$

□

* * *

Exercises.

Exercise 12.11. Show that the following identities hold for a function f and a vector field X

$$div_{\Omega}(fX) = X(f) + f\,div_{\Omega}X\,, \tag{12.53a}$$

$$div_{f\Omega}(X) = div_{\Omega}X + \frac{1}{f}X(f)\,. \tag{12.53b}$$

In the second identity it is assumed that f vanishes nowhere; in this case we see that

$$div_{f\Omega}(X) = \frac{1}{f}div_{\Omega}(fX)\,. \tag{12.54}$$

Exercise 12.12. Show that

$$i_X\eta = *X^{\flat}\,, \tag{12.55}$$

where $X^{\flat}$ is the 1-form belonging to the vector field X, i.e. $X^{\flat}(Y) = g(X,Y)$ for all $Y \in \mathcal{X}(M)$. Conclude from this

$$div\,X = \delta X^{\flat}\,. \tag{12.56}$$

Therefore, the theorem of Gauss in the form (12.48) can be written as

$$\int_D \delta X^{\flat}\eta = \int_{\partial D} *X^{\flat}\,. \tag{12.57}$$

13 Affine Connections

In this section we introduce an important additional structure on differentiable manifolds, thus making it possible to define a "covariant derivative" which transforms tensor fields into other tensor fields.

13.1 Covariant Derivative of a Vector Field

The following definition abstracts certain properties of the directional derivative in $\mathbb{R}^n$.

Definition 13.1. *An* ***affine (linear) connection*** *or* ***covariant differentiation*** *on a manifold M is a mapping ∇ which assigns to every pair X, Y of C^∞ vector fields on M another C^∞ vector field $\nabla_X Y$ with the following properties:*

(i) $\nabla_X Y$ is $\mathbb{R}$-bilinear in X and Y;
(ii) If $f \in \mathcal{F}(M)$, then

$$\nabla_{fX} Y = f\nabla_X Y,$$
$$\nabla_X (fY) = f\nabla_X Y + X(f)Y.$$

We show that ∇ can be localized.

Lemma 13.1. *Let ∇ be an affine connection on M and let U be an open subset of M. If either X or Y vanishes on U, then $\nabla_X Y$ also vanishes on U.*

Proof. Suppose Y vanishes on U. Let $p \in U$. Choose a function $h \in \mathcal{F}(M)$ with $h(p) = 0$ and $h = 1$ on $M \setminus U$. From $hY = Y$ it follows that $\nabla_X Y = \nabla_X (hY) = X(h)Y + h\nabla_X Y$. This vanishes at p. The statement about X follows similarly. □

This lemma shows that an affine connection ∇ on M induces a connection on every open submanifold U of M. In fact, let $X, Y \in \mathcal{X}(U)$ and $p \in U$. Then there exist vector fields $\tilde{X}, \tilde{Y} \in \mathcal{X}(M)$ which coincide with X, respectively Y on an open neighborhood V of p (see the continuation Lemma 12.1 of Sect.12.3). We then set

$$(\nabla|U)_X Y|_q = (\nabla_{\tilde{X}}\tilde{Y})|_q\,, \quad \text{for} \quad q \in V\,.$$

By Lemma 13.1, the right-hand side is independent of the choice of $\tilde{X}$, $\tilde{Y} \in \mathcal{X}(M)$. $\nabla|U$ is obviously an affine connection on U.

We show that $(\nabla_X Y)_p$ depends on X only via X_p, whence $\nabla_v Y$ is well-defined for $v \in T_pM$.

Lemma 13.2. *Let $X, Y \in \mathcal{X}(M)$. If X vanishes at p, then $\nabla_X Y$ also vanishes at p.*

Proof. Let U be a coordinate neighborhood of p. On U, we have the representation $X = \xi^i\,\partial/\partial x^i$ with $\xi^i \in \mathcal{F}(U)$, where $\xi^i(p) = 0$. Then $(\nabla_X Y)_p = \xi^i(p)(\nabla_{\partial/\partial x^i} Y)_p = 0$. □

Definition 13.2. *Set, relative to a chart $(U, x^1, \ldots, x^n)$,*

$$\nabla_{\partial/\partial x^i}\left(\frac{\partial}{\partial x^j}\right) = \Gamma^k{}_{ij}\left(\frac{\partial}{\partial x^k}\right). \tag{13.1}$$

The n^3 functions $\Gamma^k{}_{ij} \in \mathcal{F}(U)$ are called the ***Christoffel symbols*** *(or* ***connection coefficients****) of the connection ∇ in the given chart.*

The connection coefficients are not the components of a tensor. Their transformation properties under a coordinate transformation to the chart $(V, \bar{x}^1, \ldots, \bar{x}^n)$ are obtained from the following calculation: On the one hand we have

$$\nabla_{\partial/\partial \bar{x}^a}\left(\frac{\partial}{\partial \bar{x}^b}\right) = \bar{\Gamma}^c{}_{ab}\frac{\partial}{\partial \bar{x}^c} = \bar{\Gamma}^c{}_{ab}\frac{\partial x^k}{\partial \bar{x}^c}\frac{\partial}{\partial x^k} \tag{13.2}$$

and on the other hand the axioms imply

$$\begin{aligned}
\nabla_{\partial/\partial \bar{x}^a}\left(\frac{\partial}{\partial \bar{x}^b}\right) &= \nabla_{\frac{\partial x^i}{\partial \bar{x}^a}\frac{\partial}{\partial x^i}}\left(\frac{\partial x^j}{\partial \bar{x}^b}\frac{\partial}{\partial x^j}\right) \\
&= \frac{\partial x^i}{\partial \bar{x}^a}\left[\frac{\partial x^j}{\partial \bar{x}^b}\Gamma^k{}_{ij}\frac{\partial}{\partial x^k} + \frac{\partial}{\partial x^i}\left(\frac{\partial x^j}{\partial \bar{x}^b}\right)\frac{\partial}{\partial x^j}\right] \\
&= \frac{\partial x^i}{\partial \bar{x}^a}\frac{\partial x^j}{\partial \bar{x}^b}\Gamma^k{}_{ij}\frac{\partial}{\partial x^k} + \frac{\partial^2 x^j}{\partial \bar{x}^a \partial \bar{x}^b}\frac{\partial}{\partial x^j}\,.
\end{aligned}$$

Comparison with (13.2) results in

$$\frac{\partial x^k}{\partial \bar{x}^c}\bar{\Gamma}^c{}_{ab} = \frac{\partial x^i}{\partial \bar{x}^a}\frac{\partial x^j}{\partial \bar{x}^b}\Gamma^k{}_{ij} + \frac{\partial^2 x^k}{\partial \bar{x}^a \partial \bar{x}^b} \tag{13.3}$$

or

$$\bar{\Gamma}^c{}_{ab} = \frac{\partial x^i}{\partial \bar{x}^a}\frac{\partial x^j}{\partial \bar{x}^b}\frac{\partial \bar{x}^c}{\partial x^k}\Gamma^k{}_{ij} + \frac{\partial^2 x^k}{\partial \bar{x}^a \partial \bar{x}^b}\frac{\partial \bar{x}^c}{\partial x^k}\,. \tag{13.4}$$

Conversely, if for every chart there exist n^3 function $\Gamma^k{}_{ij}$ which transform according to (13.4) under a change of coordinates, then it is straightforward to show that there exists a unique affine connection ∇ on M which satisfies (13.1).

For every vector field X we can introduce the tensor field $\nabla X \in \mathcal{T}^1_1(M)$, defined by

$$\nabla X(Y,\omega) := \langle \omega, \nabla_Y X\rangle\,, \tag{13.5}$$

where ω is a 1-form. ∇X is the **covariant derivative** (or **absolute derivative**) of X. In a chart $(U, x^1, \ldots, x^n)$ let

$$X = \xi^i\,\partial_i \quad \text{with} \quad \partial_i := \partial/\partial_i\,,$$
$$\nabla X = \xi^i{}_{;j}\, dx^j \otimes \partial_i\,.$$

Then one has

$$\begin{aligned}\xi^i{}_{;j} &= \nabla X(\partial_j, dx^i) = \langle dx^i, \nabla_{\partial_j}(\xi^k \partial_k)\rangle \\ &= \left\langle dx^i, \xi^k{}_{,j}\partial_k + \xi^k \Gamma^s{}_{jk}\partial_s\right\rangle \\ &= \xi^i{}_{,j} + \Gamma^i{}_{jk}\xi^k\,,\end{aligned}$$

so that

$$\xi^i{}_{;j} = \xi^i{}_{,j} + \Gamma^i{}_{jk}\xi^k\,. \tag{13.6}$$

Definition 13.3. *An affine connection is called* ***symmetric*** *if*

$$\nabla_X Y - \nabla_Y X = [X,Y]\,. \tag{13.7}$$

This is easily seen to be equivalent to the symmetry property $\Gamma^k{}_{ij} = \Gamma^k{}_{ji}$ for any coordinate system.

13.2 Parallel Transport Along a Curve

Definition 13.4. *Let $\gamma : J \longrightarrow M$ be a curve in M with velocity field $\dot\gamma(t)$, and let X be a vector field which is defined on an open neighborhood of $\gamma(J)$. X is said to be* ***autoparallel*** *along γ if*

$$\nabla_{\dot\gamma} X = 0 \tag{13.8}$$

on γ. The vector $\nabla_{\dot\gamma} X$ is sometimes denoted by $\frac{DX}{dt}$ or $\frac{\nabla X}{dt}$ ***(covariant derivative along*** *γ).*

In terms of coordinates, we have

$$X = \xi^i\,\partial_i\,, \qquad \dot\gamma = \frac{dx^i}{dt}\,\partial_i\,,$$
$$\nabla_{\dot\gamma} X = \left(\frac{d\xi^k}{dt} + \Gamma^k{}_{ij}\frac{dx^i}{dt}\xi^j\right)\partial_k\,. \tag{13.9}$$

This shows that $\nabla_{\dot\gamma}X$ depends only on the values of X along γ. In terms of coordinates, (13.8) reads

$$\frac{d\xi^k}{dt} + \Gamma^k{}_{ij}\frac{dx^i}{dt}\xi^j = 0\,. \tag{13.10}$$

In connection with (13.9) we can state more precisely:

Proposition 13.1. *For a given curve $\gamma : J \longrightarrow M$ there is unique map from the set $\mathcal{X}(\gamma)$ of smooth vector fields along γ into $\mathcal{X}(\gamma)$, $X \longmapsto \nabla X/dt$, called the* ***induced covariant derivative****, such that*

(i) $\nabla(a_1X_1 + a_2X_2)/dt = a_1\nabla X/dt + a_2\nabla X/dt$, with a_1, $a_2 \in \mathbb{R}$;
(ii) $\nabla(fX)/dt = (df/dt)X + f\nabla X/dt$, with $f \in \mathcal{F}(J)$;
(iii) For $X \in \mathcal{X}(M)$ and $\bar X = X|\gamma$ we have

$$\frac{\nabla \bar X}{dt} = \nabla_{\dot\gamma}X$$

along γ.

Proof. The easy proof is left to the reader. □

We shall generalize this in Sect. 14.3.

For a given curve $\gamma(t)$ with $X_0 \in T_{\gamma(0)}(M)$ there exists a unique autoparallel field $X(t)$ along γ with $X(t) \in T_{\gamma(t)}(M)$ and $X(0) = X_0$. Since the equation for $X(t)$ is linear, there is no limitation on t. For every curve γ and any two points $\gamma(s)$ and $\gamma(t)$, there is then a linear isomorphism

$$\tau_{t,s} : T_{\gamma(s)}(M) \longrightarrow T_{\gamma(t)}(M)\,,$$

which transforms a vector v at $\gamma(s)$ into the parallel transported vector $v(t)$ at $\gamma(t)$ (see Fig. 13.1). The mapping $\tau_{t,s}$ is the **parallel transport** along γ from $\gamma(s)$ to $\gamma(t)$. We have, as a result of the uniqueness theorem for ordinary differential equations,

$$\tau_{t,s} \circ \tau_{s,r} = \tau_{t,r} \qquad \text{and} \qquad \tau_{s,s} = Id\,. \tag{13.11}$$

We can now give a geometrical interpretation of the covariant derivative that will later be generalized to tensor fields.

Theorem 13.1. *Let X be a vector field along γ. Then*

$$\nabla_{\dot\gamma}X(\gamma(t)) = \left.\frac{d}{ds}\right|_{s=t} \tau_{t,s}X(\gamma(s))\,. \tag{13.12}$$

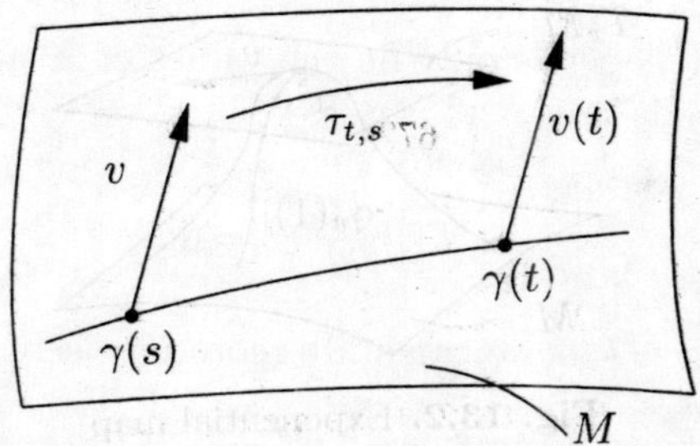

Fig. 13.1. Parallel transport.

Proof. We shall work in a chart. By construction, $v(t) = \tau_{t,s}v_0$ with $v_0 \in T_{\gamma(s)}(M)$ satisfies the equation

$$\dot{v}^i + \Gamma^i{}_{kj}\,\dot{x}^k\,v^j = 0\,.$$

If we write $(\tau_{t,s}v_0)^i = (\tau_{t,s})^i_j v_0^j$, then

$$\left.\frac{d}{dt}\right|_{t=s} (\tau_{t,s})^i_j = -\Gamma^i{}_{kj}\,\dot{x}^k\,.$$

Since $\tau_{t,s} = (\tau_{s,t})^{-1}$ and $\tau_{s,s} = Id$, it follows that

$$\begin{aligned}\left.\frac{d}{ds}\right|_{s=t} \left[\tau_{t,s}X(\gamma(s))\right]^i &= \left.\frac{d}{ds}\right|_{s=t} \left[(\tau_{s,t})^{-1}X(\gamma(s))\right]^i \\ &= -\left.\frac{d}{ds}\right|_{s=t} (\tau_{s,t})^i_j X^j + \left.\frac{d}{ds}\right|_{s=t} X^i(\gamma(s)) \\ &= \Gamma^i{}_{kj}\,\dot{x}^k\,X^j + \left.\frac{d}{ds}\right|_{s=t} X^i(\gamma(s)) = \left(\frac{DX}{dt}\right)^i.\end{aligned}$$

□

13.3 Geodesics, Exponential Mapping and Normal Coordinates

Definition 13.5. *A curve γ is a **geodesic** if $\dot{\gamma}$ is autoparallel along γ, i.e.,* $\nabla_{\dot{\gamma}}\dot{\gamma} = 0$.

In local coordinates, (13.10) implies that along a geodesic

$$\ddot{x}^k + \Gamma^k{}_{ij}\,\dot{x}^i\dot{x}^j = 0 \tag{13.13}$$

For given $\gamma(0)$, $\dot{\gamma}(0)$ there exists a unique maximal geodesic $\gamma(t)$; this is an immediate consequence of the existence and uniqueness theorems of

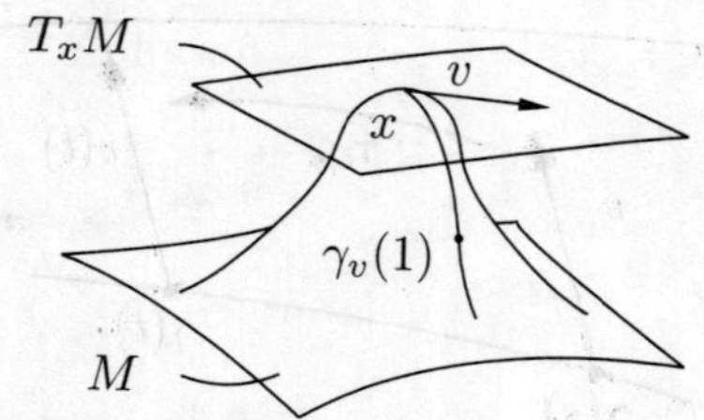

Fig. 13.2. Exponential map.

ordinary differential equations. If $\gamma(t)$ is a geodesic, then $\gamma(at)$, $a \in \mathbb{R}$, is also a geodesic, with initial velocity $a\dot{\gamma}(0)$. Hence, for some neighborhood V of $0 \in T_x(M)$, the geodesics γ_v, with $\gamma_v(0) = x$ and $\dot{\gamma}_v(0) = v \in V$, exist for all $t \in [0,1]$. We introduce the **exponential mapping** at x by (see Fig. 13.2)

$$\exp_x : v \in V \longrightarrow \gamma_v(1)\,.$$

Obviously $\gamma_{tv}(s) = \gamma_v(ts)$. If we set $s = 1$, we obtain $\exp_x(tv) = \gamma_v(t)$. The mapping $\exp_x$ is differentiable[1] at v and the last equation implies $T\exp_x(0) = Id$, since

$$T\exp_x(0)v = \left.\frac{d}{dt}\right|_{t=0} exp_x(tv) = \dot{\gamma}_v(0) = v$$

for all $v \in T_x(M)$. As an application of the implicit function theorem we then obtain

Theorem 13.2. *The mapping* $\exp_x$ *is a diffeomorphism from a neighborhood of* $0 \in T_x(M)$ *to a neighborhood of* $x \in M$.

This theorem permits the introduction of special coordinates. If we choose a basis $e_1, \ldots, e_n$ of $T_x(M)$, then we can represent a neighborhood of x uniquely by $\exp_x(x^i e_i)$. The functions $(x^1, \ldots, x^n)$ are known as **normal coordinates**. Since $\exp_x(tv) = \gamma_v(t)$, the curve $\gamma_v(t)$ has normal coordinates $x^i = v^i t$, with $v = v^i e_i$. In terms of these coordinates, (13.13) becomes

$$\Gamma^k{}_{ij}\, v^i v^j = 0\,;$$

hence we have $\Gamma^k{}_{ij}(0) + \Gamma^k{}_{ji}(0) = 0$. If the connection is symmetric, it follows that $\Gamma^k{}_{ij}(0) = 0$. This fact will turn out to be important.

13.4 Covariant Derivative of Tensor Fields

We now extend the parallel transport $\tau_{s,t}$ of Sect. 13.2 to arbitrary tensors. If $\alpha \in T^*_{\gamma(s)}(M)$, we define $\tau_{t,s}\alpha \in T^*_{\gamma(t)}(M)$ by

[1] Since $\gamma_v(t)$ depends on the initial conditions in differentiable manner.

$$\langle \tau_{t,s}\alpha, \tau_{t,s}v\rangle = \langle \alpha, v\rangle \qquad \text{for all} \quad v \in T_{\gamma(s)}(M)\,,$$

or

$$\langle \tau_{t,s}\alpha, w\rangle = \langle \alpha, \tau_{t,s}^{-1}w\rangle \qquad \text{for all} \quad w \in T_{\gamma(t)}(M)\,. \tag{13.14}$$

For a tensor $S \in (T_{\gamma(s)}(M))^p_q$ we naturally set

$$\begin{aligned}(\tau_{t,s}S)(\,\alpha_1\,,\ldots,\alpha_p, v_1,\ldots,v_q) \\ = S\big(\tau_{t,s}^{-1}\alpha_1,\ldots,\tau_{t,s}^{-1}\alpha_p, \tau_{t,s}^{-1}v_1,\ldots,\tau_{t,s}^{-1}v_q\big)\,,\end{aligned} \tag{13.15}$$

where $\alpha_i \in T^*_{\gamma(t)}(M)$ and $v_i \in T_{\gamma(t)}(M)$.

Now let X be a vector field and $\gamma(t)$ be an integral curve of X, starting at p, so that $p = \gamma(0)$. If $S \in \mathcal{T}^r_s(M)$ is a tensor field, we define the covariant derivative in the direction of X by

$$(\nabla_X S)_p = \frac{d}{ds}\bigg|_{s=0} \tau_s^{-1} S_{\gamma(s)}\,, \tag{13.16}$$

where we have used the shorthand notation $\tau_s := \tau_{s,0}$. This expression is a generalization of (13.12). If $X(p) = 0$ we set $(\nabla_X S)_p = 0$. For a function $f \in \mathcal{F}(M)$ we put $\nabla_X f = Xf$. Finally ∇_X is extended to a linear mapping of $\mathcal{T}(M)$.

Proposition 13.2. *∇_X is a derivation of the tensor algebra $\mathcal{T}(M)$.*

Proof. Obviously, $\tau_s(S_1 \otimes S_2) = (\tau_s S_1) \otimes (\tau_s S_2)$. From this we obtain the Leibniz rule for derivations as follows:

$$\begin{aligned}(\nabla_X(S_1 \otimes S_2)) &= \frac{d}{ds}\bigg|_{s=0} \tau_s^{-1}\Big(S_1(\gamma(s)) \otimes S_2(\gamma(s))\Big) \\ &= \frac{d}{ds}\bigg|_{s=0} \Big(\tau_s^{-1}S_1(\gamma(s)) \otimes \tau_s^{-1}S_2(\gamma(s))\Big) \\ &= \frac{d}{ds}\bigg|_{s=0} \tau_s^{-1}S_1(\gamma(s)) \otimes S_2(p) + S_1(p) \otimes \frac{d}{ds}\bigg|_{s=0} \tau_s^{-1}S_2(\gamma(s)) \\ &= (\nabla_X S_1)_p \otimes S_2(p) + S_1(p) \otimes (\nabla_X S_2)_p\,.\end{aligned}$$

□

Proposition 13.3. *∇_X commutes with contractions.*

Proof. We give a proof for the special case $S = Y \otimes \omega$, $Y \in \mathcal{X}(M)$, $\omega \in \mathcal{X}^*(M)$. The general case is completely analogous. If C denotes the contraction operation, we have by (13.14)

$$\begin{aligned}C\tau_s^{-1}(Y \otimes \omega)_{\gamma(s)} &= C\left(\tau_s^{-1}Y_{\gamma(s)} \otimes \tau_s^{-1}\omega_{\gamma(s)}\right) \\ &= \left\langle \tau_s^{-1}\omega_{\gamma(s)}, \tau_s^{-1}Y_{\gamma(s)}\right\rangle \\ &= \left\langle \omega_{\gamma(s)}, Y_{\gamma(s)}\right\rangle.\end{aligned}$$

Hence,

$$C\Big(s^{-1}\big(\tau_s^{-1}(Y\otimes\omega)_{\gamma(s)}-(Y\otimes\omega)_p\big)\Big)=s^{-1}\big(\langle\omega,Y\rangle_{\gamma(s)}-\langle\omega,Y\rangle_p\big)\,.$$

In the limit $s\to 0$ we obtain, with the definition (13.16),

$$C\big(\nabla_X(Y\otimes\omega)\big)=\nabla_X\big(C(Y\otimes\omega)\big)\,.$$

□

Application

Since, according to Proposition 13.2,

$$\nabla_X(Y\otimes\omega)=(\nabla_XY)\otimes\omega+Y\otimes\nabla_X\omega$$

we have, after contraction, and use of Proposition 13.3,

$$\nabla_X(\omega(Y))=\omega(\nabla_XY)+(\nabla_X\omega)(Y)$$

or

$$(\nabla_X\omega)(Y)=X\omega(Y)-\omega(\nabla_XY)\,. \tag{13.17}$$

Equation (13.17) gives an expression for the covariant derivative of an 1-form. Note that $\nabla_X\omega$ is $\mathcal{F}(M)$-linear in X. Because of the derivation property, ∇_X is then $\mathcal{F}(M)$-linear on the entire tensor algebra:

$$\nabla_{fX}S=f\nabla_XS\qquad\text{for all}\quad f\in\mathcal{F}(M)\,,\ \ S\in\mathcal{T}(M)\,.$$

Hence we can define a mapping

$$\nabla:\mathcal{T}_p^q(M)\longrightarrow\mathcal{T}_{p+1}^q(M)$$

by

$$\begin{aligned}(\nabla S)(X_1,\ldots,X_{p+1},\omega_1,\ldots,\omega_q)&\\ =(\nabla_{X_{p+1}}S)(X_1,\ldots,X_p,\omega_1,\ldots,\omega_q)\,.&\end{aligned} \tag{13.18}$$

∇S is the **covariant derivative** of the tensor field S. We easily obtain a general expression for ∇S, $S\in\mathcal{T}_p^q(M)$, by generalizing the previous discussion: If ω_i are 1-forms and Y_j are vector fields, then

$$\begin{aligned}&\nabla_X(\,Y_1\otimes\ldots\otimes Y_q\otimes\omega_1\otimes\ldots\otimes\omega_p\otimes S)\\ &\quad=\nabla_XY_1\otimes Y_2\otimes\ldots\otimes S+\ldots+Y_1\otimes\ldots\otimes Y_q\otimes\nabla_X\omega_1\otimes\ldots\otimes S+\ldots+\\ &\quad\quad+Y_1\otimes\ldots\otimes\omega_p\otimes\nabla_XS\,.\end{aligned}$$

Taking the complete contraction gives

$$\nabla_X(S(Y_1,\ldots,Y_q,\omega_1,\ldots,\omega_p))$$
$$= S(\nabla_X Y_1,\ldots,\omega_p)+\ldots+S(Y_1,\ldots,\nabla_X\omega_p)+(\nabla_X S)(Y_1,\ldots,\omega_p)\,.$$

From this we find the following expression for $\nabla_X S$,

$$(\nabla_X S)(Y_1,\ldots,Y_q,\omega_1,\ldots,\omega_p) \tag{13.19}$$
$$= X\big(S(Y_1,\ldots,Y_q,\omega_1,\ldots,\omega_p)\big) - S(\nabla_X Y_1,Y_2,\ldots,\omega_p) - \ldots - $$
$$-S(Y_1,\ldots,\nabla_X\omega_p)\,. \tag{13.20}$$

Local Coordinate Expression for the Covariant Derivative

Finally we derive a local coordinate expression for the covariant derivative. Let $(U,x^1,\ldots,x^n)$ be a chart and let $S\in\mathcal{T}_p^q(U)$, with

$$S = S^{i_1\ldots i_p}_{j_1\ldots j_q}\,\partial_{i_1}\otimes\ldots\otimes\partial_{i_p}\otimes dx^{j_1}\otimes\ldots\otimes dx^{j_q}\,,\qquad X=X^i\,\partial_i\,.$$

We use

$$X\,S^{i_1\ldots i_p}_{j_1\ldots j_q} = X^k\,S^{i_1\ldots i_p}_{j_1\ldots j_q,k} \tag{13.21}$$

and

$$\nabla_X(\partial_i) = X^k\nabla_{\partial_k}(\partial_i) = X^k{\Gamma^l}_{ki}\partial_l\,. \tag{13.22}$$

In addition we have, using (13.17),

$$(\nabla_X dx^j)(\partial_i) = X\langle dx^j,\partial_i\rangle - \langle dx^j,\nabla_X\partial_i\rangle = -X^k{\Gamma^l}_{ki}\,,$$

or

$$\nabla_X dx^j = -X^k\,{\Gamma^l}_{ki}\,dx^i\,. \tag{13.23}$$

If we use (13.21)-(13.23) in (13.19) for $\omega_j = dx^j$ and $Y_i=\partial_i$, we obtain an expression for the components of ∇S , denoted by $S^{i_1\ldots i_p}_{j_1\ldots j_q;k}\equiv\nabla_k S^{i_1\ldots i_p}_{j_1\ldots j_q}$,

$$S^{i_1\ldots i_p}_{j_1\ldots j_q;k} = S^{i_1\ldots i_p}_{j_1\ldots j_q,k} + {\Gamma^{i_1}}_{kl}\,S^{li_2\ldots i_p}_{j_1\ldots j_q}+\ldots$$
$$-{\Gamma^l}_{kj_1}\,S^{i_1\ldots i_p}_{lj_2\ldots j_q}-\ldots\,. \tag{13.24}$$

This implies that ∇S is differentiable. In particular, we have for contravariant and covariant vector fields

$$\xi^i_{\ ;k} = \xi^i_{\ ,k} + {\Gamma^i}_{kl}\,\xi^l\,,$$
$$\eta_{i;k} = \eta_{i,k} - {\Gamma^l}_{ki}\,\eta_l\,. \tag{13.25}$$

As also implied by Proposition 13.3, the covariant derivative of the Kronecker tensor vanishes:

$$\delta^i_{\ j;k} = 0\,.$$

Covariant Derivative and Exterior Derivative

Let $\omega \in \bigwedge^p(M)$ and let $\nabla\omega$ be the covariant derivative of ω with respect to a symmetric affine connection. If we apply the alternation operator $\mathcal{A}$ to $\nabla\omega$, then

$$\mathcal{A}(\nabla\omega) = \frac{(-1)^p}{p+1}\, d\omega\,. \tag{13.26}$$

Proof. According to (13.18) and (13.19) we have

$$\begin{aligned}(-1)^p\nabla\omega(X_1,\dots,X_{p+1})\\ &= X_1\omega(X_2,\dots,X_{p+1}) - \sum_j \omega(X_2,\dots,\nabla_{X_1}X_j,\dots,X_{p+1})\,.\end{aligned}$$

Hence,

$$\begin{aligned}(-1)^p\,\mathcal{A}\,(\nabla\omega)(X_1,\dots,X_{p+1})\\ &= \frac{1}{p+1}\Big\{\sum_i(-1)^{i+1}\,X_i\,\omega(X_1,\dots,\hat{X}_i,\dots,X_{p+1})\\ &\quad + \sum_{i<j}(-1)^{i+j}\,\omega([\nabla_{X_i}X_j - \nabla_{X_j}X_i],X_1,\dots,\hat{X}_i,\dots,\hat{X}_j,\dots,X_{p+1})\Big\}\,.\end{aligned}$$

Using (13.7) in the last sum, the comparison with (12.14) for $d\omega$ shows that (13.26) is correct. □

13.5 Curvature and Torsion of an Affine Connection, Bianchi Identities

We now introduce two important tensor fields, associated to an affine connection.

Definition 13.6. *Let ∇ be an affine connection on M. The* ***torsion*** *is defined as the mapping $T:\mathcal{X}(M)\times\mathcal{X}(M)\longrightarrow\mathcal{X}(M)$,*

$$T(X,Y) = \nabla_X Y - \nabla_Y X - [X,Y]\,, \tag{13.27}$$

and the ***curvature*** *as the mapping $R:\mathcal{X}(M)\times\mathcal{X}(M)\times\mathcal{X}(M)\longrightarrow\mathcal{X}(M)$,*

$$R(X,Y)Z = \nabla_X(\nabla_Y Z) - \nabla_Y(\nabla_X Z) - \nabla_{[X,Y]}Z\,. \tag{13.28}$$

Note that

$$T(X,Y) = -T(Y,X)\,, \qquad R(X,Y) = -R(Y,X)\,. \tag{13.29}$$

One can easily verify that

$$T(fX,gY) = fg\,T(X,Y)\,, \qquad R(fX,gY)\,hZ = fgh\,R(X,Y)Z$$

for all $f,g,h \in \mathcal{F}(M)$. The mapping

$$\mathcal{X}^*(M) \times \mathcal{X}(M) \times \mathcal{X}(M) \longrightarrow \mathcal{F}(M)\,,$$
$$(\omega, X, Y) \longmapsto \langle \omega, T(X,Y)\rangle$$

is thus a tensor field in $\mathcal{T}^1_2(M)$ and is known as the **torsion tensor**. Similarly, the mapping $(\omega, Z, X, Y) \longmapsto \langle \omega, R(X,Y)Z\rangle$ is a tensor field in $\mathcal{T}^1_3(M)$. This **curvature tensor** plays a central role in general relativity.

In local coordinates the components of the torsion tensor are given by

$$T^k_{ij} = \langle dx^k, T(\partial_i, \partial_j)\rangle = \langle dx^k, \nabla_{\partial_i}\partial_j - \nabla_{\partial_j}\partial_i\rangle\,,$$

and hence from (13.1)

$$T^k{}_{ij} = \Gamma^k{}_{ij} - \Gamma^k{}_{ji}\,. \tag{13.30}$$

If the torsion vanishes, we have $\Gamma^k{}_{ij} = \Gamma^k{}_{ji}$ in every chart. Then $\Gamma^k{}_{ij}(x_0) = 0$ in normal coordinates with origin x_0, as mentioned at the end of Sect. 13.3.

The components of the curvature tensor are (the ordering of the indices is important):

$$\begin{aligned} R^i{}_{jkl} &= \langle dx^i, R(\partial_k, \partial_l)\partial_j\rangle = \langle dx^i, (\nabla_{\partial_k}\nabla_{\partial_l} - \nabla_{\partial_l}\nabla_{\partial_k})\partial_j\rangle \\ &= \langle dx^i, \nabla_{\partial_k}\Gamma^s{}_{lj}\partial_s - \nabla_{\partial_l}\Gamma^s{}_{kj}\partial_s\rangle \\ &= \Gamma^i{}_{lj,k} - \Gamma^i{}_{kj,l} + \Gamma^s{}_{lj}\Gamma^i{}_{ks} - \Gamma^s{}_{kj}\Gamma^i{}_{ls}\,. \end{aligned} \tag{13.31}$$

Definition 13.7. *The* ***Ricci tensor*** *is the following contraction of the curvature tensor:*

$$R_{jl} = R^i{}_{jil} = \Gamma^i{}_{lj,i} - \Gamma^i{}_{ij,l} + \Gamma^s{}_{lj}\Gamma^i{}_{is} - \Gamma^s{}_{ij}\Gamma^i{}_{ls}\,. \tag{13.32}$$

In order to formulate the next theorem, we need the following preliminaries. An $\mathcal{F}(M)$-multilinear mapping

$$K : \underbrace{\mathcal{X}(M) \times \ldots \times \mathcal{X}(M)}_{\text{p-times}} \longrightarrow \mathcal{X}(M)$$

can be regarded as a tensor field $\tilde{K} \in \mathcal{T}^1_p(M)$:

$$\tilde{K}(\omega, X_1, \ldots, X_p) = \langle \omega, K(X_1, \ldots, X_p)\rangle\,. \tag{13.33}$$

The covariant derivative of K is naturally defined by

$$\langle \omega, (\nabla_X K)(X_1, \ldots, X_p)\rangle = (\nabla_X \tilde{K})(\omega, X_1, \ldots, X_p)\,.$$

According to (13.19), the right-hand side is

$$\begin{aligned}(\nabla_X \tilde{K})(\,\omega\,, X_1, \ldots, X_p) &= X\,\tilde{K}(\omega, X_1, \ldots, X_p) - \tilde{K}(\nabla_X \omega, X_1, \ldots, X_p) \\ &\quad -\tilde{K}(\omega, \nabla_X X_1, \ldots, X_p) - \ldots - \tilde{K}(\omega, X_1, \ldots, \nabla_X X_p)\,.\end{aligned}$$

Since $\nabla_X \omega(Y) = X\omega(Y) - \omega(\nabla_X Y)$, the right-hand side becomes

$$\Big\langle \omega, \nabla_X \big(K(X_1, \ldots, X_p)\big) - \sum_i K(X_1, \ldots, \nabla_X X_i, \ldots, X_p)\Big\rangle .$$

We may now drop the argument ω and obtain

$$(\nabla_X K)(X_1, \ldots, X_p) = \nabla_X \big(K(X_1, \ldots, X_p)\big) - \sum_i K(X_1, \ldots, \nabla_X X_i, \ldots, X_p)\,. \tag{13.34}$$

Theorem 13.3. *Let T and R be the torsion and curvature of an affine connection ∇. If X, Y and Z are vector fields, then we have*

$$\sum_{cyclic} \{R(X,Y)Z\} = \sum_{cyclic} \Big\{T(T(X,Y),Z) + (\nabla_X T)(Y,Z)\Big\} \tag{13.35}$$

(1st Bianchi identity),

$$\sum_{cyclic} \Big\{(\nabla_X R)(Y,Z) + R(T(X,Y),Z)\Big\} = 0 \tag{13.36}$$

(2nd Bianchi identity).

Proof. By (13.34)

$$(\nabla_X R)(Y,Z) = \nabla_X (R(Y,Z)) - R(\nabla_X Y, Z) - R(Y, \nabla_X Z) - R(Y,Z)\nabla_X\,.$$

The cyclic sum of the two middle terms is

$$\begin{aligned}&-\{R(\nabla_X Y, Z) + R(Y, \nabla_X Z) + R(\nabla_Y Z, X) + R(Z, \nabla_Y X) \\ &\quad + R(\nabla_Z X, Y) + R(X, \nabla_Z Y)\} \\ &= -\{R(\nabla_X Y, Z) + R(Z, \nabla_Y X) + \text{cyclic permutations}\} \\ &= -R(T(X,Y),Z) - R([X,Y])Z) + \text{cyclic permutations}\,,\end{aligned}$$

and hence the left-hand side of (13.36) is equal to

$$\begin{aligned}&\nabla_X(\,R\,(Y,Z)) - R([X,Y])Z) - R(Y,Z)\nabla_X + \text{cyclic permutations} \\ &= \nabla_X(\nabla_Y \nabla_Z - \nabla_Z \nabla_Y - \nabla_{[Y,Z]}) \\ &\quad -(\nabla_Y \nabla_Z - \nabla_Z \nabla_Y - \nabla_{[Y,Z]})\nabla_X \\ &\quad -(\nabla_{[X,Y]}\nabla_Z - \nabla_Z \nabla_{[X,Y]} - \nabla_{[[X,Y],Z]}) + \text{cyclic permutations}\,.\end{aligned}$$

If one uses the Jacobi identity, the last term in the cyclic sum drops out. The remaining terms transform in pairs into each other under cyclic permutation, except for the sign, and thus cancel in pairs in the cyclic sum. This proves the second Bianchi identity.

We now prove the first Bianchi identity for $T = 0$. In that case the left-hand side of (13.35) is

$$\begin{aligned}(\nabla_X \nabla_Y - &\nabla_Y \nabla_X)Z + (\nabla_Y \nabla_Z - \nabla_Z \nabla_Y)X + (\nabla_Z \nabla_X - \nabla_X \nabla_Z)Y \\ &- \nabla_{[X,Y]}Z - \nabla_{[Y,Z]}X - \nabla_{[Z,X]}Y \\ &= \nabla_X [Y,Z] - \nabla_{[Y,Z]}X + \text{cyclic permutations} \\ &= [X,[Y,Z]] + \text{cyclic permutations} \\ &= 0\,.\end{aligned}$$

The general case ($T \neq 0$) is left as an exercise. □

13.6 Riemannian Connections

Definition 13.8. *Let (M, g) be a pseudo-Riemannian manifold. An affine connection is a* **metric connection** *if parallel transport along any smooth curve γ in M preserves the inner product: For autoparallel fields $X(t), Y(t)$ along γ, $g_{\gamma(t)}(X(t), Y(t))$ is independent of t.*

Proposition 13.4. *An affine connection ∇ is metric if and only if*

$$\nabla g = 0\,. \tag{13.37}$$

Proof. By definition an affine connection is metric if, for all γ, Y,Z

$$\frac{d}{dt} g_{\gamma(t)}\left(\tau_{t,s} Y_{\gamma(s)}, \tau_{t,s} Z_{\gamma(s)}\right) = 0\,. \tag{13.38}$$

Because of the group property of $\tau_{t,s}$ this is equivalent to the same condition for $t = s$. By (13.15) and (13.16), this is equivalent to $\nabla_{\dot\gamma} g = 0$. □

Remark. As a result of (13.19), $\nabla g = 0$ is equivalent to the so-called **Ricci-identity**

$$X g(Y, Z) = g(\nabla_X Y, Z) + g(Y, \nabla_X Z)\,. \tag{13.39}$$

The following result is often called the miracle of pseudo-Riemannian geometry. Although not very deep, it is of prime importance.

Theorem 13.4. *For every pseudo-Riemannian manifold (M, g), there exists a unique affine connection ∇ such that*

(i) ∇ has vanishing torsion (∇ is symmetric);
(ii) ∇ is metric.

Proof. From the Ricci-identity (13.39) and (i) we obtain

$$Xg(Y,Z) = g(\nabla_Y X, Z) + g([X,Y],Z) + g(Y, \nabla_X Z). \tag{13.40}$$

From this we obtain through cyclic permutation

$$Yg(Z,X) = g(\nabla_Z Y, X) + g([Y,Z],X) + g(Z, \nabla_Y X), \tag{13.41a}$$

$$Zg(X,Y) = g(\nabla_X Z, Y) + g([Z,X],Y) + g(X, \nabla_Z Y). \tag{13.41b}$$

Taking the linear combination (13.41*a*) + (13.41*b*) − (13.40) results in the **Koszul formula**

$$\begin{aligned} 2g(\nabla_Z Y, X) = &-Xg(Y,Z) + Yg(Z,X) + Zg(X,Y) \\ &-g([Z,X],Y) - g([Y,Z],X) + g([X,Y],Z). \end{aligned} \tag{13.42}$$

The right-hand side is independent of ∇. Since g is non-degenerate, the *uniqueness* of ∇ follows from (13.42).

Now, we prove the *existence*: Define the mapping $\omega : \mathcal{X}(M) \longrightarrow \mathcal{F}(M)$, $2\omega(X)$ equal to the right-hand side of (13.42). The mapping ω is clearly additive, and is also $\mathcal{F}(M)$-homogeneous, $\omega(fX) = f\omega(X)$, as can be demonstrated by a short computation (exercise). To the 1-form ω there corresponds a unique vector field $\nabla_Z Y$ with $g(\nabla_Z Y, X) = \omega(X)$. The mapping $\nabla : \mathcal{X}(M) \times \mathcal{X}(M) \longrightarrow \mathcal{X}(M)$ defined in this manner satisfies the defining properties of an affine connection (see Definition 13.1): additivity in Y and Z is obvious, and homogenity in Z is easily demonstrated by a short calculation. We now verify the derivation property, writing $\langle X,Y\rangle$ instead of $g(X,Y)$:

$$\begin{aligned} 2\langle \nabla_Z fY, X\rangle &= -X\langle fY, Z\rangle + fY\langle Z,X\rangle + Z\langle X, fY\rangle \\ &\quad -\langle [Z,X], fY\rangle - \langle [fY,Z],X\rangle + \langle [X,fY],Z\rangle \\ &= 2\langle f\nabla_Z Y, X\rangle - (Xf)\langle Y,Z\rangle + (Zf)\langle X,Y\rangle \\ &\quad +(Zf)\langle Y,X\rangle + (Xf)\langle Y,Z\rangle \\ &= 2(\langle f\nabla_Z Y, X\rangle + (Zf)\langle Y,X\rangle) \\ &= 2\langle f\nabla_Z Y + (Zf)Y, X\rangle . \end{aligned}$$

This implies

$$\nabla_Z(fY) = f\nabla_Z Y + (Zf)Y .$$

The affine connection constructed in this manner has vanishing torsion. In fact, (13.42) implies that

$$\langle \nabla_Z Y, X\rangle = \langle \nabla_Y Z, X\rangle + \langle [Z,Y],X\rangle .$$

Furthermore, summation of the right hand sides of (13.42) for

$$2\langle \nabla_Z Y, X\rangle + 2\langle \nabla_Z X, Y\rangle$$

shows that the Ricci identity is satisfied. □

Definition 13.9. *The unique connection on (M,g) from Theorem 13.4 is called the* ***Riemannian*** *or* ***Levi-Civita connection****.*

Local Expressions

We determine the Christoffel symbols of the Riemannian connection in a chart $(U, x^1, \ldots, x^n)$. For this purpose take $X = \partial_k$, $Y = \partial_j$ and $Z = \partial_i$ in (13.42) and use $[\partial_i, \partial_j] = 0$, as well as $\langle \partial_i, \partial_j \rangle = g_{ij}$. The result is

$$2\langle \nabla_{\partial_i}\partial_j, \partial_k \rangle = 2\Gamma^l_{\ ij} g_{lk} = -\partial_k \langle \partial_j, \partial_i \rangle + \partial_j \langle \partial_i, \partial_k \rangle + \partial_i \langle \partial_k, \partial_j \rangle$$

or

$$g_{mk}\Gamma^m_{\ ij} = \frac{1}{2}\left(g_{jk,i} + g_{ik,j} - g_{ij,k}\right). \tag{13.43}$$

If (g^{ij}) denotes the matrix inverse to (g_{ij}), we obtain from (13.43)

$$\Gamma^l_{\ ij} = \frac{1}{2} g^{lk}\left(g_{ki,j} + g_{kj,i} - g_{ij,k}\right). \tag{13.44}$$

Note also that for coordinates which are normal at x_0, the Ricci identity

$$\partial_k g_{ij} = \langle \nabla_{\partial_k}\partial_i, \partial_j \rangle + \langle \partial_i, \nabla_{\partial_k}\partial_j \rangle$$

implies that the first derivatives of g_{ij} vanish at x_0.

Proposition 13.5. *The curvature tensor of a Riemannian connection has the following additional symmetry properties:*

$$\langle R(X,Y)Z, U \rangle = -\langle R(X,Y)U, Z \rangle, \tag{13.45a}$$
$$\langle R(X,Y)Z, U \rangle = \langle R(Z,U)X, Y \rangle. \tag{13.45b}$$

Proof. It suffices to prove these identities for vector fields with pairwise vanishing Lie brackets (e.g., for the basis fields ∂_i of a chart). Equation (13.45a) is equivalent to

$$\langle R(X,Y)Z, Z \rangle = 0. \tag{13.46}$$

Since ∇ is a Riemannian connection,

$$\langle \nabla_X \nabla_Y Z, Z \rangle = X\langle \nabla_Y Z, Z \rangle - \langle \nabla_Y Z, \nabla_X Z \rangle$$

and

$$\langle \nabla_Y Z, Z \rangle = \frac{1}{2} Y \langle Z, Z \rangle.$$

Hence,

$$2\langle R(X,Y)Z, Z \rangle = XY\langle Z, Z \rangle - YX\langle Z, Z \rangle = 0,$$

thus proving (13.45a).

The first Bianchi identity (13.35) with $T = 0$ implies

$$\begin{aligned}\langle R(X,Y)Z, U \rangle &= -\langle R(Y,X)Z, U \rangle \\ &= \langle R(X,Z)Y, U \rangle + \langle R(Z,Y)X, U \rangle.\end{aligned}$$

In addition, (13.45a) and the first Bianchi identity imply

$$\begin{aligned}\langle R(X,Y)Z,U\rangle &= -\langle R(X,Y)U,Z\rangle \\ &= \langle R(Y,U)X,Z\rangle + \langle R(U,X)Y,Z\rangle .\end{aligned}$$

Adding these last two equations results in

$$\begin{aligned}2\langle R(X,Y)Z,U\rangle = &\langle R(X,Z)Y,U\rangle + \langle R(Z,Y)X,U\rangle \\ &+\langle R(Y,U)X,Z\rangle + \langle R(U,X)Y,Z\rangle .\end{aligned}$$

If we interchange the pairs X,Z and Y,U, we obtain

$$\begin{aligned}2\langle R(Z,U)X,Y\rangle = &\langle R(Z,X)U,Y\rangle + \langle R(X,U)Z,Y\rangle \\ &+\langle R(U,Y)Z,X\rangle + \langle R(Y,Z)U,X\rangle .\end{aligned}$$

Use of (13.45a) and $R(X,Y) = -R(Y,X)$ then shows that the right-hand sides of both equations agree. □

Remarks.

a) The metric tensor field g permits us to map $\mathcal{X}(M)$ uniquely onto $\mathcal{X}^*(M)$ by $X \longmapsto X^\flat$, with

$$X^\flat(Z) = g(X,Z) \qquad \text{for all} \quad Z \in \mathcal{X}(M) .$$

This operator can be applied to tensors to produce new ones. For example, we may assign to every field $t \in \mathcal{T}_2^0(M)$ a unique field $\tilde{t} \in \mathcal{T}_1^1(M)$:

$$\tilde{t}(X, Y^\flat) = t(X,Y) .$$

In local coordinates, this corresponds to raising and lowering indices with the metric tensor. The inverse of the map $\flat$ is denoted by $\sharp$.

b) Since $g_{ij}g^{jk} = \delta_i^k$, it follows that $g^{ij}{}_{;k} = 0$ (remember that $\delta_i{}^j{}_{;k} = 0$).

We now give coordinate expressions for the identities satisfied by the curvature tensor of a Riemannian connection. These read, if $\sum_{(ijk)}$ denotes the cyclic sum and $R_{ijkl} := g_{is}R^s{}_{jkl}$,

$$\sum_{(jkl)} R^i{}_{jkl} = 0 \qquad \textbf{(1st Bianchi identity)}, \tag{13.47a}$$

$$\sum_{(klm)} R^i{}_{jkl;m} = 0 \qquad \textbf{(2nd Bianchi identity)}, \tag{13.47b}$$

$$R^i{}_{jkl} = -R^i{}_{jlk} , \tag{13.47c}$$

$$R_{ijkl} = -R_{jikl} , \tag{13.47d}$$

$$R_{ijkl} = R_{klij} . \tag{13.47e}$$

Proof. From (13.31) we have

$$R^i{}_{jkl} = \langle dx^i, R(\partial_k, \partial_l)\partial_j \rangle\,, \quad R_{ijkl} = \langle \partial_i, R(\partial_k, \partial_l)\partial_j \rangle\,. \tag{13.48}$$

Equation (13.47c) follows as a result of $R(X,Y) = -R(Y,X)$. Equation (13.47a) and (13.47b) follow from the first (resp. second) Bianchi identities (13.35) and (13.36) with $T = 0$. Equation (13.47d) and (13.47e) are a consequence of (13.45a) and (13.45b). □

Contracted Bianchi Identity

If, as in (13.32), we denote the Ricci tensor by R_{ik} and the **scalar curvature** by

$$R = g^{ik} R_{ik}\,. \tag{13.49}$$

then the **contracted Bianchi identity**

$$\left(R_i{}^k - \frac{1}{2}\delta_i{}^k R\right)_{;k} = 0 \tag{13.50}$$

holds. Furthermore, the Ricci tensor is symmetric

$$R_{ik} = R_{ki}\,. \tag{13.51}$$

Proof. The symmetry of R_{ik} follows from (13.32), i.e.

$$R_{jl} = g^{ik} R_{ijkl} \tag{13.52}$$

and (13.47e). Now consider

$$R_j{}^m{}_{;m} = g^{ml} R_{jl;m} = g^{ml} g^{ik} R_{ijkl;m} \overset{(13.47e)}{=} g^{ml} g^{ik} R_{klij;m}\,.$$

Use of the second Bianchi identity (13.45b) results in

$$R_j{}^m{}_{;m} = -g^{ml} g^{ik} (R_{kljm;i} + R_{klmi;j})\,.$$

According to (13.47c), (13.47d) and (13.52), the first term on the right hand side is equal to

$$-g^{ik} R_{kj;i} = -R_j{}^l{}_{;l}\,.$$

From (13.47d) and (13.52), the second term is equal to $g^{ik} R_{ki;j} = R_{,j}$. Hence,

$$R_i{}^k{}_{;k} = \frac{1}{2} R_{,i} = \frac{1}{2} (\delta_i{}^k R)_{;k}\,.$$

□

Definition 13.10. *The **Einstein tensor** is defined by*

$$G_{ik} = R_{ik} - \frac{1}{2} g_{ik} R\,. \tag{13.53}$$

By (13.50), it satisfies the contracted Bianchi identity

$$G_{i;k}^{k} = 0 . \tag{13.54}$$

This is extremely important in connection with Einstein's field equation.

* * *

Exercise 13.1. Let ϕ be an isometry of a pseudo-Riemannian manifold (M, g), $\phi^* g = g$. Show that for two vector fields X and Y

$$\nabla_{\phi_* X}(\phi_* Y) = \phi_*(\nabla_X Y) ,$$

and conclude from this that parallel transport is preserved under ϕ.

Hint. Let

$$\tilde{\nabla}_X Y := \phi_*^{-1}(\nabla_{\phi_* X}(\phi_* Y))$$

and show that $\tilde{\nabla}$ satisfies all the properties of a Riemannian connection.

13.7 The Cartan Structure Equations

Let M be a differentiable manifold with an affine connection. Let $(e_1, \ldots, e_n)$ be a basis of C^∞ vector fields[2] defined on an open subset U (which might be the domain of a chart, for example), and let $(\theta^1, \ldots, \theta^n)$ denote the corresponding dual basis of 1-forms. We define **connection forms** $\omega^i{}_j \in \bigwedge^1(U)$ by

$$\nabla_X e_j = \omega^i{}_j(X) e_i . \tag{13.55}$$

We may generalize the definition of the Christoffel symbols (relative to the basis $\{e_i\}$) as follows:

$$\nabla_{e_k} e_j = \Gamma^i{}_{kj} e_i = \omega^i{}_j(e_k) e_i .$$

Thus,

$$\omega^i{}_j = \Gamma^i{}_{kj} \theta^k . \tag{13.56}$$

Since ∇_X commutes with contractions, we have, according to (13.55),

$$\begin{aligned} 0 = \nabla_X \langle \theta^i, e_j \rangle &= \langle \nabla_X \theta^i, e_j \rangle + \langle \theta^i, \nabla_X e_j \rangle \\ &= \langle \nabla_X \theta^i, e_j \rangle + \underbrace{\langle \theta^i, \omega^k{}_j(X) e_k \rangle}_{\omega^i{}_j(X)} . \end{aligned}$$

That is,

$$\nabla_X \theta^i = -\omega^i{}_j(X) \theta^j , \tag{13.57}$$

[2] Such a set is called a **moving frame** (**vierbein**, or **tetrad** for $n = 4$).

or

$$\nabla\theta^i = -\theta^j \otimes \omega^i{}_j\,. \tag{13.58}$$

If α is a general 1-form, $\alpha = \alpha_i\,\theta^i$, with $\alpha_i \in \mathcal{F}(U)$, then (13.57) implies

$$\begin{aligned}\nabla_X\alpha &= (X\alpha_i)\theta^i + \theta^i\nabla_X\theta^i \\ &= \langle d\alpha_i - \alpha_l\,\omega^l{}_i, X\rangle\,\theta^i\,,\end{aligned}$$

so that

$$\nabla\alpha = \theta^i \otimes (d\alpha_i - \omega^k{}_i\alpha_k)\,. \tag{13.59}$$

In an analogous manner, we find for a vector field $X = X^i\,e_i$

$$\nabla X = e_i \otimes (dX^i + \omega^i{}_k X^k)\,, \tag{13.60}$$

or more explicitly

$$\nabla_Y X = \langle dX^i + \omega^i{}_k X^k, Y\rangle\, e_i\,. \tag{13.61}$$

The torsion and the curvature define differential forms Θ^i (**torsion forms**) and $\Omega^i{}_j$ (**curvature forms**) by

$$T(X,Y) = \Theta^i(X,Y)e_i\,, \tag{13.62}$$

$$R(X,Y)e_j = \Omega^i{}_j(X,Y)e_i\,. \tag{13.63}$$

Theorem 13.5. *The torsion forms and curvature forms satisfy the* ***Cartan structure equations***

$$\Theta^i = d\theta^i + \omega^i{}_j \wedge \theta^j\,, \tag{13.64}$$

$$\Omega^i{}_j = d\omega^i{}_j + \omega^i{}_k \wedge \omega^k{}_j\,. \tag{13.65}$$

Proof. According to the definition and (13.55), we have

$$\begin{aligned}\Theta^i(X,Y)e_i &= \nabla_X Y - \nabla_Y X - [X,Y] \\ &= \nabla_X(\theta^j(Y)e_j) - \nabla_Y(\theta^j(X)e_j) - \theta^j([X,Y])e_j \\ &= \big(X\theta^j(Y) - Y\theta^j(X) - \theta^j([X,Y])\big)e_j \\ &\quad + \big(\theta^j(Y)\omega^i{}_j(X) - \theta^j(X)\omega^i{}_j(Y)\big)e_i \\ &= d\theta^i(X,Y)e_i + (\omega^i{}_j \wedge \theta^j)(X,Y)e_i\,.\end{aligned}$$

Equation (13.64) follows from this. The derivation of (13.65) is similar: From the definition and (13.55), we have

$$\begin{aligned}\Omega^i{}_j(X,Y)e_i &= \nabla_X\nabla_Y e_j - \nabla_Y\nabla_X e_j - \nabla_{[X,Y]}e_j \\ &= \nabla_X(\omega^i{}_j(Y)e_i) - \nabla_Y(\omega^i{}_j(X))e_i - \omega^i{}_j([X,Y])e_i \\ &= \big(X\omega^i{}_j(Y) - Y\omega^i{}_j(X) - \omega^i{}_j([X,Y])\big)e_i \\ &\quad + \big(\omega^i{}_j(Y)\omega^k{}_i(X) - \omega^i{}_j(X)\omega^k{}_i(Y)\big)e_k \\ &= d\omega^i{}_j(X,Y)e_i + (\omega^i{}_k \wedge \omega^k{}_j)(X,Y)e_i\,.\end{aligned}$$

A comparison of the components results in (13.65). □

We can expand $\Omega^i{}_j$ as follows

$$\Omega^i{}_j = \frac{1}{2} R^i{}_{jkl}\, \theta^k \wedge \theta^l\,, \qquad R^i{}_{jkl} = -R^i{}_{jlk}\,. \tag{13.66}$$

Since

$$\langle \theta^i, R(e_k, e_l)e_j \rangle = \langle \theta^i, \Omega^s{}_j(e_k, e_l)e_s \rangle = \Omega^i{}_j(e_k, e_l) = R^i{}_{jkl}\,,$$

we have

$$R^i{}_{jkl} = \langle \theta^i, R(e_k, e_l)e_j \rangle\,. \tag{13.67}$$

As a result of (13.48), $R^i{}_{jkl}$ agree with the components of the Riemann tensor for the special case $e_i = \partial/\partial x^i$. Equation (13.67) defines these components for an arbitrary basis $\{e_i\}$. Thus the expansion coefficients of the curvature form in (13.66) are the components of the Riemann tensor.

In an analogous manner we have

$$\Theta^i = \frac{1}{2} T^i_{\ kl}\, \theta^k \wedge \theta^l\,, \qquad T^i_{\ kl} = \langle \theta^i, T(e_k, e_l) \rangle\,. \tag{13.68}$$

Proposition 13.6. *An affine connection ∇ is metric if and only if*

$$\omega_{ik} + \omega_{ki} = dg_{ik}\,, \tag{13.69}$$

where $\omega_{ik} = g_{ij}\, \omega^j{}_k$ and $g_{ik} = g(e_i, e_k)$.

Proof. As a result of Proposition 13.4, a connection is metric if and only if the Ricci identity holds:

$$\begin{aligned} dg_{ik}(X) &= X g_{ik} = X\langle e_i, e_k\rangle \\ &= \langle \nabla_X e_i, e_k\rangle + \langle e_i, \nabla_X e_k\rangle \\ &\overset{(13.55)}{=} \omega^j{}_i(X)\langle e_j, e_k\rangle + \omega^j{}_k(X)\langle e_i, e_j\rangle \\ &= g_{jk}\, \omega^j{}_i(X) + g_{ij}\, \omega^j{}_k(X)\,. \end{aligned}$$

□

Thus, for a Riemannian connection the following equations hold:

$$\begin{aligned} \omega_{ij} + \omega_{ji} &= dg_{ij}\,, \\ d\theta^i + \omega^i{}_j \wedge \theta^j &= 0\,, \\ d\omega^i{}_j + \omega^i{}_k \wedge \omega^k{}_j &= \Omega^i{}_j = \frac{1}{2} R^i{}_{jkl}\, \theta^k \wedge \theta^l\,. \end{aligned} \tag{13.70}$$

Solution of the Structure Equations

We expand $d\theta^i$ in terms of the basis θ^i:

$$d\theta^i = -\frac{1}{2}C^i{}_{jk}\,\theta^j \wedge \theta^k\,, \qquad C^i{}_{jk} = -C^i{}_{kj}\,. \tag{13.71}$$

Comparison with the first structure equation gives

$$\Gamma^i{}_{kl} - \Gamma^i{}_{lk} = C^i{}_{kl}\,. \tag{13.72}$$

Remark. For $\theta^i = dx^i$, we have $C^i{}_{kl} = 0$ and the $\Gamma^i{}_{kl}$ are symmetric in k and l.

In addition we set

$$dg_{ij} = g_{ij,k}\,\theta^k\,, \qquad g_{ij,k} = e_k(g_{ij})\,. \tag{13.73}$$

Since, according to (13.56), $\omega_{ij} = g_{is}\Gamma^s{}_{lj}\theta^l$, the first equation in (13.70) implies

$$g_{ij,k} = g_{is}\Gamma^s{}_{kj} + g_{js}\Gamma^s{}_{ki}\,. \tag{13.74}$$

We take cyclic permutations to obtain

$$g_{jk,i} = g_{js}\Gamma^s{}_{ik} + g_{ks}\Gamma^s{}_{ij}\,, \tag{13.75}$$

$$g_{ki,j} = g_{ks}\Gamma^s{}_{ji} + g_{is}\Gamma^s{}_{jk}\,. \tag{13.76}$$

Forming the combination (13.74)+(13.75)−(13.76), and using (13.72) results in

$$g_{ij,k} + g_{jk,i} - g_{ki,j} = g_{is}C^s{}_{kj} + g_{ks}C^s{}_{ij} + g_{js}(\Gamma^s{}_{ki} + \Gamma^s{}_{ik})\,.$$

If we contract this equation with g^{lj}, we obtain

$$\Gamma^l{}_{ki} + \Gamma^l{}_{ik} = g^{lj}(g_{ij,k} + g_{jk,i} - g_{ki,j}) - g^{lj}g_{is}C^s{}_{kj} - g^{lj}g_{ks}C^s{}_{ij}\,.$$

We now use (13.72) again and obtain

$$\Gamma^l{}_{ki} = \frac{1}{2}\left(C^l{}_{ki} - g_{is}g^{lj}C^s{}_{kj} - g_{ks}g^{lj}C^s{}_{ij}\right) + \frac{1}{2}g^{lj}\left(g_{ij,k} + g_{jk,i} - g_{ki,j}\right). \tag{13.77}$$

The second term vanishes for an orthonormal basis.

According to (13.56), we have

$$d\omega^i{}_j = d\Gamma^i{}_{lj} \wedge \theta^l + \Gamma^i{}_{lj}d\theta^l\,.$$

We now set $d\Gamma^i{}_{lj} = \Gamma^i{}_{lj,s}\theta^s = e_s(\Gamma^i{}_{lj})\theta^s$ and use (13.71). The result is

$$d\omega^i{}_j = \Gamma^i{}_{lj,s}\,\theta^s \wedge \theta^l - \frac{1}{2}\Gamma^i{}_{lj}C^l{}_{ab}\,\theta^a \wedge \theta^b\,,$$

or

$$d\omega^i{}_j = \frac{1}{2}(\Gamma^i{}_{bj,a} - \Gamma^i{}_{aj,b} - \Gamma^i{}_{lj}C^l{}_{ab})\theta^a \wedge \theta^b .$$

As a consequence, the second structure equation (the third equation in (13.70)) becomes

$$\begin{aligned} d\omega^i{}_j + \omega^i{}_l \wedge \omega^l{}_j &= \frac{1}{2}(\Gamma^i{}_{bj,a} - \Gamma^i{}_{aj,b} - \Gamma^i{}_{lj}C^l{}_{ab})\theta^a \wedge \theta^b \\ &\quad + \frac{1}{2}(\Gamma^i{}_{al}\Gamma^l{}_{bj} - \Gamma^i{}_{bl}\Gamma^l{}_{aj})\theta^a \wedge \theta^b \\ &= \frac{1}{2}R^i{}_{jab}\,\theta^a \wedge \theta^b = \Omega^i{}_j . \end{aligned}$$

Thus, the components of the curvature tensor are given by

$$R^i{}_{jab} = \Gamma^i{}_{bj,a} - \Gamma^i{}_{aj,b} - \Gamma^i{}_{lj}C^l{}_{ab} + \Gamma^i{}_{al}\Gamma^l{}_{bj} - \Gamma^i{}_{bl}\Gamma^l{}_{aj} . \tag{13.78}$$

Remark. Cartan's structure equations are also useful for curvature calculations, as we shall see in many instances. The reader is invited to study at this point the application to "warped products" of pseudo-Riemannian manifolds in the Appendix B.

13.8 Bianchi Identities for the Curvature and Torsion Forms

In the following we again consider an arbitrary affine connection ∇ on a manifold M. As before, let $\{e_i\}$ and $\{\theta^i\}$ be mutually dual bases and let $\omega^i{}_j$ denote the connection forms.

Proposition 13.7. *Under a change of basis*

$$\bar{\theta}^i(x) = A^i{}_j(x)\theta^j(x) , \tag{13.79}$$

$\omega = (\omega^i{}_j)$ *transforms in matrix notation as follows*

$$\bar{\omega} = A\omega A^{-1} - dA\,A^{-1} . \tag{13.80}$$

Proof. In an obvious matrix notation we have, according to (13.57),

$$\begin{aligned} \nabla_X\bar{\theta} = -\bar{\omega}(X)\bar{\theta} &= \nabla_X(A\theta) = dA(X)\theta + A\nabla_X\theta \\ &= dA(X)\theta - A\omega(X)\theta = dA(X)A^{-1}\bar{\theta} - A\omega(X)A^{-1}\bar{\theta} . \end{aligned}$$

From this the conclusion follows immediately. □

We emphasize that the $\omega^i{}_j$ are not the components of a tensor valued form in the sense of the

Definition 13.11. *A* ***tensor valued*** ***p-form*** *of type (r,s) is a skew symmetric p-multilinear mapping*

$$\phi : \underbrace{\mathcal{X}(M) \times \mathcal{X}(M) \times \ldots \times \mathcal{X}(M)}_{p\text{-times}} \longrightarrow \mathcal{T}^r_s(M).$$

The p-forms $\phi^{i_1\ldots i_r}_{j_1\ldots j_s} = \phi(\theta^{i_1},\ldots,\theta^{i_r},e_{j_1},\ldots,e_{j_s})$ are the ***components*** *of the tensor valued p-form relative to the basis θ^i.*

Proposition 13.8. *For every tensor valued p-form ϕ of type (r,s) there exists a unique tensor valued $(p+1)$-form $D\phi$, also of type (r,s), which has the following components relative to the basis θ^i:*

$$\begin{aligned}(D\phi)^{i_1\ldots i_r}_{j_1\ldots j_s} = d\,\phi^{i_1\ldots i_r}_{j_1\ldots j_s} &+ \omega^{i_1}{}_l \wedge \phi^{l i_2\ldots i_r}_{j_1\ldots j_s} + \ldots \textit{for all upper indices}\\ &- \omega^l{}_{j_1} \wedge \phi^{i_1\ldots i_r}_{l j_2\ldots j_s} - \ldots \textit{for all lower indices}.\end{aligned} \tag{13.81}$$

Proof. It suffices to show that the right-hand side of (13.81) transforms like the components of a tensor of type (r,s) under a change of basis $\bar{\theta}(x) = A(x)\theta(x)$. Since every pair of upper and lower indices in (13.81) behaves in the same manner, we may consider a tensor valued p-form ϕ^i_j of type $(1,1)$. The transformed components $\bar{\phi} = (\bar{\phi}^i_j)$ are given by $\bar{\phi} = A\phi A^{-1}$. In addition,

$$D\bar{\phi} = d\bar{\phi} + \bar{\omega} \wedge \bar{\phi} - (-1)^p \bar{\phi} \wedge \bar{\omega},$$

or, using (13.80) and $dA\,A^{-1} = -A\,dA^{-1}$,

$$\begin{aligned}D\bar{\phi} &= d(A\phi A^{-1}) + (A\omega A^{-1} - dA\,A^{-1}) \wedge A\phi A^{-1}\\ &\quad -(-1)^p A\phi A^{-1} \wedge (A\omega A^{-1} - dA\,A^{-1})\\ &= dA \wedge \phi A^{-1} + A\,d\phi A^{-1} + (-1)^p A\phi \wedge dA^{-1} + (A\omega - dA) \wedge \phi A^{-1}\\ &\quad -(-1)^p A\phi \wedge (\omega A^{-1} + dA^{-1})\\ &= A\big(d\phi + \omega \wedge \phi - (-1)^p \phi \wedge \omega\big)A^{-1}\\ &= A(D\phi)A^{-1}.\end{aligned}$$

The tensor valued $(p+1)$-form $D\phi$ of Proposition 13.8 is known as the **absolute exterior differential** of the tensor valued p-form ϕ. □

Special Cases

a) A "usual" p-form is a tensor valued form of type $(0,0)$ and for this case we have $D = d$.

b) A tensor field $t \in \mathcal{T}^r_s(M)$ is a tensor valued 0-form of type (r,s) and

$$Dt = \nabla t \tag{13.82}$$

Indeed, for a coordinate basis $\theta^i = dx^i$, (13.81) reduces with the help of (13.56) to (13.24).

It follows trivially from (13.81) that D satisfies the antiderivation rule for two tensor valued forms:

$$D(\phi \wedge \psi) = D\phi \wedge \psi + (-1)^p \phi \wedge D\psi,$$

where p is the degree of ϕ. Here, $\wedge$ denotes exterior multiplication of the components[3].

Remarks. 1. A connection is metric if and only if $Dg = 0$.

2. The basis θ^i is a tensor valued 1-form of type $(1,0)$. The first structure equation can be written in the form $D\theta^i = \Theta^i$, as one easily infers from (13.81).

We may now write the Bianchi identities in a very compact form. Obviously, Θ^i and $\Omega^i{}_j$ are tensor valued 2-forms of type $(1,0)$ and $(1,1)$, respectively (see the defining equations (13.62), (13.63)).

Proposition 13.9. *The torsion and curvature forms satisfy the following identities :*

$$D\Theta^i = \Omega^i{}_j \wedge \theta^j \qquad \textit{(1st Bianchi identity)}, \tag{13.83a}$$

$$D\Omega^i{}_j = 0 \qquad \textit{(2nd Bianchi identity)}. \tag{13.83b}$$

Proof. Using (13.81) and Cartan's structure equations we have, in matrix notation,

$$\begin{aligned} D\Theta &= d\Theta + \omega \wedge \Theta \\ &= d(d\theta + \omega \wedge \theta) + \omega \wedge (d\theta + \omega \wedge \theta) \\ &= d\omega \wedge \theta - \omega \wedge d\theta + \omega \wedge d\theta + \omega \wedge \omega \wedge \theta = \Omega \wedge \theta\,. \end{aligned}$$

In an analogous manner,

$$\begin{aligned} D\Omega &= d\Omega + \omega \wedge \Omega - \Omega \wedge \omega = d\Omega + \omega \wedge d\omega - d\omega \wedge \omega\,, \\ d\Omega &= d(d\omega + \omega \wedge \omega) = d\omega \wedge \omega - \omega \wedge d\omega \end{aligned}$$

and hence $D\Omega = 0$. □

We now show that (13.83a) and (13.83b) are equivalent to the previous formulations of the Bianchi identities (13.35), (13.36). In a natural basis $\theta^i = dx^i$, (13.83a) has the form (see also (13.68))

---

[3] If ϕ is a tensor valued p-form and ψ a tensor valued q-form, then $(\phi \wedge \psi)(X_1, \ldots, X_{p+q}) = \frac{1}{p!\,q!} \sum_{\sigma \in \mathcal{S}_{p+q}} sgn(\sigma)\, \phi(X_{\sigma(1)} \ldots X_{\sigma(p)}) \otimes \psi(X_{\sigma(p+1)} \ldots X_{\sigma(p+q)})$.

$$\begin{aligned} D\Theta^i &= d\Theta^i + \omega^i{}_j \wedge \Theta^j \\ &= d\left(\frac{1}{2} T^i{}_{kl}\, dx^k \wedge dx^l\right) + \omega^i{}_j \wedge \frac{1}{2} T^j{}_{kl}\, dx^k \wedge dx^l \\ &= \Omega^i{}_j \wedge \theta^j = \frac{1}{2} R^i{}_{jkl}\, dx^k \wedge dx^l \wedge dx^j \end{aligned}$$

or

$$\left(T^i{}_{kl,j} + \Gamma^i{}_{js} T^s{}_{kl}\right) dx^j \wedge dx^k \wedge dx^l = R^i{}_{jkl}\, dx^j \wedge dx^k \wedge dx^l \,.$$

Thus,

$$\sum_{(jkl)} T^i{}_{kl,j} = \sum_{(jkl)} \left(R^i{}_{jkl} - \Gamma^i{}_{js} T^s{}_{kl}\right).$$

We now use (13.30) and obtain

$$\begin{aligned} \sum_{(jkl)} T^i{}_{kl;j} &= \sum_{(jkl)} \left(T^i{}_{kl,j} + \Gamma^i{}_{js} T^s{}_{kl} - \Gamma^s{}_{jk} T^i{}_{sl} - \Gamma^s{}_{jl} T^i{}_{ks}\right) \\ &= \sum_{(jkl)} \left(T^i{}_{kl,j} + \Gamma^i{}_{js} T^s{}_{kl} + T^s{}_{jl} T^i{}_{ks}\right). \end{aligned}$$

Hence,

$$\sum_{(jkl)} R^i{}_{jkl} = \sum_{(jkl)} \left(T^i{}_{kl;j} - T^s{}_{jl} T^i{}_{ks}\right). \tag{13.84}$$

This equation is valid in any basis $\{\theta^i\}$ and is simply (13.35) written in terms of components.

Similarly one finds from (13.83b)

$$\sum_{(klm)} R^i{}_{jkl;m} = \sum_{(klm)} T^s{}_{km} R^i{}_{jsl}\,, \tag{13.85}$$

which is (13.36) in terms of components.

Let ϕ be a tensor valued p-form. If we apply the absolute exterior differential twice and use (13.81) as well as the second structure equation, we obtain immediately

$$(D^2\phi)^{i_1 \dots i_r}_{j_1 \dots j_s} = \Omega^{i_1}{}_l \wedge \phi^{l i_2 \dots i_r}_{j_1 \dots j_s} + \ldots - \Omega^l{}_{j_1} \wedge \phi^{i_1 \dots i_r}_{l j_2 \dots j_s} - \ldots \,. \tag{13.86}$$

Examples.

1. In particular, we have for a function f

$$D^2 f = 0\,.$$

Now $Df = df = f_{,i}\, dx^i = f_{;i}\, dx^i$ and, using the antiderivation rule,

$$D^2 f = D f_{;i} \wedge dx^i + f_{;i} D\, dx^i\,.$$

By the first structure equation $D\, dx^i = \Theta^i = \frac{1}{2} T^i{}_{kl}\, dx^k \wedge dx^l$. Hence, with (13.82), we have

$$D^2 f = f_{;i;k}\, dx^k \wedge dx^i + \frac{1}{2} f_{;s}\, T^s_{\ kl}\, dx^k \wedge dx^l \,,$$

so that

$$f_{;i;k} - f_{;k;i} = T^s_{\ ik} f_{;s} \,. \tag{13.87}$$

2. For a vector field ξ^i Eq. (13.86) becomes

$$D^2 \xi^i = \Omega^i_{\ l} \xi^l = \frac{1}{2} R^i_{\ jkl}\, \xi^j\, dx^k \wedge dx^l \,.$$

On the other hand,

$$\begin{aligned} D^2 \xi^i &= D(\xi^i_{\ ;l} dx^l) = D\xi^i_{\ ;l} \wedge dx^l + \xi^i_{\ ;s} \Theta^s \\ &= \xi^i_{\ ;l;k}\, dx^k \wedge dx^l + \xi^i_{\ ;s} \frac{1}{2} T^s_{\ kl}\, dx^k \wedge dx^l \,. \end{aligned}$$

Direct comparison results in

$$\xi^i_{\ ;l;k} - \xi^i_{\ ;k;l} = R^i_{\ jkl} \xi^j + T^s_{\ lk} \xi^i_{;s} \,. \tag{13.88}$$

In particular, if $\Theta = 0$, we have

$$\begin{aligned} f_{;i;k} &= f_{;k;i} \,, \\ \xi^i_{\ ;l;k} - \xi^i_{\ ;k;l} &= R^i_{\ jkl} \xi^j \qquad \textbf{(Ricci identity)} \,. \end{aligned} \tag{13.89}$$

13.9 Locally Flat Manifolds

Definition 13.12. *Let (M, g) and (N, h) be two pseudo-Riemannian manifolds. A diffeomorphism $\phi : M \longrightarrow N$ is an* ***isometry*** *if $\phi^* h = g$. Every pseudo-Riemannian manifold which is isometric to $(\mathbb{R}^n, \bar{g})$, with*

$$\bar{g} = \sum_{i=1}^{n} \varepsilon_i\, dx^i\, dx^i \,, \qquad \varepsilon_i = \pm 1 \tag{13.90}$$

is said to be ***flat****. If a space is locally isometric to a flat space, then it is said to by* ***locally flat****.*

Theorem 13.6. *A pseudo-Riemannian space is locally flat if and only if the curvature of the Riemannian connection vanishes.*

Proof. If the space is locally flat, the curvature vanishes. In order to see this, choose a chart in the neighborhood of a point for which the metric has the form (13.90). In this neighborhood the Christoffel symbols vanish and hence also the curvature.

In order to prove the converse, we use Theorem 13.7 below. According to this theorem, parallel transport is locally independent of the path when

$\Omega^i{}_j = 0$. Hence it is possible to parallel displace a basis of vectors in $T_p(M)$, $p \in M$, to all points in some neighborhood of p. In this manner we obtain local basis fields e_i which have vanishing covariant derivative. For the basis $\{\theta^i\}$ dual to $\{e_i\}$ we always have

$$d\theta^i(e_j, e_k) = -\theta^i([e_j, e_k]).$$

But for a symmetric connection

$$[e_j, e_k] = \nabla_{e_j} e_k - \nabla_{e_k} e_j = 0.$$

Consequently $d\theta^i = 0$. According to the Lemma of Poincaré 12.2 there exist local functions $x^i : U \longrightarrow \mathbb{R}$, $p \in U$, for which $\theta^i = dx^i$. If we choose these x^i as coordinates, then $e_i = \partial/\partial x^i$ and

$$0 = \nabla_{e_i} e_j = \Gamma^k{}_{ij} e_k .$$

Thus the $\Gamma^k{}_{ij}$ vanish identically in the coordinate system x^i in a neighborhood of p. If the connection is pseudo-Riemannian, then the metric coefficient g_{ik} are constant in the neighborhood U (see, e.g., (13.69)). In this case, they can be transformed to the normal form (13.90) by a suitable choice of coordinates. □

Theorem 13.7. *For an affine connection, parallel transport is independent of the path if and only if the curvature tensor vanishes.*

We prove this theorem after the following heuristic discussion. Let $\gamma : [0,1] \longrightarrow M$ be a closed path, with $\gamma(0) = p$. We displace an arbitrary vector $v_0 \in T_p(M)$ parallel along γ and obtain the field $v(t) = \tau_t v_0$, $v(t) \in T_{\gamma(t)}(M)$. We assume that the closed path is sufficiently small that we can work in the image of some particular chart. Then $\dot{v}^i = -\Gamma^i{}_{jk} \dot{x}^j v^k$. We are interested in

$$\begin{aligned} \Delta v^i = v^i(1) - v^i(0) &= \int_0^1 \dot{v}^i \, dt = -\int_0^1 \Gamma^i{}_{jk}(\gamma(t)) v^k \dot{x}^j \, dt \\ &= -\int_\gamma \Gamma^i{}_{kj} v^j \, dx^k = -\int_\gamma \omega^i{}_j v^j . \end{aligned} \tag{13.91}$$

Let us choose the coordinate system to be normal at the point $p = \gamma(0)$. We now evaluate the path integral (13.91) for an infinitesimal loop. For this purpose we use

$$\omega^i{}_j v^j = \omega^i{}_j (v^j - v^j_0) + \omega^i{}_j v^j_0$$

and replace $\omega^i{}_j$ by $\Gamma^i{}_{kj}(p) dx^k = 0$ in the first term of the right hand side. Using the Theorem of Stokes and the second structure equation, we obtain, when σ denotes a surface enclosed by γ,

$$\Delta v^i \simeq -\int_\gamma \omega^i{}_j\, v_0^j = -\int_\sigma d\omega^i{}_j\, v_0^j$$
$$\simeq -\frac{1}{2} R^i{}_{jkl}(p) v_0^j \int_\sigma dx^k \wedge dx^l. \tag{13.92}$$

Equation (13.92) implies heuristically Theorem 13.7. We now give a rigorous proof of this theorem.

Proof (of Theorem 13.7). Consider the 1-parameter family of curves $H(t,s)$, $(\alpha \le t \le \beta)$, $s \in J$, with $H(\alpha, s) = p$ and $H(\beta, s) = q$, for all $s \in J$. Let $v \in T_p(M)$ and let $Y(t,s)$ be the set of vectors which one obtains by parallel displacement of v along $t \longmapsto H(t,s)$ for every fixed s. Let V be the field tangent to the curves $t \longmapsto H(t,s)$ and T be the field tangent to the curves $s \longmapsto H(t,s)$ (for fixed t). With this notation, we have $Y(\alpha, s) = v, s \in J$, and

$$(\nabla_V Y)_{t,s} = 0\,, \qquad (\nabla_T Y)_{\alpha,s} = 0\,. \tag{13.93}$$

Naturally, $[V,T] = 0$ since differentiation with respect to s and t is commutative (see the Remark below). Now suppose that the curvature vanishes. It follows from (13.93) that

$$\nabla_V \nabla_T Y|_{t,s} = \nabla_T \nabla_V Y|_{t,s} = 0\,,$$

which means that the family $t \longmapsto \nabla_T Y$ is autoparallel along $t \longmapsto H(t,s)$. Together with the second equation of (13.93) this implies

$$\nabla_T Y|_{t,s} = 0\,. \tag{13.94}$$

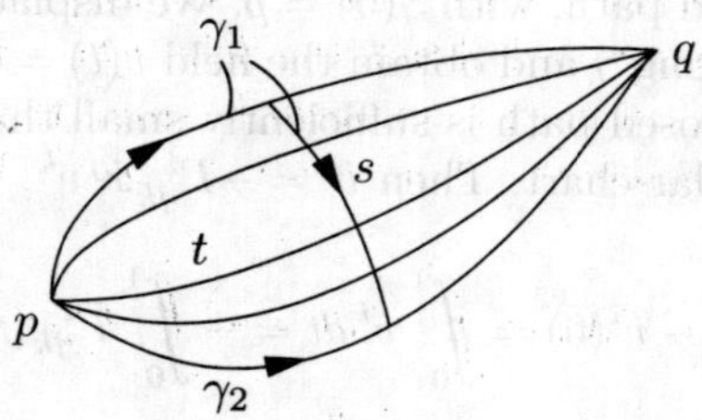

Fig. 13.3. Parallel transport along a one-parameter family of paths between two points.

One thus obtains the same result whether v is transported parallel along the path segment γ_1 or γ_2 in Fig. 13.3. By going to the limit, one sees that parallel displacement to q is independent of the path.

Conversely, assume that parallel displacement is independent of the path. Then (13.94) is valid, and from this we conclude that

$$(\nabla_V \nabla_T - \nabla_T \nabla_V) Y = R(V,T) Y = 0$$

for all t, s and hence that $\Omega = 0$. □

Remark. Above we have been working somewhat intuitively with "vector fields along maps" and their Lie brackets and induced covariant derivatives. For precisions we refer to Sect. 14.3 and 14.4. (A special case was discussed in the remark at the end of Sect. 13.1.)

* * *

Exercises.

Definition 13.13. *The* **Laplacian** *on differential forms is defined by*

$$\Delta = d \circ \delta + \delta \circ d . \tag{13.95}$$

Exercise 13.2. Show that

$$*\Delta = \Delta * , \quad \delta\Delta = \Delta\delta , \quad d\Delta = \Delta d . \tag{13.96}$$

Exercise 13.3. Prove that Δ is self-adjoint, in the sense that

$$(\Delta\alpha, \beta) = (\alpha, \Delta\beta) , \tag{13.97}$$

where

$$(\alpha, \beta) = \int_M \langle \alpha, \beta \rangle \eta = \int_M \alpha \wedge *\beta , \tag{13.98}$$

by showing that d and δ are adjoint to each other, up to a sign:

$$(d\alpha, \beta) = -(\alpha, \delta\beta) , \qquad \alpha \in \bigwedge^{p-1}(M) , \; \beta \in \bigwedge^{p}(M) . \tag{13.99}$$

Exercise 13.4. Show that on a proper Riemannian manifold the form ω is **harmonic**, i.e. $\Delta\omega = 0$, if and only if $d\omega = \delta\omega = 0$.

Exercise 13.5. Derive a coordinate expression for $\Delta\omega$.

Exercise 13.6. Prove that a harmonic function with compact support on a Riemannian manifold is constant.

Exercise 13.7. Use (12.56) and show that for a vector field X

$$div\, X = \delta X^\flat = X^k{}_{;k} . \tag{13.100}$$

Exercise 13.8. Let ω be a 1-form and K the corresponding vector field relative to the metric g ($K = \omega^\sharp$). Prove the following identity

$$\nabla\omega = \frac{1}{2}\left(-d\omega + L_K\, g\right) , \tag{13.101}$$

where ∇ is the Levi-Cività connection.

Solution. From (13.18) and (13.19) we have

$$\nabla\omega(X,Y) = (\nabla_Y\omega)(X) = Y\omega(X) - \omega(\nabla_Y X)$$

and, interchanging X with Y,

$$\nabla\omega(Y,X) = X\dot{\omega}(Y) - \omega(\nabla_X Y)\,.$$

We consider the difference and the sum of the two equations. According to (12.14) the difference is equal to $-d\omega(X,Y)$. For the sum we get

$$\begin{aligned} &X\omega(Y) + Y\omega(X) - \omega(\nabla_X Y + \nabla_Y X) \\ &\quad = Xg(K,Y) + Yg(K,X) - g(K, \nabla_X Y + \nabla_Y X) = (L_K g)(X,Y)\,. \end{aligned}$$

The last equality sign is obtained as follows. From general formulas we get

$$\begin{aligned} (L_K g)(X,Y) &= Kg(X,Y) - g(\underbrace{[K,X]}_{\nabla_K X - \nabla_X K},Y) - g(X,\underbrace{[K,Y]}_{\nabla_K Y - \nabla_Y K}) \\ &= Kg(X,Y) - g(\nabla_K X, Y) - g(X, \nabla_K Y) + g(\nabla_X K, Y) \\ &\quad + g(X, \nabla_Y K)\,. \end{aligned}$$

The first three terms add up to $\nabla g(X,Y) = 0$. Applying the Ricci identity to the last two terms gives the claimed equality. Together we obtain the identity (13.101).

Exercise 13.9. As a generalization of (13.100) show that the codifferential of a p-form ω is given by

$$\delta\omega = \frac{1}{(p-1)!}\nabla^j \omega_{jk_1\ldots k_{p-1}}\, dx^{k_1}\wedge\ldots\wedge dx^{k_{p-1}}\,. \tag{13.102}$$

Hint. Use normal coordinates, as well as (12.35).

Exercise 13.10. On a two-dimensional Riemannian manifold (M,g) one can always introduce coordinates such that the metric is conformally flat:

$$g = \lambda^2(x,y)\left(dx^2 + dy^2\right)\,. \tag{13.103}$$

(The existence of such **isothermal coordinates** is difficult to prove, except in the real analytic case which was already treated by Gauss; see, e.g., [43].)

a) Determine with the help of the structure equations the curvature scalar R, which is twice the Gaussian curvature K. If $L = 1/\lambda$, $L_x = \partial L/\partial x$, etc., one finds for K

$$K = L(L_{xx} + L_{yy}) - (L_x^2 + L_y^2)\,. \tag{13.104}$$

b) Choose $L(x,y) = y$ and conclude that **Poincaré's half-plane** $M = \{(x,y) \in \mathbb{R}^2 \,|\, y > 0\}$ with the Riemannian metric

$$g = \frac{dx^2 + dy^2}{y^2} \tag{13.105}$$

has Gaussian curvature $K = -1$.

c) Set $\lambda = \mathrm{e}^{\varphi}$ and derive the differential equation for φ such that $K = \pm 1$. The result is the **Liouville equation**

$$\Delta\varphi \pm \mathrm{e}^{2\varphi} = 0\,, \tag{13.106}$$

where Δ is the flat Laplace operator.

13.10 Weyl Tensor and Conformally Flat Manifolds

For a pseudo-Riemannian manifold of dimension $n \geq 3$ the **Weyl tensor** C_{ijkl} is defined as that part of the Riemann tensor with vanishing contractions

$$\begin{aligned} R_{ijkl} =: C_{ijkl} &+ \frac{2}{n-2}\left(g_{i[k}R_{l]j} - g_{j[k}R_{l]i}\right) \\ &- \frac{2}{(n-1)(n-2)} R\, g_{i[k}g_{l]j}\,. \end{aligned} \tag{13.107}$$

As always, a square bracket enclosing a group of indices denotes antisymmetrization in these indices, while a round bracket means symmetrization.

The Weyl tensor C_{ijkl} has the same symmetries as R_{ijkl}. It behaves very simply under conformal transformations $\tilde{g} = e^{2\phi} g$:

$$\tilde{C}_{ijkl} = e^{2\phi} C_{ijkl}\,. \tag{13.108}$$

For this reason it is often called the **conformal tensor**.

An important property of the Weyl tensor is that for $n \geq 4$ it vanishes if and only if the metric is locally conformally flat. (For a proof, see, e.g., [44].)

14 Some Details

In this section we give detailed proofs of some of the theorems formulated in previous sections. In addition, we clarify the notion of vector fields along maps and their induced covariant derivatives, because this is used at various places in the book. For a convenient formulation we introduce the tangent bundle of a manifold, the prototype of a vector bundle. Applications to variations of curves will illustrate the usefulness of the concepts.

14.1 Proofs of Some Theorems

Proof (of Theorem 11.6).

a) First we show the following: Let D_X and D_Y be the derivations of $\mathcal{F}(M)$ belonging to the vector fields X and Y. Then X and Y are ϕ-related if and only if $\phi^* \circ D_Y = D_X \circ \phi^*$. To see this we note

$$D_X \circ \phi^*(f)|_p = D_X(f \circ \phi)|_p = X_p(f \circ \phi) = (T_p\phi \cdot X_p)f$$

and

$$\phi^* \circ D_Y(f)|_p = (D_Y f)|_{\phi(p)} = Y_{\phi(p)} f \,.$$

The left-hand sides of these two equations are obviously equal if and only if X and Y are ϕ-related.

b) With a straightforward calculation one establishes the following: Suppose that derivations D_i and D'_i, for $i = 1, 2$, of $\mathcal{F}(M)$ are related as

$$\phi^* \circ D'_i = D_i \circ \phi^* \,,$$

then $\phi^* \circ [D'_1, D'_2] = [D_1, D_2] \circ \phi^*$.

c) Using the notation of Theorem 11.6 b) implies that

$$\phi^* \circ \underbrace{[D_{Y_1}, D_{Y_2}]}_{D_{[Y_1,Y_2]}} = \underbrace{[D_{X_1}, D_{X_2}]}_{D_{[X_1,X_2]}} \circ \phi^*$$

and hence a) that $[X_1, X_2]$ and $[Y_1, Y_2]$ are ϕ-related.

□

Proof (of Theorem 11.8). Only the last statement $L_X Y = [X, Y]$ is not obvious. To show this the following remark is useful. Let $f : (-\varepsilon, \varepsilon) \times M \longrightarrow \mathbb{R}$ be smooth (C^∞) and $f(0, p) = 0$ for all $p \in M$, then there is a C^∞ function $g : (-\varepsilon, \varepsilon) \times M \longrightarrow \mathbb{R}$ with

$$f(t,p) = t\,g(t,p)\,, \qquad \frac{\partial f}{\partial t}(0,p) = g(0,p)\,. \tag{14.1}$$

A function with these properties is

$$g(t,p) = \int_0^1 f'(s\,t,p)\,ds\,,$$

where the prime denotes the derivative of f with respect to the first argument.

Now let $f \in \mathcal{F}(M)$ and ϕ_t be the flow of X, $|t| < \varepsilon$. From what has just been said we know that there is a family of C^∞ functions g_t on M such that

$$f \circ \phi_t = f + t\,g_t\,, \qquad g_0 = Xf\,. \tag{14.2}$$

The last proposition of Theorem 11.8 is equivalent to

$$[X,Y]_p = \lim_{t \to 0} \frac{1}{t}\left[Y_p - T_p\phi_t \cdot Y_q\right]\,, \quad q = \phi_{-t}(p)$$

or

$$[X,Y] = \lim_{t \to 0} \frac{1}{t}\left[Y - \phi_{t*}Y\right].$$

For calculating the right-hand side we determine its action on f. We have, using Equation (14.2),

$$\left(\phi_{t*}Y\right)_p(f) = \left(T_q\phi_t \cdot Y_q\right)(f) = Y_q(f \circ \phi_t) = Y_{\phi_{-t}(p)}(f + t\,g_t)\,,$$

so

$$\begin{aligned}
&\lim_{t \to 0} \frac{1}{t}\left[Y_p - (\phi_{t*}Y)_p\right](f) \\
&= \lim_{t \to 0} \frac{1}{t}\Big[(Yf)(p) - (Yf)\big(\phi_{-t}(p)\big)\Big] - \lim_{t \to 0}(Y g_t)\big(\phi_{-t}(p)\big) \\
&= \big(L_X(Yf)\big)(p) - (Y g_0)(p) = \big(X(Yf) - Y(Xf)\big)(p)\,.
\end{aligned}$$

This proves $L_X Y = [X, Y]$. □

Proof (of Theorem 11.9). Both statements will immediately follow from the

Lemma 14.1. *Two derivations of the algebra of tensor fields $\mathcal{T}(M)$, which preserve the type and commute with all contractions, are equal if they agree on $\mathcal{F}(M)$ and $\mathcal{X}(M)$.*

Before we prove this, we show that the Lemma implies Theorem 11.9. For the first statement this is obvious. The second statement follows from

$$[L_X, L_Y]f = XYf - YXf = [X,Y]f = L_{[X,Y]}f\,,$$

for $f \in \mathcal{F}(M)$, and, using the Jacobi identity,

$$[L_X, L_Y]Z = [X,[Y,Z]] - [Y,[X,Z]] = [[X,Y],Z]\,,$$

for $Z \in \mathcal{X}(M)$. □

Proof (of Lemma 14.1). First one shows, as in the proof of Theorem 12.4, that derivations can be localized. The Lemma 14.1 is equivalent to the statement that for a derivation D with the stated properties, $D|\mathcal{F}(M) = 0$ and $D|\mathcal{X}(M) = 0$ imply that $D = 0$. Now, $K \in \mathcal{T}^r_s(M)$ can locally be represented on the domain of a chart as

$$K = K^{i_1 \dots i_r}_{j_1 \dots j_s} \partial_{i_1} \otimes \dots \otimes \partial_{i_r} \otimes dx^{j_1} \otimes \dots \otimes dx^{j_s}\,.$$

The functions $K^{i_1 \dots i_r}_{j_1 \dots j_s}$, the vector fields ∂_i and the 1-forms dx^j can be extended to M such that equality holds in a smaller neighborhood (see Lemma 12.1). Clearly, $DK = 0$ once we show that $D\omega = 0$ for every 1-form on M. For this, let Y be a vector field and C the contraction from $\mathcal{T}^1_1(M)$ to $\mathcal{F}(M)$. Then

$$0 = D\big(\underbrace{C(Y \otimes \omega)}_{=\,\omega(Y)}\big) = C\big(D(Y \otimes \omega)\big) = C(\underbrace{DY}_{=\,0} \otimes \omega) + C(Y \otimes D\omega) = (D\omega)(Y)\,.$$

□

As a preparation for the proof of Theorem 11.10 we need the

Proposition 14.1. *Let φ be a diffeomorphism of M and $X \in \mathcal{X}(M)$ with local flow ϕ_t. Then $\varphi \circ \phi_t \circ \varphi^{-1}$ is a local flow of $\varphi_* X$.*

Proof. Obviously, $\varphi \circ \phi_t \circ \varphi^{-1}$ is a local one-parameter group of local transformations. Let $p \in M$ be an arbitrary point and $q = \varphi^{-1}(p)$. The vector $X_q \in T_q(M)$ is tangential to the curve $\gamma(t) = \phi_t(q)$ at $q = \gamma(0)$. Hence $(\varphi_* X)_p = T_q\varphi \cdot X_q \in T_p(M)$ is tangential to the curve $\tilde{\gamma} = \varphi \circ \gamma$, $\tilde{\gamma}(t) = \big(\varphi \circ \phi_t \circ \varphi^{-1}\big)(p)$. □

Corollary 14.1. *The vector field $X \in \mathcal{X}(M)$ is invariant under φ (i.e., $\varphi_* X = X$) if and only if*

$$\varphi \circ \phi_t = \phi_t \circ \varphi\,.$$

Proof (of Theorem 11.10). The Corollary 14.1 implies, that from $\phi_t \circ \psi_s = \psi_s \circ \phi_t$ follows the invariance of Y under ϕ_t and thus $L_X Y = 0$, i.e., $[X,Y] = 0$. Furthermore, using the definition of the Lie derivative we conclude from

$(\phi_t \circ \psi_s)^* = \psi_s^* \circ \phi_t^* = (\psi_s \circ \phi_t)^* = \phi_t^* \circ \psi_s^*$ that $L_X \circ L_Y = L_Y \circ L_X$. Conversely, let $[X, Y] = 0$. Then Proposition 14.2 below implies

$$\frac{d}{dt}\phi_{t*}Y = 0\,,$$

for every t in the domain of definition. Thus $\phi_{t*}Y$ is constant in any point $p \in M$, so Y is invariant under ϕ_t. By the Corollary ψ_s commutes with every ϕ_t. As above this implies $L_X \circ L_Y = L_Y \circ L_X$. Finally, let us assume the second property of Theorem 11.10: $L_X \circ L_Y = L_Y \circ L_X$. By the second statement of Theorem 11.9 we then have $L_{[X,Y]} = 0$, and thus $[X, Y] = 0$. □

Proposition 14.2. *With the previous notation the following equality holds*

$$\left.\frac{d}{dt}\right|_{t=s} \phi_{t*}Y = -\phi_{s*}[X, Y]\,. \tag{14.3}$$

Proof. For a fixed s we consider the vector field $\phi_{s*}Y$. By the last statement of Theorem 11.8 we know that

$$[X, \phi_{s*}Y] = \lim_{t\to 0}\frac{1}{t}\big[\phi_{s*}Y - \underbrace{(\phi_{t*}\circ\phi_{s*})}_{(\phi_{t+s})_*}Y\big]\,.$$

On the other hand the Corollary 14.1 implies $\phi_{s*}X = X$. Since ϕ_{s*} preserves the Lie bracket (see Theorem 11.6), we obtain

$$\phi_{s*}[X, Y] = [X, \phi_{s*}Y]\,.$$

By comparison with the previous equation we find

$$\phi_{s*}[X, Y] = \lim_{t\to 0}\frac{1}{t}\big[\phi_{s*}Y - (\phi_{t+s})_*Y\big] = -\left.\frac{d}{dt}\right|_{t=s}\phi_{t*}Y\,.$$

□

Proof (of Poincaré's Lemma 12.2). It suffices to show that a closed p-form ω on open ball $U = \{x \in \mathbb{R}^n : |x| < 1\}$ of $\mathbb{R}^n$ is exact. (This is then also true for any domain diffeomorphic to U.)

Cartan's formula (12.11) implies for any vector field X

$$L_X\omega = d\,i_X\omega\,.$$

We use the radial field $X = x^i\partial_i$ and construct a map $H : \bigwedge^p(U) \longrightarrow \bigwedge^p(U)$ that satisfies the two equations

$$\begin{array}{ll} a) & H \circ L_X = id\,, \\ b) & d \circ H = H \circ d\,. \end{array}$$

Once we have shown this we are done, because we then have

$$\omega = H(L_X\omega) = H\circ d(i_X\omega) = d(H\, i_X\omega)\,.$$

We claim that the following map H has the desired properties. Let

$$H(f\, dx^{i_1}\wedge\ldots\wedge dx^{i_p}) = \left(\int_0^1 t^{p-1} f(tx)\, dt\right) dx^{i_1}\wedge\ldots\wedge dx^{i_p}$$

and extend H to all p-forms by linearity. We verify the two properties by direct calculations.

Verification of a). We have

$$\begin{aligned}
&H\circ L_X\big(f\, dx^{i_1}\wedge\ldots\wedge dx^{i_p}\big)\\
&= H\Big(x^i\partial_i f\, dx^{i_1}\wedge\ldots\wedge dx^{i_p} + f\, L_X(dx^{i_1}\wedge\ldots\wedge dx^{i_p})\Big)\\
&= H\Big(x^i\partial_i f\, dx^{i_1}\wedge\ldots\wedge dx^{i_p} + p\, f\, dx^{i_1}\wedge\ldots\wedge dx^{i_p}\Big)\\
&= \left[\left(\int_0^1 t^{p-1} t\, x^i\partial_i f(tx)\, dt\right) + \left(\int_0^1 p\, t^{p-1} f(tx)\, dt\right)\right] dx^{i_1}\wedge\ldots\wedge dx^{i_p}\\
&= \left(\int_0^1 \frac{d}{dt}\big(t^p f(tx)\big) dt\right) dx^{i_1}\wedge\ldots\wedge dx^{i_p}\\
&= f(x)\, dx^{i_1}\wedge\ldots\wedge dx^{i_p}\,.
\end{aligned}$$

Verification of b). We have

$$\begin{aligned}
H\circ d\big(f\, dx^{i_1}\wedge\ldots\wedge dx^{i_p}\big) &= H\big(\partial_i f\, dx^{i_1}\wedge\ldots\wedge dx^{i_p}\big)\\
&= \left[\left(\int_0^1 t^p\, \partial_i f(tx)\, dt\right) dx^i\right]\wedge dx^{i_1}\wedge\ldots\wedge dx^{i_p}\\
&= d\left(\int_0^1 t^{p-1} f(tx)\, dt\right)\wedge dx^{i_1}\wedge\ldots\wedge dx^{i_p}\\
&= d\left[\left(\int_0^1 t^{p-1} f(tx)\, dt\right) dx^{i_1}\wedge\ldots\wedge dx^{i_p}\right]\\
&= d\circ H\big(f\, dx^{i_1}\wedge\ldots\wedge dx^{i_p}\big).
\end{aligned}$$

□

* * *

Exercise 14.1. Show that the Proposition 14.2 can be generalized to an arbitrary tensor field $T\in\mathcal{T}(M)$:

$$\frac{d}{dt}\phi_{t*}T = -\phi_{t*}(L_X T)\,. \tag{14.4}$$

Hints. From the group property of ϕ_t one deduces readily

$$\frac{d}{dt}\phi_{t*}T = -L_X(\phi_{t*}T)\,.$$

It remains to show that $L_X \circ \phi_{t*} = \phi_{t*} \circ L_X$. For this consider the action of the derivations L_X and $(\phi_{t*})^{-1} \circ L_X \circ \phi_{t*}$ on $\mathcal{F}(M)$ and $\mathcal{X}(M)$, and use the Lemma in the proof of Theorem 11.9.

Remark. Formula (14.4) implies, in particular, the statement of Theorem 11.11.

14.2 Tangent Bundles

Definition 14.1. *If M is an n-dimensional differentiable manifold, the* **tangent bundle** *of M, denoted by TM, is the disjoint union of all tangent spaces of M:*

$$TM = \bigcup_{p\in M} T_pM\,. \tag{14.5}$$

Thus elements (points) of TM are tangent vectors to M at some point $p \in M$. The **natural projection** $\pi_M : TM \longrightarrow M$ maps a tangent vector $v \in T_pM$ to the point $p \in M$ at which the vector v is attached. Obviously, the inverse image $\pi_M^{-1}(p)$ is the tangent space T_pM, also called the **fiber** of the tangent bundle over the point $p \in M$.

The differentiable structure of M naturally induces one of TM, turning this set into a differentiable manifold. Note first that each chart (U, h) of M with coordinates $\{x^i\}$ defines a **natural bundle chart** with domain $TU = \pi_M^{-1}(U)$ and the following mapping $Th : TU \longrightarrow \mathbb{R}^{2n}$. The image of $v \in T_pM$, with $v = v^i\partial_i$, is the point $x^1(p), \ldots, x^n(p), v^1, \ldots, v^n$ in $\mathbb{R}^{2n}$. If this is done for each chart of an atlas of M, it is easy to see that the corresponding set of natural bundle charts satisfies the conditions in the Remark on p. 534 of Sect. 9, and thus defines a differentiable structure (and a unique Hausdorff topology) of TM. Moreover, π_M is then a smooth map.

If $\phi : M \longrightarrow N$ is a smooth map between two differentiable manifolds, then $T\phi : TM \longrightarrow TN$ denotes the mapping $v \in T_pM \longmapsto T_p\phi \cdot v \in T_{\phi(p)}N$. The following diagram is commutative

$$\begin{array}{ccc} TM & \xrightarrow{\;T\phi\;} & TN \\ \Big\downarrow{\scriptstyle \pi_M} & & \Big\downarrow{\scriptstyle \pi_N} \\ M & \xrightarrow[\;\phi\;]{} & N\,. \end{array}$$

Moreover, we have the **chain rule**

$$T(\psi \circ \phi) = T\psi \circ T\phi . \tag{14.6}$$

It should also be clear that a differentiable vector field $X \in \mathcal{X}(M)$ can equivalently be defined as a smooth map $X : M \longrightarrow TM$ such that the diagram

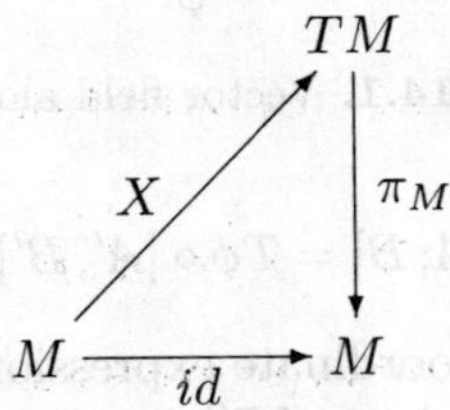

is commutative: $\pi_M \circ X = id$. Such maps are also called smooth **sections** of the tangent bundle.

14.3 Vector Fields Along Maps

We now use this convenient language to generalize the notion of vector fields.

Definition 14.2. *Let $\phi : N \longrightarrow M$ be a smooth mapping. A* **vector field along** *ϕ is a smooth map $A : N \longrightarrow TM$ such that the diagram*

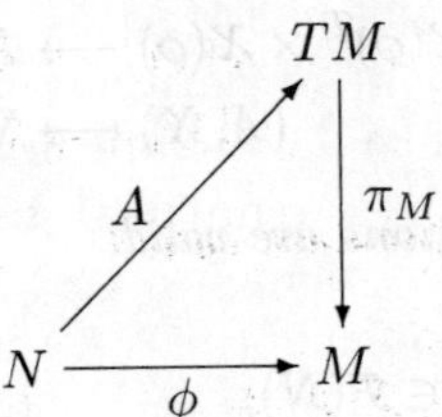

commutes.

The set $\mathcal{X}(\phi)$ of such vector fields is naturally a module over $\mathcal{F}(N)$: For $A \in \mathcal{X}(\phi)$ and $f \in \mathcal{F}(N)$ we define $f\,A$ by $(f\,A)(p) = f(p)A(p)$. $\mathcal{X}(\phi)$ is also an $\mathcal{F}(M)$-module: If $g \in \mathcal{F}(M)$ then $g \circ \phi \in \mathcal{F}(N)$ and thus $(g \circ \phi)A \in \mathcal{X}(\phi)$. Each vector field $Y \in \mathcal{X}(M)$ defines the vector field $Y \circ \phi \in \mathcal{X}(\phi)$.

Special vector fields along ϕ are of the form

$$A = T\phi \circ A' , \qquad A' \in \mathcal{X}(N) , \tag{14.7}$$

called the **tangential vector fields along** ϕ (see Fig. 14.1). We denote the set of such fields by $\mathcal{X}(\phi)^T$.

For two tangential vector fields $A = T\phi \circ A'$ and $B = T\phi \circ B'$ we define the bracket naturally by

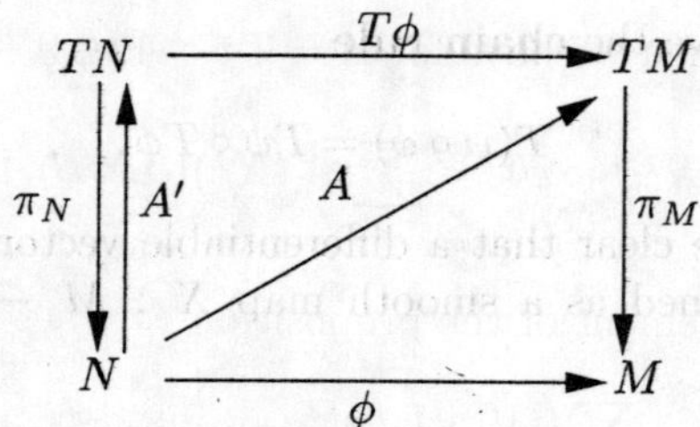

Fig. 14.1. Vector field along ϕ.

$$[A, B] = T\phi \circ [A', B'] . \tag{14.8}$$

This is again tangential. A coordinate expression is given in (14.12) below.

In practice N is often a subset of $\mathbb{R}^2$ (see Sect. 14.4). Commuting tangential vector fields are then given by the partial derivatives in $\mathbb{R}^2$.

Induced Covariant Derivatives

Let now ∇ be a connection on M. We show below that this defines an associated covariant derivative $\tilde{\nabla}$ of vector fields along ϕ in the direction of tangential vector fields along ϕ.

Proposition 14.3. *Let $\phi : N \longrightarrow M$ be a smooth map and ∇ an affine connection on M. Then there is a unique map*

$$\tilde{\nabla} : \mathcal{X}(\phi)^T \times \mathcal{X}(\phi) \longrightarrow \mathcal{X}(\phi) ,$$
$$(A, X) \longmapsto \tilde{\nabla}_A X ,$$

such that the following equations are valid:

(i) $\tilde{\nabla}_{A+B} X = \tilde{\nabla}_A X + \tilde{\nabla}_B X$;
(ii) $\tilde{\nabla}_{fA} X = f\,\tilde{\nabla}_A X$ *with* $f \in \mathcal{F}(N)$;
(iii) $\tilde{\nabla}_A (X + Y) = \tilde{\nabla}_A X + \tilde{\nabla}_A Y$;
(iv) $\tilde{\nabla}_A (fX) = f\,\tilde{\nabla}_A X + (A'f)X$ *for* $A = T\phi \circ A'$ *and* $f \in \mathcal{F}(N)$;
(v) For the special case $X = Y \circ \phi$, $Y \in \mathcal{X}(M)$ *the following holds*

$$\tilde{\nabla}_A (Y \circ \phi) = (\nabla_A Y) \circ \phi ,$$

where $\big((\nabla_A Y) \circ \phi\big)(p) = \nabla_{A(p)} Y \in T_{\phi(p)} M$ *is the covariant derivative (in* M*) of* Y *in the direction* $A(p)$.

Proof. We first show the *uniqueness*. Assume that the image $\phi(U)$ of an open set $U \subset N$ is in the domain of a chart of M with coordinates $\{x^1, \ldots, x^n\}$. An $X \in \mathcal{X}(\phi)$ can be expanded in a neighborhood of a point $\phi(p)$, $p \in U$, as

$$X = \sum_{j=1}^{n} X^j (\partial_j \circ \phi) ,$$

with $X^j \in \mathcal{F}(U)$. Then we have for $A = T\phi \circ A' \in \mathcal{X}(\phi)^T$

$$\tilde{\nabla}_A X = \sum_j \tilde{\nabla}_A (X^j \partial_j \circ \phi) \overset{(iv)}{=} \sum_j A'(X^j)(\partial_j \circ \phi) + \sum_j X^j \tilde{\nabla}_A (\partial_j \circ \phi)$$

or with the last statement of the proposition

$$\tilde{\nabla}_A X = \sum_j \Big(A'(X^j)(\partial_j \circ \phi) + X^j (\nabla_A \partial_j) \circ \phi \Big) . \tag{14.9}$$

The right-hand side is of this equation is uniquely determined by ∇.

Next, we show the *existence*. Conversely, define $\tilde{\nabla}_A$ by Equation (14.9) for $U \subset M$, such that $\phi(U)$ is in the domain of a chart of M. A direct calculation shows, that all five properties of the Proposition 14.3 are satisfied. Because of the uniqueness of the local definition of $\tilde{\nabla}_A X$, a vector field in $\mathcal{X}(\phi)$ is defined. □

Definition 14.3. *For X and $Y \in \mathcal{X}(\phi)$ we define the* ***torsion*** $\tilde{T}$ ***along*** ϕ *by $\tilde{T}(X,Y)(p) = T(X_p, Y_p)$. The right-hand side is well-defined since the torsion T of ∇ is a tensor field. Correspondingly, we define the* ***curvature*** $\tilde{R}(X,Y)Z$ ***along*** ϕ, *for $X, Y, Z \in \mathcal{X}(\phi)$.*

As expected, the following holds

Proposition 14.4. *For $A, B \in \mathcal{X}(\phi)^T$ and $X \in \mathcal{X}(\phi)$ the following equations are valid*

$$\tilde{T}(A,B) = \tilde{\nabla}_A B - \tilde{\nabla}_B A - [A,B] , \tag{14.10a}$$

$$\tilde{R}(A,B)X = (\tilde{\nabla}_A \tilde{\nabla}_B - \tilde{\nabla}_B \tilde{\nabla}_A)X - \tilde{\nabla}_{[A,B]} X . \tag{14.10b}$$

Proof. It suffices to establish these equations locally. Then we can use for the covariant derivative $\tilde{\nabla}$ expression (14.9). For calculating $[A,B]$ we need an explicit expression for $A = T\phi \circ A'$. We can write

$$(T\phi \circ A')(p) = T\phi(A'(p)) = T_p\phi \cdot A'(p) = \sum_i (T_p\phi \cdot A'(p))(x^i) \partial_i|_{\phi(p)} ,$$

so

$$A = T\phi \circ A' = \sum_i A'(x^i \circ \phi)(\partial_i \circ \phi) . \tag{14.11}$$

This gives with Equation (14.8)

$$[A,B] = T\phi \circ [A', B'] = \sum_i [A', B'](x^i \circ \phi) \partial_i \circ \phi .$$

Thus (dropping from now on the summation symbol) we have

$$[A,B] = \Big(A'\big(B'(x^i \circ \phi)\big) - B'\big(A'(x^i \circ \phi)\big) \Big) \partial_i \circ \phi . \tag{14.12}$$

By definition of $\tilde{T}$, we obtain with (14.11)

$$\tilde{T}(A,B) = A'(x^i \circ \phi) B'(x^j \circ \phi) T(\partial_i \circ \phi, \partial_j \circ \phi)\,.$$

The last factor is equal to $T(\partial_i, \partial_j)\circ\phi = (\nabla_{\partial_i}\partial_j - \nabla_{\partial_j}\partial_i)\circ\phi$. But by Equation (14.11)

$$A'(x^i \circ \phi)(\nabla_{\partial_i}\partial_j)\circ\phi = (\nabla_A \partial_j)\circ\phi\,.$$

Hence,

$$\tilde{T}(A,B) = B'(x^i\circ\phi)(\nabla_A\partial_i)\circ\phi - A'(x^i\circ\phi)(\nabla_B\partial_i)\circ\phi\,. \tag{14.13}$$

For the two remaining terms in (14.10a) we use

$$\tilde{\nabla}_A B \overset{(14.11)}{=} \tilde{\nabla}_A\big(B'(x^i\circ\phi)\partial_i\circ\phi\big) \overset{(14.9)}{=} A'\big(B'(x^i\circ\phi)\big)\partial_i\circ\phi + B'(x^i\circ\phi)(\nabla_A\partial_i)\circ\phi\,.$$

From this, as well as (14.12) and (14.13) we obtain Equation (14.10a). The energetic reader may similarly verify (14.10b). □

Now, we assume that M is a pseudo-Riemannian manifold.

Proposition 14.5. *If M is a pseudo-Riemannian manifold with metric g and corresponding Levi-Cività connection ∇, then*

$$\tilde{T}(X,Z) = 0\,, \tag{14.14a}$$

$$A'\langle X,Z\rangle = \langle\tilde{\nabla}_A X, Z\rangle + \langle X, \tilde{\nabla}_A Z\rangle \tag{14.14b}$$

for X, $Z \in \mathcal{X}(\phi)$ and $A = T\phi\circ A'$.

Proof. Equation (14.14a) is obvious. For establishing (14.14b) it is convenient to introduce a pseudo-orthonormal basis of (local) vector fields $e_1,\ldots,e_n$ with $\langle e_i, e_j\rangle = \pm\delta_{ij}$ on M. Let $E_j = e_j\circ\phi$ be the corresponding vector fields along ϕ, and expand $X = X^i E_i$, $Z = Z^j E_j$. Then

$$A'\langle X,Z\rangle = A'(X^i Z_i) = (A'X^i)Z_i + X^i(A'Z_i)\,.$$

For the right-hand side of (14.14b) we note, using the characterizing properties of $\tilde{\nabla}_A X$ in Proposition 14.3,

$$\begin{aligned}\langle\tilde{\nabla}_A X, Z\rangle &= \big\langle (A'X^i)E_i + X^i \underbrace{\tilde{\nabla}_A E_i}_{\nabla_A e_i\circ\phi}, Z\big\rangle \\ &= (A'X^i)Z_i + X^i Z^j\langle\nabla_A e_i, e_j\rangle\circ\phi\,.\end{aligned}$$

From this one sees that the Ricci identity for ∇ on M implies (14.14b). □

Remarks.

1. The formalism of this subsection will be used at various places, in particular in the next subsection. It is also useful for the treatment of bosonic strings.
2. We shall often drop the tildes in $\tilde{\nabla}_A X$, etc.. In books on differential geometry the notation $\nabla_{A'}X$ for $A = T\phi\circ A'$ is often used.
3. A special case has already be treated before, namely vector fields along parameterized curves and their covariant derivatives (see Proposition 13.2).

14.4 Variations of a Smooth Curve

Definition 14.4. *A* ***variation*** *of a smooth curve* $c : [a, b] \longrightarrow M$ *is a smooth map*

$$H : [a, b] \times (-\varepsilon, \varepsilon) \longrightarrow M\,, \quad (s, t) \longmapsto H(s, t)\,,$$

such that $H(s, 0) = c(s)$. *For fixed t the curve* $s \longmapsto H(s, t)$ *will be denoted by* c_t.

We introduce the tangential fields along the map H belonging to $\partial/\partial s$ and $\partial/\partial t$

$$T = TH \circ \frac{\partial}{\partial s} \quad \text{and} \quad V = TH \circ \frac{\partial}{\partial t}\,.$$

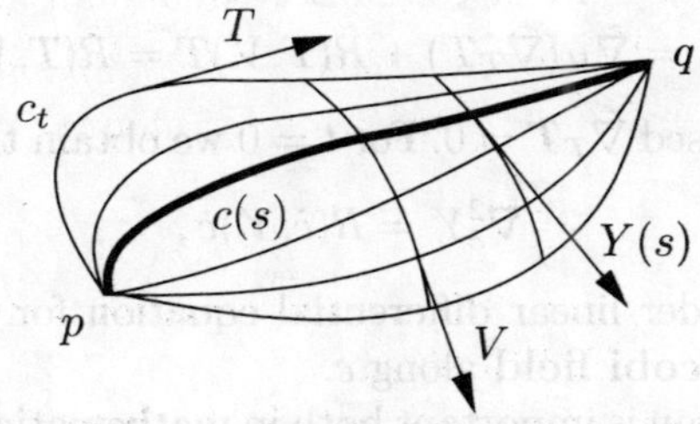

Fig. 14.2. Variation of a smooth curve.

Note that $[T, V] = 0$. The restriction of V to $t = 0$ will be denoted by Y, thus $Y(s) = T_{(s,0)}H \cdot \partial/\partial t$. For the restriction of T we shall also write $\dot{c}$. If we take H with fixed ends ($H(a, t) = p$, $H(b, t) = q$), then $Y(a) = Y(b) = 0$ (see Fig. 14.2).

First Variation Formula

With this notation we study the first variation of the **energy functional** for a pseudo-Riemannian manifold (M, g)

$$E(c_t) = \frac{1}{2}\int \langle T, T\rangle\, ds\,. \tag{14.15}$$

We assume that either all curves c_t are **spacelike**, i.e. $\langle T, T\rangle > 0$ or all curves are **timelike**, i.e. $\langle T, T\rangle < 0$. Using the tools developed in Sect. 14.3 the first variation can easily be determined. Consider

$$\begin{aligned}\frac{1}{2}\frac{\partial}{\partial t}\langle T, T\rangle &= \langle \tilde{\nabla}_V T, T\rangle = \langle \tilde{\nabla}_T V, T\rangle \\ &= \frac{\partial}{\partial s}\langle V, T\rangle - \langle V, \tilde{\nabla}_T T\rangle\,.\end{aligned} \tag{14.16}$$

Integrating this over s at $t = 0$, we obtain the **first variation formula**

$$\left.\frac{d}{dt}\right|_{t=0} E(c_t) = \langle Y, \dot{c}\rangle|_a^b - \int_a^b \langle Y, \nabla_{\dot{c}}\dot{c}\rangle\, ds\,. \tag{14.17}$$

Consider variations with fixed ends and require that for all such variations the first variation of the energy functional (left-hand side of (14.17)) vanishes, then the curve c is a geodesic, $\nabla_{\dot{c}}\dot{c} = 0$, because Y vanishes at the ends, and is otherwise arbitrary. Note that instead of $E(c_t)$ one can use the length functional $L(c_t)$.

Jacobi Equation

With the previous notation, we now assume that the curves c_t are all geodesics and investigate the propagation of Y along c. Consider the second derivative

$$\begin{aligned}\tilde{\nabla}_T^2 V &= \tilde{\nabla}_T(\tilde{\nabla}_T V) = \tilde{\nabla}_T(\tilde{\nabla}_V T)\\ &= \tilde{\nabla}_V(\tilde{\nabla}_T T) + \tilde{R}(T, V)T = \tilde{R}(T, V)T\,,\end{aligned}$$

where we have also used $\tilde{\nabla}_T T = 0$. For $t = 0$ we obtain the **Jacobi equation**

$$\nabla_{\dot{c}}^2 Y = R(\dot{c}, Y)\dot{c}\,, \tag{14.18}$$

which is a second order linear differential equation for Y. A field satisfying this equation is a **Jacobi field** along c.

The Jacobi equation is important both in mathematics and physics. Jacobi fields are used in the celebrated Morse theory (see, e.g., [41]). We use the Jacobi equation (equation for geodesic deviation) in Chap. 2 (Sect. 2.1) to interprete tidal forces in terms of the curvature of spacetime.

* * *

Exercises.

Exercise 14.2.

1. Use the language of Sect. 14.2 to show that $X \in \mathcal{X}(M)$ and $Y \in \mathcal{X}(N)$ are ϕ-related relative to the smooth map $\phi : M \longrightarrow N$, if and only if
$$T\phi \circ X = Y \circ \phi\,.$$
2. If ϕ is a diffeomorphism prove that
$$\phi_* X = T\phi \circ X \circ \phi^{-1}.$$
In particular, X is invariant under ϕ, i.e., $\phi_* X = X$, if and only if
$$T\phi \circ X = X \circ \phi\,.$$

Exercise 14.3. Do the necessary calculations in the existence proof of Proposition 14.3.

A Fundamental Equations for Hypersurfaces

In this appendix we consider submanifolds with codimension 1 of a pseudo-Riemannian manifold and derive, among other things, the important equations of Gauss and Codazzi-Mainardi. These have many useful applications, in particular for a $3+1$ splitting of the Einstein equations (see Sect. 2.9).

For definiteness, we consider a spacelike hypersurface Σ of a Lorentz manifold (M, g). Other situations, e.g., timelike hypersurfaces can be treated in the same way, with sign changes here and there. The formulation can also be generalized to submanifolds of codimension larger than 1 of arbitrary pseudo-Riemannian manifolds (see, for instance, Vol.II of [43]), a situation that occurs in string theory.

Let $\iota : \Sigma \longrightarrow M$ denote the embedding and let $\bar{g} = \iota^* g$ be the **induced metric** on Σ. The pair $(\Sigma, \bar{g})$ is a three-dimensional Riemannian manifold. Other quantities which belong to Σ will often be indicated by a bar. The basic results of this appendix are most efficiently derived with the help of **adapted** moving frames and the structure equations of Cartan. So let $\{e_\mu\}$ be an 'orthonormal' frame on an open region of M, with the property that the e_i, for $i = 1, 2, 3$, in points of Σ are tangent to Σ. The dual basis of $\{e_\mu\}$ is denoted by $\{\theta^\mu\}$. On Σ the $\{e_i\}$ can be regarded as a triad of Σ that is dual to the restrictions $\bar{\theta}^i := \theta^j | T\Sigma$ (i.e., the restrictions to tangent vectors of Σ). Indices for objects on Σ always refer to this pair of dual basis.

Formulas of Gauss and Weingarten

Consider the restriction of the first structure equation

$$d\theta^\mu + \omega^\mu{}_\nu \wedge \theta^\nu = 0 \tag{A.1}$$

to $T\Sigma$. Since $\theta^0 | T\Sigma = 0$ we obtain

$$d\bar{\theta}^i + \omega^i{}_j \wedge \bar{\theta}^j = 0 \qquad \text{on } T\Sigma, \tag{A.2a}$$

$$\omega^0{}_k \wedge \bar{\theta}^k = 0 \qquad \text{on } T\Sigma\,. \tag{A.2b}$$

Because the connection forms $\bar{\omega}^i{}_j$ of $(\Sigma, \bar{g})$ also satisfy (A.2a), and have the same symmetry properties, we conclude (see Sect. 13.7) that

$$\bar{\omega}^i{}_j = \omega^i{}_j \qquad \text{on } T\Sigma\,. \tag{A.3}$$

This has a simple geometrical meaning. Let ∇ be the Levi-Cività connection on (M, g) and $\bar{\nabla}$ that of $(\Sigma, \bar{g})$, then we get from (A.3)

$$\langle \nabla_X e_j, e_i \rangle = \omega_{ij}(X) = \bar{\omega}_{ij}(X) = \langle \bar{\nabla}_X e_j, e_i \rangle \tag{A.4}$$

for $X \in T\Sigma$. This shows that for any $X \in T\Sigma$ and any vector field Y of M tangent to Σ, the tangential projection of $\nabla_X Y$ on $T\Sigma \subset TM$ is equal to $\bar{\nabla}_X Y$.

From (A.2b) we obtain information about the component of $\nabla_X Y$ normal to $T\Sigma$. This equation tells us that the $\omega^0_{\ i} = \omega^i_{\ 0}$ satisfy the hypothesis of the following

Lemma A.1 (Cartan). *If $\alpha^1, \ldots, \alpha^n$ are linearly independent 1-forms on a manifold M of dimension $n' \geq n$, and $\beta_1, \ldots, \beta_n$ are one forms on M satisfying*

$$\sum_{i=1}^{n} \alpha^i \wedge \beta_i = 0\,, \tag{A.5}$$

then there are smooth functions f_{ij} on M such that

$$\beta_i = \sum_{j=1}^{n} f_{ij}\, \alpha^j\,; \tag{A.6}$$

moreover

$$f_{ij} = f_{ji}\,. \tag{A.7}$$

Proof. In a neighborhood of any point we can choose 1-forms $\alpha^{n+1}, \ldots, \alpha^{n'}$ so that $\alpha^1, \ldots, \alpha^{n'}$ are everywhere linearly independent. Then there are smooth functions f_{ij} $(i \leq n, j \leq n')$ with

$$\beta_i = \sum_{j=1}^{n'} f_{ij}\, \alpha^j\,.$$

Now Equation (A.5) implies

$$0 = \sum_{i=1}^{n} \sum_{j=1}^{n'} f_{ij}\, \alpha^i \wedge \alpha^j = \sum_{1 \leq i \leq j \leq n} (f_{ij} - f_{ji}) \alpha^i \wedge \alpha^j + \sum_{i=1}^{n} \sum_{j>n} f_{ij}\, \alpha^i \wedge \alpha^j\,.$$

Since the $\alpha^i \wedge \alpha^j$ for $i < j$ are linearly independent, we conclude $f_{ij} = f_{ji}$ for $i, j \leq n$ and $f_{ij} = 0$ for $j > n$. □

According to the lemma and (A.2b) there are functions K_{ij} on Σ such that

$$\omega^0_{\ i} = -K_{ij}\, \theta^j \qquad \text{on } T\Sigma\,, \tag{A.8a}$$

$$K_{ij} = K_{ji}\,. \tag{A.8b}$$

From this and $\nabla_{e_\mu} e_\nu = \omega^\lambda{}_\nu(e_\mu)e_\lambda$ we obtain

$$\langle \nabla_{e_i} e_j, e_0 \rangle = -\omega^0{}_j(e_i) = K_{ij} = K_{ji} = \langle \nabla_{e_j} e_i, e_0 \rangle . \tag{A.9}$$

So the bilinear form $K(X,Y)$ on $T\Sigma$ belonging to K_{ij} gives the normal component of $\nabla_X Y$:

$$\nabla_X Y = \bar{\nabla}_X Y - K(X,Y)N \tag{A.10}$$

for $X \in T\Sigma$, $Y \in \mathcal{X}(M)$ tangent to Σ, where $N = e_0$ (normalized normal vector on Σ, $\langle N, N \rangle = -1$).

Equation (A.10) is the **Gauss formula**. The symmetric bilinear form K, called the **second fundamental form** or **extrinsic curvature**, satisfies according to (A.9) the **Weingarten equations**

$$K(X,Y) = \langle N, \nabla_X Y \rangle = -\langle \nabla_X N, Y \rangle . \tag{A.11}$$

Remarks. The reader should be warned that some authors use as extrinsic curvature the negative of our K. The linear transformation of the tangent spaces of Σ corresponding to K is known as the **Weingarten map** of the second fundamental tensor. Its real eigenvalues are called **principle curvatures**.

Equations of Gauss and Codazzi-Mainardi

Next, we look for relations between the curvature forms $\Omega^\mu{}_\nu$ and $\bar{\Omega}^i{}_j$ of (M,g), respectively $(\Sigma, \bar{g})$. For this we restrict the second structure equation

$$\Omega^\mu{}_\nu = d\omega^\mu{}_\nu + \omega^\mu{}_\lambda \wedge \omega^\lambda{}_\nu \tag{A.12}$$

to Σ. Consider first the restriction of

$$\Omega^i{}_j = d\omega^i{}_j + \omega^i{}_k \wedge \omega^k{}_j + \omega^i{}_0 \wedge \omega^0{}_j$$

to $T\Sigma$. Using (A.3) and (A.8a) this gives

$$\Omega^i{}_j = \bar{\Omega}^i{}_j + K^i{}_k K_{jl}\, \bar{\theta}^k \wedge \bar{\theta}^l \quad \text{on } T\Sigma . \tag{A.13}$$

The other components give on $T\Sigma$

$$\begin{aligned} \Omega^0{}_j &= d\omega^0{}_j + \omega^0{}_i \wedge \omega^i{}_j = -d(K_{ij}\bar{\theta}^j) - K_{ik}\, \bar{\theta}^k \wedge \bar{\omega}^i{}_j \\ &= -dK \wedge \bar{\theta}^j - K_{ij} d\bar{\theta}^i - K_{ik}\, \bar{\theta}^k \wedge \bar{\omega}^i{}_j \end{aligned}$$

or, using the first structure equation on $(\Sigma, \bar{g})$,

$$\Omega^0{}_j = -\bar{D} K_{ij} \wedge \bar{\theta}^i \quad \text{on } T\Sigma . \tag{A.14}$$

Here $\bar{D}$ denotes the absolute exterior differential of the tensor field K.

The relations (A.13) and (A.14) are the famous **equations of Gauss** and **Codazzi-Mainardi** in terms of differential forms. We want to rewrite them in a more useful form. Consider for tangential X, Y

$$\begin{aligned}\langle R(X,Y)e_j, e_i\rangle &= \Omega_{ij}(X,Y)\\ &= \bar{\Omega}_{ij}(X,Y) + K_{ik}K_{jl}(\bar{\theta}^k \wedge \bar{\theta}^l)(X,Y)\\ &= \langle \bar{R}(X,Y)e_j, e_i\rangle +\\ &\quad K(e_i,X)K(e_j,Y) - K(e_i,Y)K(e_j,X)\,.\end{aligned}$$

This shows that for tangent vectors X, Y, Z and W of Σ we have

$$\begin{aligned}\langle R(X,Y)Z,W\rangle = &\langle \bar{R}(X,Y)Z,W\rangle -\\ &-K(X,Z)K(Y,W) + K(Y,Z)K(X,W)\,. \qquad \text{(A.15)}\end{aligned}$$

One calls this relation often **Gauss' Theorema Egregium**. (Note that some of the signs are different from the ones for the Riemannian case, that is usually treated in mathematics books.)

Similarly, we obtain from (A.14)

$$\begin{aligned}\langle R(X,Y)e_j, N\rangle &= \Omega_{0j}(X,Y) = (\bar{D}K_{ij} \wedge \bar{\theta}^i)(X,Y)\\ &= \bar{\nabla}_k K_{ij}(\bar{\theta}^k \wedge \bar{\theta}^i)(X,Y)\\ &= X^k Y^i \bar{\nabla}_k K_{ij} - (X \longleftrightarrow Y)\\ &= (\bar{\nabla}_X K)(Y, e_j) - (X \longleftrightarrow Y)\end{aligned}$$

for $X, Y \in T\Sigma$. Hence, Eq. (A.14) is equivalent to the following standard form of the Codazzi-Mainardi equation:

$$\langle R(X,Y)Z, N\rangle = (\bar{\nabla}_X K)(Y,Z) - (\bar{\nabla}_Y K)(X,Z)\,, \qquad \text{(A.16)}$$

for $X, Y, Z \in T\Sigma$.

The basic equations of Gauss and Codazzi-Mainardi allow us to obtain interesting and useful expressions for the components R_{0i} and G_{00} of the Ricci and Einstein tensors on Σ. We first note that for any frame

$$i_{e_\alpha}\Omega^\alpha{}_\beta = \frac{1}{2}\,R^\alpha{}_{\beta\rho\sigma} i_{e_\alpha}(\theta^\rho \wedge \theta^\sigma) = R_{\beta\sigma}\,\theta^\sigma\,,$$

so that

$$R_{\beta\sigma} = \Omega^\alpha{}_\beta(e_\alpha, e_\sigma)\,. \qquad \text{(A.17)}$$

Especially, we obtain

$$R_{0i} = \Omega^j{}_0(e_j, e_i)\,. \qquad \text{(A.18)}$$

Relative to our adapted basis we need $\Omega^j{}_0$ on the right-hand side only on $T\Sigma$, and can thus use (A.14) to get

$$R_{0i} = \bar{\nabla}_i K^j{}_j - \bar{\nabla}_j K^j{}_i\,. \qquad \text{(A.19)}$$

Next, we consider

$$G_{00} = R_{00} + \frac{1}{2}(-R_{00} + R^i{}_i) = \frac{1}{2}(R_{00} + R^i{}_i) .$$

From (A.17) we get

$$R_{00} = \Omega^j{}_0(e_j, e_0) ,$$
$$R^i{}_i = \Omega^{ji}(e_j, e_i) + \Omega^{0i}(e_0, e_i) ,$$

and thus

$$G_{00} = \frac{1}{2}\Omega^{ij}(e_i, e_j) . \tag{A.20}$$

Here we can use (A.13) to obtain

$$2G_{00} = \bar{R}^i{}_i + K^i{}_i K^j{}_j - K^j{}_j K^j{}_i ,$$

or

$$G_{00} = \frac{1}{2}\bar{R} + \frac{1}{2}(K^i{}_i K^j{}_j - K^i{}_j K^j{}_i) . \tag{A.21}$$

This form of Gauss' equation is particular useful in GR.

Consider, for illustration, a static metric as in Sect. 2.1. From (2.23) we see that the second fundamental form for any time slice vanishes, thus (A.21) gives

$$G_{00} = G_{\mu\nu}N^\mu N^\nu = \frac{1}{2}\bar{R} ,$$

in agreement with (2.30). In the exercises we consider time-dependent metrics.

These formulas simplify if Σ is totally geodesic in the sense of

Definition A.1. *A hypersurface Σ is called* **totally geodesic** *if every geodesic $\gamma(s)$ of (M, g) with $\gamma(0) \in \Sigma$ and $\dot{\gamma}(0) \in T_{\gamma(0)}\Sigma$ remains in Σ for s in some open interval $(-\varepsilon, \varepsilon)$.*

This is equivalent to the property that $\nabla_X Y$ is tangent to Σ whenever X and Y are (see Exercise A.3). From (A.10) we conclude that $\Sigma \subset M$ is totally geodesic if and only if the second fundamental form K of Σ vanishes.

* * *

Null Hypersurfaces

Null hypersurfaces of Lorentz manifolds play an important role in GR.

Definition A.2. *A hypersurface Σ of a Lorentz manifold (M, g) is* **null**, *if the induced metric on Σ is degenerate.*

The reader may easily show that this is equivalent to the property that at any point $p \in \Sigma$ there exists a non-vanishing null vector in T_pM which is perpendicular to $T_p\Sigma$. As a consequence of property c) in Exercise 7.5, this null vector is unique up to a scale factor (up to a positive scale factor if the null vector is future directed). Hence, there exists locally a smooth non-vanishing null vector field l along Σ which is tangent to Σ and whose normal space at any point $p \in \Sigma$ coincides with T_pM. From the quoted exercise it also follows that tangent vectors of Σ not parallel to l are spacelike.

The null vector field l, which is unique up to rescaling, has the following remarkable property:

Proposition A.1. *The integral curves of l, when suitably parameterized, are null geodesics.*

Proof. We show below that $\nabla_l l$ is at each $p \in \Sigma$ perpendicular to T_pM. This implies that $\nabla_l l$ is proportional to l, which is equivalent to the proposition.

To establish the stated property, let $u \in T_pM$ and extend u locally to a vector field along Σ which is invariant under the flow belonging to l (show that this is possible). Then $0 = L_l u = [l, u] = \nabla_l u - \nabla_u l$. From $\langle l, u\rangle = 0$ we obtain

$$0 = l\,\langle l, u\rangle = \langle \nabla_l l, u\rangle + \langle l, \nabla_l u\rangle\,,$$

thus

$$\langle \nabla_l l, u\rangle = -\langle l, \nabla_u l\rangle = -\frac{1}{2}\,u\,\langle l, l\rangle = 0\,.$$

(These formulas should be interpreted in the sense of Sect. 14.3 for vector fields along the embedding map of Σ into M.) Since u is arbitrary in p, $\nabla_l l$ is indeed perpendicular to $T_p\Sigma$. □

The integral curves of l are called **null geodesic generators** of Σ.

* * *

Exercises.

Exercise A.1. Consider a metric of the form

$$g = -\varphi^2(t, x^k)\,dt^2 + g_{ij}(t, x^k)\,dx^i dx^j \tag{A.22}$$

and show that the second fundamental form of the slices $\{t = \text{const}\}$ is given by

$$K_{ij} = -\frac{1}{2\varphi}\partial_t g_{ij}\,. \tag{A.23}$$

Solution. The Ricci identity and $[\partial_t, \partial_i] = 0$ give

$$\begin{aligned}\partial_t g_{ij} &= \langle \nabla_{\partial_t}\partial_i, \partial_j\rangle + \langle \partial_i, \nabla_{\partial_t}\partial_j\rangle \\ &= \langle \nabla_{\partial_i}\partial_t, \partial_j\rangle + \langle \partial_i, \nabla_{\partial_j}\partial_t\rangle\,.\end{aligned}$$

Here we use $N = \frac{1}{\varphi}\partial_t$ and (A.11) to obtain

$$\begin{aligned}\partial_t g_{ij} &= \langle \nabla_{\partial_i}(\varphi N), \partial_j \rangle + \langle \partial_i, \nabla_{\partial_j}(\varphi N) \rangle \\ &= \varphi\big(\langle \nabla_{\partial_i} N, \partial_j \rangle + \langle \partial_i, \nabla_{\partial_j} N \rangle\big) \\ &= -\varphi(K_{ij} + K_{ji}) .\end{aligned}$$

Exercise A.2. Use (A.23), the Gauss equation (A.21), and (2.78) to show that for the **Friedmann metric**

$$g = -dt^2 + a^2(t) h_{ij}\, dx^i \wedge dx^j \tag{A.24}$$

on $\mathbb{R} \times S^3$, where $h = h_{ij}\, dx^i \wedge dx^j$ is the standard metric on S^3, the time-time component of the Einstein tensor is given by

$$G_{00} = 3\left(\frac{\dot{a}^2}{a^2} + \frac{1}{a^2}\right) . \tag{A.25}$$

Hence, Einstein's field equation implies the following **Friedmann equation**

$$\left(\frac{\dot{a}}{a}\right)^2 + \frac{1}{a^2} = 8\pi G(\rho + \rho_\Lambda) , \tag{A.26}$$

where $\rho_\Lambda := \frac{\Lambda}{8\pi G}$ and ρ is the matter energy density.

Exercise A.3. Prove that a hypersurface Σ of (M, g) is totally geodesic if and only if $\nabla_X Y$ is tangent to Σ whenever X and Y are.

Solution. We already know that the second property holds if and only if the second fundamental form K of Σ vanishes. So let us show that $K = 0$ if and only if Σ is totally geodesic.

Let $\gamma : s \longmapsto \gamma(s)$ be a geodesic in (M, g), with $\gamma(0) \in \Sigma$, $\dot{\gamma}(0) \in T_{\gamma(0)}\Sigma$, and let $\bar{\gamma}$ be a geodesic in $(\Sigma, \bar{g})$ with the same initial conditions. From (A.10) we get

$$\nabla_{\dot{\bar{\gamma}}}\dot{\bar{\gamma}} = \bar{\nabla}_{\dot{\bar{\gamma}}}\dot{\bar{\gamma}} - K(\dot{\bar{\gamma}}, \dot{\bar{\gamma}})N = -K(\dot{\bar{\gamma}}, \dot{\bar{\gamma}})N .$$

Hence, if K vanishes we see that $\bar{\gamma}$ is also a geodesic of (M, g) with the same initial conditions as γ, thus $\bar{\gamma} = \gamma$ on their common domain. This proves that a vanishing K implies that Σ is totally geodesic.

Conversely, let Σ be totally geodesic. Then γ stays in Σ for some open interval $(-\varepsilon, \varepsilon)$. From (A.10) we now conclude

$$0 = \nabla_{\dot{\gamma}}\dot{\gamma} = \bar{\nabla}_{\dot{\gamma}}\dot{\gamma} - K(\dot{\gamma}, \dot{\gamma})N .$$

Therefore, $\bar{\nabla}_{\dot{\gamma}}\dot{\gamma} = 0$ and $K(\dot{\gamma}, \dot{\gamma}) = 0$. In particular, γ is also a geodesic of $(\Sigma, \bar{g})$. Moreover, since $\gamma(0) \in \Sigma$ and $\dot{\gamma}(0) \in T_{\gamma(0)}\Sigma$ are arbitrary, we conclude that $K = 0$ (use the symmetry of K and polarization).

B Ricci Curvature of Warped Products

In GR one often encounters so-called warped products (see, e.g., Appendix 3.10). By definition, a pseudo-Riemannian manifold (M, g) is a **warped product** of two pseudo-Riemannian manifolds $(\tilde{M}, \tilde{g})$ and $(S, \hat{g})$ with an everywhere positive **warp function** $f : \tilde{M} \longrightarrow \mathbb{R}$, written as

$$M = \tilde{M} \times_f S\,, \tag{B.1}$$

if M is the product manifold $\tilde{M} \times S$ and the metric has the form

$$g = \pi^*(\tilde{g}) + (f \circ \pi)^2 \sigma^*(\hat{g})\,, \tag{B.2}$$

where π and σ are the projections from M onto $\tilde{M}$ and S, respectively. In the following we shall drop the pull-backs π^*, σ^*.

The calculation of the curvature quantities of (M, g) in terms of those of $(\tilde{M}, \tilde{g})$ and $(S, \hat{g})$ is best done within the framework of Cartan's calculus. We shall work with the orthonormal tetrad fields $\{\theta^\alpha\}$ of (M, g), which are adapted as follows: Let $\alpha = (a, A)$ and let

$$\theta^a = \tilde{\theta}^a\,, \qquad \theta^A = f\,\hat{\theta}^A\,, \tag{B.3}$$

where $\tilde{\theta}^a$, for $a = 1, \ldots, \dim(\tilde{M})$, and $\hat{\theta}^A$, for $A = \dim(\tilde{M}) + 1, \ldots, \dim(M)$, are orthonormal bases of $(\tilde{M}, \tilde{g})$ and $(S, \hat{g})$, respectively. The connection forms of the various spaces are denoted by $\omega^\alpha{}_\beta$, $\tilde{\omega}^a{}_b$ and $\hat{\omega}^A{}_B$, and for the corresponding curvature forms we use the symbols $\Omega^\alpha{}_\beta$, $\tilde{\Omega}^a{}_b$ and $\hat{\Omega}^A{}_B$.

To start, we determine the connection forms $\omega^\alpha{}_\beta$ from Cartan's first structure equation

$$d\theta^\alpha + \omega^\alpha{}_\beta \wedge \theta^\beta = 0\,. \tag{B.4}$$

We need

$$\begin{aligned} d\theta^A &= d(f\,\hat{\theta}^A) = f\,d\hat{\theta}^A - \hat{\theta}^A \wedge df \\ &= -f\,\hat{\omega}^A{}_B \wedge \hat{\theta}^B - f_b\,\hat{\theta}^A \wedge \tilde{\theta}^b \\ &= -\hat{\omega}^A{}_B \wedge \theta^B - \frac{f_b}{f}\theta^A \wedge \theta^b\,, \end{aligned}$$

where we have set $df = f_a\,\tilde{\theta}^a$. Comparing this result with (B.4) leads to the guess

$$\omega^A_{\ B} = \hat{\omega}^A_{\ B}, \qquad \omega^A_{\ b} = \frac{f_b}{f}\,\theta^A. \tag{B.5}$$

In addition, we also have to satisfy

$$\begin{aligned} d\theta^a &= -\omega^a_{\ B}\wedge\theta^B - \omega^a_{\ b}\wedge\theta^b \\ &= d\tilde{\theta}^a = -\tilde{\omega}^a_{\ b}\wedge\tilde{\theta}^b = -\tilde{\omega}^a_{\ b}\wedge\theta^b . \end{aligned}$$

Taking

$$\omega^a_{\ b} = \tilde{\omega}^a_{\ b}, \tag{B.6}$$

we see that the structure equation (B.4) is fulfilled for the connection forms (B.5) and (B.6). Therefore, these equations provide the correct (and unique) result.

Using the second structure equation,

$$\Omega^\alpha_{\ \beta} = d\omega^\alpha_{\ \beta} + \omega^\alpha_{\ \gamma}\wedge\omega^\gamma_{\ \beta}, \tag{B.7}$$

and the corresponding equations for $\tilde{\Omega}^a_{\ b}$ and $\hat{\Omega}^A_{\ B}$ as well, we find after a short routine calculation the curvature forms

$$\Omega^a_{\ b} = \tilde{\Omega}^a_{\ b}, \tag{B.8a}$$

$$\Omega^A_{\ B} = \hat{\Omega}^A_{\ B} - \frac{\langle df, df\rangle}{f^2}\,\theta^A\wedge\theta_B, \tag{B.8b}$$

$$\Omega^a_{\ B} = -\frac{1}{f}\tilde{\nabla}_b\tilde{\nabla}^a f\,\theta^b\wedge\theta_B, \tag{B.8c}$$

from which we can now extract the Ricci tensor. This is best done by taking advantage of the formula

$$R_{\beta\gamma} = \Omega^\alpha_{\ \beta}(e_\alpha, e_\gamma), \tag{B.9}$$

which follows from $\Omega^\mu_{\ \nu} = R^\mu_{\ \nu\alpha\beta}\,\theta^\alpha\wedge\theta^\beta$ (e_α denotes the orthonormal frame dual to θ^α). One finally obtains

$$R_{ab} = \tilde{R}_{ab} - \frac{\dim(S)}{f}\tilde{\nabla}_b\tilde{\nabla}_a f, \tag{B.10a}$$

$$R_{aB} = 0, \tag{B.10b}$$

$$\begin{aligned} R_{AB} &= \hat{R}_{AB} \\ &\quad -\big(f\tilde{\Delta}f + \langle df, df\rangle(\dim(S)-1)\big)\hat{g}_{AB}. \end{aligned} \tag{B.10c}$$

This result obviously holds also for non-orthonormal adapted frames. (Note that all indices refer to the frame $\theta^\alpha = (\theta^a, \theta^A)$.)

Application: Friedmann Equations

Important examples of warped products are **Friedmann-Lemaître (-Robertson-Walker) spacetimes** in cosmology. For these $\tilde{M}$ is an open interval of $\mathbb{R}$ with the metric $-dt^2$, $(S, \hat{g})$ is a 3-dimensional space of constant curvature $k = \pm 1, 0$, and the warp function is the scale factor $a(t)$. The spacetime metric is thus

$$g = -dt^2 + a^2(t)\, \hat{g}\,. \tag{B.11}$$

We use the general results (B.10a)–(B.10c) to compute the Ricci tensor belonging to (B.11).

Since $(\tilde{M}, \tilde{g})$ is now a 1-dimensional Euclidean space, (B.10a) gives

$$R_{00} = -3\,\frac{\ddot{a}}{a}\,. \tag{B.12}$$

According to (B.10b) we have

$$R_{0i} = 0\,, \tag{B.13}$$

and (B.10c) implies

$$R_{ij} = \hat{R}_{ij} + \left[2\left(\frac{\dot{a}}{a}\right)^2 + \frac{\ddot{a}}{a}\right] a^2\, \hat{g}_{ij}\,.$$

According to (2.79), which holds for all Riemannian spaces of constant curvature, we have

$$\hat{R}_{ij} = 2\,k\,\hat{g}_{ij}\,, \tag{B.14}$$

hence

$$R_{ij} = \left[2\left(\frac{\dot{a}}{a}\right)^2 + 2\,\frac{k}{a^2} + \frac{\ddot{a}}{a}\right] a^2\, \hat{g}_{ij}\,. \tag{B.15}$$

For an orthonormal tetrad $\{\theta^\alpha\}$ we have $a^2\, \tilde{g}_{ij} = \delta_{ij}$ (because of (B.3)). Then we find for the Einstein tensor

$$\begin{aligned} G_{00} &= 3\left[\left(\frac{\dot{a}}{a}\right)^2 + \frac{k}{a^2}\right], \\ G_{0i} &= 0\,, \\ G_{ij} &= -\left[2\frac{\ddot{a}}{a} + \left(\frac{\dot{a}}{a}\right)^2 + \frac{k}{a^2}\right] \delta_{ij}\,. \end{aligned} \tag{B.16}$$

In order to satisfy the field equations, the energy-momentum tensor must have the form

$$(T_{\mu\nu}) = \mathrm{diag}(\rho, p, p, p)\,. \tag{B.17}$$

From (B.16) and (B.17) we obtain the independent Friedmann equations

$$\left(\frac{\dot{a}}{a}\right)^2 + \frac{k}{a^2} = \frac{8\pi G}{3}\,\rho + \frac{\Lambda}{3}, \tag{B.18a}$$

$$-\,2\frac{\ddot{a}}{a} - \left(\frac{\dot{a}}{a}\right)^2 - \frac{k}{a^2} = 8\pi G\,p - \Lambda\,. \tag{B.18b}$$

Note also, that (B.12) gives the (dependent) equation

$$\ddot{a} = -\frac{4\pi G}{3}(\rho + 3p) + \frac{\Lambda}{3}\,a\,. \tag{B.19}$$

This shows that $\ddot{a}$ depends on the combination $\rho + 3p$, and that a *positive* Λ leads to an **accelerated expansion**.

* * *

Exercise B.1. Use the connection forms (B.5) and (B.6) to show that $T^{\alpha\beta}{}_{;\beta} = 0$ implies for $\alpha = 0$:

$$\dot{\rho} + 3\,\frac{\dot{a}}{a}(\rho + p) = 0\,. \tag{B.20}$$

Derive this equation also from (B.18a) and (B.18b).

C Frobenius Integrability Theorem

In this appendix we develop an important differential geometric tool of GR that is also one of the pillars of differential topology and the calculus on manifolds. The following problem is a natural generalization of finding integral curves for a vector field. Instead of a vector field we assign to each point p of a manifold M a subspace E_p of T_pM which depends on p in a smooth manner and ask: When is it possible to find for each point $p_0 \in M$ a submanifold N with $p_0 \in N$, which is tangent to E_p, i.e., such that $T_pN = E_p$ for each point $p \in N$? In general, such integral manifolds do not exist, even locally. For the local problem the so-called **Frobenius theorem** (due to A. Clebsch and F. Deahna) gives necessary and sufficient criteria for integral manifolds to exist.

We begin with some precisions.

Definition C.1. *A* ***k-dimensional distribution*** *(**k-plane, k-direction field**) on M is a map $p \longmapsto E_p$, where E_p is a k-dimensional subspace of T_pM. We call this distribution* ***smooth*** *if the following condition is satisfied: For each $p_0 \in M$ there is a neighborhood V and k differentiable vector fields $X_1, \ldots, X_k$ such that $X_1(p), \ldots, X_k(p)$ form a basis of E_p for all $p \in V$.*

Definition C.2. *A k-dimensional submanifold N of M is called an* ***integral manifold*** *of the distribution E if for every $p \in N$ we have*

$$T_p\,\iota(T_pN) = E_p\,,$$

where $\iota : N \hookrightarrow M$ is the inclusion map. We say that the distribution E on M is ***integrable*** *if, through each $p \in M$, there passes an integral manifold of E.*

We want to find out when E is integrable. For this we consider vector fields which belong to E: We say that the vector field X **belongs to** E if $X_p \in E_p$ for all p. Suppose now that N is an integrable manifold of E, and $\iota : N \hookrightarrow M$ is the inclusion map. If X, Y are two vector fields which belong to E, then for all $p \in N$ there are unique $\bar{X}_p, \bar{Y}_p \in T_pN$ such that

$$X_p = T_p\,\iota \cdot \bar{X}_p\,, \qquad Y_p = T_p\,\iota \cdot \bar{Y}_p\,.$$

It is not difficult to show[1] that the vector fields $\bar{X}, \bar{Y}$ are C^∞. In other words $\bar{X}$ and X, as well as $\bar{Y}$ and Y, are ι-related. Then we know (Theorem 11.6) that $[\bar{X}, \bar{X}]$ and $[X, Y]$ are ι-related. Thus

$$T_p\iota \cdot [\bar{X}, \bar{X}]_p = [X, Y]_p\,.$$

Since $[\bar{X}, \bar{X}]_p \in T_pN$ we conclude that $[X, Y]_p \in E_p$. We thus have shown: If E is integrable then $[X, Y]$ also belongs to E. This suggests to introduce the following concept:

Definition C.3. *We say that the k-distribution E on M is* ***involutive*** *if $[X, Y]$ belongs to E whenever the vector fields X and Y (defined on open sets of M) belong to E.*

Using this terminology, we have shown that if E is integrable then E is involutive. It turns out that the reverse is also true.

Theorem C.1 (Frobenius' Theorem (first version)). *Let E be a C^∞ k-dimensional distribution on M which is involutive. Then E is integrable. More precisely the following holds: For each point $p \in M$ there is a coordinate system $\{x^1, \ldots, x^n\}$ in a neighborhood U of p with*

$$x^i(p) = 0\,, \quad x^i(U) \subset (-\varepsilon, \varepsilon) \quad (\varepsilon > 0)\,,$$

such that for each $(a^{k+1}, \ldots, a^n)$ with all $|a^i| < \varepsilon$ the set

$$\{q \in U \,:\, x^{k+1}(q) = a^{k+1}, \ldots, x^n(q) = a^n\}$$

is an integral manifold of E. Any connected integral manifold of E restricted to U is contained in one of these sets.

We postpone the proof of this important theorem and first reformulate it in a useful manner in terms of differential forms. For this we need some concepts.

To each k-dimensional distribution E we associate the set $\mathcal{I}(E) \subset \bigwedge(M)$ of differential forms ω with the property that each homogeneous component $\omega^l \in \bigwedge^l(M)$ of ω ($\omega = \sum_l \omega^l$) annihilates the sets $(X_1, \ldots, X_l)$ of vector fields belonging to E:

$$\omega^l(X_1, \ldots, X_l) = 0\,.$$

The set $\mathcal{I}(E)$ is called the **annihilator** of E. Clearly, $\mathcal{I}(E)$ is an ideal of the algebra $\bigwedge(M)$ of exterior forms: With $\omega_1, \omega_2 \in \mathcal{I}(E)$ the sum $\omega_1 + \omega_2$ also belongs to E, and $\eta \wedge \omega \in \mathcal{I}(E)$ if $\omega \in \mathcal{I}(E)$ and $\eta \in \bigwedge(M)$. Locally, the ideal $\mathcal{I}(E)$ is generated by $n-k$ independent 1-forms $\omega^{k+1}, \ldots, \omega^n$. In fact, around each point $p \in M$ we can choose coordinates $\{x^i\}$ so that $\{\partial_1|_p, \ldots, \partial_k|_p\}$ span

[1] Since ι is an immersion one can use coordinates as in Definition 9.7. For details, see [43], Vol. I, p.258.

E_p. Then $dx^1(p) \wedge \ldots \wedge dx^k(p)$ is non-zero on E_p and by continuity also in a small neighborhood of p. This implies that $\{dx^1(q), \ldots, dx^k(q)\}$ are linearly independent in E_q. Therefore, if we restrict $dx^\alpha(q)$ for $\alpha = k+1, \ldots, n$ to E_q we have the expansion

$$dx^\alpha(q) = \sum_{\beta=1}^{k} f^\alpha_\beta(q)\, dx^\beta(q)\,.$$

Therefore

$$\omega^\alpha = dx^\alpha - \sum_{\beta=1}^{k} f^\alpha_\beta\, dx^\beta\,, \quad \alpha = k+1, \ldots, n$$

are linearly independent 1-forms which belong to $\mathcal{I}(E)$. These generate the ideal. To show this we complete $\omega^{k+1}, \ldots, \omega^n$ to a basis $\omega^1, \ldots, \omega^n$ of 1-forms and consider the dual basis $X_1, \ldots, X_n$ of vector fields. Then $X_1, \ldots, X_k$ span the k-dimensional distribution E. Suppose an l-form $\omega \in \mathcal{I}(E)$ contains in the expansion with respect to the basis $\omega^1, \ldots \omega^n$,

$$\omega = \sum_{i_1 < i_2 < \ldots < i_l} c_{i_1 \ldots i_l}\, \omega^{i_1} \wedge \ldots \wedge \omega^{i_l}\,,$$

a term $c_{j_1 \ldots j_l}\, \omega^{j_1} \wedge \ldots \wedge \omega^{j_l}$ which contains no factor from $\omega^{k+1}, \ldots, \omega^n$. Then we have $\omega(X_{j_1}, \ldots, X_{j_l}) \neq 0$, in contradiction to $\omega \in \mathcal{I}(E)$. This proves:

Proposition C.1. *Let $\mathcal{I}(E)$ be the ideal of $\bigwedge(M)$ belonging to the k-dimensional distribution E. Then $\mathcal{I}(E)$ is* ***locally generated*** *by $n-k$ independent 1-forms: For each point of M there is a neighborhood U and $n-k$ pointwise linearly independent 1-forms $\omega^{k+1}, \ldots, \omega^n \in \bigwedge^1(U)$ such that for each $\omega \in \mathcal{I}(E)$*

$$\omega | U = \sum_{i=k+1}^{n} \theta_i \wedge \omega^i$$

for some $\theta_i \in \bigwedge(U)$.

Now we can reformulate the condition in the Frobenius theorem.

Proposition C.2. *A distribution E on M is involutive if and only if $\mathcal{I}(E)$ is a* ***differential ideal****: $d\mathcal{I}(E) \subset \mathcal{I}(E)$.*

Proof. We use the same notation as above. It is easy to see that E is involutive if and only if there exist smooth functions ${C^l}_{ij}$ such that

$$[X_i, X_j] = \sum_{l=1}^{k} {C^l}_{ij}\, X_l$$

for $i, j = 1, \ldots k$. Now we have (see (12.14))

$$d\omega^\alpha(X_i, X_j) = X_i(\omega^\alpha(X_j)) - X_j(\omega^\alpha(X_i)) - \omega^\alpha([X_i, X_j]).$$

For $1 \le i, j \le k$ and $\alpha > k$ the first two terms on the right vanish. So $d\omega^\alpha(X_i, X_j) = 0$ if and only if $\omega^\alpha([X_i, X_j]) = 0$. But the latter equation holds for all i, j if and only if each $[X_i, X_j]$ belongs to E (i.e., if E is involutive), while $d\omega^\alpha(X_i, X_j) = 0$ holds exactly when $d\omega^\alpha \in \mathcal{I}(E)$. □

Next we establish some equivalent conditions that a locally (finitely) generated ideal is a differential ideal.

Proposition C.3. *Let $\mathcal{I}$ be an ideal of $\bigwedge(M)$ locally generated by $n-k$ linearly independent forms $\omega^{k+1}, \ldots, \omega^n \in \bigwedge^1(U)$. Furthermore let $\omega \in \bigwedge^{n-k}(U)$ be the form $\omega = \omega^{k+1} \wedge \ldots \wedge \omega^n$. Then the following statements are equivalent:*

(i) $\mathcal{I}$ is an differential ideal.
(ii) $d\omega^\alpha = \sum_{\beta=k+1}^n \omega^\alpha_\beta \wedge \omega^\beta$ for certain $\omega^\alpha_\beta \in \bigwedge^1(U)$.
(iii) $\omega \wedge d\omega^\alpha = 0$ for $\alpha = k+1, \ldots, n$.
(iv) There exists $\theta \in \bigwedge^1(U)$ such that $d\omega = \theta \wedge \omega$.

Proof. The equivalence of (i) and (iii) follows immediately from the definitions. The same is true for the implication (i) $\Longrightarrow$ (iv). For the proof of (iv) $\Longrightarrow$ (iii) note that the condition (iv) means that

$$\sum_{\alpha=k+1}^{n} (-1)^\alpha \, d\omega^\alpha \wedge \omega^{k+1} \wedge \ldots \wedge \hat{\omega}^\alpha \wedge \ldots \wedge \omega^n = \theta \wedge \omega^{k+1} \wedge \ldots \wedge \omega^n .$$

Multiplying this equation with ω^α we get (iii). It remains to show that (iii) $\Longrightarrow$ (ii): Again, let $\omega^1, \ldots, \omega^n$ be a basis of $\bigwedge^1(U)$ such that $\omega^{k+1}, \ldots, \omega^n$ generate $\mathcal{I}$ over U. Then

$$d\omega^i = \sum_{j<l} f^i_{jl} \, \omega^j \wedge \omega^l , \tag{C.1}$$

where $f^i_{jl} \in \mathcal{F}(U)$. But

$$0 = d\omega^\alpha \wedge \omega = \sum_{1 \le j < l \le k} f^\alpha_{jl} \, \omega^j \wedge \omega^l \wedge \omega^{k+1} \wedge \ldots \wedge \omega^n .$$

Thus $f^\alpha_{jl} = 0$ for $\alpha = k+1, \ldots, n$ and $1 \le j < l \le k$. Hence the sum in (C.1) is of the form (ii). □

Now we assemble the preceding results in the following version of the Frobenius theorem.

Theorem C.2 (Frobenius' Theorem). *Let M be an n-dimensional manifold and E a k-dimensional distribution on M, and $\mathcal{I}(E)$ the associated ideal. The following statements are all equivalent:*

(i) *E is integrable.*
(ii) *E is involutive.*
(iii) *$\mathcal{I}(E)$ is a differential ideal locally generated by $n-k$ linearly independent 1-forms $\omega^{k+1},\ldots,\omega^n \in \bigwedge^1(U)$.*
(iv) *For every point of M there is a neighborhood U and $\omega^{k+1},\ldots,\omega^n \in \bigwedge^1(U)$ generating $\mathcal{I}(E)$ such that*

$$d\omega^\alpha = \sum_{\beta=k+1}^{n} \omega^\alpha_\beta \wedge \omega^\beta \tag{C.2}$$

for certain $\omega^\alpha_\beta \in \bigwedge^1(U)$.
(v)

$$d\omega^\alpha \wedge \omega^{k+1} \wedge \ldots \wedge \omega^n = 0 \tag{C.3}$$

(in the notation of (iv)).
(vi) *There exists a $\theta \in \bigwedge^1(U)$ such that*

$$d\omega = \theta \wedge \omega\,, \qquad \omega := \omega^{k+1} \wedge \ldots \wedge \omega^n\,. \tag{C.4}$$

Applications

Consider first a single timelike vector field X on a Lorentz manifold (M,g). The orthogonal complement of X_p in every $T_p(M)$ defines a 3-dimensional distribution E. Obviously, X is hypersurface orthogonal if and only if E is integrable. Since $\omega = X^\flat$ generates the ideal $\mathcal{I}(E)$ the Frobenius theorem tells us that E is integrable if and only if

$$\omega \wedge d\omega = 0\,. \tag{C.5}$$

(We have used the equivalence of (i) and (v).) This shows that (C.5) is necessary and sufficient for X being locally hypersurface orthogonal.

Our main application in GR will be the introduction of adapted coordinates if there are several Killing fields on (M,g), in particular, if there exist two commuting Killing fields. Let us, however, first introduce adapted coordinates for more general situations.

Let again, for a k-dimensional distribution E, $E|U = \text{span}\{X_1,\ldots,X_k\}$ and $\omega^{k+1},\ldots,\omega^n$ the generating 1-forms of the ideal $\mathcal{I}(E)$. We use the following notation: Indices running from $1,2,\ldots,n$ are denoted by Greek letters; for the first k numbers we use small Latin letters, while for indices with values $k+1,\ldots,n$ capital letters will be used.

Assume that E is involutive and use adapted coordinates $\{x^\mu\}$ as in the first version of the Frobenius theorem. The (local) integral manifolds are given by $\{x^A = \text{const}\}$. For the basis of vector fields X_a belonging to E we have

$$X_a(x^A) = 0 \qquad (X_a = X_a{}^b \frac{\partial}{\partial x^b}\,,\ X_a{}^A = 0). \tag{C.6}$$

Since $\langle \omega^A, X_a \rangle = 0$ we obtain, setting $\omega^A = \omega^A{}_\mu \, dx^\mu$,

$$0 = X_a{}^b \omega^A{}_\mu \langle dx^\mu, \partial_b \rangle = X_a{}^b \omega^A{}_b = 0$$

for all $a = 1, \ldots, k$. Hence $\omega^A_b = 0$ and thus

$$\omega^A = \omega^A_B \, dx^B \,. \tag{C.7}$$

Clearly, the matrices $(X_a{}^b)$ and (ω^A_B) are non-singular.

Let us specialize this to commuting fields X_a. We show that one can choose adapted coordinates such that

$$X_a = \frac{\partial}{\partial x^a} \qquad (a = 1, \ldots, k)\,. \tag{C.8}$$

To prove this, consider the flows ϕ^a_t of X_a. We know that these commute (Theorem 11.10). Now choose the coordinates x^a for a point p of an integral manifold $\{x^A = \text{const}\}$ according to

$$p = \left(\phi^1_{x^1} \circ \ldots \circ \phi^k_{x^k}\right)(p_0)\,,$$

where p_0 is a fixed "origin". With this choice the representation (C.8) obviously holds.

Returning to the general case we assume now that M has a distinguished pseudo-Riemannian metric g. Keeping the previous notation we can then also introduce the 1-forms $\omega^a = X^\flat_a$. These generate an ideal, denoted by $\mathcal{I}(E^\perp)$, which is the annihilator of a distribution $E^\perp$. Let $E^\perp$ be spanned by the vector fields X_A, $E^\perp = \text{span}\{X_A\}$. Then we have

$$\langle \omega^a, X_A \rangle = 0 \Longleftrightarrow g(X_a, X_A) = 0 \, : \, E \perp E^\perp \,. \tag{C.9}$$

In what follows we assume that the restrictions $g|E$ and $g|E^\perp$ are nonsingular. Then $E \cap E^\perp = \{0\}$, so

$$T_pM = E_p \oplus E_p^\perp \,. \tag{C.10}$$

Dually we have

$$T^*_pM = \text{span}\{\omega^a\} \oplus^\perp \text{span}\{\omega^A\}\,, \tag{C.11}$$

where $\omega^A = X^\flat_A$. We have thus constructed a basis of vector fields $\{X_\mu\}$ and the basis $\{\omega^\mu = X^\flat_\mu\}$ of 1-forms such that $\{\omega^A\}$ generates the ideal $\mathcal{I}(E)$, and $\{\omega^a\}$ the ideal $\mathcal{I}(E^\perp)$.

Let us now assume that E and $E^\perp$ are involutive. We use coordinates $\{u^\mu\}$ which are adapted to E and also coordinates $\{v^\mu\}$ that are adapted to $E^\perp$. Then we have according to (C.6) and (C.7)

$$X_a(u^A) = 0\,, \quad \omega^A = \omega^A_B \, du^B \,, \tag{C.12a}$$

$$X_A(v^a) = 0\,, \quad \omega^a = \omega^a_b \, dv^b \,. \tag{C.12b}$$

Now we define coordinates $\{x^\mu\}$ by

$$x^a = v^a \,, \qquad x^A = u^A \,. \tag{C.13}$$

The x^μ indeed form a coordinate system. For this we have to show that the dx^μ are linearly independent. Suppose there would be a linear dependence

$$f_a\, dx^a = g_A\, dx^A \,,$$

i.e., $f_a\, dv^a = g_A\, du^A$, then applying this to X_c leads to

$$f_a \langle dv^a, X_c \rangle = g_A \langle du^A, X_c \rangle = g_A X_c(u^A) = 0 \,,$$

i.e., $\langle dv^a, f_a X_c \rangle = 0$ or, by (C.12b), $\langle \omega^a, f_a X_c \rangle = 0$. Thus $f_a X_c \in E \cap E^\perp$, whence $f_a X_c = 0$ thus $f_a = 0$, implying $g_A = 0$.

The coordinates $\{x^\mu\}$ have, as a result of (C.12a) and (C.12b), the following properties:

(i) $X_a(x^A) = 0$,
(ii) $g_{aA} = 0$.

The last property is obtained as follows: Let $\omega^a = \omega^a{}_\mu\, dx^\mu$ and use (C.12b) to conclude that $\omega^a{}_A = 0$. But by (i)

$$\omega^a{}_A = g_{A\mu} X^\mu{}_a = g_{Ab} X^b{}_a$$

which implies (ii).

Summarizing, we have the result:

Proposition C.4. *Let $E = span\{X_1, \ldots, X_k\}$ be a k-dimensional distribution which is involutive. Assume in addition that the ideal generated by $\{\omega^a = X^b_a\}$ is differential. If this is the annihilator of $E^\perp$, then under the assumption that $E \cap E^\perp = \{0\}$ we can introduce coordinates $\{x^\mu\}$ with $(\mu = (a, A))$*

$$g_{aA} = 0 \,, \qquad X_a(x^A) = 0 \,. \tag{C.14}$$

Assume next, that the generating vector fields $X_1, \ldots, X_k$ in this Proposition are *Killing fields*, $L_{X_a} g = 0$. In adapted coordinates this becomes

$$X^b{}_a\, g_{\mu\nu,b} + g_{b\nu}\, X^b{}_{a,\mu} + g_{\mu b}\, X^b{}_{a,\nu} = 0 \,. \tag{C.15}$$

If we apply this for $\mu = A$, $\nu = B$, and use (C.14), we obtain $g_{AB,b} = 0$. Thus g_{AB} *depends only on* x^C. Choosing, on the other hand, $\mu = A$, $\nu = c$ in (C.15) we obtain

$$X^b{}_a\, g_{Ac,b} + g_{bc}\, X^b{}_{a,A} + g_{Ab}\, X^b{}_{a,c} = 0 \,. \tag{C.16}$$

Since $g_{aA} = 0$ the first and the last term vanish, and we conclude that $X^b{}_{a,A} = 0$. Thus $X^b{}_a$ *depends only on* x^c. Summarizing, we have the useful

Proposition C.5. *If in addition to the assumptions of the Proposition C.4 it is assumed that X_a are Killing fields of the pseudo-Riemannian manifold (M, g), then there are local coordinates $\{x^\mu\}$ such that (in the notation introduced above)*

$$g = g_{ab}(x^\mu)\, dx^a dx^b + g_{AB}(x^C)\, dx^A dx^B, \tag{C.17a}$$

$$X_a = X^b{}_a\, \partial_b, \tag{C.17b}$$

where $X^b{}_a$ are only functions of x^c.

Note, as a result of this, that the metric $g_{ab}\, dx^a dx^b$ on the integral manifolds $\{x^A = \text{const}\}$ has k Killing fields X_a, whence this submanifolds are homogeneous spaces.

Finally, we consider the important special case when the Killing fields X_a commute. We can then choose the adapted coordinates such that (C.8) holds. If we use this in (C.15) for $\mu = c$, $\nu = d$, we get

$$X^b{}_a g_{cd,b} + g_{bd} X^b{}_{a,c} + 0 = 0\,.$$

Since the second term vanishes we see that $g_{cd,b} = 0$. Hence g_{cd} depends only on x^A. In this special case we can, therefore, find coordinates x^μ such that

$$g = g_{ab}(x^C)\, dx^a dx^b + g_{AB}(x^C)\, dx^A dx^B, \tag{C.18a}$$

$$X_a = \frac{\partial}{\partial x^a}\,. \tag{C.18b}$$

This is used as the starting point of the derivation of the Kerr solution in Chap. 7.

Proof of Frobenius' Theorem (in the first version)

Proof (of Theorem C.1). Since this theorem is a local statement, we can work in $\mathbb{R}^n$ and choose $p = \{0\}$. Moreover, we can assume that $E_0 \subset T_0\mathbb{R}^n$ is spanned by the basis vectors

$$\left.\frac{\partial}{\partial t^1}\right|_0, \ldots, \left.\frac{\partial}{\partial t^k}\right|_0, \tag{C.19}$$

where $(t^1, \ldots, t^n)$ are the standard coordinates of $\mathbb{R}^n$. Let $\pi : \mathbb{R}^n \longrightarrow \mathbb{R}^k$ be the projection onto the first k factors. The corresponding tangential map $d\pi$ ($= T\pi$) maps E_0 isomorphically on $T_0\mathbb{R}^k$. By continuity, $d\pi$ is bijective on E_q for q near 0. So near 0, we can choose unique $X_1(q), \ldots, X_k(q) \in E_q$ so that

$$d\pi \cdot X_i(q) = \left.\frac{\partial}{\partial t^i}\right|_{\pi(q)} \qquad (i = 1, \ldots, k)\,. \tag{C.20}$$

Then (on a neighborhood U of $0 \in \mathbb{R}^n$) the vector fields X_i and $\frac{\partial}{\partial t^i}$ (on $\mathbb{R}^k$) are π-related. This then also holds for their Lie brackets:

$$d\pi \cdot [X_i, X_j]_q = \left[\frac{\partial}{\partial t^i}, \frac{\partial}{\partial t^j}\right]_{\pi(q)} = 0\,. \tag{C.21}$$

Since E was assumed to be involutive, we have $[X_i, X_j]_q \in E_q$, thus with (C.21) $[X_i, X_j] = 0$ ($d\pi$ is on E_q injective). According to the proof of (C.8) we can find coordinates $\{x^i\}$ such that

$$X_a = \frac{\partial}{\partial x^a}\,, \quad a = 1, \ldots, k\,. \tag{C.22}$$

The sets $\{q \in U \mid x^{k+1}(q) = a^{k+1}, \ldots, x^n(q) = a^n\}$ are then obviously integral manifolds of E, since their tangent spaces are spanned by (C.22).

Conversely, if N is a connected integral manifold restricted to U, with the inclusion $\iota : N \hookrightarrow U$, then we have for $x^\alpha \circ \iota$ with $\alpha = k+1, \ldots, n$ and $X_q \in T_qN$:

$$d(x^\alpha \circ \iota)(X_q) = X_q(x^\alpha \circ \iota) = d\iota \cdot X_q(x^\alpha) = 0\,,$$

since $d\iota \cdot X_q \in E_q$, which is spanned by the vectors (C.22) taken in q. Thus $d(x^\alpha \circ \iota) = 0$, implying that $x^\alpha \circ \iota$ is constant on the connected manifold N. □

* * *

Exercise C.1. Consider a 1-form ω on an open set $U \subset \mathbb{R}^n$. Use the Frobenius Theorem to show that the $(n-1)$-distribution $E = \{v \in TU \mid \omega(v) = 0\}$ is integrable if and only if ω is of the form

$$\omega = g\, df\,. \tag{C.23}$$

The smooth function g is called an **integrating factor**.

Since by the Frobenius Theorem integrability is equivalent to

$$\omega \wedge d\omega = 0\,, \tag{C.24}$$

we see that (C.24) is necessary and sufficient for the existence of an integrating factor. Note that (C.24) is always satisfied for $n = 2$, so that ω has always an integrating factor.

Solution. If (C.23) is satisfied, then $\{f = \text{const}\}$ are integral manifolds of E. Conversely, if E is integrable then according to the first version of the Frobenius theorem, the integral manifolds of E are given by $\{f = \text{const}\}$ for a smooth function, so that df annihilates E. Thus the ideal $\mathcal{I}(E)$ generated by ω is also generated by df, whence $\omega = g\, df$ for some smooth function g on U.

D Collection of Important Formulas

The following list summarizes some of the important formulas which have been obtained in the foregoing chapters and will constantly be used throughout this book.

Vector Fields, Lie Brackets

The set of smooth vector fields $\mathcal{X}(M)$ on a manifold M with the commutator product $[X, Y]$ form a $\mathbb{R}$-Lie algebra and a module over the associative algebra $\mathcal{F}(M)$ of C^∞-functions. In local coordinates we have

$$X = \xi^i \frac{\partial}{\partial x^i}, \qquad Y = \eta^i \frac{\partial}{\partial x^i},$$

$$[X, Y] = \left(\xi^i \frac{\partial \eta^j}{\partial x^i} - \eta^i \frac{\partial \xi^j}{\partial x^i} \right) \frac{\partial}{\partial x^j}.$$

Under a diffeomorphism $\varphi \in \mathrm{Diff}(M)$ the Lie bracket transforms naturally:

$$\varphi_* [X, Y] = [\varphi_* X, \varphi_* Y].$$

Differential Forms

The differential forms $\bigwedge(M)$ on M form a real associative algebra with the exterior product $\wedge$ as multiplication; $\bigwedge^p(M)$ denotes the $\mathcal{F}(M)$-module of p-forms. The exterior product satisfies,

$$\alpha \wedge \beta = (-1)^{kl} \beta \wedge \alpha, \qquad \alpha \in \bigwedge^k(M), \quad \beta \in \bigwedge^l(M).$$

For a differentiable map φ,

$$\varphi^*(\alpha \wedge \beta) = \varphi^* \alpha \wedge \varphi^* \beta.$$

Exterior Differential

The exterior differential d is an antiderivation of degree $+1$; in particular

$$d(\alpha \wedge \beta) = d\alpha \wedge \beta + (-1)^k \alpha \wedge d\beta\,, \qquad \text{for } \alpha \in \bigwedge^k(M);\ \beta \in \bigwedge(M)\,.$$

Furthermore

$$d \circ d = 0\,.$$

If $\alpha \in \bigwedge^p(M)$ and $X_1, \ldots, X_{p+1} \in \mathcal{X}(M)$, then

$$d\alpha(X_1, \ldots, X_{p+1}) = \sum_{1 \le i \le p+1} (-1)^{i+1} X_i\, \alpha(X_1, \ldots, \hat{X}_i, \ldots, X_{p+1}) + \sum_{i<j} (-1)^{i+j}\, \alpha\big([X_i, X_j], X_1, \ldots, \hat{X}_i, \ldots, \hat{X}_j, \ldots, X_{p+1}\big)\,.$$

For a map φ,

$$\varphi^* d\alpha = d\varphi^* \alpha$$

Poincaré Lemma

If $d\alpha = 0$, then α is locally exact; that is, there is a neighborhood U of every $x \in M$ on which $\alpha = d\beta$.

Interior Product

The interior product i_X is an antiderivation of degree -1; in particular

$$i_X(\alpha \wedge \beta) = (i_X\alpha) \wedge \beta + (-1)^p \alpha \wedge i_X\beta\,, \qquad \text{for } \alpha \in \bigwedge^p(M)\,,\ \beta \in \bigwedge(M)\,.$$

We obviously have

$$i_{fX}\alpha = f\, i_X\alpha = i_X f\, \alpha\,, \qquad \text{for } f \in \mathcal{F}(M)\,,$$

and

$$i_X \circ i_X = 0\,.$$

For a diffeomorphism φ (see Exercise D.1),

$$\varphi^* i_X \alpha = i_{\varphi^* X} \varphi^* \alpha\,.$$

Lie Derivative

The Lie derivative L_X with respect to the vector field X is a derivation of degree 0 of the tensor algebra $\mathcal{T}(M)$; it is $\mathbb{R}$-linear in X. For a diffeomorphism φ and $T \in \mathcal{T}(M)$

$$\varphi^* L_X T = L_{\phi^* X} \varphi^* T\,.$$

For a covariant tensor field $T \in \mathcal{T}^0_q(M)$ and $X_1, \ldots, X_q \in \mathcal{X}(M)$

$$(L_X\, T)(X_1, \ldots, X_q) = X\big(T(X_1, \ldots, X_q)\big) - \sum_{k=1}^{q} T\big(X_1, \ldots, [X, X_k], \ldots, X_q\big).$$

In local coordinates the components of $L_X T$ for a tensor field $T \in \mathcal{T}^r_s(M)$ are given by

$$\begin{aligned}(L_X T)^{i_1\ldots i_r}_{j_1\ldots j_s} &= X^i\, T^{i_1\ldots i_r}_{j_1\ldots j_s, i} \\ &\quad - T^{k i_2\ldots i_r}_{j_1\ldots j_s} \cdot X^{i_1}_{,k} - \text{all upper indices} \\ &\quad + T^{i_1\ldots i_r}_{k j_2\ldots j_s} \cdot X^k_{,j_1} + \text{all lower indices}.\end{aligned}$$

Relations Between L_X, i_X and d

The following identities hold for differential forms

$$\begin{aligned} L_X\alpha &= d\, i_X\alpha + i_X d\alpha\,, \qquad \textbf{(Cartan's formula)}\\ L_{fX}\alpha &= f\, L_X\alpha + df \wedge i_X\alpha\,,\\ L_{[X,Y]} &= [L_X, L_Y] = L_X \circ L_Y - L_Y \circ L_X\,,\\ i_{[X,Y]} &= [L_X, i_Y] = L_X \circ i_Y - i_Y \circ L_X\,,\\ L_X \circ d &= d \circ L_X\,,\\ L_X \circ i_X &= i_X \circ L_X\,. \end{aligned}$$

Volume Form

For an oriented n-dimensional pseudo-Riemannian manifold (M, g), with $\theta^1, \ldots, \theta^n$ an oriented basis of one-forms, the canonical volume form can be represented as

$$\eta = \sqrt{|g|}\, \theta^1 \wedge \cdots \wedge \theta^n\,, \quad g = g_{ik}\, \theta^i \otimes \theta^k\,, \quad |g| = |\det(g_{ik})|\,.$$

Hodge-Star Operation

The Hodge-star operator for (M, g) is a linear isomorphism $* : \bigwedge^k(M) \longrightarrow \bigwedge^{n-k}(M)$ and satisfies for $\alpha, \beta \in \bigwedge^k(M)$:

$$\begin{aligned} \alpha \wedge *\beta &= \beta \wedge *\alpha = \langle\alpha, \beta\rangle\eta\,,\\ *(*\alpha) &= (-1)^{k(n-k)} sgn(g)\alpha\,,\\ \langle *\alpha, *\beta\rangle &= sgn(g)\langle\alpha, \beta\rangle\,. \end{aligned}$$

In local coordinates

$$*(\theta^{i_1} \wedge \ldots \wedge \theta^{i_p}) = \frac{\sqrt{|g|}}{(n-p)!}\, \varepsilon_{j_1\ldots j_n}\, g^{j_1 i_1} \ldots g^{j_p i_p}\, \theta^{j_{p+1}} \wedge \ldots \wedge \theta^{j_n}\,.$$

Codifferential

The codifferential for (M, g) is a linear isomorphism $\delta : \bigwedge^k(M) \longrightarrow \bigwedge^{k-1}(M)$ defined by

$$\delta := sgn(g)(-1)^{np+n} * d* ,$$

and satisfies

$$\delta \circ \delta = 0 .$$

In local coordinates, for $\alpha \in \bigwedge^k(M)$,

$$(\delta\alpha)^{i_1 \dots i_{p-1}} = \frac{1}{\sqrt{|g|}} \left(\sqrt{|g|}\, \alpha^{k i_1 \dots i_{p-1}} \right)_{,k} .$$

Covariant Derivative

The covariant derivative ∇_X in the X-direction of an affine connection on M is a derivation of the tensor algebra $\mathcal{T}(M)$, which commutes with all contractions. For $S \in \mathcal{T}^p_q(M)$,

$$\begin{aligned}(\nabla_X S)(Y_1 , \dots , Y_q, \omega_1, \dots , \omega_p) \\ &= X\big(S(Y_1, \dots , Y_q, \omega_1, \dots , \omega_p)\big) - S(\nabla_X Y_1, Y_2, \dots , \omega_p) - \dots - \\ &\quad - S(Y_1, \dots , \nabla_X \omega_p) .\end{aligned}$$

In local coordinates, the components of $\nabla_X S$ are given by

$$\begin{aligned} S^{i_1 \dots i_p}_{j_1 \dots j_q ; k} = S^{i_1 \dots i_p}_{j_1 \dots j_q , k} &+ \Gamma^{i_1}_{\ kl}\, S^{l i_2 \dots i_p}_{j_1 \dots j_q} + \dots \\ &- \Gamma^{l}_{\ k j_1}\, S^{i_1 \dots i_p}_{l j_2 \dots j_q} - \dots .\end{aligned}$$

Connection Forms

Let $(e_1, \dots , e_n)$ be a moving frame with the dual basis $(\theta^1, \dots , \theta^n)$ of one-forms. The connection forms $\omega^i_{\ j}$ are given by

$$\nabla_X e_j = \omega^i_{\ j}(X) e_i .$$

or

$$\nabla_X \theta^i = -\omega^i_{\ j}(X) \theta^j .$$

Curvature Forms

The expansion coefficients of the curvature forms $\Omega^i_{\ j}$ are the components of the Riemann tensor relative to a moving frame:

$$\Omega^i_{\ j} = \frac{1}{2} R^i_{\ jkl}\, \theta^k \wedge \theta^l , \qquad R^i_{\ jkl} = -R^i_{\ jlk} .$$

Cartan's Structure Equations

The torsion forms Θ^i and curvature forms $\Omega^i{}_j$ satisfy Cartan's structure equations

$$\Theta^i = d\theta^i + \omega^i{}_j \wedge \theta^j\,,$$
$$\Omega^i{}_j = d\omega^i{}_j + \omega^i{}_k \wedge \omega^k{}_j\,.$$

Riemannian Connection

For the Riemannian connection with metric $g = g_{ik}\,\theta^i \otimes \theta^k$

$$\omega_{ik} + \omega_{ki} = dg_{ik}\,, \qquad \omega_{ik} = g_{ij}\,\omega^j{}_k\,.$$

Coordinate Expressions

Riemann tensor

$$R^i{}_{jkl} = \Gamma^i{}_{lj,k} - \Gamma^i{}_{kj,l} + \Gamma^s{}_{lj}\Gamma^i{}_{ks} - \Gamma^s{}_{kj}\Gamma^i{}_{ls}\,,$$

Ricci tensor

$$R_{jl} = R^i{}_{jil} = \Gamma^i{}_{lj,i} - \Gamma^i{}_{ij,l} + \Gamma^s{}_{lj}\Gamma^i{}_{is} - \Gamma^s{}_{ij}\Gamma^i{}_{ls}\,,$$

Scalar curvature

$$R = g^{ik} R_{ik}\,,$$

Einstein tensor

$$G_{ik} = R_{ik} - \frac{1}{2} g_{ik}\, R\,,$$

Christoffel symbols

$$\Gamma^l{}_{ij} = \frac{1}{2} g^{lk}\left(g_{ki,j} + g_{kj,i} - g_{ij,k}\right).$$

Absolute Exterior Differential

The absolute exterior differential of a tensor valued p-form ϕ of type (r, s) has the components

$$(D\phi)^{i_1\ldots i_r}_{j_1\ldots j_s} = d\,\phi^{i_1\ldots i_r}_{j_1\ldots j_s} + \omega^{i_1}{}_l \wedge \phi^{l i_2\ldots i_r}_{j_1\ldots j_s} + \ldots \text{ for all upper indices}$$
$$- \,\omega^l{}_{j_1} \wedge \phi^{i_1\ldots i_r}_{l j_2\ldots j_s} - \ldots \text{ for all lower indices}\,.$$

The second derivative is given by

$$(D^2\phi)^{i_1\ldots i_r}_{j_1\ldots j_s} = \Omega^{i_1}{}_l \wedge \phi^{l i_2\ldots i_r}_{j_1\ldots j_s} + \ldots - \Omega^l{}_{j_1} \wedge \phi^{i_1\ldots i_r}_{l j_2\ldots j_s} - \ldots\,.$$

Bianchi Identities

In terms of the coordinate components of the Riemann tensor the Bianchi identities read for a symmetric affine connection:

$$\sum_{(jkl)} R^i{}_{jkl} = 0 \qquad \textbf{(1st Bianchi identity)}\,,$$

$$\sum_{(klm)} R^i{}_{jkl;m} = 0 \qquad \textbf{(2nd Bianchi identity)}\,.$$

In terms of the torsion and curvature forms they are

$$D\Theta^i = \Omega^i{}_j \wedge \theta^j$$

$$D\Omega^i{}_j = 0\,.$$

* * *

Exercise D.1. Let $\varphi : M \longrightarrow N$ be a differentiable mapping between the manifolds M and N, $\alpha \in \bigwedge(N)$, $X \in \mathcal{X}(N)$, $Y \in \mathcal{Y}(M)$ such that Y and X are φ-related. Show that

$$i_Y \varphi^* \alpha = \varphi^* i_X \alpha\,.$$

In particular, if φ is a diffeomorphism, then

$$i_{\varphi^* X}\, \varphi^* \alpha = \alpha^* i_X \alpha\,.$$

Remark. This naturalness property of the interior product follows directly from the definitions.

References

Textbooks on General Relativity

Classical Texts

1. W. Pauli, *Theory of Relativity.* Pergamon Press 1958; Relativitätstheorie, In: *Enzyklopädie der Mathematischen Wissenschaften*, Band 5, Teil 2, Teubner, Leipzig 1921; neu herausgegeben und kommentiert von D. Giulini, Springer-Verlag 2000
2. H. Weyl, *Space-Time-Matter.* Dover, New York 1952 (translated from the 3rd German edition by H.L. Brose); *Raum-Zeit-Materie*, Vorlesungen über allgemeine Relativitätstheorie, Springer, Berlin 1918; neu herausgegeben und ergänzt von J. Ehlers, Springer-Verlag 1993 (8th edition)
3. A.S. Eddington, *The Mathematical Theory of Relativity.* Chelsea Publishing Company 1975
4. V. Fock, *The Theory of Space Time and Gravitation.* Pergamon Press, Oxford 1959
5. J.L. Synge, *Relativity, The General Theory.* North-Holland 1971

Selection of (Graduate) Textbooks

6. C.M. Will, *Theory and Experiment in Gravitational Physics.* Cambridge University Press, Cambridge 1993 (2nd edition)
7. R.M. Wald, *General Relativity.* The University of Chicago Press, Chicago 1984
8. G. Ellis and S. Hawking, *The Large Scale Structure of Space-Time.* Cambridge Univeristy Press 1973
9. J.L. Anderson, *Principles of Relativity Physics.* Academic Press 1967
10. L.D. Landau and E.M. Lifshitz, *The Classical Theory of Fields.* Addison Wesley 1987 (4th edition)
11. R. Adler, M. Bazin and M. Schiffer, *Introduction to General Relativity.* McGraw-Hill 1975 (2nd edition)
12. B. Schutz, *A First Course in General Relativity.* Cambridge University Press 1985
13. C.W. Misner, K.S. Thorne and J.A. Wheeler, *Gravitation.* Freeman 1973
14. N. Straumann, *General Relativity and Relativistic Astrophysics*, Texts and Monographs in Physics, Springer-Verlag Berlin 1984
15. F. de Felice and C. Clarke, *Relativity on Curved Manifolds.* Cambridge University Press 1990

General Physics and Astrophysics Books

16. L.D. Landau and E.M. Lifshitz, *Fluid Mechanics.* Pergamon Press 1987 (2nd edition)
17. G. Wentzel, *Quantum Theory of Fields.* Interscience Publishers 1949
18. J.D. Jackson, *Classical Electrodynamics.* John Wiley and Sons 1999 (3rd edition)
19. P. Schneider, J. Ehlers and E.E. Falco, *Gravitational Lenses.* Springer Heidelberg 1992
20. J. Binney and S. Tremaine, *Galactic Dynamics.* Princeton University Press, Princeton 1987
21. S.L. Shapiro and S.A. Teukolsky, *Black Holes, White Dwarfs and Neutron Stars.* John Wiley and Sons, Inc. 1983
22. N.K. Glendenning, *Compact Stars.* Springer 1997
23. F. Weber, *Pulsars as Astrophysical Laboratories for Nuclear and Particle Physics*, Institute of Physics Publishing. Bristol and Philadelphia 1999
24. S. Chandrasekhar, *Structure and Evolution of the Stars.* University of Chicago Press 1939
25. P.A. Charles and F.D. Seward, *Exploring the X-ray Universe.* Cambridge University Press 1995
26. P. Mészáros, *High-Energy Radiation from Magnetized Neutron Stars.* University of Chicago Press 1992
27. J. Frank, A. King and D. Raine, *Accretion Power in Astrophysics.* Cambridge University Press 2002 (3rd edition)
28. V.I. Arnold, *Mathematical Methods of Classical Mechanics.* Graduate Texts in Mathematics, Vol. 60, Springer-Verlag 1989
29. N. Straumann, *Relativistische Quantentheorie.* Springer-Verlag 2004

Mathematical Tools

Modern Treatments of Differential Geometry for Physicists

30. W. Thirring, *A Course in Mathematical Physics I and II: Classical Dynamical Systems and Classical Field Theory.* Springer-Verlag 1992 (2nd edition)
31. Y. Choquet-Bruhat, C. De Witt-Morette and M. Dillard-Bleick, *Analysis, Manifolds and Physics.* North-Holland 1982 (rev. edition)
32. G. von Westenholz, *Differential Forms in Mathematical Physics.* North-Holland 1978
33. R. Abraham and J.E. Marsden, *Foundations of Mechanics.* Benjamin 1978 (2nd edition)
34. T. Frankel, *The Geometry of Physics.* Cambridge University Press 1997

Selection of Mathematical Books

35. S. Kobayashi and K. Nomizu, *Foundations of Differential Geometry I and II.* Interscience Publishers 1963/69
36. Y. Matsushima, *Differentiable Manifolds.* Marcel Dekker, New York 1972

37. R. Sulanke and P. Wintgen, *Differentialgeometrie und Faserbündel.* Birkhäuser 1972
38. R. L. Bishop and R. J. Goldberg, *Tensor Analysis on Manifolds.* McMillan, New York 1968
39. B. O'Neill, *Semi-Riemannian Geometry with Applications to Relativity.* Academic Press, Inc. 1983
40. S. Lang, *Differential and Riemannian Manifolds.* Graduate Texts in Mathematics 160, Springer 1995
41. J. Jost, *Riemannian Geometry and Geometric Analysis.* Springer 2002 (3rd edition)
42. B. O'Neill, *Elementary Differential Geometry.* Academic Press 1997
43. M. Spivak, *A Comprehensive Introduction to Differential Geometry I-V*, Publish or Perish 1979
44. P.K. Raschewski, *Riemannsche Geometrie und Tensoranalysis.* Deutscher Verlag der Wissenschaften 1959
45. L.H. Loomis and S. Sternberg, *Advanced Calculus.* Addison-Wesley 1968
46. L.C. Evans, *Partial Differential Equations.* Graduate Studies in Mathematics, Vol. 19, American Mathematical Society 1998
47. F. John, *Partial Differential Equations.* Springer-Verlag 1982 (4th edition)
48. M.E. Taylor, *Partial Differential Equations.* Springer Verlag 1996

Historical Sources

49. G. Nordström, *Phys. Zeitschrift* **13**, 1126 (1912); *Ann. Phys. (Leipzig)* **40**, 856 (1913); *Ann. Phys. (Leipzig)* **42**, 533 (1913)
50. A. Einstein and A.D. Fokker, *Ann. Phys. (Leipzig)* **44**, 321 (1914)
51. L. O'Raifeartaigh and N. Straumann, *Rev. Mod. Phys.* **72**, 1 (2000)
52. E. Cartan, *Annales Scientifiques de l'École Normale Supérieure* **40**, 325 (1923), and **41**, 1 (1924)
53. K. Friedrichs, *Mathematische Annalen* **98**, 566 (1928)
54. H. Hertz, *Über die Constitution der Materie.* Springer Verlag 1999
55. A. Pais, *Subtle is the Lord, The Science and the Life of Albert Einstein.* Oxford University Press 1982
56. J. Stachel, *Einstein from "B" to "Z".* Einstein Studies, Vol. 9, Birkhäuser 2002
57. A. Einstein, *Collected Papers.* Vol. 4, Doc. 10
58. A. Einstein, *Collected Papers.* Vol. 6, Doc. 25, Doc. 24, Doc. 32
59. A. Einstein, *Collected Papers.* Vol. 7, Doc. 9, Doc. 1
60. A. Einstein, *Collected Papers.* Vol. 8
61. L. Corry, J. Renn and J. Stachel, *Science* **278**, 1270-1273 (1997)
62. D. Hilbert, *Nach. Ges. Wiss. Göttingen*, 395 (1916)
63. A. Einstein, *Sitzungsber. Preuss. Akad. Wiss.* Phys.-math. Klasse VI, 142 (1917)
64. J. Norton, *General Covariance and the Foundations of General Relativity: Eight Decades of Dispute.* Reports on Progress in Physics **56**, 791-858 (1993)
65. M. Friedmann, *Foundations of Space-Time Theories, Relativistic Physics and Philosophy of Science.* Princeton University Press 1983
66. A. Einstein, *The Meaning of Relativity.* Princeton 1956 (5th edition)
67. W. Israel, Dark Stars: The Evolution of an Idea. In: S. Hawking and W. Israel (eds.), *300 Years of Gravitation.* Cambridge University Press 1987

Recent Books on Cosmology

Although cosmology is not treated in this book, we give references to some useful recent textbooks.

68. P.J.E. Peebles, *Principles of Physical Cosmology.* Princeton University Press 1993
69. J.A. Peacock, *Cosmological Physics.* Cambridge University Press 1999
70. A.R. Liddle and D.H. Lyth, *Cosmological Inflation and Large Scale Structure.* Cambridge University Press 2000
71. S. Dodelson, *Modern Cosmology.* Academic Press 2003
72. G. Börner, *The Early Universe.* Springer-Verlag 2003 (4th edition)

Research Articles, Reviews and Specialized Texts

Chapter 1

73. J. Ehlers and W. Rindler, *Gen. Rel. Grav.* **29**, 519 (1997)
74. C.M. Will, *Living Reviews in Relativity* **4** (2001); gr-qc/9811036
75. N. Straumann, *Fundamental Physics in Space & Related Topics.* ESA-CERN Workshop 5-7 April 2000, European Space Agency, SP-469, 55 (2001); astro-ph/0006423
76. W. Pauli and M. Fierz, *Helv. Phys. Acta* **12**, 287 (1939); *Proc. Roy. Soc. (London) Ser. A* **173**, 211 (1939)
77. S. Deser, *Gen. Rel. Grav.* **1**, 9 (1970)

Chapter 2

78. D. Lovelock, *J. Math. Phys.* **13**, 874 (1972)
79. N. Straumann, On the Cosmological Constant Problems and the Astronomical Evidence for a Homogenous Energy Density with Negative Pressure. In: B. Duplantier and V. Rivasseau (eds.), *Vacuum Energy-Renormalization*, Séminaire Poincaré (2002). Birkhäuser Verlag 2003; astro-ph/0203330; and The History of the Cosmological Constant. In: *On the Nature of Dark Energy*, Proceedings of the 18th IAP Astrophysics Colloquium, Frontier Group 2002; gr-qc/0208027
80. D.E. Rowe, *Phys. Perspect.* **3**, 379-424 (2001)
81. A.E. Fischer and J.E. Marsden, The Initial Value Problem and the Dynamical Formulation of General Relativity. In: S.W. Hawking and W. Israel (eds.), *General Relativity, an Einstein Centenary Survey.* Cambridge University Press 1979
82. Y. Choquet-Bruhat and J.W. York, The Cauchy Problem. In: A. Held (ed.), *General Relativity and Gravitation*, Vol. 1, Plenum Press 1980
83. S. Klainerman and F. Nicolò, *The Evolution Problem in General Relativity.* Birkhäuser 2002
84. R. Schoen and S.T. Yau, *Commun. Math. Phys.* **65**, 45 (1976); *Phys. Rev. Lett.* **43**, 1457 (1979); *Commun. Math. Phys.* **79**, 231 (1981); *Commun. Math. Phys.* **79**, 47 (1981)
85. L. Rosenfeld, *Mem. Roy. Acad. Belg. Cl. Sci.* **18**, 6 (1940)

86. D. Christodoulou and S. Klainermann, *The Global, Nonlinear Stability of the Minkowski Space*. Princeton University Press, Princeton 1993
87. M. Shibata and T. Nakamura, *Phys. Rev.* D **52**, 5428 (1995)
88. T.W. Baumgarte and S.L. Shapiro, *Phys. Rev.* D **59**, 024007 (1999)
89. M. Alcubierre et al., *Phys. Rev.* D **62**, 044034 (2000)
90. S. Frittelli and O. Reula, *Commun. Math. Phys.* **166**, 221 (1994)
91. C. Bona, J. Massó, E. Seidel and J. Stela, *Phys. Rev. Lett.* **75**, 600 (1995)
92. A. Abrahams, A. Anderson, Y. Choquet-Bruhat and J.W. York Jr., *Phys. Rev. Lett.* **75**, 3377 (1996)
93. M.H.P.M. van Putten and D.M. Eardley, *Phys. Rev.* D **53**, 3056 (1996)
94. H. Friedrich, *Class. Quantum Grav.* **13**, 1451 (1996)
95. A. Anderson, Y. Choquet-Bruhat, and J.W. York,Jr., *Top. Methods Nonlinear Anal.* **10**, 353 (1997)
96. http://cargese.univ-course.fr/Pages/Ecoles2002.html

Chapter 3

97. R.A. Hulse, *Rev. Mod. Phys.* **66**, 699 (1994)
98. J.H. Taylor, *Rev. Mod. Phys.* **66**, 711 (1994)
99. P. Coles, Einstein, Eddington and the 1919 Eclipse. In: V.J. Martinez, V. Trimble and M.J. Pons-Borderia (eds.), Proceedings on International School on *The Historical Development of Modern Cosmology*, Valencia 2000, ASP Conference Series (to appear); astro-ph/0102462
100. M. Froeschlé, F. Miguard and F. Arenon, in Proceedings of the Hipparcos Venice 1997 Symposium, to appear
101. I. Shapiro et al., *Phys. Rev. Lett.* **26**, 1137 (1971)
102. R.D. Reasonberg et al., *Astrophys. J.* **234**, L 219 (1979)
103. J.G. Williams et al., *Phys. Rev.* D **53**, 6730 (1996)
104. J.M. Weisberg and J.H. Taylor, *Astrophys. J.* **576**, 942 (2002)
105. S. Chandrasekhar, *The Mathematical Theory of Black Holes*. Clarendon Press, Oxford University Press 1983
106. J.B. Hartle, *Phys. Reports* **46**, 201 (1978)
107. C.V. Vishveshwara, *Phys. Rev.* D **1**, 2870 (1970)
108. F.J. Zerilli, *Phys. Rev. Lett.* **24**, 737 (1970)
109. R. Schödel et al., *Nature* **419**, 694 (2002); astro-ph/0210426
110. B. Bertotti, L. Iess and P. Tortora, *Nature* **425**, 374-376 (2003)

Chapter 4

111. S. Rowan and J. Hough, *Gravitational Wave Detection by Interferometry*; http://www.livingreviews.org/Articles/Volume3/2000-3hough/.
112. *System and Technology Study Report* ESA/SCI (2000) 11, by the LISA Science Team
113. B.F. Schutz, *Gravitational Wave Astronomy*, Class. Quant. Grav. **16**, A131-156 (1999).
114. C. Cutler and K.S. Thorne, gr-qc/0204090
115. N. Straumann, Lectures on Gravitational Lensing. In: R. Durrer and N. Straumann (eds.), *New Methods for the Determination of Cosmological Parameters*. Troisième Cycle de la Physique en Suisse Romande (1999)

116. M.J. Irvin et al., *Astrophys. J.* **98**, 1989 (1989)
117. A. Milsztajn, The Galactic Halo from Microlensing. In: *Matter in the Universe*, Space Science Series of ISSI, Vol. 14 (Reprinted in from *Space Science Reviews*, Vol. 100, Nos. 1-4, 2002)
118. N. Kaiser and G. Squires, *Astrophys. J.* **404**, 441 (1993)
119. M. Bartelmann and P. Schneider, *Weak Gravitational Lensing*, Physics Reports **340**, 291-472 (2001)
120. N. Kaiser, *Astrophys. J.* **439** (1995)
121. R.A. Isaacson, *Phys. Rev.* **166**, 1263 (1968); *Phys. Rev.* **166**, 1272 (1968)
122. S. Seitz and P. Schneider, *Astron. Astrophys.* **305**, 383 (1996)

Chapter 5

123. T. Damour, The Problem of Motion in Newtonian and Einsteinian Gravity. In: S.W. Hawking and W. Israel (eds.), *300 Years of Gravitation.* Cambridge University Press, London 1987
124. R. Penrose, The Light Cone at Infinity. In: L. Infeld (ed.), *Relativistic Theories of Gravitation.* Pergamon Press, Oxford 1964
125. H. Friedrich, Conformal Einstein Evolution. In: J. Frauendiener and H. Friedrich (eds.), *The Conformal Structure of Spacetime.* Springer-Verlag, Berlin Heidelberg 2002
126. J. Frauendiener, *Conformal Infinity*, Living Reviews in Relativity, **3**, 2000; http://www.livingreviews.org/Articles/Volume3/2000-4frauendiener/.
127. H. Friedrich , *Proc. Roy. Soc. London* A**378**, 401 (1981)
128. H. Bondi, M.G.J. Van der Burg and A.W.K. Metzner, *Proc. Roy. Soc. London* A**289**, 21 (1962)
129. J. Nester and W. Israel, *Phys. Lett.* **85A**, 259 (1981)
130. A. Ashtekar and A. Magnon-Ashtekar, *Phys. Rev. Lett.* **43**, 181 (1979)
131. L. Blanchet, T. Damour and G. Schäfer, *MNRAS* **242**, 289 (1990)
132. S. Chandrasekhar, *Astrophys. J.* **142**, 1488 (1965)
133. T. Damour, Gravitational Radiation and the Motion of Compact Bodies. In: N. Deruelle and T. Piran (eds.), *Gravitational Radiation.* North-Holland, Amsterdam 1983
134. L. Blanchet, *Gravitational Radiation from Post-Newtonian Sources and Inspiralling Compact Binaries*, Living Reviews in Relativity, **5**, 2002; http://www.livingreviews.org/Articles/Volume/2002-3Blanchet/.
135. M.E. Pati and C.M. Will, *Phys. Rev.* D **62**, 124015 (2000)
136. M.E. Pati and C.M. Will, *Phys. Rev.* D **65**, 104008 (2002); gr-qc/0201001
137. G. Schäfer, *Gen. Relativ. Gravit.* **18**, 255 (1986)
138. T. Damour, P. Jaranowski and G. Schäfer, *Pys. Rev.* D **62**, 021501 (2000)
139. T. Damour, Gravitational Radiation Reaction in the Binary Pulsar and the Quadrupole Formula Controversy. In: S. Benenti, M. Ferraris and M. Francavighia (eds.), *Proceedings of Journées Relativistes 1983.* Pitagora Editrice 1985; *Phys. Rev. Lett.* **51**, 1019 (1983)
140. R.A. Hulse and J.H. Taylor, *Astropys. J.* **195**, L 51 (1975)
141. J.H. Taylor, *Millisecond pulsars: Natur's mos stable clocks* Proc. IEEE **79**, 1054 (1991)
142. C.M. Will, *Ann. Phys.* **155**, 133 (1984)

143. R. Blandford and S.A. Teukolsky, *Astrophys. J.* **205**, 580 (1976)
144. T. Damour and N. Deruelle, *Ann. Inst. H. Poincaré (Physique Théorique)* **43**, 107 (1985)
145. T. Damour and N. Deruelle, *Ann. Inst. H. Poincaré (Physique Théorique)* **44**, 263 (1986)
146. T. Damour and J.H. Taylor, *Phys. Rev.* D **45**, 1840 (1992)
147. J.H. Taylor, *Phil. Trans. R. Soc.* **A 341**, 117 (1992)
148. J.M. Weisberg and J.H. Taylor, The Relativistic Binary Pulsar B 1913+16. In: D.J. Nice and S.E. Thorsett (eds.), Proceedings of *Binary Pulsars*, Chania, Crete 2002 ASP Conference Series; astro-ph/0211217
149. T. Damour and J.H. Taylor, *Astrophys. J.* **366**, 501 (1991)
150. L.L. Smarr and R. Blandford, *Astrophys. J.* **207** 574 (1976)
151. A. Wolszczan, *Nature* **350**, 688 (1991)
152. I.H. Stairs, S.E. Thorsett, J.H. Taylor and A. Wolszczan, *Astrophys. J.* **581**, 501 (2002); astro-ph/0208357
153. I.H. Stairs, E.M. Splaver, S.E. Thorsett, D.J. Nice and J.H. Taylor, *MNRAS* **314**, 459 (2000)
154. J.H. Taylor and J.M. Cordes, *Astrophys. J.* **411**, 674 (1993)
155. T. Damour and G. Esposito-Farèse, *Phys. Rev* D **54**, 1474 (1996); D **58**, 042001 (1998)
156. A.G. Lyne, et al., *Science* **303**, 1153 (2004)
157. M. Burgay, et al., *Nature* **426**, 531 (2003)

Chapter 6

158. J.L. Provencal, H.L. Shpiman, E. Hog and P. Thejll, *Astrophys.* **494**, 759 (1998)
159. L.D. Landau, *Phys. Z. Sowjetunion* **1**, 285 (1932)
160. N. Straumann, Physics of Type II Supernova Explosions. In: H. Mitter and F. Widder (eds.), *Particle Physics and Astrophysics, Current Viewpoints.* Springer-Verlag 1983
161. H.A. Bethe, *Rev. Mod. Phys.* **62**, 801 (1990)
162. J.R. Oppenheimer and G.M. Volkoff *Phys. Rev* **55**, 374 (1939)
163. H. Heiselberg and V. Pandharipande, *Ann. Rev. Nucl. and Part. Science* **50**, 481-524 (2000)
164. S. Balberg and S.L. Shapiro, *The Properties of Matter in White Dwarfs and Neutron Stars*; astro-ph/0004317
165. R.B. Wiringa, V.G.J. Stoks and R. Schiavilla, *Phys. Rev.* **C53**, R1483 (1996)
166. A. Akmal and V.R. Pandharipande, *Phys. Rev.* C **56**, 2261 (1997)
167. J.L. Forest, V.R. Pandharipande and J.L. Friar, *Phys. Rev.* C **52**, 568 (1995)
168. A. Akmal, V.R. Pandharipande and R. Ravenhall, *Phys. Rev.* C **58**, 1804 (1998)
169. H. Heiselberg and M. Hjorth-Jensen, *Phys. Rep.* **328**, 237 (2000)
170. F.M. Walter and L.D. Mathews, *Nature* **389**, 358 (1997)
171. D. Kramer, H. Stephani, M. MacCallum and E. Herlt, *Exact Solutions of Einstein's Field Equations.* Cambridge Monographs on Mathematical Physics 2003
172. M. Ansorg, A. Leinwächter and R. Meinel, *Astron. Astrophys.*, **381** L49 (2002)

173. D.G. Yakovlev, A.D. Kaminker, O.Y. Gnedin and P. Haensel, *Neutrino Emission from Neutron Stars*, Phys. Rept. **354**, 1 (2001); astro-ph/0012122
174. H.-Th. Janka, K. Kifonidis and M. Rampp, Supernova Explosions and Neutron Star Formation. In: (eds.) Proc. of International Workshop on *Physics of Neutron Star Interiors*, Trento 2000. Springer Lecture Notes in Physics, Vol. 578, Springer Berlin 2001
175. K.S. Thorne, The General Relativistic Theory of Stellar Structure and Dynamics. In: C. DeWitt, E. Schatzmann and P. Veron (eds.), *High Energy Astrophysics.* Gardon and Breach 1967
176. G. Baym et al., *Nucl. Phys.* **A 175**, 225 (1971); *Astrophys. J.* **170**, 99 (1972)
177. E. Flowers and N. Itoh, *Astrophys. J.* **206**, 218 (1976)
178. S. Ayasli and P.C. Joss, *Astrophys. J.* **256**, 637 (1982)
179. P. Giannone and A. Weigert, *Zs. f. Ap.* **67**, 41 (1967)
180. T.M. Tauris and E.P.J. van den Heuvel, Formation and Evolution of Compact Stellar X-ray Sources. In: W.H.G. Lewin and M. van der Klis (eds.), *Compact Stellar X-Ray Sources.* Cambridge University Press (in preparation); astro-ph/0303456
181. W.H.G. Lewin and M. van der Klis, *Compact Stellar X-Ray Sources.* Cambridge University Press (in preparation)
182. E.P.J. van den Heuvel, M. Livio and S.N. Shore, *Interacting Binaries*, Saas-Fee Advanced Course 22 (1994), Swiss Society for Astrophysics and Astronomy
183. S. Chandrasekhar, *Phil. Mag.* **11**, 592 (1931); *Astrophys. J.* **74**, 81 (1931)
184. S. Chandrasekhar, *Observatory* **57**, 373 (1934)
185. R.H. Fowler, *Mon. Not. R. Astron. Soc.*, **87**, 114 (1926)
186. B. Paczynski, *Ann. Rev. Astron. Astrophys.* **9**, 183 (1971)
187. R. Schoen and S.-T. Yau, *Comm. Math. Phys.* **90**, 575 (1983)

Chapter 7

188. W. Israel, *Phys. Rev.* **164**, 1776 (1967); *Commun. Math. Phys.* **8**, 245 (1968)
189. M. Heusler, *Black Hole Uniqueness Theorems.* Cambridge University Press 1996
190. M. Heusler, *Stationary Black Holes: Uniqueness and Beyond*, Living Reviews, **1**, 1998; http://www.livingrews.org
191. M.S. Volkov and D.V. Gal'tsov, *Phys. Rep.* **319**, 1-83 (1999)
192. R.P. Kerr, *Phys. Rev. Lett.* **11**, 237 (1963)
193. E.T. Newman and A.I. Janis, *J. Math. Phys.* **6**, 915 (1965)
194. E.T. Newman et al., *J. Math. Phys.* **6**, 918 (1965)
195. G.C. Debney et al., *J. Math. Phys.* **10**, 1842 (1969)
196. R.D. Blandford and R.L. Znajek, *Mon. Not. R. Astron. Soc.* **179**, 433 (1977)
197. K.S. Thorne, R.H. Price and D.A. MacDonald, *Black Holes: The Membrane Paradigm.* Yale University Press, New Haven and London (1986)
198. N. Straumann, The Membrane Model of Black Holes and Applications. In: F.W. Hehl, C. Kiefer and R.J.K. Metzler (eds.), *Black Holes: Theory and Observation.* Springer-Verlag 1998
199. H.K. Lee, R.A. Wijers and G.E. Brown, *Phys. Rep.* **325**, 83-114 (2000)
200. B. Carter, *Commun. Math. Phys.* **10**, 280 (1968)
201. J.M. Bardeen, W.H. Press and S.A. Teucholsky, *Astrophys. J.* **178**, 347 (1972)

202. M. Heusler and N. Straumann, *Class. Quantum Grav.* **10**, 1299 (1993); *Phys. Lett. B* **315**, 55 (1993)
203. V. Iyer and R.M. Wald, *Phys. Rev. D* **50**, 846 (1994)
204. R.M. Wald, Black Holes and Thermodynamics. In: R.M. Wald (ed.), *Black Holes and Relativistic Stars.* University of Chicago Press 1998
205. B. Carter, *J. Math. Phys.* **10**, 70 (1969)
206. D. Giulini, *J. Math. Phys.* **39**, 6603 (1998)
207. D.R. Gies and C.T. Bolton *Astrophys. J.* **304**, 371 (1986)
208. A.P. Cowley et al. *Astrophys. J.* **272**, 118 (1983)
209. J.E. McClintock and R.A. Remillard *Astrophys. J.* **308**, 110 (1986)
210. J.A. Orosz, Inventory of Black Hole Binaries. In: K.A. van der Hucht, A. Herraro and C. Esteban (eds.), IAU Symposium No. 212, *A Massive Star Odyssey, from Main Sequence to Supernova*, Lanzarote 2002, ASP Conference Series (to appear); astro-ph/0209041
211. R.M. Wagner et al. *Astrophys. J* **556**, 42 (2001)
212. J.E. McClintock et at. *Astrophys. J.* **551**, L147 (2001)
213. M. Miyoshi, J. Moran, J. Herrnstein, L. Greenhill, N. Nakai, P. Diamonnd and M. Inone *Nature* **373**, 127 (1995)
214. T. Ott et al. *ESO Messenger* **111**, 1 (2003); astro-ph/0303408
215. M. Kozlowski, M. Jaroszynski and M.A. Abramowicz, *Astron. Astrophys.* **63**, 209 (1978); M. Sikore, M. Jaroszynski and M.A. Abramowicz, *Astron. Astrophys.* **63**, 221 (1978)
216. D. Giulini, *J. Math. Phys.* **39**, 6603 (1998)
217. P.T. Chrusciel, E. Delay, G. Galloway and R. Howard, *Annales Henri Poincaré* **2**, 109 (2001)
218. P.T. Chrusciel, *Helv. Phys. Acta* **69**, 529 (1996)
219. J.H. Krolik, *Active Galactic Nuclei.* Princeton University Press 1999
220. Ch.J. Willott, R.J. McLure and M. Jarvis, *Astrophys. J.* **587**, L15 (2003)
221. R. Penrose, Structure of Spacetime. In: C. DeWitt and J.A. Wheeler (eds.), *Battelle Rencontres: 1967 Lectures in Mathematics and Physics.* W.A. Benjamin New York 1968
222. R. Genzel et al., *Nature* **425**, 934 (2003)

Chapter 8

223. E. Witten, *Commun. Math. Phys.* **80**, 381 (1981)
224. J. Stewart, *Advanced General Relativity.* Cambridge University Press, 2003
225. J. Nester, *Phys. Lett.* **83A**, 241 (1981)
226. R. Penrose and W. Rindler, *Spinors and Space-Time.* Cambridge University Press 1984
227. R.P. Geroch, *J. Math. Phys.* **9**, 1739 (1968)
228. O. Reula, *J. Math. Phys.* **23**, 810 (1982)
229. T. Parker and C.H. Taubes, *Comm. Math. Phys.* **84**, 223 (1982)
230. G.W. Gibbons, S.W. Hawking, G.T. Horowitz and M.J. Perry, *Comm. Math. Phys.* **88**, 295 (1983)
231. R. Penrose, *Ann. N.Y. Acad. Sci.* **224**, 125 (1973)
232. G. Huisken and T. Ilmanen, *J. Diff. Geom.* **59**, 353 (2001)

Index

Texts and Monographs in Physics

Essential Relativity Special, General, and Cosmological Revised 2nd edition By W. Rindler

The Elements of Mechanics By G. Gallavotti

Generalized Coherent States and Their Applications By A. Perelomov

Quantum Mechanics II By A. Galindo and P. Pascual

Geometry of the Standard Model of Elementary Particles By. A. Derdzinski

From Electrostatics to Optics A Concise Electrodynamics Course By G. Scharf

Finite Quantum Electrodynamics The Causal Approach 2nd edition By G. Scharf

Path Integral Approach to Quantum Physics An Introduction 2nd printing By G. Roepstorff

Supersymmetric Methods in Quantum and Statistical Physics By G. Junker

Relativistic Quantum Mechanics and Introduction to Field Theory By F. J. Ynduráin

Local Quantum Physics Fields, Particles, Algebras 2nd revised and enlarged edition By R. Haag

The Mechanics and Thermodynamics of Continuous Media By M. Šilhavý

Quantum Relativity A Synthesis of the Ideas of Einstein and Heisenberg By D. R. Finkelstein

Scattering Theory of Classical and Quantum *N*-Particle Systems By. J. Derezinski and C. Gérard

Effective Lagrangians for the Standard Model By A. Dobado, A. Gómez-Nicola, A. L. Maroto and J. R. Peláez

Quantum The Quantum Theory of Particles, Fields, and Cosmology By E. Elbaz

Quantum Groups and Their Representations By A. Klimyk and K. Schmüdgen

Multi-Hamiltonian Theory of Dynamical Systems By M. Błaszak

Renormalization An Introduction By M. Salmhofer

Fields, Symmetries, and Quarks 2nd, revised and enlarged edition By U. Mosel

Statistical Mechanics of Lattice Systems Volume 1: Closed-Form and Exact Solutions 2nd, revised and enlarged edition By D. A. Lavis and G. M. Bell

Statistical Mechanics of Lattice Systems Volume 2: Exact, Series and Renormalization Group Methods By D. A. Lavis and G. M. Bell

Conformal Invariance and Critical Phenomena By M. Henkel

The Theory of Quark and Gluon Interactions 3rd revised and enlarged edition By F. J. Ynduráin

Quantum Field Theory in Condensed Matter Physics By N. Nagaosa

Quantum Field Theory in Strongly Correlated Electronic Systems By N. Nagaosa

Information Theory and Quantum Physics Physical Foundations for Understanding the Conscious Process By H.S. Green

Magnetism and Superconductivity By L.-P. Lévy

The Nuclear Many-Body Problem By P. Ring and P. Schuck

Perturbative Quantum Electrodynamics and Axiomatic Field Theory By O. Steinmann

Quantum Non-linear Sigma Models From Quantum Field Theory to Supersymmetry, Conformal Field Theory, Black Holes and Strings By S. V. Ketov

Texts and Monographs in Physics

The Statistical Mechanics of Financial Markets 2nd Edition By J. Voit

Statistical Mechanics A Short Treatise
By G. Gallavotti

Statistical Physics of Fluids
Basic Concepts and Applications
By V. I. Kalikmanov

Many-Body Problems and Quantum Field Theory An Introduction 2nd Edition
By Ph. A. Martin and F. Rothen

Foundations of Fluid Dynamics
By G. Gallavotti

High-Energy Particle Diffraction
By E. Barone and V. Predazzi

Physics of Neutrinos
and Applications to Astrophysics
By M. Fukugita and T. Yanagida

Relativistic Quantum Mechanics
By H. M. Pilkuhn

The Geometric Phase in Quantum Systems
Foundations, Mathematical Concepts, and Applications in Molecular and Condensed Matter Physics
By A. Bohm, A. Mostafazadeh, H. Koizumi, Q. Niu and J. Zwanziger

The Atomic Nucleus as a Relativistic System
By L.N. Savushkin and H. Toki

The Frenkel–Kontorova Model
Concepts, Methods, and Applications
By O. M. Braun and Y. S. Kivshar

Aspects of Ergodic, Qualitative and Statistical Theory of Motion
By G. Gallavotti, F. Bonetto and G. Gentile

Quantum Entropy and Its Use
By M. Ohya and D. Petz

General Relativity
With Applications to Astrophysics
By N. Straumann

Coherent Dynamics of Complex Quantum Systems By V. M. Akulin